ICL Contribution to Landslide Disaster Risk Reduction

Series Editor

Kyoji Sassa, The International Consortium on Landslides, ICL, Kyoto, Japan

The ICL Contribution to Landslide Disaster Risk Reduction book-series publishes integrated research on all aspects of landslides. The volumes present summaries on the progress of landslide sciences, disaster mitigation and risk preparation. The contributions include landslide dynamics, mechanisms and processes; volcanic, urban, marine and reservoir landslides; related tsunamis and seiches; hazard assessment and mapping; modeling, monitoring, GIS techniques; remedial or preventive measures; early warning and evacuation and a global landslide database.

More information about this series at http://www.springer.com/series/16332

Fausto Guzzetti • Snježana Mihalić Arbanas •
Paola Reichenbach • Kyoji Sassa •
Peter T. Bobrowsky • Kaoru Takara
Editors

Understanding and Reducing Landslide Disaster Risk

Volume 2 From Mapping to Hazard and Risk Zonation

Springer

See next page

ISSN 2662-1894 ISSN 2662-1908 (electronic)
ICL Contribution to Landslide Disaster Risk Reduction
ISBN 978-3-030-60226-0 ISBN 978-3-030-60227-7 (eBook)
https://doi.org/10.1007/978-3-030-60227-7

Cover illustration: Aratozawa landslide triggered by the Iwate-Miyagi inland earthquake on 14 June 2008 in Miyagi, Japan. (Tohoku Regional Forest Office, The Forestry Agency JAPAN. All Rights Reserved).

This Springer imprint is published by the registered company Springer Nature Switzerland AG
The registered company address is: Gewerbestrasse 11, 6330 Cham, Switzerland

Editors
Fausto Guzzetti
Department of Civil Protection
Rome, Italy

Paola Reichenbach
Research institute for Geo-Hydrological Protection
(Italian National Research Council)
Perugia, Italy

Peter T. Bobrowsky
Geological Survey of Canada
Sidney, BC, Canada

Snježana Mihalić Arbanas
Faculty of Mining,
Geology and Petroleum Engineering
University of Zagreb
Zagreb, Croatia

Kyoji Sassa
International Consortium
on Landslides
Kyoto, Japan

Kaoru Takara
Graduate School of Advanced Integrated Studies in Human Survivability (Shishu-kan)
Kyoto University
Kyoto, Japan

Associate Editors
Hiromitsu Yamagishi
Hokkaido Research Center
of Geology (HRCG)
Sapporo, Japan

Dario Peduto
Department of Civil Engineering
University of Salerno
Fisciano, Italy

Mike Winter
Winter Associates Limited
Kirknewton, UK

Olga Mavrouli
Faculty of Geo-Information Science
and Earth Observations (ITC)
University of Twente
Enschede, The Netherlands

Paola Salvati
Research Institute for Geo-Hydrological Protection
(Italian National Research Council)
Perugia, Italy

Ping Lu
College of Surveying and Geo-Informatics
Tongji University
Shanghai, China

Binod Tiwari
California State University Fullerton
Fullerton, CA, USA

Alessandro C. Mondini
Research Institute for Geo-Hydrological Protection
(Italian National Research Council)
Perugia, Italy

Daniele Giordan
Research Institute for Geo-Hydrological Protection
(Italian National Research Council)
Torino, Italy

Oded Katz
Geological Survey of Israel
Jerusalem, Israel

D. P. Kanungo
Geo-Hazard Risk Reduction (GHRR) Group
CSIR—Central Building Research Institute (CBRI)
Roorkee, Uttarakhand, India

Veronica Pazzi
Department of Earth Science
University of Florence
Florence, Italy

Yifei Cui
Department of Hydraulic Engineering
Beijing Tsinghua University
Beijing, China

Elias Garcia-Urquia
Civil Engineering Department
National Autonomous University of Honduras
Tegucigalpa, Honduras

ICL and Springer created a new book series “ICL Contribution to Landslide Disaster Risk Reduction” in 2019 which is registered as ISSN 2662-1894 (print version) and ISSN 2662-1908 (electronic version). The first books in this series are six volume of books “Understanding and Reducing Landslide Disaster Risk” containing the recent progress of landslide science and technologies from 2017 to 2020.

Editor-in-Chief: Kyoji Sassa
Assistant Editor-in-Chief: Željko Arbanas

Organizational Structure of the Fifth World Landslide Forum

Organizers

International Consortium on Landslides (ICL)
Global Promotion Committee of International Programme on Landslides (IPL-GPC), including: United Nations Educational, Scientific and Cultural Organization (UNESCO), World Meteorological Organization (WMO), Food and Agriculture Organization (FAO), United Nations Office for Disaster Risk Reduction (UNDRR), United Nations University (UNU), International Science Council (ISC), World Federation of Engineering Organizations (WFEO), International Union of Geological Sciences (IUGS), International Union of Geodesy and Geophysics (IUGG)
Kyoto University (KU), Japan Landslide Society (JLS), Japanese Geotechnical Society (JGS), Japan Society for Natural Disaster Science (JSNDS) and Japan Association for Slope Disaster Management (JASDiM)

Co-sponsors

Cabinet Office (Disaster Management Bureau) of Japan; Ministry of Foreign Affairs of Japan (MOFA); Ministry of Education, Culture, Sports, Science and Technology-Japan (MEXT); Ministry of Land Infrastructure, Transport and Tourism (MLIT); Ministry of Agriculture, Forestry and Fisheries (MAFF); Science Council of Japan (SCJ); Japan International Cooperation Agency (JICA); Japan Society of Civil Engineers (JSCE); Japanese Society of Irrigation, Drainage and Rural Engineering (JSIDRE); Japan Society of Erosion Control Engineering; Japan Society of Engineering Geology.

Supporting Organizations with Finance

Tokyo Geographical Society
International Union of Geological Sciences (IUGS)
Association for Disaster Prevention Research, Kyoto, Japan

Organizing Committee

Honorary Chairpersons
Audrey Azoulay, Director-General of UNESCO*
Mami Mizutori, Special Representative of the United Nations Secretary-General for Disaster Risk Reduction*

Petteri Taalas, Secretary-General of WMO*
Qu Dongyu, Director-General of FAO*
David Malone, Under-Sectretary General of the Unitred Nations and Rector of UNU
Daya Reddy, President of ISC
Gong Ke, President of WFEO
Qiuming Cheng, President of IUGS
Kathryn Whaler, President of IUGG
Qunli Han, Executive Director of Integrated Research on Disaster Risk (IRDR)
Walter Ammann, President and CEO of Global Risk Forum GRF Davos, Switzerland
Juichi Yamagiwa, President of Kyoto University, Japan
Angelo Borrelli, Head of the National Civil Protection Department, Italian Presidency of the Council of Ministers, Italy
Darko But, Director General of the Administration for Civil Protection and Disaster Relief of the Republic of Slovenia, Slovenia
Akifumi Nakao, Director, International Cooperation Division, Disaster Management Bureau, Cabinet Office, Japan
Kazuyuki Imai, Director General of Sabo Department, Ministry of Land Infrastructure, Transport and Tourism, Japan*
Chungsik Yoo, President of the International Geosynthetics Society
Rafig Azzam, President of the International Association for Engineering Geology and the Environment
(*to be confirmed)

Chairpersons
Kyoji Sassa, Professor Emeritus, Kyoto University; Secretary General of ICL
Peter T. Bobrowsky, Geological Survey of Canada; President of ICL
Kaoru Takara, Kyoto University, Japan; Executive Director of ICL

Members
Željko Arbanas (University of Rijeka, Croatia)
Snježana Mihalić Arbanas (University of Zagreb, Croatia)
Nicola Casagli (University of Firenze, Italy)
Fausto Guzzetti (Department of Civil Protection, Italy)
Matjaž Mikoš (University of Ljubljana, Slovenia)
Paola Reichenbach (Research Institute for Geo-Hydrological Protection, National Research Council, Italy)
Shinji Sassa (Port and Airport Research Institute, Japan)
Alexander Strom (Geodynamics Research Center LLC, Russia)
Binod Tiwari (California State University, Fullerton, USA)
Veronica Tofani (University of Firenze, Italy)
Vít Vilímek (Charles University in Prague, Czech Republic)
Fawu Wang (Tongji University, China)

Chairpersons of Local Organizing Committee
Kaoru Takara (Kyoto University)
Daisuke Higaki (Japan Landslide Society)
Ikuo Towhata (Japanese Geotechnical Society)

Secretary Generals
Ryosuke Uzuoka (Disaster Prevention Research Institute, Kyoto University)
Kazuo Konagai (International Consortium on Landslides)
Khang Dang (International Consortium on Landslides)

International Scientific Committee

Beena Ajmera, North Dakota State University, USA
Snježana Mihalić Arbanas, University of Zagreb, Croatia
Željko Arbanas, Faculty of Civil Engineering, University of Rijeka, Croatia
Amin Askarinejad, Technische Universiteit Delft, Delft, The Netherlands
Peter T. Bobrowsky, Geological Survey of Canada, Sidney, Canada
Michele Calvello, University of Salerno, Italy
Giovanna Capparelli, Universita degli Studi della Calabria, Rende, Italy
Nicola Casagli, University of Florence, Italy
Yifei Cui, Tsinghua University, Beijing, China
Sabatino Cuomo, University of Salerno, Fisciano, Italy
Khang Dang, International Consortium on Landslides, Kyoto, Japan
Elias Garcia-Urquia, National Autonomous University of Honduras, Tegucigalpa, Honduras
Stefano Luigi Gariano, Research Institute for Geo-Hydrological Protection, CNR, Perugia, Italy
Daniele Giordan, Research Institute for Geo-Hydrological Protection, CNR, Italy
Fausto Guzzetti, Department of Civil Protection, Italy
Baator Has, Asia Air Survey, Tokyo, Japan
Hans-Balder Havenith, Universite de Liege, Liege, Belgium
D. P. Kanungo, Central Building Research Institute (CBRI), Roorkee, Uttarakhand, India
Oded Katz, Geological Survey of Israel, Jerusalem, Israel
Kazuo Konagai, International Consortium on Landslides, Kyoto, Japan
Doan Huy Loi, International Consortium on Landslides, Kyoto, Japan
Ping Lu, Tongji University, Shanghai, China
Olga Mavrouli, University of Twente, Enschede, The Netherlands
Matjaž Mikoš, Faculty of Civil and Geodetic Engineering, University of Ljubljana, Slovenia
Alessandro C. Mondini, Research Institute for Geo-Hydrological Protection, CNR, Italy
Veronica Pazzi, Department of Earth Science, University of Florence, Florence, Italy
Dario Peduto, Department of Civil Engineering, University of Salerno, Fisciano, Italy
Paola Reichenbach, Research Institute for Geo-Hydrological Protection, CNR, Italy
Paola Salvati, Research Institute for Geo-Hydrological Protection, CNR, Italy
Katsuo Sasahara, Kochi University, Japan
Kyoji Sassa, International Consortium on Landslides, Kyoto, Japan
Shinji Sassa, Port and Airport Research Institute, Japan
Andrea Segalini, University of Parma, Italy
Hendy Setiawan, Universitas Gadjah Mada, Yogyakarta, Indonesia
Alexander Strom, Geodynamics Research Center LLC, Moscow, Russia
Kaoru Takara, Kyoto University, Japan
Faraz Tehrani, Deltares, Delft, The Netherlands
Binod Tiwari, California State University, Fullerton, California, USA
Veronica Tofani, University of Florence, Italy
Ryosuke Uzuoka, Kyoto University, Kyoto, Japan
Vít Vilímek, Faculty of Science, Charles University, Prague, Czech Republic
Fawu Wang, College of Civil Engineering, Tongji University, Shanghai, China
Gonghui Wang, Kyoto University, Kyoto, Japan
Mike Winter, Winter Associates Limited, Kirknewton, UK
Hiromitsu Yamagishi, Hokkaido Research Center of Geology (HRCG), Sapporo, Japan

Local Organizing Committee

Shinro Abe, Okuyama Boring Co., Ltd.
Kiminori Araiba, Fire and Disaster Management College
Shiho Asano, Forestry and Forest Products Research Institute
Has Baator, Asia Air Survey Co., Ltd.

Hiromu Daimaru, Forestry and Forest Products Research Institute
Khang Dang, International Consortium on Landslides
Mitusya Enokida, Japan Conservation Engineers & Co., Ltd.
Kumiko Fujita, International Consortium on Landslides
Kazunori Hayashi, Okuyama Boring Co., Ltd.
Daisuke Higaki, The Japan Landslide Society
Kiyoharu Hirota, Kokusai Kogyo Co., Ltd.
Kazuo Konagai, International Consortium on Landslides
Taketoshi Marui, MARUI & Co., Ltd.
Satoshi Nishikawa, Nagoya University
Keisuke Oozone, OYO Corporation
Katsuo Sasahara, Kochi University
Kyoji Sassa, International Consortium on Landslides
Shinji Sassa, Port and Airport Research Institute
Go Sato, Teikyo Heisei University
Nobuyuki Shibasaki, Nippon Koei Co., Ltd.
Nobuo Sugiura, Japan Association for Slope Disaster Management
Kaoru Takara, Kyoto University
Keisuke Takimoto, GODAI KAIHATSU Corporation
Yoko Tomita, Public Works Research Institute
Ikuo Towhata, The Japanese Geotechnical Society
Kenichi Tsukahara, Kyushu University
Ryosuke Tsunaki, Sabo & Landslide Technical Center
Taro Uchida, Saitama University
Mie Ueda, International Consortium on Landslides
Ryosuke Uzuoka, Kyoto University
Fawu Wang, Tongji University
Hiroshi Yagi, Yamagata University
Hiromitsu Yamagishi, Shin Engineering Consultants Co., Ltd.
Maki Yano, OSASI Technos Inc.

Foreword by Mami Mizutori

More landslides can be expected as climate change exacerbates rainfall intensity. The long-term trend of the last 40 years has seen the number of major recorded extreme weather events almost double, notably floods, storms, landslides, and wildfires.

Landslides are a serious geological hazard. Among the host of natural triggers are intense rainfall, flooding, earthquakes or volcanic eruption, and coastal erosion caused by storms that are all too often tied to the El Niño phenomenon. Human triggers including deforestation, irrigation or pipe leakage, and mine tailings, or stream and ocean current alteration can also spark landslides. Landslides can also generate tsunamis, as Indonesia experienced in 2018.

Globally, landslides cause significant economic loss and many deaths and injuries each year. Therefore, it is important to understand the science of landslides: why they occur, what factors trigger them, the geology associated with them, and where they are likely to happen.

Landslides with high death tolls are often a result of failures in risk governance, poverty reduction, environmental protection, land use and the implementation of building codes. Understanding the interrelationships between earth surface processes, ecological systems, and human activity is the key to reducing landslide risk.

The Sendai Framework for Disaster Risk Reduction, the global plan to reduce disaster losses adopted in 2015, emphasizes the importance of tackling these risk drivers through improved governance and a better understanding of disaster risk.

One important vehicle for doing that is the Sendai Landslide Partnerships 2015–2025 for global promotion of understanding and reduction of landslide risk facilitated by the International Consortium on Landslides (ICL) and signed by the leaders of 22 global stakeholders, including the UN Office for Disaster Risk Reduction (UNDRR), during the Third UN World Conference on Disaster Risk Reduction in Sendai, Japan.

The Sendai Landslide Partnerships—featured on the Sendai Framework Voluntary Commitments online platform—helps to provide practical solutions and tools, education, and capacity building, to reduce landslide risks.

The work done by the Sendai Partnerships can be of value to many stakeholders including civil protection, planning, development and transportation authorities, utility managers, agricultural and forest agencies, and the scientific community.

UNDRR fully supports the work of the Sendai Landslide Partnerships and ICL and looks forward to an action-oriented outcome from the 5th World Landslide Forum to be held in November 2020 in Kyoto, Japan. Successful efforts to reduce disaster losses are a major contribution to achieving the overall 2030 Agenda for Sustainable Development.

Mami Mizutori
United Nations Special Representative of the
Secretary-General for Disaster Risk Reduction

Foreword by the Assistant Director-General for the Natural Sciences Sector of UNESCO for the Book of the 5th *World Landslide Forum*

As the world slowly recovers from the COVID-19 global pandemic, and looking back at the way this crisis developed, it becomes evident that as a global community we were not prepared for an event of this scale. Although not commonly perceived as such, biological hazards such as epidemics are included in the Sendai Framework for Disaster Risk Reduction 2015–2030. In that sense, the preparedness approach for a pandemic is very similar to that of a geophysical natural hazard such as landslides.

Although natural hazards are naturally occurring phenomena, the likelihood of their occurrence and of associated disasters is rising. Climate change, urban pressure, under-development and poverty and lack of preparedness are increasingly transforming these natural hazards into life-threatening disasters with severe economic impacts. Therefore, Disaster Risk Reduction (DRR) is gaining momentum on the agenda of the UN system of Organizations including UNESCO. While the Sendai Framework for Disaster Risk Reduction 2015–2030 is the roadmap for DRR, other global agendas including the Sustainable Development Goals, the Paris Climate Agreement and the New Urban Agenda have targets which cannot be attained without DRR.

In shaping its contribution to those global agendas, UNESCO is fully committed in supporting its Member States in risk management, between its different mandates and disciplines and with relevant partners. The International Consortium on Landslides (ICL) is UNESCO's key partner in the field of landslide science. The Organization's support to the Consortium is unwavering. Since ICL was established in 2002, the two organizations have a long history of cooperation and partnership and UNESCO has been associated with almost all of ICL activities. I am very glad that ICL and UNESCO are mutually benefitting from their collaboration.

The 5th World Landslide Forum (WLF5) is expected to represent a milestone in the history of landslide science particularly for scientists and practitioners. One of the major outcomes of WLF5 will be the Kyoto 2020 Commitment for global promotion of understanding and reducing landslide disaster risk (KLC2020). This commitment is expected to strengthen and expand the activities of the Sendai Landslide Partnership 2015–2025. With UNESCO already engaged as a partner, the adoption of this international commitment will raise global awareness on landslide risk and mobilize wider partnerships that draw together stakeholders from all levels of society, across different regions, sectors and disciplines.

It is my great pleasure to congratulate the organizers for holding this event and assure you that UNESCO is fully committed in contributing to its success. As part of that contribution, our Organization is proud to host a session on landslides and hazard assessment at UNESCO-designated sites such as natural World Heritage sites, biosphere reserves and UNESCO Global Geoparks. This session aims to assess landslide impacts on our shared cultural and natural heritage, providing the best opportunity to generate public awareness and capacity development for landslide disaster reduction.

I am confident that WLF5 will contribute to further advance the knowledge of both scientists and practitioners regarding landslide disaster risk reduction. This book paves the way for the science, knowledge and know-how which will feature in the deliberations of the Forum. UNESCO commends all of the contributors to this publication. I look forward to an enhanced collaboration between UNESCO and ICL in future activities and undertakings.

Shamila Nair-Bedouelle
Assistant Director-General for Natural Sciences
UNESCO

Preface I

Understanding and Reducing Landslide Disaster Risk

Book Series: ICL Contribution to Landslide Disaster Risk

The International Consortium on Landslides (ICL) was established in pursuance of the 2002 Kyoto Declaration "Establishment of an International Consortium on Landslides," with its Statutes adopted in January 2002. The Statutes define the General Assembly of ICL as follows: in order to report and disseminate the activities and achievements of the Consortium, a General Assembly shall be convened every 3 years by inviting Members of the International Consortium on Landslides, individual members within those organizations, and all levels of cooperating organizations and individual researchers, engineers, and administrators. The General Assembly developed gradually prior to, during and after its first meeting in 2005. In the light of the 2006 Tokyo Action Plan, the Assembly was further facilitated at, and following the First World Landslide Forum held in November 2008. On the occasion of each of its triennial forums, ICL publishes the latest progress of landslide science and technology for the benefit of the whole landslide community including scientists, engineers, and practitioners in an understandable form. Full color photos of landslides and full color maps are readily appreciated by those from different disciplines. We have published full color books on landslides at each forum. In 2019, ICL created a new book series "ICL Contribution to Landslide Disaster Risk Reduction" ISSN 2662-1894 (print version) and ISSN 2662-1908 (electronic version). Six volumes of full color books *Understanding and Reducing Landslide Disaster Risk* will be published in 2020 as the first group of books of this series.

The Letter of Intent 2005 and the First General Assembly 2005

The United Nations World Conference on Disaster Reduction (WCDR) was held in Kobe, Japan, 18–22 January 2005. At this Conference, ICL organized session 3.8 "New international Initiatives for Research and Risk Mitigation of Floods (IFI) and Landslides (IPL)" on 19 January 2005 and adopted a "Letter of Intent" aimed at providing a platform for a holistic approach in research and learning on 'Integrated Earth System Risk Analysis and Sustainable Disaster Management'. This Letter was agreed upon and signed, during the first semester of 2005, by heads of seven global stakeholders including the United Nations Educational, Scientific and Cultural Organization (UNESCO), the World Meteorological Organization (WMO), the Food and Agriculture Organization of the United Nations (FAO), the United Nations International Strategy for Disaster Risk Reduction (UNISDR-currently UNDRR), the United Nations University (UNU), the International Council for Science (ICSU-Currently ISC), and the World Federation of Engineering Organizations (WFEO).

The first General Assembly of ICL was held at the Keck Center of the National Academy of Sciences in Washington D.C., USA, on 12–14 October 2005. It was organized after the aforementioned 2005 World Conference on Disaster Reduction (WCDR). ICL published the first full color book reporting on Consortium activities for the initial 3 years, 2002–2005 titled "Landslides-Risk analysis and sustainable disaster management". In the preface of this book, the Letter of Intent for Integrated Earth System Risk Analysis and Sustainable Disaster Management was introduced. Results of the initial projects of the International Programme on Landslides (IPL) including IPL C101-1 Landslide investigation in Machu Picchu World Heritage, Cusco, Peru and previous agreements and MoU between UNESCO, ICL and the Disaster Prevention Research Institute of Kyoto University including UNESCO/KU/ICL UNITWIN Cooperation programme were published as well in this book.

The 2006 Tokyo Action Plan and the First World Landslide Forum 2008

Based on the Letter of Intent, the 2006 Tokyo Round-Table Discussion—"Strengthening Research and Learning on Earth System Risk Analysis and Sustainable Disaster Management within UN-ISDR as Regards Landslides"—towards a dynamic global network of the International Programme on Landslides (IPL) was held at the United Nations University, Tokyo, on 18–20 January 2006. The 2006 Tokyo Action Plan—Strengthening research and learning on landslides and related earth system disasters for global risk preparedness—was adopted. The ICL exchanged Memoranda of Understanding (MoUs) concerning strengthening cooperation in research and learning on earth system risk analysis and sustainable disaster management within the framework of the United Nations International Strategy for Disaster Reduction regarding the implementation of the 2006 Tokyo action plan on landslides with UNESCO, WMO, FAO, UNISDR (UNDRR), UNU, ICSU (ISC) and WFEO, respectively in 2006. A set of these MoUs established the International Programme on Landslides (IPL) as a programme of the ICL, the Global Promotion Committee of IPL to manage the IPL, and the triennial World Landslide Forum (WLF), as well as the concept of the World Centres of Excellence on Landslide Risk Reduction (WCoE).

The First World Landslide Forum (WLF1) was held at the Headquarters of the United Nations University, Tokyo, Japan, on 18–21 November 2008. 430 persons from 49 countries/regions/UN entities were in attendance. Both Hans van Ginkel, Under Secretary-General of the United Nations/Rector of UNU who served as chairperson of the Independent Panel of Experts to endorse WCoEs, and Salvano Briceno, Director of UNISDR who served as chairperson of the Global Promotion Committee of IPL, participated in this Forum. The success of WLF1 paved the way to the successful second and third World Landslide Forum held in Italy and China respectively.

The Second World Landslide Forum 2011 and the Third World Landslide Forum 2014

The Second World Landslide Forum (WLF2)—Putting Science into Practice—was held at the Headquarters of the Food and Agriculture Organization of the United Nations (FAO) on 3–9 October 2011. It was jointly organized by the IPL Global Promotion Committee (ICL, UNESCO, WMO, FAO, UNDRR, UNU, ISC, WFEO) and two ICL members from Italy: the Italian Institute for Environmental Protection and Research (ISPRA) and the Earth Science Department of the University of Florence with support from the Government of Italy and many Italian landslide-related organizations. It attracted 864 participants from 63 countries.

The Third World Landslide Forum (WLF3) was held at the China National Convention Center, Beijing, China, on 2–6 June 2014. A high-level panel discussion on an initiative to create a safer geoenvironment towards the UN Third World Conference on Disaster Risk Reduction (WCDRR) in 2015 and forward was moderated by Hans van Ginkel, Chair of Independent Panel of Experts for World Centers of Excellence (WCoE). In a special address to this high-level panel discussion, Irina Bokova, Director-General of UNESCO, underlined that countries should be united to work against natural disasters and expressed commitment that UNESCO would like to further deepen cooperation with ICL. Ms. Bokova awarded certificates to 15 World Centres of Excellence.

The Sendai Landslide Partnerships 2015 and the Fourth World Landslide Forum 2017

The UN Third World Conference on Disaster Risk Reduction (WCDRR) was held in Sendai, Japan, on 14–18 March 2015. ICL organized the Working Session "Underlying Risk Factors" together with UNESCO, the Japanese Ministry of Land, Infrastructure, Transport and Tourism (MLIT) and other competent organizations. The session adopted ISDR-ICL Sendai Partnerships 2015–2025 (later changed to Sendai Landslide Partnerships) for global promotion of understanding and reducing landslide disaster risk as a Voluntary Commitment to the World Conference on Disaster Risk Reduction, Sendai, Japan, 2015 (later changed to Sendai Framework for Disaster Risk Reduction). After the session on 16 March 2015, the Partnerships was signed by Margareta Wahlström, Special Representative of the UN Secretary-General for Disaster Risk Reduction, Chief of UNISDR (UNDDR), and other representatives from 15 intergovernmental, international, and national organizations. Following the Sendai Landslide Partnerships, the Fourth World Landslide Forum was held in Ljubljana, Slovenia from 29 May to 2 June in 2017. On that occasion, five volumes of full color books were published to disseminate the advances of landslide science and technology. The high-level panel discussion on 30 May and the follow-up round table discussion on 31 May adopted the 2017 Ljubljana Declaration on Landslide Risk Reduction. The Declaration approved the outline of the concept of "Kyoto 2020 Commitment for global promotion of understanding and reducing landslide disaster risk" to be adopted at the Fifth World Landslide Forum in Japan, 2020.

The Fifth World Landslide Forum 2020 and the Kyoto Landslide Commitment 2020

The Fifth World Landslide Forum was planned to be organized on 2–6 November 2020 at the National Kyoto International Conference Center (KICC) and the preparations for this event were successfully ongoing until the COVID-19 pandemic occurred over the world in early 2020. The ICL decided to postpone the actual Forum to 2–6 November 2021 at KICC in Kyoto, Japan. Nevertheless, the publication of six volumes of full color books *Understanding and Reducing Landslide Disaster Risk* including reports on the advances in landslide science and technology from 2017 to 2020 is on schedule. We expect that this book will be useful to the global landslide community.

The Kyoto Landslide Commitment 2020 will be established during the 2020 ICL-IPL Online Conference on 2–6 November 2020 on schedule. Joint signatories of Kyoto Landslide Commitment 2020 are expected to attend a dedicated session of the aforementioned Online Conference, scheduled on 5 November 2020 which will also include and feature the Declaration of the launching of KLC2020. Landslides: Journal of the International Consortium on Landslides is the common platform for KLC2020. All partners may contribute and publish news and reports of their activities such as research, investigation, disaster reduction administration in the category of News/Kyoto Commitment. Online access or/and hard copy of the Journal will be sent to KLC2020 partners to apprise them of the updated information from other partners. As of 21 May 2020, 63 United Nations, International and national organizations have already signed the KLC2020.

Call for Partners of KLC2020

Those who are willing to join KLC2020 and share their achievements related to understanding and reducing landslide disaster risk in their intrinsic missions with other partners are invited to inform the ICL Secretariat, the host of KLC2020 secretariat (secretariat@iclhq.org). The ICL secretariat will send the invitation to the aforementioned meeting of the joint signatories and the declaration of the launching of the KLC2020 on 5 November 2020.

Eligible Organizations to be Partners of the KLC2020

1. ICL member organizations (full members, associate members and supporters)
2. ICL supporting organization from UN, international or national organizations and programmes
3. Government ministries and offices in countries having more than 2 ICL on-going members
4. International associations /societies that contribute to the organization of WLF5 in 2021 and WLF6 in 2023
5. Other organizations having some aspects of activities related to understanding and reducing landslide disaster risk as their intrinsic missions.

Kyoji Sassa
Chair of WLF5/
Secretary-General of ICL
Kyoto, Japan

Peter T. Bobrowsky
President of ICL
Sidney, Canada

Kaoru Takara
Executive Director of ICL
Kyoto, Japan

Appendix: World Landslide Forum Books

WLF	Place/participants	Title	Editors	Publisher/pages
WLF0 (1st General Assembly) 2005	Washington D.C., USA 59 from 17 countries/UNs	Landslides-Risk Analysis and Sustainable Disaster Management	Kyoji Sassa, Hiroshi Fukuoka, Fawu Wang, Goghui Wang	Springer/377 pages ISBN: 978-3-540-2864-6
WLF1 2008	Tokyo, Japan 430 from 49 countries/regions/UNs	Landslides-Disaster Risk Reduction	Kyoji Sassa, Paolo Canuti	Springer/649 pages ISBN: 978-3-540-69966-8
WLF2 2011	Rome, Italy 864 from 63 countries	Landslide Science and Practice Vol. 1 Landslide inventory and Sustainability and Hazard Zoning	Claudia Margottini, Paolo Canuti, Kyoji Sassa	Springer/607 pages ISBN: 978-3-642-31324-0
		Vol. 2 Early Warning, Instrumentation and Monitoring		Springer/685 pages ISBN: 978-3-642-31444-5
		Vol. 3 Spatial Analysis and Modelling		Springer/440 pages ISBN: 978-3-642-31309-7
		Vol. 4 Global Environmental Change		Springer/431 pages ISBN: 978-3-642-31336-3
		Vol. 5 Complex Environment		Springer/354 pages ISBN: 978-3-642-31426-1
		Vol. 6 Risk Assessment, Management and Mitigation		Springer/789 pages ISBN: 978-3-642-31318-9
		Vol. 7 Social and Economic Impact and Policies		Springer/333 pages ISBN: 978-3-642-31312-7
WLF3 2014	Beijing, China 531 from 45 countries/regions/UNs	Landslide Science for a Safer Geoenvironment Vol. 1 The International Programme on Landslides (IPL)	Kyoji Sassa, Paolo Canuti, Yueping Yin	Springer/493 pages ISBN: 978-3-319-04998-4
		Vol. 2 Methods of Landslide Studies		Springer/851 pages ISBN: 978-3-319-05049-2
		Vol. 3 Targeted Landslides		Springer/717 pages ISBN: 978-3-319-04995-3
WLF4 2017	Ljubljana, Slovenia 588 from 59 countries/regions/UNs	Advancing Culture of Living with Landslides Vol. 1 ISDR-ICL Sendai Partnerships 2015-2025	Kyoji Sassa, Matjaž Mikoš, Yueping Yin	Springer/585 pages ISBN: 978-319-53500-5

(continued)

WLF	Place/participants	Title	Editors	Publisher/pages
		Vol. 2 Advances in Landslide Science	Matjaž Mikoš, Binod Tiwari, Yueping Yin, Kyoji Sassa	Springer/1197 pages ISBN: 978-319-53497-8
		Vol. 3 Advances in Landslide Technology	Matjaž Mikoš, Željko Arbanas, Yueping Yin, Kyoji Sassa	Springer/621 pages ISBN: 978-3-319-53486-2
		Vol. 4 Diversity of Landslide Forms	Matjaž Mikoš, Nicola Casagli,Yueping Yin, Kyoji Sassa	Springer/707 pages ISBN: 978-3-319-53484-8
		Vol. 5 Landslides in Different Environments	Matjaž Mikoš,Vít Vilímek,Yueping Yin, Kyoji Sassa	Springer/557 pages ISBN: 978-3-319-53482-4
WLF5	2020 (publication) 2021 (Forum)	Understanding and Reducing Landslide Disaster Risk Vol. 1 Sendai Landslide Partnerships and Kyoto Landslide Commitment	Kyoji Sassa, Matjaž Mikoš, Shinji Sassa, Peter T. Bobrowsky, Kaoru Takara, Khang Dang	Springer In Process
		Vol. 2 From mapping to hazard and risk zonation	Fausto Guzzetti, Snježana Mihalić Arbanas, Paola Reichenbach, Kyoji Sassa, Peter T. Bobrowsky, Kaoru Takara	
		Vol. 3 Monitoring and early Warning	Nicola Casagli, Veronica Tofani, Kyoji Sassa, Peter T. Bobrowsky, Kaoru Takara	
		Vol. 4 Testing, modelling and risk assessment	Binod Tiwari, Kyoji Sassa, Peter T. Bobrowsky, Kaoru Takara	
		Vol. 5 Catastrophic landslides and Frontier of Landslide Science	Vit Vilimek, Fawu Wang, Alexander Strom, Kyoji Sassa, Peter T. Bobrowsky, Kaoru Takara	
		Vol. 6 Specific topics in landslide science and applications	Željko Arbanas, Peter T. Bobrowsky, Kazuo Konagai, Kyoji Sassa, Kaoru Takara	

Preface II

Volume 2 *From Mapping to Hazard and Risk Zonation*

Landslides are known to occur in all continents and in the seas and oceans, where they play a key role in shaping landscapes and in producing sediments. In many areas of the world, landslides also represent a severe threat. It is therefore not surprising that landslide mapping and modelling, landslide hazard assessment, and landslide risk evaluation are increasingly popular among scientists, practitioners and decision makers. However, due to their inherent natural variability, landslides are difficult to predict, and this limits our collective ability to mitigate landslide risk and to reduce the landslide societal, economic and environmental consequences, and particularly the direct consequences on the population.

From Mapping to Hazard and Risk Zonation is Volume 2 of the book *Understanding and Reducing Landslide Disaster Risk*, prepared to collect selected contributions presented to the Fifth World Landslide Forum (WLF5), to be held in Tokyo, Japan, 2–6 November 2021. A contribution of the International Consortium of Landslides (ICL) to landslide disaster risk reduction, the volume is divided in six parts, and comprises an editorial introduction, two keynote articles, and 52 articles covering five main topics related to (i) landslide detection, recognition and mapping techniques and methods tested at various geographical scales and in different morphological, geological and climatic settings, (ii) landslide susceptibility assessment and spatial landslide modelling at different geographical scales, using consolidated and innovative techniques, (iii) landslide size statistics and landslide temporal modelling, which are key, yet inadequately investigated to landslide hazard assessment, (iv) landslide data and information collection, organization, and sharing for improved disaster mitigation, and (v) vulnerability to landslides of people, communities and the built environment, key but poorly understood components of landslide risk assessment.

Overall, the 54 articles presented in the volume cover study areas in 23 Nations in all continents, except Antarctica. Collectively, the articles provide a perspective of the "science and art" of landslide mapping, and of the current capabilities the international landslide community has to model landslide susceptibility and to prepare susceptibility zoning maps, to model and evaluate landslide hazards, to assess the vulnerability to landslides of various elements at risk, and to estimate and—hopefully—mitigate landslide risk.

A collective volume like the one we were fortunate to edit is the result of team work. We acknowledge the more than 180 authors who decided to contribute to the volume, sharing their data, information and experience. All articles published in the volume were peer reviewed, and were accepted with at least one positive review decision. A special appreciation goes to the handling editors and to the many referees who dedicated their time and effort to read the manuscripts, review them, and provide constructive suggestions to improve the quality of the articles. Everyone benefited from their objective, careful and fair comments. We are indebted to the colleagues who accepted our invitation to convene the sessions of the WL5 where the contributions presented in the volume, and several others, will be present in November 2020.

Ultimately, we wish to thank the International Consortium of Landslides (ICL), the organizers of the Fifth World Landslide Forum (WLF5), and particularly Prof. Kyoji Sassa, for giving us the opportunity to edit the volume. We are grateful to Dr. Khang Dang for his enduring professional support during the preparation of the volume.

Perugia, Italy — Paola Reichenbach
Zagreb, Croatia — Snježana Mihalić Arbanas
Rome, Italy — Fausto Guzzetti

Contents

Introduction to the Volume 'From Mapping to Hazard and Risk Zonation'

Paola Reichenbach, Snježana Mihalić Arbanas, and Fausto Guzzetti

Abstract

"From mapping to hazard and risk zonation" is Volume 2 of the book "Understanding and Reducing Landslide Disaster Risk". The volume collects 54 articles covering five main general topics, including (i) landslide detection, recognition and mapping, (ii) landslide susceptibility assessment and spatial landslide modelling, (iii) landslide size statistics and landslide temporal modelling, (iv) data and information for landslide disaster mitigation, and (v) vulnerability to landslides of people, communities and the built environment. More than 180 authors contributed to the volume, and presented the results of their research conducted in 23 Nations in all continents, except Antarctica. Overall, the 54 articles illustrate a variety of consolidated and innovative methods and techniques, and together represent a contribution of the international landslide community to the long-term effort towards landslide disaster risk reduction.

Keywords

Landslide • Map • Susceptibility • Hazard • Vulnerability • Risk • Model

From mapping to hazard and risk zonation is Volume 2 of the book **Understanding and Reducing Landslide Disaster Risk**, a contribution to the International Consortium of Landslides (ICL) to landslide disaster risk reduction.

The volume collects 54 articles covering five main general topics, including, landslide detection, recognition and mapping, landslide susceptibility assessment and spatial landslide modelling, landslide size statistics and landslide temporal modelling, data and information for landslide disaster mitigation, and vulnerability to landslides of people, communities and the built environment. Overall, the articles are presented by more than 180 authors, and cover study areas in 23 Nations in all continents, except Antarctica. Based on the prevalent content of the individual contributions, we loosely grouped the articles in six sections, namely,

- keynotes,
- landslide detection, recognition and mapping,
- landslide susceptibility assessment and modelling,
- landslide size statistics and temporal modelling,
- data and information for landslide disaster mitigation, and
- vulnerability to landslides of people, communities and the built environment.

The keynotes section begins with an article by *Miyagi* who provides an overview of landslide mapping methods in Japan, focusing on the production of geomorphological inventory maps prepared through the visual interpretation of stereoscopic aerial photographs of different vintages, and on the application of the Analytical Hierarchical Process (AHP) classification method for the preparation of landslide reactivation potential maps and of landslide susceptibility maps. The authors also discuss the identification and mapping of landslides using digital elevation models (DEMs) obtained from remote sensing sources, including Unmanned Aerial Vehicles (UAVs), Structure from Motion (SfM) and 3D visualisation technologies.

P. Reichenbach (✉)
Consiglio Nazionale delle Ricerche, Istituto di Ricerca per la Protezione Idrogeologica, via della Madonna Alta 126, 06128 Perugia, Italy
e-mail: paola.reichenbach@irpi.cnr.it

S. Mihalić Arbanas
Faculty of Mining, Geology and Petroleum Engineering, University of Zagreb, Zagreb, Croatia
e-mail: smihalic@rgn.hr

F. Guzzetti
Department of Civil Protection, Italian Presidency of the Council of Ministers, via Vitorchiano 2, 00189 Rome, Italy
e-mail: fausto.guzzetti@protezionecivile.it

F. Guzzetti et al. (eds.), *Understanding and Reducing Landslide Disaster Risk*,
ICL Contribution to Landslide Disaster Risk Reduction,
https://doi.org/10.1007/978-3-030-60227-7_1

In the second keynote article, *Yagi and co-workers* illustrate the production of landslide inventory maps for susceptibility modelling, with examples from Japan and Honduras. The authors show an application of the AHP method to determine the morphological and land cover settings that favour landslide reactivation, and provides examples of medium-scale landslide susceptibility maps and of the application of stereoscopic aerial photogrammetry and AHP for the large-scale susceptibility assessment of the 1998 El Eden landslide triggered by Hurricane Mitch in Honduras, and susceptibility mapping using high-resolution DEM in the Akatani study area, Nara Prefecture, Japan.

The section on landslide detection, recognition and mapping collects eleven articles that present examples, discuss problems, and propose solutions for improved cartographic representations of landslides of different types for a variety of applications, from site-specific engineering geological mapping to small-scale regional landslide mapping.

The section opens with five articles that illustrate techniques for the production of landslide inventory maps in different geological and geomorphological settings. Four studies exploit high resolution, bare-earth DEMs obtained by airborne LiDAR for landslide detection and mapping, whereas the fifth study uses satellite images for landslide mapping. *Sikora and Wojciechowski* present a geomorphological inventory of 2,444 landslides in the Sudetes Mountains, in the Bohemian Massif, Poland, which the authors use to investigate the size and spatial distributions of landslides in relation to the geological and geomorphological settings. *Jagodnik and co-workers* show a geomorphological inventory of landslides and gullies in the flysch deposits of the Dubračina River basin, Croatia. The inventory was compiled through the visual interpretation of maps showing a number of topographic and morphometric variables obtained from a high-resolution DEM. Working in the Polish Carpathians, *Pawluszek-Filipiak and Borkowski* compare a visually interpreted landslide inventory to the results of a classification model developed using a Support Vector Machine applied to seven topographic and morphometric variables. Results reveal the potential for automated landslide detection in complex landscapes. *Kasai* exploits multiple, very-high resolution LiDAR DEMs from 2006 to 2010 to prepare landslide roughness maps which are used to investigate the activity of the large Nagatono landslide, Japan, which was reactivated in 2011. Lastly, *Dagdelenler and co-workers* use a multi-resolution segmentation process for the semi-automatic mapping of deep-seated rotational landslides in the Ulus District, along the Western Black Sea Region, Turkey.

The next two articles illustrate examples of landslide detection and mapping using high resolution DEMs obtained by processing images taken by UAVs, combined with SfM photogrammetric range imaging techniques. *Damjanović and co-workers* use a high-resolution DEM and SfM, in combination with field surveys, to map in detail a fossil landslide in the Pannonian Basin, Croatia, and to perform a slope stability analysis of the landslide. *Thanh and co-workers* also use high resolution DEMs and SfM to prepare a large-scale landslide inventory for the Lao Cai Province, Vietnam, that shows more than 40 landslides, rock falls, and debris flows.

The following two articles discuss the use of Synthetic Aperture Radar (SAR) differential interferometry for the characterization and quantification of landslide activity. *Greif and co-workers* use ESA Sentinel-1 images taken between December 2014 and May 2017 to estimate the velocity of 182 landslides in the Kosice basin, eastern Slovakia; and *Pawluszek-Filipiak and Borkowski* use ESA Sentinel-1 images taken in 2017 to evaluate the activity of 205 landslides shown in an existing inventory for the Rożnów Lake area, in Polish Carpathians range.

Stefanelli and co-workers use a geomorphological landslide inventory for the Arno River basin, central Italy, and calculate the Morphological Obstruction Index—which depends on the valley floor morphometry and on the estimated landslide volumes—to assess the potential predisposition to river obstruction of the 27,500 landslides shown in the inventory.

In the last article of the section, *Pham and co-workers* discuss the preparation of a landslide event inventory map for a section of the Halong-Vandon expressway, in the Quang Ninh Province, Vietnam. The inventory shows rainfall induced landslides occurred in the period from June 2017 to August 2018. The study includes results of soil testing on the landslide materials and an analysis of rainfall thresholds for possible landslide initiation.

The section on landslide susceptibility assessment and modelling comprises 18 articles that give examples of landslide susceptibility modelling and associated zonings, including heuristic approaches, statistical methods, neural network analyses, and machine learning methods, in a variety of morphological and geological settings.

The section begins with two articles describing slope instability settings conditioned by single factors. *Barykina and co-authors* use a detailed geological map and data obtained from boreholes to identify landslide types, and to detect the sliding surface of landslides affecting the Vorobyovy Gory, along the right side of the Moskva River, in Russia. *Borboni and co-workers* analyse the effects of agronomical practices on soil properties, grapevine root systems, and the proneness to shallow landslides in the northern Italian Apennines. Results reveal that soil hydraulic conductivity is strongly influenced by soil management practices, and that vineyards with inter-rows alternation management exhibit the highest root density and the strongest root reinforcement.

In the next four articles, authors present susceptibility analyses based on multiple factors. For two catchments in the La Ciénega River basin, Mexico, *García Reyes and co-workers* use aerial photographs, field surveys, and geomorphological evaluation to map landslides, to define landform units, and to assess landslide susceptibility using a heuristic approach. Working in India, *Singh and co-workers* propose a susceptibility evaluation based on an expert-based weighting schema of seven environmental variables. Adopting slope units as the mapping unit of reference, *Lin and co-workers* use discriminant analysis to assess the potential for deep-seated landslides in four catchments in the Namasia area, Kaoshiung, Taiwan. *Nolasco-Javier and Kumar* use binary logistic regression to model landslide susceptibility in a study area in the Philippines. The authors discuss variables multicollinearity, which can reduce the ability of the model to identify statistically significant independent variables.

The next five articles present applications of different machine learning methods for landslide susceptibility zonation. *Huqqani and co-workers* illustrate a landslide susceptibility map for the Penang Island, Malaysia, prepared using an artificial neural network model. The study evaluates the effect of the number of hidden layers on the susceptibility zonation confronting metrics of model accuracy and computation time. *Wang and co-workers* use the "deep belief network" to evaluate landslide susceptibility in a large river catchment in China, and evaluate the model performance confronting the model output with two different susceptibility zonations prepared using the back propagation neural network and logistic regression. Working in Japan, *Dou and co-workers* confront two advanced artificial intelligence models to evaluate landslide susceptibility in a study area. For the Cinque Terre national park, Northern Italy, *Di Napoli and co-workers* discuss ensembles of single susceptibility models prepared using an Artificial Neural Network, Generalized Boosting, and Maximum Entropy algorithms. Lastly, *Kalantar and co-workers* describe three machine learning methods—the Generalized Linear Model, Boosted Regression Trees, and Support Vector Machine—and their ensemble, to model landslide susceptibility in the Sajadrood region, Iran.

The next two articles discuss some of the limits of landslide susceptibility models. *Braun and co-workers* present a landslide susceptibility zonation prepared using Neural Networks and Decision Trees, and they analyse how data quality, quantity, complexity, and preparation can have major effects on the outcomes of the modelling. In addition, they discuss the necessity of standards for input data, modelling output, and result communication to improve the usability of landslide susceptibility models in urban planning. *Torizin and co-workers* focus on biases and sampling errors in landslide susceptibility modelling, they provide examples from the Lanzhou, China, case study, and offer practical suggestions on how to account for uncertainties in data-driven landslide susceptibility assessments.

The next three articles deal with physically-based landslide susceptibility modelling. Working in the Serra do Mar mountain range (Brazil), *Cabral and Reis* compare the performances of SHALSTAB and SINMAP, two well-known physically-based, numerical models for shallow landslide susceptibility assessment. *Marjanović and co-workers* compare heuristic and deterministic landslide susceptibility assessment models, in North Macedonia. In China, *Chen and co-workers* evaluate the seismic landslide susceptibility after a major seismic event that triggered more than 4,834 landslides applying the Newmark model.

The last two articles of the section consider landslide risk. *Doshida* exploits LiDAR data taken by a UAV immediately after the Yabakei landslide, Japan, to evaluate residual risk associated to parts of the landslide, for rescue operations. *Vega and Hidalgo* evaluate the probability that a sliding mass reaches the La Liboriana River, Colombia, and the probability that the river is dammed by the landslide, causing problems to the water supply of nearby rural communities.

The section discussing landslide size statistics and landslide temporal modelling encompasses eleven articles that focus on methods and techniques for the estimation of the frequency and the probability density functions of landslide size, and the characteristics of landslides and debris flows at different scales.

The section opens with the article by *Akgun and co-workers* who examine the frequency and the probability distributions of the area of individual landslides in flysch deposits along the Western Black Sea Region of Turkey, using the Double Pareto and the Inverse Gamma distribution models. In the next paper, *Steger and co-workers* investigate statistical associations between landslides of the slide type and a set of environmental variables. The authors highlight the need to consider the source of the landslide information to avoid erroneous inferences. *Paulin and co-workers* describe an inventory of more than 600 landslides in volcanic and sedimentary terrain. Using measurements from representative landslides, they establish an empirical power law relationship linking landslide area to volume, and use the relationship to estimate the potential contribution of landslide material delivered from volcanic and sedimentary landform units.

The following eight articles deal with debris flows, in various ways. *Brideau and co-workers* propose an empirical approach to estimate the potential runout distance of debris flows, and the associated impact areas. *Baum* proposes an approach for narrowing the parameter uncertainty for debris flow initiation models, and to assess the likely extent of the potential debris flow source areas. *Nakatani and co-workers* use information gathered for the severe debris flow event that hit Hiroshima, Japan, in 2014, to investigate the impact of DEM resolution on debris flow simulations. *Hu and co-workers* propose a physically-based, distributed, debris flow peak discharge model to evaluate debris flow routing and to estimate debris flow peak discharge in areas previously affected by earthquakes. *Watanabe and co-workers* use a simple model to assess the stability of the eastern slopes of the Takumi village, Japan, during Typhoon Hagibis, in October 2019. *Koi and co-workers* study the magnitude and timing of lahars related to recent volcanic eruptions in Japan. *Li and Wang* evaluate the impact of the new Sichuan-Tibet Railway, in China, on geological hazards, using different models. In the last article of the session, *Rouai and co-authors* illustrate a methodology to map slope instabilities and to identify landslide prone areas in the Volubilis archaeological site, Morocco.

The section dealing with data and information for landslide disaster management and mitigation consists of five articles that discuss topics related to the collection and use of data and information for landslide risk and crisis management, at various scales. *Ramli and Ahmad* describe the Slope Hazard and Risk Mapping Project aimed at producing hazard and risk maps in Malaysia. A practical and evidence-based guide supports implementation, ensures engagement by stakeholders, and strengthens accountability in disaster risk reduction. Working in the Cameron Highlands, the most populated and economically developed mountain area in Malaysia, *Abdul Kadir and co-workers* analyse long-term trends in land use changes, and investigate the influence of human-induced activities on landslides for better, risk-informed land planning. Working in Italy, *Tarolli and co-workers* combine remote sensing imagery, digital terrain analysis, and modelling techniques, to prepare accurate landslide maps of agricultural landscapes. The obtained landslide maps support land owners and planners in managing complex and fragile landscapes, and help preserve cultural heritage, ecosystem services and food safety, maintaining economic and environmental sustainability. *Del Soldato and co-workers*, also working in Italy, propose an operational framework for the continuous, real-time monitoring of hazardous scenarios—including the acceleration of landslide or subsidence phenomena—over large and very large areas, through the systematic collection and repeated processing of satellite imagery. Lastly, *Miyagi and co-workers* show examples of landslide maps at different scales in Japan and Vietnam, and their possible application for crisis management.

The last section of the volume includes seven articles that discuss aspects related to landslide vulnerability of people, communities and the built environment. The section begins with two articles dealing with the human consequences of landslides. *Salvati and her co-workers* use a catalogue on 1039 landslide fatalities in Italy in the 50-year period 1970–2019 to show that, in Italy, vulnerability to landslides depends on age and gender. *Goto and co-workers* use information collected by interviewing citizens and governmental representatives three months after a series of deadly debris flows that impacted the community of Montecito, California, in 2018, to study the factors that influenced the vulnerability of people to debris flows in this community.

The next two articles discuss applications of satellite imagery for the assessment of the vulnerability to landslides of structures and infrastructures. *Peduto and co-workers* present two case studies of the use of DInSAR data obtained by processing SAR images taken by COSMO-SkyMed satellites to investigate the vulnerability of buildings and roads to slow-moving landslides in Calabria, southern Italy. *Bianchini and co-workers* exploit DInSAR data obtained by processing ESA Sentinel-1 imagery for the evaluation of landslide hazard and landslide impact in Valle d'Aosta, northern Italy, a mountain region where landslides are abundant and more than 50% of the territory is above 2000 m of elevation.

The last three articles tackle issues related to the use of high-resolution topographic data, aided by photogrammetric imaging techniques and GIS, for vulnerability and risk assessment of structures threatened by landslides. *Garnica-Peña and co-workers* describe the preparation of a map showing landslide exposure for the La Gloria neighbourhood of Teziutlán, in the Sierra Norte de Puebla Region, Mexico, which was affected by a severe landslide disaster in October 1999. *Goto and Clarke* use Ordinal Logistic Regression to evaluate shallow landslide risk for the city of Sao Paulo, Brazil, based on geomorphological, lithological, hydrological, and landslide information. The authors evaluate the performance of their model by comparing training and validation sets. Ultimately, *Singh and co-workers* illustrate the assessment of rock fall risk in the Chamoli District, Garhwal Himalayas, India, as a function of physical vulnerability and proximity of buildings to site-specific rock fall zones and drainage channels using a semi-quantitative approach in a GIS.

Overall, the 54 articles in the Volume illustrate and discuss a variety of consolidated and innovative methods and techniques, and together represent a contribution of the international landslide community to the long-term effort towards landslide disaster risk reduction.

Perugia, Zagreb, Rome, June 29, 2020.

Acknowledgements We thank the many authors who have decided to share their findings with the articles presented in this volume. We are indebted to the many reviewers who have given us—and the authors—some of their time. We thank the organizers of the Fifth World Landslide Forum (WLF5) and the International Consortium of Landslides (ICL) for giving us the opportunity to edit this volume of the book Understanding & Reducing Landslide Disaster Risk. We hope to see them all from 2 to 6 November 2021 in Kyoto, Japan.

Part I
Keynotes

Landslide Recognition and Mapping for Slope Disaster Risk Reduction and Management–Keynote Speech

2020 Kyoto Japan

Toyohiko Miyagi

Abstract

Understanding of landslide phenomena and mapping has advanced technically, as have the tools and related theory. The change began during the international decade for Natural Disaster Reduction launched by the United Nations in 1990. The utmost importance of these issues worldwide has clarified business solutions for disaster prevention by the United Nations World Conference on Disaster Reduction in 2015. Here, an improved understanding of landslide recognition and mapping serve as fundamental ways to both reduce and prevents disasters. Naturally for engineers and researchers, recognition and mapping must benefit in both capacities. However, a question remains as to the ability for everyone to understand the results, such as with a map. Whenever a disaster occurs, disaster victims claim due to "not understand that a disaster could occur here". During such times, engineers, clients, and local residents must aim to increase the understanding and use of maps for slope disaster risk reduction (SLOPE DRR). This keynote lecture introduces examples in connection with the circumstances of understanding and mapping.

Keywords

Landslide mapping • Risk evaluation • Sensing tool • Site prediction • Slope DRR

Introduction

This keynote lecture begins with a discussion of how to understand landslides, followed by a short review of the current status of mapping that includes the fundamental approach to visualizing and understanding landslides. This keynote lecture concludes with an explanation of its importance through discussions and visualizations.

The overall image of a landslide is defined by the gravitational topography and geologic actions of the ground surface. First, one must understand the fundamental characteristics of the landslide processes in order to understand landslides. One must consider what locations to emphasize and also consider purposes and modes by which these locations should be examined. This discussion clarifies and reconfirms the following: (1) Landslides occur when a sloping portion of land with immobile surroundings and clear borders moves simultaneously as a mass. (2) The main landslide component is the moving landslide body. An unmoving base called the slip surface is created beneath it. (3) In some landslides, the landslide body itself might remain on the slope. In such cases, the terrain and geological characteristics change autonomously becoming factors in further destruction and future movements. (4) Movement results from latent vulnerabilities of the land on which a landslide occurs when a distributive factor is triggered. The landslide potential of sloping areas can be understood based on these factors. This information can be used for disaster prevention contributing to earth science.

Landslides, which have the characteristics described in (1) above, occur in a clearly defined area, examining which enables the mapping of landslide phenomena. This examination supports landslide topography maps that differentiate areas with changes in landslide probability. In addition, because sequence changes in (3) are caused by those in the materials and surfaces, which can be mapped and visualized, combining the investigation, experimentation, and consideration of (2) and (4) improves the speed at which mapping can be achieved. Three-dimensional visualization is often useful to comprehensively understand landslides.

Landslides are common. They are a fundamental phenomenon caused by earth surface movements. Based on this understanding, comprehensive, rational, and strategic

T. Miyagi (✉)
Advantechnology Co. Ltd., 1-4-8 Kakyoin Aobaku, Sendai, 980-0013, Japan
e-mail: c1934009@mail.tohoku-gakuin.ac.jp

F. Guzzetti et al. (eds.), *Understanding and Reducing Landslide Disaster Risk*, ICL Contribution to Landslide Disaster Risk Reduction,
https://doi.org/10.1007/978-3-030-60227-7_2

responses should be considered. Currently, multiple organizations are using construction information modeling and management (CIM). Evolution of diversification through CIM tools such as ADCALC 3D, UAV/SfM, Lasers, and other sensing tools, is the ultimate goal of this presentation. Hot topics that contribute to the development of logical disaster prevention and reduction policies are introduced to bring about a comprehensive understanding of landslides.

Landslide Recognition

Simple Recap of Mapping

Landslides can cause death and property destruction, and thus, measures for addressing this phenomenon are crucially important for society. Every economically developed country worldwide has areas prone to landslides. Technicians and government officials are concerned about where landslides might occur during heavy precipitation or an intensive earthquake. In spite of these conditions, research and understanding of landslides has emphasized scientific and technical fields to date. If it says the main points, scientific and technical research has progressed with the accumulation of field experience as research materials for actual landslide areas, improvement of technology for implementing disaster prevention and control, and implementation of monitoring and countermeasures. As a starting point, this approach requires a disaster to occur first. The coping paradigm is inescapable, but knowledge and analytical techniques to ascertain the fundamental nature of landslide phenomena (both theoretical and mechanical) have developed through these methods.

This session title is “Recognition and Mapping.” In recent years, the fundamental understanding of landslide phenomena which supports the need to map has been applied to related issues such as identifying areas in danger of landslides, evaluating danger levels of landslide-prone areas, and grasping the latent potential of landslide disasters. As a result, mapping is regarded as more important than ever. This trend has been manifested not through the post-disaster activities discussed above but rather through specific examination of where landslides will occur requiring the investigation of the latent potential for landslides: (1) the geographical characteristics of landslides to be visually understood, (2) landslide movement areas to be determined through wide-area sensing, and (3) expansion to additional challenges such as detecting disaster risk through detailed observation of movement areas is also possible.

The main points for understanding mapping landslide phenomena are explained below.

1. Interface science: From an earth science perspective, mass movements occur on the earth surface. Inner and outer forces have fixed material characteristics such as weathering, deformation, and geological structure. This interface causes changes that make up one field of interface science (Fig. 1).
2. Main components of landslides: As a factor related directly to the mechanism of landslide activity, the clump of land that causes landslide movements is called the landslide body. The slip surface is a boundary surface between the landslide body and the stable surroundings. The landslide body at the top of the slip surface moves, differentiates, flows, spreads, and stops. The landslide body boundary with immobile surroundings can be broken up into distinctive component areas such as main scarps, frank scarps, and separation scarps (Fig. 2, Varnes 1978).
3. Landslide sizes: Surface destruction occurs only from movements of the soil layers: small landslides, surface landslides, and slope failures (1 m thick or less with surface areas less than several square meters). Various other landslide classifications exist. For example, massive landslides are designated when thicknesses are up to 100 m with areas on the order of several square kilometers.
4. Styles of movement: Various types of movements have been classified in to falls, rolling, toppling, rotational slides, translational slides, spreads, flows, and complex slides (Varnes 1978).
5. Sequential destruction process: Through the occurrence of initial destruction in sloping areas not prone to landslides, landslide movements generate factors for additional destruction within the landslide body, which in turn generate further destructive factors in a sequential

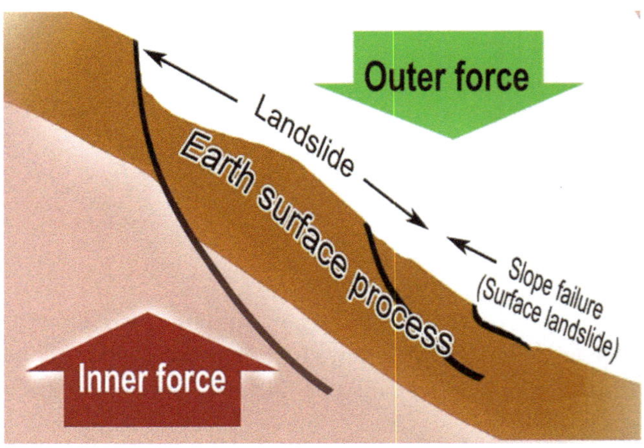

Fig. 1 Basic landslide pross components as interface science

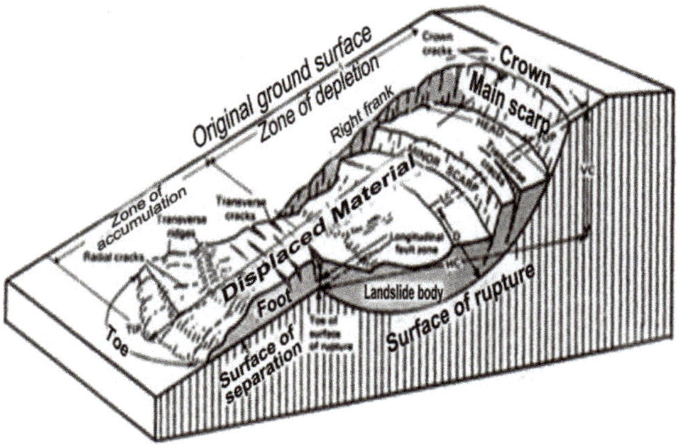

Fig. 2 Illustration of style and the related form and structure of landslide (Style: Slump) (Varnes 1978)

fashion (control under autonomous destruction processes). In this way, the process of change can be regarded as autonomous sequence (Fig. 3).

6. The following is a summary of points that characterize the above landslide motivations or which directly relate to their evaluation. Landslides are a type of mass movement with extremely strong unification. Cliff faces are differentiated from their surroundings. With the slip surface as the boundary, a chain of deformation occurs in the landslide body at this boundary. Accordingly, understanding landslide movement requires an understanding of the surroundings and grasping the movement style, movement orientation, and slip surface presence for a landslide. These aspects can only be understood in three dimensions.

Descriptions of Landslides, Body Material Characteristics, Movements, and Topography

How exactly does the landslide surface topography reflect its movement and internal structure? By peeling back layer-by-layer and examining the internal structure of the ground surface and landslide body all the way down to the sliding surface, appropriate methods can be found. In addition to determining the location of landslide movement, topography is useful to estimate the paths which a landslide is likely to take in certain scenarios. This can also be used to bolster a discussion of mapping, as explained in the next section. To date, we have conducted investigations related to the link between the ground surface topography and internal structure, specifically examining the Okamizawa Landslide (Miyagi et al. 2004). This lecture presents an overview of results from these investigations, as shown in Figs. 4 and 5. Furthermore, the following are examples of the relation between landslide appearance and its internal contents (Miyagi and Hamasaki 2016).

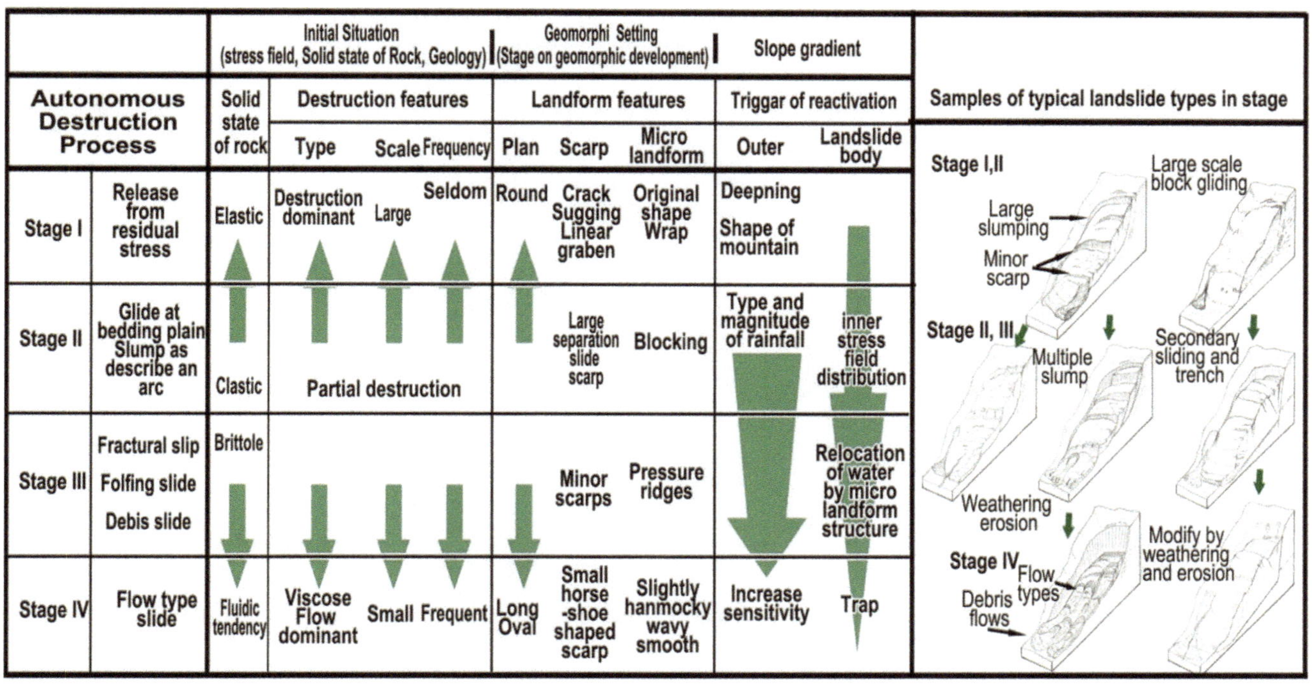

Fig. 3 Autonomous landslide destruction process and typical features

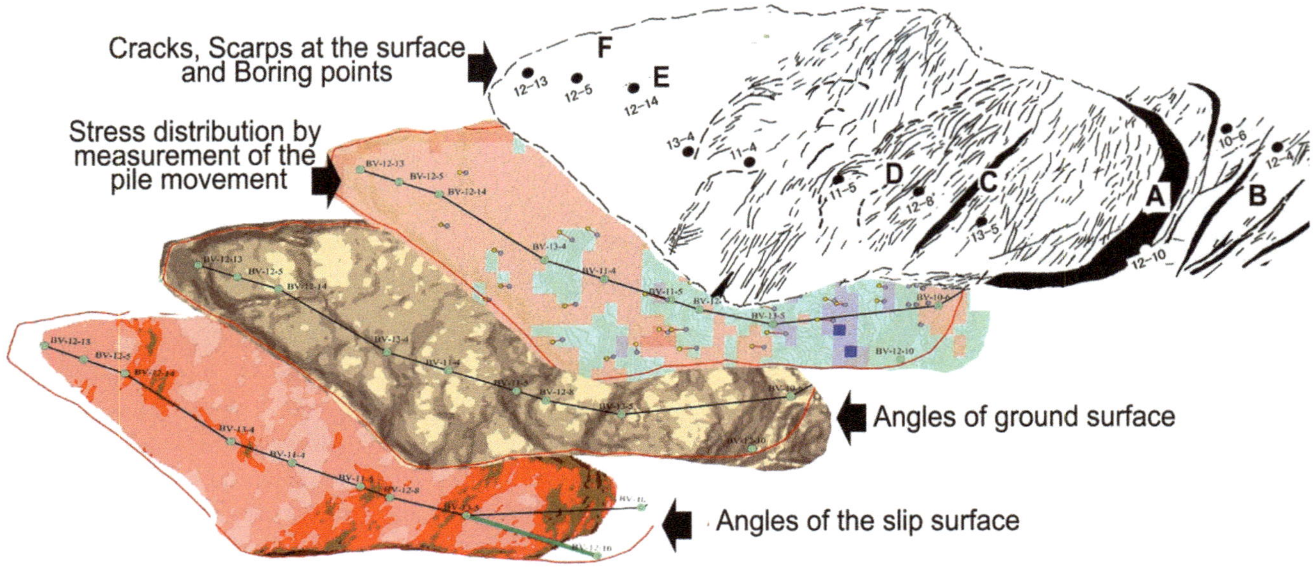

Fig. 4 Field investigation data of the Okamizawa landslide area, Akita Prefecture, Northeastern Japan

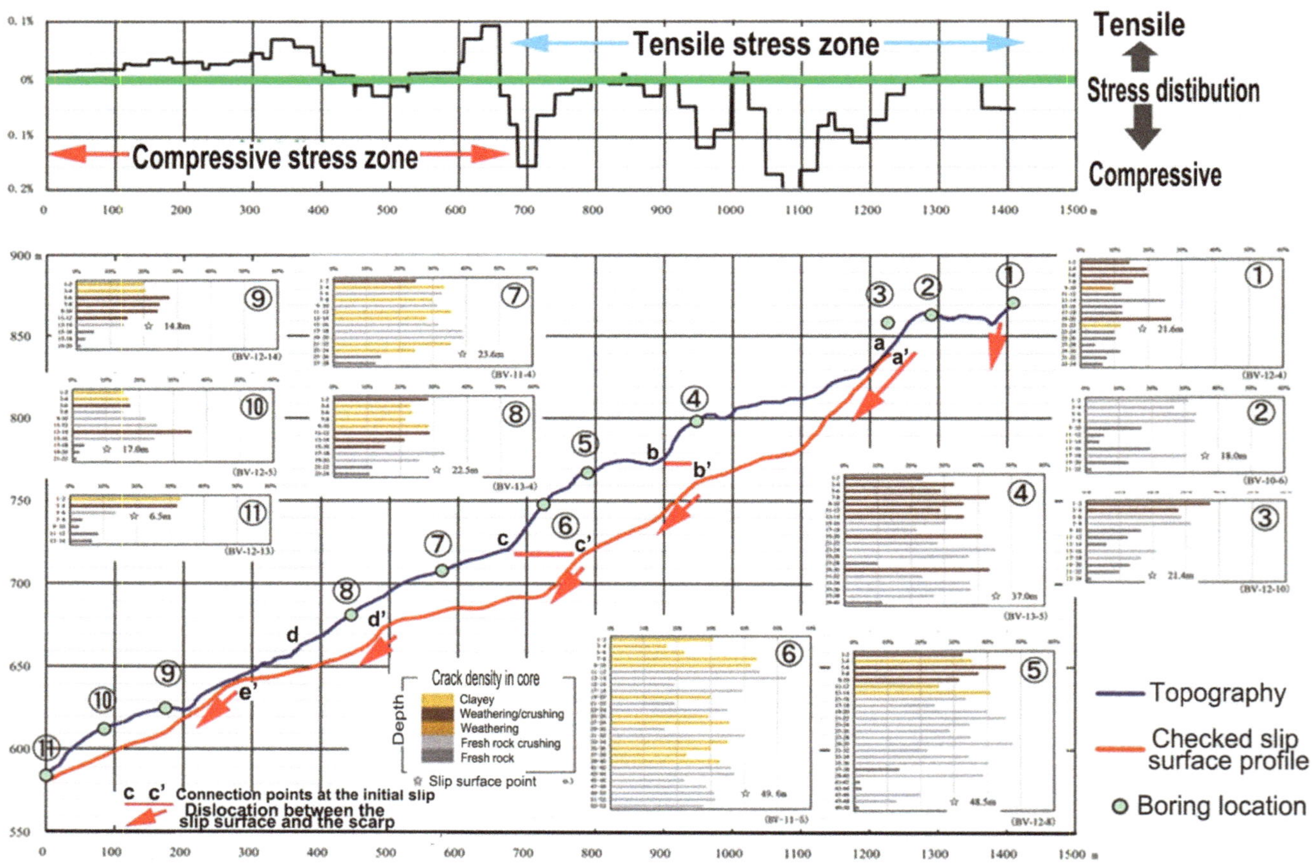

Fig. 5 Cross profile of the landslide surface, slip surface, crack density graphs and drilling core materials, and the distribution of stress at surface at the Okamizawa landslide area, Akita Prefecture, Northeastern Japan

Okamizawa Landslide: The Okamizawa Landslide is among the few in Japan which is being investigated in detail. It is a large-scale landslide approximately 1.5 km long and approximately 0.7 km wide at its maximum point. It is currently under the jurisdiction of the Forestry Agency. Its geology is hard shale and sandstone from the middle Miocene age, interbeded with tuffs and other alternating sedimentary rocks. The geological bedding and the homoclinal structures both are 10–25° westerly dipping. Overall, it has the characteristics of a bedding plane dip-slip landslide. With the landslide body as the focus, a ground surface strain distribution map was developed based on maps containing detailed contours, land form classifications, surface crack distributions, deduced slip surface contours, surface dislocations and vector mapping. Drilling core photography was used for data conversion of crack density and lithofacies texture. These data were used to elucidate the landslide movement and topography handling.

Landslide Movements Reflected in Surface Structures
Some of the distribution map data described above is depicted in Fig. 4. Using the distribution of features such as the developing cracks from the landslide body and the main scarp, the following characteristics can be understood. In the upper portion of the Okamizawa Landslide, the large main scarp (A) shows a horseshoe shape. Huge slabs behind it have cracks formed in the direction of movement (B). Within the downward slope of the main scarp, minor scarps C and D can be found. Further down in Zone E zone, neither crack nor minor topography of the toe of the landslide body has scarps. The surface instead shows undulating and hummocky features.

Some of the distribution map data described above is depicted in Fig. 4. Using the distribution of features such as the developing cracks from the landslide body and the main scarp, the following characteristics can be understood. In the upper portion of the Okamizawa Landslide, the large main scarp (a) shows a horseshoe shape. Huge slabs behind it have cracks formed in the direction of movement (b). Within the downward slope of the main scarp, minor scarps C and D can be found. Further down in Zone E zone, neither crack nor minor topography of the toe of the landslide body has scarps. The surface instead shows undulating and hummocky features.

The undulations gradually become smaller. At the lower part, a gentle slope (F) is reached toward the mountain stream. The difference in height between the landslide body toe and the mountain stream is small. The toe area has no noticeable small landslides.

Based on this micro-topography, the landslide direction can be predicted as described below. In the initial stage, it takes the form of a bedding plane dip-slip landslide and rock slide (B) as suggested by the large, parallel open cracks behind main scarp A. The main scarp with its clear horseshoe shape and the cracks in front of it (A), show a large-scale slump type slip, which represents the sliding of a full-blown landslide. Several hundred meters in front of A is a large scarp (C). It is not immediately clearly understood why it is there. Moreover, in areas such as near the toe, which has no cracks (F), numerous questions still exist about what movements may have caused these results. Figure 5 was created based on drilling core sampling shown in the crack classification diagram. The cross-profile figure shows the ground surface, the estimated slip surface, and the strain distribution along the ground surface. Points a, b, and c are located on the ground surface, whereas a′, b′, c′, d′, and e′ are on the slip surface with a steeper slope. The scarp with the horseshoe shape denoted as a in Fig. 5 continues from a’ on the steep slope of the slip surface. In the case of a slump type slip, the main scarp is always exposed by the slip surface as shown in Fig. 2. The cross-section clearly exhibits the slump type slip.

If b in the steep sloping section is a scarp, then this extension should also appear in Fig. 5, but there is no continuation with the slip surface in the cross-section. However, if c in the steep part is regarded as the main scarp, then there is no agreement between the steep part and the slip surface at this location. At this point, the findings suggest that this landslide is retrogressive starting from the slope bottom. If this is the case, then although steep scarps b and c are already known to be connected to the steep part of the slip surface, they were pushed down the slope by a new landslide body formed above. However, because the slip surface has a downward-sloping structure, it remained in the same position. The steep parts of the two cross-sections are thought to have been formed by the resulting dislocation and slippage. These deviations can be regarded as shown by the movement history of the intermittent retrogressive landslides. In other words, the central portion of the current landslide is A. Through movements of A, this section is pushed into C before they displace together. Scarp B is the main scarp of an earlier landslide. It formed a scarp and slip surface series with the underground steep part (b) of the slip surface. However, with sliding of A, the foot of scarp C and slip surface b’ were pushed forward approximately 60 m. Similarly, the distance between c and slip surface c’ is even larger at approximately 120 m. In addition, the distance is approximately 200 m for the steep surface up to d. With reference to the strain distribution, the cracks show a tensional trend. The lower half of the landslide body is in compression. Two places exist in which tension resulted in a particularly large strain corresponding to b and c. This result suggests that even though A moves close to C and D, each is moving autonomously. Based on this finding, C and D are regarded as functioning as a single zone corresponding to the slip surface.

Numerous drilling investigations have been conducted for the Okamizawa landslide. The materials and properties of the landslide body have been investigated comprehensively. Figure 5 confirms changes in the surface that have been observed in the landslide body material structure, as discussed above. Using eleven core photographs arranged along the topographic cross-section, crack density imaging and material properties were used to generate data (Fig. 51–11). Here, crack density is found for all cracks 1 m or longer created by crushing forces and appearing in the core images. Furthermore, the core lithological facies were categorized as fresh rock, fresh sheared fragments, weathered rock, weathered and crushed rock, and clay soil. Based on these results, the landslide body composition changed from fresh rock to weathered and crushed rock in a B-A-C-D order. Corresponding changes in crack density were remarkable. However, the cracks were found to terminate they reached the clay soil. These lithological facies also correspond to strain rate. The facts for the Okamizawa Landslide can be summarized as follows: (1) Repeated landslides of approximately 150 m long are combined to create the landslide at its current scale. These can be traced back over time. (2) Movement of the upper area is in direct agreement to movement of the lower area because of pressure from above. (3) Movement which is mirrored around the slip surface itself is striking in the vicinity of D, distinct cracks, lithological facies, and strain showing tendencies toward tension. (5) The lower half of the landslide body is free of cracks because of the clay soil. The form of the clay soil changes to an overall cleaving extending from the slip surface to the landslide body. (6) Moving from the lithological facies layers to the clay soil layer of the landslide requires three to four repetitions of large movements. (7) Through the process described in (6), the main scarp seems to disappear gradually.

Excellent correspondence can be found between the landslide's external appearance and internal contents. By observing the micro-topography, the movement orientation of the landslide was understood.

Landslide Recognition by Mapping

Surface movements in landslides are always the result of topography. However, even though this phenomenon is dramatic, technology and tools developed to date have been insufficient to understand the surface movements. As a result, it has remained unclear what position it occupies in relation to a landslide event. The notion of grasping potential risk of landslides was not explored thoroughly. It cannot be said that that curated landslide distribution maps have not existed until now. However, such maps mostly consist merely of dots placed in the locations where landslide disasters have occurred. There are no feature visual representations of the actual landslide topography. Mapping locations at risk for landslides has not even been considered. Perhaps the main reason is that viewing topography in three dimensions was not deemed to be necessary.

Numerous reports have described the importance of mapping landslide phenomena worldwide. Many reports on such landslide mapping will also be presented as part of this lecture. Landslide mapping opens up some avenues for expansion. The reports are classifiable into several categories.

1. Reports linked directly to achieving "site prediction," one of three basic propositions of disaster prevention research. For disaster prevention, the two other basic propositions are "time prediction" and "size prediction." However, because of the improvements to surface sensing technology and the progression of research related to landslide phenomena and the resulting topography after they occur over the last 50 years, almost all the tools and technology necessary for site prediction are still now available.
2. Predicting locations is now possible. Therefore, information can be shared with potentially affected residents nearby. Most municipalities in Japan produce hazard maps. Data related to the potential for landslides and other slope disasters are prepared nationally. An example is the Landslide Distribution Maps prepared by the National Research Institute for Earth Science and Disaster Resilience (NIED).
3. Nevertheless, it is also true that these slope-related disasters occur so frequently that counting their numbers is difficult. However, even with advances in technology and information preparation combined with increased public reporting, most people living in disasters areas have continued say, "I did not know." Residents seem to have insufficient understanding of the importance of achieving "a framework for evacuation from dangerous areas" through information disclosure and visualization.

Actual Mapping

Landslide Topographic Area Mapping

In Japan, the explanation of topography classifications within the Fundamental Land Classification Survey report issued in 1967 state slope stability items such as landslide topography, knick point and varied point of slope in the other classification items (Economic Planning Agency Japan 1967). Aerial photographs taken by the US military in 1946–1947 were used for this mapping. At that time, understanding of landslide phenomena was insufficient with only simple markings of landslide disaster sites. Hatano (1972) first interpreted and mapped out actual landslide distribution slopes over a wide area.

Landslide topography was ascertained from interpretation of aerial photography. Landslide topography distribution maps were released by Chida et al. (1971), Miyagi (1979), and Shimizu (1984) in various areas across Japan. These studies presented characteristic landslide topographies of areas where landslide phenomena caused movements. In addition, most landslide disasters which actually occurred were shown to occur within this landslide topography. At disaster prevention research laboratories in 1982–2015, great effort was spent creating and issuing landslide topography distribution maps for all of Japan. A seamless nationwide landslide topography database was completed with these data.

Figure 6 depicts a sample map of river terraces in small basin areas between mountains. The occurrence of landslide topography movements and when these basins will be destroyed have been understood using aerial photography (Miyagi et al. 2004). Landslides occur on a yearly basis in many areas in Japan. Therefore, Hatano conducted extensive interpretation of aerial photography in different areas. Through his interpretations, he determined that the landslide mechanism is one of the key processes for mountain slope developments in humid areas.

Figure 7 presents a aerial photograph and map of an area where landslides frequently occur. In addition, Fig. 8 portrays a topographic distribution map that we prepared using aerial photography from 1978. Topography such as the lava flows of the volcanic slopes in the northwest of the map and the basin alluvial fans in the eastern area are spread throughout in this map. Although the volcanic slopes dissect rivers to create valleys, numerous landslide topography sites are distributed along these valleys heading toward the main rivers. Land material movements caused by landslides are apparent throughout. This classification map includes landslide movement patterns and vulnerability classification evaluation information. This information enables an understanding of several aspects such as places where landslide-type slope disasters might occur, the scale at which they could occur, and the types of landslides which can be currently measured. The characteristics of neighboring areas which might suffer from damage are compiled in a visual format. This information is used for multiple applications such as river basin management and disaster prevention measures in local areas as well as for drafting evacuation plans. This accumulated information is becoming a fundamental resource for preventing disasters and reducing their damage.

Evaluating Landslide Reactivation Potential

Among natural slopes, there are numerous topographical regions that originated from landslides. Most landslide movements unfolding in front of our eyes represent the

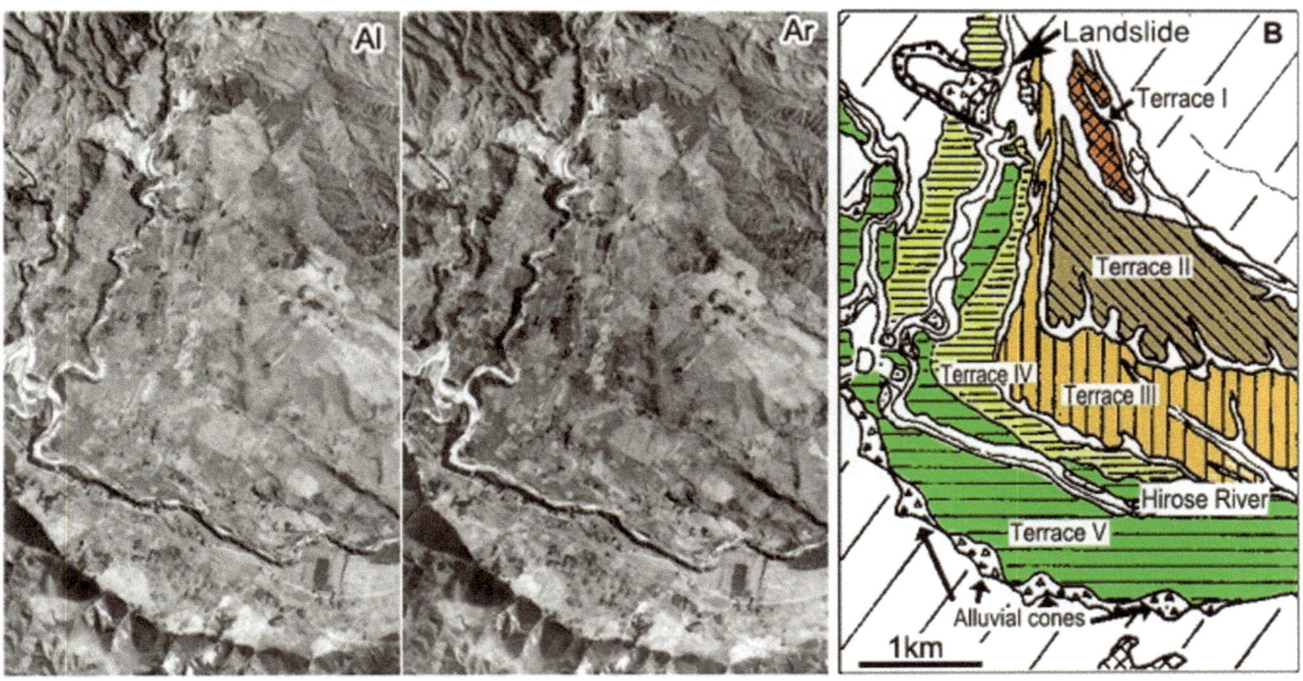

Fig. 6 Sample of the stereo pair aerial photographs (Al and Ar); for 3D photo interpretation. **b** The result as the landform classification map (Air photograph taken by US Army at 1947, 1/40,000, Monochrome) (B is mapped by Ohuchi 1973, and modifyed by Miyagi et al. 2004)

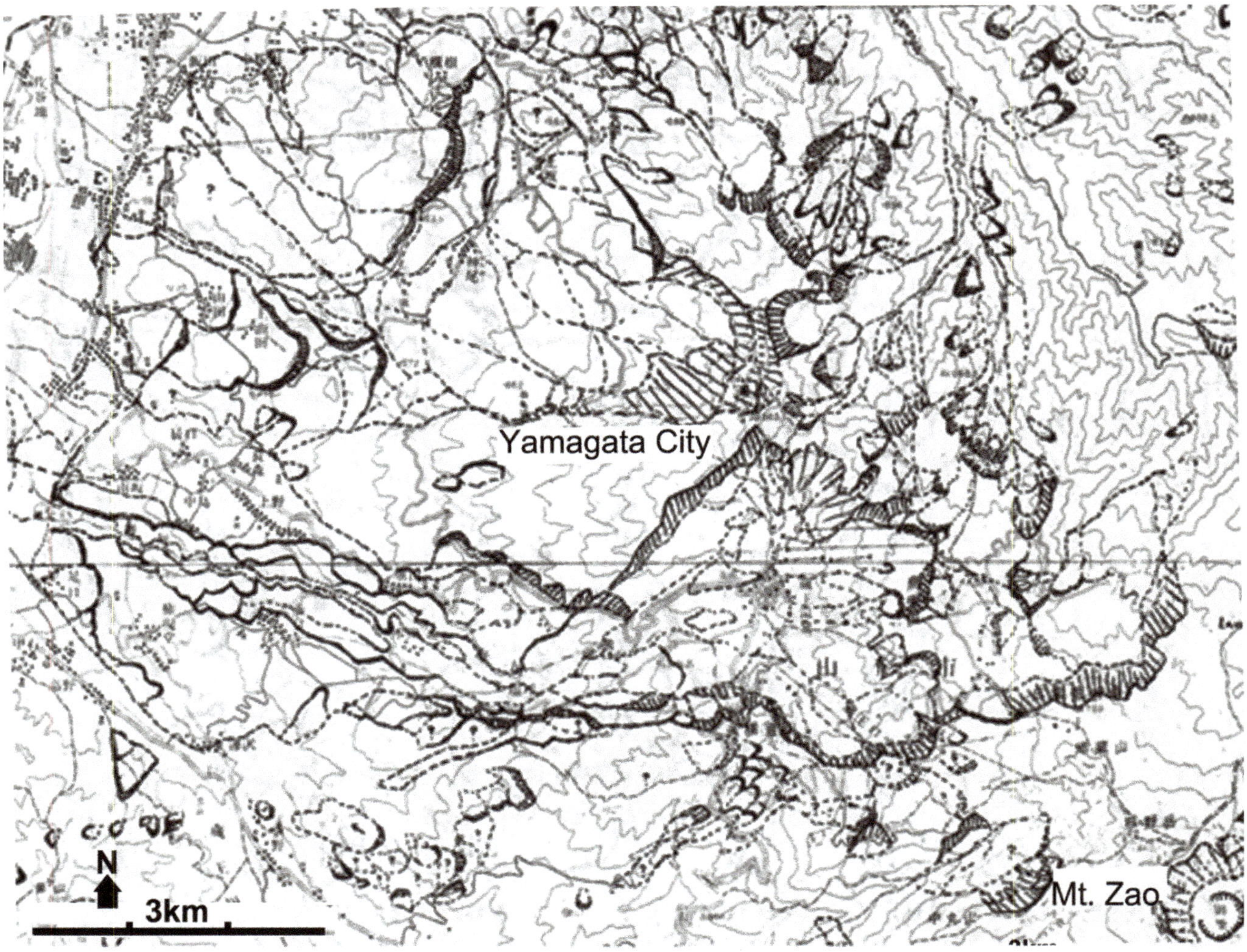

Fig. 7 First wider area landslide topography distribution map (Based on aerial photograph interpretation (API), Yamagata prefecture Japan, 1/50,000, drawn by Hatano 1972)

reactivation of some existing feature of landslide topography. Accordingly, the need to evaluate which landslide topographies have high probabilities for reactivation is clear. As shown in Fig. 3, the possibility for landslide reactivation is higher in slope areas that have been destroyed by past landslides rather than those which have not. Construction works such as dams and roads can exaggerate the occurrence of landslide disasters in the greatly disturbed soil in these areas. As shown in Figs. 4 and 5, it is true that landslide movements and material composition correspond to the topography of the applicable area. Accordingly, methods have been developed to evaluate effectively the possibility of landslide movement based on topography (Miyagi et al. 2004). By particularly addressing the distribution of micro-topographic elements which make up the landslide topography, an attempt to evaluate the possibility of landslide movement as suggested by these elements was conducted using AHP. Figure 9 was created to convert each confirmed landslide topography unit into target landslide units to evaluate the possibility of their reactivation. By marking what and how was observed, this sheet is used for such evaluations. This was further simplified by the Tohoku Branch of the Japan Landslide Society (Miyagi et al. 2004) recreated to show a stronger visual component. The application of AHP should be considered from the perspective of quantitative evaluations of landslide areas where countermeasures have been implemented throughout Japan (such as Fig. 10). Figure 11 portrays the landslide topography distribution for three levels of reactivation possibility levels.

Landslide Risk for Areas with Artificial Land Reclamation Areas

In recent years, large-scale land development for roads, railways, and main transportation lines is being undertaken all over the world. This artificial modification of topography generates land reclamation areas on a large-scale. Land

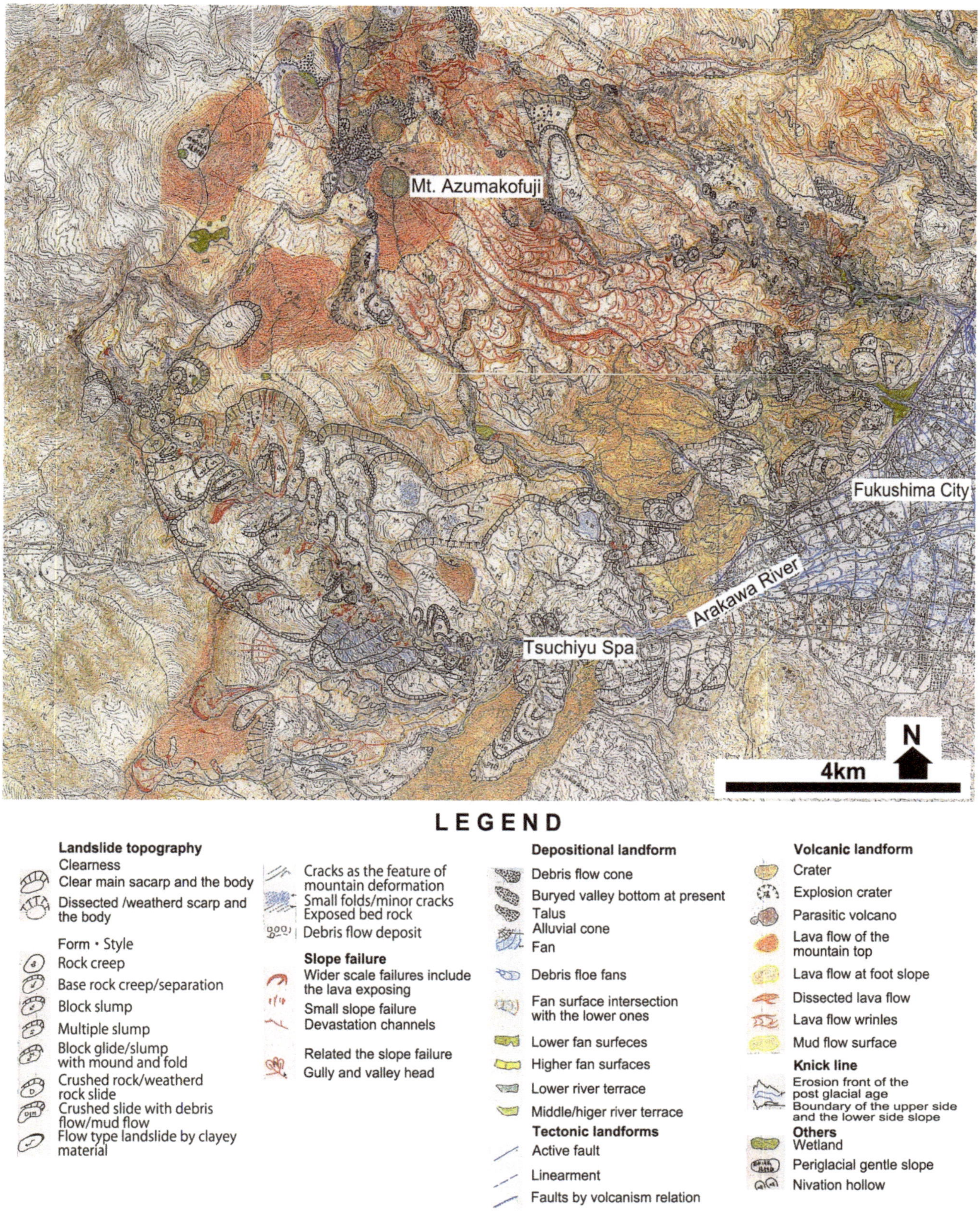

Fig. 8 Typical landform classification map in high landslide concentration area Northwest of Fukushima City, in Fukushima prefecture, Japan. Style and microfeatures of landslide topography were identified using high-resolution aerial photographs (Based on photographs taken in 1978, 1/15,000, full color, drawn by Miyagi 1995)

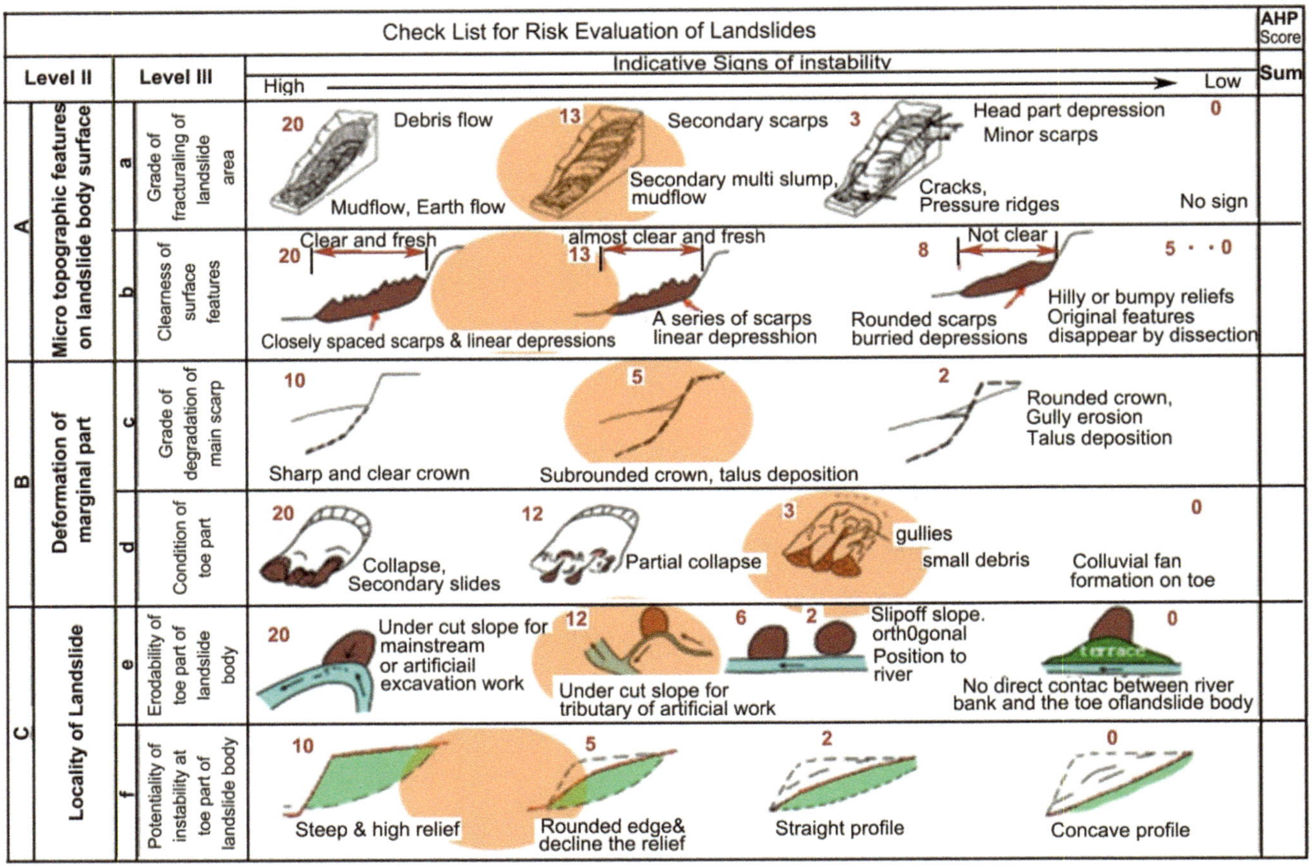

Fig. 9 Inspection sheet for AHP score evaluation for landslide risk (Hamasaki, 2106 modified from Miyagi et al. 2004)

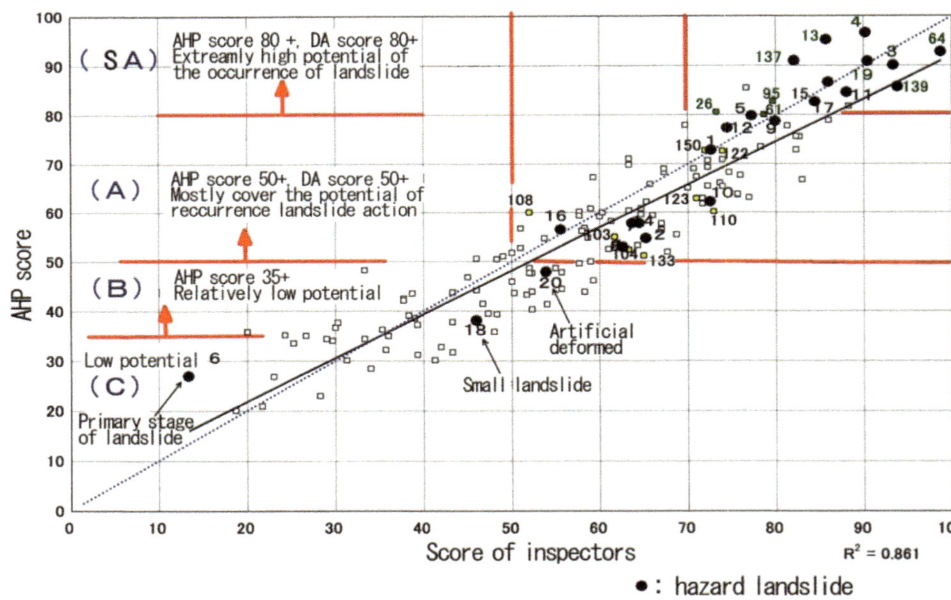

Fig. 10 Validity of the AHP model by inspector evaluation (Miyagi et al. 2004)

formerly made up of slope and valley areas is filled by artificial soils until it is nearly flat. However, this process disturbs the natural soil condition severely, which can be expected to reduce its foundation strength drastically. In these artificially modified areas, although changes in the topography are not readily noticed to the naked eye, the conditions can be determined clearly by comparing topographic data before and after the modifications. Maps that

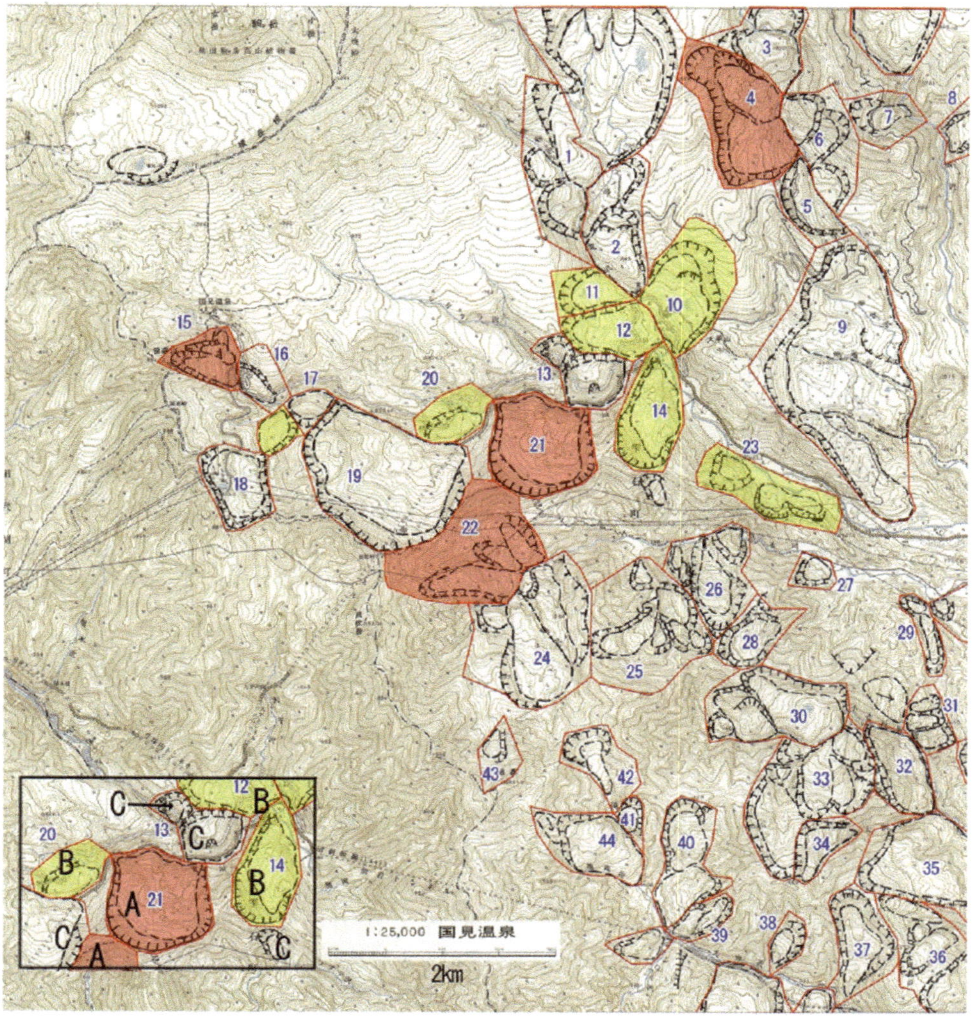

Fig. 11 Landslide topography map with three landslide hazard reactivation levels (Miyagi et al. 2004). Explanation of the simbols of map: blue color figure: number of inspected landslide unit. A/B/C: reactivation risk levels by AHP score refered in Fig. 10

organize and present this information are being prepared. According to verification data from Japan, land reclamation areas almost never have "progressive consolidation and increased foundation strength," although time has passed since these modifications were made (Asada 2004). Land reclamation areas have high potential risk for landslide reactivation disasters. Hamasaki et al. (2010) used a trial spherical slip model to calculate safety ratios for such land reclamation areas. Earthquakes occur frequently in Japan. Strong earthquakes often damage the slopes of land reclamation areas. These slopes can also move during typhoons, during heavy rains of the rainy season, and during heavy snow melt season also. Figure 12 presents one evaluation example. In hilly areas, cut and fill land modifications have been extensively conducted (Fig. 12a) for making the artificial flat areas (Fig. 12b) for residential and public use. In such cases, although the land appears to be flat, the modifications have produced major differences in the soil properties. The results of safety factor calculations for such land reclamation areas are shown in Fig. 12c. At this site, small landslides occurred in three places after the construction was completed. Two landslides were triggered by the Great East Japan Earthquake in 2011. The third was caused by some heavy rainfall in the area.

Mapping Landslide Potential

As discussed above, unstable slopes are formed all over Japan by landslide action. These areas are showing landslide topography through the "Autonomous destruction process," as discussed in Sect. 2. However, not all slopes are always of this type. It is clear that small-scale surface and shallow landslides can occur and disperse in mere moments, but cannot be traced accurately. Information that is useful to evaluate the potential for landslide occurrence on the entire slope must be prepared. A landslide susceptibility map mean exactly this. The slopes of target areas are arranged in a grid with an appropriate mesh size. The form and related factors are quantified and displayed as indicators. These data are

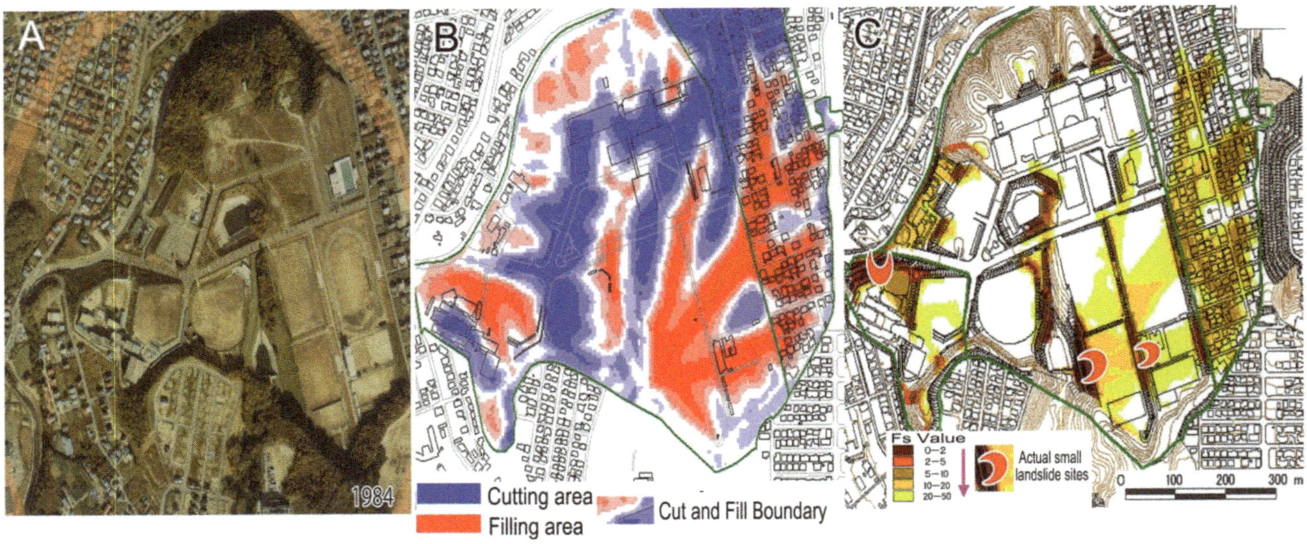

Fig. 12 Sample of simulated landslide potential evaluation and the relations of the artificial land deformation area: **a** just before the land deformation (1988); **b** current condition of the showing cutting and filling area; **c** calculation result and the zones of small landslide hazards

compared to those of reports from actual landslide disasters. Modeling is used for optimization to improve accuracy, and contribute to the prediction of future disasters. Whether risky topography exists or not, modeling is based on the notion of topography as a continuum. This method is becoming increasingly advanced worldwide. Expressing susceptibility levels through the classification of each grid is the standard method. Figure 13 contains an example of a slope disaster susceptibility classification along a national route in Vietnam. On this map, 81% of the zone is categorized as having a high or very high risk of slope disaster (Fig. 13).

New Approach for Landslide Susceptibility Mapping

The following is an example of the use of AHP method buffer and blunder probability analyses for the predictive evaluation of the possibility of landslides and collapse occurring over a wide area (Fig. 14). The process has three stages: (1) implement buffer movement analysis for factor extraction and evaluation of AHP slope movement prediction; (2) apply blunder probability analysis to find the applicability and accuracy of the collapse and landslide disaster model created based on results of Step (1); and (3) establish the optimal target variable.

The AHP is conducted using layering and weighting of factors based on factor analysis and experimental knowledge of a high-level engineer. Using this approach, the AHP calculation total score is the slope movement evaluation score for topography factors applied to each grid in the mesh area. Buffer movement analysis is a technique for skipping through a circular section of the GIS surface with radius R at a certain interval and collecting general data related to the landslide and collapse factors. The AHP calculations from the extracted factor values are applied to the mesh area (Fig. 14a). Here, R and the skip interval are modified on a generated scale based on the object variable. This lecture examines the 2008 Iwate–Miyagi Nairiku earthquake, which caused over 3,500 collapses (slope failures) and landslides.

As shown in Fig. 14b, c, landslides and slope failures were set respectively to $R = 250$ m with a skip interval = 150 m and $R = 100$ m with a skip interval = 50 m, based on the average size of the area affected by each type of phenomenon.

The blunder probability analysis method finds the movement prediction evaluation model suitability and its success or failure through comparison with actual landslide and collapse occurrences. Areas with high AHP evaluation scores where landslides or collapses occurred are regarded as having high occurrence probability, whereas those with few actual cases have low AHP evaluation scores. Lower probability duplication quantity indicates a model with high applicability.

The Monte Carlo Method is used for analysis. The determination items acquired from AHP are analyzed using the Blunder Probability Method, while modifying their weights as the process progresses. Through repeated analysis, the model gives the most accurate understanding of the actual scenario. Results of these analyses showed favorable compatibility with the actual conditions at the sites of these earthquake-induced landslides and collapses.

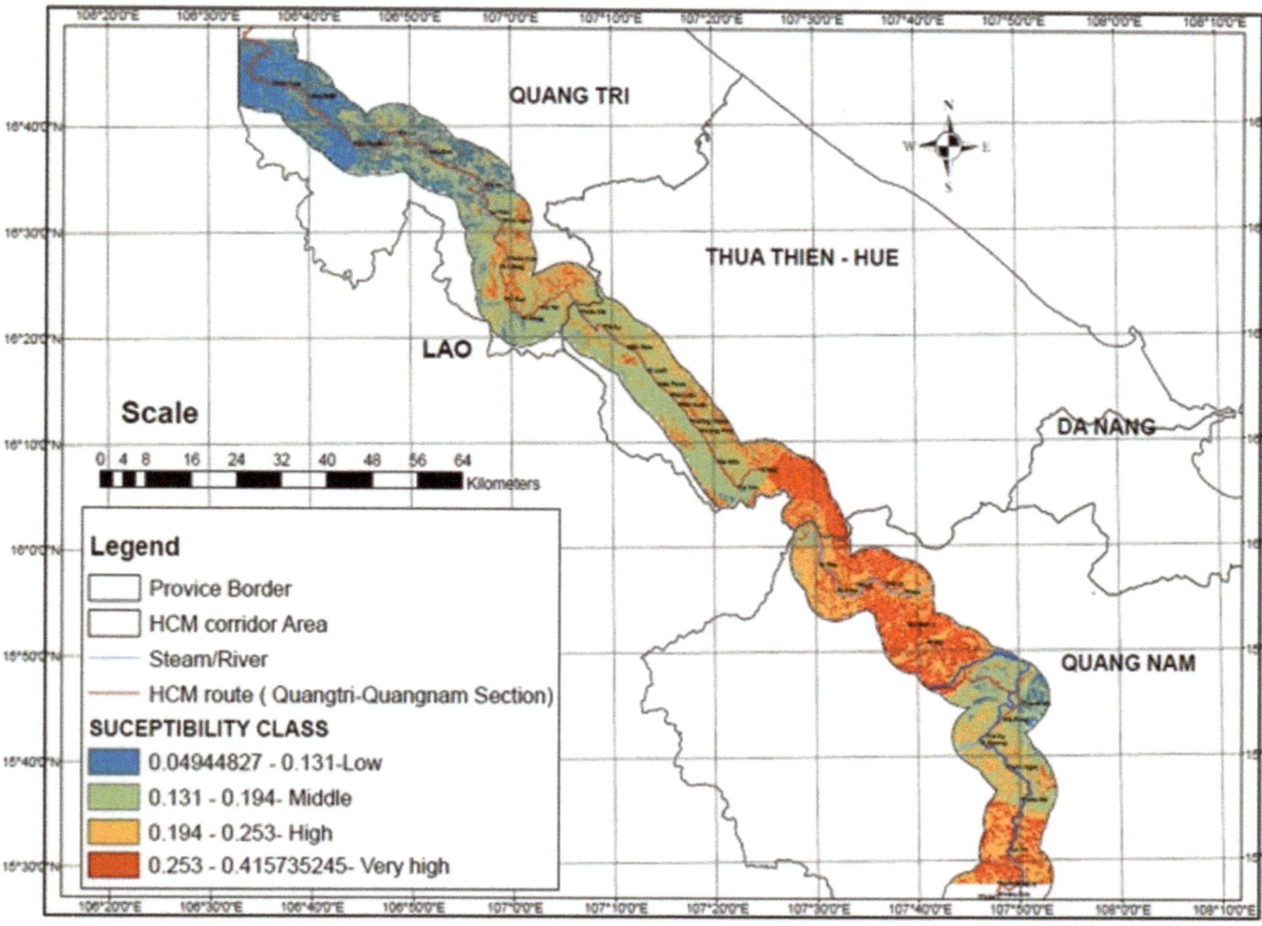

Fig. 13 Susceptibility map of small scale landslides along Ho Chi Minh Route, Central Vietnam (Tien et al. 2016)

Advanced Information Mapping for Landslide Countermeasures and Management

In recent years, sensing tools are becoming increasingly advanced. Innovation of mapping technology is also progressing.

The mapping concept remains unchanged. The maps mark locations at risk in advance enabling countermeasures to be implemented. Although aerial photographic interpretation has been conducted for the past 50 years, there are many recent changes. These changes are a result of the following: (1) The creation of three-dimensional coordinates has simplified in case of using UAV data and application soft. As a result, anyone can clearly visualize both the specific details and the overall conditions of landslides. (2) Data collection is drastically cheaper than in the past. For example, with visual images such as those available through Google Earth, the entire world can be observed as if it were in the palm of one's hand. With NASA's SRTM or AW3D (World Digital 3D), 30 m DEM (Digital Elevation Model) is useful and free of charge with a multi-flexible scale and orthographic imaging. Even if the grid is used at 5 m or finer, information can be secured at an extremely low price. (3) Stakeholders such as residents and government officers, along with technicians, can easily looked the sites and, process the data, and convert it into a mapping format. In this way, management points of interest can be confirmed. Combining UAVs (Drones) with three dimensional data processing techniques has also undergone remarkable development in the recent years, including decreases in costs. Consequently, the parties burdened with disaster risks and those proposing disaster prevention measures can mutually communicate for implementing these measures to reduce damage in the event of a disaster. In this modern era, mapping information is opening up new avenues of application (Fig. 15). The following section shows several examples.

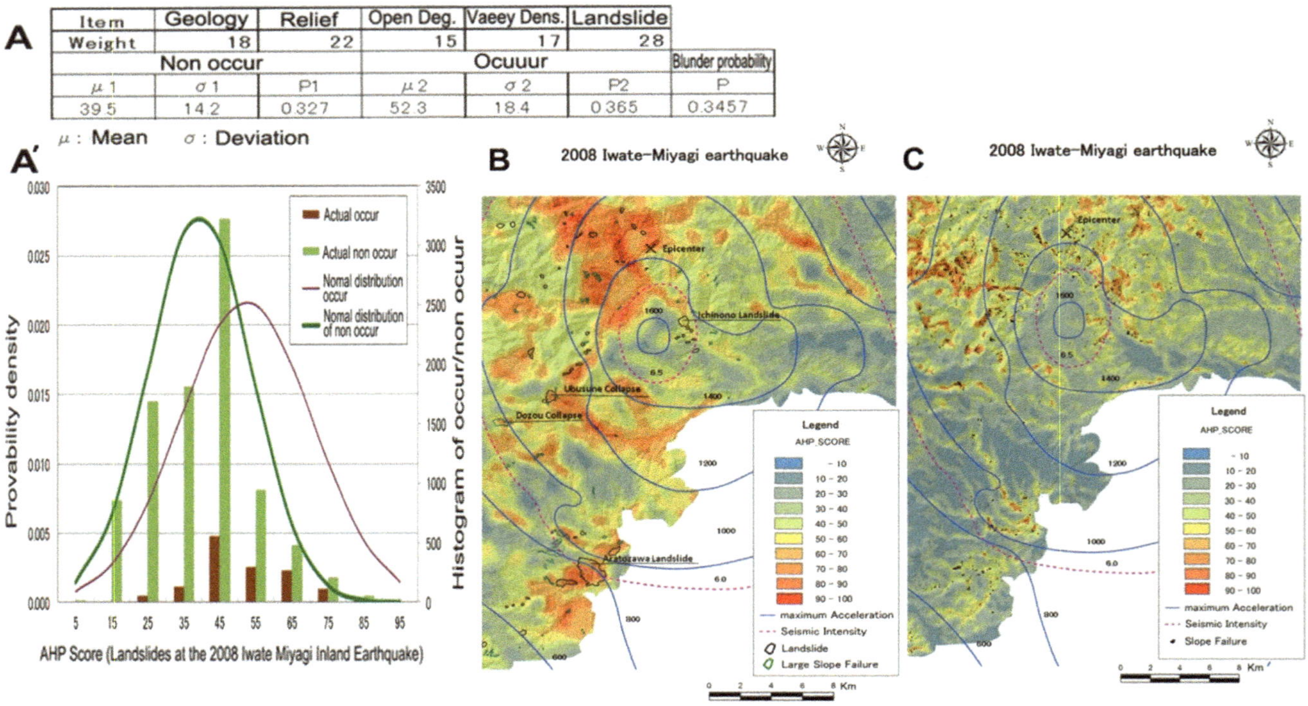

A

Item	Geology	Relief	Open Deg.	Vaeey Dens.	Landslide	
Weight	18	22	15	17	28	
Non occur			Ocuuur			Blunder probability
μ 1	σ 1	P1	μ 2	σ 2	P2	P
39.5	14.2	0.327	52.3	18.4	0.365	0.3457

μ : Mean σ : Deviation

Fig. 14 Susceptibility map from buffer analysis and probability analysis using AHP method for landslide/slope failure susceptibility compared with the 2008 Iwate–Miyagi inland earthquake (Hamasaki et al. 2015): **a** Values and A', Distribution results; **b** Susceptibility for large-scale landslide; **c** Susceptbility for collapses such as slope failures

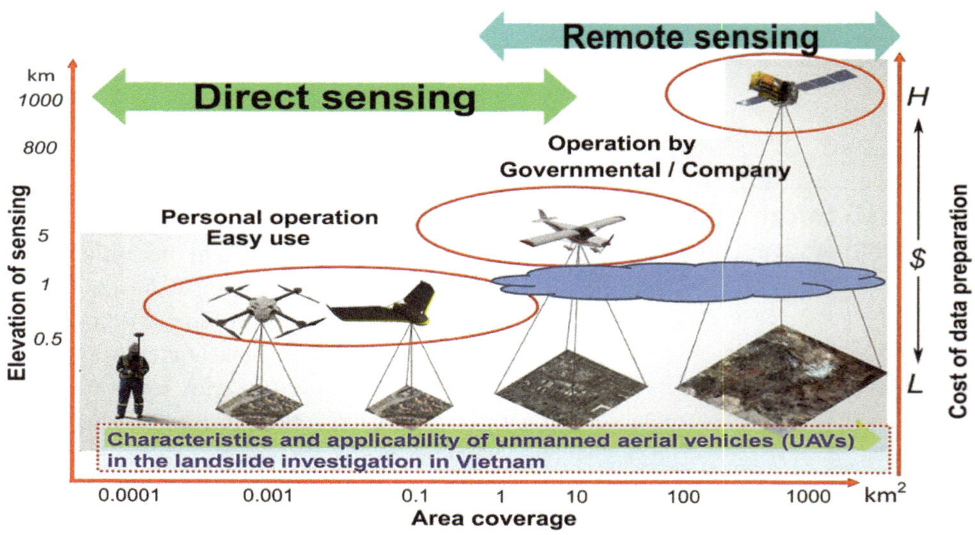

Fig. 15 Variety of sensing tools for various purposes (Modified after Thanh et al. 2020)

Landslide Mapping Through World Digital 3D Mapping (AW3D) and Google Earth

When the "Daichi" Earth surveying satellite was launched, JAXA, NTT data, and RESTEC took 4 million photographs of conditions worldwide. Digital three-dimensional coordinate information based on these images is widely available. This global three-dimensional information itself can already be used at no charge through services such as NASA's SRTM. However, the ability to use it at a low cost in various formats such as 5 and 2.5 m grids in combination with orthographic images. The 5 m grid size is useful for observations roughly equivalent to the 1/20,000 scale aerial photography which was being used up until now. Accordingly, even in areas where securing wide-ranging image information through traditional means such as aerial photography is difficult, it is now possible to gather detailed data from sources such as Google Earth. The free availability of high-resolution bird's-eye-view images for observation can be considered as standard. Through the combination of this information, preliminary exploration of inaccessible areas has also become possible. Figure 16 shows the landslide topography distribution conditions for the 10 m interval contour topography map of Paektu/Changbai Mountain, which lies on the border of China and North Korea. Within the decipherable scope, a total of 41 landslide locations (Ls 01–41) were recognized. The middle section of Fig. 16 shows Google Earth images for three especially noteworthy locations of landslides (Fig. 16 Right). In rural areas, position information is available through landslide topography locations extracted from these 41 locations. Observation of these points shows that Ls 09a and 09b in the right is a large-scale landslide that branched off from the left bank of the river which flows from "Heaven Lake" crater of Paektu/Changbai Mountain. Accordingly, it can be inferred that soil and sand were pushed out into the mountain stream as a debris flow, causing the flow path to curve east. Ls20a and 20b in the right is probably a landslide originating from the slope outside of the crater. Based on the contour lines, the scope of this landslide has 300 m wide, 700 m long, and 30 m deep. When combined with Google Earth data, the movement seems to be reminiscent of a slow-moving debris avalanche. The Paektu/Changbai Mountain zone is in a periglacial environment. Therefore, this landslide could be a solifluction agent. The only peninsula jutting out into Heaven Lake is Ls23a and 23b in the right. Signs of a landslide are apparent. The peninsula itself does not appear having undergone marked changes resulting from the landslide movement. However, closer inspection reveals numerous cracks which have developed breaking through the ridge, such as groove-like depression. Observation from a long distance reveals distinct oxide contamination on the slope related to this depression. This result suggests the possibility of a future landslide occurrence.

The AW3D has a level of detail almost equivalent to a 1/20,000 scale aerial photograph. In rural areas, this information is available only through DSM (Digital Surface Model). This lack of availability has been criticized. "Since it is not DEM, this is not an expression of topography itself." As one might expect, this criticism reveals a lack of understanding of the objective.

Aerial photographs are analog images. Interpreting topographic information from images with vegetation coverage has been done ever since aerial imaging first came into use as an information collection method. Large-scale landslide topography is apparent along

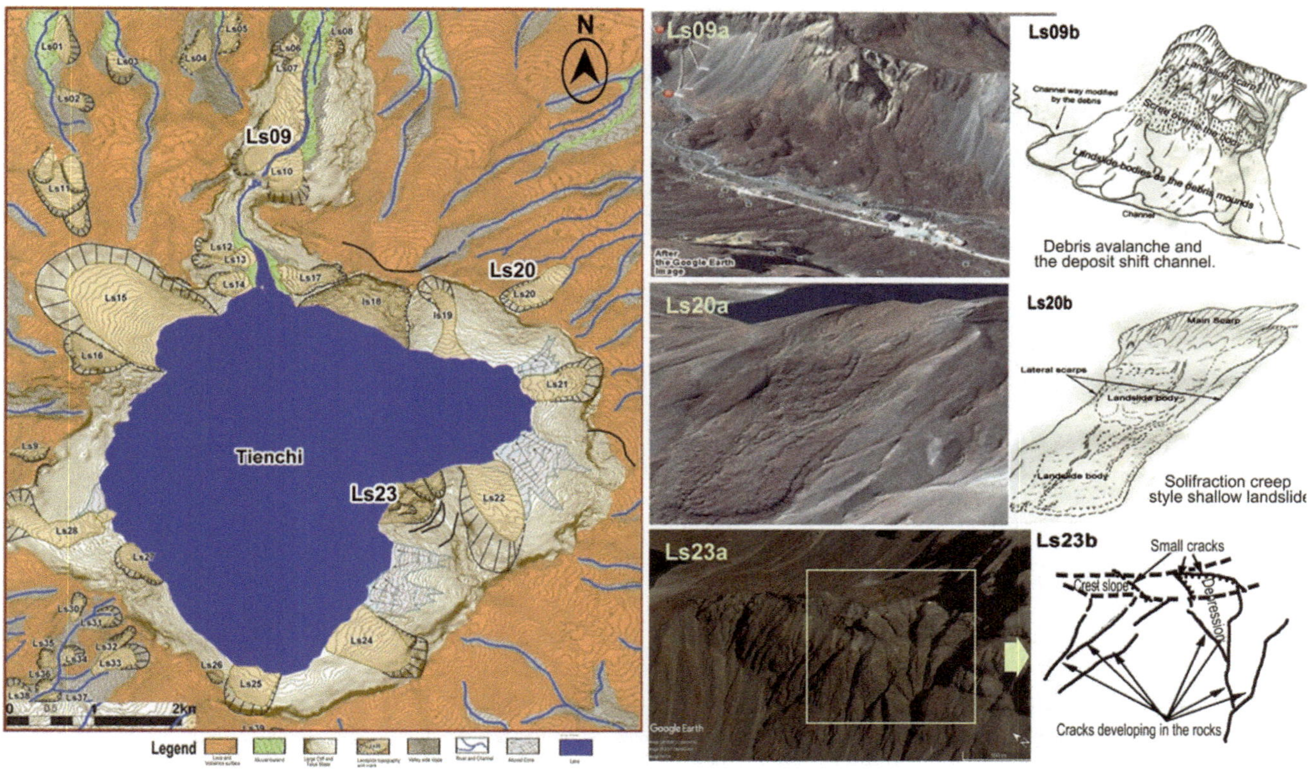

Fig. 16 Landslide topography map by AW3D 5 m DSM around Mt. Chanbai, Tienchi Lake at the China—North Korea border. Typical landslide samples from Google images (Miyagi 2017)

Fig. 17 Sample of landslide topography map by AW3D 5 m DSM along the National Route No.7 central Vietnam: **a**, overview of the landslide; **b**, B1 is the grand view from P1 point. B2 is the close up photographs of the landslide site and the rock from P2 (Dung et al. 2016)

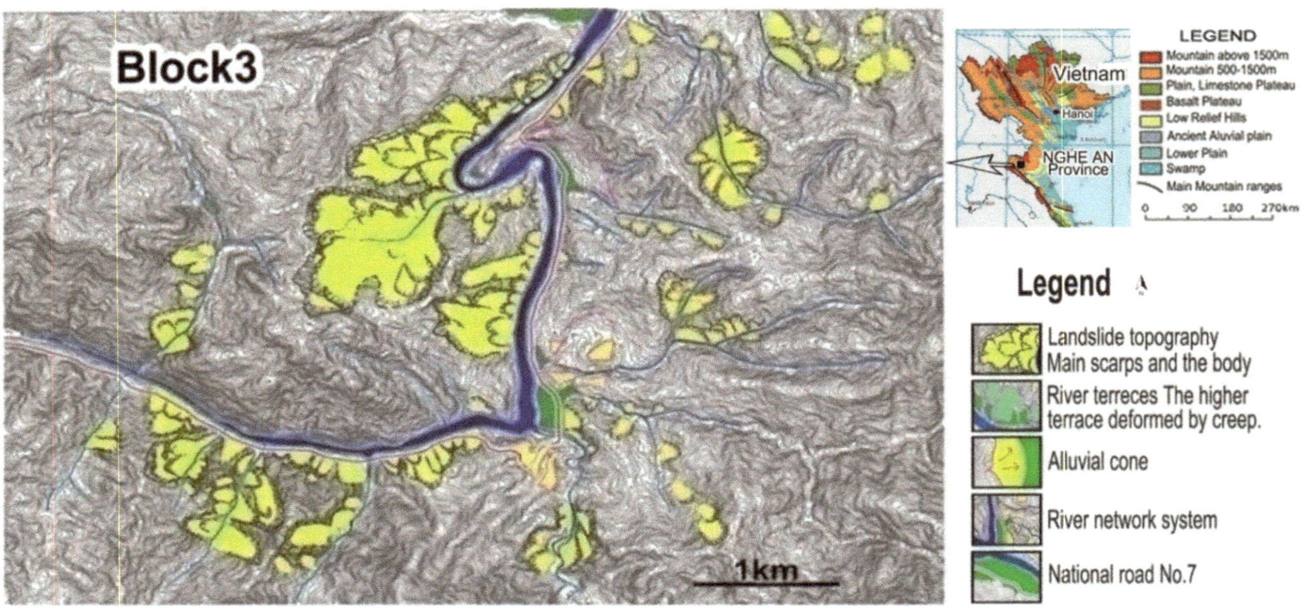

Fig. 18 Sample of large scale landslide topography distribution map by AW3D 5 m DSM along National Route No. 7 Central Vietnam (Dung et al. 2016)

Vietnam's National Highway 7. It can be interpreted to have separated into smaller-scale landslides inside as shown in Fig. 17. Field reconnaissance has been a powerful tool in this case. Figure 18 portrays a partial interpretation of a 70 km stretch of mountainous area along Vietnam's National Highway 7. This interpretation reflects 1,000 small and large landslide topography locations.

Landslide Topography Identification by Airborne Laser Measurement Data

During the past 15 years, the development of airborne laser measurement equipment for acquisition of topographical information has been active leading to useful technology in the field. Unome (2020) describes characteristics of the technique, as explained hereinafter. Airborne laser measurements use a dedicated system mounted to an aircraft. The position and height of the reflection off of the surface are used to generate precise three-dimensional measurement data. Information can be acquired not only for buildings and trees, but also for their underlying ground surfaces. Measurement can be done in areas such as underneath trees and in shadows that were difficult to view stereoscopically with traditional aerial photography. Through advances in data processing technology over the last 20 years, data accuracy and stability have improved tremendously. As a result of the progress in data management nationwide resulting from public measures instituted about 15 years ago, the technology is also being applied to various fields such as forestry sabo, river, urban engineerings, and disaster prevention in the recent years. As a means of understanding the properties of massive landslides, this section introduces the example of the magnitude 7.4 Inland Earthquake which occurred in June 2008 (Fig. 19). Aerial photography in Japan has a 60 year history. These historical images are being used. In the event of a large-scale disaster such as Iwate-Miyagi Nairiku Earthquake, wide-area photographs covering the disaster area were taken immediately. Accordingly, the information included in these aerial photographs used to obtain an understanding before and after disaster is a fundamental operation.

Figure 19a contains an aerial photograph from 1977 that shows the dam that had just been completed. It also shows the trees that were cut down along the forest management boundary. Figure 19b is an aerial photograph taken the day after the earthquake (June 15, 2008). In two places, heavy landslide damage is visible. Other aspects, such as muddy water, inside the dam lake can be confirmed as well. Figure 19c shows a contour map with 2 m intervals created based on the laser data acquired in October 2008. This is digital three-dimensional data. Therefore, it can be combined with precision ground control points to acquire extremely accurate three-dimensional data. The degree of displacement and ground movement can be understood from these points, leading to the acquisition of data values directly linked to disaster prevention proposals. Micro-topography is easily captured and its distribution map as shown in Fig. 19d can also be created. In Japan, the evaluation for inundation depth to assess the risk of flooding in at-risk river areas is progressing throughout the country, an especially wide-ranging

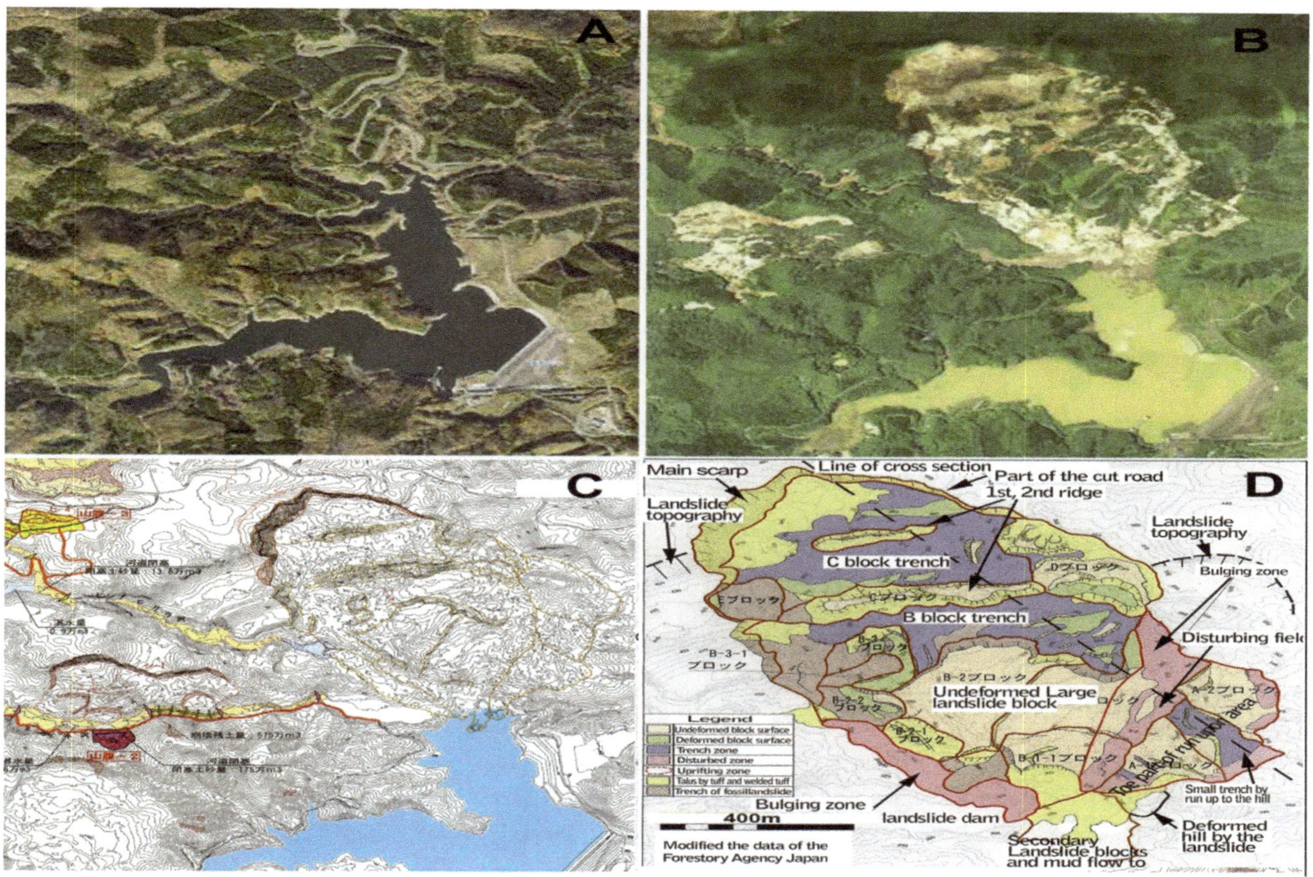

Fig. 19 Massive Aratozawa landslide: **a** Aerial photograph taken in 1977; **b** Aerial photograph taken the day after the earthquake; **c** 2 m contour map established based on laser data collected in Nov. 2008; D, Land classification map of the massive landslide for countermeasures, monitoring, etc.

target area. For landslides, because of it the movement volume and other information can be collected repeatedly as high-precision three-dimensional data, this method is frequently used when repeated data acquisition is necessary.

Combining Landslide Recognition and Mapping for Future Disaster Prevention and Mitigation Projects

During the last half decade, sensing technologies for the acquisition of high-precision topographic information have developed to near perfection. Combining UAV imaging information acquisition with SfM imaging and measurement technology, three-dimensional DSM data or orthographic image acquisition and generation for a specified time, place, and sensitivity can be achieved inexpensively, even by individual researchers.

In Japan, GIS management and publication of geographic maps have been progressing up to the same level. For example, 5 m DEMs covering the some of the country are now available with no charge. As described above, free and low cost information availability has expanded through various avenues such as Google Earth, SRTM, AW3D, and World Imagery. At this point, useful data for preventing and mitigating landslides and other slope disasters in various local areas can be accessed and used freely. The following section will introduce two examples of high-precision data utilization that we attempted.

Repeated UAV and SfM Data Collection for Landslide Dislocation Monitoring

As described above, a large scale landslide was triggered by the 2008 Iwate–Miyagi Nairiku Earthquake. Slope failures and landslides also occurred. Therefore, the possibility of grasping recent trends in three dimensions was explored. As Fig. 20 shows, imaging was conducted from October 2015 through June 2016. UAV (Drones) were used to take photographs at 2 s intervals. Landmarks with high visibility such as roads and plazas with central white lines near them were used for GCP (Ground Control Points). Five lines were set within the photographed area. Two period surface dislocation was confirmed. Measurement line A Phantom 2 and Phantom 3 were used for the data acquisition. A RICOH GR and its 4 K camera is one of the five sets and is superimposed across the cross-section profile clarifying no marked changes in the landslide were observed. However, the toe of the landslide body was confirmed to have been disturbed as suggested by deformed trees along the cross-section. As a result, the accuracy of the data in shaded areas was reduced. Precise changes in the landslide were recognized sufficiently through the modification of the data acquisition target characteristics. This system is highly accurate and its ease of use in terms of both time and expense are important benefits. For applications such as monitoring and construction management, this tool will start a new era. It also has an extensive range of applications for mapping.

The report by Yanagiba (2019) specifically reports activities completed when the 2014 Hiroshima landslide disaster occurred. This report had already identified the possible application of UAV and SfM. At the time the disaster occurred, DSM data were created using SfM processing for aerial photographs from the past (2008). DSM data for the day after the disaster was created using UAV and SfM processing. The differences between the two DSMs (before and after disaster) were analyzed. The erosion depth and sedimentation thickness which resulted from the debris flows were quantified. These operations were completed within a day and a half. The results were then distributed immediately to the people (rescue team, etc.) involved in the search efforts.

For the people searching for the numerous missing persons just after the disaster, these data reported the areas in the sea of mud and debris with especially thick accumulations that were the most important data for their efforts. Here, speed was crucially important. In addition, techniques of multi-view photographic measurement are applicable to aerial photographs for map creation taken in the past. The results can be compared between recent maps and those from the past. Currently, the use of this system is rapidly expanding to the other cases.

Modification of this kind is exactly what is needed. As on-site data from an actual area can also be created by users, we are currently collaborating with local users in areas in Vietnam to create and use disaster risk maps.

Visualization and Mapping of the Massive Landslide Using 5 m DEM

At the Geospatial Information Authority of Japan (GSI) maintenance of 5 m DEM covering the coastal area of the whole country is advanced. Figure 21 is a trial visualization of the topographic features of a large-scale landslide area. The micro-feature classification figure is

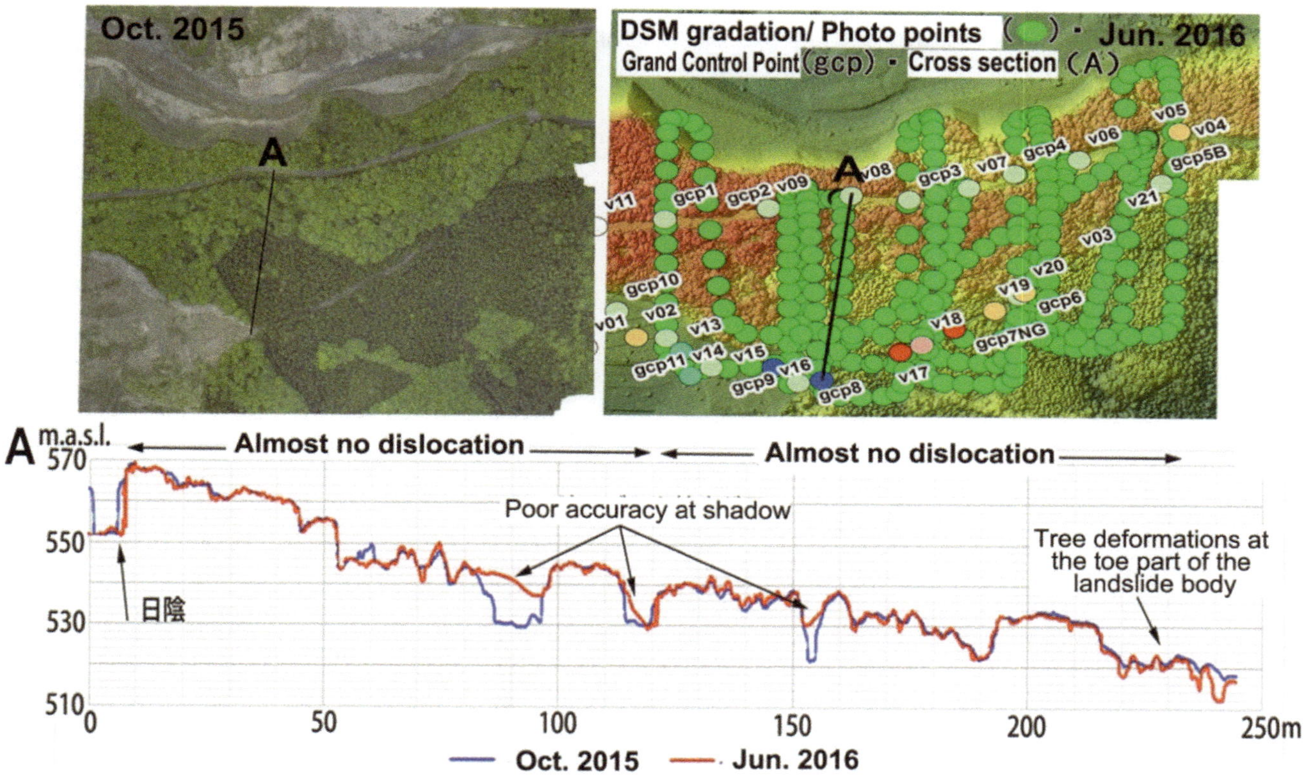

Fig. 20 Monitoring of the landslide deformation based on multi-period data collection by UAV/SfM analysis (Data collection and analysis was carried out by Uchiyama and Miyagi 2015–2016)

deciphered and created based on this. Figures 21a, b are a stereo pair. Both of them are digital data. As shown in Fig. 21c, a contour map can be generated from the 5 m DEM. The data in Figs. 21a–c describe the geographical features of landslides. It is also important for people who see the real objects, but cannot do the deciphering since shade processing is performed to elucidate topographic features intuitively. Although three dimensional aerial photographs are difficult for many people to grasp topographic features, such imaging is increasingly popular.

From the feature classification in Fig. 21d, the gully erosion region of the landslide of two A1 and A2, B1 and B2, and C1 can be identified. A1 A2 A3 is 2 km wide across a single huge landslide extending for 5 km. Here A1 is the main scarp, A2 is the landslide body and movable body front toe part of the body can be set as A3. From shading in the figure, the fine micro-features can be checked. In addition, the main slip scarp (A1) received the severe gully erosion as (C1).

Landslide B1 B2 B3 occurred in order of B3, B-2, and B1. This landslide can be regarded as a large scale slump type retrogressive landslide, such as the Okamizawa landslide introduced earlier in Sect. 2.2. As described above, using 5 m DEM, it was possible to reveal the fine geographical features of the landslide.

After Visualization: Time of Treating the Data of Scales Free

As Sect. 2 described, preliminary mapping has relied on the best efforts of various concerned contributors: “Can an engineer understand a disaster factor called a landslide, and grasp the features?”“Has almost all map information been visualized and released?” “Have the policy makers, engineers, and researchers tried hard enough?” Nevertheless, local residents who are the actual disaster victims meet with a scene claiming “not having known information about the area itself” until a disaster occurs. Even if many administrative officers in charge also complete their tasks, did they fully understand the “landslide phenomenon and taken measures of rational conviction?” Indeed, few among even concerned professionals have firm beliefs, and can be characterized as “right.” Especially in the case of catastrophic massive landslide disasters, the costs related to measures become immense. The landslide disasters are not to be understood the contents of earlier work.

After the construction finishes, subsequent maintenance management is a business service of the governments’ subordinate agencies. A person in charge understands the complicated and advanced business, and

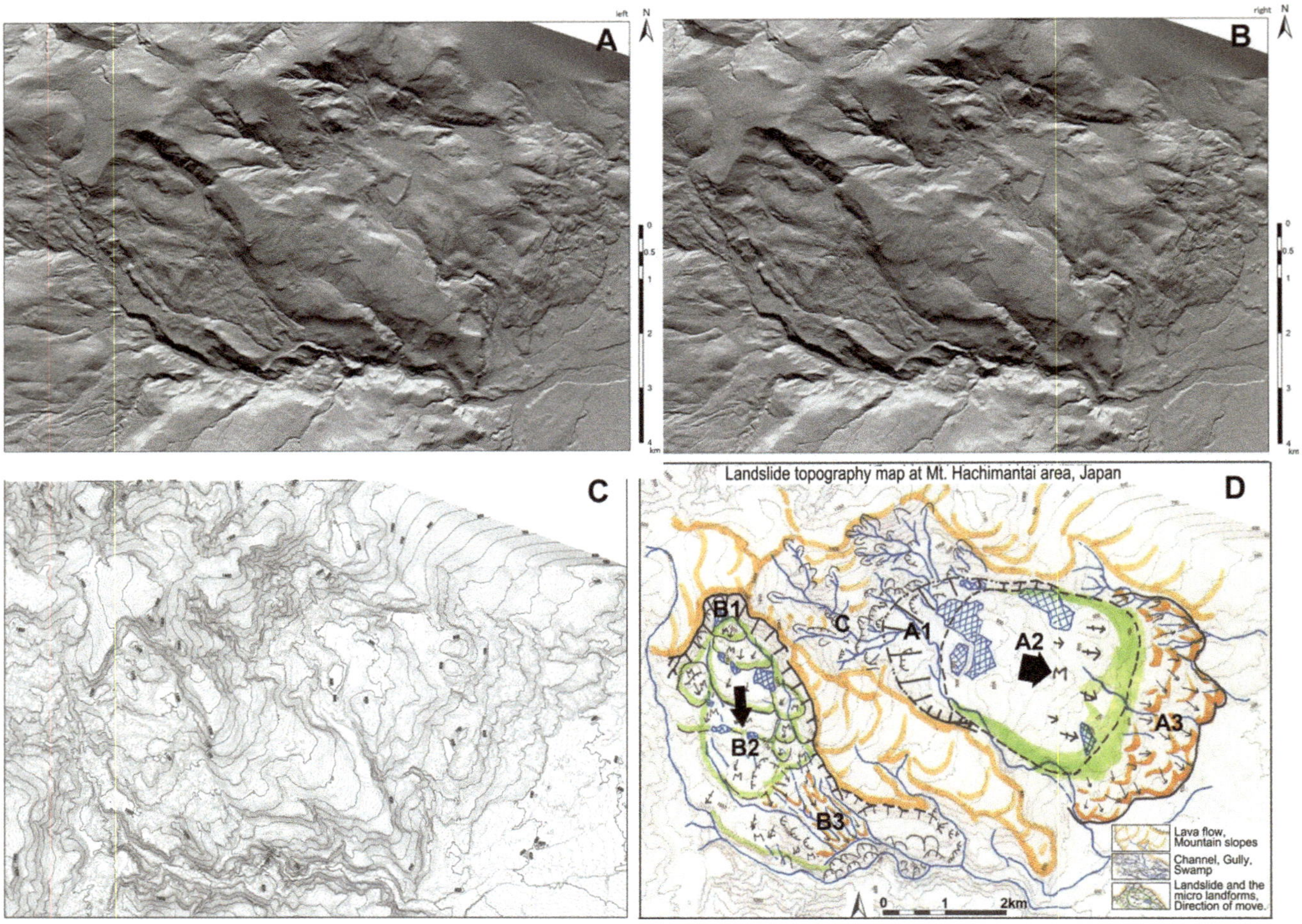

Fig. 21 Landform data of Hachimantai, Iwate Prefecture Northeast Japan, processing results and landslide classification at Iwate prefecture, Japan based on the 5 m DEM data by Japanese Government: **a** Relief map left, **b** Relief map right for stereo pair; **c** Establishment of the 2 m contour map; **d** landslide distribution map

he/she performs the sufficient management. The capability to perceive signs of a disaster promptly, to ascertain the meanings of the signs, and to build a rationale early is important as a person in charge. The basic structure and the change tendencies (direction, speed, and style) of the landslide are usually understood. As the end change signs being to appear, predicting their results quickly will be necessary.

Map information, a result of analysis of the earth surface situation becomes important to ascertain how changes occur inside a landslide.

Micro feature Mapping Using Laser Data and 3D Model Validation

The Dozangawa landslide (Fig. 22) is a massive landslide which is moving as a toe unit. Such massive landslides have some difficulties of movement mechanism recognition. Signs show that it moved toward the Dozangawa River through a movable body based on 3D analysis of laser data. Apparently, this landslide moves northward as a whole.

However, to the north near the main scarp, it slows in the toe from a water-induced landslide. The deformation tendency of the landslide is toward the west. Moreover, the geological map by Geological Survey of Japan suggests that the geological structure of the upper body inclines eastward. Consequently, applying the 3D simulation for movement tendency modeling is difficult. But, the change direction can be presumed from micro features restored using laser data (Fig. 22b). Results show that the movable body is classified into the Upper, Middle, and Lower portions. The order of movement is clear: the Lower portion is the beginning, moving westward to the river followed by the Middle portion moving the same direction causing a stress release in the front and a release toward the direction of a river by movement of the Lower portion. The Upper portion slides to the north as a result of the front thereafter. This movement agrees with GPS measurement results and reinforces the basis of change reproduced in the simulation (Fig. 22c).

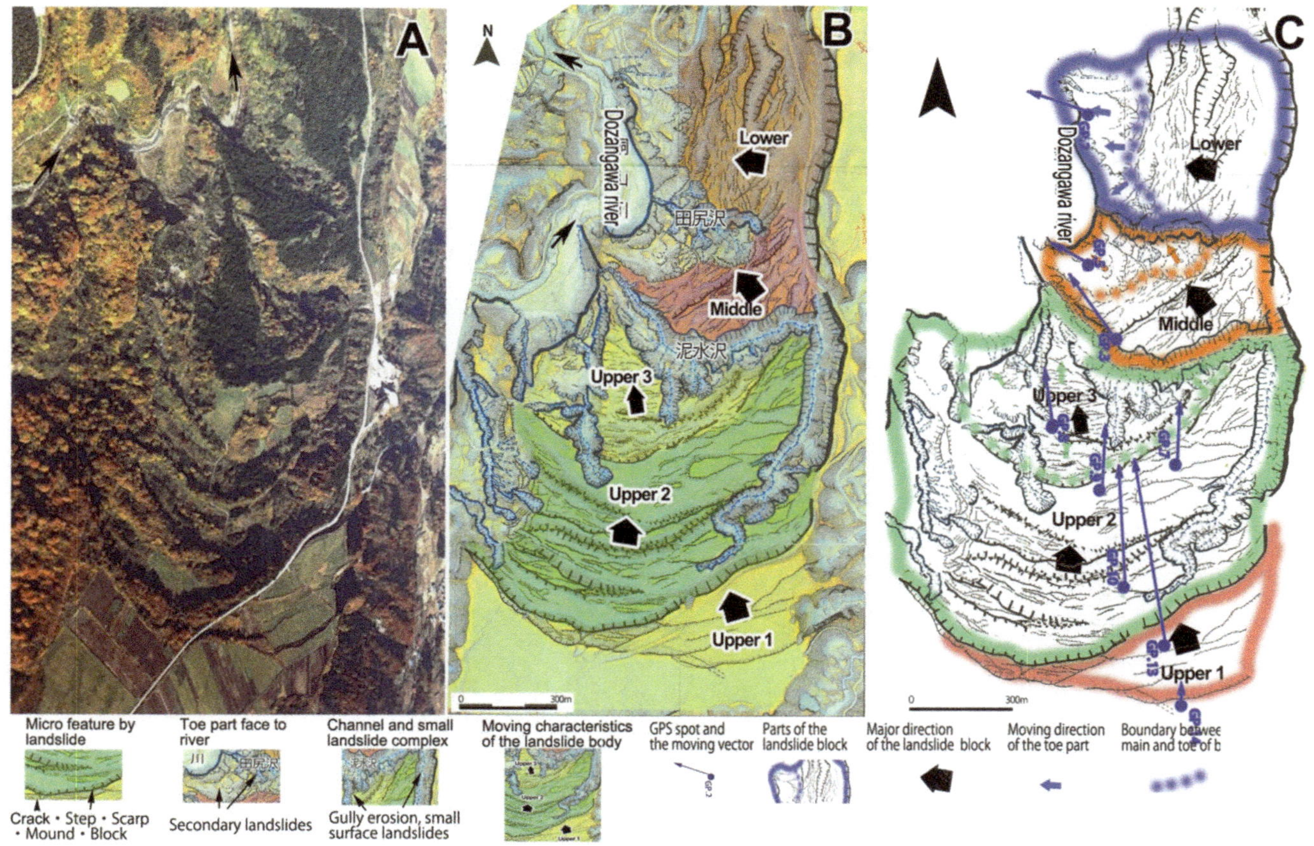

Fig. 22 Sample of microfeature landslide mapping to evaluate movement conditions related to a massive landslide in case of the Dozangawa Landslide, Yamagata prefecture, Japan. **a** Bird's–eye photo of the Dozangawa Lanslide area. **b** Deduced major movement direction based on the micro landforme feature analysys by using the 1 m DEM of Laser data. **c** Overlap the GPS measuring data and map B (Original data: collected by Tohoku Regional Forest offce, Mapping by Miyagi 2019)

Integrated Consideration by Innovation and Breakthrough to Common Recognition for Stake Holders from 3D Mapping

Development of visualization demands common understanding and rational measures. Understanding a phenomenon by development of three dimensions from sensing technology contributes greatly to disaster reduction and field management of a slope disaster. However, dealing with a phenomenon using three-dimensional mapping conventionally entailed high costs. There is no need associated with understanding all landslides in three dimensions. However, it is considered as the common base at the time of visualizing combined information necessary for a three-dimensional map for understanding a complicated and large object. The Ministry of Land, Infrastructure, Transport and Tourism (MILTT) report called "A system of considering landslide management as a visualization base" used Construction Information Management (CIM) corresponding to a landslide. The co-author of this keynote lecture, E. Hamasaki, developed and improved a comprehensive visualization system, Adcalc 3D, which carries out information control for land, geology, measurement, groundwater, etc. using three-dimensional image processing with cheap and sufficient operation. The epoch-making reality will pull scales from the eyes for professionals in this area. This way of thinking also leads to a useful concept in the field of sensing which is advancing in the recent years. "A specialist, administrator, official responsible for administrative affairs, local residents, etc. of those who look at it, can check the same place similarly." With an on-site sense, even for landslides works, perception is easy. In exploring a spot, it is unclear "how it is necessary to see what is now here." Normalization through CIM will lead to the realization of a rational argument, inducing all stakeholders' for a common understanding. This will fill a large gap for common understanding between the mapping sides and the using sides, namely between who are promoting mapping, and the side receiving it (Fig. 23).

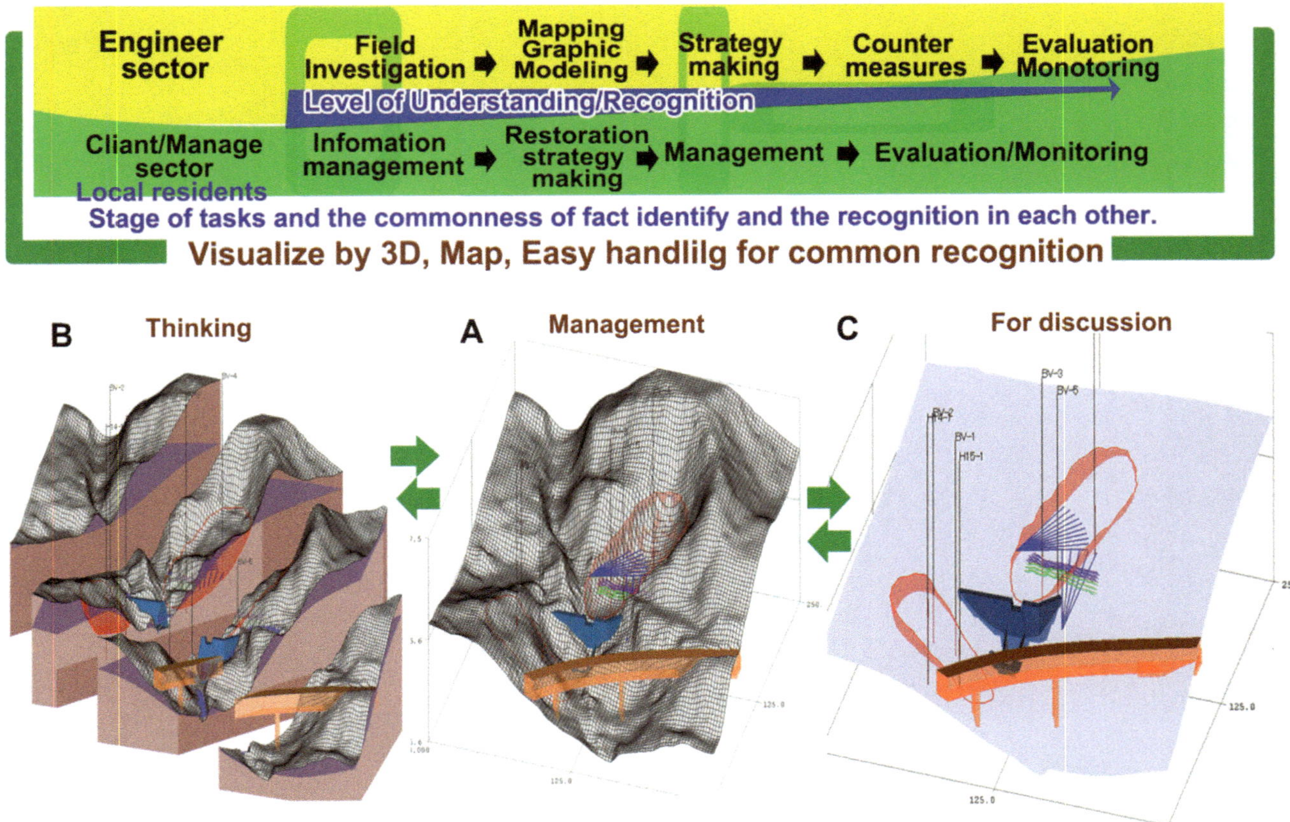

Fig. 23 Examples of 3D, flex layer, and flex purpose data management as CIM using Adcalc 3D (coordinated by Hamasaki). Explanation of the breakthrough possibility for the common recognition beyond the bottleneck among the Client, Person and Engineers

Conclusion

An outline of mapping necessary for a landslide to serve for its recognition can be summarized as following points:

(1) Landslide geographical feature to obtain an understanding of the portions connected with subsequent changes. (2) An understanding and mapping connected with susceptibility zoning of landslide generation. (3) Depending on public information to acquire and map at a low price and independently using available resources. (4) Setting up a report from viewpoints such as importance and mapping them intelligibly, and increasingly combining them.

Sendai Framework 2015–30 was proposed in the Third WCDRR held in Sendai in 2014. There, rather than taking disaster measures after a disaster occurs, prior investment for prevention of a disaster is pursued when it can be attained at a low price. Disaster prevention, actually disaster risk reduction, has currently become the mainstream of management of landslide disasters. All stakeholders can understand the slope disaster characteristics of their areas. Today, a map can be made that is useful towards realizing suitable management. That is an important reason that mapping of slope disasters suitable for disaster reduction (Slope DRR) is called upon under these circumstances.

Acknowledgements We would like to say thanks to Prof. H. Yamagishi, Prof. K. Sassa and Prof. S. Mihalic Arbanas for their useful guidance and comments for our lecture. Also many thanks to Dr. Ajmera, Mr. Doan Huy Loi, graduate student of Kyoto University for kind support in the finalization of this manuscript.

References

Asada A (2004) Hazard prediction and the measure at the land reclamation residential area for future earthquake in Miyagi prefecture, Japan. 197ps (Manuscript paper)

Chida N, Sugawara K, Miura O (1971) Landslides at the Northeast foot slope of Mt. Funagata Volcanic area. Ann Assoc Tohoku Geogr 23:175

Dung ND, Miyagi T, Luong LH, Hamasaki E, Hayashi K, Tien DV, Daimaru H, Abe S (2016) Trial of landslide topography mapping using ALOS W3D data—case study along the National Road No. 7 in central Vietnam—. Trans. Japanese Geomorph Union 33–1:127–140

Economic Planning Agency (1967) Fundamental land classification survey, geomorphology, subsurface geology and soil. SENDAI National Land Survey, 70 ps. and maps

Hamasaki E (2019) Simple instruction of the CIM (construction information modeling) and ADCALC 3D. 10 ps

Hamasaki E, Higaki D, Hayashi K (2015) Buffer movement analysis and BLUNDER PROBABILITY analysis for GIS based landslide susceptibility mapping—a case study of the 2008 Iwate-Miyagi Nairiku Earthquake, Japan. Landslide, J Japan Landslide Soc 52:51–19

Hamasaki E, Miyagi T, Takeuchi N, Ohnishi Y (2007) Risk evaluation of the earthquake triggered landslide on the land reclamation slope by the three dimensional instability analysis of simplified RBSM. Landslide, J Japan Landslide Soc 48:251–258

Hatano S (1972) 1/50000 landslide mapping at Yamagata 1/200000 area. Summary for Japan Geographical Association

Landslide Society Tohoku Branch (n.d.) Landslide and Landslide topography "Authors Committee for landslide mapping", 140 ps, 100 sheets of maps and geological maps

Miyagi T (2012) International meeting on regional disasters and environmental vulnerability. Proceedings of Kick off Meeting by Monka-sho aid, pp 1–87

Miyagi T (2016) Drill of hazard mapping as the practice of physical geography and disaster prevention study in the Eastern Japan great Earthquake 2011, "crustal disaster reduction and the academic/education society". Japan Sci Council Series No 22:88–99

Miyagi T (2017) Preliminary study of the landslide risk identification by landform classification using the AW3D data at around Tienchi Lake, Chanbaishan, China—North Korea border. Proceedings of research report of NE China, pp 16–21

Miyagi T, Hamasaki H (2016) Evaluation the potential of landslide reactivation and the micro landform features. In: Fujimoto K, Miyagi T, Saojo K, Takeuchi Y (eds) Micro landforms—as the tie between human and the nature (pp 160–181) Kokon-Shoin, Japan

Miyagi T, Thanh NK, Tien DV, Luong LH, Viet QV (2020) Slope disaster risk reduction map as a communication tool for community based DRR in Japan and Vietnam

Miyagi T, Prasada BG, Tanavud C, Potichan A, Hamasaki E (2004) Landslide risk evaluation and mapping—manual of aerial photo interpretation for landslide topography and risk management—. . Report National Res Instit Earth Sci Disaster Prevent No 66:75–137

Miyagi T (1979) Landslides in Miyagi prefecture. Sci Rept Tohoku University Ser.7(Geogr.) 31:1–14

Nguyen KT, Miyagi T, Shinobu S, Tien D, Luong LH, Ha DN (2020) Developing recognition and simple mapping by UAV for local resident in mountainous area in Vietnam (Submitting to WLF 5)

Ohuchi Y (1973) River terraces and their displacements along the Hirose river, Miyagi Prefecture. Ann Tohoku Geogr Assoc 25:84–90

Saaty TL (2008) Decision making with the analytical hierarchy process. Int J Serv Sci 1:83–98

Sato G, Yagi H, Urquia EG, Mullings M, Kuwano T, Isono K (2018) Landslide mapping using a 2 m DEM based on AW3D digital topographic data in Tegucigalpa, Honduras. In: 2nd Central America and Caribbean Landslide Congress, pp 18–20

Technical committee for Agriculture and Fishery (1964) Manual for Land use planning. Statistical Institute for Agriculture and Forestry, 432 ps

Tien DV, Miyagi T, Abe S, Hamasaki E, Yoshimatsu H (2016) Landslide susceptibility mapping along the Ho Chi Minh route in central Vietnam: AHP approach applied to a humid tropical area. . Trans Japanese Geomorph Union 33–1:79–104

Uchiyama S, Miyagi T (2016) Acquisition and utilization of high-definition digital surface models through aerial photography using a small unmanned aerial system: an example of typhoon damage in Iriomote Island mangrove forests. . Trans Japanese Geomorph Union 33–1:159–174

Unome S (2020) Current status of Airborne LiDAR system with a case study in Amami Oshima Island and introduction of the latest LiDAR technology. Mangrove Sci 11:27–35

Varnes DJ (1978) Slope movement types and processes. In: Schuster RL, Krizek RJ (eds) Special Report 176. Landslide analysis and control, pp 11–33. TRB, National Research Council, Washington, D.C.

Yamagishi H, et al (2012) Hokkaido digital landslide map (with DVD). Hokkaido University Press, 100 p

Yanagiba S (2019) Quantification of surface environmental changes by high resolution photogrammetry method at multiple times. Doctor thesis of The University of Tokyo, 171 ps

Landslide Susceptibility Mapping by Interpretation of Aerial Photographs, AHP and Precise DEM

2020 Kyoto Japan

Hiroshi Yagi, Kazunori Hayashi, and Go Sato

Abstract

Aerial photo interpretation which make us realize the nature of landforms is strong tool to detect landslide prone area, even though precise DEM becomes to be available. Combination of above two methods is important to prepare landslide susceptibility map (LSM). Scale of landslide susceptibility map (LSM) varies according to the objectives of the projects. Landslide inventory of small scale is useful for nation wide planning. While AHP method is suitable for LSM of middle scale, 1/20,000–1/50,000 by aerial photo interpretation to nominate landslide susceptible area. Landslide susceptibility mapping of large scale for implementation of landslide prevention work or installation of observation equipment requires ground truth and comprehensive evaluation combined with AHP. Bell-Shape Index provide us the convenient criteria to evaluate landslide susceptibility in high relief mountain region, assessing overburdened mountain profile in high relief area.

Keywords

Landslide susceptibility mapping • Aerial photo interpretation • AHP • Precise DEM • Bell-shape index

H. Yagi (✉)
Yamagata University, Kojirakawa, 990-8560, Yamagata, Japan
e-mail: yagi@e.yamagata-u.ac.jp

K. Hayashi
Okuyama Boring Co.,Ltd., Yokote, 013-0046, Japan
e-mail: k.hayashi@okuyama.co.jp

G. Sato
Teikyo Heisei University, Nakano, Tokyo, 170-8445, Japan
e-mail: gsato@thu.ac.jp

Introduction

Landslides are one of the severest natural hazard that has to be addressed in relation to the development activities. However, planning, designing and construction of development projects are often implemented without proper assessment of potential hazards, such as triggering or causing new landslides due to slope instabilities during & after badly managed construction work or urbanization without proper planning. In order to minimize risk to infrastructure and human lives, Landslide Susceptibility Mapping (LSM) is required so that safe or least hazardous areas could be selected for urbanization or infra-development. Landslide susceptibility mapping is based on the assumption that landslides occur as a result of similar geological, geomorphological and hydrological conditions that caused in the past.

Understanding Geomorphological Processes from Landform

Landforms directly reflect the formative processes related their geomorphological evolution. These formative processes include tectonic movement, volcanic, fluvial and colluvial processes. Gravitational process such as fall, slide, flow, creep and spread are also included. Evolution of terrain is ongoing in several scales of space such as large scale, medium scale, small scale and micro scale ranging from several ten kilometers to one meter with different time scale, respectively. In other words, terrain consists of several layer sheets of different time and space scale. We can select the suitable layer sheet to understand the nature of the objective landform. Realizing formative process of landform is alternate-repeatedly overviewing wide area and focusing on the small area, detecting overall topographic features, i.e. relief, roundness, concavity, convexity and micro topography allocated on ground surface, i.e. cracks, steps, shrinks,

F. Guzzetti et al. (eds.), *Understanding and Reducing Landslide Disaster Risk*,
ICL Contribution to Landslide Disaster Risk Reduction,
https://doi.org/10.1007/978-3-030-60227-7_3

scarplets, depressions. However, we can not overview the terrain as long as we are on the ground level. Topographic map provides us relief information of the terrain, by reading pattern of contour lines (Fig. 1). But a contour map is two dimensional information that is projected from three dimensional information of the terrain. Furthermore, relief data is rounded based on contour interval. Subsequently, the contour map view is a rough image, neglecting smaller surface ruptures than the contour interval. Elaborate contour reading also takes much time, depending on an observer's capability of 3D imaging for a wide area.

Aerial Photograph Interpretation Focused on Landslide

Stereoscopic aerial photogrammetry allows one to obtain an overall and detail view of the terrain in a three-dimensional image quickly, though that depends much on scale of photographs and skill and experience of the observer.

What can we detect on the ground through aerial photo interpretation? The answers are as follows; aerial photo interpretation facilitate us to detect synoptic information of the terrain such as drainage system, ridges, slopes, mounds, hollows, terraces and dis-continuity of geomorphological surfaces and also three dimensional distribution of the surfaces at different levels. Geomorphological surface was formed under a specific climatological, hydrological and tectonic conditions at a particular area and a geological age of the bedrock and structure, just like a terrace with spacial continuity rimmed by breaks of slopes (Fig. 2). And the geomorphological surfaces develop at the different levels, icising the former geomorphological surfaces due to change of the environmental conditions. Therefore, various geomorphological surfaces of the different ages and levels are allocated on the ground. Subsequently aerial photo interpretation facilitates us to understand the processes and evolution of the terrain. Evolution of mountain slopes is also developed by erosion and mass movement, forming discontinuity on slope such as gullies, cracks, landslide crowns and other micro topography with various scale ranging from a few ten centimeters to a few meters.

Landslides occur under the similar geological, geomorphological and hydrological conditions that caused the past landslides. Dormant landslides are often reactivated when they are disturbed their equilibrium of slope stability by some causualities as follows; rise of ground water level due to heavy rain or snow melt, increase of acceleration due to earthquake, human activities due to excavation works such as removing earth at the toe parts of the dormant landslides or due to increasing in weight such as building on a sliding mass as driving force.

The past is a key to predict future landslide hazard. The first step for reducing human and economic losses due to landslide is to prepare landslide maps that inventory dormant landslide sites. Landslide inventory mapping (LM) project all over Japan was carried out in the 80s and 90s for the sake of realizing landslide prone area in Japan by NIED (National Institute for Natural Science and Disaster Resilience) and

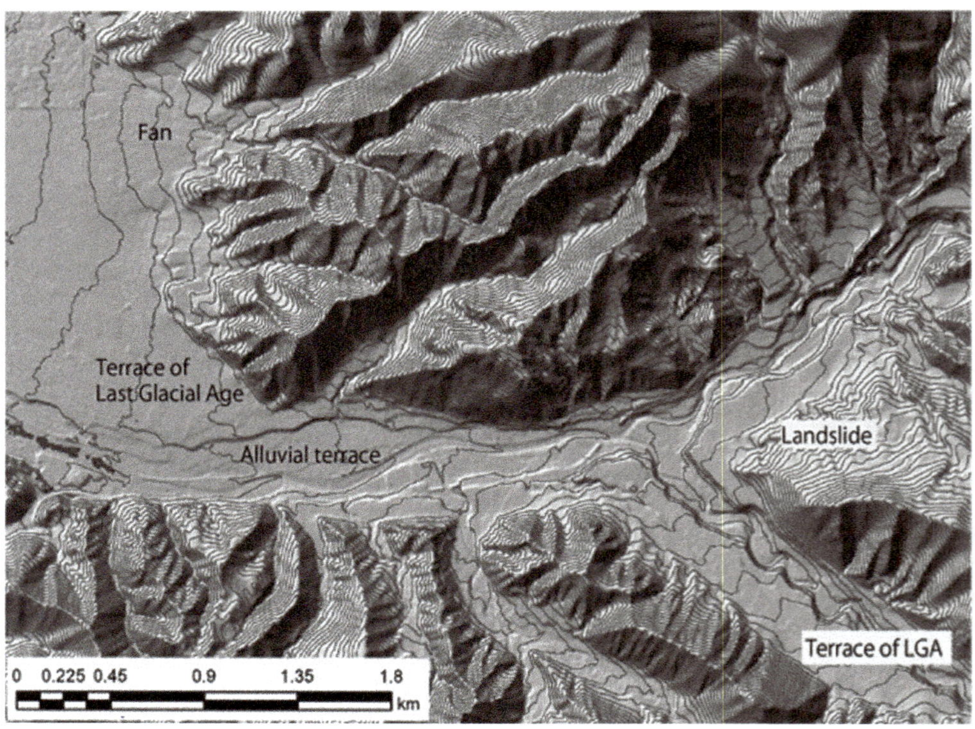

Fig. 1 Landform view of medium-small scale, Yamadera area, Yamagata, Japan Contour interval: 10 m

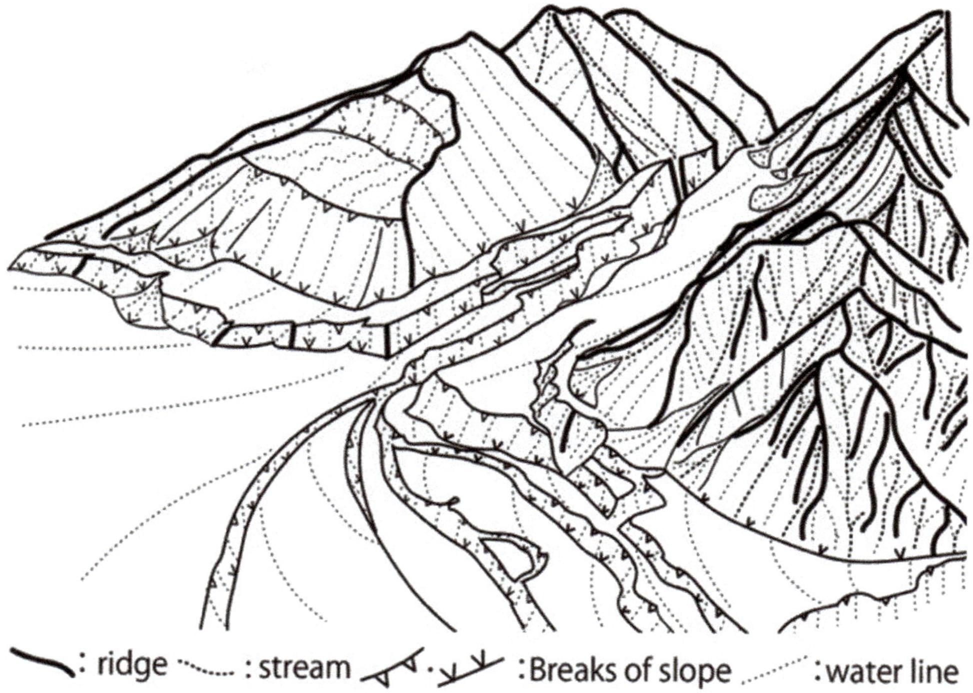

Fig. 2 Schematic birds-eye view of distribution of geomorphological surfaces fringed by break of slopes

was completed at the beginning of 21th Century. This inventory mapping was implemented by aerial photo interpretation of 1/40,000 and topographic map of 1/25,000 (Fig. 3; NIED 1998). The achievement of LM is used for raising awareness on landslide disaster, superimposing on the other thematic maps such as geology and relief maps. New project to re-utilize LM combining high resolution DEM created by LiDAR technology is going to launch to prepare Landslide Susceptibility Map (LSM).

Example of Landslide Interpretation by Aerial Photo on Different Scale of Topographic Map

We try to show image difference of two inventory maps of landslide crown detected by interpretation of aerial photos of 1/40,000 on 10 m contour map (Fig. 4) and by that of 1/20,000 on 5 m contour interval map originated from high resolution DEM (Fig. 5). It is apparent that the latter one can express much more micro slope breaks as secondary crown due to slight movement of the landslide mass. However, detailed interpretation by the high resolution map takes much time and is not good to get outline of geomorphological evolution for a wide area. Therefore, screening of landslides for the wide area is the first step. The second step is elaborate detection of micro displacement in the target area to prepare proper observation of the movement at a field site, using more precise topo-maps.

Mapping Scale, Method and Targets of LSM Work

Before taking up a landslide susceptibility mapping, it is first necessary to clarify mapping scale, considering objectives of LSM work. This is because the method adopted for LSM should be determined according to a geographic extent depending on the intended mapping scale. Basic concept of identification of the scale & objectives of the susceptibility mapping is generally categorized as follows:

- National scale 1:200,000–1:1,000,000, nation wide development planning of infrastructure
- Regional scale 1:50,000 - 1:200,000, regional development planning of infrastructure
- Medium scale 1:20,000–1:50,000, urban & road development planning
- Large scale 1:5,000–1:20,000, practical urban desgin & implementation of disaster prevention work.

Landslide inventory map is enough for national or regional level planner to know the landslide susceptibility of the targeted areas and it is also known as landslide zonation map in scale of 1:100,000–1/1,000,000. It is usually compiled from the inventory maps of the middle scale LSM prepared by aerial photo interpretation. However, hazard susceptibility mapping of small scale less than 1:50,000 is

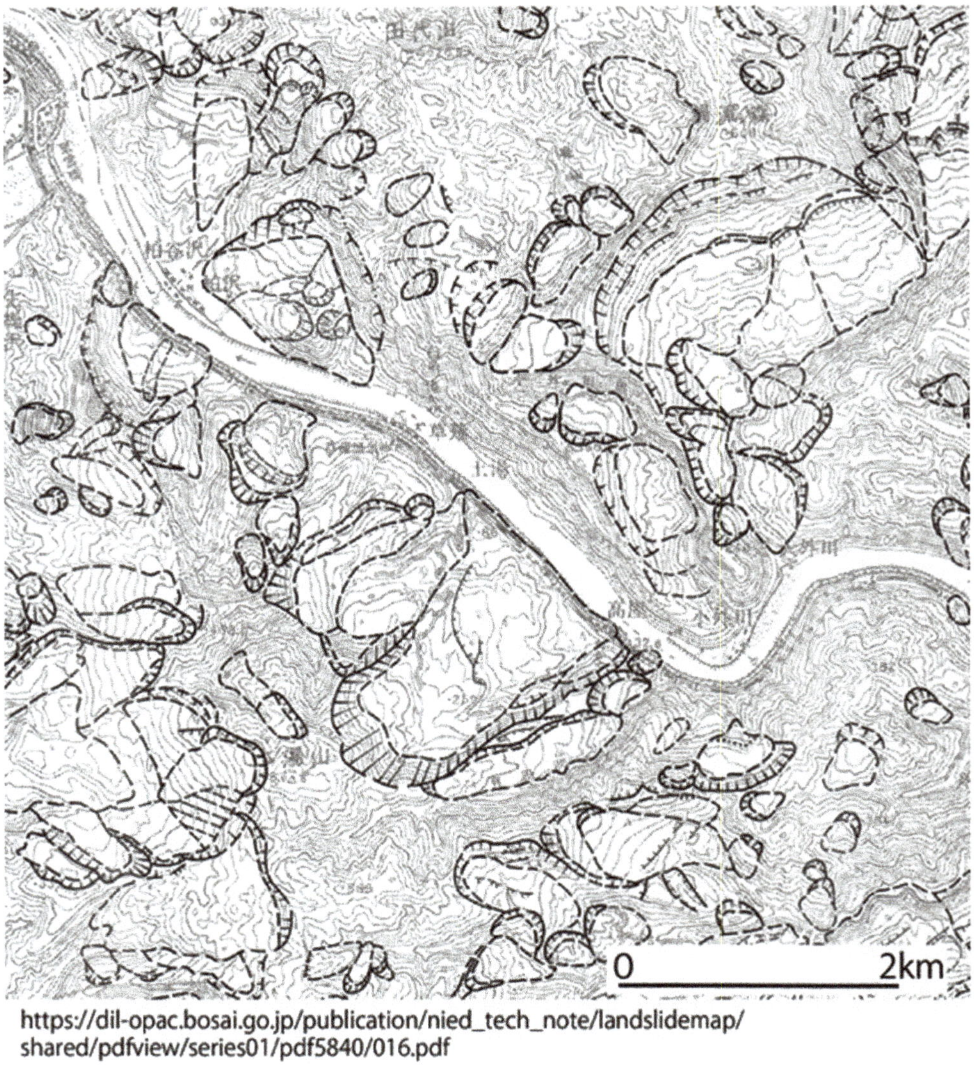

Fig. 3 Landslide map of Kiyokawa area, Japan, issued by NIED

not feasible for engineering projects or disaster prevention work. If data on slope gradient and geological information are superimposed onto the inventory maps in scale of 1/50,000 or more, processed by GIS using wide grid size of DEM, i.e. 10–30 m, the planner can realize the susceptible zone to be avoided when they make the development plan. Because it can be prepared quickly and conveniently, covering wider area (Yagi 2016). For example, SRTM of 30 m grid size originated from satellite image is available freely. It is easy to find a susceptible landslide of which toe part angle is steeper than 30° and of which relative height at the toe part is higher than 10 m, using SRTM data. Landslide susceptibility maps in the range of about 1:20,000–1:50,000 should be prepared for urban development planning or feasibility assessment for road alignment decision using GIS method. The medium scale LSM is much practical because most of study materials such as topographic maps, aerial photographs and engineering geological maps of this scale are available even in under developing countries as Nepal (Yagi et.al. 2018). LSM of the middle scale refers to an arithmetic method of portraying the spatial variation in the susceptibility of slope to fail, assessing landslide factors such as micro topography relating to landslide reactivation through aerial photo interpretation. And it is prepared on the basis of criteria system based on ranking of each slope facet. To quantify susceptibility ranking of slope, a weighting or rating system is introduced. Range of landslide susceptibility rate obtained is to be categorized into some qualitative levels in terms of relative instability, high, moderate or low etc. GIS is a good tool for such proceedures.

Large scale LSM is desirable, if high resolution digital elevation model (DEM) of 5–10 m grid is available and field investigation at a study site is implementable. Objectives of the large scale LSM is detection of high hazardous area from the aspect as follows; designing of countermeasure work for slope disaster, prioritization of prevention work for high risky infrastructures and determining the residential areas where need further monitoring for the landslide.

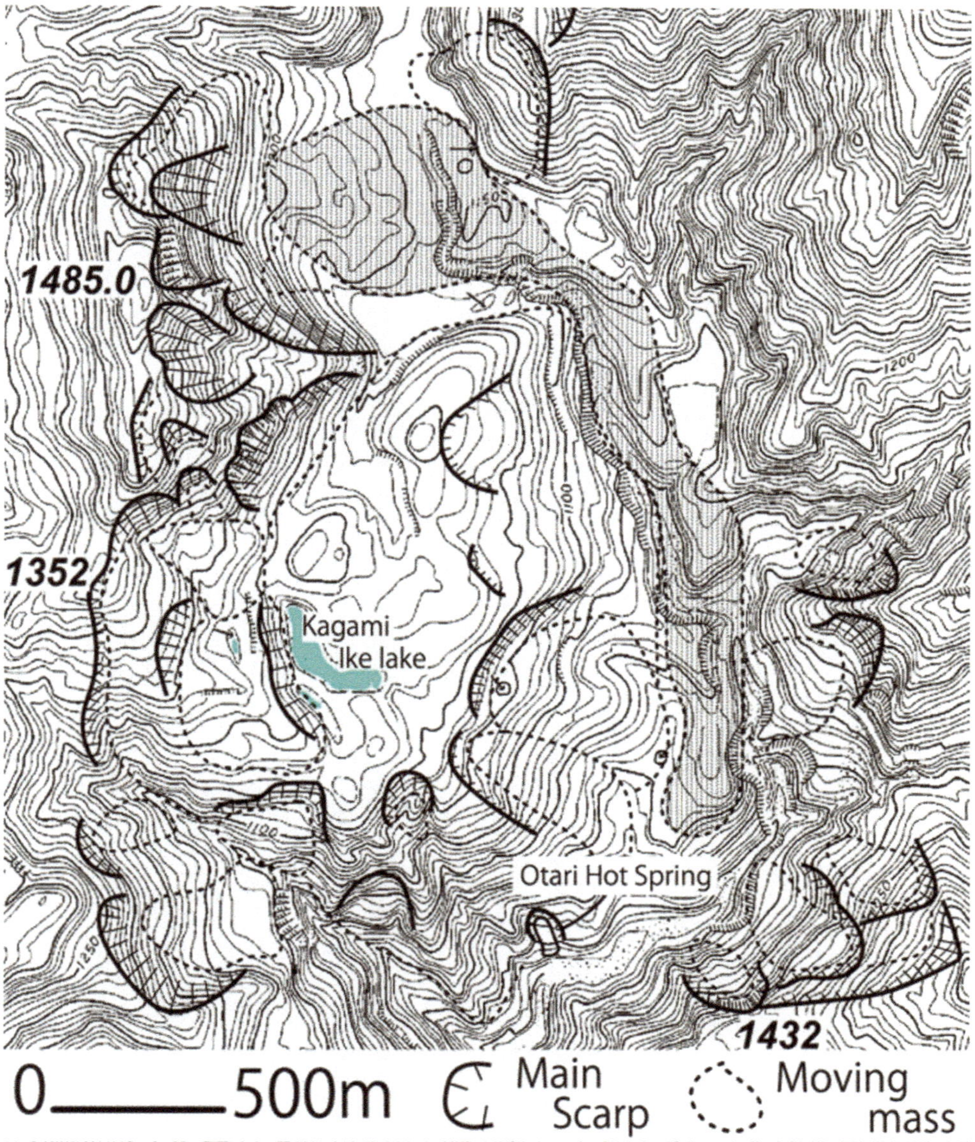

Fig. 4 Landslide inventory map of 10 m contour interval the study area locates behind Otari hot spring, Japan

Middle Scale Landslide Susceptibility Mapping Combined with Aerial Photo Interpretation and AHP

Landslide inventory map of the middle scale prepared by interpretation of aerial photographs is the most basic way to know landslide prone area spatially, because old landslides are often reactivated by equilibrium change of slope stability due to erosion or artificial cutting. However, it is a matter of discussion for geotechnical engineers concerning to landslide management and control to presume which landslides will be reactivated in the near future. There are too much landslides to be checked in the field (Fig. 6: GMDTT 2014). So more detail qualification system to evaluate the slope stability in larger scale is required, using aerial photo interpretation, setting a field study site (Yagi 2016). Aerial photo interpretation is actually a cheap and effective method for landslide susceptibility assessment study. However, it requires for the geotechnical engineers much experience and formalization of evaluation, keeping quantitative identification of landslide. In this context, an expert system using AHP method to assess landslide susceptibility of the middle scale by aerial photo interpretation is adopted based on the interview and brain-storming among experts on the photo interpretation and reconnaissance study on landslides (Yagi et al. 2009; Hamasaki and Miyagi 2018), focussing on topographic features of landslide and location of landslide affecting its stability of slope. Photo scale used in the study is 1/20,000.

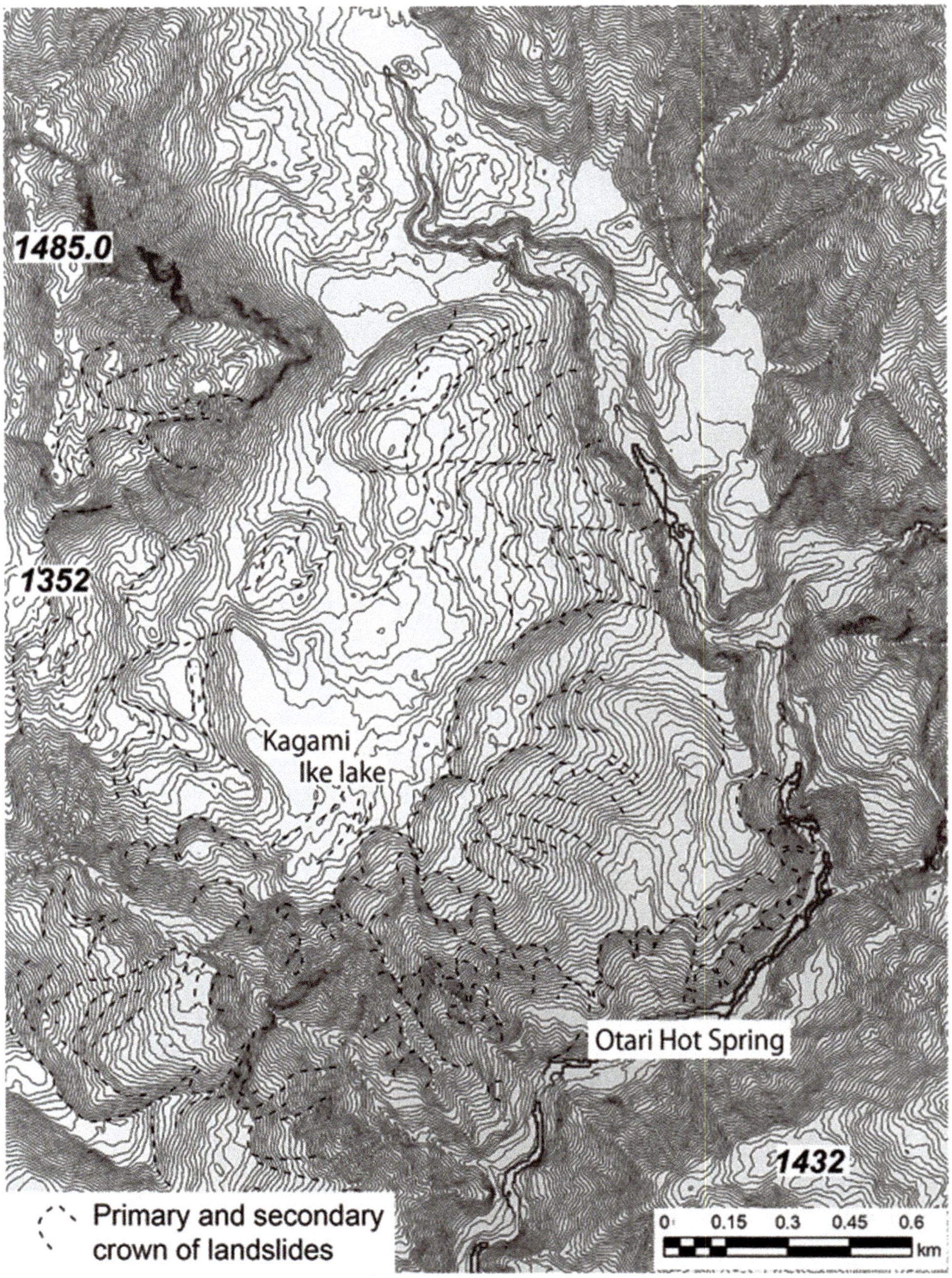

Fig. 5 Landslide inventory map of 5 m contour interval the study area is almost same as that shown in Fig. 4

What is Analytical Hierarchical Process (AHP)?

The Analytical Hierarchical Process (AHP) method, developed by Satty (1980) decomposes the process of subjective decision of a person into a hierarchical structure of simple and independent factors and expresses the process qualitatively. It decides relative importance between two factors as a pair-wise comparison as follows; equal, weakly, strongly, much strongly. And then it decides the weight for each factor and assigns the weight to each factor by each landslide. Consequently, AHP is a method for formalizing decision-making based on attributing factors that vary personally. Therefore, a key point is to select which factors are important and attributing to your decision.

For landslide hazard susceptibility assessment, the first step is to nominate important factors causing landslides from geomorphological aspects.

The second step is to decide the relative importance of the factors. Subsequently, it can evaluate the relative landslide potentiality of reactivation assigning weight for each factor.

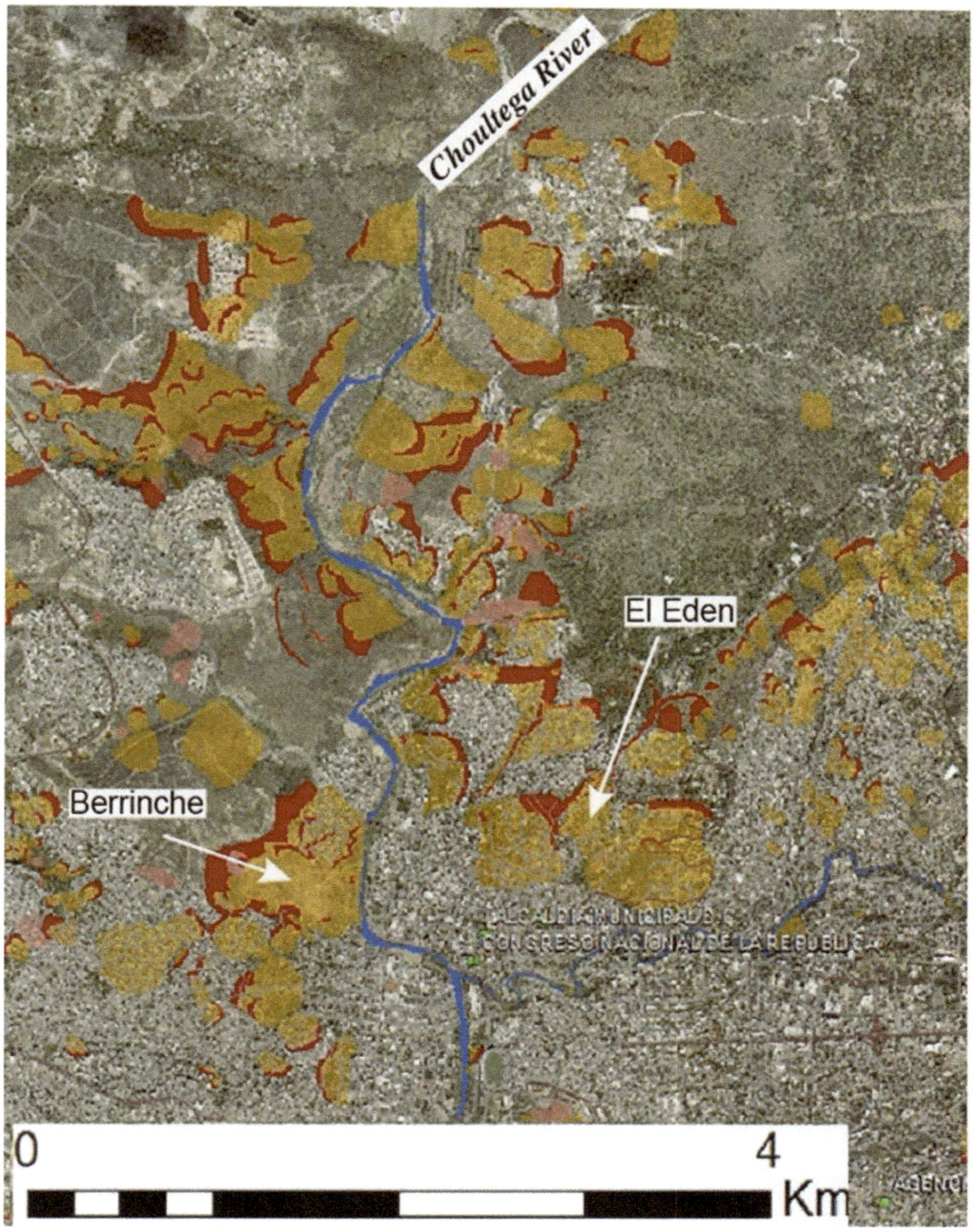

Fig. 6 Landslide map in northern Tegucigarpa metropolitan area interpreted by 1/20,000 aerial photographs after GMDTT (2014) red are: main scarp, yellow area: moved zone

Topographic Features as Attributing Factors

Landslide inventory mapping is the most usual step of the hazard assessment (Soeters and Westen 1996). Micro topography on a landslide mass had been also focused (Kienholz 1978). Six geomorphological factors were selected as evaluating criteria in this study from the view points of geomorphological evolution, landslide activity and destabilizing possibility, based on the brain-storming among the landslide researchers of Japan Landslide Society (Yagi et. al. 2009). They are as follows:

1. Sharpness and clearness of micro-topography formed by landslide,
2. Fragmentation of a primary block into sub-blocks
3. Profile of landslide mass and toe part
4. Erodibility of toe part of landslide mass
5. Water collectability from upper slope of landslide crown
6. Land cover, artificial change and habitation on landslide mass.

Criteria of (1) and (2) represent the stability due to landslide evolution, and those of (3) and (4) are the stability factors attributed from conditions of the landslide toe part, respectively.

Sharpness and Clearness of Micro-topography Formed by Landslide

Sharpness and clearness of such micro-topography on a moving block and a main scarp are checked as the important factor of the susceptibility evaluation. Micro-topographies such as scarps, cracks, steps, hollows and ridges formed in and around a crown or landslide mass represent its recent ground motion from the aspect of geomorphological evolution, because the micro-topographic features are easily subject to erosion and become rounded and unclear with the lapse of time (Fig. 7). For example, main scarps of old landslides are usually eroded and show rounded because the old landslides are mostly dormant. If the landslide is active, those micro-topographies will be densely distributed and will be very sharp and clear.

Fragmentation of a Primary Block into Sub-blocks

Rock mass of a landslide is subjected to be fractured with retrogressive or progressive movement inside the original rock mass. Subsequently it is prone to be divided into the secondary or tertiary sub-blocks (Fig. 8). Such fragmentation of the primary landslide mass also indicates recent activity and continuity of gravitational movement from the views point of geomorphological evolution. Consequently, such fragmentation is adopted as an evaluation criterion of the landslide susceptibility.

Profile of Landslide Mass and Toe Part

Geotechnical engineer usually checks how much driving force is remained within a landslide mass. Topographic profile of a landslide mass and toe part directly reflects its stability low or high. Thus, the profile of landslide mass and toe part is chosen as one of the criteria of the susceptibility evaluation (Fig. 9). If a steep and top-heavy profile at the toe part suggests high instability of landslide wholly, high score should be allocated to it.

Erodibility of Toe Part of Landslide Mass

Under-cutting off a toe part of landslide easily breaks stability of a dormant landslide. If a river course shows incised meander and a toe part of the dormant landslide is located in an undercut slope, it is subject to sever erosion and is prone to be reactivated (Fig. 10). So erodibility of toe part of landslide mass is nominated as the evaluation criterion of landslide susceptibility.

Water Collectability from Upper Slope of Landslide Crown

Movement of landslide is prone to be controlled by a ground water condition. Landslide occurs if driving force derived from gravity of slope materials becomes bigger than resistant force along a slope. Resistant force is reduced by the ground water pressure, therefore, the stability of a slope much depends on inflow of water to the landslide mass. Then water collectability from upper slope of landslide is considered as slope concavity of a water catchment (Fig. 11). It is chosen as the attributing factor for landslide reactivation.

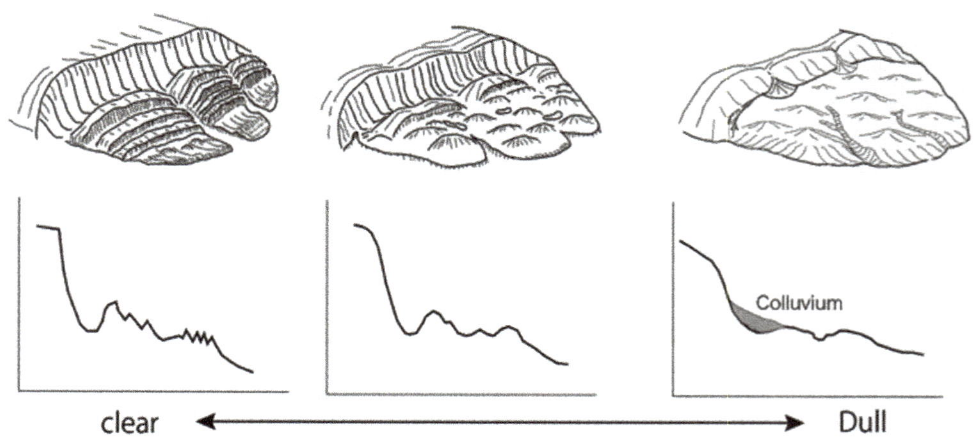

Fig. 7 Schematic view of sharpness and clearness of micro-topography as evaluation criteria of AHP

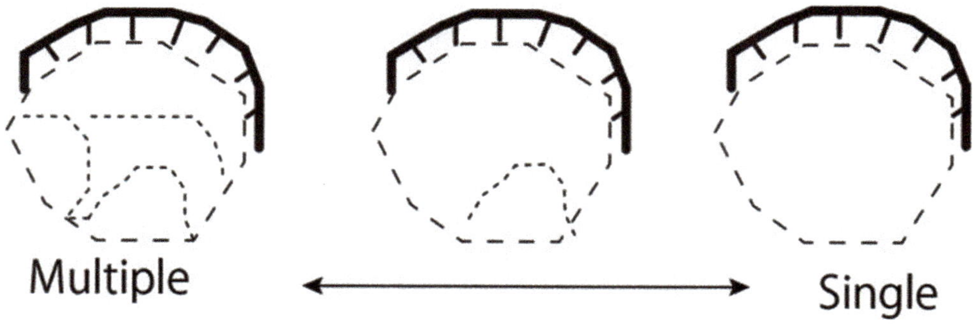

Fig. 8 Schematic planer view of fragmentation of landslide block

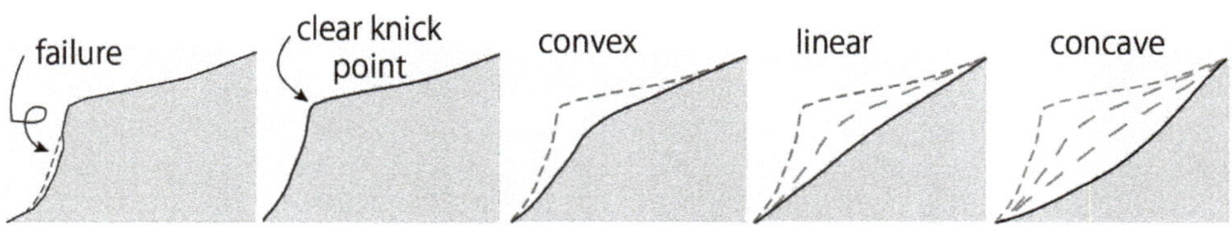

Fig. 9 Schematic profile of landslide mass and toe part

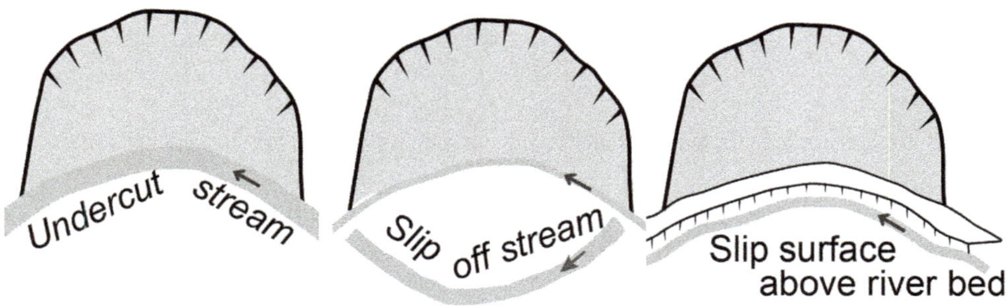

Fig. 10 Schematic planer view of erodibility of toe part

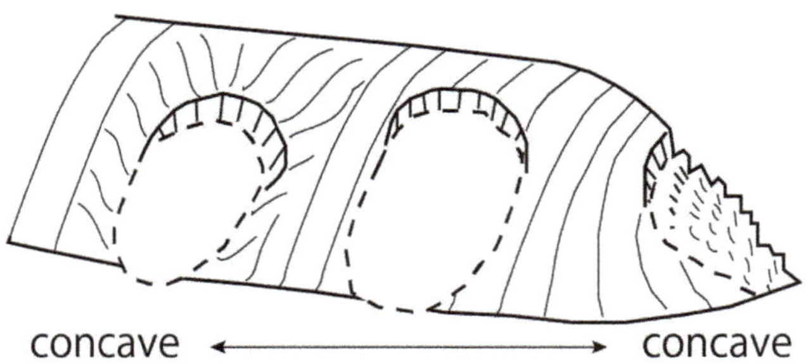

Fig. 11 Schematic view of water collectability

Fig. 12 Schematic view of land cover and habitation

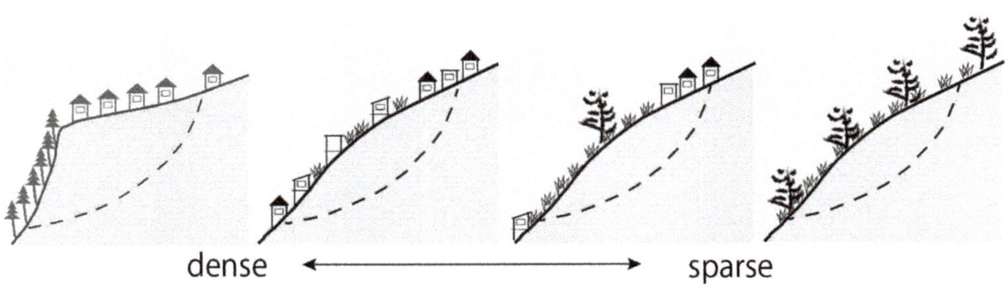

Land Cover, Artificial Change and Habitation on Landslide Mass

Human activity or habitation on a landslide slope sometimes affects stability of the slope due to irregular cutting of slope or infiltration of living drainage if proper urban planning and designing is not done before development. Integrated attention and observation to such land-cover pattern shown in Fig. 12 should be paid as the contributing factors for landslide reactivation.

Weighting System for Susceptibility Assessment of Landslide

As mentioned above, the AHP method decomposes the process of personal decision into a hierarchy of simple and independent criteria such as Level II and Level III in descending order, and expresses the process qualitatively by each hierarchical level. That of Level I is the comprehensive final decision. We show the evaluation process in hierarchical Level II as below.

Hierarchical Level II of Landslide Susceptibility

Matrix of attributing factors as the weighting criteria of the hierarchical Level II, which consists of: sharpness and clearness of micro-topography formed by landslide, fragmentation of a primary block into sub-blocks, profile of landslide mass and toe part, erodibility of toe part of landslide mass, water collectability from upper slope of landslide crown, land cover, artificial change and habitation on landslide mass was decided by a series of pair-wise comparisons (Table 1). Each number of a cell in the matrix (A*ij*) shows the relative important level of the objective *i*th row to the objective *j*th column, allocating odd number according to the relative importance by brain storming among skilled engineers with much experiences about landslide interpretation of aerial photos. They are as follows:

1: Objectives *i*th row and *j*th column are of equal importance.

3: Objective *i*th row is important than objective *j*th column.

1/3: Objective *i*th row is less important than objective *j*th column (inverse number of 3).

Erodibility of the toe part shows big number for other attributing factors, because engineers usually pay much attention to condition of the toe part.

Hierarchical Level III

Example of an important factor at a hierarchical level III in a case of toe part erodibility is shown in Table 2. This case depends on combination of stream scale and slope type, undercut slope or slip-off slope, considering position of a slip surface. Where a big stream is facing to the undercut slope, higher score will be given, compared to the slip-off slope facing to the relatively small stream (Table 2).

Weighting System

The weight coefficient for the hazard susceptibility is calculated, multiplying the weight of hierarchical level II and level III. The score in Table 3 is percentage of each weight coefficient for the total, respectively. Among the attributing factors of level II, erodibility of the toe part is highly weighted. And higher score is given to the undercut slope facing to the big stream in level III. Consequently, landslides of high potentiality of reactivation usually occur along main streams of big rivers, where the landslides have undergone toe erosion due to under cutting by the strong streams. Checking each decision criterion of level III for each attributing factor of level II by aerial photograph interpretation for each landslide block is implemented in order to assess the landslide susceptibility by AHP. Figure 13 is an illustrated weighting system at the hierarchical level III, allocating score to each criterion according to its slope condition. Such simplified image evaluation system based on micro-topography related to landslides is conceived as the expert system. And it is accessible for common geotechnical engineers to detect highly susceptible area among many coherent and dormant landslides.

Improved Weighting System for Hilly Area in Tegucigarpa

There are many active landslides in Tegucigalpa, Honduras (see Fig. 6). Two third of them are distributed along Choluteca River and its tributaries (Grupo de Mapeo de Deslizamientos de Tierra Tegucigalpa; GMDTT, 2014). The most known one is Berrinche landslide that is located on the left bank of Choluteca River (Fig. 14).and was triggered by Hurricane Mitch in 1998. Meanwhile, Tegucigalpa valley is surrounded by steep wall composed of thick welded ignimbrites or basalt sheets overlying the Cretaceous sedimentary soft sediments. Such geological/geomorphological feature is called "cap rock structure" that is prone to cause landslides due to a top-heavy profile

Table 1 Matrix of a series of pair-wise comparison of attributing factors of level II for landslides distributed along river

		Sharpness & clearness of landslide micro-topography	Fragmentation of a primary block into sub-blocks	Profile of landslide mass & toe part	Erodibility of toe part of landslide mass	Water collectability from upper slope of landslide	land cover, artificial change and habitation on landslide mass	Geometric Mean	weight	Score
Evolution of landslide	Sharpness & clearness of landslide micro-topography	1	1	0.333	0.333	1	1	0.693	0.105	11
	Fragmentation of a primary block into sub-blocks	1	1	1	1	1	3	1.201	0.181	18
Instability of toe part	Profile of landslide mass & toe part	3	1	1	1	3	3	1.732	0.261	26
	Erodibility of toe part of landslide mass	3	1	1	1	3	3	1.732	0.261	26
Water condition	Water collectability from upper slope of landslide	1	1	0.333	0.333	1	1	0.693	0.105	11
Landcover and landuse	land cover, artificial change and habitation on landslide mass	1	0.333	0.333	0.333	1	1	0.577	0.087	9
								6.629	1.000	101.0

1: Objective *i th* row and *j th* column are of equal importance.
3: Objective i th row is important than objective j th column.
0.333 : Objective i th row is less important than objective j th column.

Table 2 Example of pair-wise comparison of attributing factors of level III, Erodibility of toe part

	undercut ·big river	undercut ·small river	slip off· big stream	slip off· small/higher position·slide surface			core
undercut·big river	1.00	3.00	5.00	7.00	3.20	0.589	59.00
undercut·small river	0.33	1.00	1.00	3.00	1.00	0.184	18.00
slip off·big stream	0.20	1.00	1.00	3.00	0.88	0.162	16.00
slip off·small/higher position·slide surface	0.14	0.33	0.33	1.00	0.35	0.065	7.00
					5.44	1.0	100

and viscous/fluid property of underlying stratum due to water infiltration through joint or fault system in welded tuff. Such type of landslide is called as block glide and spread. And infiltrated water through the upper pyroclastics sheet usually spring out along the discontinuity between the ignimbrite as a cap rock and sediments (Fig. 15).

Pediment slope develops on the soft sediments along the foot of the volcanic rock wall and many dormant landslides are also detected on the slope of the pediment hill. A landslide such as El Eden site (Fig. 6) is very active though they are not facing to major streams. Then all water derived through the cap rock sheet is presumed to make all

Table 3 Weight and score of AHP for landslide susceptibility assessment

Score List for Susceptibility Evaluation				Block No.			
	Level II	Level III	Score				
Evolution of landslide	Sharp and clearness of landslide microtopography	Clear	9				
		Slightly clear	2				
		dull	0				
	Fragmentation of primary block into sub-blocks	multiple	11				
		partly	6				
		non	1				
Erodibility and unstability of toe part	Topographic feature of landslide mass or toe part profile	steep slope where slope failure occurs	15				
		steep slope with clear nick point	8				
		convex slope with rounded nick point	2				
		linearly declined	1				
		concave slope	0				
	Erodibility of toe part of landslide mass VS river scale and slip surface	Undercut Slope/ Main stream	16				
		Undercut Slope/ Subsidary stream Slip off Slope/ Main stream	5				
		Slip surface on terrace	4				
		slip surface at high position	1				
Water conditi on	Water collectability from upper slope of	Vallley head	7				
		Linearly declined	3				
		Ridge	1				
Landcover and landuse	Land cover, artifitial change and urbanization	urbanized/forest on steep slope	5				
		densely urbanized on rarther steep slope	3				
		partly urbanized on rather steep slope	1				
		wild land or forest on rather steep slope	0				
		total weight	101				

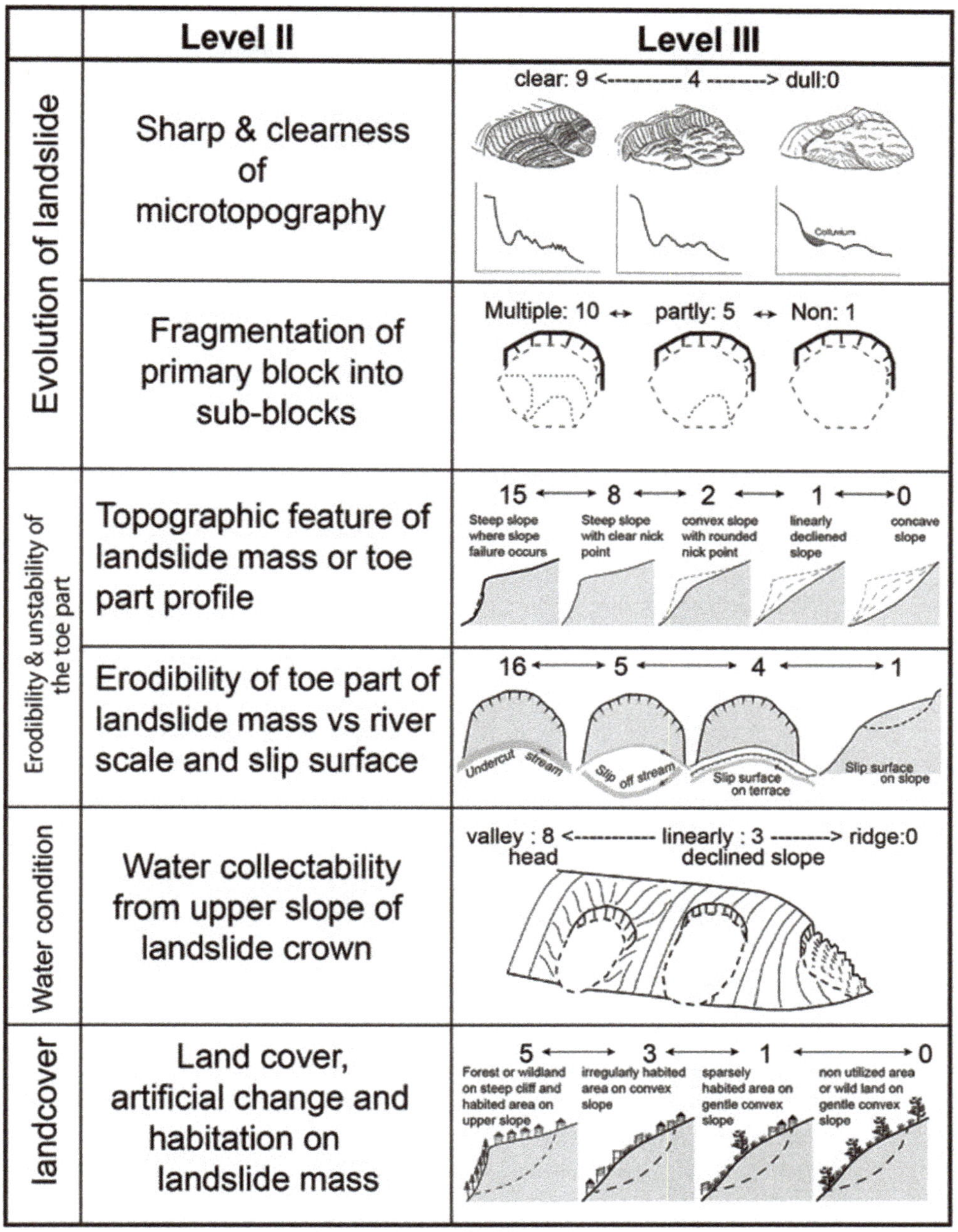

Numeral is a weight for each criteria.

Fig. 13 Schematic illustration of weighting system of Level II and III for landslides locating along the river course

landslides active. Furthermore, non-planned urbanization is proceeding on the hilly area where the capacity of living drainage is not enough. Subsequently, another set of evaluation factor is prepared for such hilly area (Table 4). Then, a geological and topographic setting is selected as an attributing factor in a new list of weighting criteria of the hierarchical Level II replacing the erodibility of the toe part of landslide mass.

According to change of the attributing factor, a series of weight and score is also alternated because relative importance among the factors is different on a new chart. They are schematically illustrated in Fig. 16.

Landslide Susceptibility Mapping of Large Scale Combining Ground Truth and Aerial Photo Interpretation Based on AHP

Middle scale landslide susceptibility mapping in scale of 1/20,000–1/50,000 using aerial photo interpretation is suitable to distinguish landslides which are active or dormant. However, it is difficult to detect a landslide block under a critical condition causing a slight movement only by aerial photo interpretation. Concrete planning or designing of prevention work for landslide blocks or installation of

Fig. 14 Berrinche landslide in Tegucigarpa (shot by H.Yagi)

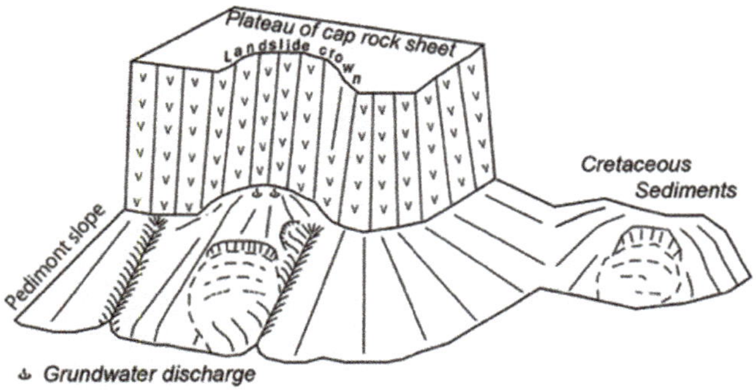

Fig. 15 Schematic geological and geomorphological model of hilly land in marginal part of Tegucigalpa city

equipments to observe initial movement of landslides should be done for critical landside blocks. To solve this matter and to make LSM more practical at the landslide site, landslide susceptibility map of 1/5,000–1/20,000 in scale combining ground truth data and aerial photo interpretation is proposed, if high resolution DEM more precise than 5 m grid or aerial photo in scale of approx. 1/10,000 in scale are available. Contour map super-imposed on orthophotograph is preferable to plot in situ ground deformation (Fig. 17). Figure 17 shows sub-block distribution of landslides at El Eden which was activated by Hurricane Mitch in 1998. Detail interpretation of large scale aerial photographs clarified that the El Eden landslide consists of several sub-blocks. White arrow in Fig. 17 shows the activated part by the torrential rain. Surrounding sub-blocks have been moving after the Hurricane Mitch. Many houses and huts are locating on the sub-blocks. Based on the large scale landslide inventory map, field investigation on recent displacement and plotting onto the map were carried out. To allocate susceptibility to each sub-block, total score of each was calculated according to the result of aerial study and ground truth.

Weighting and score system of AHP for the landslide susceptibility evaluation of the level II consists of micro topography (Micro-topo) caused by landslide movement, slope profile of landslide mass, geological strata and structure consisting of landslide mass (Geo-setting), water condition and land use and vegetation cover. Information on the micro-topography, geology, water condition such as water spring & land-cover are observed by ground truth at the site. Aerial photo interpretation can also detect the micro-topographic features such as steps, opening crack and profile of landslide mass such as bulging.

Table 5 shows a matrix of a pair-wise comparison between attributing factors of the level II above mentioned.

Table 6 shows an evaluation list by each factor. Score of evaluation for each factors of the level II is allocated the highest number among those of the level III consisting of the factors of the level II, based on the result of ground truth.

Result of the susceptibility mapping is shown in Fig. 18 (UNAH and JICA 2016), indicating that marginal blocks of EL Eden landslide are active because they are remanet of the core part shown by the white arrow that caused slush movement due to Hurricane Mitch in 1998.

Susceptibility Mapping Using Precise DEM

Interpretation of aerial photograph and AHP for susceptibility mapping deeply depend on observation skill and experience of engineers or researchers. We tried to eliminate human sense for the susceptibility mapping in a high relief mountain area, using precise DEM for numerical calculation of slope distortion. It is called Bell Shape Index as mentioned below.

Double ridges or up-hill facing scarplets distributed on mountain ridges of high relief (Figs. 19, 20 and 21) are empirically known as indicators that the mountain bodies are slowly undergoing gravitational creep deformation and distortion of rock mass has been accumulated for long geological period (Yagi 1981; Chigira 1983 etc.). They are also considered as precursory signs of large landslides.

Such micro topography on ridgetops of Japan Alps have developed since 30 ka (Fusejima 1988). That is presumably attributed to para-glacial phenomena and have intermittently developed in a time scale of ca. 10 ka. However, it is quite gradual movement. Therefore, it is very difficult to find which mountain slope is under critical condition to collapse soon. Earthquake tremor will affect stability of high relief mountains, though dense distribution of the up-hill facing scarplets do not always indicate degree of stringent to collapse in near future.

We focused on the characteristics of profile of mountain in this lecture. Up-hill facing scarplets usually occur on gentle or flat ridge tops fringed with distinct breaks of slopes with high relief. That means much loading as causative factor of deformation of mountain remains on the ridge tops.

Table 4 Matrix of a series of pair-wise comparison of attributing factors of level II for landslides in the marginal part of Tegucigalpa valley

		Sharp & clearness of landslide micro-topography	Fragmentation of primary block into sub-blocks	Topographic feature of landslide mass	G/G setting	Water collectability from upper slope of landslide	Land cover, artifitial change and unbanization	Geometric Mean	weight	Score
Evolution of landslide	Sharp & clearness of landslide micro-topography	1	1	0.333	1	1	1	0.833	0.129	13
	Fragmentation of primary block into sub-blocks	1	1	1	1	0.333	3	1.000	0.155	16
Instability of toe part	Topographic feature of landslide mass	3	1	1	3	3	3	2.080	0.323	32
Loc. Condition	Geological/topographic setting	1	1	0.333	1	1	1	0.833	0.129	13
Water condition	Water collectability from upper slope of landslide	1	3	0.333	1	1	1	1.000	0.155	16
Landcover and landuse	Land cover, artifitial change and unbanization	1	0.333	0.333	1	1	1	0.693	0.108	11
								6.438	1.000	101

1: Objective *i th* row and *j th* column are of equal importance.
3: Objective i th row is important than objective j th column.
0.333 : Objective i th row is less important than objective j th column.

Cross sections of high relief mountains sometimes show distinct contrast between sharp lower slopes steeper than 40° and gentle ridge top less than 15° and subsequently they are similar to bell-shaped profiles. We analysed the topographic features of mountain ridge in the Akaishi and Hida Ranges (the Southern and Northern Japan Alps respectively), using DEM of 10 m grid size which was made before 2011, paying attention to how the ridge top shows a top heavy profile. Then we defined Bell-Shap Index to evaluate degree of the overburdened profile as follows; area ratio of a mountain profile per a subterranean aperture area from a ridge top to a valley bottom (Fig. 22), averaging those of eight directions from the top by every grid (Fig. 23). Data mining process and a flow chart of calculation are shown in Figs. 23 and 24. In other words, the Bell-Shape Index (BSI) is defined as one kind of the convexity of ridge including effect of neighbouring grid as a whole.

Example of calculation of BSI was carried out for Akatani landslide area, Nara prefecture, Japan where was affected by torrential rain caused by Typhoon 12th in 2011 (Fig. 25). Figure 25 implies that high relief slope with high BSI are distributed around the ridge and some of them slushed down due to the heavy rain and other high BSI slopes along the ridge top remain as the landslide susceptible slope. We think Bell-Shape Index is a convenient criteria to evaluate degree of gravitational rock creep and subsequent deformation of mountain body and that it can be used for susceptibility mapping in mountainous regions which are prone to suffer torrential rain or epicentral earthquake.

Concluding Remarks

Aerial photo interpretation which make us realize the nature of landforms is a strong tool to detect landslide prone area, even though precise DEM becomes to be available at the present. Combination of such method is important to prepare landslide susceptibility mapping (LSM). Scale of LSM varies according to the objectives of the projects. LSM of small scale such as 1/200,000 is useful for nation wide planning, avoiding landslide prone area. Using AHP method is suitable for LSM of middle scale, 1/20,000–1/50,000 combined with aerial photo interpretation to nominate landslide susceptible area. Landslide susceptibility mapping of large scale, 1/5,000–1/20,000 is feasible for implementation of the landslide prevention work or installation of observation equipment, combined with the ground truth and AHP. Bell-Shape Index (BSI) can show the degree of overburdened mountain body which has been undergoing gradual gravitational deformation, using DEM of which grid size is ca. 10 m. BSI provide us the convenient criteria to evaluate the landslide susceptibility in high relief mountain region where big earthquakes are expected in near future.

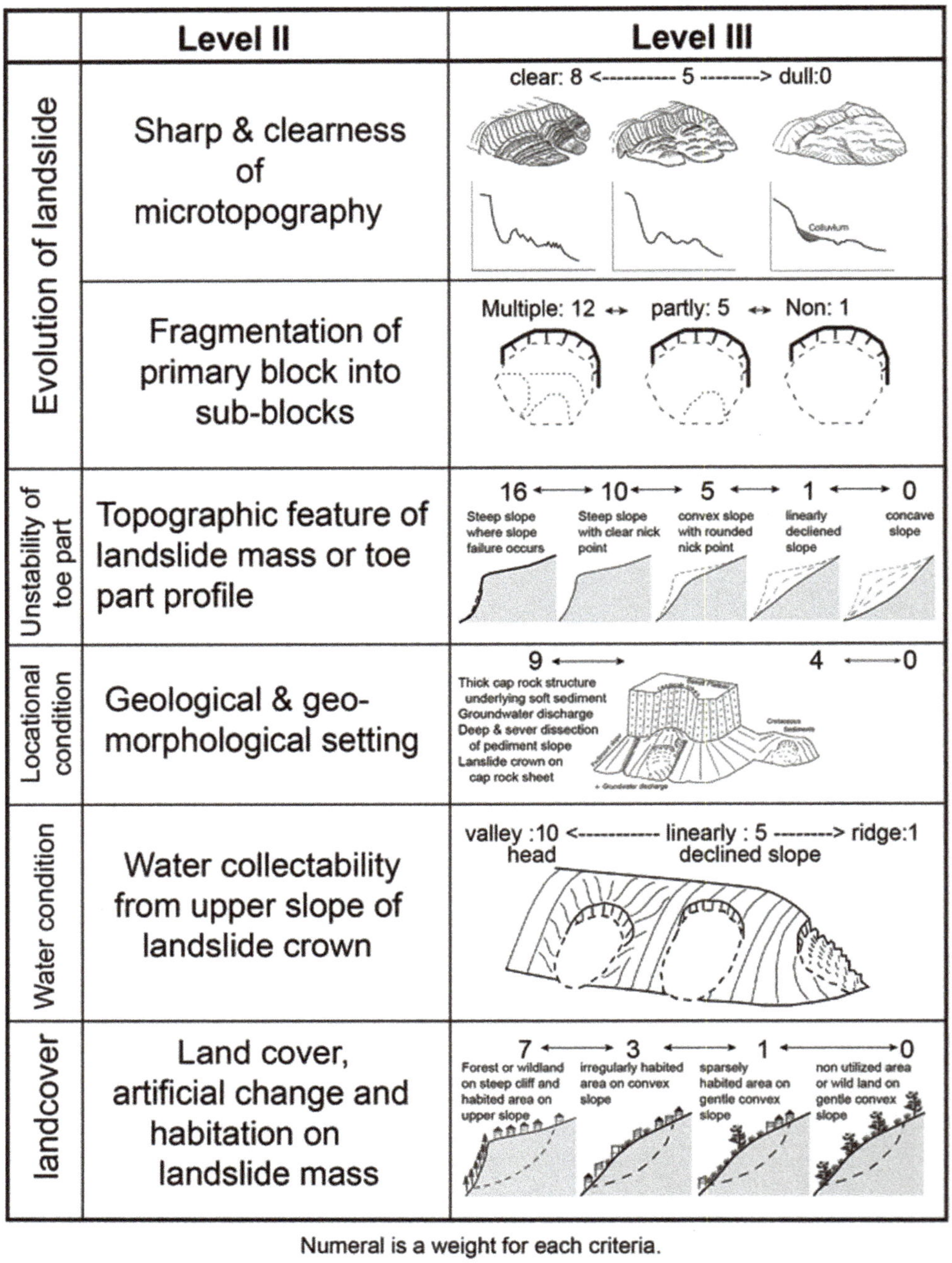

Fig. 16 Schematic illustration of weighting system for landslides locating on hilly slope in Tegucigarpa

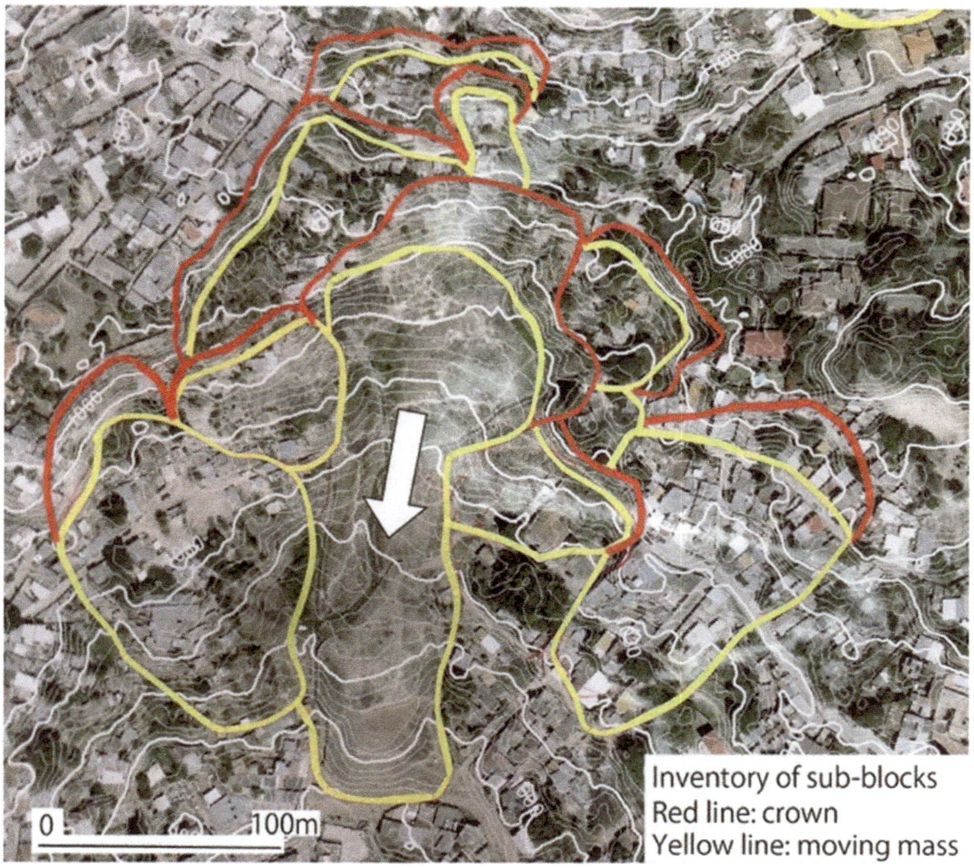

Fig. 17 Large scale landslide inventory map projected on a orthophoto map contoured by 2 m at EL Eden, Tegucigarpa

Table 5 Matrix of a series of pair–wise comparison of attributing factors of Level II for landslide assessment of large scale combining aerial photograph and ground truth

	Micro-topo	slope profile	Geo_setting	water condition	land use/ cover	Geo-Mean	Weight	Score
Micro-topo	1	1	1	1	3	1.25	0.24	24
slope profile	1	1	3	1	3	1.55	0.30	30
Geo-setting	1	0.333	1	1	1	0.80	0.15	15
water condition	1	1	1	1	1	1.00	0.19	19
land use/ cover	0.333	0.333	1	1	1	0.64	0.12	12
						5.24	1.00	100

Table 6 Weight and score of AHP for landslide susceptibility assessment of large scale

Score List for Susceptibility Evaluation				Block No.			
Level II		Level III		1	2	3	····i
Micro-topography	24	tilting of houses or ground	24				
		step on ground	20				
		crack on ground or wall	18				
		fissure on ground or wall	15				
		no signs	0				
Profile of landslide mass	30	steep slope with slope failure	30				
		steep slope with clear nick poin	20				
		convex slope	10				
		linerly decline slope	5				
		concave slope	0				
Geolo-gical setting	15	Collvium(Landslide debris)	15				
		Tertiary sediments	10				
		Hard rock	5				
Water condition	19	Spring water on slope	19				
		Spring water at the toe part	15				
		piping on the slope	15				
		natural darin of surface water	10				
Land-cover	12	sprawl urbanization	12				
		scattered spral	8				
		wild land or forest	5				
total	100						

Score of evaluation for each criterion of Level II is allocated the highest number among those of Level III consisting level II.

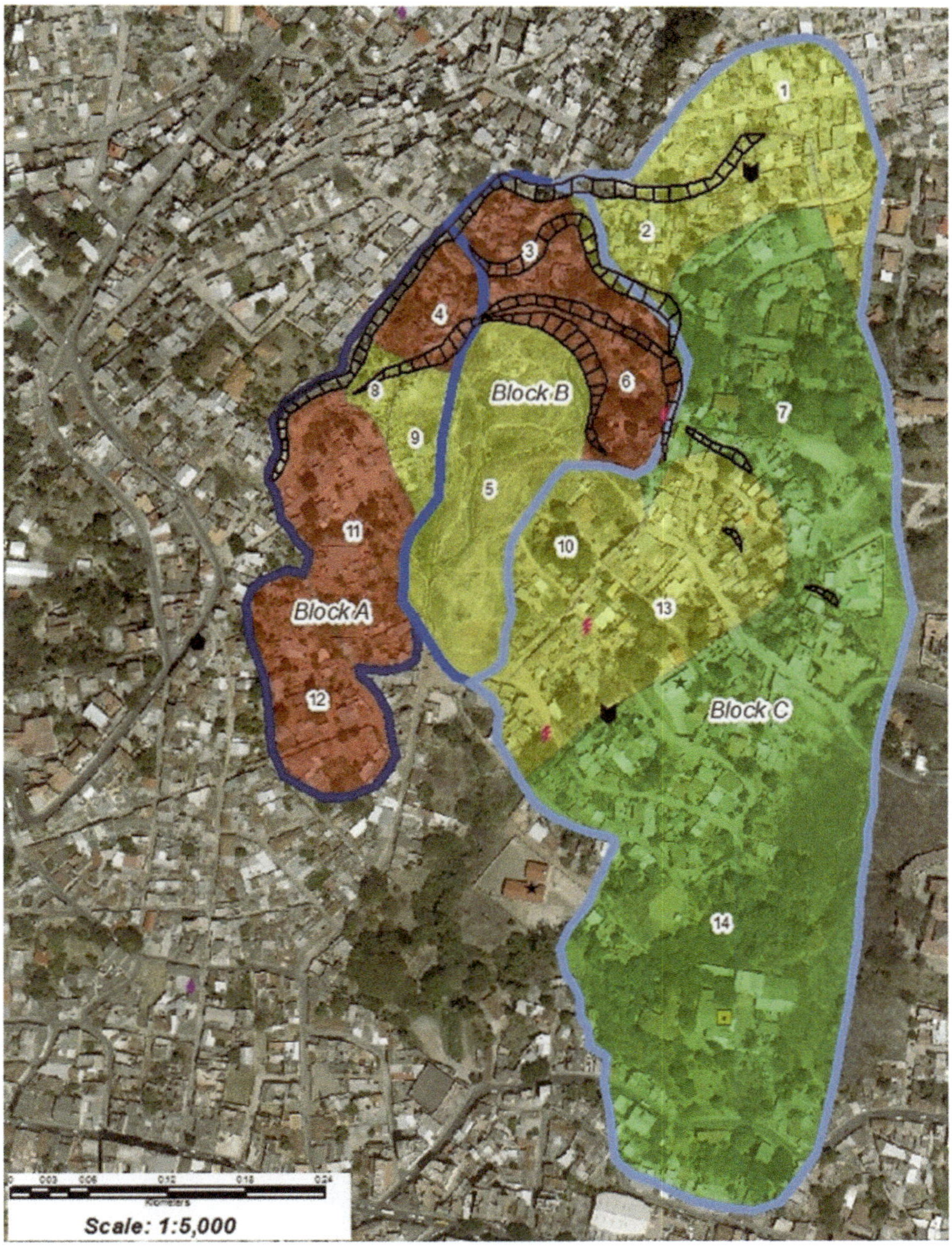

Fig. 18 Large scale landslide susceptibility map of El Eden landslide, Tegucigarpa after UNAH and JICA (2016) red colored area: active, green colored area: potential yellow colored: moderate, numerals: number of sub-block

Fig. 19 Wedge like subsidence of flat ridge top at Mt. Shichimenzan, Yamanashi, Japan (pictured by H. Yagi) It shows top heavy profile of the mountain body

Fig. 20 Multiple ridges causing subsidence of the ridge top, Mt. Kamigochi-Dake, southern Japan Alps, Shizuoka (pictured by H. Yagi)

Fig. 21 Multiple ridges formed on Hyakken daira (200 yards flat ridge top), Mt. Akaishi-dake (3121 m a.s.l.), southern Japan Alps, Shizuoka (pictured by H. Yagi)

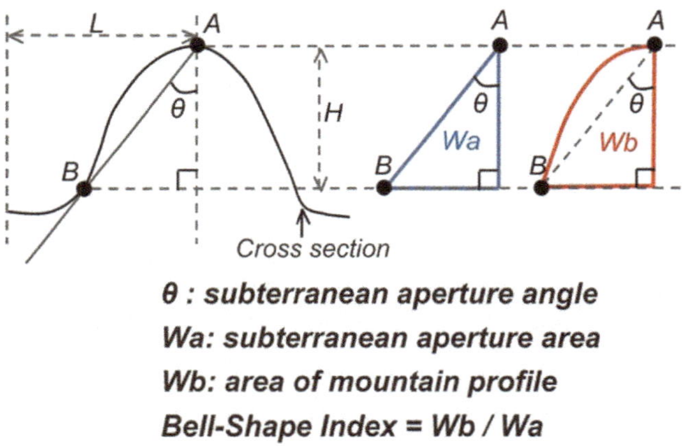

Fig. 22 Concept of *Bell-Shape Index*

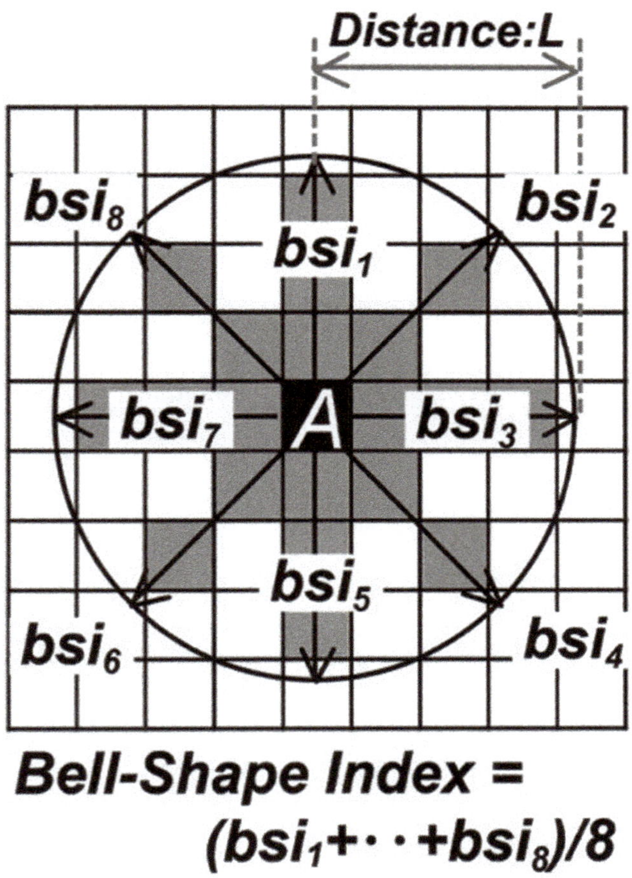

Fig. 23 Concept of data mining for BSI averaging

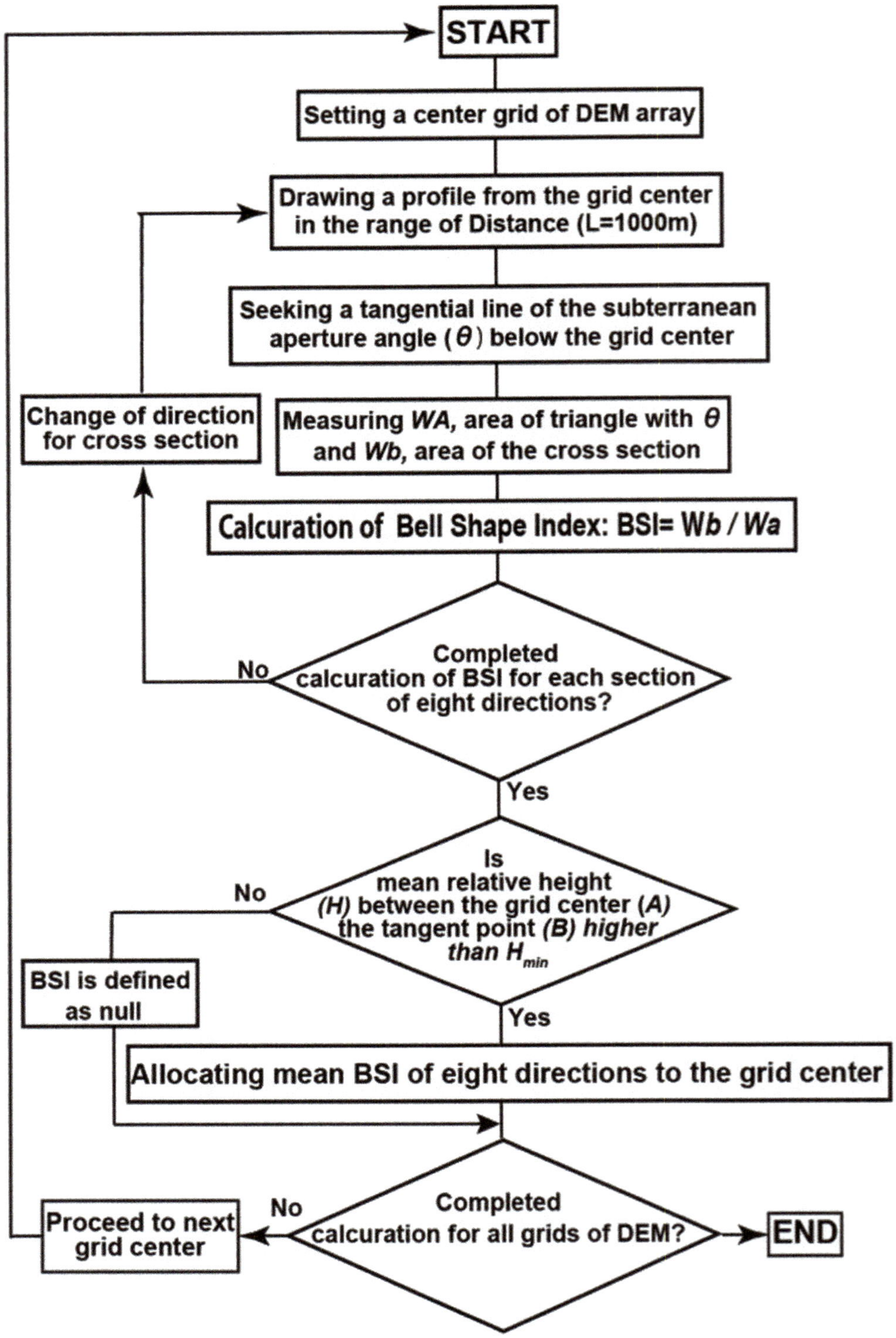

Fig. 24 Flow chart of BSI calculation

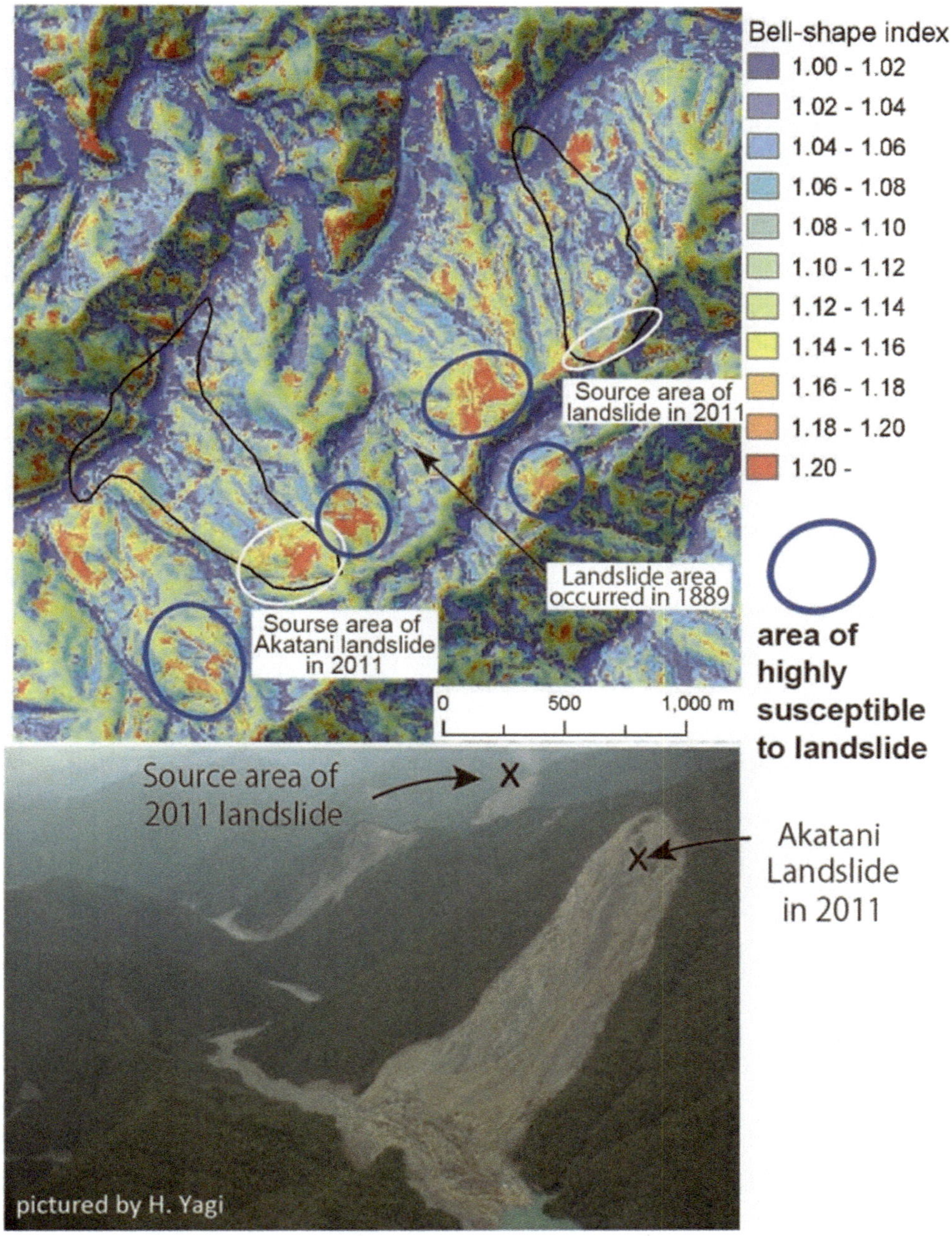

Fig. 25 Upper: BSI map in Akatani area, Nara Pref, Japan lower: bird's eye view of Akatani landslide which occurred in 2011

Acknowledgements We deeply appreciate Prof. K. Sassa giving us a chance to write this paper. We also thank Prof. H. Yamagishi for many suggestive advices. Part of this study depends much on the LSM programme in Tegucigarpa carried by JICA and Honduras organizations, eg. UPI, UNAH & CODEM from 2012–2016. H. Yagi & G. Sato of the authors participated in the programme. We express much appreciation for all personnel and organizations concerned.

References

Chigira M (1983) Long term gravitational deformation of rocks by mass rock creep. Eng Geol 32(3):157–184

Fusejima Y (1988) Mulptiple ridges developed on Mt. Chogatake, Northern Japan Alps. Meeting Abstract Japanese Geograph. (*ChiriYo*), (33): 112–113 (in Japanese)

Grupo de Mapeo de Deslizamientos de Tierra Tegucigalpa; GMDTT (2014) Manual para elaboracion de mapa de inventario de deslizamientos de tierra. Casa de aplicacion ciudad de Tegucigalpa/Grupo de Mapeo de Deslizamientos de Tierra de Tegucigalpa. Agencia de Cooperacion Internacional de Japon (JICA)/[Tegucigalpa]:[Publigraficas], p 80

Hamasaki K, Miyagi T (2018) Risk evaluation using the analytic Hierachy process (AHP)—introduction to the process concept. In: Sassa k, Tiwali B, Liu K, McSaveney M, Strom A, Setiawan H (eds) Landslide Dynamics, ISDR-ICL Landslide Interactive Teaching Tools, pp 461–474

Kienholz H (1978) Maps of geomorphology and natural hazards of Grindelward, Switzerland, scale of 1:10,000. Arct Alp Res 10:169–184

NIED (1998) Landslide maps, series 1 “Shinjo and Sakata” and explanations for the landslide maps. Technical note of the National Research Institute for Earth Science and disaster prevention No. 69 (in Japanese and English abstract)

Saaty TL (1980) The analytic hierarchy process. McGraw-Hill Book Company, New York, NY, p 265

Soeters R, Westen CJ (1996) Slope instability recognition, analysis and zonation. In: Turner AK, Schuster RL (eds) Landslides: investigation and mitigation, transportation research board, national research council, pp 129–177. National Academy Press, Washington D.C.

Yagi H (1981) Origin of uphill-facing scarplets distributed on the high mountain slopes of Japan. Geograph Rev Japan 54(5):272–280 ((in Japanese and English abstract))

Yagi H (2016) Landslide susceptibility mapping adopting AHP method. In: Special issue of 2nd central America and Caribbean landslide Congress, pp 177–182. https://iplhq.org/icl/wp-content/uploads/2018/02/Proceeding-of-the-2nd-Central-America-and-Caribbean-Landslide-Congress-in-English.pdf

Yagi H, Higaki D, Japan Landslide Society (2009) Methodological study on landslide hazard assessment by interpretation of aerial photographs combined with AHP in the middle course area Agano River, Central Japan. J Japan Landslide Soc 45(5):8–16 ((in Japanese and English abstract))

Yagi H, Hayashi K, Higaki D, Tsou C, Sato G (2018) Dormant landslides distributed in upper course of Sun Kosi watershed and landslides induced by Nepal Gorkha earthquake 2015. J Nepal Geol Soc 55(Special Issue):61–67

UNAH-JICA (2016) El Eden landslides' susceptibility map. Project of strenghtening landslide controle and mitigation capacities.

Part II

Landslide Recognition and Mapping

New Landslide Inventory Map of the Sudetes Mountains (South-Western Poland)

2020 Kyoto Japan

Rafał Sikora and Tomasz Wojciechowski

Abstract

The paper presents the new landslide inventory map in the Sudetes Mountains (NE part of Bohemian Massif) in south-western Poland. The inventory is based on the remote sensing method—analysis of the digital elevation model (from LiDAR data) and topographic maps. The information from published papers, geological maps and field works were also used. The results contain a characterization of the study area by the number of landslides and index of landslide occurrence for each mesoregion. Moreover, a preliminary analysis of landslide distribution in relation to the geology and major tectonic zones of the Sudetes was made.

Keywords

Landslide • LiDAR • Lower Silesia • Bohemian Massif

Introduction

In the second decade of the twenty-first century it was possible to use the digital elevation model performed on Light Detection and Ranging (LiDAR-DEM) to identify landslides in Poland. Application of the DEM to landslide investigation is most advanced in the Polish part of the Carpathians. In the Sudetes, south-western Poland, this method was used to investigate landslides only in selected areas. The paper presents the first LiDAR-DEM-based landslide inventory map for the entire Sudetes.

Our study of occurrence of landslides in the Polish part of the Sudetes, including their regional distribution analysis, was undertaken to plan the implementation of the Landslide Counteracting System project (SOPO in Polish) in this area. A part of the project is currently compiled as the *Map of Landslides and Mass Movements Hazard Terrains* (*MOTZ* in Polish), which is based on detailed mapping of landslides at the scale 1:10,000. The results of the research presented in this paper will be utilized for setting the scope and the schedule of the SOPO project.

Landslide Maps in the Polish Part of the Sudetes

In contrast to the Polish segment of the Carpathian Mountains, were tens of thousands of landslides have been documented (Wójcik et al. 2020), the number of landslides in the Sudetes is underestimated. Moreover, there was much less interest and low awareness of the landslide problem before the start of the SOPO project. In addition, this unconsciousness occurred among both locals and many researchers.

Various authors gave different estimates of the number of landslides in the Sudetes. According to Bażyński and Küchn (1971), there were 141 landslides. Lemberger (2005) suggest occurrence of only 24 landslides in the whole Sudetes, although in some areas about 25 have been identified (Synowiec 2003). On the archival map, Grabowski (2008) marked 174 landslides in the Sudetes. Consequently, the Sudetes were thought to be poorly affected by landslides. Furthermore, the ELSUS data (Wilde et al. 2018) shows that the Sudetes area is classified mainly as moderately or low susceptible to landslides. According to ELSUS data, only the uppermost parts of the mountains are highly susceptible for landslides. The first calculations of landslide susceptibility in the Sudetes were made only for some areas (Kasprzak and Traczyk 2012; Sikora et al. 2017). By the latest research

R. Sikora (✉) · T. Wojciechowski
Polish Geological Institute—National Research Institute, Geohazards Center, Skrzatów 1, 31-560 Cracow, Poland
e-mail: rafal.sikora@pgi.gov.pl

T. Wojciechowski
e-mail: tomasz.wojciechowski@pgi.gov.pl

F. Guzzetti et al. (eds.), *Understanding and Reducing Landslide Disaster Risk*,
ICL Contribution to Landslide Disaster Risk Reduction,
https://doi.org/10.1007/978-3-030-60227-7_4

(Wojciechowski 2019) the Sudetes are a second-ranked region among areas susceptible to landslides. Due to the estimated low number of landslides, no broad and comprehensive study of mass movements problem has been undertaken in the Sudetes.

Our previous study detected a total of 2444 landslides and landslide groups, based on new geospatial data (Sikora and Wojciechowski 2019). These landslides were recognized on natural slopes or are indicated by a human in mining areas. The following sections also present a preliminary analysis of geological factors regarding the distribution of landslides in the Sudetes on natural slopes.

Study Area

Geographical Settings

The Sudetes constitute the NE edge of the Bohemian Massif. The research area covers the Polish part of the ridge of the Sudetes Mountains (Fig. 1). The area is 6397 km^2 and stretches between SW Poland's border and the mountain front related to the Sudetic Marginal Fault (also known as the Sudetic Boundary Fault). The Sudetes are the second most extended mountain range in Poland. They can be classified as a medium-altitude range, half of which reaches a level of 400–600 a.s.l. (Placek 2011). There are 25 mesoregions distinguished within four macroregions of the Sudetes (Sołon et al. 2018).

Geological Settings

Geologically, the area of the Sudetes is characterized by a map-view "mosaic-like" geological structure, where rock formations of various ages, ranging from Neoproterozoic to Cenozoic, are exposed. The pre-Permian rocks are in most part affected by Variscan low- to high-grade metamorphism, igneous intrusions, folding and thrusting (e.g. Aleksandrowski and Mazur 2002; Żelaźniewicz and Aleksandrowski 2008). They constitute the Variscan basement in areas overlain by remains of the Permo-Mesozoic cover, heavily faulted and, locally, folded at the end of Cretaceous times (e.g. Oberc 1977). The Cenozoic sedimentary cover with sporadic basaltic intrusions occurs mainly in the western part of the Sudetes.

Material and Methods

The identification of landslides was achieved through a detailed analysis (1:5,000 scale) of the LiDAR-DEM, 1:10,000 scale topographical charts, and the orthophotomap. The LiDAR-DEM and topographical data were available in the *National Geodetic and Cartographic Data Base of Poland* (PZGiK in Polish).

The LiDAR-DEM display parameters (vertical exaggeration, lighting, hill shading, shader colors) were customized considering the objectives of the research and the illumination direction modified to suit the strike of the ridges and river valleys. In some areas, where many landslides were too small to be analyzed individually within the assumed scale, the landslides were combined into groups. Moreover, due to the assumed scale of the analysis, it was also not possible to identify very small forms, especially small-sized debris flow and earth flow.

Landslide distribution analysis concerning lithology and tectonic zones of the geological basement was based on the regional maps by Sawicki (1995) and Cymerman (2004).

Results

Comparison with Archival Landslide Maps

The analysis confirmed the existence of only 36 landslides out all landslides, which were marked on the map compiled by Grabowski (2008). Hundred and twenty-three landslides have not been confirmed on the LiDAR-DEM, and 17 require verification of location data. All landslides described in publications over the last decade have been verified positively.

Landslides in Mining Areas

Out of 2444 landslides and landslides groups recognized in the Sudetes, 707 (28.92%) were identified in different mining areas. Most of the landslides were induced by human activity in the Turów Mining Area (Żytawa Basin). The morphology of excavation slopes and pile slopes are very variable over time, and in this case, the LiDAR-DEM does not show the current topography. Despite this, our analysis shows a huge impact of human mining activity on landslides triggering.

Updated Landslide Inventory Map

Finally, our analysis has been limited to natural slopes of the Sudetes which are covered by 1737 landslides (Fig. 1). Among all landslides 80.90% (1405 in total) do not exceed 1 ha, 14% (243) are in the range of 1–5 ha, 3.5% (61) landslides are in the range of 5–10 ha, and only 1.6% (28) landslides are larger than 10 ha (Fig. 2). The inventory map shows the distribution of landslides across the Sudetes and enables the identification of areas most affected by them.

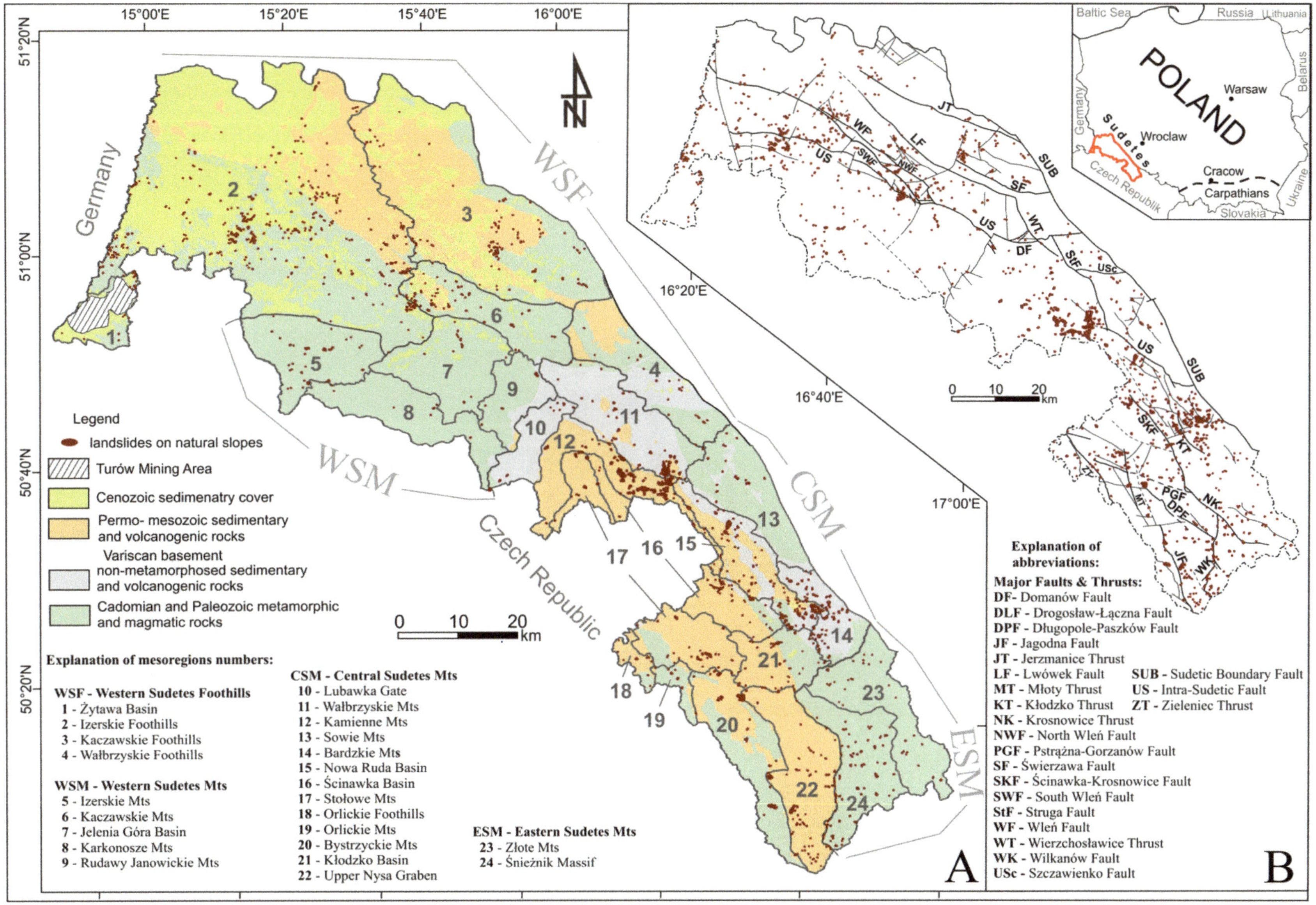

Fig. 1 Landslide inventory map for natural slopes of the Sudetes Mts. **a**—distribution map of landslides by geographical division and lithology of basement. **b**—Distribution of landslides versus major tectonic zones

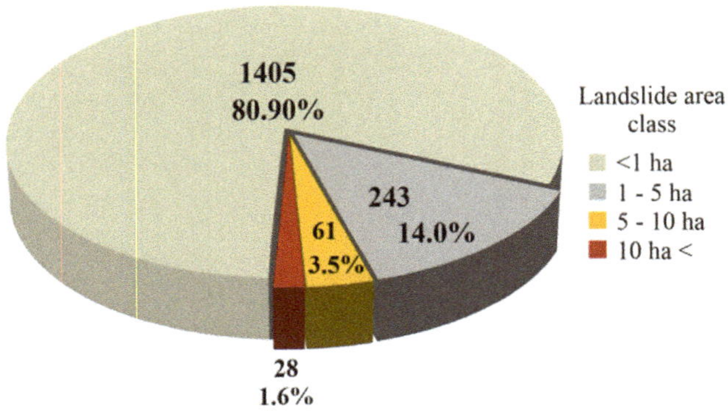

Fig. 2 Landslide size distribution on natural slopes of the Sudetes

Noticeable is a considerable in the number of landslides between the individual geographical units (mesoregions). The numbers of landslides vary from 5 (Rudawy Janowickie Mts) to 495 (Izerskie Foothills) per unit (Fig. 3). However, the analysis that is based on their total area per unit is more reasonable (Fig. 4). In this respect, the Kamienne Mountains (4.62 km^2), Nowa Ruda Basin (2.33 km^2), Izerskie Foothills (2.25 km^2), Bystrzyckie Mountains (1.93 km^2) and Wałbrzyskie Mountains (1.62 km^2) are most affected by the landslides. This aspect is better described by the index of landslide occurrence (ILO) calculated for each geographical unit (Bober 1984), i.e. the ratio between the total area of landslides to the total area of mesoregion (Fig. 4). The highest ILO was calculated for the regions of the Kamienne Mountains (2.46%), Nowa Ruda Basin (1.22%), Bystrzyckie Mountains (0.84%) and Bardzkie Mountains (0.81%).

Geological Factors of Landslides Occurrence

A preliminary evaluation suggests an important role of geological factors in landslide development. The distribution of landslides on the inventory map closely relates to the variability of the geological basement (Fig. 1a). The slopes of the Kamienne Mts, Nowa Ruda Basin, Wałbrzyskie Mts and Bystrzyckie Mts, which are most affected by landslides (including the largest ones), are composed of sedimentary or volcanogenic rocks of the Late Carboniferous—Mesozoic succession of the Intra-Sudetic Synclinorium and Upper Nysa Kłodzka Graben. The slopes of the Bardzkie Mts and the southern part of the Nowa Ruda Basin are built of strongly deformed and cataclastic Lower Carboniferous rocks (Kulm facies) of the Bardo Structural Unit. On the other hand, landslides are rare in the areas where the

Fig. 3 Landslides distribution by geographical division

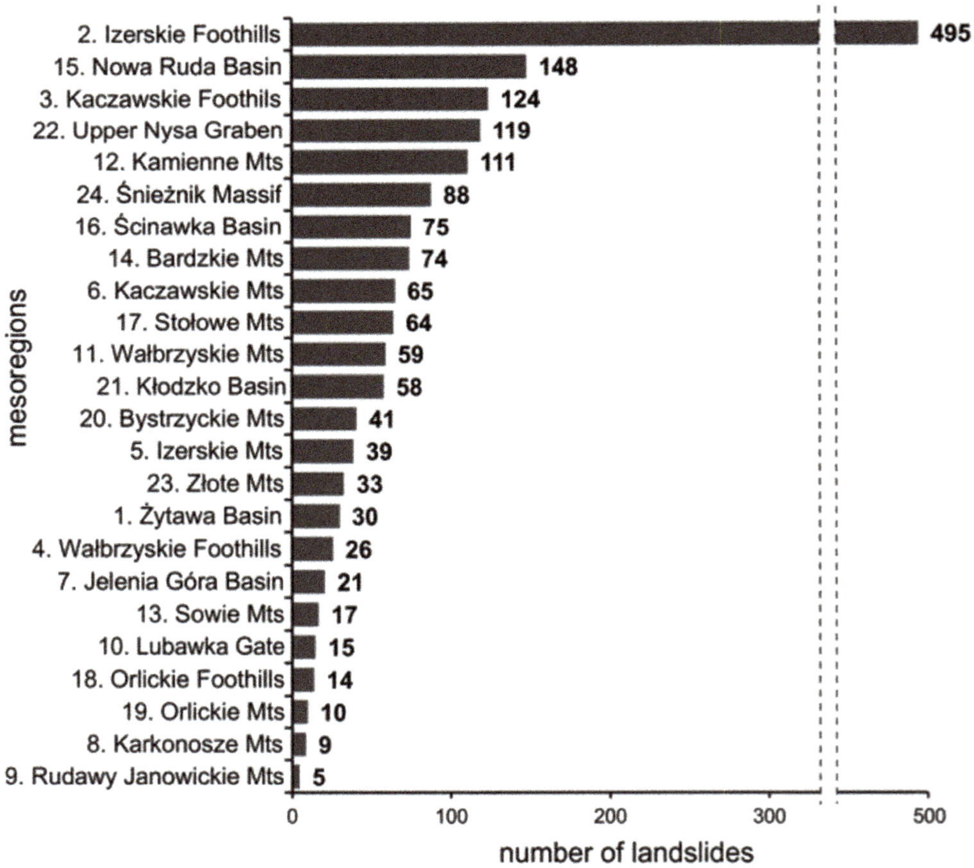

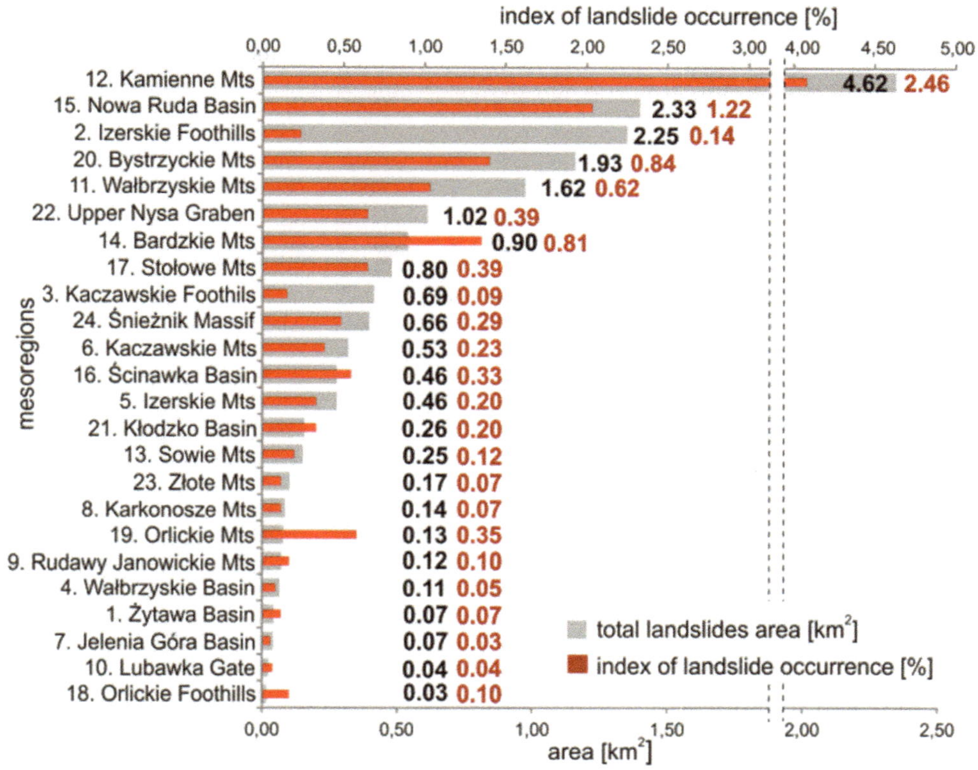

Fig. 4 Total landslide area vs ILO index

underlying Variscan basement is represented by metamorphic or magmatic rocks.

Numerous, generally small landslides in the western part of the Sudetes (Izerskie Foothills) are related to the Cenozoic sedimentary cover. Most of these landslides developed within Pleistocene glaciogenic rocks. The analyzed map shows that many landslides are located near major faults or thrust zones (e. g. Intra Sudetic Fault, Wilkanów Fault, Ścinawka-Krosnowice Fault, Jagodna Fault; Fig. 1b).

Discussion

The new inventory map shows that the Sudetes are more affected by landslides than previously studied. There are over 10 times more than estimated from an earlier report (Grabowski 2008). The discovery of more landslides was possible due to analyzing of the high-resolution DEM from LiDAR data that has not been used in the regional-scale mapping of landslides in the whole Sudetes. Moreover, their distribution is very variable, and large concentration of landslides are limited to specific areas. Our results have shown even more landslides in the Kamienne Mts, Wałbrzyskie Mts and Bystrzyckie Mts than indicated by Migoń et al. (2014, 2017), Osika et al. (2018) and Różycka et al. (2015). We have also confirmed earlier data on the large number of landslides in the Bardzkie Mts found in our previous works (Sikora et al. 2017). The occurrence of landslides in the western part of the Sudetes (e. g. Izerskie Foothills) had not been investigated previously. Despite the identification of a much larger number of landslides than previously recognized, the Sudetes should be generally considered as a mountain range poorly covered by landslides. This fact is confirmed by the ILO, calculated, to be in the order of magnitude lower than for the nearby Carpathians (Kaczorowski and Kułak 2019).

The preliminary evaluation of landslide distribution suggests that the geological factor is significant for mass movements development in the whole Sudetes. Furthermore, the relief in the Sudetes is controlled by lithological and structural factors of bedrock geology (Migoń and Palcek 2014). Several authors demonstrated the impact of the geological factor on the occurrence of landslides in a different part of Sudetes (e. g. Duszyński et al. 2017; Kowalski 2017; Migoń et al. 2017; Sikora et al. 2017). In this aspect, the development of landslides, controlled by the basement structure, is interesting and was discussed in some articles (e. g. Kasprzak et al. 2016; Kowalski 2017; Sikora et al. 2016; Sikora and Piotrowski 2017; Kowalski et al. 2019). A relationship between landslides scarps and tectonic discontinuity has been confirmed in many cases by field works in the Bardzkie Mts (Sikora and Piotrowski 2017).

Many faults are visible as morphological escarpments which are prone to landslides (e.g. Jagodna Fault, Wilkanów Fault), but some faults zones are not visible in the topography of study area (e.g. Intra-Sudetic Fault).

Seismic activity and earthquakes are presently rare in the Sudetes. Since the fifteenth century, there have been 16 earthquakes reported. Most of them were of 4 to 5 in the EMS scale, and only four were of 6–7 (Guterch, 2009). Some landslides in the Sudetes were probably induced by more intense earthquakes, but this problem requires detailed analysis. In our study, the contemporary effect of the tectonic zones on the formation and activity of the landslides is of passive character. The rocks present near the fault zones are intensely fractured and prone to mass movements. Unfortunately, cataclastic zones are often inadequately examined and documented. After all, the lack of landslides on the morphological edge of the Sudetic Marginal Fault is surprising.

The presented results were used to develop the Landslide Susceptibility Map of the Sudetes (Sikora and Wojciechowski 2019). The areas that are most affected by landslides are confirmed as most vulnerable for mass movements. This map shows that 38% of the Polish part of the Sudetes is susceptible for landslides. Only about 8% of the study area is classified as highly or very highly susceptible.

Conclusions

The new inventory map shows the spatial distribution of 1737 landslides on natural slopes of the Polish part of the Sudetes. This map is the result of the first regional-scale study of the LiDAR-DEM for the identification of mass movements identification in south-western Poland. With the exception of some areas, the Sudetes are poorly covered by landslides. The Susceptibility Map of Poland (Wojciechowski 2019) shows them as one of most susceptible area in Poland. On the other hand, high and very high susceptibility was calculated only for selected areas.

We present a preliminary analysis on geological conditions of landslide distribution in the study area. This problem should be investigated in detail in the future. Moreover, the presented map can be useful to study the relation of mass movements to different controlling factors in the Sudetes on a regional scale.

In the SOPO project, the presented map can be used to plan the detailed mapping of landslides and to evaluate preliminarily the landslide hazards for buildings and roads in the Sudetes.

In particular, the occurrence of landslides in mining areas in the Sudetes should be studied in the future. Determination of the dynamics and the scale of landslide formation in this area is crucial for ensuring the safety of exploitation and later reclamation of the area.

References

Aleksandrowski P, Mazur S, (2002) Collage tectonics in the northeasternmost part of the Variscan Belt: the Sudetes, Bohemian Massif. In: Winchester J, Pharaoh T, Verniers J (eds) Paleozoic Amalgamation of Central Europe, vol 201, pp 237–277. Geological Society, London

Bażyński J, Kühn A (1971) Landslides registration in Poland. Instytut Geologiczny, Warszawa (in Polish)

Bober L (1984) Landslides areas in the Polish Flysh Carpathians and Their connection with the geological structure of the region. Biuletyn Instytutu Geologicznego 340: 115–158 (in Polish, with English summary)

Cymerman Z (2004) Tectonic map of the Sudetes and the Fore-Sudetic Block: 1:200 000. PIG–PIB. Warszawa (ISBN_ 8373727388_)

Duszyński F, Jancewicz K, Kasprzak M, Migoń P (2017) The role of landslides in downslope transport of caprock-derived boulders in sedimentary tablelands, Stołowe Mts, SW Poland. Geomorphology 295:84–101

Grabowski D (2008) Outline map of landslides and areas prone to mass movements, scale 1:50 000 in Poland without Carpathians. PIG-PIB, Warszawa (in Polish)

Guterch B (2009) Seismicity in Poland in the light of historical records. Przegląd Geologiczny 57(6):513–520 (in Polish, with English abstract)

Kaczorowski J, Kułak M (2019) Landslides in the Polish Carpathians in statistical terms on the basis of the SOPO project. In: II Polish Landslide Conference, pp 14–17. Wieliczka, Poland (in Polish)

Kasprzak M, Traczyk A (2012) Conditioning of landslides development in central part of the Kamienne mountains. Landform Anal 20:65–77

Kasprzak M, Duszyński F, Jancewicz K, Michniewicz A, Różycka M, Migoń P (2016) The Rogowiec landslide complex (central Sudetes, SW Poland)—a case of a collapsed mountain. Geol Quarterly 60 (3):695–713

Kowalski A (2017) Distribution and origin of landslide forms in the Bóbr river valley near Wleń (Western Sudetes). Przegląd Geologiczny 65(10):629–641 (in Polish, with English abstract)

Kowalski A, Kasza D, Wajs J (2019) Structural control of mass movements on slopes formed of magmatic and metamotphic rocks: the case study of Wielisławka Mt. (SW Poland, Sudetes Mts.). Geol Quarterly 63(3):460–477

Lemberger M (ed) (2005) Registration and inventory of natural hazards (especially landslides and other geodynamic phenomena) throughout the country. National Geological Archives, Warszawa (in Polish)

Migoń P, Placek A (2014) Lithological and structural control on the relief of the Sudetes. Przegląd Geologiczny 64:36–43 (in Polish, with English abstract)

Migoń P, Jancewicz K, Kasprzak M (2014) The extend of landslide-affected areas in the Kamienne mountains (Middle Sudetes)—a comparision of geological maps and LiDAR based digital elevation model. Przegląd Geologiczny 64:463–471 (in Polish, with English abstract)

Migoń P, Jancewicz K, Różycka M, Duszyński F, Kasprzak M (2017) Large-scale slope remodelling by landslides—geomorphic diversity and geological controls, Kamienne Mts Central Europe. Geomorphology 289:134–151

Oberc J (1977) The late Alpine epoch in south-west Poland. In: Książkiewicz M,. Oberc J, Pożaryski W (eds) Geology of Poland, 4, Tectonics, pp 451–475. Wydawnictwa Geologiczne, Warszawa

Osika A, Wistuba M, Malik I (2018) Relief evolution of landslide slopes in the Kamienne Mts (Central Sudetes, Poland)—analysis of high-resolution DEM from airbone LiDAR. Contemp Trends Geosci 7:1–20

Placek A (2011) Structural morphology of the Sudetes in the light of rock strength measurements and DEM analysis. Rozprawy Naukowe Instytutu Geografii i Rozwoju Regionalnego 16. Uniwersytet Wrocławski, p 190. (ISBN _9788362673032_) (in Polish)

Różycka M, Michniewicz A, Migoń P, Kasprzak M (2015) Identification and morphometric properties of landslides in the Bystrzyckie Mountains (Sudetes, SW Poland) based on data derived from airborne LiDAR. Geomorphometry Geosci 1:247–250

Sawicki L (1995) Geological map of Lower Silesia with adjacent Czech and German territories (without Quaternary deposits): 1:100 000 PIG, Warszawa

Sikora R, Kowalski A, Piotrowski A (2016) Implications of landslides development and basament geology diversity along the Bóbr Valley in the Izerskie and Kaczawskie Highlands area (Western Sudetes). In: Wojewoda J (ed) 3rd Polish Geological Congress "Challenges of Polish Geology", pp. 348–350. PTG, Wrocław (ISBN_9788394230425_) (in Polish)

Sikora R, Kowalski A, Badura J, Gotowała R, Piotrowski A, Różański P, Urbański K (2016) Selected landslides of the Lower Silesia and their relationships with geological structure. In: Wojewoda J, Kowalski A (eds) 3rd Polish Geological Congress "Challenges of Polish Geology". Field trip guide, pp 44–60. PTG, Wrocław (ISBN_9788394230432_) (in Polish)

Sikora R, Wojciechowski T, Tomaszczyk M, Piotrowski A (2017) Geological condition of landslides occurrence in the Bardzkie Mountains and adjacent areas (Sudetes, SW Poland). WLF4 Local Proceedings with Programme, p 222. Ljubljana, Slovenia

Sikora R, Piotrowski A (2017) Relationships of selected landslide forms with structure of the basement in the ravined Nysa Kłodzka river valley in the Bardzkie Moutains. In: Abstracts of the 21th Meeting of Polish Geomorphologist, p 137. Warszawa (in Polish)

Sikora R, Wojciechowski T (2019) Landslides in the Sudetes. Przegląd Geologiczny 67(5):360–368 (in Polish, with English abstract)

Solon J, Borzyszkowski J, Bidłasik M et al (2018) Physico-geographical mesoregions of Poland: verification and adjustment of boundaries on the basis of contemporary spatial data. Geographica Polonica 2(91):143–170

Synowiec G (2003) Structural landslides in the Kamienne Góry Mts., Sudetes, SW Poland. In: Rybař J, Stemberk J, Wagner G (eds) Landslides Swets and Zeitlinger Lisse, pp 311–314

Wilde M, Günther A, Reichenbach P, Malet J-P, Hervás J (2018) Pan-European landslide susceptibility mapping: ELSUS Version 2. J Maps 14(2):97–104

Wojciechowski T (2019) Landslide susceptibility of Poland. Przegląd Geologiczny 67(5):320–325 (in Polish, with English abstract)

Wójcik A, Wojciechowski T, Wódka M, Kamieniarz S, Kaczorowski J, Sikora R, Kułak M, Karwacki K, Warmuz B, Perski Z (2020) Development of landslide research in Polish Geological Institute. Przegląd Geologiczny 68(5):356–363 (in Polish, with English abstract)

Żelaźniewicz A, Aleksandrowski P, (2008) Tectonic subdivision of Poland: southwestern Poland. Przegląd Geologiczny 56(10):904–2011 (in Polish, with English abstract)

Gullies as Landforms for Landslide Initiation —Examples from the Dubračina River Basin (Croatia)

Petra Jagodnik, Vedran Jagodnik, Željko Arbanas, and Snježana Mihalić Arbanas

Abstract

This paper presents the general characteristics and spatial distribution of the gullies formed in flysch deposits in the Dubračina River Basin, and the associated landslide phenomena initiated in gullies, based on the visual interpretation of 1 × 1 m LiDAR imagery. Two types of landslides are identified: debris slides and debris slide-debris flows. Specific topographic locations of landslides have influenced the possibilities for landslide identification and mapping on the LiDAR imagery, as it is presented for the representative examples of identified landslides located within one of the largest gullies identified in the study area.

Keywords

Gully erosion • Historical landslide inventory • Historical gully inventory • LiDAR derivatives • Visual interpretation • Dubračina River Basin • Vinodol Valley

P. Jagodnik (✉)
Faculty of Civil Engineering, University of Rijeka, Radmile Matejčić 3, 51000 Rijeka, Croatia
e-mail: petra.jagodnik@gradri.uniri.hr

V. Jagodnik · Ž. Arbanas
Faculty of Civil Engineering, University of Rijeka, 51000 Rijeka, Croatia
e-mail: vedran.jagodnik@gradri.uniri.hr

Ž. Arbanas
e-mail: zeljko.arbanas@gradri.uniri.hr

S. M. Arbanas
Faculty of Mining, Geology and Petroleum Engineering, University of Zagreb, 10000 Zagreb, Croatia
e-mail: snjezana.mihalic@rgn.unizg.hr

F. Guzzetti et al. (eds.), *Understanding and Reducing Landslide Disaster Risk*, ICL Contribution to Landslide Disaster Risk Reduction,
https://doi.org/10.1007/978-3-030-60227-7_5

Introduction

Gully erosion is one of the major causes of land degradation (Valentin et al. 2005). Formation of the permanent gullies is particularly considered as an important factor in environmental changes and landscape evolution, producing more sediment loss than other forms of soil erosion (Martínez-Casasnovas et al. 2003). Gullies may form at any break of slope or vegetation cover when the geological substrate is mechanically weak; varying from different types of superficial deposits to weak sedimentary rocks (Selby 1993). High sediment amounts discharged from gullies can further be transported to river systems and water reservoirs (Poesen et al. 2003).

The morphological evolution of gully landforms and thus their more destructive impact on an environment are accelerated by initiation of landslides in the gully channel walls (Desta and Adugna 2012). Different types of movement (e.g., sliding, falling or even toppling) determine the way in which the gully channel geometry is modified (Ezechi and Okagbue 1989), as well as the total sediment amount transported from a catchment (Borelli et al. 2015). The destructive impact of associated erosional and gravitational processes will continue until the gully is stabilized by the effective control measures (Poesen 1993). Thus, determination of the spatial distribution and general characteristics of gully erosion phenomena becomes an important element for making rational decisions when managing hazardous processes in an area (Poesen et al. 2003), in particular because the erosion processes are one of the main landslide preparatory causal factors (Popescu 2002). A key step for this purpose is the production of a landform inventory (e.g., Conforti et al. 2013; Santangelo et al. 2015; Bernat Gazibara et al. 2019), which provide information on the types and locations of individual hazardous phenomena.

The Dubračina River Basin (43.22 km^2), situated in the Vinodol Valley (64.57 km^2) in the NW coastal part of the Republic of Croatia, is specific by the active geomorphological processes influencing the overall topography. This area is rural, with numerous small settlements being connected with dense road network, and it has a long tradition of agricultural activities. Material damages on private and public properties caused by landslides and gully erosion are often significant and continuously cause economic losses. Other important hazardous phenomena in this area are the periodical transports of the large sediment amounts discharged from the tributaries to the Dubračina riverbed, which are further being transported to the river mouth into the Adriatic Sea.

For the Vinodol Valley, the historical landslide inventory (Guzzetti et al. 2012), as well as the historical erosion inventory, are produced based on the visual interpretation of 1-m digital terrain model (DTM) computed from the airborne LiDAR (Light Detection and Ranging) data, in the previous author's work (Đomlija 2018). Numerous gullies of diverse sizes and geometries have been identified in the flysch bedrock and superficial deposits, as well as more than 600 hundred landslides. Given that landslides were already previously known to be associated with erosion processes (e.g., Jurak et al. 2005), but not exactly to what extent, the preparation of the geomorphological inventories enabled an analysis of the interrelation and spatial distribution of landslides and their types with respect to the gully landforms in the study area.

This paper in brief presents: (i) the general characteristics and spatial distribution of gullies formed in flysch deposits in the Dubračina River Basin; (ii) the landslide types (Hungr et al. 2014) initiated within the gully landforms; (iii) the locations, dimensions and spatial extent of individual landslides along the gully channel walls; and (iv) the possibilities for identification and mapping of landslides initiating in gullies based on the visual interpretation of high resolution (HR) LiDAR imagery, for the representative examples of different landslide types.

Study Area

The Dubračina River basin (43.22 km^2) encompasses the north-western and the central parts of the Vinodol Valley (64.57 km^2) (Đomlija 2018). It is about 15 km long and 1–4 km wide. It has an elongated shape (Fig. 1a), which is narrower in the north-western (NW) Basin part from Križišće to Pećca settlements, and wider in the south-eastern (SE) Basin part from Pećca to Kričina settlements. The inner Basin area is surrounded by the relatively steep slopes and cliffs (Fig. 1a).

Elevations range from 0 to 923 m a.s.l., with 44% of the area being lower than 200 m a.s.l. (Bernat et al. 2014). The prevailing slope angles (58%) are between 5° and 20°.

The Basin flanks are built of carbonate rocks, while the lower parts and the bottom of the Basin are built of siliciclastic rocks, i.e., flysch deposits (Blašković 1999). Flysch bedrock is mostly covered by various types of superficial deposits (Đomlija 2018; Jagodnik et al. 2020), formed by geomorphological processes active in both the carbonate and flysch rock mass.

The climate is maritime with mean annual precipitation between 300 and 700 mm. The rainy period lasts from November to May, with precipitation at its maximum in November. The predominant land covers are forests and schrubs.

The main watercourse is Dubračina River, with the length of about 12 km (Đomlija 2018). The river mouth into the Adriatic Sea is in the City of Crikvenica (Fig. 1a). In the inner parts of the Basin, the surface hydrological network is formed due to the impermeable flysch bedrock. The catchment widens south of the Tribalj settlement and it encompasses numerous periodical surface flows that form along gullies during the intense or prolonged rainfall events (Fig. 1a). There are several larger tributary flows of the torrential type in the SE Basin part: Ričina Tribaljska, Pećca, Kostelj, Slani Potok, Mala Dubračina and Kučina, which periodically transport large amounts of sediment to the Dubračina riverbed.

Erosion and sliding processes have significantly influenced the topography and land degradation in the study area. Numerous landslides continuously cause direct damages on private and public properties, such as the landslides shown in Fig. 1b–e. Namely, the certain landslide phenomena often reactivates, e.g., the landslide shown in Fig. 1e, which has reactivated on the County Road CR 5064 several times in the past few years, despite the performed remedial measures.

Materials and Methods

The altitude data were acquired in March 2012, using the multi-return LiDAR system. The DTM of 1 × 1 m resolution was produced from the bare ground returns acquired at the point density of 4.03 points/m^2, with an average point distance of 0.498 m.

Eight types of topographic datasets were derived from the DTM using standard operations in *ArcGIS 10*: (i) the hillshade maps with different illumination parameters (e.g., Jagodnik et al. 2020); (ii) the slope map, (iii) the contour line maps with 1, 2 and 5 m contour intervals; (iv) the aspect map; (v) the topographic roughness map; (vi) the profile curvature map; (vii) the planform curvature map; and the (viii) stream power index map. DTM derivatives were visually analysed both singularly and in combinations, for topographic characteristics indicative for gullies and

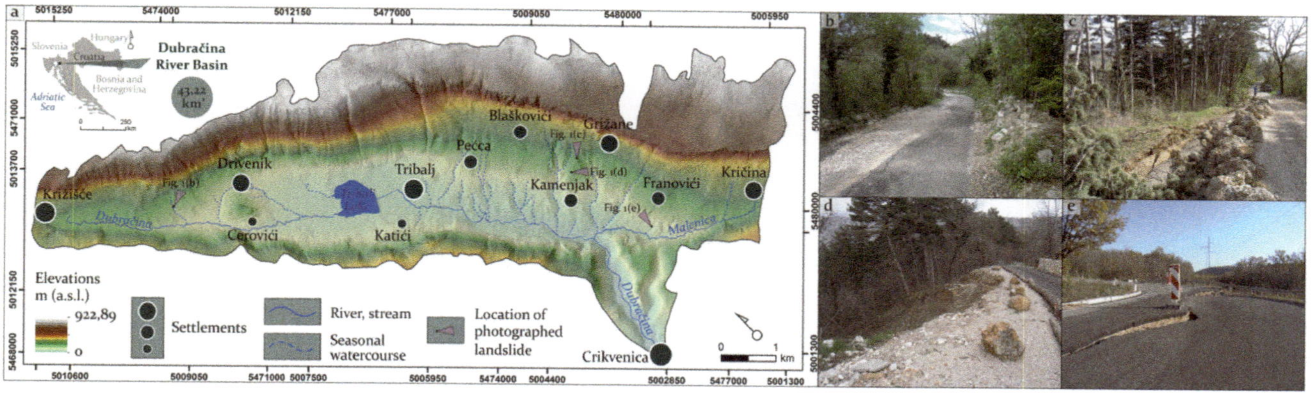

Fig. 1 Geographical location and relief map of the Dubračina River Basin (**a**), with examples of landslides initiated along the gully channel margins, which have caused material damages on the local (**b–d**) and county (**e**) roads passing along gullies

landslides (Đomlija 2018). Thereby, the study concerns only the gully forms wider than >3 m, given that initiation of landslides in the study area is not associated to narrower gullies (Đomlija 2018).

The detailed description of the methodology for identification and mapping of gully erosion phenomena in the study area is given in Đomlija (2018) and Đomlija et al. (2019). Whole procedure of identification and precise delineation of landslides within a gullied landform using the visual interpretation of LiDAR topographic derivatives is presented in Jagodnik et al. (2020). Both the gullies and landslides are mapped with polygons, according to established mapping criteria which imply that certain recognition features (e.g., Jagodnik et al. 2020) are visually analysed on the most effective LiDAR map (Đomlija 2018).

The precise delineation was performed in scales ranging from 1:100 to 1:300 for landslides, and from 1:500 to 1:5000 for gullies, depending on their sizes. Landslides are classified according to Hungr et al. (2014). The type of movement is determined based on the shape of the delineated polygon, while the type of material is determined based on the reconnaissance geological mapping, as well as the general knowledge of geological conditions in the study area (Blašković 1999; Đomlija 2018). Landslide and gully dimensions are calculated in *ArcGIS 10.0*. Descriptive landslide statistics are calculated on only the total number of complete landslides whose polygons depict the entire landslide body.

Field verification was conducted during the winter and spring of 2015 and 2016. Most of the gullies are identified in the field, given their sizes and formation within the populated areas. In contrast, field identification of landslides was limited, due to the complex gully topography and gully dimensions, and dense vegetation. However, portions of individual landslides are identified where it was possible to recognize mainly the landslide crowns and main scarps, and the landslide-forming material.

Results

There are 107 gullies identified in the Dubračina River Basin (Table 1), with the total area of 1.63 km^2. Most of them are formed in the NW Basin part. Gullies are significantly larger in the SE Basin part (Table 1; Fig. 2), with 75% of the phenomena smaller than 35,568 m^2. The largest gully in the study area (0.48 km^2) is the Slani Potok gully (Fig. 2b).

Totally 444 landslides are identified in gullies (Table 1), of two types (Hungr et al. 2014): (i) debris slide (DS), and (ii) debris slide-debris flow (DSDF). DS phenomena prevail (313 phenomena), and are mostly located in the gullies formed in the SE Basin part (Fig. 2b). Only one DSDF phenomenon is identified in the NW Basin part (Table 1). Total area of landslides located in gullies is 0.78 km^2 (Table 1): area of 0.48 km^2 of DS, and area of 0.30 km^2 of DSDF. Most of the landslides are located in the Slani Potok, Mala Dubračina and Kričina gullies (Fig. 2b). There are 181 landslides (107 DS, and 74 DSDF) identified in the Slani Potok gully only (Jagodnik et al. 2020). As in the other gully forms in the study area, landslides mostly initiate along the gully channel margins and extend to the channel bottom. Landslide are predominantly successive, hence the portions of individual landslides are often overlapped with zones of accumulations of adjacent landslides.

The smallest DS has an area of 65 m^2, and the largest is 43,151 m^2 (Table 2). Both phenomena are located in gullies in the SE Basin part. In this area, 75% of DS have an area <1,755 m^2. DSDF phenomena are generally larger than DS (Table 2), with 75% of landslides that are smaller than 2,958 m^2.

Examples of identified landslides are presented for the Kučina gully (0.15 km^2) formed in the SE Basin part, on certain LiDAR derivatives used for the visual analysis (Fig. 3).

The main scarps were mainly recognized on the slope map already during the identification and mapping of gullies

Table 1 Descriptive statistics for gullies identified in the Dubračina River Basin (DRB), with general data on landslides initiated within the gullies

Part of DRB	Nr. of gullies	A_{tot} (m^2)	A_{min} (m^2)	A_{Q1} (m^2)	A_{Q2} (m^2)	A_{Q3} (m^2)	A_{max} (m^2)	Nr. of landslides in gullies		Total area of landslides in gullies (m^2)	
								DS	DSDF	DS	DSDF
NW	76	198,526	96	281	889	2,623	25,759	21	1	9,400	534
SE	31	1,435,093	317	1,723	6,699	35,568	476,879	292	130	469,716	296,753

Abbreviations are used for the landslide types (DS—debris slide; DSDF—debris slide-debris flow)

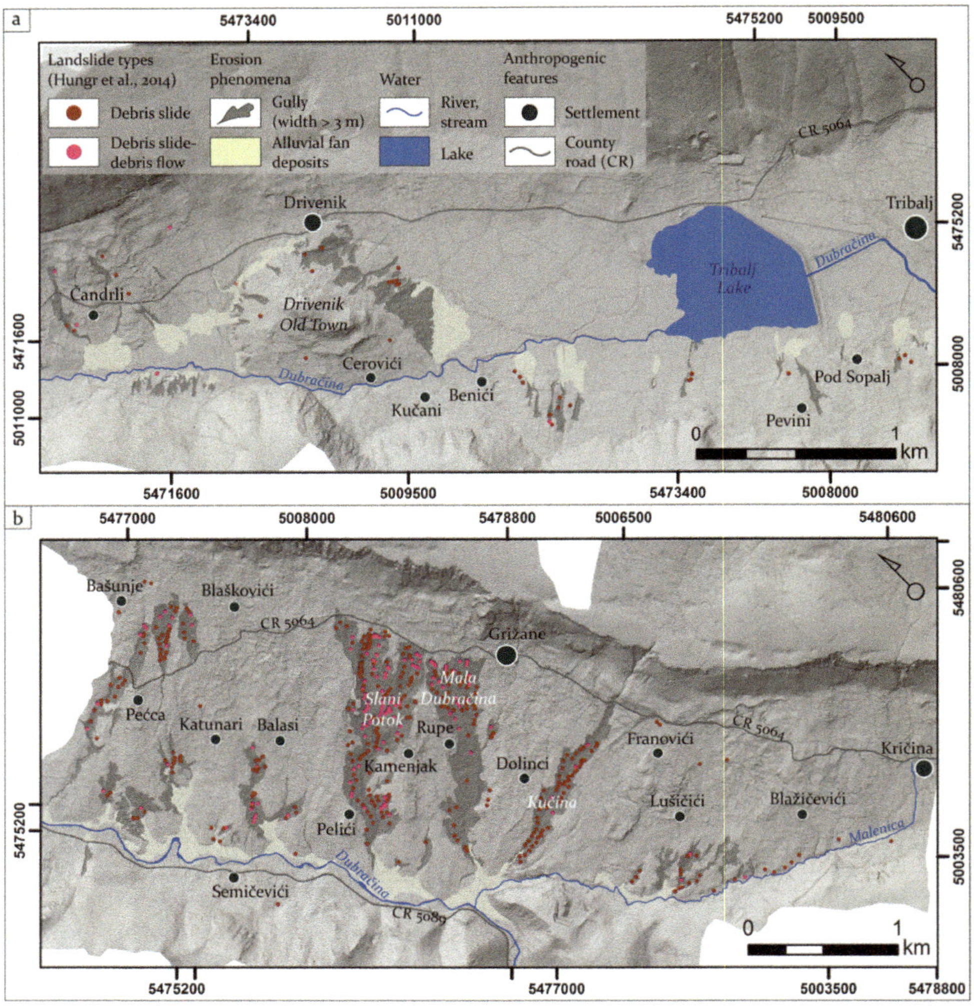

Fig. 2 Details from the historical landslide inventory and historical erosion inventory of the Vinodol Valley (Đomlija 2018) for: **a** the northwestern part, and **b** the southeastern part of the Dubračina River Basin

Table 2 Descriptive statistics for landslides identified within the gullies in the Dubračina River Basin, calculated according to the number of completely identified landslides

Landslide type	Part of DRB	Nr. of complete landslides	A_{tot} (m^2)	A_{min} (m^2)	A_{Q1} (m^2)	A_{Q2} (m^2)	A_{Q3} (m^2)	A_{max} (m^2)
Debris slide	NW	21	9,400	98	212	413	646	1,235
	SE	236	388,491	65	450	978	1,755	43,151
Debris slide-debris flow	NW	1	534	–	–	–	–	–
	SE	104	257,413	154	848	1,611	2,958	19,923

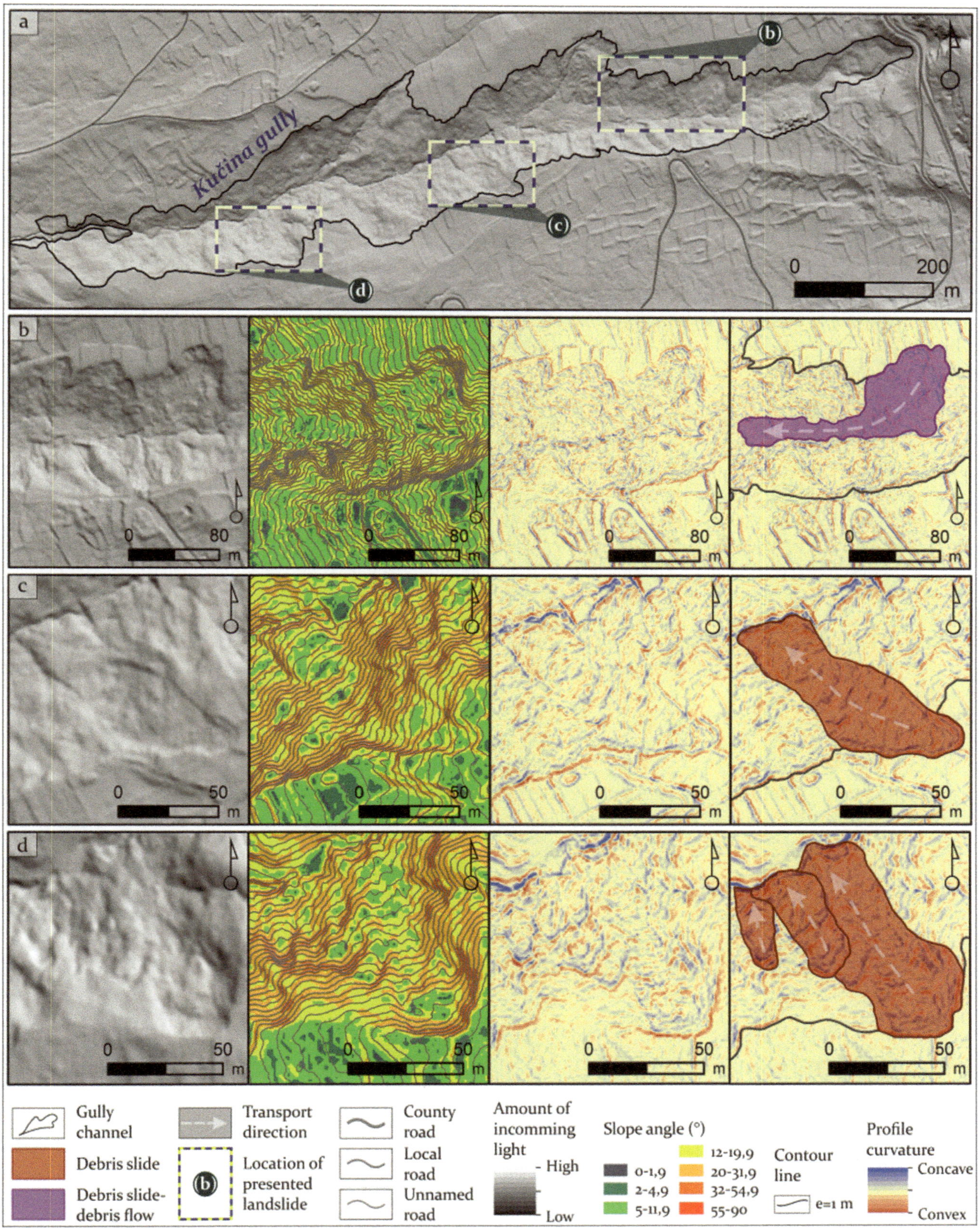

Fig. 3 Examples of landslides initiated in the Kučina gully in the SE part of the Dubračina River Basin: **a** the gully channel presented on the hillshade map, with depicted locations of landslides presented in the figure; **b** debris slide-debris flow, and **c**, **d** debris slides presented on (from left to right) the hillshade map, the contour line map over the slope map, the profile curvature map without and with delineated landslide polygons

(Đomlija et al. 2019), given that most of the landslide crowns coincide with the channel margins where there are abrupt changes in slope angles between the channel and the surrounding slopes. Both the crowns and main scarps can also be easily recognized on the profile curvature map and the contour line map, such in the cases of the landslides presented in Fig. 3. However, the landslide mapping in this study has revealed that the contour line map, even with the 1-m contour interval, does not clearly reflect the shape of the crown of very small landslides (Đomlija 2018). The effectiveness of this map for mapping the landslides initiated in gullies is also considered to be reduced for identification and precise delineation of the landslide flanks.

Most of the landslides initiate along the gully channel margins and extend to the gully channel bottom (Fig. 3). Thus, the profile curvature, as well as the planform curvature and the stream power index maps (not presented), are highly effective for delineation of the landslide toes, given that these landslide features coincide with the gully thalwegs that are, in most cases, clearly expressed on these maps. The profile curvature map is particularly effective for identification of the zones of accumulation of DSDF phenomena, clearly reflecting its convex lobate topography. The planform curvature and the stream power index maps are highly effective for precise delineation of the flanks and zones of accumulations of successive landslides located in gullies, due to the small erosional channels that form along their boundaries.

There are significant differences in the appearance of landslide topography on the hillshade map. Namely, DS presented in Fig. 3c can be easily recognized, as well as the zone of depletion of DSDF shown in Fig. 3b that is clearly expressed. In contrast, the landslide topography is difficult to recognize on the hillshade map for the examples of DS phenomena presented in Fig. 3d. Only the overall disturbed appearance of the gully channel wall and the smoothed appearance of the largest landslide phenomenon can be determined, while the recognition of the two adjacent smaller landslides is almost impossible without visually analysing other LiDAR derivatives.

Discussion and Conclusions

In total, 444 landslides are identified in in total 107 gullies formed in flysch deposits in the Dubračina River Basin, based on the visual interpretation of 1-m LiDAR DTM. Two types of landslides (Hungr et al. 2014) initiate within the gullies in the study area: (i) debris slides, (313 identified phenomena); and (ii) debris slide-debris flows (131 identified phenomena). Landslides in total cover 48% of all identified gully landforms. There is the pronounced diversity of gully forms along the study area (Fig. 2): although more gullies have been formed in the NW Basin part, gully forms are significantly larger in the SE part of the Basin and include the largest number of identified landslides. Landslides are mostly successive in distribution in both small and large gullies, and they generally initiate along the gully channel margins and extend to the gully bottom (Đomlija 2018; Jagodnik et al. 2020). The initiation of landslides in gully walls, which is predetermined by the similar geological conditions along the study area, results in the same or similar landslide type, while the spatial arrangement and extent of landslides within the gullies results in similar sizes of individual landslide phenomena.

The topographic location have influenced the possibilities for landslide identification and mapping on the LiDAR imagery, in both positive and negative sense. On the one hand, erosion processes have modified either the entire landslide topography or the topography of only individual landslide features, which made it difficult to recognize a certain number of landslides on the hillshade map that is commonly used as the first map in studies when visually searching for landslides (e.g., Van den Eeckhaut 2005). On the other hand, the knowledge that landslides occur mainly in gullies facilitated the detection of landslides as it directed the researcher to locations in the study area where landslide occurrences are more probable. Moreover, the spatial extent of most identified landslides from the gully channel margins to the channel bottoms facilitated the delineation of landslide features coinciding with the gully channel elements, even in the case of landslides with topography that is not clearly expressed on LiDAR derivatives.

Finally, gully landforms may distribute considerably due to the recent climate changes and thus even more influence the landscape evolution in an area, in relatively short time (Kukemilks and Saks 2013). Increase in precipitation can trigger new landslides and/or reactivate the landslides already existing in gullies, which results in the exposure of underlying material and further triggers erosional processes. The methodology used in this study has resulted in identification of portions of the studied area that are most affected by hazardous processes, as well as in knowledge of their types and interrelation. Such information can certainly facilitate planning of mitigation measures, as well as forecasting the locations and types of future sliding and erosion phenomena.

In future research, a more detailed statistical analysis of landslide-gully relationship will be carried out for each identified gully, focusing on the total number of landslides and their frequency, spatial distribution and geometric features of landslide phenomena of different type in different types of gullies. Due to great diversity of gully forms in the study area, classification of gully types is necessary for this type of analysis. Also, a quantitative analysis of potential of LiDAR derivatives for visual landslide mapping within gullied topography will be performed, in order to determine the optimal types of LiDAR derivatives and their combinations that are the most effective for identification and precise delineation of different landslide features, given their topographic location. In this way, the efficiency of visual analysis of LiDAR DTM for landslide identification and mapping in erosional environments will be also quantified, and more detailed understanding of the relationship between identified landslides and different types of gullies will enable spatial prediction of these processes in the study area.

Acknowledgements The results presented herein have been obtained in the framework of the Croatian-Japanese bilateral project Risk Identification and Land-Use Planning for Disaster Mitigation of Landslides and Floods in Croatia, funded by the Japan Agency for Science and Technology (JST) and Japan International Cooperation Agency (JICA) through the Science and Technology Research Partnership for Sustainable Development (SATREPS) Program.

References

Bernat S, Đomlija P, Mihalić Arbanas S (2014) Slope movements and erosion phenomena in the Dubračina river Basin: a geomorphological approach. Landslide and flood hazard assessment. In: Proceedings of 1st regional symposium on landslides in the Adriatic-Balkan Region, pp 79–84. Zagreb, Croatia (6–9 April 2013)

Bernat Gazibara S, Krkač M, Mihalić Arbanas S (2019) Landslide inventory mapping using LiDAR data in the City of Zagreb (Croatia). J Maps 15(2):773–779

Blašković I (1999) Tectonics of part of the Vinodol Valley within the model of the continental crust subduction. Geologia Croatica 52:153–189

Borelli L, Cofone G, Coscarelli R, Gullà G (2015) Shallow landslides triggered by consecutive rainfall events at Catanzaro strait (Calabria–Southern Italy). J Maps 11(5):730–744

Conforti M, Pascale S, Pepe M, Sdao F, Sole A (2013) Denudation processes and landforms map of the Camastra River catchment (Basilicata–South Italy). J Maps 9(3):444–455

Desta L, Adugna B (2012) A field guide on gully prevention and control. Nile Basin Initiative Eastern Nile Subsidiary Action Program, Addis Ababa, Ethiopia, p 67

Đomlija P (2018) Identification and classification of landslides and erosion phenomena using the visual interpretation of the Vinodol valley digital elevation model. PhD thesis, Faculty of Mining, Geology and Petroleum Engineering, University of Zagreb, Zagreb, Croatia (in Croatian)

Đomlija P, Bernat Gazibara S, Arbanas Ž, Mihalić Arbanas S (2019) Identification and mapping of soil erosion processes using the visual interpretation of LiDAR imagery. ISPRS Int J Geo-Inf 8(10):438

Ezechi JI, Okagbue CO (1989) A genetic classification of gullies in Eastern Nigeria and its implications on control measures. J Afr Earth Sci 9(3/4):711–718

Guzzetti F, Mondini AC, Cardinali M, Fiorucci F, Santangelo M, Chang KT (2012) Landslide inventory maps: new tools for an old problem. Earth Sci Rev 112:42–66

Hungr O, Leroueil S, Picarelli L (2014) The Varnes classification of landslide types, an update. Landslides 11:167–194

Jagodnik P, Jagodnik V, Arbanas Ž, Mihalić Arbanas S (2020) Landslide types in the Slani Potok gully Croatia. Geologia Croatica 73(1):13–28

Jurak V, Slovenec D, Mileusnić M (2005) An excessive flysch erosion—Slani Potok. In: Excursion guide book of 3rd geological congress, pp 51–55. Zagreb, Croatia (September 01–October 2005)

Kukemilks M, Saks T (2013) Landslides and gully erosion on banks of the Guaja River between the towns of Sigulda and Līgatne. Estonian J Earth Sci 62(4):231–243

Martínez-Casasnovas JA, Antón-Fernández C, Ramos MC (2003) Sediment production in large gullies of the Mediterranean area (NE Spain) from high-resolution digital elevation models and geographical information system analysis. Earth Surf Proc Land 28:443–456

Poesen J (1993) Gully typology and gully control measures in the Europeal loess belt. In: Wicherek S (ed) Farm land erosion in temperate plains environment and hills, pp 221–239. Elsevier Science, Amsterdam

Poesen J, Nachtergaele J, Verstraeten G, Valentin C (2003) Gully erosion and environmental change: importance and research needs. Catena 50:91–133

Popescu ME (11–12 July, 2002) Landslide causal factors and landslide remediatial options. In: Proceedings of the 3rd international conference on landslides, slope stability and safety of infra-structures, pp 61–81. Singapore

Santangelo M, Gioia D, Cardinali M, Guzzetti F, Schiattarella (2015) Landslide inventory map of the upper Sinni River valley Southern Italy. J Maps 11(3):444–453

Selby MJ (1993) Hillslope materials and processes, 2nd edn., p 451. Oxford University Press, New York, USA. (ISBN 0–19–874183–9)

Valentin C, Poesen J, Yong L (2005) Gully erosion: Impacts, factors and control. Catena 63:132–153

Van den Eeckhaut M, Poesen J, Verstraeten G, Vanacker V, Moeyersons J, Nyssen J, Van Beek LPH (2005) The effectiveness of hillshade maps and expert knowledge in mapping old deep-seated landslides. Geomorphology 67:351–363

Object-Oriented Automatic Landslide Detection from High Resolution Digital Elevation Model—Opportunities and Challenges Based on a Case Study in the Polish Carpathians

2020 Kyoto Japan

Kamila Pawluszek-Filipiak and Andrzej Borkowski

Abstract

One of the remedies for reducing the negative effects of landslide activity is landslide mapping. Landslide detection, carried out by using historical data analysis, stereoscopic photo interpretation and/or field works, is expensive, time-consuming and requires expert knowledge and experience. Automatic approaches for landslide detection can provide benefits such as increased efficiency and reduced costs and time. Many attempts have been made to automate the process of landslide identification but the key information for this process is provided by the high-resolution Digital Elevation Model (DEM) delivered from Airborne Laser Scanning (ALS) data. Having considered this, the objective of this study is to utilise the Object-Oriented Approach (OOA) and DEM for the detection of landslides. In this study, we use the results archived from Pawluszek et al. (ISPRS Int J Geo-Inf 8:321, 2019). The challenges and opportunities of automatic approaches are discussed, based on an investigation conducted in an area heavily affected by landslides. The study area is located close to Rożnów Lake, in Poland and stands out by various land uses. The automatic detection results achieved (OA = 85% and K = 0.6) indicate that there is a huge potential in automatic approaches. However, these approaches face difficulties in landslide detection due to the smoothing of typical landslide features. This situation appears for old landslides and landslides located in areas of active agricultural treatments. Besides the fuzzy delineation of the landslide extent, landslide amalgamation in the OOA results can be observed. Thus, automatic approaches still need to be developed and improved. At the current stage of the development, automatic approaches cannot replace validation based on field reconnaissance but can support an interpreter in their work.

Keywords

Landslide • Landslide detection • Object-oriented approach • Classification • Airborne laser scanning

Introduction

A landslide is a sudden mass movement occurring on a slope under the influence of gravity (Pawluszek and Borkowski 2017). Landslides are one of the most common natural hazards that occur in different parts of the world. The fundamental step to reduce the negative effects of landslide activity, such as economic and environmental losses, is landslide mapping and susceptibility assessment (Petschko et al. 2016). Effective landslide hazard and risk assessment can only be conducted when landslide identification and susceptibility mapping are continuously updated. Many countries in the world have already created (or are in the process of creating) landslide databases. Van Den Eeckhaut and Hervás (2012) reported that in 2012 the completeness of landslide databases for many European countries was below 25%. Therefore, it is crucial to complete national landslide-related databases.

Landslide identification is carried out by exploiting historical data analysis, stereoscopic photographic interpretation and/or field works but these are expensive, time-consuming and require expert knowledge and experience in the people performing the landslide detection (Guzzetti et al. 2012). In the case of forested and urbanised areas, it is also challenging to correctly identify the landslide extent because the slope cannot be observed by the

K. Pawluszek-Filipiak (✉) · A. Borkowski
Institute of Geodesy and Geoinformatics, Wroclaw University of Environmental and Life Sciences, ul. Grunwaldzka 53, 50-357 Wrocław, Poland
e-mail: kamila.pawluszek-filipiak@upwr.edu.pl

A. Borkowski
e-mail: andrzej.borkowski@upwr.edu.pl

F. Guzzetti et al. (eds.), *Understanding and Reducing Landslide Disaster Risk*, ICL Contribution to Landslide Disaster Risk Reduction,
https://doi.org/10.1007/978-3-030-60227-7_6

interpreter integrally, only partially. This can lead to incorrect landslide identification (Wojciechowski et al. 2012).

Airborne Laser Scanning (ALS) technology captures very detailed information about terrain with a high accuracy and high point density (significantly more than 10 points/m^2), including forested areas (Jaboyedoff et al. 2012; Wojciechowski et al. 2012). A Digital Elevation Model (DEM) created from ALS data represents detailed terrain forms without the vegetation coverage. Based on a highly accurate, detailed DEM, it is possible to identify a landslide and its features with very high reliability, even if it is covered by forests (Tarolli et al. 2012; Wojciechowski et al. 2012; Guzetti et al. 2012; Pawluszek 2019). Advantages of ALS data and its increasing availability have led to a lot of investigations aimed at the automation of the landslide identification process, based on high-resolution DEM created from ALS data. Automatic approaches for landslide detection can provide benefits in the future, such as increased efficiency and reduced costs and time.

Many methods for automatic landslide detection from DEM can be found in the literature (McKean and Roering 2004; Chen et al. 2014; Leshchinsky et al. 2015; Gaidzik et al. 2017; Knevels et al. 2019; Pawluszek et al. 2017, 2018, 2019). These are mainly pixel-based, machine learning approaches (PBA) and object-oriented approaches (OOA). The basic information unit in PBA is the single-pixel while in OOA this unit is represented by an image object (Li et al. 2015). Both approaches have their advantages and disadvantages however, many recent researchers report that OOA outperforms PBA (Li et al. 2015; Pawluszek et al. 2019).

Having considered this fact, the basic objective of this study is to utilise OOA for landslide detection in a complex environment. In this study we use some of the results from the already published investigation by Pawluszek et al. (2019), conducted in an area located close to Rożnów Lake, Poland. The area is characterised by various land uses and is heavily affected by landslides. In this paper, the challenges and opportunities for automatic approaches have to be discussed in correlation with various land use at this location. Many other aspects related to OBIA applications for landslide detection in this study area are documented in the base publication (Pawluszek et al. 2019).

Study Area

The study area is located in the Małopolskie municipality in Poland (Fig. 1). The area covers latitudes 49° 43′ N to 49° 46′ N and longitudes 20° 38′ E to 20° 43 ′E and is located in close proximity to Rożnów Lake, in the central part of the Polish Carpathians. The area in question is mainly used for agriculture (42% of the total area) and, despite the

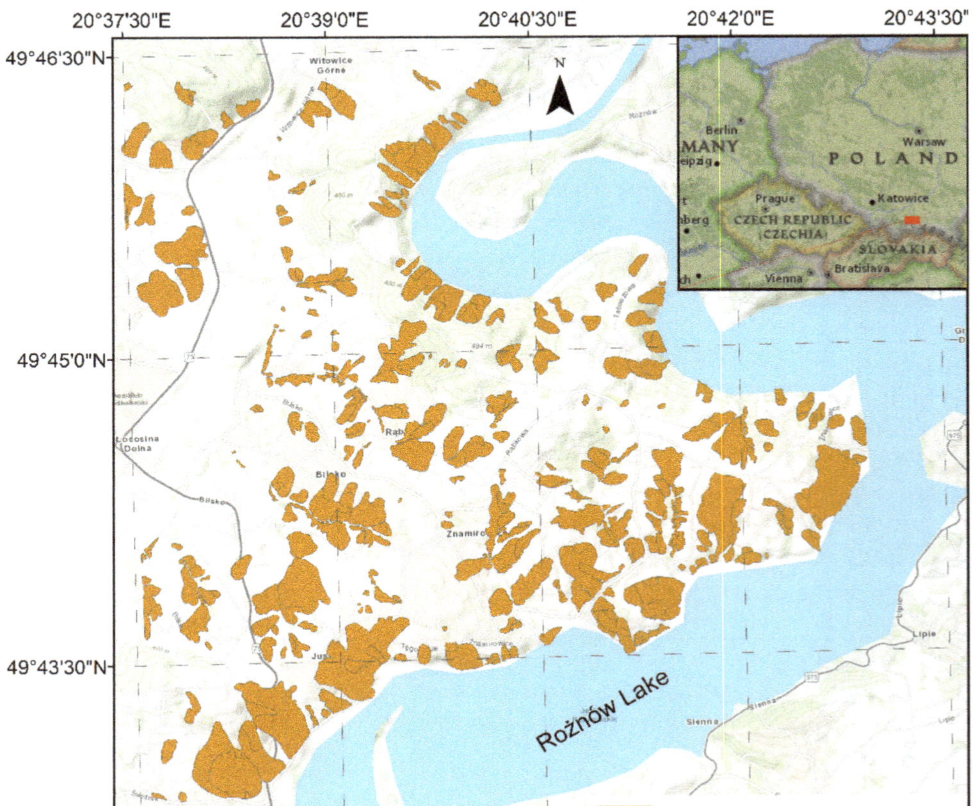

Fig. 1 Location of the study area

existence of landslides, the landslide affected areas are still extensively used for economic purposes (Kroh 2017). Forested areas account for 34% (Kroh 2017). Landslide identification in this region is additionally challenging due to large land management such as farmland, grasslands, and pasture as well as continuing construction and other developments.

Data Used

In the present work, we acquired the reference landslide inventory records from the Landslide Counteracting System, called SOPO (Table 1). ALS data was acquired in the framework of the Country's Protection System (called ISOK project) covering the whole country. Point clouds were acquired in 2012 by means of the Riegl LiteMapper 6800i system. The point cloud density is from 4 to 6 points/m^2. The DEM interpolated from ALS data is presented in Fig. 2. The road layer was captured by the OpenStreetMap platform and the river and streams network was extracted from the DEM according to the approach presented in Pawluszek et al. (2019). The map in Fig. 3 shows the land use over the study area.

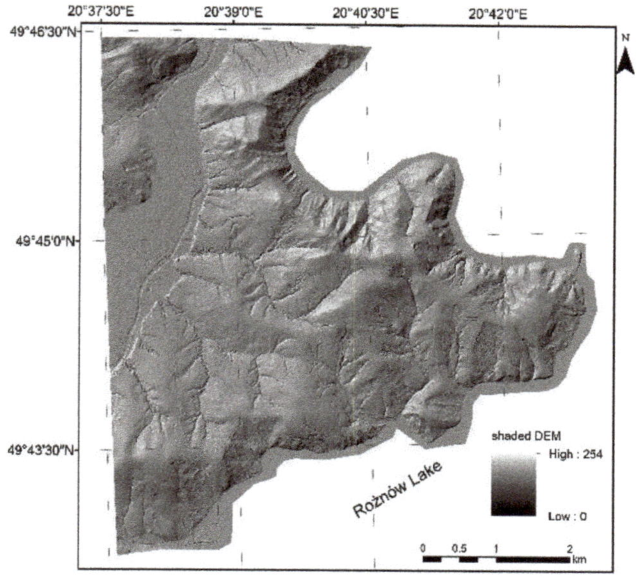

Fig. 2 DEM generated from ALS data

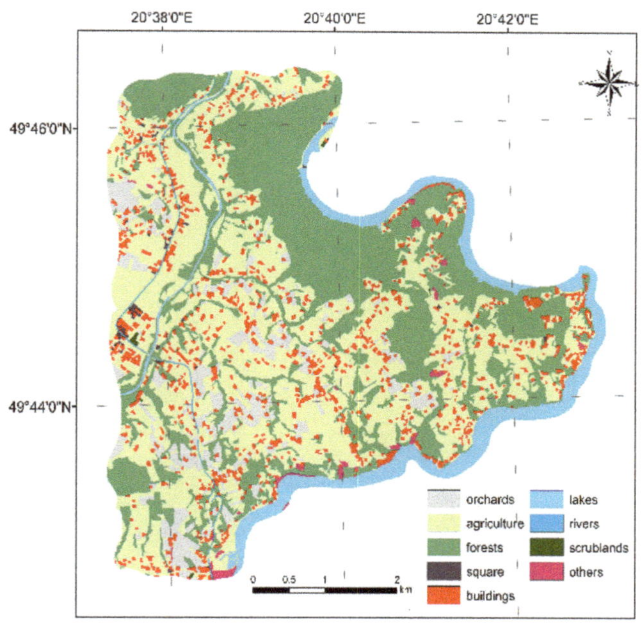

Fig. 3 Land use over the study area

Methodology

The methodology flowchart is presented in Fig. 4. From the ALS data, a DEM with a resolution of 2 m was generated (Fig. 2). Afterwards, based on DEM, various topographical factors were computed: aspect, curvature, slope, roughness, hillshade, openness and Sky View factor, in particular. These topographical layers were calculated with the optimum window size according to Pawluszek et al. (2018) and then utilised for supervised classification. The Support Vector Machine method was used as a classifier. For the training of the classifier, we used randomly selected landslide objects which cover 20% of the total study area. Before the classification, a segmentation process had to be performed. A multi-resolution segmentation approach,

Table 1 Data used in presented study

Data	Source	References
ALS data	ISOK project	https://www.isok.gov.pl
Landslide database	Landslide Counteracting System (SOPO)	https://geoportal.pgi.gov.pl/portal/page/portal/SOPO/Wyszukaj3
River network	Extracted using DEM	Pawluszek et al. (2019)
Road network	Open street map	https://www.openstreetmap.org/#map=5/51.500/-0.100

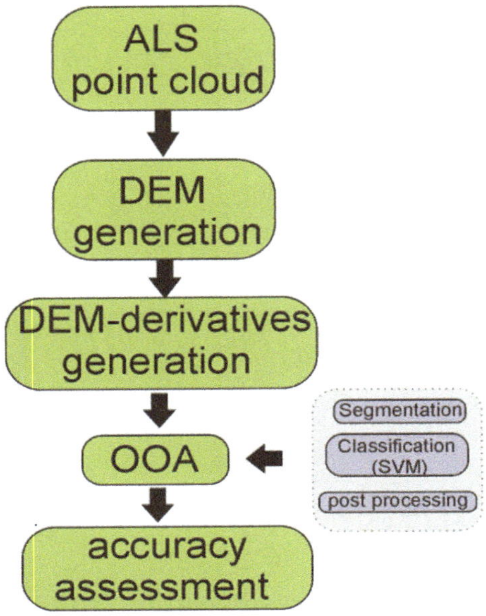

Fig. 4 Research methodology

exploiting all available topographical layers and the road and river networks, was utilised for this purpose. The results from the SVM classification were filtered by the thresholding of the length/width ratio, the minimum object size and the tool called assign class by thematic layers. Various classification parameters, such as segmentation scale, were selected based on a trial and error basis. The broader description of various parameters influencing the final results is provided in Pawluszek et al. (2019). Finally, the results that were achieved were evaluated. As validation indexes, we utilised Overall Accuracy (OA) and Kappa (K) index. The OA represents the sum of correctly classified pixels divided by the total number of pixels. In contrast, K reflects the agreement between the classification results and the true values. Kappa equal to 1 represents a perfect agreement, while kappa close to 0 indicates non-agreement. For all calculations, eCognition software was utilised.

Results and Discussion

The landslides detected using the presented methodology are shown in Fig. 5. The OA and K-index are 85.5% and 0.6 respectively. These results demonstrate the huge potential for automatic approaches to landslide detection. Figure 6 presents examples of various landslides visualised on the DEM background. Figures 6a–c represent landslide extents extracted by OOA, while Figs. 6d–f represent the reference landslide extents from the SOPO database. The reference landslides have been mapped by geologists in the field and by various photographic interpretations. Therefore, this data can be considered as being the ground truth data. It can be easily observed that the extent of the landslides in the case of

Fig. 5 Results for landslides detected using OOA

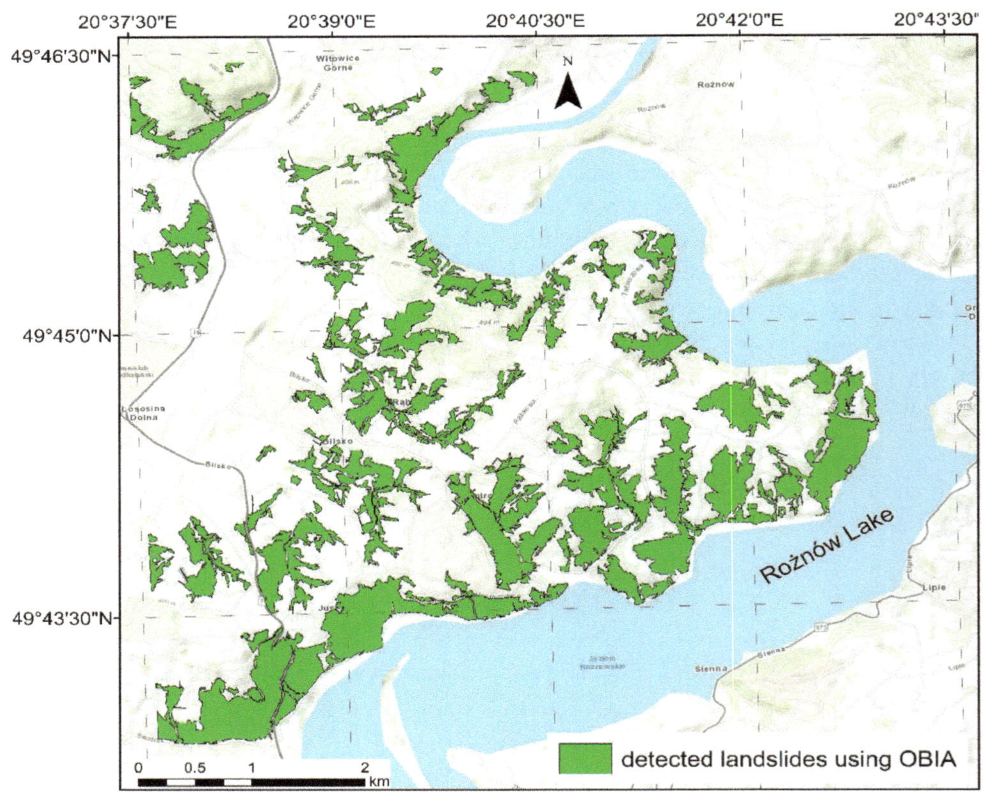

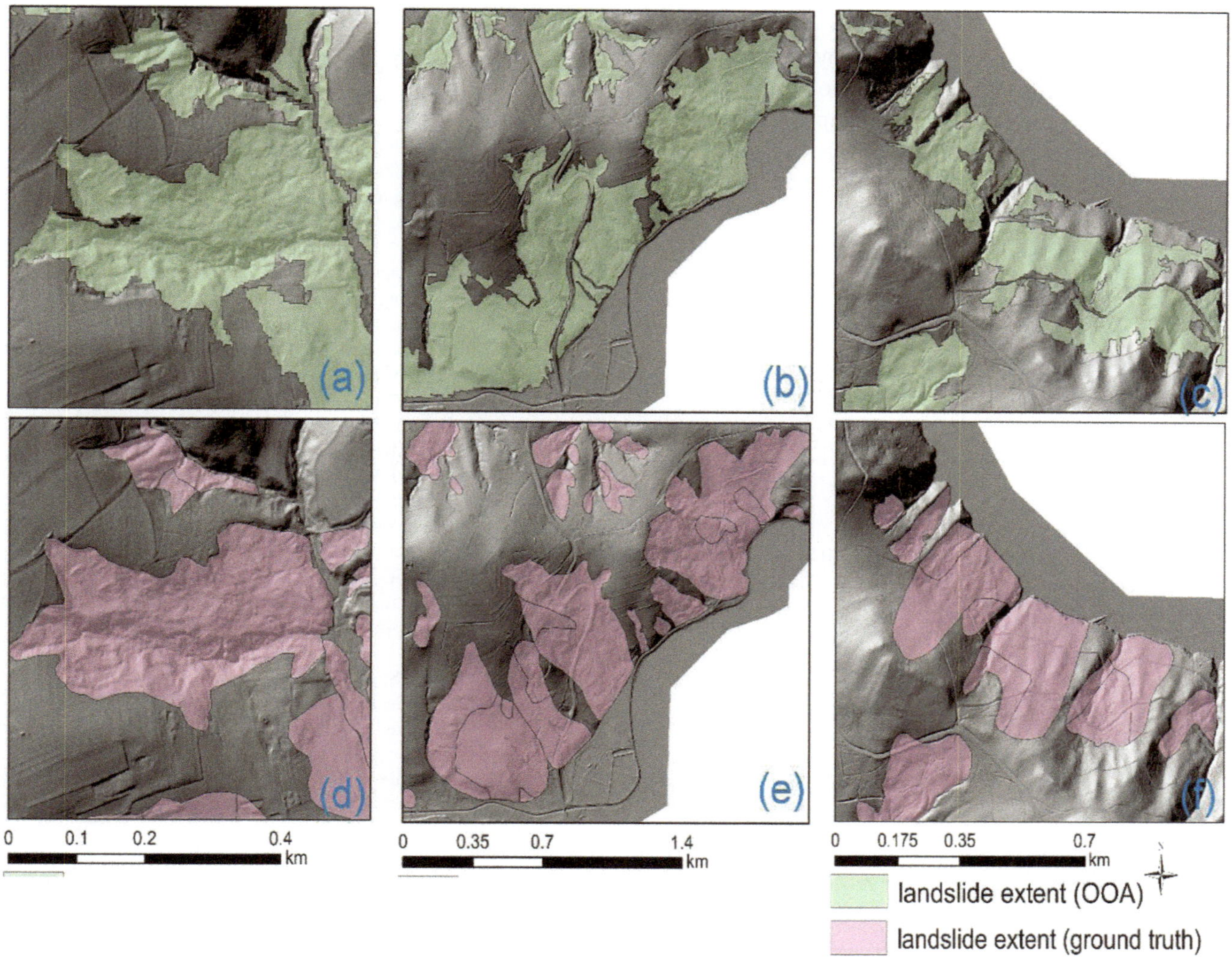

Fig. 6 Landslides detected using OBIA (**a**, **b**, **c**) and ground-truth landslide extent from SOPO database

Fig. 6a, d, correspond with each other well. An opposite situation can be observed in the case of the landslides presented in Fig. 6b, e, c, f. Besides the deviations in the landslide extent delineation, we can also observe landslide amalgamation. In Fig. 6e, many separate landslides are visible while in the case of the automatic detection (Fig. 6b), a lot of these landslides have been merged into one object. Moreover, OOA found it difficult to correctly extract landslide boundaries. This is probably due to the loss of the distinctive landslide morphology. The areas presented in these figures (Fig. 6b, c) are mostly urbanised and agricultural areas, therefore topography roughness (which is typical for landslides), disappeared (see Fig. 3). Figures 6c, f present very poor identification results. These landslides are totally covered by forests (compared with Fig. 3), therefore the typical landslide features should be preserved on the topography. However, the explanation for such difficulties is that these are very old landslides, whose typical landslide features have vanished due to various environmental factors, such as denudation or weathering.

Conclusions

Landslide location and extent is a fundamental source for susceptibility and hazard assessment. The development of effective automatic approaches for landslide detection can provide significant benefits. The OOA presented in this study shows great potential, especially in forested areas. However, it faces difficulties connected with the smoothing of typical landslide features. It is very challenging to identify landslides when landslide features, such as scarps or distinctive rough landslide bodies, are not preserved in the DEM. Additionally, the problem of amalgamation appears, when two landslides are located in close proximity to each other.

In summary, the presented approach can be effectively applied to young landslides with fresh morphology. Nevertheless, automatic approaches still need to be developed and improved. At the moment, automatic approaches will not replace validation based on field reconnaissance but they can support interpreters in their work. Furthermore, the adoption of machine learning methods based on deep learning can bring further progress in automatic landslide detection, as well as landsides which have suffered degradation of their typical morphology.

Acknowledgements This research has been supported by the National Science Centre, Poland (Grant No. 2018/28/T/ST10/00528).

References

Chen W, Li X, Wang Y, Chen G, Liu S (2014) Forested landslide detection using LiDAR data and the random forest algorithm: a case study of the Three Gorges, China. Remote Sens Environ 152:291–301. https://doi.org/10.1016/j.rse.2014.07.004

Gaidzik K, Ramírez-Herrera MT, Bunn M, Leshchinsky BA, Olsen M, Regmi NR (2017) Landslide manual and automated inventories, and susceptibility mapping using LIDAR in the forested mountains of Guerrero, Mexico. Geomatics, Nat Hazards Risk 8(2):1054–1079. https://doi.org/10.1080/19475705.2017.1292560

Guzzetti F, Mondini AC, Cardinali M, Fiorucci F, Santangelo M, Chang, KT (2012) Landslide inventory maps: New tools for an old problem. Earth Sci Rev 112(1–2):42–66. https://doi.org/10.1016/j.earscirev.2012.02.001

Jaboyedoff M, Oppikofer T, Abellán A, Derron MH, Loye A, Metzger R, Pedrazzini A (2012) Use of LIDAR in landslide investigations: a review. Nat Hazards 61(1):5–28. https://doi.org/10.1007/s11069-010-9634-2

Knevels R, Petschko H, Leopold P, Brenning A (2019) Geographic object-based image analysis for automated landslide detection using open source GIS software. ISPRS Int J Geo-Inf 8:551

Kroh P (2017) Analysis of land use in landslide affected areas along the Łososina Dolna Commune, the Outer Carpathians. Poland Geomat Nat Hazards Risk 8:863–875. https://doi.org/10.1080/19475705.2016.1271833

Leshchinsky BA, Olsen MJ, Tanyu BF (2015) Contour connection method for automated identification and classification of landslide deposits. Comput Geosci 74:27–38. https://doi.org/10.1016/j.cageo.2014.10.007

Li X, Cheng X, Chen W, Chen G, Liu S (2015) Identification of forested landslides using LiDAR data, object-based image analysis, and machine learning algorithms. Remote Sensing 7(8):9705–9726. https://doi.org/10.3390/rs70809705

McKean J, Roering J (2004) Objective landslide detection and surface morphology mapping using high-resolution airborne laser altimetry. Geomorphology 57(3–4):331–351. https://doi.org/10.1016/S0169-555X(03)00164-8

Pawluszek K (2019) Landslide features identification and morphology investigation using high-resolution DEM derivatives. Nat Hazards 96(1):311–330. https://doi.org/10.1007/s11069-018-3543-1

Pawluszek K, Borkowski A (2017) Automatic landslides mapping in the principal component domain. In: Mikoš M, Vilímek V, Yin Y, Sassa K (eds) WLF 2017 conference proceedings advancing culture of living with landslides, pp 421–428. Springer, Cham, Switzerland. https://doi.org/10.1007/978-3-319-53483-1_50

Pawluszek K, Borkowski A, Tarolli P (2017) Towards the optimal pixel size of DEM for automatic mapping of landslide areas. Int Arch Photogram Remote Sensing Spatial Inf Sci 42:83–90. https://doi.org/10.5194/isprs-archives-XLII-1-W1-83-2017

Pawluszek K, Borkowski A, Tarolli P (2018) Sensitivity analysis of automatic landslide mapping: numerical experiments towards the best solution. Landslides 15:1851–1865. https://doi.org/10.1007/s10346-018-0986-0

Pawluszek K, Marczak S, Borkowski A, Tarolli P (2019) Multi-aspect analysis of object-oriented landslide detection based on an extended set of LiDAR-derived terrain features. ISPRS Int J Geo-Inf 8(8):321. https://doi.org/10.3390/ijgi8080321

Petschko H, Bell R, Glade T (2016) Effectiveness of visually analyzing LiDAR DTM derivatives for earth and debris slide inventory mapping for statistical susceptibility modeling. Landslides 13(5):857–872

Tarolli P, Sofia G, Dalla Fontana G (2012) Geomorphic features extraction from high-resolution topography: landslide crowns and bank erosion. Nat Hazards 61(1):65–83. https://doi.org/10.1007/s11069-010-9695-2

Van Den Eeckhaut M, Hervás J (2012) State of the art of national landslide databases in Europe and their potential for assessing landslide susceptibility, hazard and risk. Geomorphology 139:545–558

Wojciechowski T, Borkowski A, Perski Z, Wójcik A (2012) Dane lotniczego skaningu laserowego w badaniu osuwisk-przykład osuwiska w Zbyszycach (Karpaty zewnętrzne). Przegląd Geologiczny 60(2):95–102 ([in Polish])

Can Repeat LiDAR Surveys Locate Future Massive Landslides?

Mio Kasai

Abstract

This study examined a topographic indicator of a massive landslide that occurred in 2011 in the Kii Peninsula, Japan. Assuming that the slope was deforming slowly based on surface roughening before the event, a roughness filter, the standard deviation of the slope angle for 3•3 cells on a 1 m digital elevation model (DEM), was used as a measure of change in the ground surface and calculated using light detection and ranging (LiDAR) data for 2006 and 2010. The images produced from the survey data showed that a major crack had already developed between a wide ridge and a lower undulating patch on the slope that subsequently became the head of the slide. The cells in which the standard deviation increased by 0.39–2.32 between the surveys formed a group of ripple-like features clustered in the area. The results indicate that repeat LiDAR surveys can help identify slopes in imminent danger of sliding based on the location of progressive surface roughening.

Keywords

Repeat LiDAR surveys • Surface roughness • Massive landslide

Introduction

Steep landscapes underlain by deeply weathered bedrock, such as in the crushed zone in Japan, are prone to massive landslides. Rock masses in this mountain range move downward slowly, but are constantly deforming the slope profile until an intense rainfall or earthquake triggers rapid landsliding from the ridge top. Because of their size, slides not only affect human lives, properties, and infrastructure directly but also have prolonged effects on daily lives and the economies of local communities. For disaster-mitigation planning, it is important to identify slopes in imminent danger of collapse before a triggering event occurs, by locating topographic signs of deformation. Using airborne light detection and ranging (LiDAR) surveys, previous research has demonstrated that key features are undulating patches in which cracks and troughs develop, wide, discontinuous ridges, bulging slope profiles, and small landslides or rock falls at the feet of slopes (Chigira et al. 2013). Because cracks and troughs form rough ground, a surface roughness filter can be applied to estimate the degree of slope activity (Kasai et al. 2009). In reality, however, these features can be found everywhere on topographic maps based on LiDAR surveys, probably because once they form, they remain until they erode away after the slope has stabilized. Therefore, surface roughness estimated from a single survey does not always indicate current slope activity. However, repeated LiDAR surveys might identify slopes in imminent danger of collapse. With the increasing accumulation of survey data over the last two decades, we can use the data to find dynamically changing ground surfaces associated with massive landslides.

This study examined the change in surface roughness of the slope before the 2011 Nagatono slide, on the Kii Peninsula, Japan (Fig. 1), using LiDAR survey data obtained in 2006 and 2010. The research question is, whether repeat LiDAR surveys could locate future massive landslides. The outcome of this study would help with disaster-mitigation planning.

M. Kasai (✉)
Faculty of Agriculture, Hokkaido University, Kita-9 Nishi-9 Kita-ku, Sapporo, 060-8589, Japan
e-mail: kasaim@for.agr.hokudai.ac.jp

F. Guzzetti et al. (eds.), *Understanding and Reducing Landslide Disaster Risk*, ICL Contribution to Landslide Disaster Risk Reduction,
https://doi.org/10.1007/978-3-030-60227-7_7

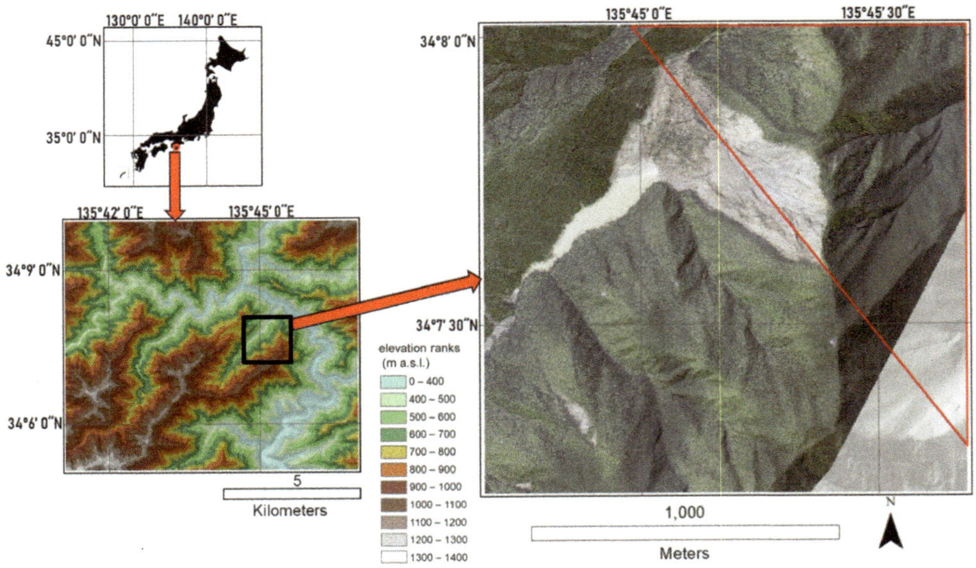

Fig. 1 Study site and the Nagatono slide. The photograph was taken in 2011. LiDAR surveys in 2006 and 2010 covered the area shown by a red triangle

Study Area

The 0.75 km^2 study area (red triangle, Fig. 1) is located in the Kii Mountains, a range underlain by crushed Cretaceous mudstones and sandstones belonging to the Shimanto Belt. The belt consisting of mildly metamorphosed sedimentary rocks extends from the Nansei Islands through Kyushu, Shikoku, Kii Peninsular, Akaishi and Kanto Mountains to the Boso Peninsula in Japan (Taira et al. 1982). The landscape is characterized by steep hillslopes averaging 37.5° covered by planted Cryptomeria japonica. An intense storm that triggered several massive slide in the region brought more than 1,300 mm of rainfall from September 1–4 in 2011 at the Kazeya metrological station 10 km from the site (Nara local meteorological office 2011). The mean annual precipitation at the station was 2,400 mm from 1978 to 2019. The Nagatono slide started from a wide ridge (indicated by a yellow arrow: Fig. 2) connected to an undulating patch underneath it (Fig. 2: encircled in black). The slope angles of the patch were largely below 35° (Fig. 2). The total area of the slide was 220,700 m^2, and it yielded 4.1 million m^3 of debris (Chigira et al. 2013). LiDAR surveys in 2006 and 2010 covered half of the slope that eventually formed the slide (Fig. 1), offering an opportunity to examine the topographic change before the event. The data resolution in 2006 was finer than that in 2010 (Fig. 3). The average point density of the former was 3.34 points/m^2, while it was 0.38 points/m^2 for the latter.

Methods

The resolution of LiDAR data usually differs between surveys, particularly in mountainous regions due to slope steepness and dense vegetation cover, both of which limit the number of ground points. To reduce the effects of the resolution difference on the analyses, the surface was first smoothened using local polynomial interpolation of the ground points on ArcGIS, with a semi-axis of 10 m and smoothing factor of 0.1. Then the 1 m DEM of the smoothened surface was used to estimate the surface roughness at the time of each survey. This study used the standard deviation of slope angle for 3•3 cells as a roughness filter, as proposed by Frankel and Dolan (2007). An increase in the value means that the location became bumpier after 4 years. After selecting the range of increase in the value that represents dynamic change in the ground surface, the spatial density of the cells in this range within a 50 m radius was calculated to find clusters of such cells. Finally, the clusters were examined to determine whether they could predict the Nagatono slide.

Results and Discussion

The appearance of a major crack running underneath the wide ridge in the 2006 image (Fig. 3) was evidence that the Nagatono slide had already started to move at least 6 years before the event. Therefore, careful examination of

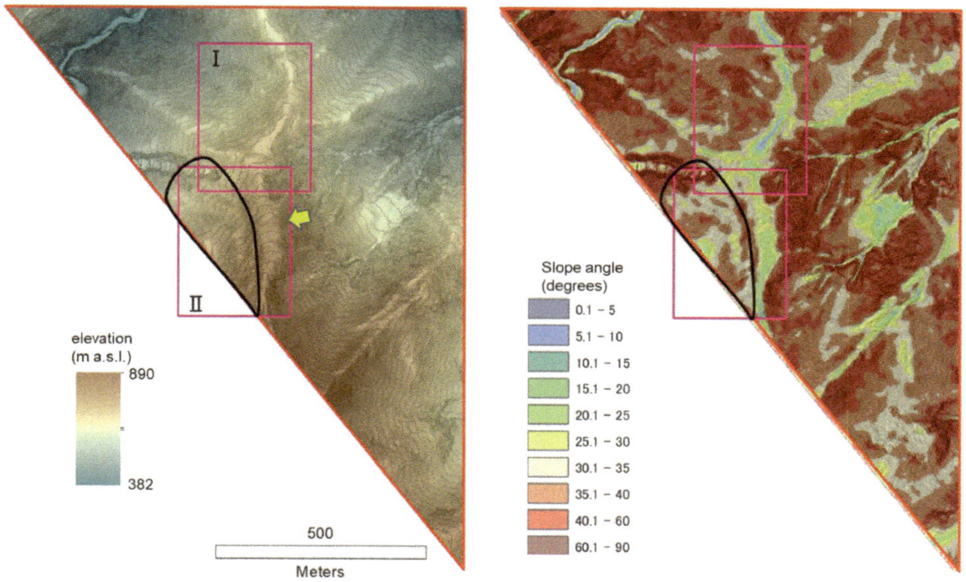

Fig. 2 Hillshade image overlain by slope gradient and elevation maps (left) and slope distribution map (right) produced by the 2006 LiDAR survey data. A wide ridge existed before the Nagatono slide (yellow arrow). A part of an undulating patch was covered by each survey in 2006 and 2010 (enclosed in black). Enlarged images of the areas of pink rectangles, I and II, are presented in Figs. 3 and 4

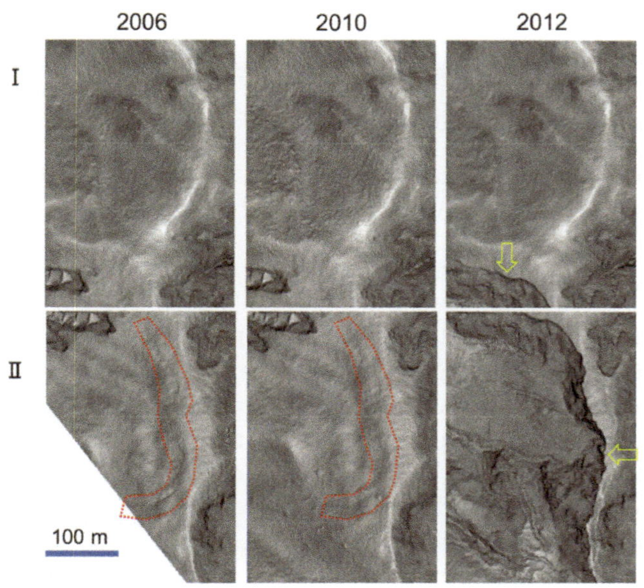

Fig. 3 Enlarged images of the I and II areas (Fig. 2), produced by the 2006, 2010 and 2012 LiDAR survey data. A crack (II, enclosed in red dotted line) was evident particularly in the 2006 image before the Nagatono slide (indicated by yellow arrow)

high-resolution topographic maps can help identify slopes that are imminent dangers. The crack was also recognizable in the 2010 image, albeit less clearly than before (Fig. 3). It is difficult to estimate the progress of slope deformation between the surveys with images only. By contrast, cells with standard deviations that increased by 0.39–2.32 during the period appeared as a group of ripple-like features along

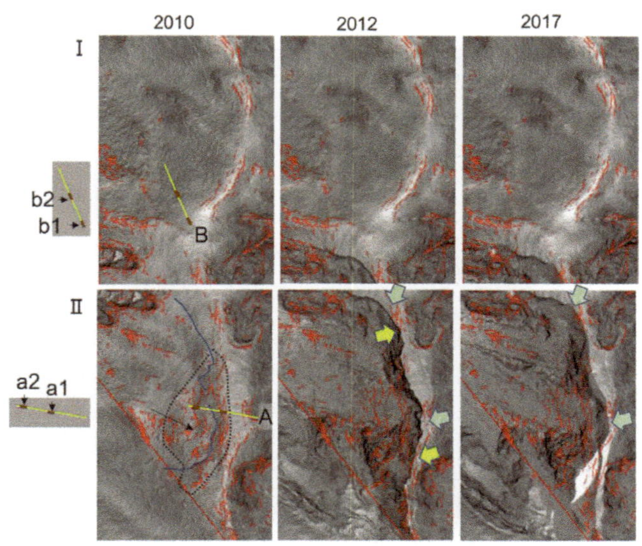

Fig. 4 Cells with standard deviations that increased by 0.39–2.32 from 2006 to 2010 (colored in red) on the topographic maps of the I and II areas, produced by the 2010, 2012 and 2017 LiDAR survey data. Longitudinal profiles of the sections from lines A and B (colored in brown) are presented in Fig. 5. The crack marked in Fig. 3 is traced in blue. A group of ripple-like features (e.g. black arrow) appeared over the undulating patch (enclosed in black dotted line). The Nagatono slide collapsed from the edges indicated by the cells (yellow arrows) in 2012 and had progressed to another edge (green arrows) by 2017

the crack and troughs (e.g. black arrow, Fig. 4) over the undulating patch (enclosed in black dot line, Fig. 4). Over the 4 years, the longitudinal profile of the smoothed surface of line A on the slope at the head of the Nagatono slide also

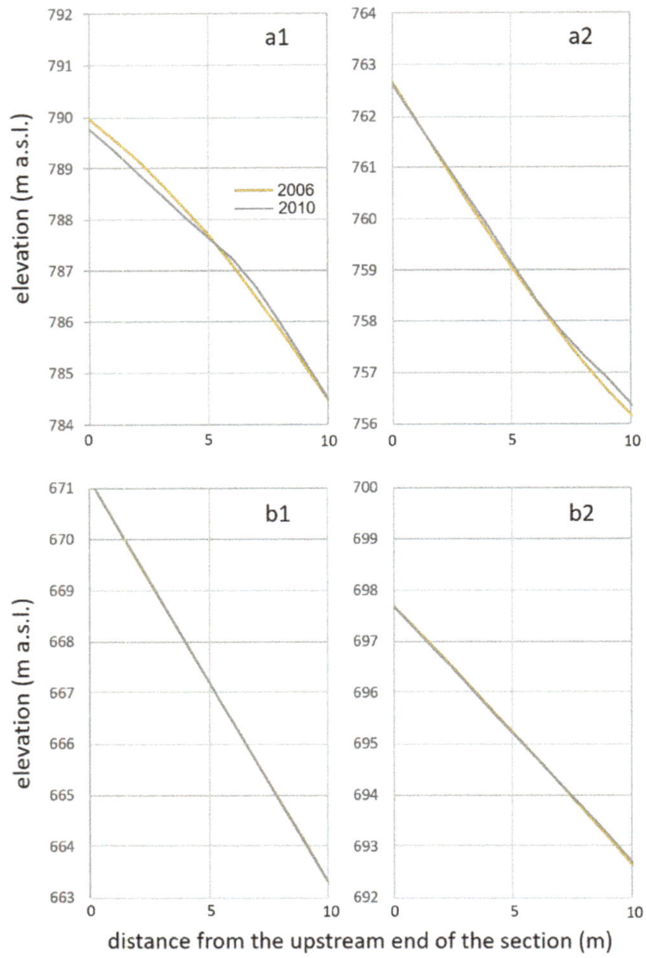

Fig. 5 Longitudinal profiles of the smoothed surfaces of 10 m sections from lines A and B (Fig. 4)

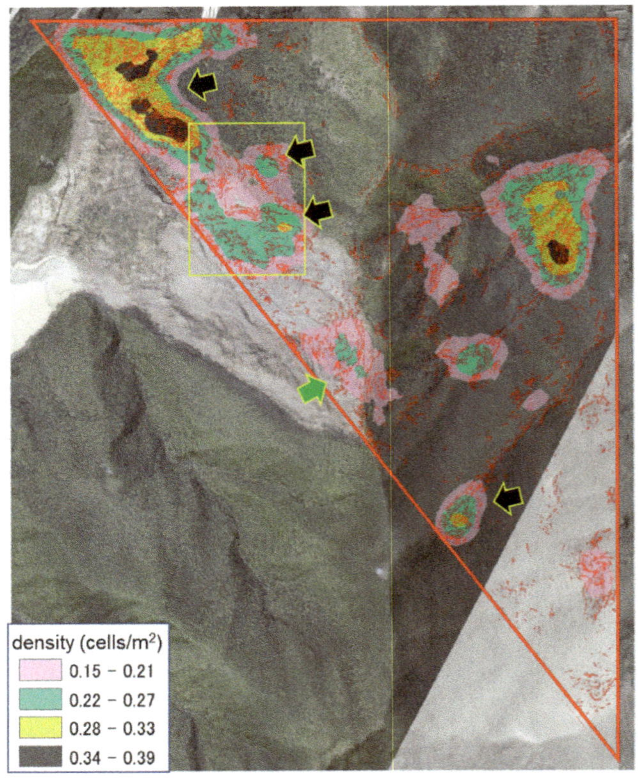

Fig. 6 Cells with standard deviations that increased by 0.39–2.32 from 2006 to 2010 (colored in red) and clusters of the cells on the 2011 photographic image. Black arrows: clusters possibly resulting from coarse ground points. Green arrow: a cluster believed to reflect the real surface change associated with the Nagatono slide. The area of a yellow rectangle is presented in Fig. 7

changed more noticeably than that of line B on the slope where little happened in 2011 (Figs. 4 and 5). In 2012, the Nagatono slide collapsed from the edges indicated by the cells (yellow arrows, Fig. 4). By 2017, the slide had progressed to another edge on the ridge top (green arrows, Fig. 4). These combined results suggest that an increase in the standard deviation of the slope angles of neighboring cells, estimated from repeat LiDAR surveys, can show the progress of slope deformation and help locate future massive landslides and determine their extent.

However, careful review of the images is required, because the wrong surface texture may be produced if there are only a few ground points. While Fig. 6 shows that clusters of cells covered the slope where the Nagatono slide occurred, the coarse resolution of the 2010 image in the area at the flank of the slide (Fig. 7) prevented the confirmation of slope activity before the event. Alternatively, the cluster may not be related to the occurrence of the slide; other clusters resulting from the coarse resolution of data are also seen on the map (black arrows, Fig. 6). Those clusters could not be eliminated after smoothing the surface, and the presented method needs improvement for dealing with coarse datasets. In comparison, the cluster at the top of the slope, which became the head of the slide (green arrow, Fig. 6), was believed to reflect the real surface change, as indicated above. Clusters were also seen around streams. These should be ignored when trying to identify the locations of massive landslides, as progressive slope deformation usually does not allow a stream network to develop.

Conclusion

This study detected topographic changes that predicted the 2011 Nagatono slide from LiDAR surveys done in 2006 and 2010. The results suggest that a cluster of cells, indicative of a roughing surface, estimated from repeat LiDAR surveys,

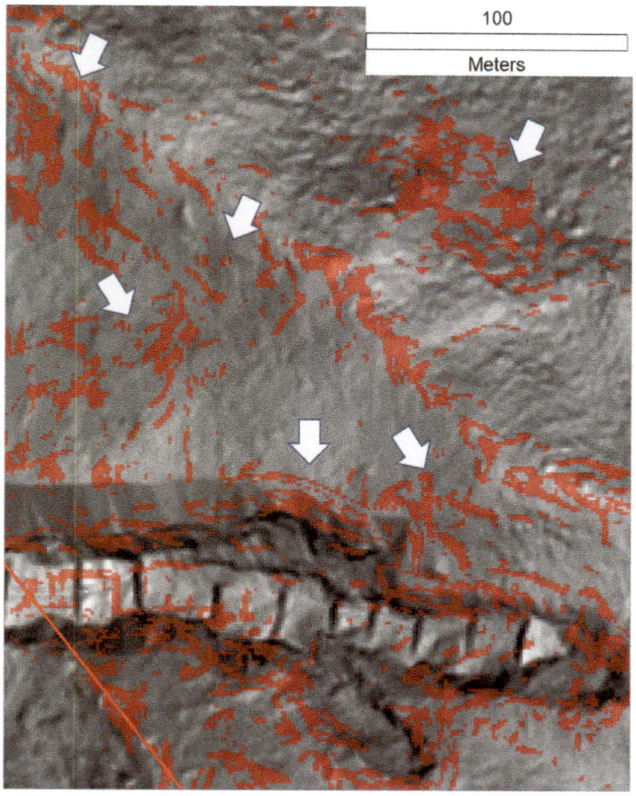

Fig. 7 Topographic map produced by the 2010 LiDAR survey data for the area shown in Fig. 6 (yellow rectangle). Sparse ground points resulted in false ground texture, affecting the value of the roughness filter (e.g. indicated by arrows)

can predict future landslides, particularly when they appear on an undulating patch on a wide ridge. The increased standard deviation of the slope angle for 3•3 cells in a 1 m DEM proved to be a useful indicator of dynamic surface change. With higher-resolution LiDAR data, slopes that are urgent concerns can be identified more correctly, helping to prepare for massive landslides during the next triggering event to mitigate associated disasters.

Acknowledgements The author thanks Kinki Regional Development Bureau of Ministry of Land, Infrastructure, Transport and Tourism, and Nara Prefecture for providing LiDAR survey data and orthophotographs.

References

Chigira M, Thou CY, Matsushi Y, Hiraishi N, Matsuzawa M (2013) Topographic precursors and geological structures of deep-seated catastrophic landslides caused by Typhoon Talas. Geomorphology 201:479–493

Frankel KL, Dolan JF (2007) Characterizing arid region alluvial fan surface roughness with airborne laser swath mapping digital topographic data. J Geophys Res Earth Surf 112:F02025

Kasai M, Ikeda M, Asahina T, Fujisawa K (2009) LiDAR-derived DEM evaluation of deep-seated landslides in a steep and rocky region of Japan. Geomorphology 113(1):57–69

Nara local meteorological office (2011) Torrential Rain and Gale Force Winds by Typhoon No. 12, Heisei 23. https://www.jma-net.go.jp/nara/kishou/pdf_files/t1112_20110908.pdf. Last accessed: 12 March, 2020 (in Japanese)

Taira A, Okada H, Whitaker JH, Smith AJ (1982) The Shimanto Belt of Japan: cretaceous-lower Miocene active-margin sedimentation. Geol Soc London Spec Publ 10:5–26

Semi-automatic Landslide Inventory Mapping with Multiresolution Segmentation Process: A Case Study from Ulus District (Bartin, NW Turkey)

Gulseren Dagdelenler, Murat Ercanoglu, and Harun Sonmez

Abstract

Landslide inventory mapping studies have been received special attention from a wide range of specialists. In accordance with this situation, the main purpose of this study was to produce a semi-automatic GIS-based inventory mapping for locating landslides by a multiresolution segmentation process, which is the first phase of Object-Oriented Analyses (OOA). Ulus district of Bartin located in the Western Black Sea Region of Turkey was chosen as the study area. For the multiresolution segmentation process, a total of 1132 objects were automatically extracted using the first 4 bands of the Landsat ETM + satellite image and three thematic maps (slope, curvature and Normalized Difference Vegetation Index) of the study area. Multiresolution segmentation process was performed with Definiens Professional Earth (DPE in Definiens professional 5 UserGuide, Definiens AG, Munchen, Germany, 2006) software. In order to determine semi-automatically landslide locations in the study area, Artificial Neural Networks (ANN) method has been applied to the study area with 8 parameters (brightness, shape index, GLDV Contrast, Length/Width, Maximum Difference, Asymmetry, GLCM Contrast and GLCM Homogeneity) as an input and landslide (1) and no landslide (0) information as an output. This semi-automatically inventory map of the area estimated 78.2% of the existing landslides accurately. The applicability of this method is easy and quick, but when applying this method to the study area, thematic, spectral, shape and textural properties of the landslides must be revealed accurately and in detail.

Keywords

Landslide • Inventory map • Object-oriented analyses (OOA) • Multiresolution segmentation • Ulus (bartin) • Artificial neural networks (ANN)

Introduction

Landslides are the most common mass movements among the natural disasters. Both in Turkey and all over the world, each year a large number of landslides occur and these landslides often cause major economic losses. Considering these losses caused by landslides, the spatial distribution of existing mass movements; in other words, landslide inventory information of the area should be determined (Van Westen et al. 2008).

The most important factor in accurate and reliable landslide susceptibility mapping (LSM) of a particular area is the need to produce landslide inventory mapping correctly and reliably. Landslide inventory maps provide a crucial contribution toward landslide susceptibility, hazard and risk evaluation (Guzzetti et al. 2006; Van Westen et al. 2008). For mapping landslides and creating landslide inventory maps, some researchers have used stereoscopic aerial photographs (Cardinali et al. 2001), some have used field observations (Santangelo et al. 2010), some have used aerial photographs (Guzzetti et al. 2012), satellite image data (Nichol et al. 2006), some researchers also used satellite-based artificial beam radar images (Bovenga et al. 2012). In recent years, many new remote sensing techniques have occured and their use in field studies has become quite common (Tarolli et al. 2009). Air-borne Laser Swath Mapping (ALSM) method using LIDAR (Light Detection and Ranging) technology has provided a remarkable advantage in terms of providing high resolution topographic data compared to other

G. Dagdelenler (✉) · M. Ercanoglu · H. Sonmez
Department of Geological Engineering, Hacettepe University, Beytepe, 06800 Ankara, Turkey
e-mail: gulsrn@hacettepe.edu.tr

M. Ercanoglu
e-mail: murate@hacettepe.edu.tr

H. Sonmez
e-mail: haruns@hacettepe.edu.tr

F. Guzzetti et al. (eds.), *Understanding and Reducing Landslide Disaster Risk*,
ICL Contribution to Landslide Disaster Risk Reduction,
https://doi.org/10.1007/978-3-030-60227-7_8

traditional methods. It is preferred by many researchers because of its success in mapping landslide morphology of LIDAR and predicting landslide activity in a particular region, whether covered or not (Van Den Eeckhaut et al. 2011). Until recently, the lack of high-resolution satellite data limited most satellite-based landslide studies to regional scale assessments. However, in recent years, landslide mapping studies using high-resolution satellite images (Quickbird, Ikonos, Landsat, SPOT-5, WorldView-1, Cartosat-1 and 2) are increasing.

Some researchers do landslide inventory mapping for a specific region, they use aerial photographs, field observations and past landslide records (Dagdelenler et al. 2015) while others use different automatically mapping techniques such as object-oriented analyses (OOA) (Aksoy and Ercanoglu 2012; Kurtz et al. 2014). OAA has a huge potential to detect landslides automatically in a better way than the pixel-based methods, by incorporating a multitude of landslide spectral, morphometric (shape) and textural features (Martha et al. 2012). In OAA analyses, the researchers use different kind of high or very high resolution satellite images to extract landslide locations automatically. However, with high resolution topographic data for example LIDAR, object-oriented image analyses (OOA), also referred to as object-based image analyses (OBIA) due to its reliable and accurate results (Van Den Eeckhaut et al. 2011). OAA has two steps, the first step is multiresolution segmentation step of images that consist of just pixels into image primitives. This process is a vital process in OOA by grouping homogenous pixels into an object (Baatz and Schape 2000). The other step is the classification of those segments, incorporating spectral, textural, morphometric (shape) landslide parameters (Blaschke 2010).

The main goal of the study is to create a database by extracting the morphological, spectral and textural properties of landslide and non-landslide objects for each segment as a result of the image segmentation process of the study area using high resolution satellite image. Therefore, the areas covered by landslide and non-landslide objects observed in the study area will be determined semi-automatically by OOA. In this study, multiresolution segmentation algorithm was selected for the analyses. Multiresolution segmentation process and extraction of spectral and textural characteristics were carried out using Definiens Professional Earth (DPE 2006). In the first stage of the segmentation process, the first 4 bands of the Landsat ETM + satellite image of 2000 of the study area were used as spectral features, and slope, curvature and NDVI (Normalized Difference Vegetation Index) maps of the study area were also used as thematic input data. As a result of this process, a total of 1132 objects were automatically sorted. Then, eight parameters (brightness, shape index, GLDV (Grey-Level Difference Vector) contrast, length/width, maximum difference, asymmetry, GLCM (Grey Level Co-occurence Matrix) contrast and GLCM homogeneity) related to landslide locations were chosen to be used as input parameters in the analyses. Thus, a GIS-based database will be created automatically on the landslide and non-landslide areas, based on the properties to be selected (spectral, figural and textural) and the landslide location. Using the parameters selected as a result of the multiresolution segmentation process, possible landslide locations were determined semi-automatically using the Artificial Neural Networks (ANN) method.

Study Area

Ulus district of Bartin, which is located in the Western Black Sea Region, was chosen as the study area due to the heavy rainfall and weak ground units (Fig. 1). The study area covers four layouts of approximately 578 km^2 within the Zonguldak E29d3, Zonguldak E29d4, Zonguldak F29a1 and Zonguldak F29a2 layouts, which are between 32,037′30″–32,045′00″ longitudes and 41,022′30″–41,037′30″ latitudes. The most important stream in the study area is Ulus Stream, and Eldeş and Gegen streams connected to this stream are the most important drainage networks. There are two dominant formations in the study area, namely, İnalti and Ulus Formation. The basement rocks of the study area are the Upper Jurassic-Lower Cretaceous aged İnalti Formation. İnalti Formation is overlain by the Cretaceous aged Ulus Formation and the current terraces/talus/alluvium units. İnalti Formation is observed in the northern part of the study area where it has a very steep topography. It consists of cream-light gray colored relatively weathered recrystallized limestones. On the other hand, Ulus formation, which consists of brown-gray-dark green, thin-medium layered, fine-medium grained sandstone, dark green-dark gray colored claystone and medium-thick layered shales, is defined as flysch deposits (Akyol et al. 1974). Most of the landslides occurred in this upper cretaceous aged Ulus Formation in other words, this formation is home to lithologies in which various types of landslides take place in the study area.

Landslide inventory mapping of the study area was made using Google Earth satellite images, aerial photographs and detailed on-site studies. A landslide inventory map of 1/500,000 scaled by the General Directorate of Mineral Exploration and Exploration was also used in the landslide inventory of the study area (Fig. 2). As a result of the landslide inventory studies carried out, a total of 195 landslide areas were determined in the study area (Dagdelenler et al. 2012) (Fig. 2). The study area covers an area of 578 km^2 on a 1:25,000 scale topographic map, and landslide areas cover an area of approximately 103 km^2. According to the Varnes (1978) landslide classification system, the landslides occurring in the study area were observed to be mostly massive deep circular landslides.

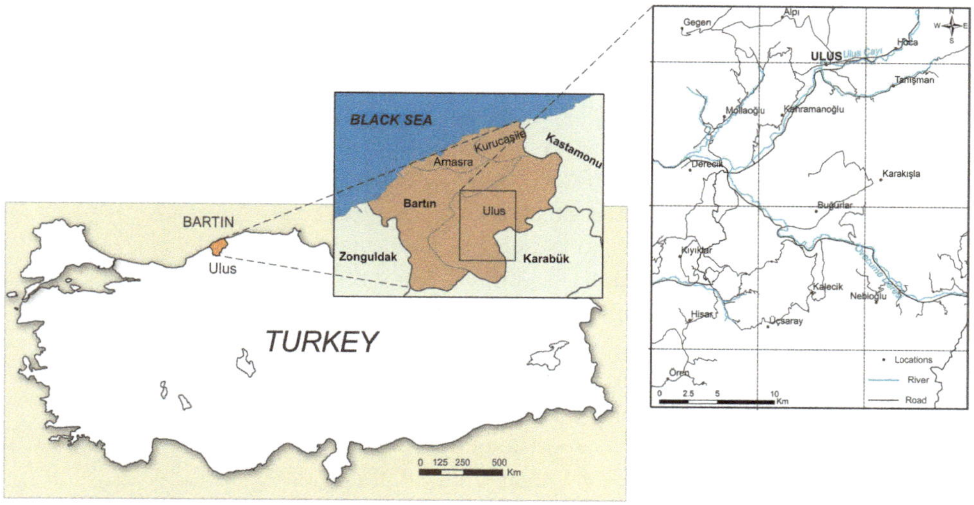

Fig. 1 Location map of the study area

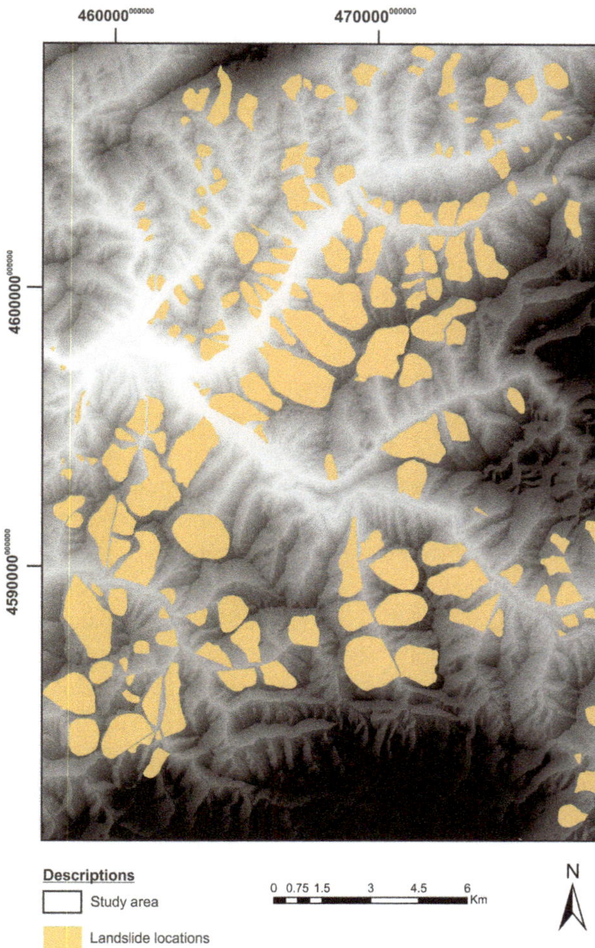

Fig. 2 Inventory map of the study area

Methodology

Producing landslide inventory maps, some researchers have traditionally relied on the analysis of aerial photographs, field investigations and the collection of historical information (Guzzetti et al. 2012). However, new techniques have emerged, one of which is the object-based image classification analysis (Martha et al. 2010). In this study, image segmentation process, the first stage of the object-oriented image analyses, was used for semi-automatic landslide inventory mapping of the study area.

Image Segmentation Process

The image segmentation process is the first stage of the Object-Based Image Classification Methods, and this method is considered as a method that takes into account the structure, textures and spectral information together (Aksoy and Ercanoglu 2012). The main purpose of this process is to place objects with similar properties in any digital image (satellite image, aerial photo, etc.) used as input. In other words, it is to create meaningful objects from the image. This means that the shape of each relevant object must be shown based on an image object (Marangoz et al. 2006). While performing image segmentation, textural, shape and spectral properties of objects are used. In the segmentation stage, scale, color, shape, smoothness and compactness parameters should be determined as realistic as possible. Scale parameter indirectly affects the average object size.

The larger the scale parameter, the larger the objects. With color and shape parameters, the effect of the homogeneity of the color and shape conflict on the object production can be corrected (Marangoz et al. 2006).

Image segmentation process takes place with different algorithms such as chessboard, quad tree, spectral difference and multiresolution segmentation. Among these algorithms, the most used in the literature is multiresolution segmentation. In this study, multiresolution algorithm has been selected in the segmentation process, and other elements (scale, color, shape, smoothness and compactness) of the segmentation process were applied by considering the values taken by Aksoy and Ercanoglu (2012) in their study (Table 1). While performing the appropriate multiresolution segmentation parameter selection, we applied the trial and error method several times. The main purpose of this method is to capture the smallest landslide area (4847 m^2) with the segment boundary. The main purpose here is to create a GIS-based database automatically in the landslide (1) and non-landslide (0) areas based on the properties to be selected (spectral, figural and textural) and these location features. With an approach followed in this way, the representation of the study area as a result of segmentation process is presented in Fig. 3. As a consequence of the segmentation process, a total of 1132 objects were automatically sorted. In this process, the first 4 bands of the Landsat ETM + satellite image of the study area as spectral properties, and as the thematic input data, slope, curvature and NDVI (Normalized Difference Vegetation Index) maps were used. In these objects obtained as a result of segmentation process, "1" is assigned to the landslide areas and "0" to the no landslide areas, in other words, an overlap was performed using 195 previously mapped landslide locations. This segmentation process used in the study was carried out with DEP (Definiens Earth Pro 2006) software.

Brightness, shape index, GLDV contrast, length/width, maximum difference, asymmetry, GLCM contrast and GLCM homogeneity) parameters, respectively, as spectral, shape and textural features in the creation of a landslide database. These parameters are the most frequently used parameters in such studies and also the only parameters that we could produce using DEP (2006). The main purpose is to examine the changes of these properties taken into account in the landslide (1) and no landslides (0) and to produce values that will be inputs to the ANN (Artificial Neural Network) analysis to be carried out in the next stage. Changes of each parameter used in the database for the landslides (1) and no landslides (0) will basically reveal the relationship values for landslide properties. For example, the basis for considering the brightness parameter is the view that the areas that are bare from the surrounding areas will give more reflection after the landslide occurs. In the parameters considered as textural features (such as GLCM, GLDV), irregularities in landslide topography are tried to be explained (Aksoy and Ercanoglu 2012). The parameters (such as Length/Width, Shape Index), which are taken into account as the formal features, are based on the principle that landslide objects will change significantly in terms of shape indexes compared to the objects around them (Aksoy and Ercanoglu 2012). As a result of this approach, each feature determined for landslide (1) and non-landslide (0) areas has been put forward based on GIS. With the help of the database created with these determined parameters and their changes, Artificial Neural Networks (ANN) method has been applied to determine possible landslides (1) and non-landslides (0).

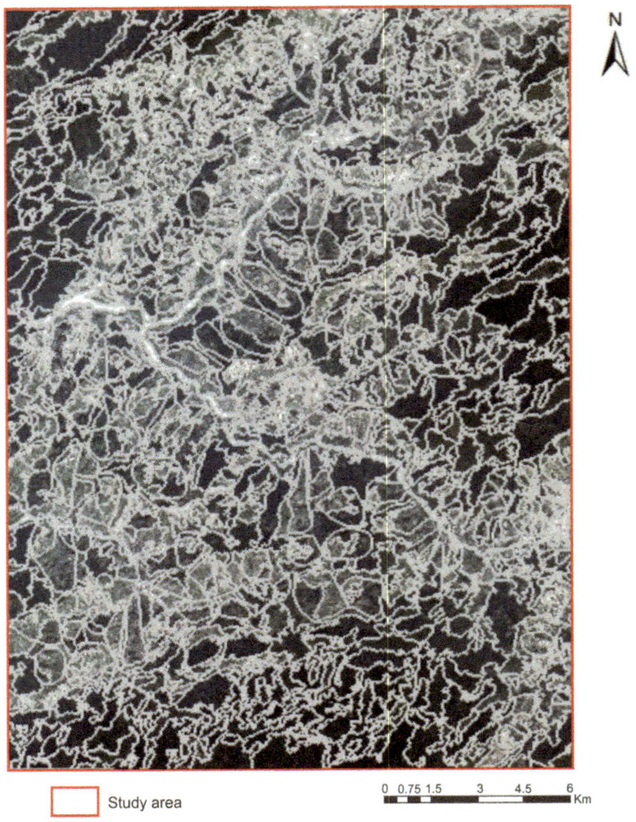

Fig. 3 Composite image of the study area obtained by segmentation process

Table 1 Selected parameters for segmentation process

Scale	Color	Shape	Smoothness	Compactness
50	0.9	0.1	0.5	0.5

ANN Analyses

Artificial neural networks (ANN) is a computer-based algorithm that is based on the human nervous system and artificially simulates this system. As it is known, the nervous system of people consists of single-cell structures called neuron and includes dentrit, dentronite, cell body, synapses and axons. From this point of view, ANN can be defined as the modelling tool of the biological nervous system in computer environment (Konar 2005).

ANN has been widely used in most of the studies on landslide susceptibility modelling in recent years (Dou et al. 2015, 2020; Pham et al. 2019). In this study, Idrisi Selva software was used to perform ANN analyses to determine possible landslide locations semi-automatically. In the ANN analyses, the multilayer perceptron (MLP) neural network was carried out. The MLP consists of at least three layers of nodes: an input layer, a hidden layer and an output layer. Except for the input nodes, each node is a neuron that uses a nonlinear activation function (Lippman et al. 1987). The most important phase of ANN is the training phase. At this stage, 849 items, 75% of the number of objects (1132) considered in the segmentation phase, were randomly selected and the training phase was carried out. The MLP algorithm is networks containing one or more layers between the input and output layers, which correspond to hidden layers. In ANN analysis, a total of 8 parameters (brightness, shape index, GLDV Contrast, Length/Width, Maximum Difference, Asymmetry, GLCM contrast and GLCM homogeneity) were used as input parameters, and landslide (1) and no landslide (0) information were used as output parameters. The best ANN network topology is determined as 8-4-1 (Dagdelenler et al. 2017). Schematic representation of this ANN architecture is presented in Fig. 4. Learning coefficient and momentum values are considered as 0.1 and error value as 0.001. As a result of these analyses ANN reached the desired error value in 3501 iteration and the training phase was ended. In addition, % importance values that express the effect of the considered parameters on the output are presented in Table 2. The values obtained as a result of all these analyses were transferred to ArcGIS (10.3) software and finally a semi-automatic landslide inventory map of the study area was produced (Fig. 5).

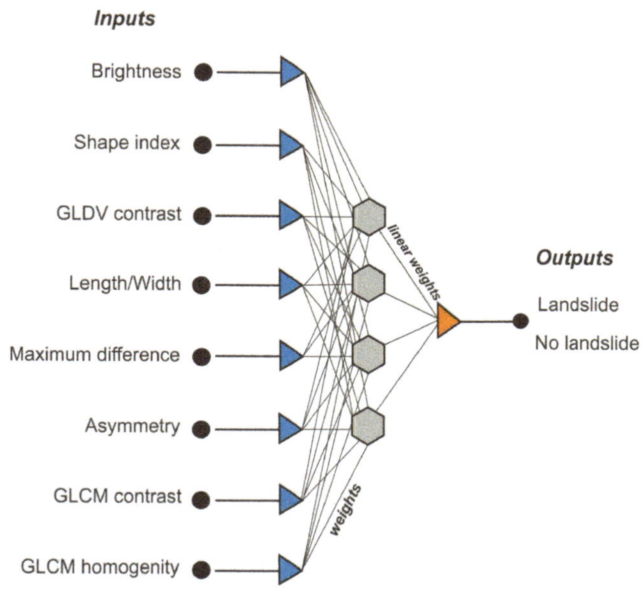

Fig. 4 ANN architecture used in landslide inventory analysis

Results and Discussion

In this study, 195 landslides were mapped using Google Earth satellite images, aerial photographs, detailed on-site studies and 1/500,000 scale landslide inventory map of the General Directorate of Mineral Research and Exploration (Dagdelenler et al. 2017). In the analysis, 25 × 25 m grid network was used. According to Varnes (1978), the landslides in the study area were mostly observed as deep massive circular failures.

In the creation of the landslide database, a total of 8 parameters, including brightness, shape index, GLDV Contrast, Length/Width, Maximum Difference, Asymmetry, GLCM Contrast and GLCM Homogeneity were taken into account. In order to determine the landslide locations automatically in the study area, these 8 parameters were used as input and landslide (1) and no landslide (0) information was applied to the study area as an output parameter. In this study, Idrisi Selva software was used to perform ANN analyses. In ANN analyses, Feed-Forward Back Propagation Neural Networks (FFBP) algorithm was used and the best

Table 2 Importance (%) values of the parameters considered

Parameter	Importance (%)
Brightness	19.9760
Shape index	6.8874
GLDV contrast	14.1773
Length/Width	8.3891
Maximum difference	0.0068
Asymmetry	16.6991
GLCM contrast	29.5576
GLCM homogeneity	4.30633

Fig. 5 Semi-automatically generated landslide inventory map of the study area

network topology is determined as 8-4-1. The values obtained as a result of the analyses were transferred to ArcGIS (10.3) software and the landslide inventory map of the study area was semi-automatically created. According to this inventory map, 78.2% of the existing landslides were estimated correctly. It should be noted that this estimate value is valid only for the study area. In other words, although this method can be applied in different areas, the areal features to be considered must be examined in a healthy way. For example, it is possible to reach wrong evaluations in applications in a field where lithology is variable because it is one of the most important parameter that directly controls the landslide mechanism.

Acknowledgements This study was supported by Hacettepe University Scientific Research Projects Coordination Unit (Project No. 014 D02 602 006-527). The author would like to thank Associate Professor Erman Ozsayin and Fatih Gulec for their valuable contribution in the data collection section of the project.

References

Aksoy B, Ercanoglu M (2012) Landslide identification and classification by object-based image analysis and fuzzy logic: an example from the Azdavay region (Kastamonu, Turkey). Comp Geosci 38:87–98

Akyol Z, Arpat E, Erdoğan B, Göğer E, Güner Y, Şaroğlu F, Şentürk İ, Tütüncü K, Uysal Ş (1974) 1/50.000 ölçekli Türkiye Jeoloji Haritası Serisi, Zonguldak E29 a, E29 b, E29 c, E29 d, Kastamonu E30 a, E30 d. MTA Yayınları, Ankara

Baatz M, Schape A (2000) Multiresolution segmentation—an optimization approach for high quality multi-scale image segmentation. In: Angewandte Geographische Informationsverarbeitung symposium, pp 12–23

Blaschke T (2010) Object based image analysis for remote sensing. ISPRS J Photogram Remote Sens 65:2–16

Bovenga F, Wasowski J, Nitti DO, Nutricato R, Chiaradia MT (2012) Using COSMO/SkyMed X-band and ENVISAT C-band SAR interferometry for landslide analysis. Remote Sens Environ 119:272–285

Cardinali M, Ardizzone F, Galli M, Guzzetti F, Reichenbach P (2001) Landslides triggered by rapid snow melting: the December 1996–January 1997 event in Central Italy. In: Proceedings of 1st EGS Plinius conference, Bios Publisher, Cosenza, 14–16 Oct 1999, pp 439–448

Dagdelenler G, Nefeslioglu HA, Gokceoglu C (2015) Modification of seed cell sampling strategy for landslide susceptibility mapping: an application from the Eastern part of the Gallipoli Peninsula (Canakkale, Turkey). Bull Eng Geol Environ 75:575–590

Dağdelenler G, Ercanoğlu M, Sönmez H, Özsayın E, Güleç F (2017) Topoğrafik haritalardan yararlanarak heyelan envanter haritalamasinin coğrafi bilgi sistemleri (CBS) ortaminda otomatik olarak elde edilmesine yönelik bir çalişma: Ulus (Bartin) havzasi. Hacettepe Üniversitesi Bilimsel Araştirma Projeleri Koordinasyon Birimi, Destek Projesi, Proje No: 014 D02 602 006–527

Definiens Professional Earth (DPE) (2006) Definiens professional 5 UserGuide. Definiens AG, Munchen, Germany, p 249

Dou J, Yamagishi H, Pourghasemi HR, Yunus AP, Song X, Xu Y, Zhu Z (2015) An integrated artificial neural network model for the landslide susceptibility assessment of Osado Island, Japan. Nat Hazards 78:1749–1776

Dou J, Yunus AP, Merghadi A, Shirzadi A, Nguyen H, Hussain Y, Avtar R, Chen Y, Pham BT, Yamagishi H (2020) Different sampling strategies for predicting landslide susceptibilities are deemed less consequential with deep learning. Sci Total Environ 720:137320

Guzzetti F, Reichenbach P, Ardizzone F, Cardinali M, Galli M (2006) Estimating the quality of landslide sus-ceptibility models. Geomorphology 81:166–184

Guzzetti F, Mondini AC, Cardinali M, Fiorucci F, Santangelo M, Chang KT (2012) Landslide inventory maps: new tools for an old problem. Earth Sci Rev 112:42–66

Konar A (2005) Computational intelligence: principles techniques and applications. Springer

Kurtz C, Stumpf A, Malet J-P, Gançarski P, Puissant A, Passat N (2014) Hierarchical extraction pf landslides from multiresolution remotely sensed optical images. ISPRS J Photogram Remote Sens 87:122–136

Lippmann RP, Gold B, Malpass ML (1987) A comparison of Hamming and Hopfield neural nets for pattern classification (Tech. Rep. TR-769). MIT Lincoln Laboratory. MIT Press, Cambridge, MA

Marangoz AM, Oruç M, Karakış S, Şahin H (2006) Comparison of pixel-based and object-oriented classification using IKONOS imagery for automatic building extraction-Safranbolu testfield. In: Proceedings of the fifth international Turkish-German joint geodetic days symposium, Berlin, Germany, pp 28–31

Martha TR, Kerle N, Jetten V, van Westen CJ, Kumar KV (2010) Characterising spectral, spatial and morphometric properties of landslides for semi-automatic detection using object-oriented methods. Geomorphology 116:24–36

Martha T, Kerle N, van Westen CJ, Jetten V, Kumar KV (2012) Object-oriented analysis of multi-temporal panchromatic images for creation of historical landslide inventories. ISPRS J Photogr Remote Sens 67:105–119

Nichol JE, Shaker A, Wong MS (2006) Application of high-resolution stereo satellite images to detailed landslide hazard assessments. Geomorphology 76(1–2):68–75

Pham BT, Prakash I, Dou J, Singh SK, Trinh PT, Tran HT, Le TM, Phong TV, Khoi DK, Shirzadi A, Bui DT (2019) A novel hybrid approach of landslide susceptibility modelling using rotation forest ensemble and different base classifiers. Geocarto Int. https://doi.org/10.1080/10106049.2018.1559885

Santangelo M, Cardinali M, Rossi M, Mondini AC, Guzzetti F (2010) Remote sensing mapping using a laser rangefinder binocular and GPS. Nat Hazards Earth Syst Sci 10:2539–2546

Tarolli P, Arrowsmith JR, Vivoni ER (2009) Understanding earth surface processes from remotely sensed digital terrain model. Geomorphology 113:1–3

Van Den Eeckhaut M, Poesen J, Gullentops F, Vandekerckhove L, Hervás J (2011) Regional mapping and characterisation of old landslides in hilly regions using LIDAR-based imagery in Southern Flanders. Quatern Res 75:721–733

Van Westen CJ, Castellanos Abella EA, Sekhar LK (2008) Spatial data for landslide susceptibility, hazards and vulnerability assessment: an overview. Eng Geol 102:112–131

Varnes DJ (1978) Slope movement types and processes. In: Schuster RL, Krizek RJ (eds) Landslides: analysis and control. National Research Council, Washington DC, Transportation Research Board, Special Report 176. National Academy Press, Washington DC, pp 11–33

Landslide Mapping Based on UAV Photogrammetry Using SfM—The Prnjavor Čuntićki Landslide Case Study, Croatia

2020 Kyoto Japan

Vedran Damjanović, Snježana Mihalić Arbanas, Sanja Bernat Gazibara, Josip Peranić, Marin Sečanj, Martin Krkač, and Željko Arbanas

Abstract

The paper presents results of identification and mapping of the Prnjavor Čuntićki landslide reactivated in March 2018 due to intensive rainfalls in Croatia. The described investigations included conventional landslide field mapping and visual mapping using high resolution digital terrain model derived from data captured by Unmanned Aerial Vehicle (UAV) and Structure from Motion (SfM). Identification of landslide contours and topography enabled back slope stability analysis that resulted in preliminary landslide model. The research results imply reactivation of large part of fossil landslide (about 25% of landslide area) and high risk due to potential damaging of ouses and local road. Preliminary interpretation of landslide mechanism will be used for planning of landslide monitoring and early warning system.

Keywords

Landslide mapping • UAV • SfM • Landslide model • Prnjavor Čuntićki

Introduction

The Prnjavor Čuntićki landslide (with the area of 7.8 ha) is large fossil landslide that possess very high risk, because it is endangering part of the village placed in the middle part of the landslide (Fig. 1). During last 60 years, the landslide area was used as agricultural area, mostly for annual crops in association with permanent crops. Only main scarp of the fossil landslide was heavily vegetated by forest. The landslide had been reactivated in spring 2014 when new cracks have been developed in upper part of the fossil landslide. Reactivation was initiated by intensive rainfall, during one of the recent MORLE (Multiple-Occurrence Regional Landslide Events) in the region (Mihalić Arbanas et al. 2017).

The fossil landslide was completely unknown to the local and regional authorities till its next reactivation in Spring 2018. Development of larger displacements was triggered by an intensive rainfall, that also caused MORLE in NW Croatia (Bernat Gazibara et al. 2019). Collapse of the reactivated landslide was prevented by urgent measures, superficial drainage by digging channels and protection of water inflow into opened cracks. After this event, landslide activity presents constant threat for 10 houses placed in the foot of the reactivated landslide or in its vicinity (Fig. 1).

This paper presents results of the identification and mapping of the Prnjavor Čuntićki landslide based on remote sensing using UAV photogrammetry in combination with Structure from Motion (SfM), followed by on-site landslide mapping. Interpretation of landslide contours of fossil and reactivated landslide was verified by slope stability back analysis. Described survey served as a base for development

V. Damjanović · S. Mihalić Arbanas (✉) · S. Bernat Gazibara · M. Sečanj · M. Krkač
Faculty of Mining, Geology and Petroleum Engineering, University of Zagreb, Pierottijeva 6, 10000 Zagreb, Croatia
e-mail: smihalic@rgn.hr

V. Damjanović
e-mail: vedran.damjanov@gmail.com

S. Bernat Gazibara
e-mail: sbernat@rgn.hr

M. Sečanj
e-mail: msecanj@rgn.hr

M. Krkač
e-mail: mkrkac@rgn.hr

J. Peranić · Ž. Arbanas
Faculty of Civil Engineering, University of Rijeka, Radmile Matejčić 3, 51000 Rijeka, Croatia
e-mail: josip.peranic@gradri.uniri.hr

Ž. Arbanas
e-mail: zeljko.arbanas@gradri.uniri.hr

F. Guzzetti et al. (eds.), *Understanding and Reducing Landslide Disaster Risk*, ICL Contribution to Landslide Disaster Risk Reduction,
https://doi.org/10.1007/978-3-030-60227-7_9

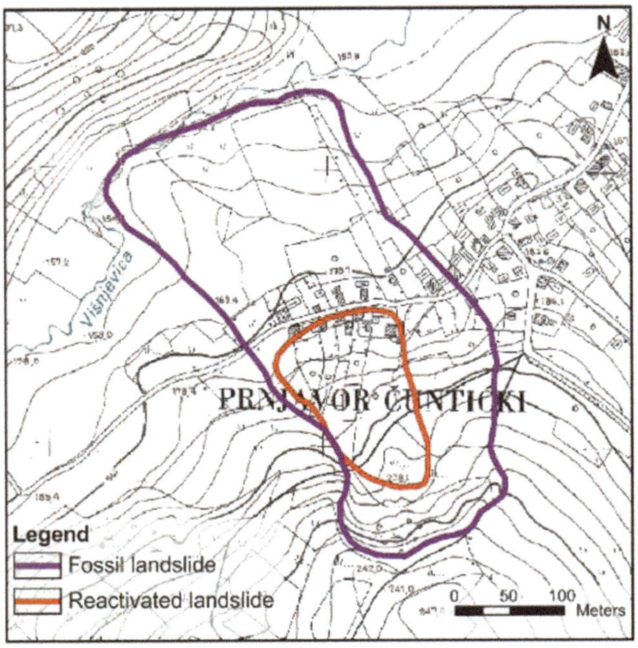

Fig. 1 Position of fossil (purple line) and reactivated landslide (red line) in Prnjavor Čuntićki village (Croatia) on topographic base map (original scale 1:7000)

of the concept of landslide monitoring for the purpose of early warning.

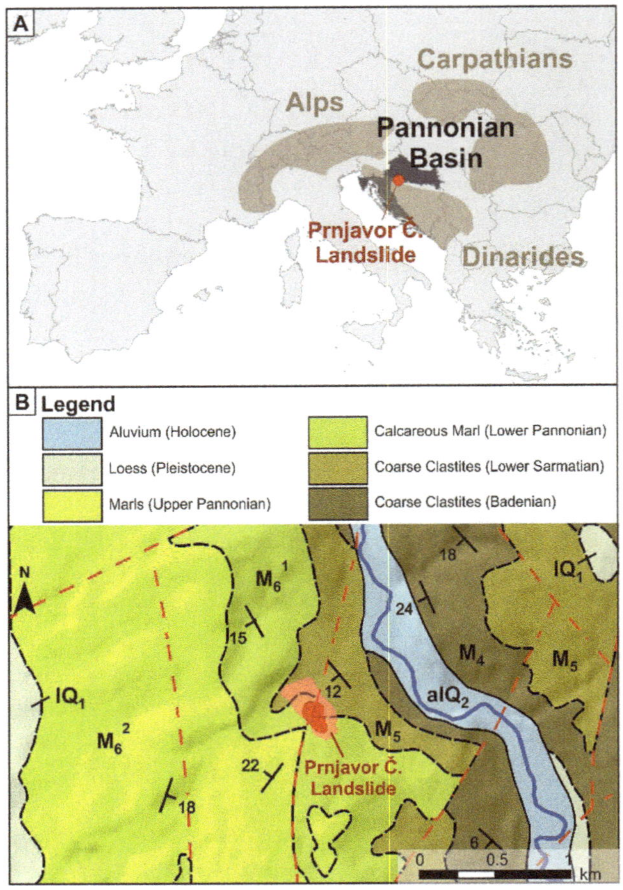

Fig. 2 Geographical location and geological settings: **a** Geographical location; **b** Basic Geological Map of Croatia in original scale 1:100,000 (Pikija 1987)

Study Area

The Prnjavor Čuntićki landslide is moderate size fossil landslide (with the total area about 80,000 m^2) that had been reactivated in the upper part of the landslide mass. Approximate dimensions of the reactivated landslide are landslide length of approx. 200 m and landslide width of approx. 100 m. The landslide is located in the village which is a part of the Petrinja City in the central part of Croatia (Fig. 2a). This area belongs to the lowlands of the Pannonian Basin, mostly built of thick Neogene and Quaternary deposits (Mihalić Arbanas et al. 2017). The landslide is developed on the slope with general dip direction towards NW and average dip angle of about 11°. Relative position of the fossil landslide is from the top to the bottom of the 500 m long slope, forming right bank of the Višnjevica stream.

Geological settings of the wider area of the Prnjavor Čuntićki landslide is complex due to several geological and fault contacts (Fig. 2b). According to Pikija (1987) the wider study area is built of clastic and carbonate deposits of Badenian, Sarmatian and Pannonian age (Middle to Upper Miocene).

Fossil landslide was formed in the Sarmatian (M_5) and Lower Pannonian ($M_6{}^1$) deposits. Sarmatian deposits consist of rhythmical alteration of thinly bedded marls and limestones with appearance of thin sandstones layers (Pikija 1987). Lower Pannonian deposits continuously overly Sarmatian deposits and they consist of rhythmical alteration of marls, clayey limestones and thin bedded limestones while layers of sandy marls and fine-grained sands appear only sporadically. Main characteristic of this stratigraphic unit is heterogeneity and anisotropy caused by significant differences in strength of marls and calcareous rocks. Generally, dominating type of rocks in the area are marls, which present very weak to weak rocks according to ISRM (2007).

According to the Pikija (1987), bedding planes of the Sarmatian and Lower Pannonian deposits at the study area are generally gently dipping toward the SW, with the dip angle of 6–15°. Bedding planes of the Upper Pannonian deposits, located near the SW landslide boundary, are dipping towards the NW, with the dip angle about 22°, which indicates that the contact between these two stratigraphic units can be erosional disconformity. Also, according to the same author, there is Quaternary fault that passes through the middle part of the fossil landslide (Fig. 2). It is characterized as right-lateral strike-slip fault, striking NNE-SSW with the tectonic transport of around 100 m (Pikija 1987). These complex structural-geological settings, together with the alteration of weak (marls) and hard (calcareous) rocks could present predisposing factors for development of the large fossil landslide.

Materials and Methods

Identification and mapping of fossil and reactivated landslides was carried out following three steps: UAV-based photogrammetric mapping of wider area of the fossil landslide in February 2020; SfM photogrammetry from images acquired by UAV; field mapping of landslide features using surface observation of deformations related to landslide movement in the period 2018–2020.

UAV Photogrammetric Mapping

UAV photogrammetric mapping was performed within one field survey on February 2020 with an extent of surveyed area of 300,000 m^2. UAV used for the purpose of landslide mapping was DJI Phantom 4 Pro, 4-motor semi-professional drone. It is equipped with 1-inch 20-megapixel CMOS sensor. Total of 10 tarpaulin markers (dimension of 1 × 1 m) with an QR code were placed on the terrain surface staggered to ensure at least two markers per image. Also, seven GCPs were placed as fixed plastic geodetic markers on randomly chosen positions inside the landslide. Their geodetic positions were determined by GNSS measurements with accuracy of 5 mm. The flight path was irregular and the total number of captured images was 1348.

SfM Photogrammetry

The photogrammetry model was created in Pix4D software. Classification of ground points in point cloud was performed using CloudCompare software. Digital Terrain Model (DTM) surface raster was generated in ArcGIS 10.6 by interpolation of points obtained from bare-earth point cloud. Technique used in Pix4D was SfM which uses computer algorithms to identify matching features in a set of overlapping digital images. Totally, 9,326,834 features were identified as a result. Seven GCPs were used for georeferencing models with geodetic locations in HTRS96/TM coordinate system for the Republic of Croatia.

Field Mapping

Reconnaissance landslide mapping of the reactivated landslide was conducted in the field in March 2018, in the period of landslide reactivation. The aim was identification of fresh landslide features and other ground deformations to identify contour of the reactivated landslide. Reconnaissance field mapping was conducted one more time in February 2020 to check newly developed landslide features. Phenomena, such as formation of springs and superficial flows in the displaced material in the backyard of the houses, including one spring in basement of one house were observed as additional indicators of landslide activity. Damages on residential buildings and other structures were poorly expressed, visible only in the form of cracks on horizontal surfaces, including local road.

Results

Results of the three described steps used in identification and mapping of fossil and reactivated landslides are shortly described here. UAV photogrammetric mapping as well as SfM photogrammetry resulted in 3D point cloud and DTM of 30-cm resolution of the area of fossil landslide. Visual interpretation of DTM enabled identification of fossil landslide contours. Field mapping of fresh landslide features enabled identification of contours of reactivated landslide. Construction of engineering geological longitudinal cross section of both landslides in combination with slope stability back analysis resulted in determination of landslide models and possible depths of landslide surfaces.

Point Cloud

In the first step of matching images and Tie Point generation, usage of markers on the ground resulted in high median of 30,045.7 matches per calibrated image (1315 of 1348 images were calibrated). High number of Tie Points from first step ensures a robust geometric link between images, important for dense point cloud generation in next step. After first step of image matching, software generated point cloud with 95,972,320 of 3D densified points, with average density of 712.55 points per cubic meter. In order to generate DTM, filtering process was applied on previously generated point cloud to separate ground and non-ground points. Cloth Simulation Filter (CSF) was used, which is surface-based method that gradually approximate the ground surface by iteratively selecting ground measurements from the original dataset. The core of this type of filtering method is to create a surface that approximates bare-earth (Zhang et al. 2016).

Table 1 Model geolocation accuracy and raster resolutions

	GCP	Orthophoto	DTM
Accuracy	0.774 m	1.46 cm/pixel	30 cm/pixel

After filtering, total number of points that represented ground points in point cloud was 64,197,239. Unfortunately, view of clear-ground surface was obstructed in parts of the terrain covered with dense vegetation, resulting in holes in final DTM after a use of CSF filter. Accuracy of model related to GCP error, as well as resolution of DTM and Orthophoto are listed in Table 1.

DTM generated from ground points of the point cloud, was used to derive slope and contour maps. Orthophoto generated from colorized point cloud (Fig. 3a) had a value of Ground Sampling Distance (GSD) of 1.46 cm/pixel which would suggest high detail image. Slope map, created using ArcMap software, is presented in Fig. 4a. Contour line map was created with contour interval of 0.5 m. Despite deformations in the DTM, caused by dense vegetation, both derivative maps express significant landslide features that

Fig. 3 Contours and features of the Prnjavor Čuntićki landslide: **a** perspective view of the orthophoto map derived by photogrammetry from March of 2020 with interpreted contours of the fossil and reactivated landslide; **b** part of the main scarp of the reactivated landslide; **c** left lateral flank of the reactivated landslide; **d** crack in the zone of accumulation of the reactivated landslide; **e** bulging related to transverse ridges in the zone of accumulation of the reactivated landslide. Photo by Edina Žuko

enabled visual identification of landslide contours of fossil landslide. Same maps were also used for verification of landslide features of the reactivated landslide during reconnaissance mapping.

Landslide Models

Field mapping of the Prnjavor Čuntićki landslide after activation of the recent landslide in 2018 as well as mapping using visual inspection of SfM DTM derivative map from 2020 enabled identification of two landslides, fossil and reactivated landslides (Fig. 3a–d). Position of sliding surfaces of both landslides were identified based on identification of landslide contours, geometry of the topographic cross-section and slope stability analyses.

Landslide models have been defined by: (i) landslide geometry depicted on landslide map (Fig. 4a) and on longitudinal engineering geological cross section (Fig. 4b); and (ii) landslide features, landslide type regarding mechanism of sliding, landslide activity (states, distributions and styles), type of landslide material and landslide causes (preparatory and triggering causal factors). The approximate positions of slip surfaces were determined by numerical analysis using strength reduction method (Matsui and San 1992). Based on presumption of homogenous soil material in the slope and assumption of initial strength parameters typical for weathered marls, numerical analysis using Rocscience RS2 software (Rocscience 2018) was conducted. The slip surface was determined at the position where the yielding is occurred in the numerical model after reduction of initial strength parameters to the values corresponding a possible resistance along the slip surface at FoS = 1.0 (Fig. 5). Slip surfaces for both, fossil and reactivated landslides are identified as relatively slightly inclined in the middle part. They are also relatively deep, comparing to the total length of landslides. This implies possibility that slip surface of fossil landslide is significantly shallower, formed along the geological contact of stratigraphical units (Sarmatian and Lower Pannonian deposits). Reactivated landslide is formed in the colluvial deposits of the fossil landslide, but the geometry of its slip surface can be only slightly changed.

The fossil landslide was activated as a result of instability of natural slope, most probably before population of this part of the Prnjavor Čuntićki village. Local road with houses was constructed through middle part of the displaced mass of the fossil landslide. Most probably, the landslide is translational sliding with sliding surface predisposed by geological contact. Based on preliminary landslide map (Fig. 4a), the total length of the fossil landslide is about 446 m, and the width of displaced mass is about 180 m. According to the preliminary landslide cross-section (Fig. 4b), approximate depth of the displaced mass and the depth of the rupture

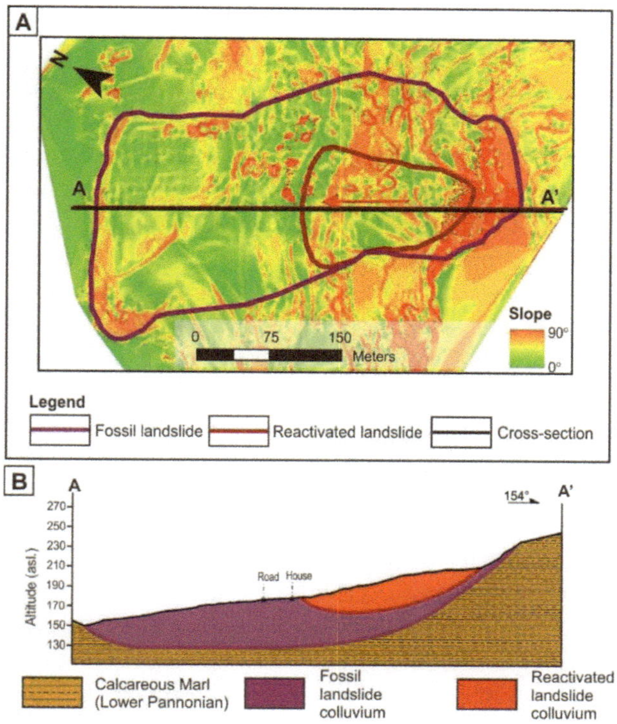

Fig. 4 Landslide map (**a**) and preliminary longitudinal engineering geological cross-section (**b**) of the fossil landslide and landslide reactivated on March 15, 2018. Original scale is 1:7000

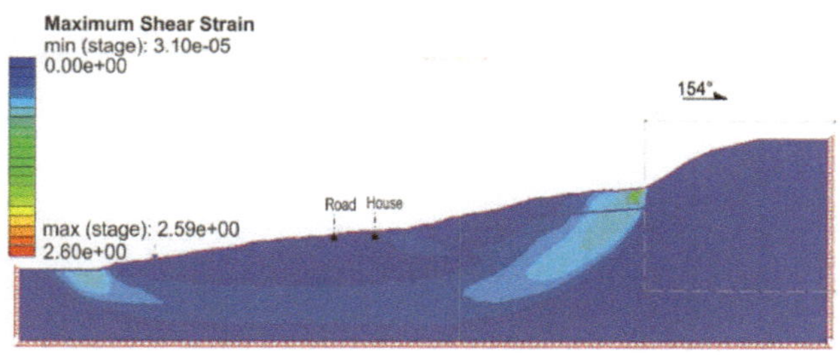

Fig. 5 Results of numerical analysis using strength reduction method in Rocscience RS2 software. The light blue zone present yielding zone and the most probable position of a slip surfaces

surface is about 53 m. Visible landslide features are landslide crown and main scarp, left landslide flank, the displaced mass and landslide foot with the toe that are clearly expressed. Right landslide flank is not clearly visible, because it is partially removed by man-made activities. Landslide materials are superficial deposits and weathered rock originating from marl and calcareous rocks. Preparatory causal factors of the historical landslide were interpreted as follows: weathered and jointed materials of the bedrock, most probably adversely oriented mass discontinuities (including bedding and joints related to faulting), fluvial erosion of the slope toe and most probably prolonged high precipitation. Triggering causal factors of the fossil landslide are not known because the landslide activated in 2018 was developed as a result of instability inside landslide mass of the fossil landslide.

In the foot part of the reactivated landslide are courtyards of six houses and in the direct vicinity of the zone of landslide toe is local road and four houses. According to the preliminary model (Fig. 4b) landslide type is identified as rotational landslide. In case of shallower slip surface of the fossil landslide, reactivated landslide can be translational, along surface of rupture of the fossil landslide. Based on preliminary landslide map (Fig. 4a), the total length of the reactivated landslide is about 182 m and the width of the displaced mass is about 105 m. According to the preliminary landslide cross-section (Fig. 4b), approximate depth of the displaced mass and the depth of the rupture surface is about 30 m. Visible landslide features are clearly expressed landslide crown and main scarp, left landslide flank, the displaced mass and landslide foot. The toe of the reactivated landslide is not clearly visible because this part is built and densely covered by houses and other structures. Landslide materials are superficial deposits derived by fossil sliding (colluvial soil). Preparatory causal factors of the landslide activated in 2018 were interpreted as follows: sheared, weathered and jointed materials of the colluvium; contrast in permeability inside colluvium and between colluvium and undisturbed ground and its effects on groundwater level; prolonged high rainfall; and vegetation removal (deforestation) at the area of the fossil landslide. Triggering causal factors of the recent landslide is rainfall event ending in May 2014.

Discussion and Conclusion

This paper presents preliminary study of the Prnjavor Čuntićki landslide in which UAV photogrammetry have been used to support other techniques of landslide identification and mapping in detailed scale. The aim of the study was the elaboration of landslide model and the assessment of the exposure of buildings and infrastructure to landslide risk. Although DTM contains errors and uncertainties in terrain surface calculations, it was possible to obtain true DTM from photogrammetric point clouds, as most of the area is not covered by dense vegetation. The analysis of this model of very high resolution allowed the detection of some scarps and other features of fossil and reactivated landslides by means of photo interpretation. Orthophotographs, contour and slope maps were used for textural analysis of the joints and other morphological features of landslides. The DTM was also used to obtain longitudinal profile of the landslide. Detection of superficial landslide features, as well as preliminary interpretation of slip surfaces, proved advantages of UAV photogrammetry and SfM in early stages of landslide investigation, because they allow fast, low cost and effective data acquisition.

Engineering geological model of the landslide, obtained as a result of the field and UAV mapping, SfM photogrammetry, as well as from back slope stability analyses, served also as a model for designing of the monitoring and Early Warning System (EWS) and for planning site investigation for the design of remediation measures. Due to high risk, protection of residents has the highest priority. The preliminary design of the monitoring and EWS includes at least two short-span extensometers, weather station, and two boreholes equipped with inclinometer casing and water level sensors. The first extensometer is planned at the toe of the reactivated landslide, location easily reachable by the residents who can be involved in monitoring activities. The purpose of the extensometer is to measure the relative displacement of the most active part of landslide, close to the nearby houses. The second extensometer is planned at the top part of the landslide, enabling measurements of the total movement of the reactivated landslide and eventual movement of fossil landslide. The weather station is planned in the stable area on the east form landslide, to be protected from the landslide movement. Finally, two boreholes are planned in the upper and in the foot part of the reactivated landslide. The planned depth is up to 55 m. The purpose of the boreholes is to confirm the depths of the sliding surfaces (reactivated and fossil landslides), to provide some data about geomechanical properties of landslide mass and to obtain data about groundwater level. Determination of borehole cores will enable verification of landslide depths as well as landslide models, and that information are necessary for EWS as well as for remediation design.

Acknowledgements The research presented in this paper was supported by EU funds, EFRR/KF/ESF Fund in the framework of the Project No. 05.1.1.02.0020 entitled "Applied landslide investigation for development of landslide mitigation measures and risk prevention" (PRI-MJER). The authors also want to thank to the anonymous Reviewer for his constructive comments.

References

Bernat Gazibara S, Cindrić Kalin K, Erak M, Krkač M, Đomlija P, Arbanas Ž, Mihalić Arbanas S (2019a) Landslide hazard analysis in national-scale for landslide risk assessment in Croatia. In: Proceedings of the 4th regional symposium on landslides in the Adriatic-Balkan region, 23–25 Oct 2019, Sarajevo, BIH, pp 175–182

ISRM (2007) The complete ISRM suggested methods for rock characterization, testing and monitoring: 1974–2006. In: Ulsay R, Hudson J (eds) Suggested methods prepared by the commission on testing methods. International Society for Rock Mechanics, Compilation Arranged by the ISRM Turkish National Group, Ankara, 628p

Matsui T, San K-C (1992) Finite element slope stability analysis by shear strength reduction technique. Soils Found 32(1):59–70

Mihalić Arbanas S, Sečanj M, Bernat Gazibara S et al (2017) Landslides in the Dinarides and Pannonian Basin—from the largest historical and recent landslides in Croatia to catastrophic landslides caused by Cyclone Tamara (2014) in Bosnia and Herzegovina. Landslides 14:1861–1876

Pikija M, Basch O, Šimunić A, Šikić K, Marinčić S (1987) Guideline for Sisak Sheet L 33–93 of the Basic Geological Map 1:100,000. Federal Geological Survey, Belgrade, Yugoslavia, 52p (In Croatian)

Pix4D (2020) Step 1. Before starting a project > 4. Getting GCPs on the field or through other sources (optional but recommended). https://support.pix4d.com/hc/en-us/articles/202557489-Step-1-Before-Starting-a-Project-4-Getting-GCPs-on-the-field-or-through-other-sources-optional-but-recommended. Last accessed 1 Apr 2020

Rocscience (2018) RS2 tutorial. www.rocscience.com/help/slide2/#t=tutorials%2FSlide_Tutorials.htm. Last accessed 01 Mar 2020

Zhang W, Qi J, Wan P, Wang H, Xie D, Wang X, Yan G (2016) An easy-to-use airborne LiDAR data filtering method based on cloth simulation. Remote Sens 8(6):501–523

Developing Recognition and Simple Mapping by UAV/SfM for Local Resident in Mountainous Area in Vietnam—A Case Study in Po Xi Ngai Community, Laocai Province

2020 Kyoto Japan

Nguyen Kim Thanh, Toyohiko Miyagi, Shinobu Isurugi, Dinh Van Tien, Le Hong Luong, and Do Ngoc Ha

Abstract

Laocai is one of the provinces of northwestern Vietnam that is at high risk of landslide. Typical natural features of this area are hilly terrain, complex geology, tropical monsoon climate with large annual average precipitation. People of ethnic minorities in the area up to now still affected by natural disasters. They desperately need appropriate tools for identifying and responding to this disaster. This study is a test fand quick landslide inventory map-making based on the application of modern technologies such as UAV/SfM to the building up of landslide identification and distribution maps in combination with checking and revising by site survey with the participation of local people for an area of Po Xi Ngai village, Lao Cai province. By recognizing the landslides on 3D images of the area in conjunction with the fieldwork checking, this method is expected to give local people the knowledge of landslide identification to develop the prevention and mitigation strategies by their capacity.

N. K. Thanh (✉) · D. Van Tien · L. H. Luong
Institute of Transport Science and Technology, 1252 Lang Street, Lang Thuong, Dong Da, Ha Noi, Vietnam
e-mail: nguyenkimthanh.itst@gmail.com

D. Van Tien
e-mail: dvtien.gbn@gmail.com

L. H. Luong
e-mail: lehongluong@gmail.com

N. K. Thanh
Ph.D. Student, Tohoku Gakuin University, Sendai, Japan

S. Isurugi
Tohoku Gakuin University, Sendai, Japan
e-mail: isurugi@mail.tohoku-gakuin.ac.jp

T. Miyagi
Advantechnolog Co. Ltd., 1-4-8 Kakyoin Abaku, Sendai, 980-0013, Japan
e-mail: c1934009@mail.tohoku-gakuin.ac.jp

D. N. Ha
University of Yamanashi, 4-4-37 Takeda, Kofu-shi, 400-8510, Yamanashi-ken, Japan
e-mail: ha.yamanashi@gmail.com

Keywords

Landslide • Simple mapping • UAV/SfM • Local resident • Mountainous area • Vietnam

Introduction

Vietnam is a country in Southeast Asia with a population of over 96 million people and 54 ethnic groups in the territory (General Statistics Office 2019). With ¾ mountainous topographic features, complex geology and monsoon tropical climate with large annual average rainfall, Vietnam is often affected by natural disasters, such as landslides. Annual landslides claim hundreds of lives and properties (Yem et al. 2006). Proactively facing natural disasters, especially landslides, is an important orientation of the government. To achieve this goal, the landslides identification at different levels of management is critical key and it is particularly significant for grassroots levels such as local communities—areas directly affected by landslide natural disasters.

Research Methodology

The traditional method for the identification of landslides to build landslide inventory maps is to use geomorphological and geological knowledge in considering of surface deformation to compare with the recognizing features of the landslide, to provide interpretation (Miyagi et al. 2004). Virtual eye-bird viewpoints to target terrain are usually set up using stereoscopic images, which were taken with high-resolution cameras on various flying vehicles. However, the use of aerial photographic data from airplanes or

F. Guzzetti et al. (eds.), *Understanding and Reducing Landslide Disaster Risk*, ICL Contribution to Landslide Disaster Risk Reduction,
https://doi.org/10.1007/978-3-030-60227-7_10

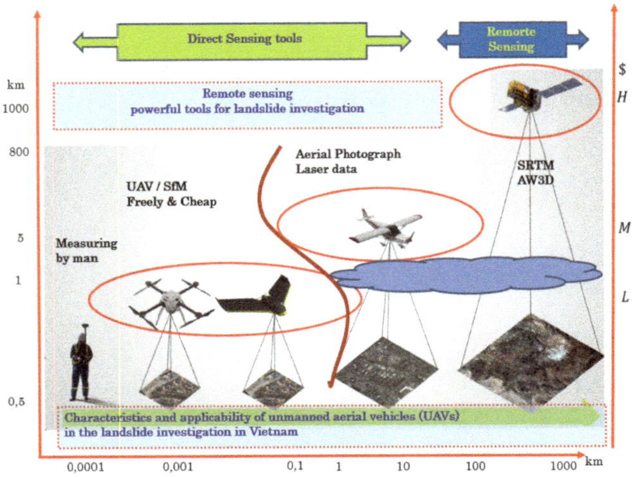

Fig. 1 Concept shift from the technology development to easy high-tech low cost

satellites is inadequate and unavailable under real-world conditions in Vietnam.

The high-resolution photogrammetry approach as direct sensing is not only meaningful to the scientist but also useful to the residents. The high-resolution camera, setting up on Unmanned aerial vehicles (UAV) are suitable for acquiring high-resolution images, and quickly. The advantages are an increased very high temporal and spatial resolution, flexible deployment, and the potential for very rapid data acquisition and processing in comparison to traditional remote-sensing methods (Fig. 1). It is also able to monitoring and analysis of active landslide involves both spatial and temporal measurements and requires a continued assessment of landslide conditions, including the extent and rate of displacement as well as changes in the surface topography. The stereoscopic image of topography is easy for recognizing and understanding as to spatial to local people.

With this study, from the terrain images taken by the UAV/SfM method, through automatic processing, a DSM model was built. The combination of topographic maps created from DSM with the achievements of "Landslide topography mapping through aerial photo interpretation" (Miyagi et al. 2013) and "Abstracting unstable slopes (landslide topography) using aerial photos and topographic maps" (Hamasaki et al. 2013), Landslides in the target area are quickly identified.

Characteristics of Research Area

To build the ability to identify potential landslides for local communities with simple, fast, low cost, appropriate proportions, the use of terrain photography was carried out. The technology of taking a topographic photo using UAV/SfM combined with the construction of contour lines to identify the characteristics of DSM model landslides has been tested for some northern areas of Vietnam, in the Lao Cai province.

The location of the study area is Po Xi Ngai village, Trung Chai commune of Sapa Town, Lao Cai province (Fig. 2), limited from latitude 22°24′58 N–22°25′50 N and longitude 103°54′23 E–103°52′57 E. This area is famous for its beautiful natural landscape with terraced fields.

Located on a natural slope with an average slope angle of 25–30° of the mountain range, running on the northwest-southeast direction, the study area has an elevation from the foot to the top of the Peak as 850–1400 m (Fig. 2). Mong Sen stream flows along the foot of mountain range, merging near perpendicularly with Ngoi Dum stream at the southeast corner of the study area creates convex terrain. The National highway 4D runs in a northeast-southwest direction parallel to the Ngoi Dum stream in this area.

Geologically, the study area is located in the area of Po Sen intrusive block (Fig. 3) with mainly Diorite, granodiorite and granite. The material on the slope is a thick, weathered Sialferite crust that creates a mixture of granite granules mixed with fine-grained sand, which has high air porosity and easy to reduce the shear resistance causing landslides during the rainy season (Van Ngoi et al. 2008; Duan et al. 2011). The average annual rainfall is 2000–3600 mm. Heavy rainfall is concentrated in June, July and August, accounting for 80–85% of the annual rainfall.

In terms of land use, the area is covered by natural vegetation interspersed with agricultural land and planted forests. The main agricultural crop is rice and maize, which are cultivated on terraced fields along the slope. The area's residents are the Dao community, with more than 38 households and more than 200 people, who have lived here for generations.

Research Results

In this study, three main areas with an area of 3.6 km^2 were conducted in 2019 using UAV—Phantom 4 pro. The study areas are shown in Fig. 4. Region A (Fig. 5) is the Mong Sen area, where many landslides have been recorded in 1998, 2002 and 2004 (Yem et al. 2006). Region B (Fig. 6) is the area of the old Po Xi Ngai village, where landslides appeared in 2016 and local authorities have decided to move the community of people living in this area since 2017. Region C (Fig. 7) is the resettlement area of Po Xi Ngai Community. This place has a narrow scope and high construction density so it will still be considered as a high risk of landslide in the rainy season.

After taking photos by UAV, the DSM data obtained from UAV images by using by Pix 4 D Mapper software with Ground Sample Distance (GSD) was developed. Orthomosaic images by UAV and simple mapping is

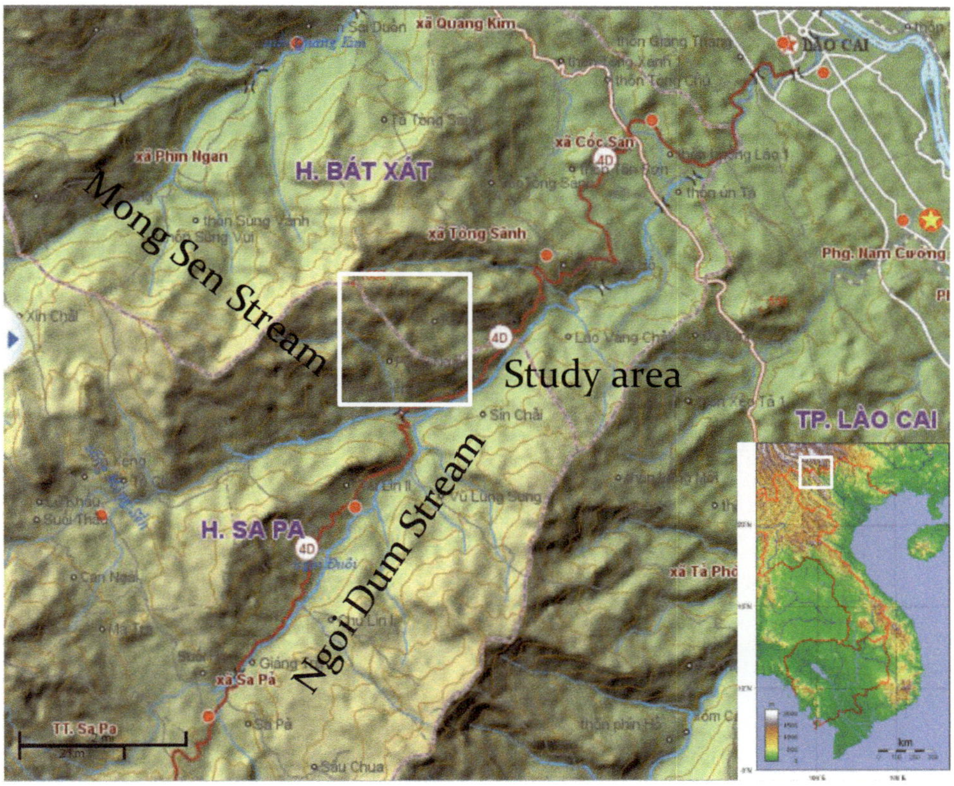

Fig. 2 Location of Lao Cai province and study area (*Source* Atlas Laocai)

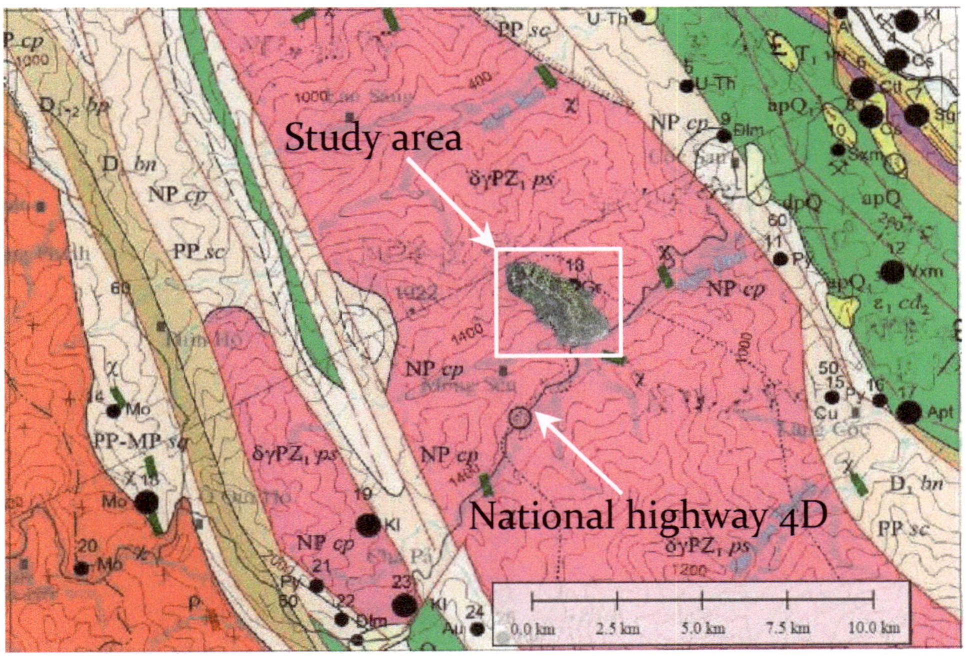

Fig. 3 Geological map (DGM 1999) and location of the case study

$\delta\gamma PZ_1$ *ps* Posen complex: diorite, granodiorite, granite NP *cp* Cha Pả Formation: quartz-sericite schist, marble. Thickness 250m

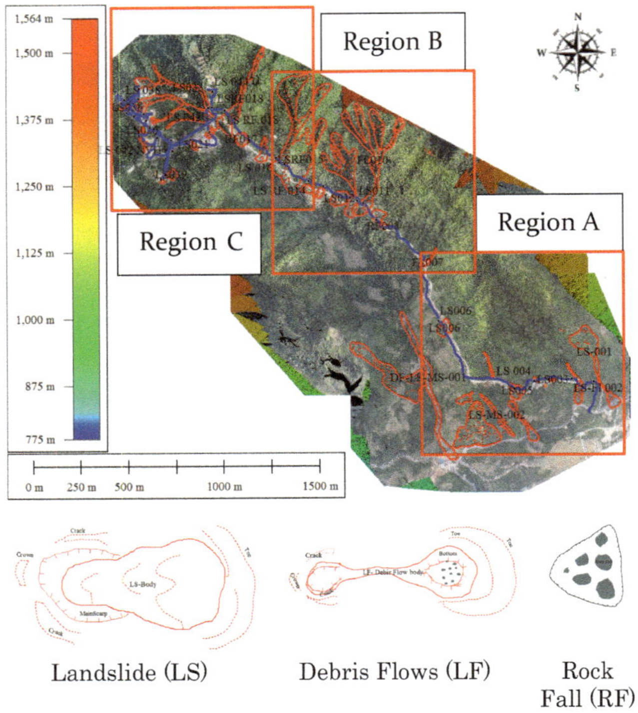

Fig. 4 Orthomosaic by UAV and simple mapping

presented in Fig. 4. The topographic maps with contour lines of 5 m were built. Using methods of image recognition (Miyagi et al. 2013) to carry out the identification of landslides from DSM topographic maps combined with 3D images. Field surveys redefining landslides to check the accuracy and re-calibration were conducted with the participation of local people.

As a result, more than 40 landslides, rocks fall and debris flows were identified along the road NH4D to Po Xi Ngai village area. Detail of results of landslide identification analysis in the study area shown in the discussion section.

Discussions

(i) Landslide recognition accuracy. Three landslide identification regions were conducted in the study area to build landslide identification and distribution map. Then the site survey was taken place for evaluation of the accuracy. The results of the survey show that the identification marks on the DSM model and Orthomosaic are completely consistent with those found in the actual terrain.

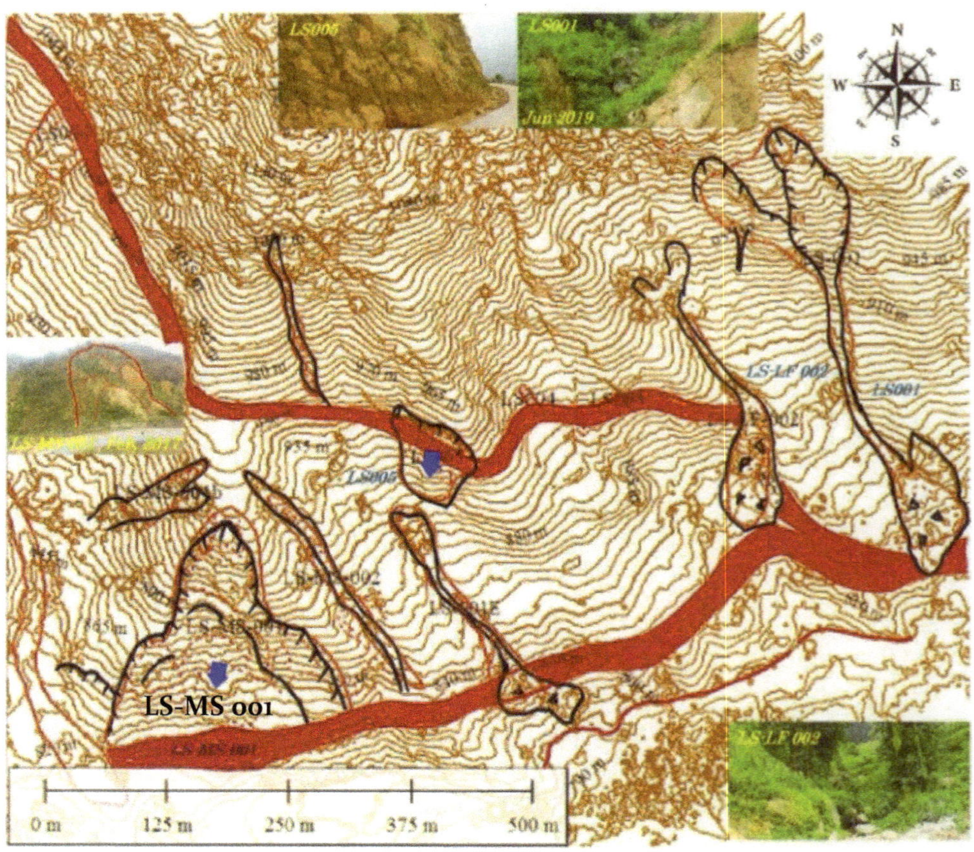

Fig. 5 Landslide recognition and simple mapping by UAV for Region A

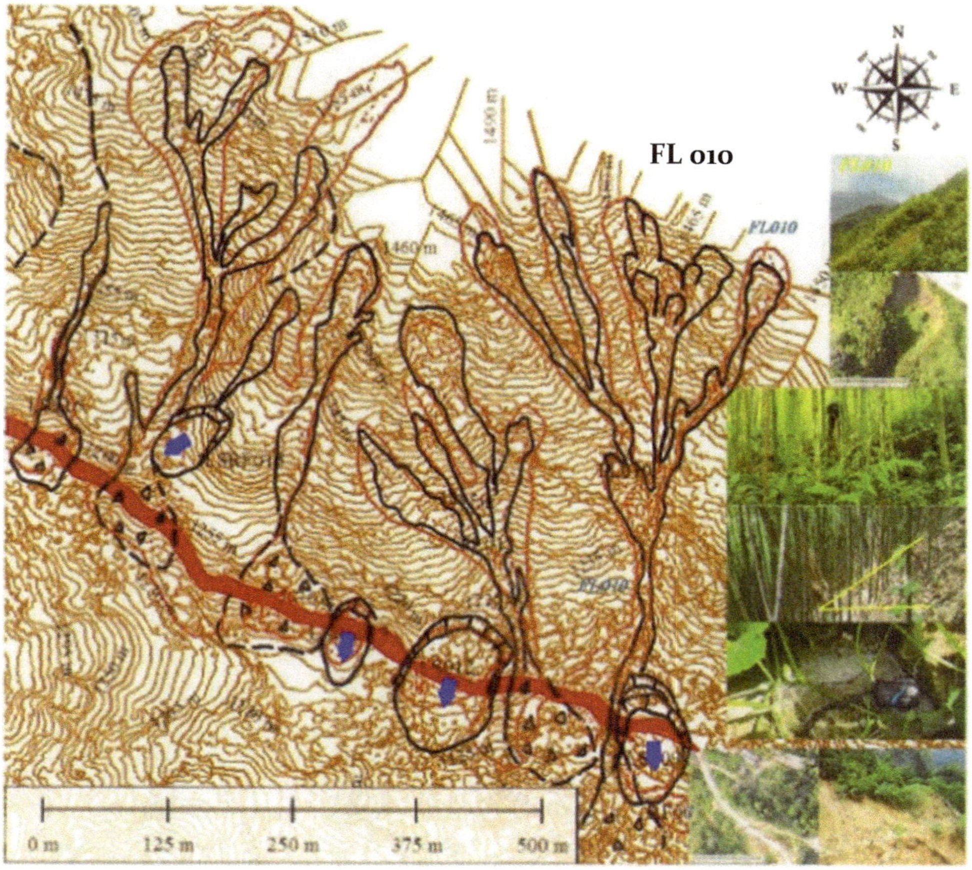

Fig. 6 Landslide recognition and simple mapping by UAV for Region B

In Region A, see the details of Fig. 5-Mong Sen Bridge area (refer to LS-MS 001), there is a typical landslide block with a scale of 250 m width, 200 m height has recognized. The reactive activities had happened in 1998, 2002 and 2004. The slip surface depth is estimated as of 10–20 m and the moved landslide body was changed form and classified into many landslide blocks. The main and lateral scarps are observed to coincide with the identification positions on the DSM model and Orthomosaic image. On upper the main scarp zone, there are some opening cracks and two of them had changed from and developed into the gullies and small debris flows.

In Region B, see the details of Fig. 6, In August 2016, from the impact of storm No. 2, there were debris flows appeared on the natural slope. The initial mechanism of landslides in this area is the appearance of the gullies on the depth weathering granite zone of the natural slope. The depth of gullies is from 1–3 m. The head of them separated but the downstream are linked together to make Fan-shaped terrain. In areas the main scarps of debris flows can be easily recognized on the Orthomosaic image (refer to FL 010). Many opening cracks were discovered on the slope of sides and the upper of debris flows. They made clear signals for identification on the surface slope. Residents have been relocated and resettled to region C, however, this area is now still used as a place, where farmers cultivate and produce.

In Region C, see the details of Fig. 7, Many translations slide mixing up with debris flows arrange on the studied slope (refer to LS-RF 018). Translation slides have average dimensions as 100–150 m width, 50–125 m height and 5–10 m depth with the main scarp in a horseshoe shape. The debris flows make depth gullies and trenches. The high-density appearance of the small and middle of these types of landslides predicts a high risk of large mass movement potentially.

(ii) Advantages and Limitations. With simple equipment and manipulation, the technology of UAV/SfM has proved to have an advantage compared to traditional air photo technology. Orthomosaic from UAV allow identification from the whole to the details of the landslides by different shooting heights. This is especially meaningful for the identification survey in the first phase of the landslide movement as well as in the process of moving (high-risk phases) without having to directly survey on site (Thanh 2018). However, due to the development of topographic

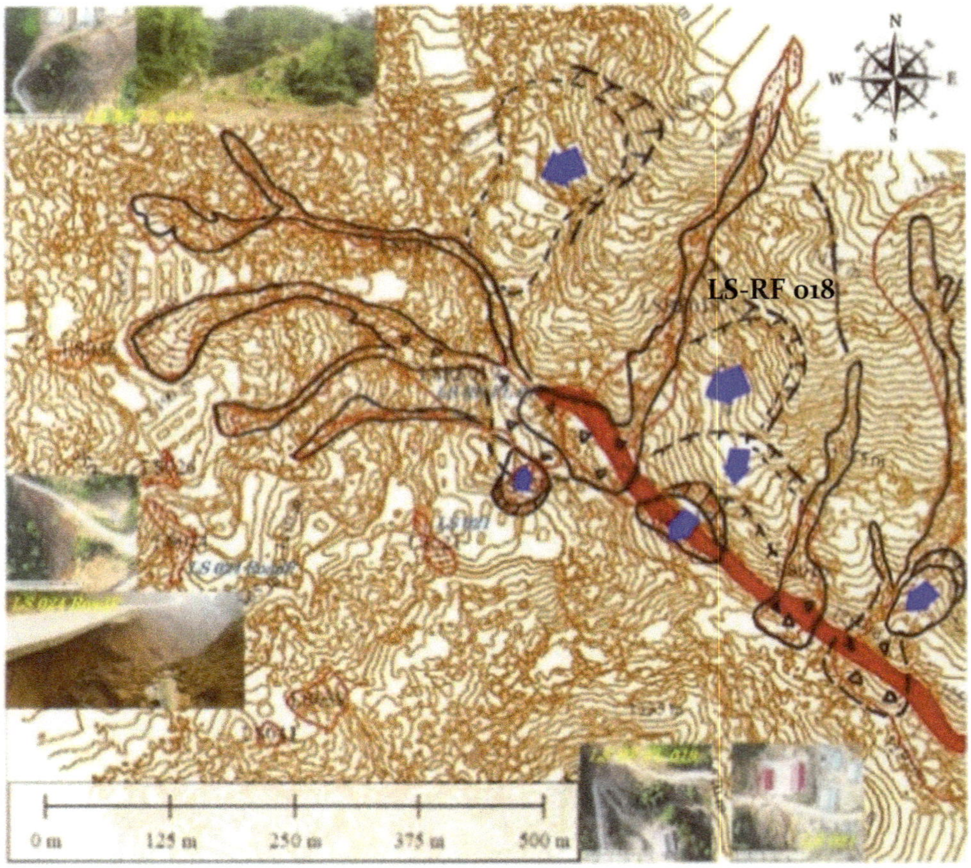

Fig. 7 Landslide recognition and simple mapping by UAV for Region C

maps of the DSM model, so for steep slopes covered with a variety of plants with different heights, simple and quick identification will be difficult.

(iii) Capacity for application: From the results of the study, the technology of UAV/SfM is effective. This is not only useful for recognize figure, boundary and the micro features of the landslide but opening the capacity for un-continuous monitoring of landslide movement by comparing and interpretation the difference phase images. The advantages of technology should be applied wider to "as the powerful tool for realistic prevention, disaster reduction and avoid the risky site of the local community area and roadside".

Conclusionsc

(i) Developing fast, simple landslide identification technology that is suitable for the local people's cognitive ability is a realistic requirement. UAV/SfM landslide identification technology has been tested based on applying basic scientific knowledge, modern technology, which made by researchers and identifying and verifying the actual scene with resident participation was conducted and proved to be effective.

(ii) Testing results show that More than 40 landslides, rocks fall and debris Flows are identified and recorded along the road from NH4D to Po Xi village through UAV/SfM photography technology. The UAV photo analysis and interpretation combined with contour lines Terrain map developed from DSM were applied. The actual site survey results the landslide inventory map, which was build up from these technologies are quite consistent.

(iii) From topography images taken by UAV, Landslide identification can be made not only science sector but also by the residents. The UAV related technology will be the tie technology between the field investigation data and realize the facts facies.

References

Chu Van Ngoi et al (2008) Assess the risk of landslides along the 4D route based on studying the relationship between geological structure and topography

DGM (1999) Geological Map of Kim Binh—Lao Cai Province, Vietnam (F-48-VIII and F-48-IV, Scale 1:200,000) published by the Department of Geology and Minerals of Vietnam

Duan NB et al (2011) Studying to determine causes of the landslide in the area of the Mong Sen Bridge, Lao Cai Province. Vietnam J Earth Sci

General Statistics Office (2019) The statistical yearbook of Vietnam. General Statistics Office of Vietnam

Hamasaki et al (2013) Abstracting unstable slopes (landslide topography) using aerial photos and topographic maps: concept and frameworks, ICL Landslide teaching tools. ISBN 978-4-9903382-2-0

Miyagi et al (2013) Landslide topography mapping through aerial photo interpretation, ICL Landslide teaching tools. ISBN 978-4-9903382-2-0

Miyagi et al (2004) Landslide risk evaluation and mapping—manual of aerial photo interpretation for landslide topography and risk management. Report of the National Research Institute for Earth Science and Disaster Prevention, No. 66; September 2004

Thanh NK (2018) Research application take photo remote sensing technology using unnamed aerial vehicles (UAV) for landslide survey along transport arteries in Viet Nam, Research under MOT

Yem NT et al (2006) Assessment of landslides and debris flows at some prone mountainous areas Vietnam and recommendation of remedial measures. Phase I: a study of the east side of the Hoang Lien Son mountainous area of Vietnam. Institute of Geological Sciences

Landslide Activity Classification Based on Sentinel-1 Satellite Radar Interferometry Data

2020 Kyoto Japan

Vladimir Greif, Jaroslav Busa, and Martin Mala

Abstract

Kosice basin located in the Eastern part of Slovakia heavily affected by landslides was studied using radar data from Sentinel-1A satellite. Existing landslide inventory activity map was reassessed using processed data from Sentinel-1A radar mission acquired between December 2014 and May 2017. PS InSAR technique was used for generation of landslide displacement permanent scatterers inside the landslide area, where average LOS velocity was applied for assessment of landslide activity in the form of thematic map. Using alternative method a LOS deformation velocity vectors were transformed into the slope direction generated from DEM resulting in k_{SLOPE} velocity data used for alternative classification map. This was possible thanks to availability of radar data from both acquisitions (ascending and descending) on the studied AOI. Comparison of both methods showed increase in number of landslides classified as active on behalf on medium active class when k_{SLOPE} transformed data were used, resulting in more comprehensive activity classification map. The transformation of velocity vector v_{LOS} must be, however, used with caution, due to variable sensitivity of radar data in different directions with regard to the satellite path.

Keywords

Sentinel-1A • Landslide activity • Monitoring • Radar interferometry • Kosice basin

V. Greif (✉) · J. Busa · M. Mala
Faculty of Natural Sciences, Department of Engineering Geology, Comenius University in Bratislava, Ilkovicova 6, Bratislava, 84215, Slovakia
e-mail: vladimir.greif@uniba.sk

J. Busa
e-mail: busa7@uniba.sk

M. Mala
e-mail: mala26@uniba.sk

Introduction

The launch of the Sentinel-1 constellation by ESA (European Space Agency) Copernicus program ensures systematic and regular acquisitions of SAR data. Equally to the regular acquisition plan, the revisiting time is a key parameter to take into account for monitoring purposes. The Sentinel-1 constellation has a shorter revisiting time (6–12 days) with respect to previous C-band radar systems that had larger temporal baselines, i.e., 35 or 24 days for ERS1/2 and ENVISAT or RADARSAT, respectively (Showstack 2014; Del Soldato et al. 2019).

Furthermore, Sentinel-1 data are characterized by high spatial coverage (approximately 250 km and 165 km in swath and azimuth, respectively) and a greater ground resolution (5 × 14 m in range and azimuth, respectively) with respect to previous satellites (ERS1/2 and ENVISAT). Sentinel-1A and Sentinel-1B, launched in April 2014 and April 2016, respectively, acquire data regularly on the same orbital plane at 180 relative inter-distance with a cumulated temporal repetition of 6 days (Torres et al. 2012).

Remote satellite techniques have been routinely used for landslide mapping and monitoring for some time, and optical imagery and GPS techniques are well established methods in studying the geomorphology and topography of landslides including monitoring slope deformation (Malet et al. 2002).

In comparison, the idea of the exploitation of satellite-born radar sensors is quite new. The practical use of radar techniques in landslide study originated in 1989 when differential interferometric synthetic aperture radar technique (DInSAR) was described by Gabriel et al. (1989). Early attempts used few interferograms which provided flexibility in assessing the ground deformations even in conditions of limited SAR data availability. The original DInSAR consisted of only two SAR data acquisitions and it contained at least two inherent errors in interferogram creation. Firstly atmospheric variations affected radar wave phase delay and secondly, the digital elevation model used for cancellation of

F. Guzzetti et al. (eds.), *Understanding and Reducing Landslide Disaster Risk*, ICL Contribution to Landslide Disaster Risk Reduction,
https://doi.org/10.1007/978-3-030-60227-7_11

the topography created from two-pass signal interference was inaccurate. Following analysis of a multitemporal and multibaseline stack of interferograms, Ferretti et al. (1999, 2001) came up with the idea of using longer acquisition sequences to overcome this limitation. This technique was called Permanent Scatterers (PS) and it mitigated the Atmospheric Phase Delay (APD) via statistical filtering of long-term radar sequences. It also provided residual topography estimates with very high accuracy on stable targets. Later intensive research and development in the field of algorithms and procedures for PS analysis was introduced (Bernardino et al. 2002; Mora et al. 2003; Wegmüller et al. 2005; Kampes and Adam 2005; Fornaro et al. 2007; Guarnieri and Tebaldini 2008).

The DInSAR technique has detailed description in scientific literature (Rott and Nagler 2006; Colesanti and Wasowski 2006), and it has been successfully used in subsidence study (Ferretti et al. 2000; Henry et al. 2004; Amelung et al. 1999; Rott et al. 2002, 2003), co-earthquake deformation analysis, volcano monitoring and for other purposes.

The first application of DInSAR in landslide study was reported from France (Fruneau et al. 1996), followed by Canada (Singhroy et al. 1998), Austria (Rott et al. 1999) and other areas (Vietmeier et al. 1999; Crosetto et al. 2005). The improved PS-InSAR is especially useful in landslide monitoring where the deformation rates are less than approx. 1.4 cm/month and long term series of SAR data is available (Colesanti et al. 2003, 2004; Hilley et al. 2004; Bovenga et al. 2006; Farina et al. 2006, 2008; Herrera et al. 2009). PS-InSAR is capable of detecting deformations of individual objects from "small" individual houses to objects only about one square meter in size. A certain density of stable objects (about 5 per km^2) is needed to enable accurate correction of APD (Singhroy 2009).

Study Area

Area of interest (AOI) is located in the eastern part of Slovakia, Kosice basin situated between Presov and Kosice cities (Fig. 1). Total area of 2271 m^2 encompasses urbanized country with large concentration of landslides.

The Kosice Basin from geological viewpoint represents north-eastern promontory of the Pannonian Basin aggregate. The Kosice Basin has a longitudinal north-south trending shape with the southern part twisted south-westwards and transiting to Hungary.

Geological structure of the Kosice Basin is relatively complex and formed during more stages. Present shape was formed during the accumulation of Neogene sedimentary formations. In the northern part of the Kosice Basin Neogene sediments are underlain by Paleogene sediments and underlying pre-Tertiary complex in the whole basin is represented by different Palaeozoic- Mesozoic rocks. N-S faults orientation in AOI played important role in the geomorphological evolution of the Kosice basin. The basin is drained by the river Hornad (flows through Kosice city) and its tributary Torysa (flows thorough Presov city).

The study area was heavily affected by landslide activity triggered by rainfalls, with daily amounts reaching up to 80 mm/day in late May and early June 2010. Together 577 new or reactivated landslides with total area of 507 ha threatening lives and property of people were mapped by Slovak Geological Survey in Slovakia in 2010. At present, there is 1038 registered landslides in the AOI covering an area 280 km^2, which account for 12.36% of total AOI extent (Simekova et al. 2006).

Methods

In order to achieve the goal of assessing the landslide activity for such large area of interest the data processing was divided into two stages. The first stage involved application of PS InSAR algorithm for radar data processing in order to obtain the velocities of ground deformation. In the second stage resulting line of sight velocities v_{los} were manipulated with regard to slope orientation and subsequently compared to existing landslide inventory database in the GIS environment.

Results were compared to ground truth data obtained from available monitoring data on existing landslides in the form of borehole inclinometric measurements or precise geodetic total station surveys.

The input radar data consisted of 67 ascending acquisitions of Sentinel-1A satellite from orbit no. 102 and 57 descending acquisitions from orbit no.153. The radar interferometric wide swath mode (IWS) SLC images were acquired between December 2014 and May 2017. The data processing was carried out using Interferometric stacking module of SARsacpe under ENVI platform distributed by HARRIS Geospatial Solutions, creating 66 interferograms from ascending orbits with master image from March 11th 2016 and 56 interferograms from descending orbits with master image from May 2nd 2016.

A Godstein filter was applied to radar images before "phase unwrapping" (Goldstein and Werner 1998).

In order to eliminate topographic and atmospheric artefacts and for the purpose of georeferencing of resulting PS a

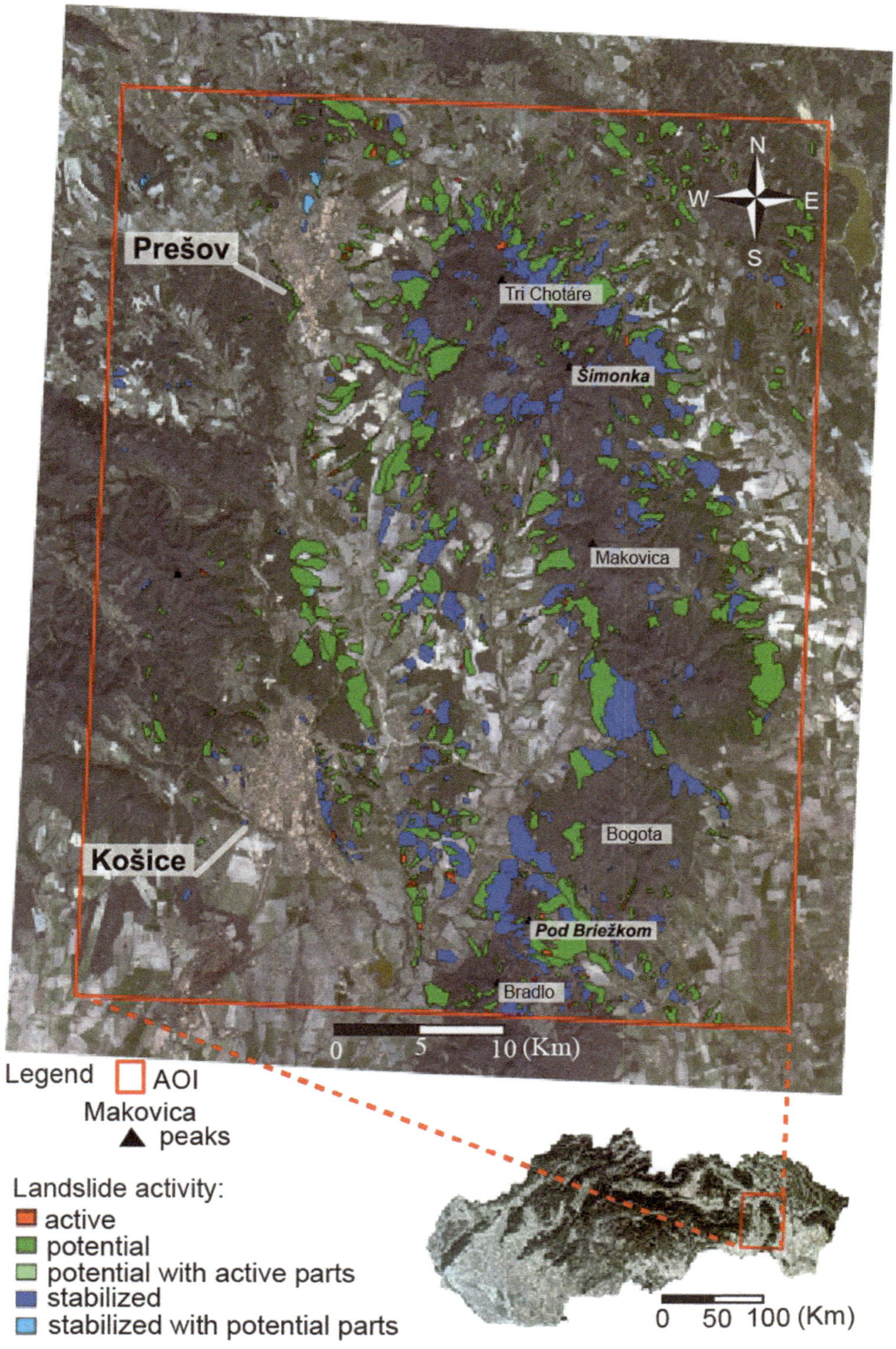

Fig. 1 Study area located in eastern Slovakia showing landslide activity classification based on the landslide inventory map

DEM from SRTM v.4 with 90 m resolution was used (Jarvis et al. 2017). Coherence value for resulting PS was set at 0.75 (Ferretti et al. 2001).

Figure 2 shows acquisition time vs relative positions for ascending and descending acquisitions of radar images.

Second stage involved the 1D LOS velocity vector transformation to the slope direction orientation k_{SLOPE}. The procedure used for the vector transformation was similar to the one used by Colesanti and Wasowski (2006), Cascini et al. (2010) or more recently by Greif and Vlcko (2012).

For the transformation of the 1D LOS displacement rates to the orientation of the landslide movement k_{SLOPE} it was necessary to know the geometry of the data acquisition and slope orientation parameters obtained from the DEM.

Fig. 2 Acquisition time vs relative positions for ascending and descending acquisitions of radar images in regard to the master image

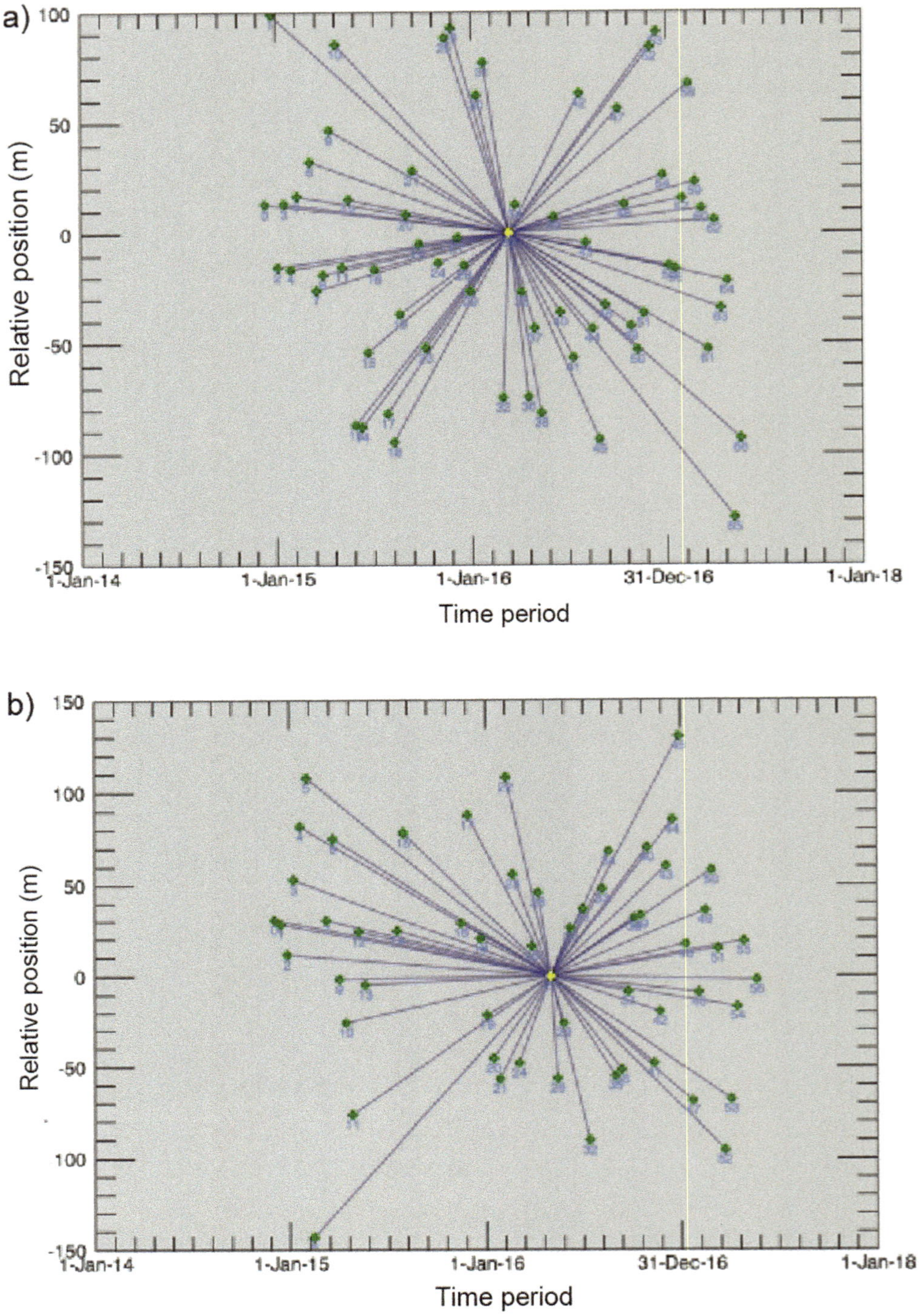

The scaling factor 1/cosβ threshold of 15 (corresponding to $\beta = 86.2°$) was adopted as the limit for acceptable projected displacement rate values, as suggested by Cascini et al. (2010). Resulting PS were overlaid by the landslide inventory map and landslide activity state was evaluated based on the v_{LOS} as well as k_{SLOPE} data. The landslide activity classification based on v_{LOS} categorized landslides into three categories with limits of 1 mm/year and 5 mm/year average velocity (Fig. 3). Similar classification was carried out for the filtered k_{SLOPE} data with limits of 3 mm/year and 10 mm/year average displacement rate in the slope direction.

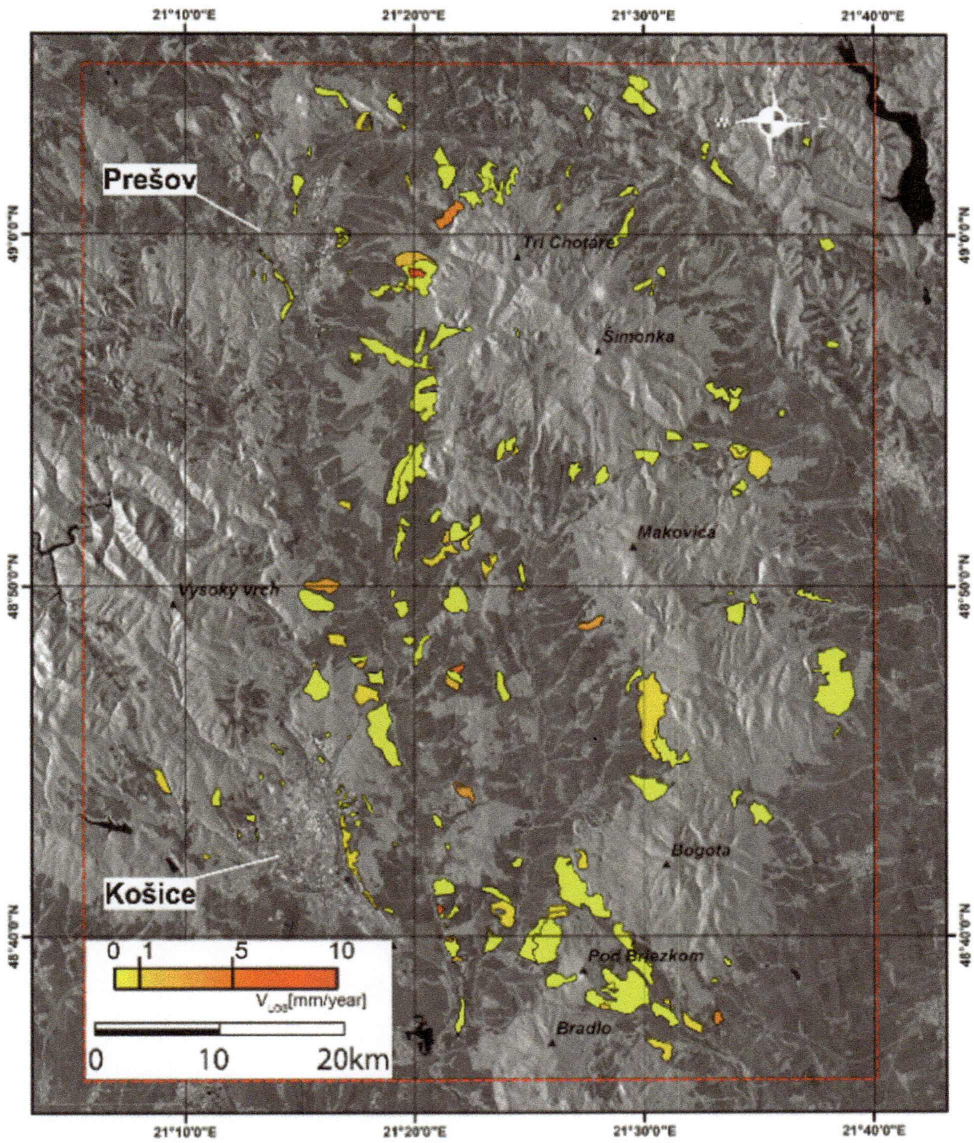

Fig. 3 Landslide activity based on the average value of displacement rate in the LOS direction v_{LOS} for the landslides with detected PS in the AOI

Results

Together 182 landslides were reclassified into three activity classes as is shown in Fig. 4. 75.82% of landslides were classified in the stable category, 22.53% as the medium activity landslides and 1.65% as active landslides based on the average absolute value of v_{LOS} displacement rate of PS located inside the boundaries of landslide inventory database within the AOI.

Further a similar procedure was carried out for the k_{slope} displacement rate and the results were as follows: Stable landslides 75.27%; medium active landslides 18.13%; active landslides 6.59%, as could be seen in Fig. 5.

Application of LOS vector displacement transformation into the k_{slope} direction resulted in the increase of the landslides classified as active on behalf of medium active landslides by 400%. This better reflects the state of the landslide activity inside the area of interest. The stable landslide category was not affected by the vector transformation.

Conclusions

Heavy rainfall in 2010 triggered many landslides in Kosice basin. Present landslide activity was reassessed using processed data acquired from Sentinel-1A radar mission between December 2014 and May 2017 in the form of

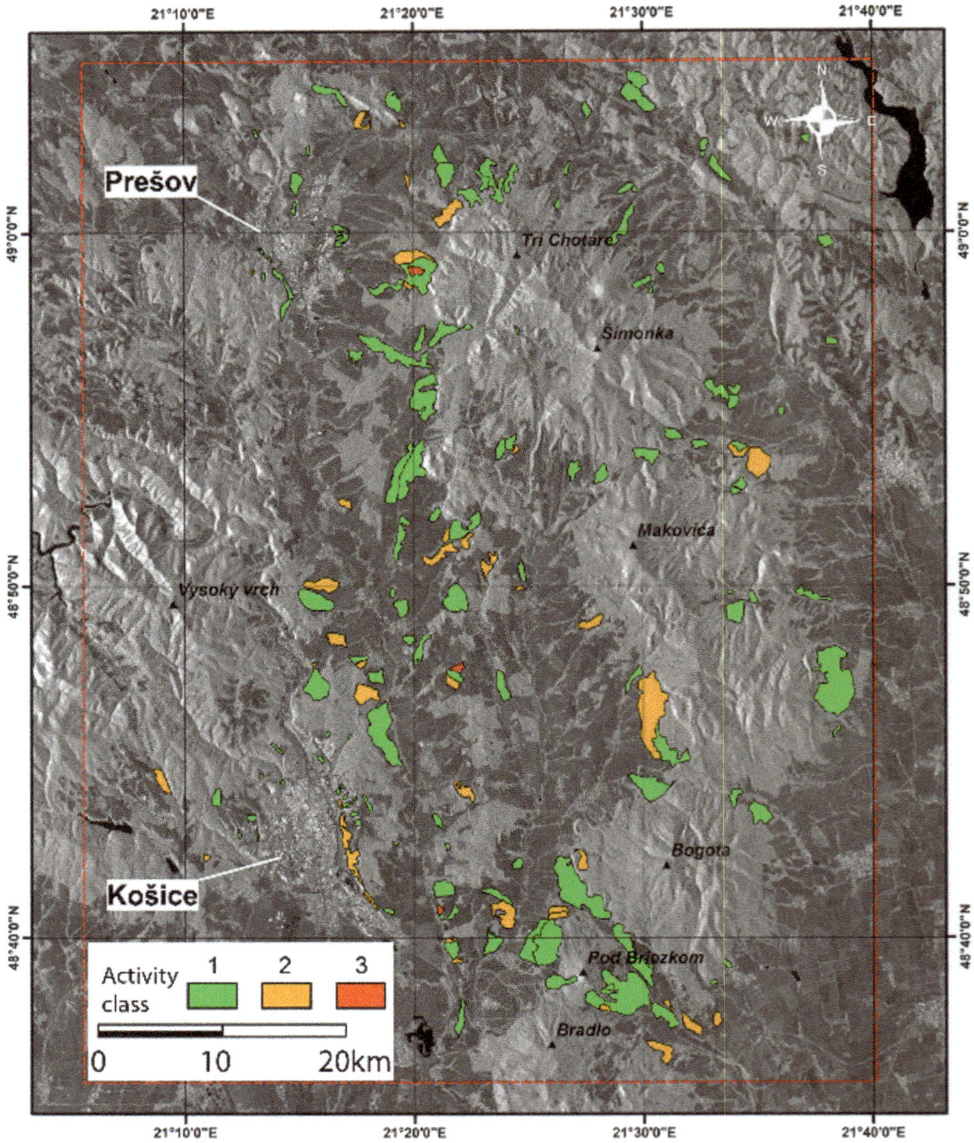

Fig. 4 Landslide activity classes derived from average PS deformation rate in the LOS direction for landslides from inventory within area of interest. 1. Stable landslides <1 mm/year; 2. Medium active > 1 < 5 mm/year; 3. Active >5 mm/year

reclassified landslide inventory map of studied area. PS InSAR technique was used for generation of landslide displacement permanent scatterers inside the landslide area, where average LOS velocity was applied for assessment of landslide activity in the form of thematic map. Using alternative method a LOS deformation velocity vectors were transformed into the slope direction generated from DEM resulting in k_{SLOPE} velocity data used for alternative classification map. This was possible thanks to availability of radar data from both acquisitions (ascending and descending) on the studied AOI.

Comparison of LOS data versus k_{SLOPE} data showed increase in number of landslides classified as active on behalf on medium active class when k_{SLOPE} transformed data were used. This showed more relevant classification of landslide activity compared to classification based on v_{LOS} displacement rate. However, transformation of velocity vector v_{LOS} must be used with caution, due to variable sensitivity of radar data in different directions with regard to the satellite path.

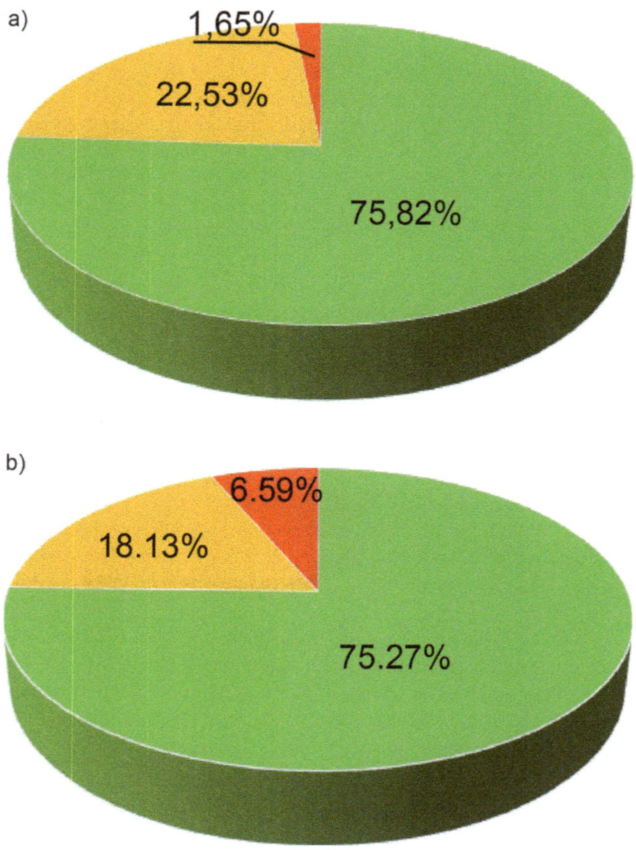

Fig. 5 Comparison of classifications based on landslide displacement rate derived from average PS deformation rate in the LOS (**a**) versus slope direction (**b**)

Acknowledgements This work was supported by a grant from the Slovak Ministry of Education VEGA (Contract no. 1/0503/19).

References

Amelung F, Galloway DL, Bell JW, Zebker HA, Laczniak RJ (1999) Sensing the ups and downs of Las Vegas: InSAR reveals structural control of land subsidence and aquifer-system deformation. Geology 27(6):483–486

Berardino P, Fornaro G, Lanari R, Sansosti E (2002) A new algorithm for surface deformation monitoring based on small baseline differential SAR interferograms. IEEE Trans Geosci Remote Sens 40(11):2375–2383

Bovenga F, Nutricato R, Refice A, Wasowski J (2006) Application of multi-temporal differential interferometry to slope instability detection in urban/peri-urban areas. Eng Geol 88:219–240

Cascini L, Fornaro G, Peduto D (2010) Advanced low- and full-resolution DInSAR map generation for slow-moving landslide analysis at different scales. Eng Geol 112(1–4):29–42

Colesanti C, Ferretti A, Prati C, Rocca F (2003) Monitoring landslides and tectonic motion with the permanent scatterers technique. Eng Geol 68:1–14

Colesanti C, Wasowski J (2004) Satellite SAR interferometry for wide-area slope hazard detection and site-specific monitoring of slow landslides. In: Proceedings Ninth Internat. Symposium on landslides, June 28–July 2, 2004, Rio de Janeiro, pp 795–802

Colesanti C, Wasowski J (2006) Investigating landslides with space-borne Synthetic Aperture Radar (SAR) interferometry. Eng Geol 88:173–199

Crosetto M, Crippa B, Biescas E (2005) Early detection and in-depth analysis of deformation phenomena by radar interferometry. Eng Geol 79(1–2):81–91

Del Soldato M, Solari L, Raspini F, Bianchini S, Ciampalini A, Montalti R, Ferretti A, Pellegrineschi V, Casagli N (2019) Monitoring ground instabilities using SAR satellite data: a practical approach. Int J Geo-Inf 8(7):307

Farina P, Casagli N, Ferretti A (2008) Radar-interpretation of InSAR measurements for landslide investigations in civil protection practices. In: Proceedings of the 1st North American landslide conference, Vail, Colorado, 3–7 June 2007, pp 272–283

Farina P, Colombo D, Fumagalli A, Marks F, Moretti S (2006) Permanent scatterers for landslide investigations: outcomes from the ESA-SLAM project. Eng Geol 88:200–217

Ferretti A, Prati C, Rocca F (1999) Multibaseline InSAR DEM reconstruction: the wavelet approach. IEEE Trans Geosci Remote Sens 37(2):705–715

Ferretti A, Prati C, Rocca F (2000) Non-linear subsidence rate estimation using permanent scatterers in differential SAR interferometry. IEEE Trans Geosci Remote Sens 38(5):2202–2212

Ferretti A, Prati C, Rocca F (2001) Permanent scatterers in SAR interferometry. IEEE Trans Geosci Remote Sens 39(1):8–20

Fornaro G, Pauciullo A, Serafino F (2007) Multipass SAR processing for urbanized areas imaging and deformation monitoring at small and large scales. Urban Remote Sens Joint Event URS 2007 Paris, 11, 13 April

Fruneau B, Achache J, Delacourt C (1996) Observation and modelling of the Saint-E' tienne-de-Tine'e landslide using SAR interferometry. Tectonophysics 265:181–190

Gabriel AK, Goldstein RM, Zebker HA (1989) Mapping small elevation changes over large areas: differential radar interferometry. J Geophys Res 94(B7):9183–9191

Greif V, Vlčko J (2012) Monitoring of post—failure landslide deformation by the PS—InSAR technique at Lubietova in Central Slovakia. Environ Earth Sci 66(6):1585–1595

Goldstein RM, Werner CL (1998) Radar interferogram filtering for geophysical applications. Geophys Res Lett 25:4035–4403

Guarnieri AM, Tebaldini S (2008) On the exploitation of target statistics for SAR interferometry applications. IEEE Trans Geosci Remote Sens 46(11):3436–3443

Hasager ChB, Jensen NO, Nielsen M, Furevik B (2002) SAR satellite image derived wind speed maps validated with in-situ meteorological observations and footprint theory for offshore wind resource mapping. In: 2002 Global Windpower Proceedings 2–5 April 2002, CNIT—La Défense—Paris—France. Available also online: https://sitecoremedia.risoe.dk/research/vea/Documents/globalwindpower2002_proceedings.pdf. Accessed 4 May 2010

Henry E, Mayer C, Rott H (2004) Mapping mining-induced subsidence from space in a hard rock mine: example of SAR interferometry application at Kiruna mine. CIM Bull 97(1083):1–5

Herrera G, Davalillo JC, Cooksley G, Monserrat O, Pancioli V, Mapping and monitoring geomorphological processes in mountainous areas using PSI data: Central Pyrenees case study. Nat Hazards Earth Syst 9:1587–1598

Hilley GE, Bürgmann R, Ferretti A, Novali F, Rocca F (2004) Dynamics of slow-moving landslides from permanent scatterer analysis. Science 304:1952–1955

Jarvis A, Reuter HI, Nelson A, Guevara E (2017) Hole-filled SRTM for the globe version 4. CGIAR-CSISRTM 90 m Database. Available online https://srtm.csi.cgiar.org. Accessed on 14 Sept 2017

Kampes BM, Adam N (2005) The STUN algorithm for persistent scatterer interferometry. In: Fringe 2005 Workshop, Frascati, Italy.

https://earth.esa.int/fringe2005/proceedings/papers/58_kampes.pdf. Accessed 19 Jan 2009

Malet JP, Maquaire O, Calais E (2002) The use of global positioning system techniques for the continuous monitoring of landslides—application to the Super-Sauze earth flow (Alpes de Haute-Provence, France). Geomorphology 43:33–54

Mora O, Mallorqui JJ, Broquetas A (2003) Linear and nonlinear terrain deformation maps from a reduced set of interferometric SAR images. IEEE Trans Geosci Remote Sens 41(10):2243–2253

Righini G, Del Ventisette C, Constantini M, Malvarosa F, Minati F (2009) Spaceborne SAR analysis for landslides mapping in the framework of the PREVIEW project. In: Sassa K, Canuti P (eds) Landslides—disaster risk reduction. Springer, pp 299–301

Rott H, Scheuchl B, Siegel A, Grasemann B (1999) Monitoring very slow slope movements by means of SAR interferometry: a case study from a mass waste above a reservoir in the Ötztal Alps, Austria. Geophys Res Lett 26:1629–1632

Rott H, Nagler T, Rocca F, Prati C, Mazzotti A, Keusen HR, Liener, Tarchi D (2002) MUSCL—monitoring urban subsidence, cavities and landslides by remote sensing, Final Report, EC Project EVG1-CT-1999-00008. Institute for Meteorology and Geophysics, University of Innsbruck, Austria

Rott H, Nagler T, Rocca F, Prati C, Mazzotti A, Keusen HR, Liener, Tarchi D (2003) InSAR techniques and applications for monitoring landslides and subsidence. In: Benes (ed) Geoinformation for European-wide integration. Proceedings of EARSeL Assembly, Prague, June 2002. Millpress, Rotterdam, pp 25–31

Rott H, Nagler T (2006) The contribution of radar interferometry to the assessment of landslide hazards. Adv Space Res 37(4):710–719

Showstack R (2014) Sentinel satellites initiate new era in earth observation. Eostrans Am Geophys Union 95:239–240

Šimeková J, Martinčeková T, Abrahám P, Gejdoš T, Grenčíková A, Grman D, Hrašna M, Jadroň D, Záthurecký A, Kotrčková E, Liščák P, Malgot J, Masný M, Mokrá M, Petro Ľ, Polaščinová E, Solčiansky R, Kopecký M, Žabková E, Wanieková D (2006) Landslide inventory of Slovakia, 1:50 000. Žilina, Ingeo ((in Slovak))

Singhroy V, Mattar KE, Gray AL (1998) Landslide characteristics in Canada using interferometric SAR and combined SAR and TM images. Adv Space Res 3:465–476

Singhroy V (2009) Satellite remote sensing applications for landslide detection and monitoring. In: Sassa K, Canuti P (eds) Landslides—disaster risk reduction. Springer, pp 143–158

Torres R, Snoeij P, Geudtner D, Bibby D, Davidson M, Attema E, Potin P, Rommen B, Floury N, Brown M (2012) GMES Sentinel-1 mission. Remote Sens Environ 120:9–24

Vietmeier J, Wagner W, Dikau R (1999) Monitoring moderate slope movements (landslides) in the southern French Alps using differential SAR interferometry. In: Proceedings of Fringe'99. Lieges, Belgium

Wegmüller U, Werner C, Strozzi T, Wiesmann A (2005) ERS—ASAR integration in the interferometric point target analysis. In: Proceedings Fringe 2005 Workshop, Frascati, Italy, 28 November–2 December. https://earth.esa.int/workshops/fringe05/proceedings/papers/196_wegmuller.pdf. Accessed 6 Mar 2009

Updating Landslide Activity State and Intensity by Means of Persistent Scatterer Interferometry

2020 Kyoto Japan

Kamila Pawluszek-Filipiak and Andrzej Borkowski

Abstract

Landslide detection and characterisation are the fundamental activities performed to reduce economic losses caused by hazardous landslide events. Since existing landslide databases have to be updated, different techniques are used for this purpose. Last decades, satellite radar interferometry has increasingly been involved in this context. In this study, a sophisticated variant of radar interferometry, namely, the persistent scatterers interferometry (PSI) technique, was used to update the landslide activity state and landslide intensity in the area of Rożnów Lake in Poland. The study area is located in the Western Carpathians. Sentinel-1 A and B images covering almost the whole year 2017 were used. Ascending and descending images were processed separately by means of the PSI approach, and afterwards, PSI velocities were projected onto the slope direction. Then the results for ascending and descending orbits were merged into the database, and the landslide activity and intensity were assessed. The landslide state assessment was performed by means of a PSI-based matrix approach. The activity state of 205 landslides has been evaluated. As a result, the majority of the landslides have been assessed as active. The results are supported by field evidence. Moreover, the results suggest the need for updating the existing landslide inventory maps.

Keywords

Landslide • Landslide state • PSI • SAR interferometry

K. Pawluszek-Filipiak (✉) · A. Borkowski
Institute of Geodesy and Geoinformatics, Wroclaw University of Environmental and Life Sciences, ul. Grunwaldzka 53, 50-357 Wrocław, Poland
e-mail: kamila.pawluszek-filipiak@upwr.edu.pl

A. Borkowski
e-mail: andrzej.borkowski@upwr.edu.pl

Introduction

Landslides are one of the most common environmental geohazards appearing worldwide (Pawluszek 2019). Landslides cause a serious threat to people, their life and existence. In Poland, landslides mainly occur in the Carpathians area. The activation of numerous landslides in 2010 as a result of intense precipitation triggered a disastrous sequence mainly destruction of buildings (Wojciechowski et al. 2012). The fundamental approach to mitigate economic and environmental losses caused by landslides is landslide identification and monitoring.

Landslide monitoring can provide knowledge about landslide behaviour and evolution over time. The knowledge about landslide displacement patterns allows an effective assessment of landslide susceptibility and hazard (Pawłuszek et al. 2019). Various geodetic, especially Global Navigation Satellite System (GNSS), and remote sensing (RS) techniques are utilised for landslide monitoring. Whilst geodetic techniques provide pointwise information, RS techniques measure terrain changes over whole areas, quantitatively. This provides dense information, which helps the understanding of landslide movement behaviour and the search for its interaction with triggering factors. This allows for a more comprehensive understanding of landslide kinematics and mechanisms. However, various RS techniques have various advantages and disadvantages, and not all RS techniques are effective in every case (Stumpf 2013).

Since Sentinel-1 A and B satellites have been launched, the six-day revisiting time of SAR satellites opens new possibilities for monitoring landslides from space. The main advantage of the SAR sensors is image acquisition. It is almost independent from natural illumination and cloud coverage. Differential synthetic aperture radar interferometry (DInSAR) is a method which analyses the phase difference between two radar images captured at distinct times (Yonezawa et al. 2012). This phase difference contains possible ground deformation and other phase components

F. Guzzetti et al. (eds.), *Understanding and Reducing Landslide Disaster Risk*, ICL Contribution to Landslide Disaster Risk Reduction,
https://doi.org/10.1007/978-3-030-60227-7_12

due to environmental factors. Thus, the DInSAR technique is limited by atmospheric, topographic and other factors (Pawluszek-Filipiak and Borkowski 2020). However, persistent scatterer interferometry (PSI) techniques, which estimate various interferometric components, allow for displacement measurements reaching even millimetre precision. PSI measurements allow assessing landslide activity state and intensity by applying a PSI-based matrix approach.

This approach has been presented by Cascini et al. (2013). The authors applied an approach based on interferometric measurements of Envisat data to update the activity state of a slow-moving landslide in southern Italy. A similar approach was presented by Sara et al. (2015) for the study area in the Setta basin in Italy. They applied ERS-1/2 and Envisat data and the PSI-based matrix approach to update the landslide activity state. The same year, Del Ventisette et al. (2014) applied the PSI matrix approach and ERS-1/2 and Envisat data to assess landslide activity within the Italian Alps. The application of Sentinel-1 images for landslide monitoring was presented by Monserrat et al. (2016) and Barra et al. (2016). They applied the PSI approach and Sentinel-1 images to update the landslide inventory map of the Molise region, in southern Italy. Recently, Kalia (2018) applied Sentinel-1 images in the descending mode for the classification of landslide activity on a regional scale, using PSI for the Moselle Valley in Germany.

Taking advantage of freely available Sentinel-1 A/B data, the aim of this study was to examine the performance of the PSI approach for another landslide-prone region. We applied this approach for landslide activity state updating for the area of Rożnów Lake in Poland. An additional objective was landslide intensity updating, which means the evaluation of landslide velocity in the study area.

Study Area and Data Used

Study Area

Landslide activity and intensity state updating have been performed for the area of Rożnów Lake in Poland. This area is located in the Western Carpathians in Małopolskie province. The study area has been selected because of the high landslide density (around 25% of the study area is covered by landslides). The reason for this is the nature of the topography and the geological structure of the Carpathians. This region is strongly prone to landslide occurrence because of the high slopes combined with flysch conditions. The flysch geology in the Western Carpathians is composed of alternating layers of permeable sandstones and poorly permeable shales, claystones and marls. There are also weathering materials, which are susceptible to landslide processes, especially close to cracks or faults (Kaczmarczyk et al. 2012). The extent of the study area is shown in Fig. 1.

Data Used

The activation of many landslides in 2010 as a result of intense precipitation brought catastrophic destruction. Many buildings and roads were destroyed. In the Polish part of the Carpathians, landslide areas often occupy about 30–40% of the province areas. This is a very important problem for local communities due to the population growing and developing properties in the landslide-prone areas (Crozier 1986). In order to mitigate the negative effects of the landslide activity, the Polish Geological Institute has created a landslide database called the Landslide Counteracting Framework (SOPO in Polish). The SOPO project is a venture implemented in 2006 from resources of the National Fund for Environmental Water Protection and Economy. The SOPO project aims at supporting the administrative and environmental protection authorities, as well as non-governmental institutions, for effective landslide hazard management (Grabowski and Wojciechowski 2015). This project provides data about the extent of all recorded landslides, their activity as well as the detailed geomorphological and geological description. Figure 1 depicts the landsides and their activity in the study area. This information has been taken from the SOPO database.

For PSI processing, we utilised 56 ascending and 52 descending Sentinel-1 A/B images. The Sentinel data has been acquired from Sentinel Data Hub, available at (https://scihub.copernicus.eu/). Additional specifications of the Sentinel-1 data are presented in Table 1.

Methodology

The methodology applied in the present work is shown in Fig. 2. Firstly, Sentinel-1 ascending and descending images were processed separately by means of the PSI approach. Afterwards, PSI velocities were projected onto the slope direction. Then results for ascending and descending orbit were merged into the one database, and the landslide activity and intensity were updated. A more detailed description of the processing steps is presented in the following subsections.

PSI Processing

Sentinel-1 data has been processed according to the PSI approach, firstly introduced by Ferretti et al. (2000, 2001). For n + 1 SAR images, n full resolution interferograms are

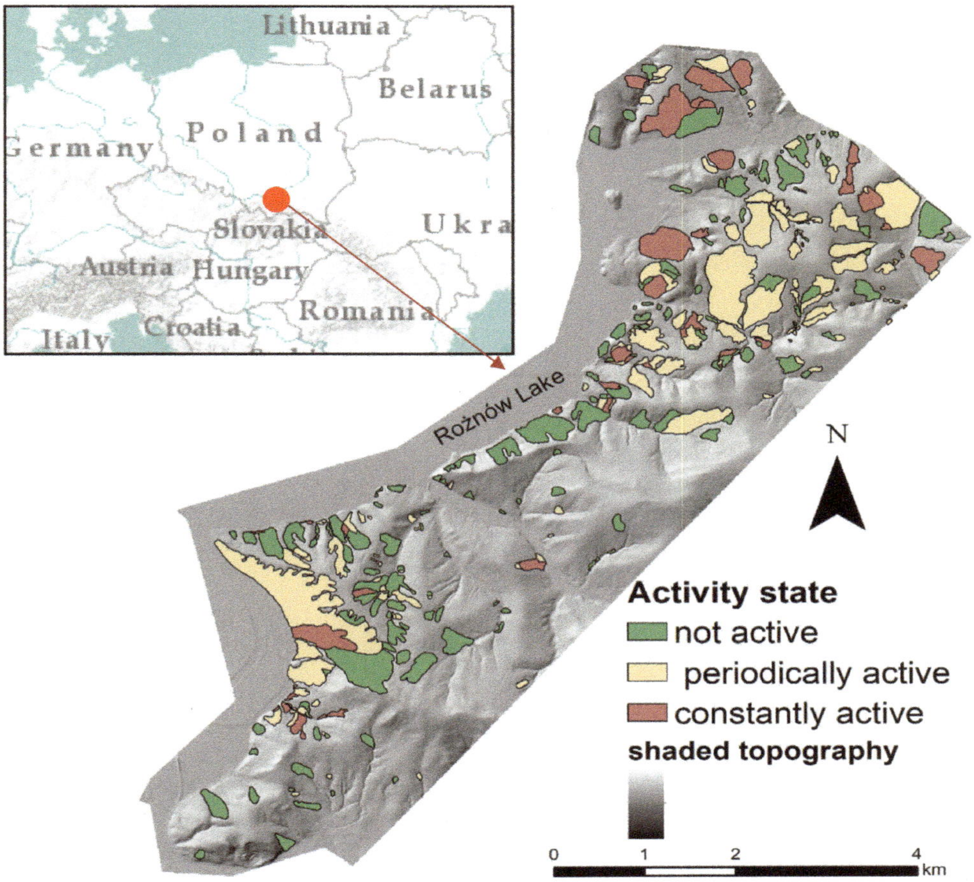

Fig. 1 Study area and location of landslides

Table 1 Specifications of the Sentinel-1 satellite data used

	Ascending images	Descending images
Incidence angle [°]	39.05	33.71
Azimuth [°]	83.82	-83.78
Time span	5/01/2017–31/12/2017	2/01/2017–22/12/2017
Polarisation	VV	VV

generated. The spatial and temporal baselines for the master and slave images are presented in Fig. 3. On the basis of Fig. 3, it can be observed that the maximum and spatial baselines for the ascending and descending images did not exceed 150 m. On the basis of the images used, 55 and 51 interferograms were generated for ascending and descending orbits, respectively. Candidates for PSI were selected on the basis of the amplitude dispersion index being greater than 0.25. The interferogram phase unwrapping was applied to sparse grid points. The Atmospheric Phase Screen (APS) was modelled using high-pass and low-pass filtering. Finally, PS points with coherence greater than 0.7 were used.

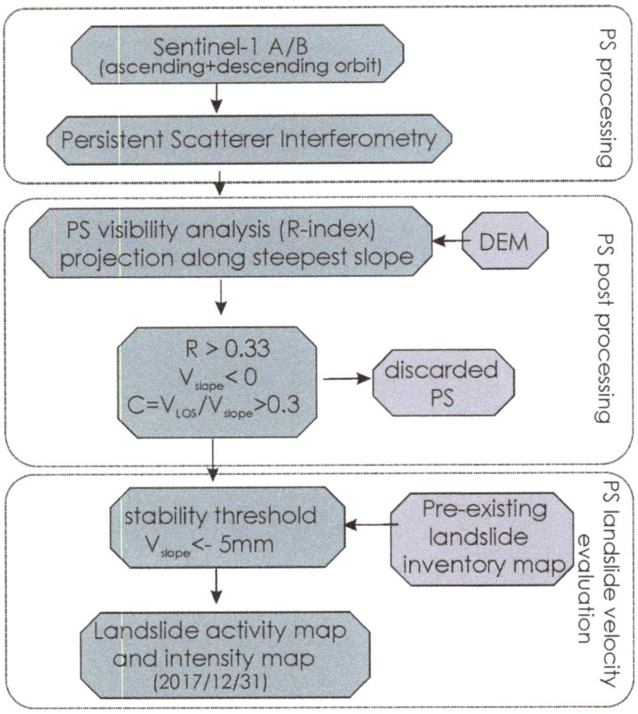

Fig. 2 Methodology flowchart

Post-processing

The results obtained from the PSI processing were post-processed in order to retrieve the most reasonable displacement estimates.

To evaluate the sensitivity of the SAR sensor in terms of the ground deformation measurement, we calculated the sensitivity index according to Notti et al. (2010) as:

$$sensitivity = -\sin(s \times \sin(a - \alpha + 90) - \theta) \quad (1)$$

where α is the satellite azimuth,θ is the incidence angle, s is the slope and a is the slope aspect. The sensitivity index takes values within the range [−1, +1]. A sensitivity index close to zero means low SAR sensitivity to measure displacements for a specific location, whilst a value close to one shows perfect sensitivity. In this study, we decided to used PS points with sensitivity index values greater than 0.33.

Additionally, in mountainous and rural regions where the PS density is low, deformation decomposition is not the best choice. Selection of the specific PS points for the decomposition and lack of the same PS points in ascending and descending geometry causes low final PS coverage. Therefore, the conversion of line of sight (LOS) deformation (V_{LOS}) into the maximum slope direction (V_{slope}) is commonly used according to the equation (Bianchini et al. 2013):

$$V_{slope} = \frac{V_{LOS}}{\cos\beta} \quad (2)$$

with β as the angle between the steepest slope and the LOS direction. When $\beta = 90°$, V_{slope} goes to infinity. Therefore, in order to discard PS with a huge velocity, which is the error of high β, we considered only PS points with $\frac{V_{LOS}}{V_{slope}} > 0.3$. Moreover, because we were interested in landslide velocity estimation, we ignored positive values of movement.

Landslide Activity State Updating

Landslide activity has been assessed on the basis of the velocity estimated using PSI. As an activity threshold,

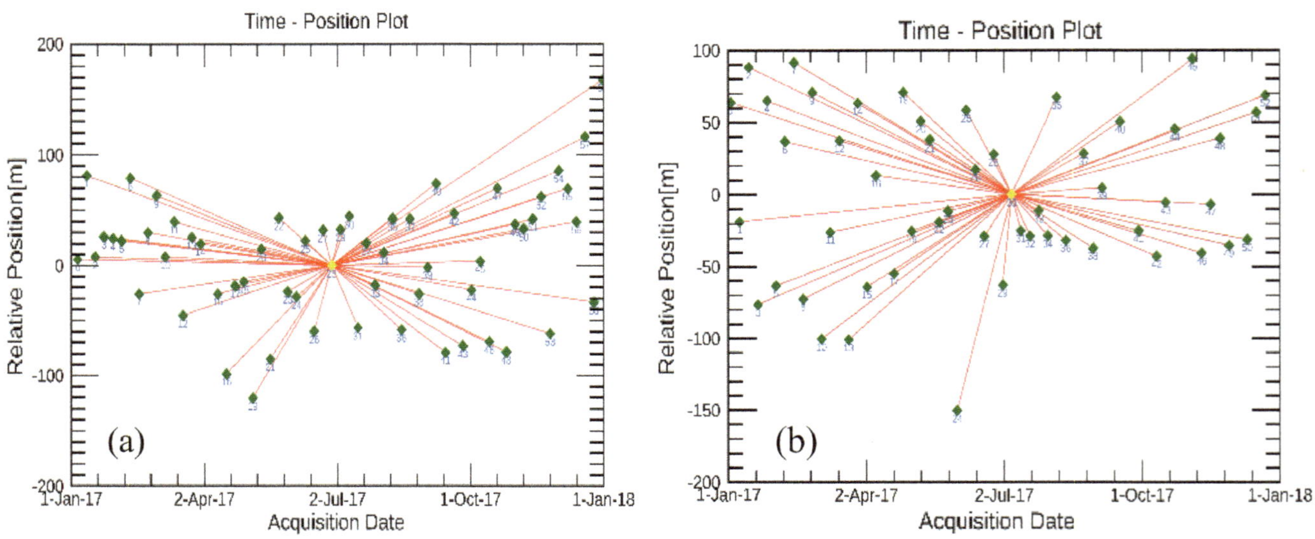

Fig. 3 Spatial and temporal relationship between master and slave images for ascending (**a**) and descending (**b**) geometry

velocity of 5 mm/yr was considered. Namely, velocity in the range of 0 and −5 mm indicate that specific landslide is stable while velocity greater than −5mm indicate that landslide is active.

Landslide Intensity State Updating

The landslide intensity has also been assessed on the basis of the PSI-based matrix approach. This PSI matrix is presented in Table 2. As the intensity threshold, an average velocity greater than −2 mm/yr was adopted. Namely, if the average velocity of PS points inside the landslide body is greater than −2 mm/yr, the velocity of the landslide is negligible. When the velocity is in the range of −2 mm/yr to −10 mm/yr, that particular landslide is considered to be extremely slow. And when the velocity is more negative than −10 mm/yr, then the landslide is considered to be very slow. Of course, it may be possible that only a few PS points exist within the landslide area. Therefore, when less than four PS points were detected within the area of a landslide, the activity and intensity state of this landslide could not be assessed (Table 2).

Table 2 Intensity PSI-based matrix

	Present PSI data	
Vs > −2 mm/yr	Vs < −2 mm/yr Vs > −10 mm/yr	Vs < −10 mm/yr
Negligible	Extremely slow	Very slow

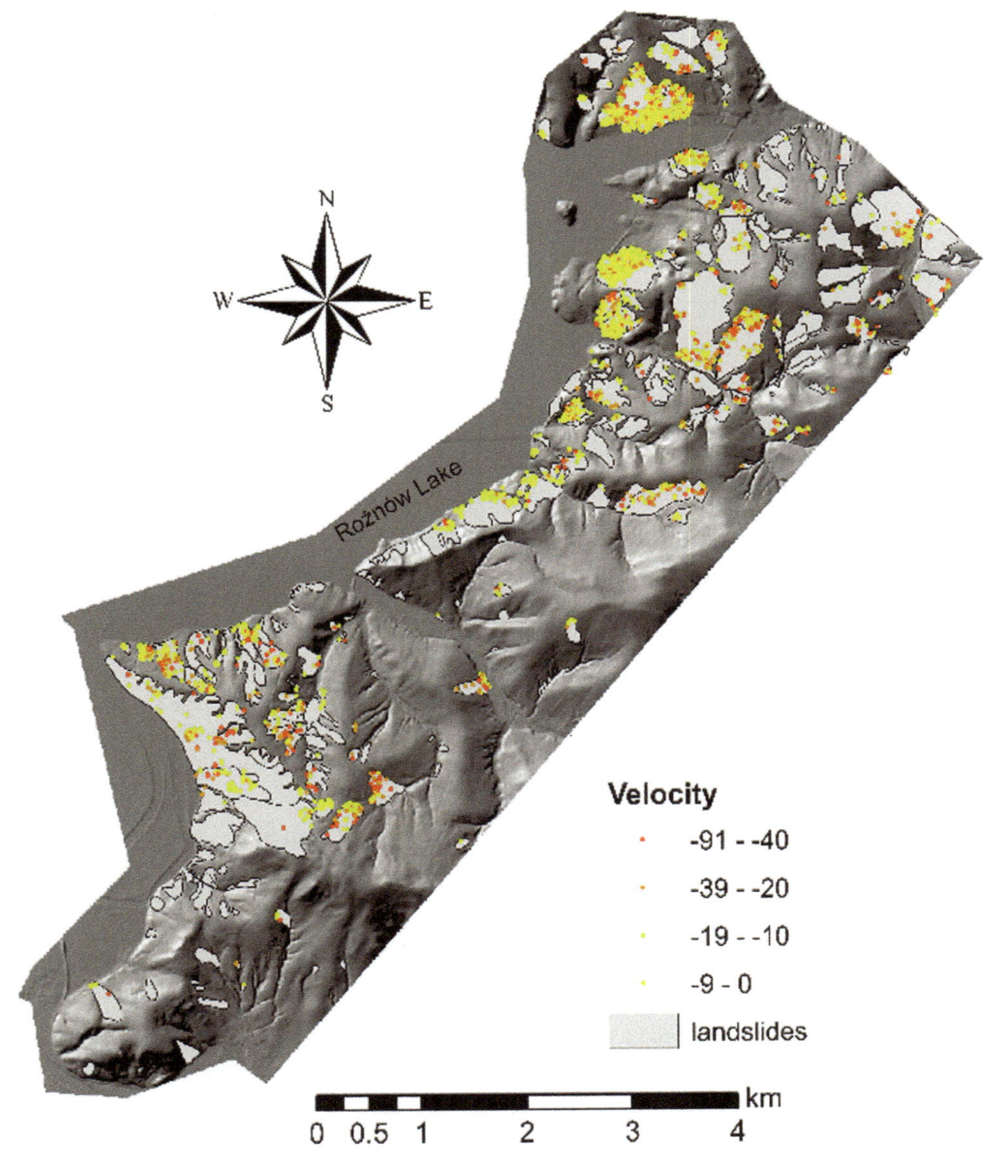

Fig. 4 Velocity [in mm] at PS points over the landslide area

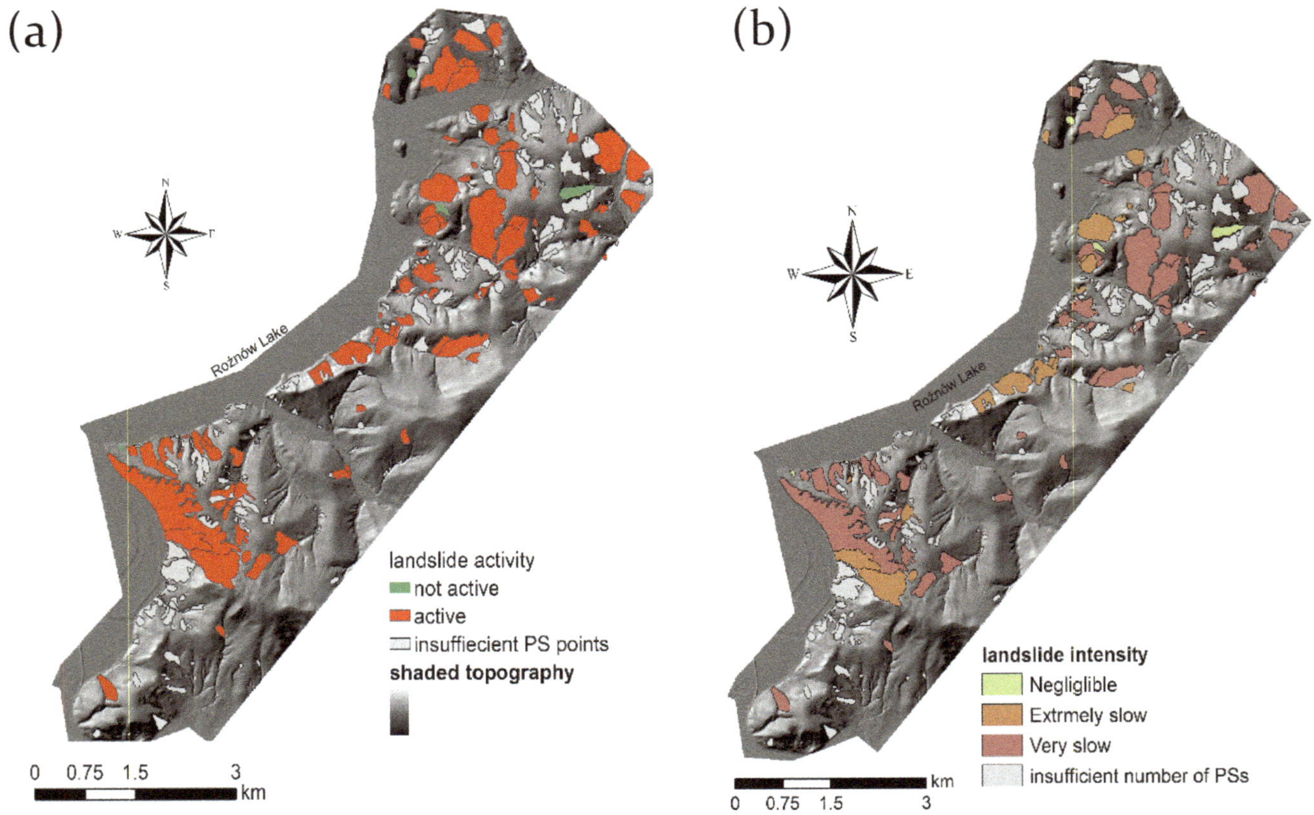

Fig. 5 Landslide activity state (**a**) and intensity (**b**) assessed on the basis of PSI for the year 2017

Results

After post-processing, the final velocity estimation on the steepest slopes was performed (Fig. 4). A maximum velocity of −91 mm/yr was detected in the study area. On the basis of obtained PS point velocities, the landslide activity state has been evaluated utilising the PSI-based matrix approach. For the landslides in which more than four PS points were detected, the activity state has been evaluated. The result is presented in Fig. 5a. It can be noticed that almost all landslides were assessed as active ones. This result agrees with the landslide activity state provided in the SOPO database (compare Fig. 1), to some extent only. Our investigation suggests that this database should be updated to present the current state of the landslides. Additionally, Fig. 5b presents the landslide intensity state. As can be seen in Fig. 5b, many landslides are extremely slow landslides.

Field Investigation

During the summer of 2018, field investigation was carried out within the study area. Many signs of activity were preserved within many landslide bodies. These signs of landslide activity are namely, building damage or cracks and road and agricultural damage. As an example, four photos taken in the study area are presented in Fig. 6. In these photos, the evidence of landslide activity can be easily noticed.

Conclusions

In this work, the landslide activity and intensity have been evaluated on the basis of persistent scatterers interferometry. The method has been performed for the area of the Polish Flysch Carpathians in Małopolskie municipality. This region has been widely affected by landslides for the last few decades.

PSI processing was performed for ascending and descending images. Velocities in LOS directions have been projected onto the steepest slopes. Thanks to this, ascending and descending results were merged into one layer. The landslide activity and intensity have been evaluated using the well-known PSI-based matrix technique. The results indicate that many landslides were active in the investigated period of 2 January–31 December 2017. Additionally, field investigation brought a great deal of evidence of landslide activity in the field, which directly confirms the PSI results.

Summarising, the PSI approach using Sentinel-1 A/B data allowed for landslide activity estimation in rural areas. For 205 landslides, the activity state has been evaluated. However, the activity of some landslides could not be

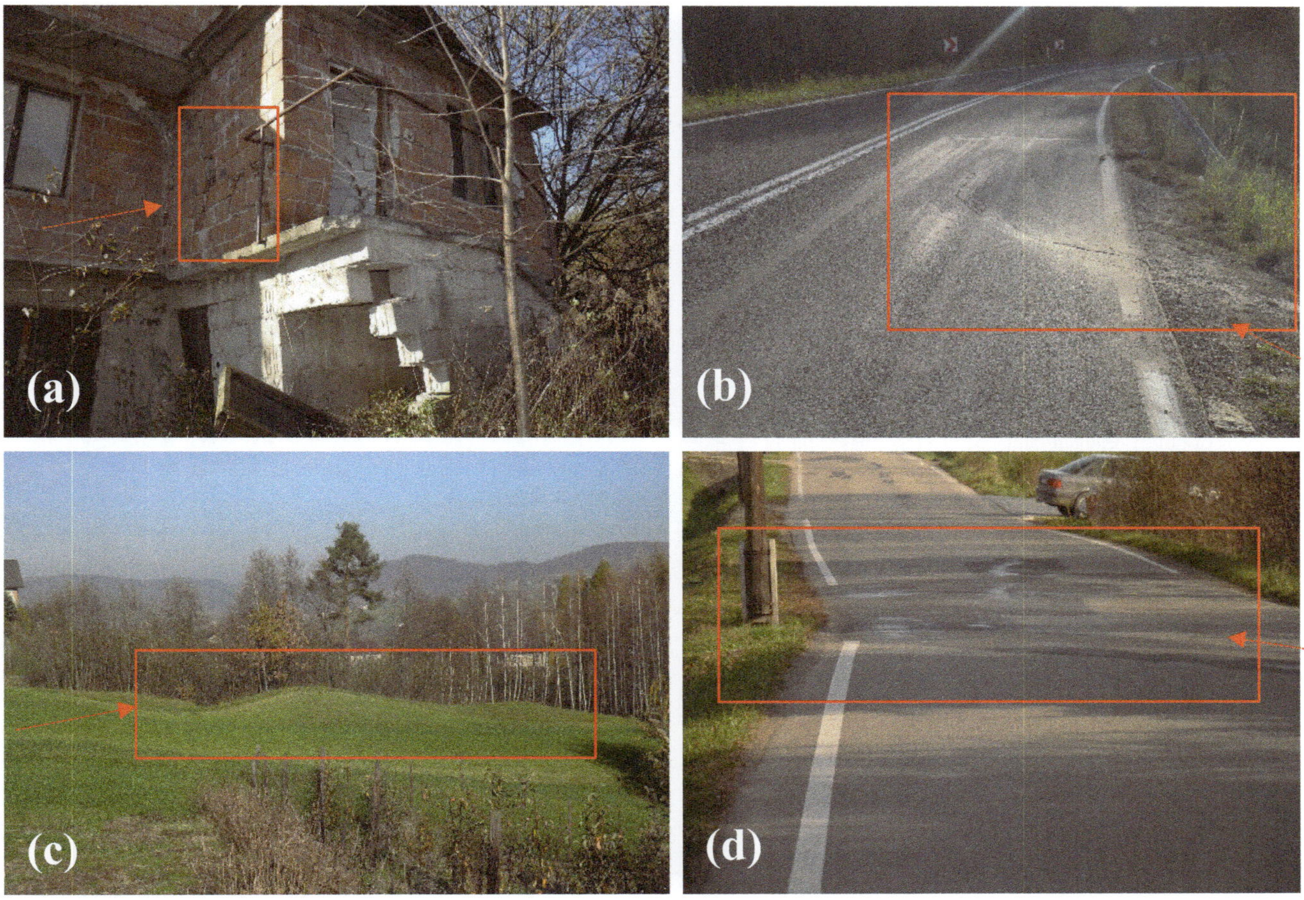

Fig. 6 Photographs taken in the field that represent damages caused by the landslide activity in the study area

assessed due to the small number of PS points within the landslide body, especially in forested areas. The results obtained suggest the need for updating the existing official landslide inventory map.

Overall, the outcomes of the work support the potential of the PSI approach for landslide studies, especially for landslide activity and intensity assessment.

Acknowledgements This work was carried out by Kamila Pawluszek-Filipiak during her research internship in the German Research Centre of Geosciences in Potsdam within the project 'Innovative Doctorate' (No. D220/0001/17), founded by Wroclaw University of Environmental and Life Science. The authors are very grateful to Prof. Mahdi Motagh for his valuable supervision given during the internship and also the Polish Geological Institute for support during the field investigations.

References

Barra A, Monserrat O, Mazzanti P, Esposito C, Crosetto M, Scarascia Mugnozza G (2016) First insights on the potential of Sentinel-1 for landslides detection. Geomatics, Nat Hazards Risk 7(6):1874–1883. https://doi.org/10.1080/19475705.2016.1171258

Bianchini S, Herrera G, Mateos RM, Notti D, Garcia I, Mora O, Moretti S (2013) Landslide activity maps generation by means of persistent scatterer interferometry. Remote Sens 5(12):6198–6222. https://doi.org/10.3390/rs5126198

Cascini L, Peduto D, Pisciotta G, Arena L, Ferlisi S, Fornaro G (2013) The combination of DInSAR and facility damage data for the updating of slow-moving landslide inventory maps at medium scale. Nat Hazards Earth Syst Sci 13(6):1527

Crozier MJ (1986) Landslides: causes, consequences & environment. Taylor & Francis

Del Ventisette C, Righini G, Moretti S, Casagli N (2014) Multitemporal landslides inventory map updating using spaceborne SAR analysis. Int J Appl Earth Obs Geoinf 30:238–246

Ferretti A, Prati C, Rocca F (2000) Nonlinear subsidence rate estimation using permanent scatterers in differential SAR interferometry. IEEE Trans Geosci Remote Sens 38:2202–2212. https://doi.org/10.1109/36.868878

Ferretti A, Prati C, Rocca F (2001) Permanent scatterers in SAR interferometry. IEEE Trans Geosci Remote Sens 39(1):8–20

Grabowski D, Wojciechowski T (2015) III etap projektu SOPO—kontynuacja i nowe trendy Ogólnopolska Konferencja O! SUWISKO 19–22 maja 2015 Wieliczka Materiały konferencyjne 21–22 [in Polish]

Kaczmarczyk R, Tchórzewska S, Woźniak H (2012) Charakterystyka wybranych osuwisk z terenu Polski południowej uaktywnionych po okresie intensywnych opadów w 2010r Nowoczesne Budownictwo Inżynieryjne wydanie lipiec-sierpień 2012 s 74–76 [in Polish]

Kalia A (2018) Classification of landslide activity on a regional scale using persistent scatterer interferometry at the Moselle Valley (Germany). Remote Sens 10(12):1880. https://doi.org/10.3390/rs10121880

Monserrat O, Crosetto M, Devanthéry N, Cuevas-González M, Barra A, Crippa B (2016) Landslide inventory and monitoring using Sentinel-1 SAR imagery. In: Living planet symposium, vol 740, p 308

Notti D, Davalillo JC, Herrera G, Mora O (2010) Assessment of the performance of X-band satellite radar data for landslide mapping and monitoring: upper Tena Valley case study. Natural Hazards Earth Syst Sci 10(9):1865

Pawluszek K (2019) Landslide features identification and morphology investigation using high-resolution DEM derivatives. Nat Hazards 96(1):311–330. https://doi.org/10.1007/s11069-018-3543-1

Pawłuszek K, Marczak S, Borkowski A, Tarolli P (2019) Multi-aspect analysis of object-oriented landslide detection based on an extended set of LiDAR-derived terrain features. ISPRS Int J Geo-Inf 8 (8):321. https://doi.org/10.3390/ijgi8080321

Pawluszek-Filipiak K, Borkowski A (2020) Integration of DInSAR and SBAS techniques to determine mining-related deformations using Sentinel-1 Data: the case study of Rydułtowy mine in Poland. Remote Sens 12(2):242. https://doi.org/10.3390/rs12020242

Sara F, Silvia B, Sandro M (2015) Landslide inventory updating by means of Persistent Scatterer Interferometry (PSI): the Setta basin (Italy) case study. Geomatics, Nat Hazards Risk 6(5–7):419–438. https://doi.org/10.1080/19475705.2013.866985

Stumpf A (2013) Landslide recognition and monitoring with remotely sensed data from passive optical sensors. Doctoral dissertation, Université de Strasbourg

Wojciechowski T, Borkowski A, Perski Z, Wójcik A (2012) Dane lotniczego skaningu laserowego w badaniu osuwisk-przykład osuwiska w Zbyszycach (Karpaty zewnętrzne). Przegląd Geologiczny 60(2):95–102

Yonezawa C, Watanabe M, Saito G (2012) Polarimetric decomposition analysis of ALOS PALSAR observation data before and after a landslide event. Remote Sens 4:2314–2328. https://doi.org/10.3390/rs4082314

Damming Predisposition of River Networks: A Mapping Methodology

Carlo Tacconi Stefanelli, Nicola Casagli, and Filippo Catani

Abstract

Landslide dams may collapse within few hours/days after their formation resulting in destructive flooding wave. Due to the limited time available since their formation, forecasting tools able to assess the damming susceptibility over large areas are more advisable for prevention and setting up mitigation measures. A semi-automated GIS-based methodology is proposed in this work to map the spatial damming predisposition over large areas, to analyse consequence and risk scenarios. The procedure is based on a morphological index that use a statistical correlation between morphometric parameters to spatially assess the chance of a river obstruction through the reactivation of an existing landslide. This damming mechanism were tested on the Arno River basin (9116 km^2) in Italy, where about 30,000 landslides are mapped. The highest mountain ridges in the Eastern part of the area resulted as the most susceptible to damming in the basin. The concentration of the historical landslide dams endorses the results for this basin.

Keywords

Landslide dams • Susceptibility • Natural hazards • GIS

Introduction

River obstructions are common geomorphologic processes in mountain regions and can cause serious hazards such as upstream backwater, catastrophic flooding, riverbed modification and channel instability (whit possible secondary landslides) (Costa and Schuster 1988; Casagli and Ermini 1999). The consequences of landslide dams may have substantial economic, social and environmental impacts in valley floors, where human activities most concentrated. It is widely documented (Ermini and Casagli 2003; Tacconi et al. 2015) that most of landslide dams have a short life, as about 40% of them fail within 24 h after their formation and less than 20% last more than one month. The available time is often not enough for a reliable stability analysis and only techniques requiring rapid data collection can be used. Planning and prevention measures (e.g. risk mapping) are essential to reduce natural hazard consequences where the expected damming probability is high.

Ancient landslides, dormant and vegetated at this time (Rosi et al. 2018), may often generate landslide dams (Crozier 2010). They were originally triggered under different climatic and environmental settings but can be reactivated by natural causes (like rainfall or earthquake). Hence, all dormant landslides should be subject to investigation if are able to reach a stream in their pathway along the slope.

Morphometrical parameters of the landslide and the river can be used to compose geomorphological indexes, according to some authors (Swanson et al. 1986; Ermini and Casagli 2003), to assess the landslide dams formation and behaviour.

The Morphological Obstruction Index (MOI) (Tacconi Stefanelli et al. 2016) is a recent index that proved a reliable capability to assess the formation and the stability of landslide dams. Farther, the required morphometrical parameters for the index are easily obtained from common Digital Elevation Model. The MOI formula, empirically achieved from 300 cases throughout Italy, associates two important

C. T. Stefanelli (✉) · N. Casagli · F. Catani
Department of Earth Sciences, University of Florence, Via La Pira 4, 50121 Firenze, Italy
e-mail: carlo.tacconistefanelli@unifi.it

N. Casagli
e-mail: nicola.casagli@unifi.it

F. Catani
e-mail: filippo.catani@unifi.it

F. Guzzetti et al. (eds.), *Understanding and Reducing Landslide Disaster Risk*,
ICL Contribution to Landslide Disaster Risk Reduction,
https://doi.org/10.1007/978-3-030-60227-7_13

parameters, the landslide volume V_l (m^3) and the valley width W_v (m):

$$\mathrm{MOI} = \log(V_l/W_v) \quad (1)$$

The index classifies landslide dams within three evolutionary domains: formed, not formed and of uncertain evolution (no assessment can be made on its evolution). The confines of these domains are marked by two lines, the "Non-formation Straight line" and the "Formation Straight line" (Fig. 1).

The "Non-formation Straight line" is expressed as follows:

$$V_l' = 1.7 \cdot W_v^{2.5} \quad (2)$$

With V_l', called "Non-formation volume", as the minimum landslide volume (m^3) potentially able to block a river valley with a width of Wv. A lower volume will not realize a complete dam.

The "Formation Straight line" is expressed as follows:

$$V_l'' = 180.3 \cdot W_v^2 \quad (3)$$

where V_l'', called the "Formation volume", is the smallest landslide volume (m^3) able to block the river valley (with 99% of confidence interval). It is the higher bound for not formed dams and the lower of the Formation domain.

In this work we present a simple GIS-based mapping methodology to verify the damming susceptibility at basin scale from inventoried landslides with geomorphological indexes. The method will be applied to the Arno River basin (Italy), where all the data needed for its application are available.

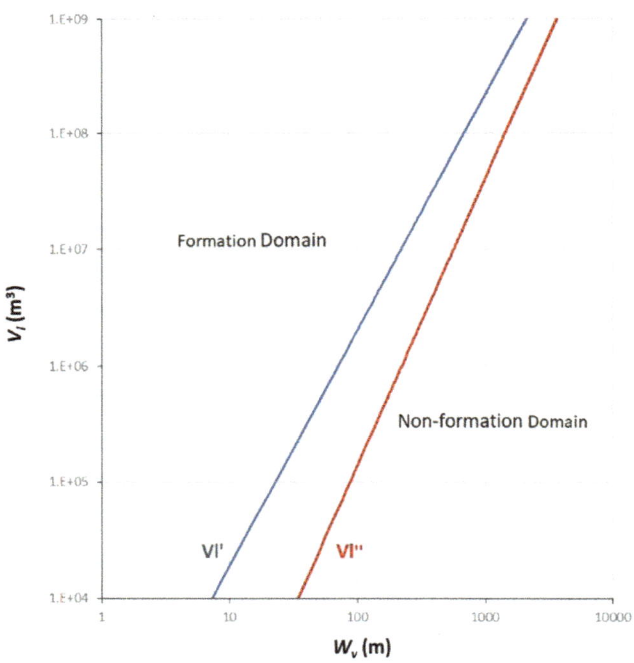

Fig. 1 Plot of the non-formation line and formation line

Study Area and Materials

The Arno River basin is in the south-eastern portion of the Northern Apennines chain in Tuscany (Central Italy) and has a surface of 9116 km^2. Here hilly and mountainous areas prevail with 78% of the total surface (Fig. 2). The essential data for the application of the method, are:

- An updated landslide inventory;
- A Digital Elevation Model (DEM);
- The river network.

The landslides database of the Tuscany region was recently updated using the persistent scatter interferometry (PSI) (Rosi et al. 2018). This database is the collection of occurrences over a large temporal scale and is an "historical inventory". About 27,500 landslides in the new inventory belong to the Arno River basin. The landslides surface ranges from 1×10^2 to 5×10^6 m^2 and most of them (about 98%) can be classified as rotational or planar slow-moving slides.

A DEM with resolution of 10 m and a vector layer of the river network are also needed.

Method

The width of each river valley can be seen as a static parameter in the MOI equation, since it does not significantly change over the time. Starting from this simplification, according to Eqs. (2) and (3), if the average river with W_v within each river stretch is evaluated two threshold values V_l' and V_l'' (Non-formation volume and Formation volume) can be computed. V_l' is the smallest volume of formation, below which a landslide is not able to realize a complete obstruction, and V_l'' is the smallest value above which an obstruction is definitely realized.

Through some assumptions and simplifications with a landslide inventory, we can estimate the landslide volumes. Landslides with volume bigger than V_l' and V_l'' for their river section are identified as potentially susceptible to dam the river. From these assumptions a "Map of the Damming Susceptibility" for reactivation of mapped landslides can be produced employing a GIS software.

As preliminary operations it is suggested to remove unnecessary data. The river obstruction happens exclusively in hilly or mountainous areas, specially along steep slopes

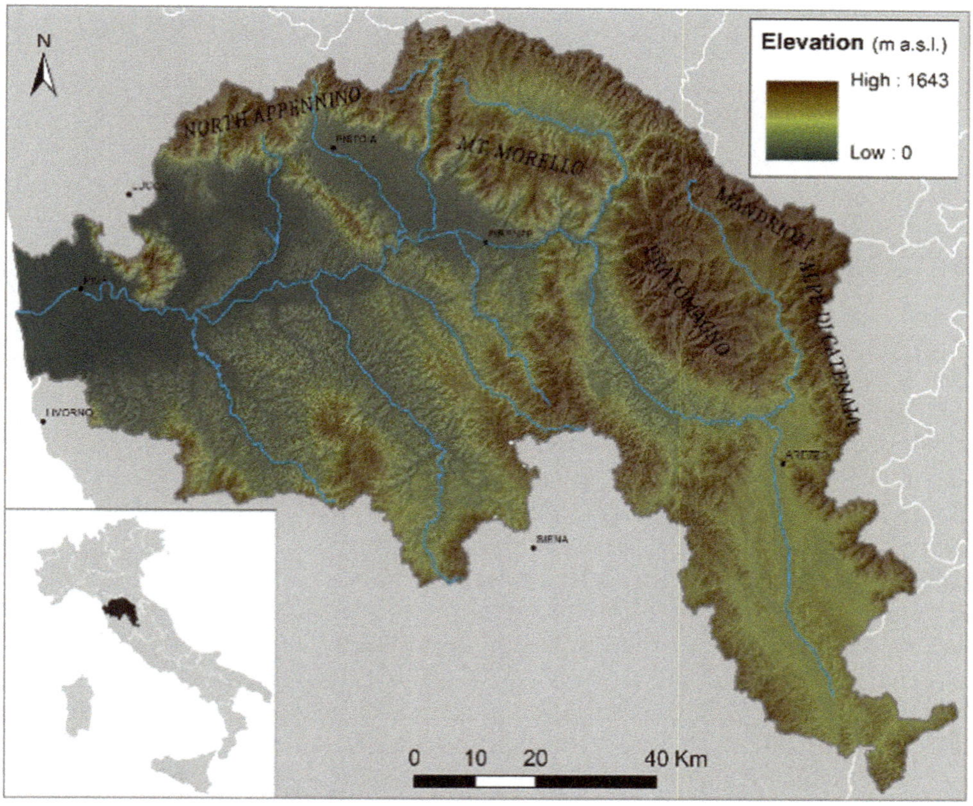

Fig. 2 Elevation distribution and river network of the Arno River basin

(Costa and Schuster 1988; Tacconi Stefanelli et al. 2015, 2018). For this reason, flat areas (slopes below 5° and elevations below 150 m a.s.l.) have been discarded from the elaborations. Short river stretches have small upstream drainage basins and the consequences of a landslide dam here would be insignificant. For these reasons, river stretches with length shorter than (arbitrary) 20 m were not considered. Furthermore, to make maps clearer, the river network has been divided in river stretches 300 m long.

With the spreading of remote sensing data and GIS applications tools that allow the extraction af landform information at basin scales are now widely used in natural science studies. Wood (1996, 2009) have realized "LandSerf" software (integrated in SAGA GIS software), able to classify landforms from DEMs, deriving land-surface parameters (curvature, slope and aspect) used in a multi-scale approach for pattern recognition and texture analysis. With this tool is possible to classify the landscape into homogeneous morphometric units (e.g. ridges, pits, channels and planes) and isolate the channels morphological units in the form of polygons, which can be used to define the valley floor limits. To associate a valley width value, W_V, to each river stretch, the distance between the two valley floor boundaries have to be measured. Lines 500 m long outdistanced by 20 m and perpendicular to the river network (hereafter "transects") are created. Then, the valley floor boundaries polygons are used to "cut" the perpendicular transects 500 m long. The median value between the transects length intersecting each river stretch is used as valley width value, W_V. At this step, the two landslide values of "Non-formation volume" and "Formation volume", V_l' and V_l'', can be easily calculated applying Eqs. (2) and (3).

It is realistic that a reactivated landslide will move downstream. Draining directions within each slope can be easily computed with a GIS software. Each landslide can then be related to the river stretch that it should reach trough the draining surfaces.

To assess the landslide volumes, according on the type of movement, two procedures have been followed, one for rotational slides and one for the remaining type of movements.

Volumes for rotational slides are computed, assuming a semi ellipsoidal sliding surface, employing the equation in WP/WLI (1990), as follows:

$$V_l = \frac{1}{6} \cdot \pi \cdot D_r \cdot L_r \cdot W_r \tag{4}$$

where D_r is the sliding surface depth (m), L_r the distance between the toe and the crown of the landslide (m) and W_r the perpendicular distance to L_r between the lateral bounds of the landslide (m). All the other landslide movement types different from rotational slides, a simplification is taken assuming a planar sliding surface with a constant depth

Table 1 Comparison between landslide volumes, V_l, with the volume of non-formation, V_l', and formation, V_l''

	$V_l > V_l'$ (V_l'')	$V_l < V_l'$ (V_l'') $< V_l \cdot 1,2$	$V_l < V_l'$ (V_l'')
P value	2	1	0

(Cruden and Varnes 1996). The landslides volume is computed with the following equation:

$$V_l = A \cdot D_r \quad (5)$$

with A (m^2) is the landslide surfaces, automatically calculated in a GIS software. In Eq. (5) constant landslide depth of 1.0 m was assumed according to the average soil thickness in the region as reported by Catani et al. (2010) and Bicocchi et al. (2019). This simplification can only be valid for specific areas and it is possible only in well-studied areas. These procedures for landslide volume calculation are based on some simplifications, since they use geometric approximations, but still give the magnitude of the process and can be useful for the study purposes.

The classification of the damming predisposition for each landslide start following Table 1: comparing the boundary volume of Non-formation or Formation (V_l' or V_l'') with the landslide volume, V_l, a P value (Predisposition value) of 2 is assigned if V_l is bigger than V_l' (or V_l''), and a value of 0 (zero) if it is smaller. If the V_l' (or V_l'') value is higher than V_l but lower than the V_l improved by 20% ($V_l \cdot 1,2$), a value of 1 is assigned. The V_l increase of 20% (an arbitrary value) is used to avoid any underestimation during parameter sampling for a cautionary principle.

Combining the two P value obtained from Table 1 for volume of Non-formation or Formation is possible to attribute to each mapped landslide the damming predisposition through the intensity matrix of Fig. 3.

The damming predisposition is divided in five (qualitative) classes: Very Low (dark green), Low (light green), Moderate (yellow), High (orange) and Very High (red). The gray boxes are not possible combinations.

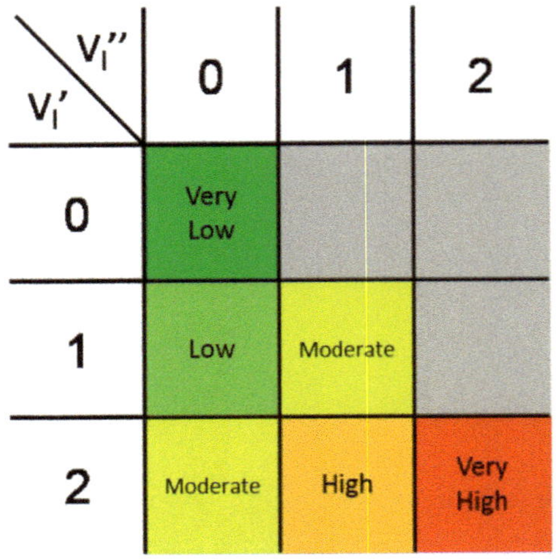

Fig. 3 Matrix used to assign the damming predisposition intensity to the mapped landslides (from Tacconi Stefanelli et al. 2020)

Results and Discussions

The assessment of damming predisposition on landslide inventory of the Arno River basin is shown in Fig. 4, where most exposed areas to damming are the Mt. Morello-Pratomagno and the Mandrioli-Alpe di Catenaia mountain ridges. The class distribution between mapped landslides is shown in the histogram of Fig. 5. The Very Low class is the most frequent, with 94.40% of the whole inventory, followed by the 4.34% of the Moderate. The remaining 1.26% is divided among Low (0.78%), Very High (0.46%) and High (0.02%) classes. The mapping method result was then compared for validation with known landslide dams in the area (from Tacconi Stefanelli et al. 2015). All censed landslide dams in Fig. 4, except by one in the southwestern part of the basin, are in area with higher concentration of landslides classified with High and Very-High damming predisposition.

This agreement between the mapping result and the censed landslide dams can be considered as a positive demonstration of the reliability of the method despite several simplifications and approximations.

Conclusions

A tool able to identify the areas with higher damming risk could radically reduce the reconstruction costs, allowing to focus monitoring works, planning activities and preventive measures.

The goal of this work was to propose a useful and easy tool to predict which areas have a higher damming susceptibility to reactivation of existing landslides. The proposed methodology was developed using the Morphological Obstruction Index and few common data and applied with some simplifications and approximations on a test area, the Arno River basin. The result is a practical and realistic map of the predisposition of the censed landslides to block the river course which agree with known past damming episodes.

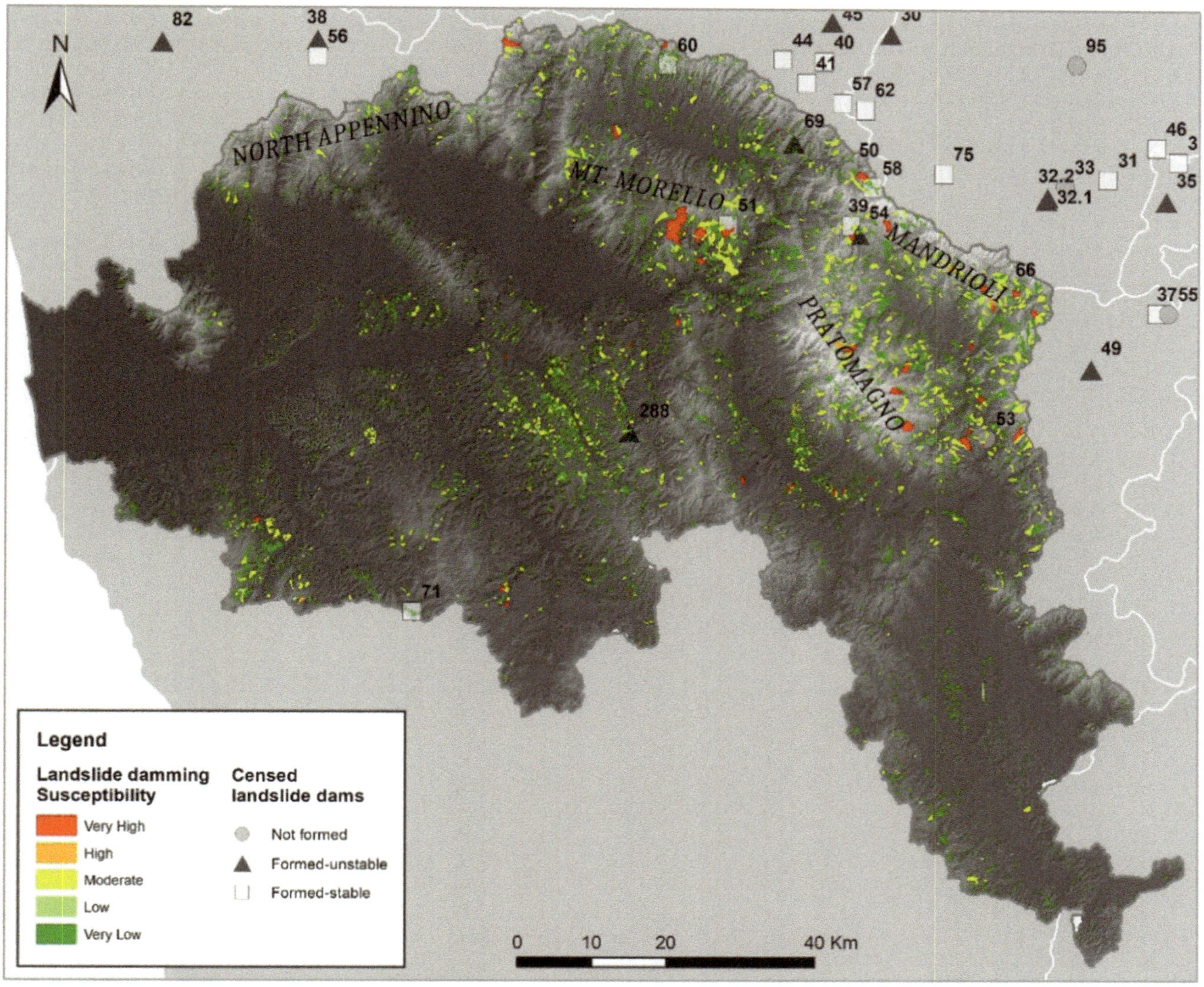

Fig. 4 Map of Damming Predisposition in the Arno River basin by reactivation of mapped landslides and censed landslide dams (from Tacconi Stefanelli et al. 2015, bold numbers are the dam IDs, with Not-formed: only a partial damming of the stream is realized; Formed-unstable: the natural dam and the lake are formed but has collapsed; Formed-stable: the dam is formed and the lake still exists or has disappeared for sediment filling)

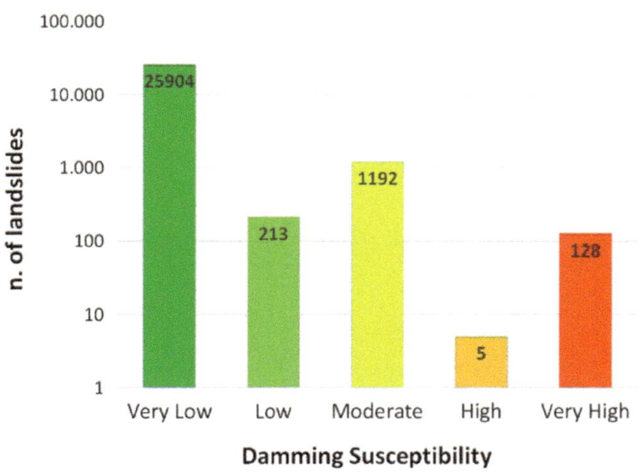

Fig. 5 Damming predisposition of mapped landslides in the Arno River basin

References

Bicocchi G, Tofani V, D'Ambrosio M, Tacconi Stefanelli C, Vannocci P, Casagli N, Lavorini G, Trevisani M, Catani F (2019) Geotechnical and hydrological characterization of hillslope deposits for regional landslide prediction modeling. B Eng Geol Environ, pp 1–17

Casagli N, Ermini L (1999) Geomorphic analysis of landslide dams in the Northern Apennine. Trans Jpn Geomorphys Union 20(3):219–249

Catani F, Segoni S, Falorni G (2010) An empirical geomorphology based approach to the spatial prediction of soil thickness at catchment scale. Water Resour Res 46(5)

Costa JE, Schuster RL (1988) Formation and failure of natural dams. Bull Geol Soc Am 100 (7):1054–1068

Crozier MJ (2010) Deciphering the effect of climate change on landslide activity: a review. Geomorphology 124(3):260–267

Cruden DM, Varnes DJ (1996) Landslides: investigation and mitigation. Chapter 3—landslide types and processes, p 247

Ermini L, Casagli N (2003) Prediction of the behavior of landslide dams using a geomorphical dimensionless index. Earth Surf Proc Land 28:31–47

Rosi A, Tofani V, Tanteri L, Tacconi Stefanelli C, Agostini A, Catani F, Casagli N (2018) The new landslide inventory of Tuscany (Italy) updated with PS-InSAR: geomorphological features and landslide distribution. Landslides 15(1):5–19

Swanson FJ, Oyagi N, Tominaga M (1986) Landslide dams in Japan. In: Schuster RL (ed) Landslide dams: processes, risk and mitigation. Geotech Sp, vol 3. ASCE, New York, pp 131–145

Tacconi Stefanelli C, Catani F, Casagli N (2015) Geomorphological investigations on landslide dams. Geoenviron Disaster 2(1):1–15

Tacconi Stefanelli C, Segoni S, Casagli N, Catani F (2016) Geomorphic indexing of landslide dams evolution. Eng Geol 208:1–10

Tacconi Stefanelli C, Vilímek V, Emmer A, Catani F (2018) Morphological analysis and features of the landslide dams in the Cordillera Blanca Peru. Landslides 15(3):507–521

Tacconi Stefanelli C, Casagli N, Catani F (2020) Landslide damming hazard susceptibility maps: a new GIS-based procedure for risk management. Landslides, 1–14

Wood J (1996) The geomorphological characterization of digital elevation models. Diss., Department of Geography, University of Leicester, U.K. online

Wood J (2009) Geomorphometry in LandSerf. In: Hengl T, Reuter HI (eds) Geomorphometry: concepts, software, applications. Dev Soil Sci, Elsevier 33:333–349

WP/WLI—International Geotechnical Societies' UNESCO Working Party on World Landslide Inventory (1990) A suggested method for reporting landslides. IAEG Bull 41:5–12

Landslides Along Halong-Vandon Expressway in Quang Ninh Province, Vietnam

2020 Kyoto Japan

Pham Van Tien, Le Hong Luong, Le Minh Nhat, Nguyen Kim Thanh, and Phuong Van Cuong

Abstract

This paper presents a preliminary study on landslides along Halong-Vandon expressway in Quang Ninh province, Vietnam through detailed site surveys, soil testing and analysis of collected data including geology, geomorphology, rainfall and project documents. The results show that landslides occurred at 44 locations from June to August in the 2017 and 2018 rainy seasons. Shallow debris slides were dominant on cut slopes along the expressway, accounting for 35 cases. Rainfall with moderate to high intensity in a short period was the landslide triggering factor while slope cutting for road construction was the main preparatory factor of the landslides. A detailed rainfall analysis presented that the cumulative event rainfall of 3 days ago was likely sufficient to saturate the soils and forming a landslide. In addition, the improper geological investigation and calculation of safety factors played a significant contributing factor, which has become a critical issue in the Halong-Vandon expressway project. These findings highly agreed with those examined in a selected landslide case at the Km 27 + 950 location.

Keywords

Landslides • Characteristics • Mechanism • Cut slopes • Rainfall • Halong-Vandon

Introduction

Landslides are one of the most frequent and serious natural hazards which can cause substantial human and economic losses in Vietnam (Tam 2001; Duc 2013). With more than ¾ of the length of national highways running through mountainous regions, landslides often occur on cut slopes in rainy seasons, particularly along national highways such as Nos. 1, 2, 3, 6, and the Ho Chi Minh (HCM) route (Tam et al. 2008; Duc 2013). Tam et al. (2008) indicated that landslides are very likely to occur during and after rainstorm events, especially the new traffic routes are highly prone to such phenomena. In recent years, landslides have been reported to cause extensive damage on the newly built roads such as the Hanoi-Laocai, Halong-Vandon (HL-VD) and Bacgiang-Langson expressways. The rapid economic development of Vietnam will be seriously hampered if the effect of landslides is not addressed or mitigated effectively.

On the 60-km HL-VD expressway—an important transport lifeline connecting the Hanoi-HaiPhong-QuangNinh economic triangle, landslides not only severely destroyed the road and its facilities but it also caused a major economic and social impact because the road operation was delayed several months. Although the landslides have been widely reported by authorities and the media, there still has a lack of understanding of mechanisms causing slidings due to less study in-depth. This study, therefore, aims to investigate characteristics, causes and mechanisms of the landslides in 2017 and 2018 rainy seasons on the HL-VD expressway (Fig. 1).

P. Van Tien (✉) · L. H. Luong · N. K. Thanh
Institute of Transport Science and Technology, Hanoi, Vietnam
e-mail: phamtiengtvt@gmail.com

L. H. Luong
e-mail: lehongluong@gmail.com

N. K. Thanh
e-mail: nguyenkimthanh.itst@gmail.com

L. M. Nhat
Viet Nam Disaster Management Authority, Ministry of Agriculture and Rural Development, Hanoi, Vietnam
e-mail: nhatkyoto@gmail.com

P. Van Cuong
Vietnam Japan Engineering Consultants Co., Ltd. (VJEC), Hanoi, Vietnam
e-mail: cuongpv@vjec.vn

F. Guzzetti et al. (eds.), *Understanding and Reducing Landslide Disaster Risk*,
ICL Contribution to Landslide Disaster Risk Reduction,
https://doi.org/10.1007/978-3-030-60227-7_14

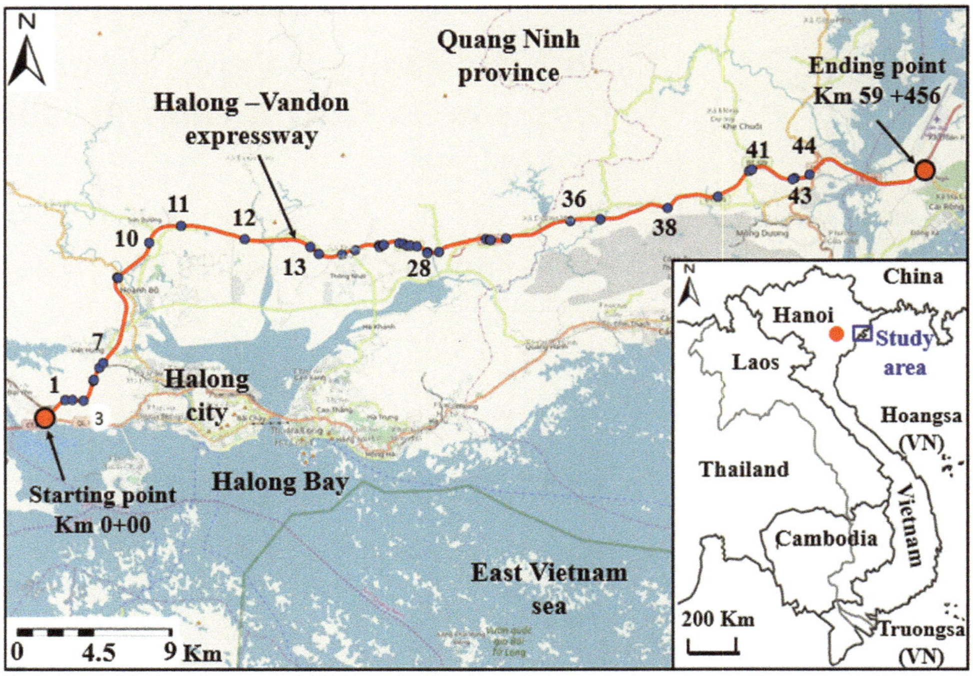

Fig. 1 Map of the Halong-Vandon expressway

Study Methods

First, we did many site investigations to study geomorphological characteristics, causes and triggering mechanism of landslides along the road. Soil samples and project documents were then collected for a further analysis. In which, several tests on the landslide samples at Km 27 + 950 were performed to investigate the physical mechanism by ring shear apparatus (Sassa et al. 2014).

To reconstruct rainfall thresholds triggering landslides, the Poisson model (Crovelli 2000) was applied in this study by using sub-daily rainfall measured at Bai Chay hydro-meteorological station in Ha Long City from 2013 to 2018. The 6-year data series of daily rainfall presents a total of 2232 days including 1.136 days with rainfall. The correlation between the amount of rainfall at the day of the landslide occurrence (P) and cumulative event rainfall at 1, 3, 5, 7, 10 and 15 days antecedent (P1, P3, P5, P7, P10, P15) is extracted and analysed to determine the number of needed rainy days before the day that landslide occurred.

Regional and Geological Settings

The 60-km HL-VD expressway is located in QuangNinh province. The route has the starting point at Halong-HaiPhong expressway and the ending point at Vandon airport. The province is adjacent to the Gulf of Tokin in the east and bordered by extended forest and mountain regions in the west. The study area is mainly across the mountainous and hilly relief, geologically characterized by Hon Gai, Binh Lieu, Bai Chay, Bac Son and Cat Ba Formations with limestone, sandstone, siltstone, conglomerates, gritstones, claystone, shale and thin lenses of coal. The faults are strongly active and mainly run in the east-west direction along the HL-VD expressway (DGM 1999). The average altitude ranges from 100 to 250 m and the highest is over 500 m with slope angles varying from 15 to 25°.

The study area experiences a tropical coastal climate with two different seasons: (1) hot and rainy in the summer from May to October and (2) cold and dry in the winter from November to April. The average annual precipitation ranges from 1.700 to 2.400 mm. There are about 70–85% of the annual average falling from June to September and about 45–50% of that dropping in July and August (Fig. 2).

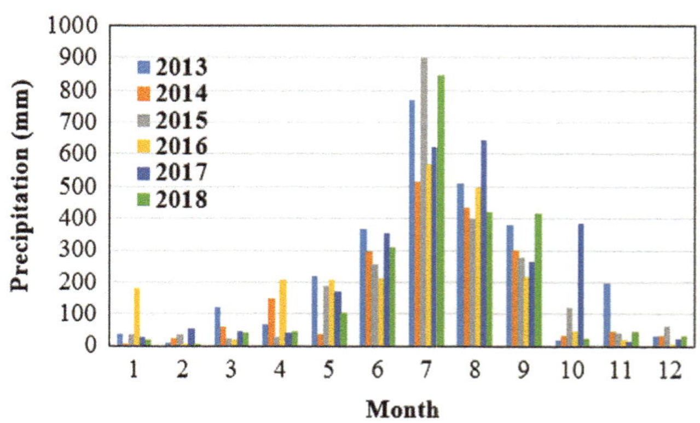

Fig. 2 Monthly rainfall from 2013 to 2018

Results and Discussions

Characteristics and Causes of Landslides

During the 2017 and 2018 rainy season, rainfall triggered a number of landslides along the HL-VD expressway. We carried out site surveys out to examine the causes and characteristics of landslides. As a result, 44 landslide locations along a length of 2411 m of the road have investigated so far and its features are listed in Table 1. All of these landslides are formed on cut slopes with regolith thickness varying from 2 up to 30 m. Based on types of movement and its materials (Varnes 1978), landslides are classified as: debris slides (Type I); debris slump (Type II); and Rockfalls (Type III). Shallow debris slide has a large majority (up to 35 cases) while there are several deep-seated landslides (6 cases). Three rockfall cases with small volumes (an average diameter less than 2 m) occurred along very steep and heavily fractured and weathered slopes. Shallow landslides failed within saturated and unconsolidated soil layers. A deep-seated landslide at Km 27 + 500 as slump type was the largest landslide along the expressway. On many slopes, landslides took place several times with an increase of extent and size. The typical slope strata consists of:

Table 1 Landslide inventory along Halong-Vandon expressway

No	Location by km	Time	Dimension (m)			Type	No	Location by km	Time	Dimension (m)			Type
			Ls	Ws	Ds					Ls	Ws	Ds	
1	1 + 90 (L)	27 Jun. 2018	30	15	1–2	I	23	27 + 950 (L)	21 Jul. 2018	140	80	24–32	II
2	1 + 360 (L)	27 Jun. 2018	30	15	1–2	I			31 Jul. 2018				
3	2 + 80 (R)	27 Jun. 2018	30	50	2–4	I			16 Aug. 2018				
4	3 + 400 (L)	27 Jun. 2018	20	40	0.5–1	I	24	28 + 200 (L)	21 Jul. 2018	30	50	2–3	I
5	3 + 460 (L)	21 Jul. 2018	32	15	0.5–1	I	25	28 + 500 (L)	21 Jul. 2018	15	50	2–3	I
6	4 + 200 (L)	27 Jun. 2018	22	15	0.5–1	I	26	28 + 600 (R)	21 Jul. 2018	75	150	6–12	II
7	4 + 550 (L)	9 Jul. 2017	38	115	1–3	I	27	29 + 480 (L)	21 Jul. 2018	15	20	0.5–1	I
		27 Jun. 2018					28	29 + 480 (R)	21 Jul. 2018	8	15	0.5–1	I
8	10 + 100 (L)	13 Jul. 2017	95	140	3–5	I	29	29 + 950 (R)	20 Jul. 2017	115	200	10–15	I
		21 July 2018					30	30 + 550 (R)	2017–2018	–	–		III
9	10 + 100 (R)	21 Jul. 2018	6	20	1–2	I	31	30 + 750 (R)	2017–2018	–	–		III
10	12 + 800 (L)	20 Jul. 2017	25	50	2–4	I	32	32 + 800 (R)	9 Jul. 2017	10	100	2–3	I
11	15 + 100 (L)	27 Jun. 2018	4	20	1–2	I			27 Jun. 2018				
12	18 + 730 (L)	21 Jul. 2018	30	40	4–5	I	33	33 + 00 (R)	13 Jul. 2017	10	100	1–2	I
13	19 + 00 (L)	2017–2018	–	–		III	34	33 + 100 (R)	27 Jun. 2018	15	10	4–5	I
14	22 + 480 (L)	20/7/2017	50	50	5–10	I	35	34 + 00 (L)	13 Jul. 2017	80	80	3–5	I
15	23 + 110 (R)	27 Jun. 2018	30	15	2–3	I			21 Jul. 2018				
16	24 + 450 (R)	9 Jul. 2017	40	50	2–4	I	36	37 + 760 (R)	30 Jun. 2017	50	100	2–3	I
		27 Jun. 2018					37	39 + 450 (R)	30 Jun. 2017	20	40	1–2	I
17	25 + 200 (R)	13 Jul. 2017	50	50	2–3	I	38	43 + 300 (R)	13 Jul. 2017	110	125	1–3	I
18	26 + 600 (L)	13 Jul. 2017	70	50	0.5–1	I	39	46 + 360 (R)	13 Jul. 2017	40	80	1–2	I
		21 Jul. 2018					40	48 + 600 (L)	5 Aug. 2017	8	18	0.5–1	I
19	26 + 600 (R)	30 Jun. 2017	6	30	1–2	I			27 Jun. 2018				
		21 Jul. 2018					41	48 + 800 (L)	27 Jun. 2018	8	18	1–2	I
20	26 + 770 (L)	30 Jun. 2017	70	50	1–2	I	42	51 + 300 (L)	13 Jul. 2017	100	215	6–10	I
		21 Jul. 2018							21 Jul. 2018				
21	26 + 950 (L)	30 Jun. 2017	70	50	2–3	I	43	51 + 300 (R)	21 Jul. 2018	53	55	8–10	I
		21 Jul. 2018					44	52 + 250 (L)	21 Jul. 2018	35	20	2–3	I
22	27 + 750 (L)	21 Jul. 2018	25	45	1–2	I							

*(R) and (L) present the right and left side of HL-VD route. Ls, Ws, and Ds are the landslide length, width and depth

- A very soft to soft or stiff clay layer containing granule, sand and grit in brownish/yellowish/whitish grey colors or brownish/reddish brown. The clay is soft and plastic when saturated, and it is stiff when dry.
- Sandstone, gritstone, siltstone, claystone in yellowish/reddish brown or yellowish/whitish/blackish grey colors. Rocks are fractured and slightly to moderately weathered.
- Coal-clay mixtures, claystone interbedded with thin layers of coaly shale.
- Strongly to moderately weathered limestone in a deep layer. Limestone are fractured in whitish or greenish grey colors.

Landslide materials are heterogeneous, loose, unstable and in low shear strength. Particularly, the silty clay layer, which has high porosity and permeability, are very susceptible to rainfall infiltration to initiate the movement. The sliding strata is mostly characterized by interbedding of claystone, siltstone, sandstone and gritstone in which the sliding surfaces are commonly formed within the claystone layers or at the geological boundary between claystone and upper layers of siltstone, sandstone and gritstone (Fig. 3a, b). In the study area, limestone rocks are likely to develop karst cavities that contribute to the build-up of groundwater table. In this regards, the geological characteristics have become important conditions for landslide occurrences.

In addition to geological causes, most of the slope surfaces after cutting are not inappropriately or sufficiently protected against water infiltration and erosion processes by slope stabilization solutions.

At rockfall locations, although the angle of cut slopes ranging from 70 to 80° are very steep, protection works had not applied. For a typical case at Km 23 + 110 (Fig. 3a), the applied remedial solution with concrete surfacing was not only too simple, but it also could not handle with groundwater factor inside the slope. As a result, landslides occurred in many locations along the road.

Another causative factor of the HL-VD landslides is the inappropriate design. There was no additional geological survey for the construction stage of the road. Soil parameters measured from a limit number of samples in the design stage was used for slope stability analysis of all engineered slopes in the construction stage. Therefore, the calculated safety factors are uncertain to meet the standards for slope stabilization design. At the Km 27 + 950, input parameters were strength values of weathered rocks, which were largely bigger than that of completely weathered sandstone and siltstone on the actual slope. Therefore, the safety factor was calculated to be larger than 1.30 whereas the actual value of the safety factor was much smaller than 1.2 (Lan et al. 2019). The use of improper soil parameters for slope stability calculation became a critical problem in this route. Consequently, the slope failed just about one month after slope cutting and under the time of the construction stage.

Physical Mechanism of the Selected Landslide Case at Km 27 + 950 by Ring Shear Test

At the Km 27 + 950 location, heavy rainfall triggered landslides in Thong Nhat commune, Hoanh Bo district in Quang Ninh province (3c). The geological structure and the strata of the slope are characterized by highly fractured and weathered layers consisting of sandstone and siltstone and underlaid limestone, which are quite typical along HL-VD expressway. The initial and subsequent slidings were triggered by heavy rainfall with continuously 5-day accumulative precipitations of 320.8 and 303.0 mm on 21 July 2018 and 31 July 2018, respectively. The entire landslide, which was characterized by rotational debris slump with a maximum depth of 32 m, consisted of the main mass movement and a subsequent sliding that slid along the bedrock of limestone (Lan et al. 2019). To understand the landslide mechanism, one soil sample (S0) taken on the left flank of the landslide (3c) was tested by ring shear apparatus ICL-2 that was equipped at Vietnam Institute of Transport Science and Technology. Grain-size distribution of the sample is shown in Fig. 4. Properties of the collected soil sample are mostly similar to those measured from boring samples. The landslide sample S0 has a high natural degree of saturation ranging from 88.56% and a high void ratio of 0.81. These parameters imply that the slope material is less unconsolidated, thus rainwater can easily infiltrate the slopes, increase the saturation degree of the soils, and reduce slope stability.

In ring shear test, the saturated sample is first consolidated to 500 kPa normal stress in the drained condition. The initial stresses (black line) are estimated from a sliding surface depth of 32 m with an inclination of 19° and soil weight of 19.1 kN/m^3. Shearing by shear displacement control mode is then applied at a designed shearing velocity (Sv) of 1 mm/s. The detailed testing procedure was presented by Sassa et al. (2014). The time-series data and stress paths of the ring shear test is indicated in Fig. 5. As seen in Fig. 5, immediately after shearing, pore water pressure was generated due to continuous grain crushing of the sample. The friction angle during motion is 31.8° while the residual strength of the sample is about 150 kPa. Less pore water pressure generation and high value of residual strength indicated a low level of landslide mobility. The test result implies the landslides moved at a low velocity that is good agreement with analysis by Lan et al. (2019).

Fig. 3 **a**, **b** Typical geological features and sliding strata of the landslides, and **c** Overview of a selected case study of deep-seated landslide with its UAV's photo

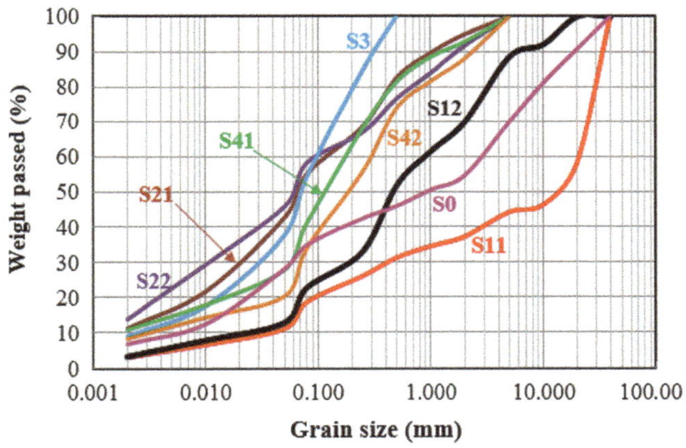

Fig. 4 Grain-size distribution of the sample S0 in comparison with samples from boring holes (BC 2018 and Lan et al. 2019)

Rainfall Analysis

The historical record indicates that the landslides have occurred in the period from June to August in 2017 and 2018, especially the majority of the events (33/44) took place in July that often has a largest amount of rainfall (Table 1). It presents that landslides were triggered by short-time rainfall events with a high accumulative amount (Fig. 6a). The 5-day accumulative precipitations triggered landslides on 21 July and 31 July, 2018 were 320.8 and 303.0 mm, respectively. In July 2018, the moderate to high heavy rainfall triggered the largest number of landslides (up to 20 landslides).

The results of rainfall threshold analysis are presented in Table 2 and Fig. 6b. It presents that daily.

Fig. 5 Time series data (left) and stress path (right) of shear displacement control test: normal stress (σ) = 500 kPa, shear displacement speed = 1 mm/s, B_D = 0.98

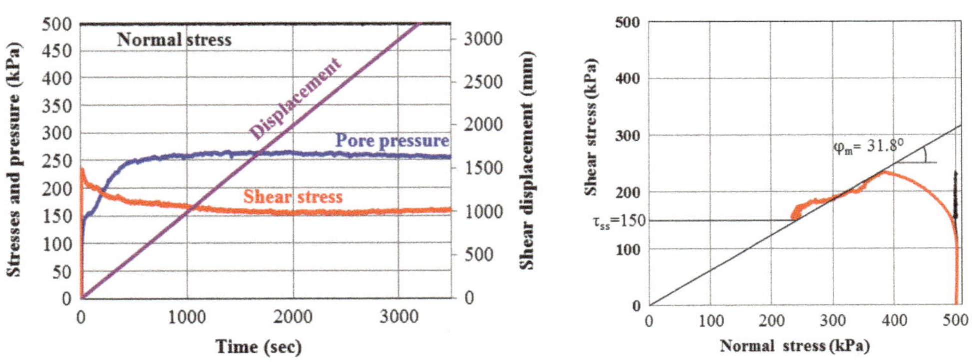

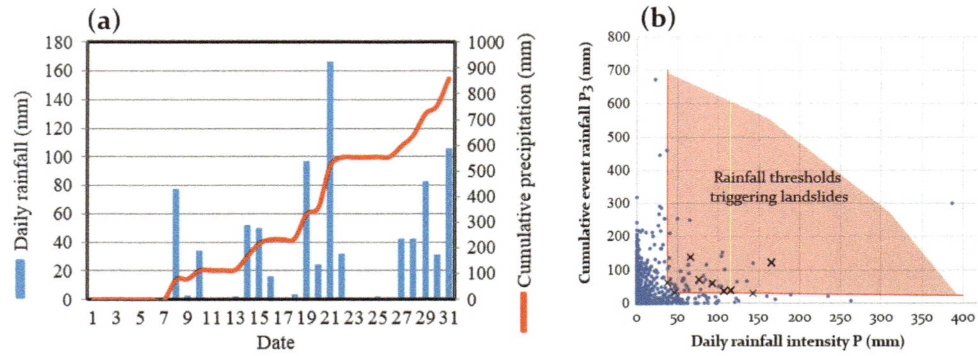

Fig. 6 **a** Daily rainfall and its cumulative precipitation in July 2018 and **b** Rainfall thresholds triggering landslide at (P3)

Table 2 Relation between rainfall thresholds and the occurrence of landslides

No	Rainfall (mm)	Rainfall thresholds triggering landslide (P) (mm)	Number of day having rainfall ≥ (P)	Number of day in which landslide occurred	Number of day in which landslide didn't occur	Probability (%)
(1)	(2)	(3)	(4)	(5)	(6)	(7) = (5)/(4)
1	P min	38.2	117	54	63	46.15
2	P3 min	29.2	55	54	1	98.18
3	P5 min	43.7	65	54	11	83.08
4	P7 min	43.7	87	54	33	62.07
5	P10 min	54.5	86	54	32	62.79
6	P15 min	79.9	91	54	37	59.34

rainfall (P) and cumulative event rainfall (P1, P3, P5, P7, P10, P15) have high correlations with landslide occurrence. Look at the Table 3 and Fig. 6b we can see minimum daily rainfall (P) triggering landslides is 38.2 mm and the percentage of the day that landslides occurred in the relation between daily rainfall and cumulative event rainfall varied from 46.15 to 98.18%. At the P3, the percentage is the largest. Thus, for the study area, it is likely that the cumulative event rainfall of 3 days ago was sufficient to saturate the soils and forming the landslides. Therefore, it is reasonable to use the threshold of 3 days earlier to identify rainfall thresholds triggering landslides and as mentioned by red colour in Fig. 6b.

Conclusions

This paper presents the preliminary results of the study on the causes, characteristics and mechanisms of landslides along the Halong-Vandon highway. The study shows that the majority of landslides were shallow debris slides on the cutting slopes and others were debris slumps and small rockfalls. Slope cutting is one of the most significant causes induced landslides while heavy rainfall is the main factor that triggered the slidings. Other contributing factors relate to the geological structure of slope strata and active west-east faults along the expressway that favored the landslide occurrence. Heterogeneous, unconsolidated, fractured and weathered materials of sandstone, siltstone, claystone and shale are prone to water saturation and shear strength reduction, leading to sliding.

In addition, this study also pointed out an inadequacy of the survey, design and construction works for slope stabilization. The HL-VD project has been not conducted additional geological surveys during the construction stage, and designs for slope remedy using parameters measured in the design stage were inappropriate with the real site conditions. Therefore, it strictly requires to conduct additional surveys to determine the actual geological data for the construction phase at landslide locations.

The detailed analysis indicated that the landslides were typically related to rainfall with moderate to high rainfall intensity during the short period. In critical conditions, the percentage of time for landslides varies between 46.15 and 98.18%. The relationship of daily rainfall with rainfall of 1, 5, 7, 10 and 15 days prior to the landslides has a lower probability to trigger the slidings in comparison with rainfall of 3 days.

For the Km 27 + 950 landslide, ring shear apparatus tests indicate that the collected landslide soil sample did not present high mobility due to less pore water pressure generation. So, the landslides characterized as rotational debris slump could not travel rapidly. This finding are incordance with site evidence. Notably, the evidence of the underground spring system and karst caves within the slope suggested a severe condition for the build-up of groundwater table to trigger the landslide.

References

BC (2018) Project documents provided by the BOT Bien Cuong Joint Stock Company

Crovelli RA (2000) Probability models for estimation of number and cost of landslides. US Geological Service. Open-file report 00-249

DGM (1999) Geological Map of Halong city (F-48-xxx, Scale 1:200.000) published by the Department of Geology and Minerals of Vietnam

Duc DM (2013) Rainfall-triggered large landslides on 15 December 2005 in Van Canh district, Binh Dinh province, Vietnam. Landslides 10:219–230

Lan CN, Tien PV, Do TN (2019) Deep-seated rainfall-induced landslides on a new expressway: a case study in Vietnam. Landslides. https://doi.org/10.1007/s10346-019-01293-6(Online)

Sassa K, Dang K, He B, Takara K, Inoue K and Nagai O (2014) A new high-stress undrained ring-shear apparatus and its application to the 1792 Unzen-Mayuyama megaslide in Japan. Landslides 11(5):827–842

Tam DM (2001) Flooding and landslides at the highways of Vietnam. In: Proceedings of the international workshop on "Saving Our Water and Protecting Our Land". Hanoi, 20–22 Oct 2001, pp 18–27

Tam DM, Hanh NH, Hung NQ, Viet NB, Hung NV, Thao PT, Anh TTV (2008) Project report on research on selection and application conditions of the new technologies for landslide risk prevention along national highways. Research project in transportation sector, Ministry of Transport, 2008, p 396

Varnes DJ (1978) Slope movement types and processes. In: Schuster RL, Krizek RJ (eds) Landslides, analysis and control, special report 176: Transportation research board. National Academy of Sciences, Washington, DC., pp 11–33

Part III

Landslide Hazard Assessment and Zonation—Susceptibility Modelling

New Data on Geological Conditions of Landslide Activity on Vorobyovy Gory (Moscow, Russia)

Olga Barykina, Oleg Zerkal, Igor Averin, and Eugene Samarin

Abstract

Landslides developing on the high slope of the Vorobyovy Gory in the right side of the Moskva River valley form one of the largest landslide massifs in Moscow (Russia). Based on the newly obtained data it was detected: firstly, the territory involved in landslide processes on the Vorobyovy Gory is characterized by much larger values, both in area and in depth, than it was previously assumed. In the head part, where the displacement zone is located at depths of 80–100 m, the deformations, confined to the lower part of the Jurassic deposits, have a block character. Secondly, we can speak of a combined mechanism for the development of a large-scale landslide massif "Vorobyovy Gory", which includes plastic flow with the formation of a ridge compression, collapse with tipping, block displacement and other types of deformations. Also it is possible to distinguish both primary and secondary displacements. According to correlation of Callovian-Oxfordian deposits the new sliding surface position is detected.

Keywords

Largest landslide • Sliding surface • Combined mechanism of displacements

O. Barykina (✉) · O. Zerkal · E. Samarin
Lomonosov Moscow State University, Moscow, 119991, Russia
e-mail: barykina@geol.msu.ru

O. Zerkal
e-mail: igzov@mail.ru

E. Samarin
e-mail: samarinen@mail.ru

I. Averin
Engineering Geology" LTD. Company, Moscow, 121351, Russia
e-mail: i.averin@mail.ru

Introduction

Landslide processes in Moscow have been studied more than a hundred and fifty years (Pavlov 1890; Nikitin 1897; Danshin 1937; Churinov 1957; Kuntzel 1965; Paretskaya 1975; Barykina et al 2017, 2019; Zerkal et al 2017). Now, more than 200 landslide sites are known on the territory of the city, including 16 sites where large-scale slope deformations develop. One of the landslide hazard areas is the Vorobyovy Gory—nature reserve within Moscow located on the right cut bank of the Moskva River.

The study area is located in the central part of the Vorobyovy Gory covering the area from the Moscow viewpoint to the Moscow metro bridge. The slope has a typical landslide landscape. The geological strata are represented by deposits of the Carboniferous, Jurassic, Cretaceous and Quaternary systems. Deep large and long-time active landslides were formed at the right side of the valley of the Moskva River in pre-Holocene time. Currently, several zones of the "Vorobyovy Gory" landslide are active. The length of the slope, within which the slope deformations develop, is several kilometers, the visible width along the axial part of the landslide is more than three hundred meters, and the total volume of soils involved in the landslide shifts in the area under consideration is estimated at several million cubic meters. The article is devoted to the description of landslide development based on new data (2015–2019).

The main features of the landslide massif structure are the following: first of all, it has been generally accepted that on the Vorobyovy Gory, the upper boundary (on the surface) of landslide deformations (head scarp) is the modern edge of the existing slope, and the main deforming horizon of the landslide massif is the clayey sediments of the Oxfordian Stage.

Additional study of the slope structure was carried out according to the axial parts of the landslide bodies—from the high surface of the "plateau" to the Moskva River. Initially, it was believed that bore-holes located on the high surface of the "plateau" are located off the area with landslide

F. Guzzetti et al. (eds.), *Understanding and Reducing Landslide Disaster Risk*,
ICL Contribution to Landslide Disaster Risk Reduction,
https://doi.org/10.1007/978-3-030-60227-7_15

deformation. The detailed study of the core of the wells on the watershed “plateau” near the slope edge, within the area of natural occurrence of rocks, as it was considered earlier, showed an another situation. In particular, in clays of the Oxfordian Stage located in the lower part of the Mesozoic-Cenozoic interval, a series of sliding surfaces was revealed.

Geological Setting

Area Characteristic

The city of Moscow is located on a naturally complex territory, which is characterized by a long geological history of development and a variety of landscapes. The valley complex of the Moskva river occupies the most part of the city (river floodplain and its three terraces); the south-western part (where the research area is located) lies within the Teplostan upland; the eastern part is the marginal part of the Mescherskaya lowland—a flat, weakly dissected swampy plain with low, absolute marks. Thus, the valley of Moskva river is the main geomorphological object of the territory, occupying a significant part and crossing the city diagonally from the north-east to south-west (Fig. 1).

The relief of Moscow inherited preglacial features and was formed because of Quaternary period glaciations, as well as erosion. In the right-sided banks, the Moskva River cuts into the valley side, forming steep landslide slopes, one of which (“Vorobyovy Gory”) studied herein.

Vorobyovy Gory are located on the right side of the valley of the Moskva River and represent a steep (Fig. 2), sometimes forested slope (up to 70 m high) with a peculiar ridge-landslide relief, stretching along the river. The main scarp is well expressed in relief—its height varies from 12 to 30 m, and steepness from 25° to 40°. Erosion forms, such as ruts, gullies, ravines, etc., are developed within the slope also. The lower part of the slope adjoining the Moskva River embankment is significantly technologically altered by anti-landslide measures.

Geology

Rocks of the Middle Pennsylvanian of Carboniferous, Bathonian-Tithonian Jurassic, Berriassian-Aptian stages of the Lower Cretaceous, and Quaternary formations represented by morainic and aquatic-glacial accumulations compose a section of the watershed part of the Vorobyovy Gory.

The deposits of the Myachkovskian age (the Moscovian stage of the Middle Pennsylvanian), are at the base of the slope. The deposits are represented by an uneven interlacing of limestones, marlstone and clays.

Limestones are white, creamy, light grey, coarse- and finely organogenic-clastic, dense, cavernous at some intervals, the caverns are filled with clay bluish-gray, solid (the size of the caverns gradually decreases down the section from 3–4 cm to 1–1.5 cm), dense, strong, layer thickness is 1.5–1.7 m.

Marlstone is white, light green, homogeneous, dense, strong, thickness from 0.3 to 1.2 m.

Clay is bluish-gray, greenish-gray, dense, soft and tight, with a thickness of 0.1–0.2 m.

At the top, limestone and dolomites are crushed to fine crushed stone 3–4 cm in size, usually in lime powder, with a crushing zone, thickness of 5–30 cm.

The top of Carboniferous deposits forms a complexly organized pre-Jurassic paleorelief. It should be emphasized that the area of the Vorobyovy Gory from the observation deck to the metro bridge is located within the “Main pre-Jurassic paleovalley”.

Above are the deposits of the Bathonian-Callovian Stage of the Middle Jurassic. In the lower part of the section, mainly in the local depressions of the pre-Jurassic relief, there are deposits of the Lublin unit, represented by black clays, soot, fat, with local ironing (bioturbation), with fine inclusions of limestone. Thickness is from 0.4 to 1.5 m, increasing from watershed to river bed.

Along the sharp border with the stratigraphic dissent, the deposits of the Velikodvorskaya unit (J_2vd) are located above the section, presented by brown clays, sandy, with interlayers of fine brown sand, with fine pebbles of carbonate rocks.

Sediment thickness of the Velikodvorskaya unit: in the watershed area—1.2 m, in the Moskva River bed—0.5 m.

With the gradual transition above the section, deposits of the Podosinkjvskaya unit (J_2pd) are found. It is presented by greenish-brown, grayish olive, greenish-gray clays, sandy, with glandular oolites and pisolites, with individual spots of carbonation, with thin (up to several centimeters) layers of oolite marlstone.

Sediment thickness of the Podosinkjvskaya unit (J_2pd): in the watershed area is 3.1 m.

Four units represent the section of the Oxfordian stage, the deposits of which overlap the Callovian units. (Unified regional stratigraphic…(2012)).

Above the section with stratigraphic dissent there are deposits of the Ratkovskaya unit (J_3rt) presented by dark brown, with light greenish interlayers, silvery, with abundant detritus of shells and foraminifera, and ammonites of middle Oxford.

Deposition thickness of the Ratkovskaya unit: in the watershed area is 1.4 m.

Sediments of the Podmoskovnaya unit (J_3pm) lie above the section with a gradual transition. Clays are brownish-dark-gray, greasy, dense, with numerous passages of silt, with fine phosphorite oolites (up to 25 mm) and Upper Oxford ammonites.

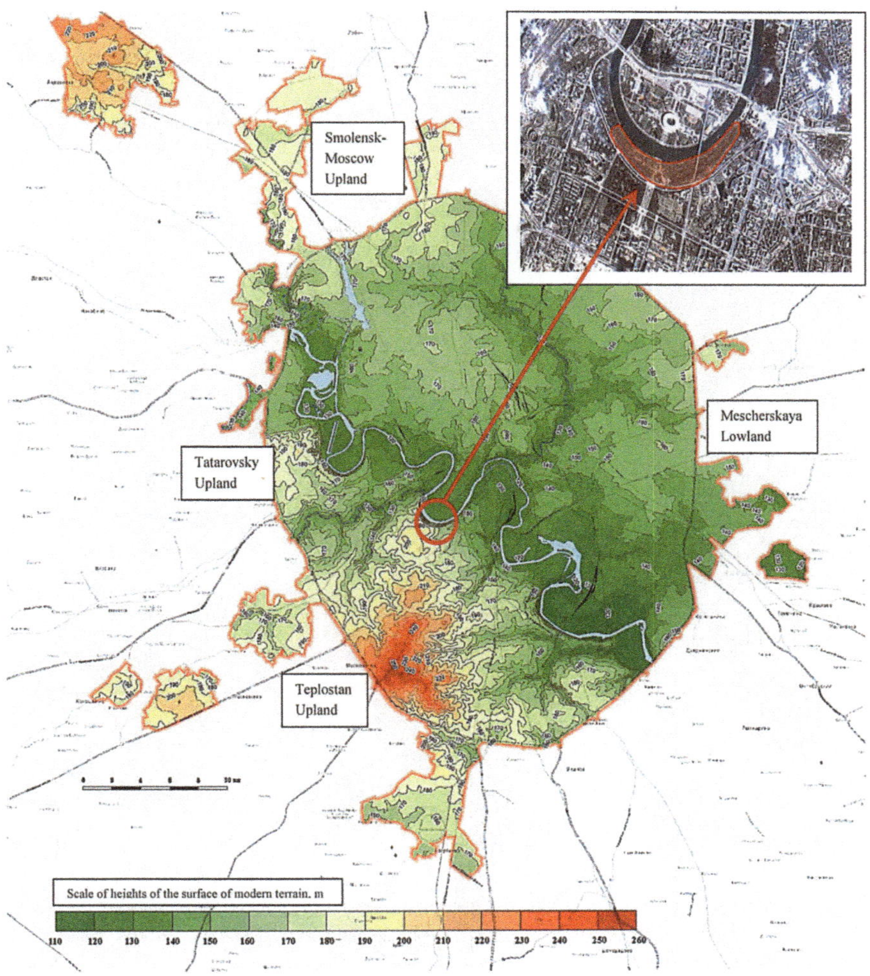

Fig. 1 Hypsometric map of Moscow area (by Geological Atlas of Moscow 2009). Research area is marked by the red circle

Sediment thickness of the Podmoskovnaya unit: in the watershed area is 5.0–6.0 m.

Sediments of the Kolomenskaya unit (J_3kl)—clays gray, dark grey, dark brownish grey, strongly siltstone, mica, limestone, with nests of powdery pyrite, below with fine phosphorites, above with nests of green glauconite (5.3–5.6 m)—are found in the section with a gradual transition.

Sediment thickness of the Kolomenskaya unit: 6.8–7.5 m near the watershed, 5.6 m in the Moskva River bed.

The displacement zone represented by a series of subhorizontal planes is described in the depth range of 50.2 and 51.2 m (Fig. 3). The depth range of 53.6 and 54.0 m describes a single sliding zone oriented at an angle of up to 20 degrees to the horizontal.

Above the section with a gradual transition lie sediments of the Makarjevskaya (Yermolinskaya) (J_3mk) unit—black clay, dense, layered, strongly micaceous, often with abundant pyrite (loose powder or dense constrictions), with small carbonate constrictions, individual layers of clay greenish and black (enriched with glauconite), solid.

Sediment thickness of the Makarjevskaya (Yermolinskaya) unit: 5.1–5.5 m near the watershed, 1.9 m in the Moskva River bed.

In the depth range of 44.5 and 45.5 m we describe the dislocation zone represented by a series of subhorizontal sliding surfaces.

The deposits of the Tithonian stage are represented by the Egoryevskaya unit (J_3eg)—glauconite green sands, dense, with phosphorites, with fragments of belemnites. Sediment thickness in the watershed area is 2.2 m.

The Filevskaya unit (J_3fl) is represented by a pack of black dense micaceous clays, layered, with constrictions of pyrite, belemnites, characteristic ammonites (Virgatites), the top greenish, silvery and sandy.

Sediment thickness in the watershed area is 5.8 m.

Deposits of the Lopatinskaya unit (J_3-K_1lp) are located above the section along the sharp rough boundary. These are

Fig. 2 The overview of the landslide slope of the Vorobyovy Gory (on the left in the background is the Moscow City business district)

green, grey and olive-green, fine sands and siltstones, with phosphorites and ammonite. In the bed of the Moskva River sediments are represented by brown-black silts, with black clay interlayers up to 0.5 cm, with a large number of grated fauna fragments and phosphorite concretions.

Sediment thickness of the Lopatinskaya unit is 4.2–4.5 m in the watershed area and 2.6 m in the Moskva River bed.

The Berriasian stage of the Lower Cretaceous is presented by gray-green, green fine-grained sands deposits.

The deposits of the Kuntsevskaya unit (K_1kn) are located above the section along the sharp rough border. The sediments are presented by sands from light grey to greenish grey are strongly micaceous, glauconite, siltstone fine-grained, with thin (up to ribbon) interlayers of black mica clay, with very rare belemnites; interlayers of green and grey-green fine and fine sands, with thin interlayers of black mica clay and dark grey siltstone.

Sediment thickness of the Kuntsevskaya unit is 13.1–13.8 m in the watershed area, and 4.8 m in the Moskva River bed.

In the borehole, passed in the Moskva River bed, the deposits of the Kuntsevskaya unit are more or less confidently identified only in the interval of the absolute markings —111.30–113.20 (lower part). The upper part of the sediments is significantly reworked by slope deformations, which is manifested by changes in color to brown-yellow (a result of glauconite decomposition), separate intervals of sand ironing, and disturbance of layering character.

The Hauterivian deposits are composed of deposits of four units. Deposits of the Diakovskaya unit (K_1dk) are located above the section along the sharp rough border. The sands are bright green, emerald-green, greenish-black, grey, fine-grained glauconite, clayey, dense, with horizontal and wavy layering, at the base with a layer of phosphorites 2–5 cm.

Sediment thickness of the Diakovskaya unit: in the watershed area—4.0–4.5 m, in the Moskva River bed–sediments are not identified.

With the gradual transition above the section, deposits of the Savel'evskaya unit (K_1sv) are found. They are presents by: dark gray to black, fine-grained sand and grey, dark gray silts, relatively homogeneous; sands swampy green, emerald green, medium grained, fine and dusty, interlayered compacted, glauconite.

Sediment thickness of the Savel'evskaya unit in the watershed area is 2.5–3.7 m. With a gradual transition above the section, deposits of the Gremyachevskaya unit (K_1gr) are found. Sands are brownish-greenish-grayish, downstream to

Fig. 3 The zone of series sliding surfaces (the depth 50.2–51.2 m)

dark grey and black, mainly fine, micaceous, with phosphorite concretions up to 20–25 mm across and rarely large phosphorite nodules.

Sediment thickness of Gremyachevskaya deposits: in the watershed area is 4.0–4.2 m.

With a gradual transition above the section there are deposits of Kotelnikovskaya unit (K_1ktl) presented by siltstones and clays—temp-gray, black, dense, tight plastic.

Sediment thickness of the Kotelnikovskaya unit is 2.3 m in the watershed.

The Barremian Stage is presented by deposits of the Butovskaya unit (K_1bt). The deposits are represented by a fine interlacing of sand of light gray, light brown, reddish brown, dusty, fine and medium coarseness, dense, micaceous with clay brown, dark gray, grey with hardness, from tight to soft plastic consistency.

Deposit thickness of the Butovskaya unit: in the watershed area is 3.9–4.4 m.

Deposits of the Aptian Stage, which crowns the thickness of the Lower Cretaceous deposits, are presented by deposits of the Ikshinskaya unit (K_1ik). The sediments are light gray, yellowish-gray, bright yellow, light brown, fine, dense, homogeneous, with single interlayers enriched with gravel material, with clay lenses, compacted at watered intervals.

Sediment thickness of the Ikshinskaya unit is in the watershed area—14.3–15.0 m.

With a gradual transition the deposits of the Vorokhobinskaya unit (K_1vrh) are found. It presented by motley-colored contrasting sandy-clayey unit with small lemonite nodules, wavy, sometimes lenticularly layered. Sands are fine grained with yellowish ferrous interlayers.

Sediment thickness of the Vorokhobinskaya unit does not exceed 3.0 m in the watershed area, as it is covered with Quaternary formations. To the west, towards the observation deck, the sediment thickness of the Vorokhobinskaya unit increases to 5.7 m.

The thickness of Quaternary sediments is composed of two horizons of moraine loam with crushed stone and gravel, separated by fluvioglacial medium- and fine-grained sands formed in the interglacial epoch (Mindel-Riss). The lower horizon of loams has a thickness of up to 7 m, and the thickness of the upper horizon (the Moscow stage of glaciation) is up to 5–6 m.

The Moskva River has a depth of about 5–7 m at the site adjacent to Vorobyovy Gory. The thickness of alluvial deposits in the channel part of the valley ranges from 7 to 10 m.

Characteristic of Landslide Accumulations

The facial analysis of Jurassic and Cretaceous deposits, described in the core of boreholes allows to draw the following conclusions.

1. The hypsometric position and thickness of the clays of the Velikodvorskaya unit (J_2vd) do not change from the watershed to the Moskva River bed.
   Deposits of the Podosinkovskaya unit (J_2pd)—greenish-brown clays—are characterized by constant thickness within the watershed part of the slope (deposits in the natural occurrence) and in the area of the landslide terrace—3.1–3.2 m. Similar to the deposits of the Ratkovskaya unit (J_3rt)—dark brown clays—at a thickness of 1.4 m the top of the layer rises by approximately 1.0 m. In the bed of the Moskva River the deposits of these units cannot be separated due to a significant clay material crumbling. The deposits of these formations lie at the depth of 40 m and have a total power of 5.5 m. The visible change in the thickness and position of the top is clearly fixed on brownish-gray clays of the Oxfordian Podmoskovnaya unit (J_3pm): if in the area of the watershed (deposits in the natural occurrence) the absolute top marks are at 89.86–90.92, and in the area of the landslide stage—91.5–91.06 with comparable thickness of 5.0–6.0 m, then in the Moskva River bed the thickness of clays of the Podmoskovnaya unit increases to 8.2 m, and the top of deposits rises on 5 m.
   Thus, the main horizon of landslide deformation is in the clays of the Podosinkovskaya and Ratskovskaya units. The thickness of landslide formations can be estimated at approximately 60.0 m.
2. In the area of high landslips, the stratigraphic volume of Jurassic and Cretaceous sediments is fully preserved, but there is a significant reduction in the thickness of individual units, especially the Mnevnikovskaya unit (Egoriyevskaya and Filevskaya units), the Lopatinskaya and the Kuntsevskaya units. Thus, the landslide has a block structure in the upper part. The increase in the thickness of sands of the Diakovskaya unit (K_1dk) from 4.0 to 4.5 m at the watershed to 6.2 m may indicate a secondary landslide with a sliding surface in the depth range of 31.0–32.0 m.
3. For deposits of the Kolomenskaya unit (J_3kl) and Makarievskaya unit (J_3mk) apply well diagnosed zones of subhorizontal displacement in depth intervals of 44.5–45.5 m, 50.2–51.2 m and 53.6–54.0 m, which may be due to subsequent after the initial block shift by horizontal clay extrusion.

Conclusions

So, according to correlation of Callovian-Oxfordian deposits the new sliding surface position is detected. The main horizon of landslide deformation is in the clays of the Podosinkovskaya and Ratkovskaya units. The thickness of landslide deposits at mid-slope can be estimated as approximately 60.0 m.

Violation of the stratigraphic volume of Jurassic and Cretaceous sediments (deposition from some horizons in the section) in the Moskva River bed, reduction of the Tithon deposits, and, conversely, increase in the Oxford-Kimmeridge interval indicates that the toe landslide part (50–100 m long) is a typical compression landslide. Taking into account the thickness of alluvial sediments, represented mainly by tight plastic sands and loams, with basal gravel and crushed stone horizon, the thickness of landslide deposits can be estimated as approximately 27.0 m. Performed analysis showed the combined mechanism of displacements (International Geotechnical Society's…(1993)) of a large-scale landslide massif "Vorobyovy Gory".

References

Barykina OS, Zerkal OV, Samarin EN, Gvozdeva IP (2017) On the development of landslide processes on Vorobyovy Gory (Moscow). Engineering-geological problems of the present and methods for their solution. In: Proceedings of conference geomarketing. Moscow, pp 111–117. (in Russian)

Barykina OS, Zerkal OV, Samarin EN, Gvozdeva IP (2019) The history of slope evolution—primary cause of its modern instability (by Example of the "Vorobyovy Gory" Landslide, Moscow). In: Natural hazards and risk research in Russia: innovation and discovery in Russian science and engineering. Springer International Publishing AG, part of Springer Nature 2019, Switzerland, pp 345–361. (ISBN 978-3-319-91832-7)

Churinov MV, (1957) Description of landslides of the right bank of the Moskva River on the site of the Leninskie Gory and the possibility of building development of this territory. Questions of hydrogeology and engineering geology. In: Proceedings of VSEGINGEO. Gosgeoltekhizdat. Moscow, pp 62–78. (in Russian)

Danshin BM (1937) The geological structure of the Leninskie Gory in connection with some issues of the stratigraphy of deposits of the Cretaceous system and landslide phenomena along the banks of the Moskva River. In: News of the Moscow geological trust, pp 3–23. (in Russian)

International Geotechnical Society's UNESCO Working Party on World Landslide Inventory (WP/WLI) (1993) A multi-lingual landslide glossary. Bitech Publ, Vancouver, p 59

Kuntzel VV (1965) On the age of deep landslides in Moscow and the Moscow region, associated with Jurassic clay deposits. Bull Moscow Soc Naturalists. Subdiv Geol XL (8):93–100. (in Russian)

Nikitin SN (1897) Les environs de Moscou. Guide des excursions du VII Congres Geologique International. St-Petersbourg, pp 1–16

Paretskaya MN (1975) Dependence of the morphology of landslides extrusion of the Moscow suburbs on the strength of Jurassic clays. Procc VSEGINGEO 81:94–97 ((in Russian))

Pavlov AP (1890) New data on the geology of the Vorobyovy Gory. Bull Nat Sci 7:301–304 ((in Russian))

Unified regional stratigraphic scheme of the Jurassic sediments of the East European Platform (2012) Explanatory letter. In: Mitta VV, Alexeev AS, Shik SM (eds) GINRAS—VNIGNI. Moscow. (in Russian)

Zerkal O, Barykina O, Samarin E, Gvozdeva I (2017) The influence of paleo-landslide activity on the modern slope stability. In: Proceedings of 2017 IPL symposium, UNESCO-ICL, Paris, pp 89–92

WLF
2020 Kyoto Japan

Impact of Agricultural Management in Vineyards to Landslides Susceptibility in Italian Apennines

Massimiliano Bordoni, Alberto Vercesi, Michael Maerker, Valerio Vivaldi, and Claudia Meisina

Abstract

Cultivation of grapevines in sloping soils is very widespread all over the world, representing also a fundamental branch of the local economy of several hilly zones. Vineyards can be managed in different ways especially in the inter-rows. These management practices may influence deeply soil properties and grapevine root development. Therefore, this work aims to analyze the effects of different agronomical practices of inter-rows on soil properties, grapevine root systems and proneness towards shallow landslides. We focused on traditional agricultural techniques of tillage and permanent grass cover as well as the alternation of these two practices between adjacent inter-rows. The research was conducted in several test-sites of the Oltrepò Pavese, one of the most important Italian zones for wine production in northern Italian Apennines. Among the examined soil properties, soil hydraulic conductivity was the most influenced one by different soil management practices. Regarding the features of the grapevine root system, vineyards with alternation management of inter-rows had the highest root density and the strongest root reinforcement. As a consequence, slopes with medium steepness were unstable if inter-rows of vineyards were tilled, while vineyards with permanent grass cover or alternation in the inter rows promoted the stability of slopes with higher steepness. The results of this study yielded important information to establish land use managements acting as mitigation measures for shallow landslides susceptibility.

Keywords

Vineyard • Soil • Root • Shallow landslides • Failure probability • Land management

Introduction

Vineyards cover currently 7.5 million ha corresponding to about 0.5% of the entire agricultural areas in the world (OIV 2017). As other human activities, viticulture has strong impacts on the environment Moreover, vineyard cultivation causes important effects in different parts of the soil system, influencing its physical, hydrological, chemical and biological properties through different management techniques, in particular of the inter-row management (Prosdocimi et al. 2016; Rodrigo-Comino 2018).

The management practices of inter-rows have also an important impact on the distribution of grapevine roots in the soil, in terms of rooting depth and, especially, of root density (Smart et al. 2006). The density of roots within the soil, together with their mechanical behavior related to shear and/or tensile forces, increases soil stability (Bischetti et al. 2009; Cohen and Schwarz 2017). Root reinforcement may have beneficial effects in preventing slope instabilities and is often used as an effective tool to decrease landslide susceptibility, in particular for shallow landslides affecting the first 2.0 m of soil (Wu 2012). Shallow landslides triggered by intense rainfall events frequently affect vineyards located in sloping terrains, causing the partial or complete destruction of the vineyards, of local structures and infrastructure and thus, creating severe economic damages. Shallow

M. Bordoni (✉) · M. Maerker · V. Vivaldi · C. Meisina
Department of Earth and Environmental Sciences, University of Pavia, Via Ferrata 1, 27100 Pavia, Italy
e-mail: massimiliano.bordoni@unipv.it

M. Maerker
e-mail: michael.maerker@unipv.it

V. Vivaldi
e-mail: valerio.vivaldi@unipv.it

C. Meisina
e-mail: claudia.meisina@unipv.it

A. Vercesi
Department of Sustainable Crop Production, Università Cattolica del Sacro Cuore, Via Emilia Parmense, 84, 29122 Piacenza, Italy
e-mail: vercesi@unicatt.it

F. Guzzetti et al. (eds.), *Understanding and Reducing Landslide Disaster Risk*,
ICL Contribution to Landslide Disaster Risk Reduction,
https://doi.org/10.1007/978-3-030-60227-7_16

landslides in vineyards are widespread in different European contexts characterized by traditional viticulture, especially, in Italy (Fonte and Masciocco 2009; Blahut et al. 2014; Bordoni et al. 2019). For this reason, a quantification of root reinforcement of grapevines in vineyards with different inter-row management is fundamental to understand the practices that might promote the stability of sloping vineyards and the ones that cause slope instability.

Root reinforcement can be also implemented in models to estimate the slope stability towards shallow failures (Chiaradia et al. 2016; Zhang et al. 2018). In the case of vineyards, different agricultural managements could determine differences in root reinforcement and, consequently, differences in modelled proneness to shallow failures.

The main aim of this paper is to analyze the effects of different inter-row management techniques on soil properties, grapevine root systems and proneness towards shallow landslides. The research was conducted in one of the most important Italian wine production areas, the Oltrepò Pavese, in Lombardy region, in north-western Italy. This area is also representative of the main geological, geomorphological and agronomical features of northern Italian Apennines (Bordoni et al. 2019).

Materials and Methods

Study Area

Oltrepò Pavese (265 km^2, Fig. 1) is characterized by a traditional viticulture, conducted on hillslopes and based on different grapevine cultivars, as Croatina, Pinot noir, Barbera, Riesling italico, Chardonnay and Moscato bianco. In the northern part of the area, bedrock materials are characterized by a Mio-Pliocenic succession consisting of sandstones, conglomerates, marls and evaporitic deposits. In this area, slopes are steep, with slope angles generally steeper than 20°. Shallow soils derived from bedrock weathering consists in sandy silts or clayey sandy silts with thickness ranging between few centimeters and 2.5 m. In the southern part of the area, bedrock is composed of Cretaceous flysch deposits and other Eocenic-Miocenic bedrocks, consisting of marls, calcareous-marls, sandstones and scaly shales. In this sector, slopes are less steep with slope angles generally between 10° and 20°. Vineyards are mainly cultivated at elevations ranging between 60 and 500 m a.s.l. and on slopes between 5° and 37°.

The entire study area is very susceptible to slope instabilities, as testified by several rainfall-induced shallow landslides occurred since 2009 (Bordoni et al. 2019). The mean density of these phenomena is of about 6 shallow landslides per km^2 (2105 phenomena since 2009). Shallow landslides are very common in cultivated vineyards of the study area (Fig. 2). In fact, 424 failures affected vineyards, occupying an area of 3.2 km^2 (2.1% of the area cultivated with vineyards). These phenomena caused the partial or the total destruction of the rows, severe damages to the roads and loss of fertile soil.

Test-Sites

29 test-sites (Fig. 1) were selected, representing the main geological, geomorphological and agronomical features of the area. 17 sites had soils with a predominantly clayey silt or clayey sandy silt texture, classified as low plastic soils (CL) according to the Unified Soil Classification System (USCS).

The other 12 sites were characterized by silt with clays or silty clays and could be considered as high plastic soils (CH), according to USCS. As regards the geomorphological features, all the test-sites were located on slopes, at elevations ranging between 115.0 and 344.1 m a.s.l. Moreover, slope angle changed between 5.0 and 37.0°, while all slope aspects apart of north directions occurred.

Most of the analyzed vineyards (25 test-sites) were characterized by a row orientation parallel to the maximum slope gradient (Par VN). Row orientation perpendicular to the maximum slope gradient (Perp VN) was less widespread (4 test-sites). The studied grapevine plants had ages between 5 and 30 years. Four different types of inter-row management, corresponding to the techniques usually adopted by the local wine-growers, were tested: (1) Total tillage (TT), which does not allow to grow grasses in the inter-rows during the entire year; (2) Tillage (TIL), where mechanical operations are conducted to limit the grow of grasses; (3) Alternating tillage-grass (ALT), in which a row is tilled while the next one is left with permanent grass cover, with alternation every 3–5 years; (4) Permanent grass cover (PGC), where grass cover is maintained in the inter-rows.

Evaluation of the Soil Properties

Soil samplings were performed in each test-site, in correspondence of a soil pit of 1.0 m × 2.0 m excavated in the inter-row. Disturbed and undisturbed soil samples were collected in the identified horizons of the soil profile for laboratory analysis. The following soil attributes, considered the most affected by inter-rows management, were analyzed (l. 2017): soil texture (especially the amount of gravel, sand, silt, and clay in the soil), dry density and porosity, soil water content, water retention curve and soil saturated hydraulic conductivity. Moreover, Atterberg limits and oedometric properties of the soils were determined to complete the characterization and to evaluate potential differences

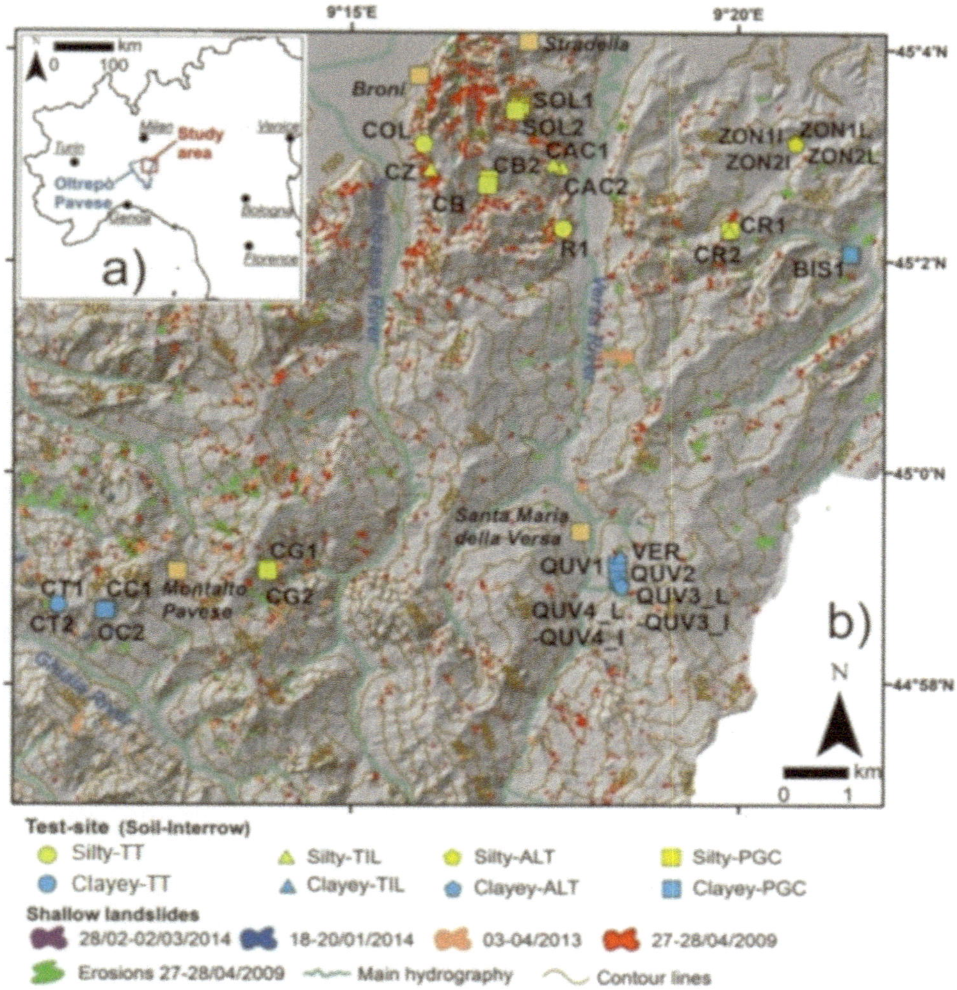

Fig. 1 Oltrepò Pavese hilly zone: **a** location of the study area; **b** location of the selected test-sites and of the shallow slope instabilities

according to different inter-row management techniques. All the laboratory tests were performed according to American Society for Testing and Materials (ASTM) procedures. Instead, saturated hydraulic conductivity (K_s) was measured in field through a constant head permeameter device. Ks was measured in at least 3 points of each test site in the first 0.2 m of soil profile and in the horizons below 0.2 m, generally between 0.5 and 1.0 m depth.

Evaluation of Root Density and Reinforcement

In most of the test-sites, root density was evaluated at three distances from the rootstock, to analyze potential variation on this parameter at different distances from the trunk: (i) one measure close to the plant, between 0.0 and 0.5 m from the stem; (ii) one measure at 0.5 and 1.0 m from the trunk; (iii) one measure between 1.0 and 1.5 m from the trunk (the middle between two adjacent rows). Root density was then quantified by counting the number of roots per root diameter class by means of the root-wall technique (Bischetti et al. 2009). The measured root amount and root density was estimated through the Root Area Ratio (RAR), which is the ratio between the cross sectional area of the roots and the soil area in the frame of known size (0.3 × 0.3 m). Data of root mechanical properties of grapevine plants of different test-sites were obtained from Bordoni et al. (2019). For calculating root reinforcement (c_r), Root Bundle Model—Weibull (RBMw) (Schwarz et al. 2013) was used, integrating the data related to root density and root mechanical properties.

Probabilistic Assessment of Failure Probability

Failure probability (P_f) was calculated for different inter-row managements by means of a probabilistic method, based on Lu and Godt's (2008) model, for the assessment of slope safety factor (FS) also under partially conditions (Eq. 1):

Fig. 2 Examples of rainfall-induced shallow landslides affecting cultivated vineyards of Oltrepò Pavese

$$FS = \left(\frac{\tan \phi'}{\tan \theta}\right) + \left(\frac{2(c' + c_r)}{\gamma z \sin 2\theta}\right) - \frac{\sigma^s}{\gamma z}[(\tan \theta + \text{cotan}\theta) \tan \phi'] \quad (1)$$

where: φ' is the soil friction angle; θ is the slope angle; c' is the soil effective cohesion; c_r is the root reinforcement, γ is the soil unit weight, z is the depth below ground level in which a potential sliding surface could develop; σ^s is the suction stress.

Because most of the involved geotechnical and root mechanical parameters are affected by natural variability, this model was applied within a probabilistic approach based on a Monte Carlo procedure (Chiaradia et al. 2016). Considering a defined range of values of the required soil and root parameters involved in Eq. 1 (φ', c', c_r, γ, σ^s), Monte Carlo was run for a total of 1000 repetitions, each considering randomly selected values of each variable parameters. The number of times in which FS was less than 1 (unstable conditions) was divided for the 1000 times that FS was calculated. In this way, the failure probability P_r was obtained as (Eq. 2):

$$P_r = P(FS < 1) \quad (2)$$

Results

Soil Properties

Dry density and porosity were similar in vineyards characterized by different inter-row management, both for sites with low plastic (CL) clayey or clayey sandy silts and high plastic (CH) clays with silts or silty clays. The average differences in dry density and porosity was, in fact, of less than 0.6 kN/m^3 and 0.05 between vineyards with different inter-row management, for both the soil. Saturation degree of dry period of the soil levels till 0.2 m from ground of test sites with different inter-row management ranged between 32 ± 3 and 37 ± 3%. In wet periods, saturation degree of the most superficial horizons were very similar concerning different inter-row management and very close to complete saturation (between 95 ± 4 and 99 ± 1%). Only K_s parameter showed an evident difference between different inter-rows management. K_s of topsoil horizons in PGC was about 4–5 times higher than TT and TIL and about 2 times higher than ALT. Instead, for the soil levels below 0.2 m from ground, K_s decreased passing from tilled vineyards to ALT and PGC ones of about 2–3 times (Fig. 3).

Root Density and Reinforcement

Grapevine root density did not show significant differences considering different distances from plant trunk, since the RAR values kept in a range of less than 0.05%. Besides the different management of the inter-row, in all the test site the highest amounts of roots were found between 0.3 and 0.6 m below ground level. Root density changed significantly according to different inter-row management (Fig. 4). ALT sites had the highest root density all along the soil profile. PGC had a root density on average 10–15% lower than ALT. While, for TT and TIL sites, the decrease of root density in respect to ALT was more evident, on average of about 51–66%.

c_r was, then, estimated at: (i) 0.3 m, in correspondence to the topsoil layer; (ii) 0.5 m, in correspondence to the highest root density; (iii) 1.0 m, where shallow landslide sliding surfaces prevalently occurred in the vineyards of the study area; (iv) 1.5 m, in correspondence to the measured highest rooting depth of all the test sites. c_r trends followed the trend of root density, with an increase at 0.5 m in respect to the topsoil layers and a consequent decrease with depth below 0.5 m (Fig. 5). Moreover, as for the root density, c_r was

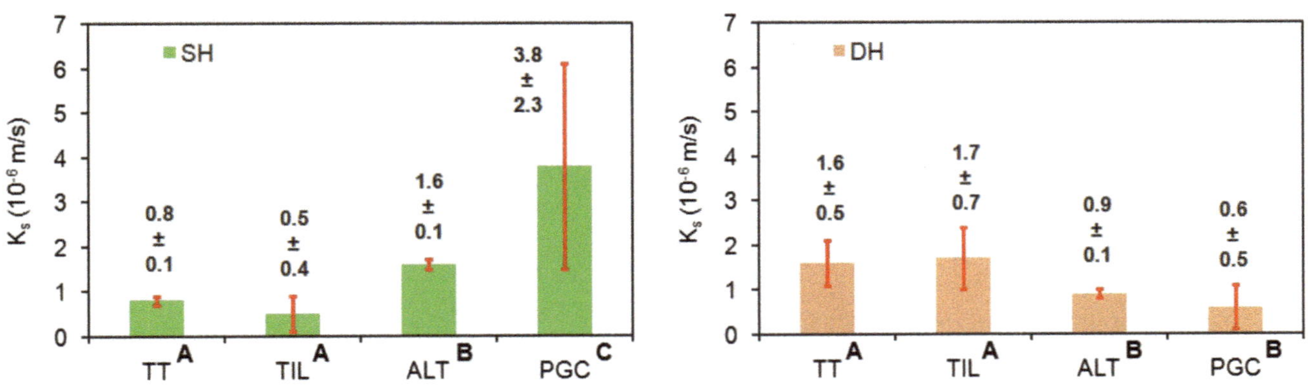

Fig. 3 Saturated hydraulic conductivity (K_s) values measured in vineyards with different management, in the first 0.2 m from the ground level (SH) and below this soil horizon (DH)

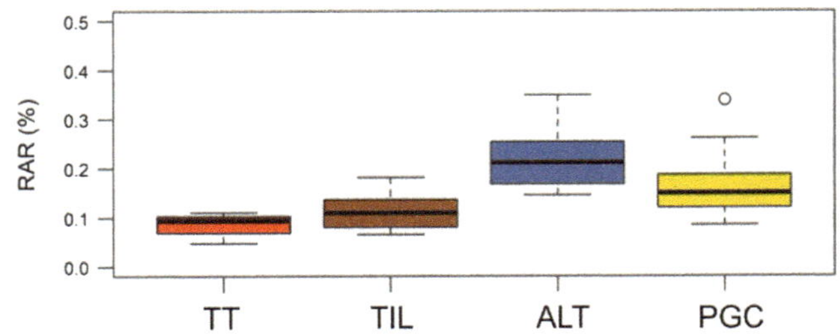

Fig. 4 RAR distribution in vineyards with different management

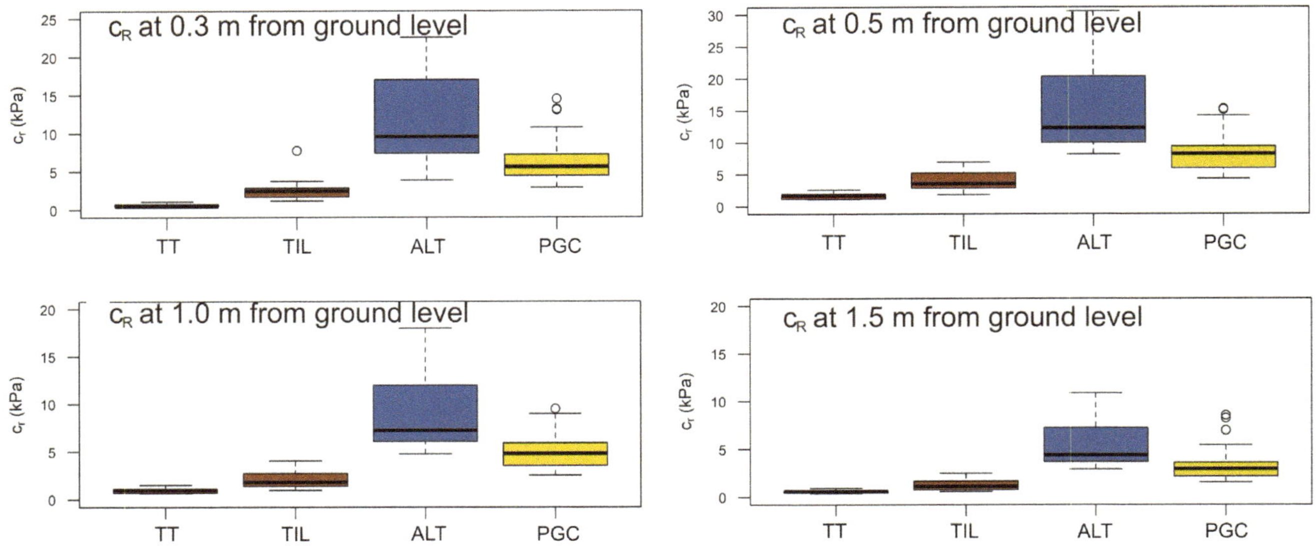

Fig. 5 Root reinforcement (c_r) distribution in vineyards with different management, at 0.3, 0.5, 1.0 and 1.5 m from the ground level

different in sites with different inter-row management (Fig. 5). ALT sites provided the highest root reinforcement, at all the considered depths (11.73 ± 2.34 kPa at 0.3 m from ground, 14.28 ± 2.59 kPa at 0.5 m from ground, 9.18 ± 1.99 kPa at 1.0 m from ground and 5.56 ± 1.59 kPa at 1.5 m from ground). Generally, PGC had c_r values 40–45% lower than ALT. While, for TT and TIL sites, a decrease of c_r in respect to ALT was on average more than 67–73%, with lowest values documented at TT sites.

Failure Probability

For the assessment of P_r through the probabilistic approach, c_r and the geotechnical parameters (φ', c', γ) were extracted from a normal distribution (Table 1). Pr was calculated for two soil types, which are the most widespread in Oltrepò Pavese area: (i) clayey silts or clayey-sandy silts; (ii) clays with silts or silty clays.

As regards σ^s, a uniform distribution between 0 and 10 kPa was considered, consistent with the values measured during past triggering events in the study area (Bordoni et al. 2015). P_r was evaluated considering slope angle ranging between 5 and 45°, which is the typical steepness of the hillslopes in the study area, and for soil depth of 1.0 m from ground, where most of the shallow landslides sliding surfaces developed (Bordoni et al. 2015). P_r obtained for different management in vineyards indicated that the probability of slopes failure for TT and TIL was the highest (Fig. 6). Regardless the type of soil, P_r decreased considering PGC and ALT (Fig. 6). Considering clayey silts or clayey-sandy silts, P_r exceeded 0.5 (50% of the simulations with FS lower than 1.0) for slope angle higher than: (i) 17° for TT; (ii) 18° for TIL; (iii) 25° for PGC; (iv) 33° for ALT. Regarding clays with silts or silty clays, P_r was higher than 0.5 at slope steepness lower for each type of land use, due to the poorer geotechnical properties than the ones of clayey silts or clayey-sandy silts. Thus, the probability of rupture exceeded 0.5 for slope angle higher than: (i) 10° for TT and TIL; (ii) 21° for PGC; (iii) 28° for ALT.

Table 1 Mean and standard deviation of the geotechnical parameters used for the assessment of the failure probability

Soil type	γ (kN/m³)	φ' (°)	c' (kPa)
Clayey silts or clayey sandy silts	18.0 ± 0.8	26 ± 4	1.8 ± 1.6
Clays with silts or silty clays	19.4 ± 0.5	12 ± 4	0.0 ± 0.0

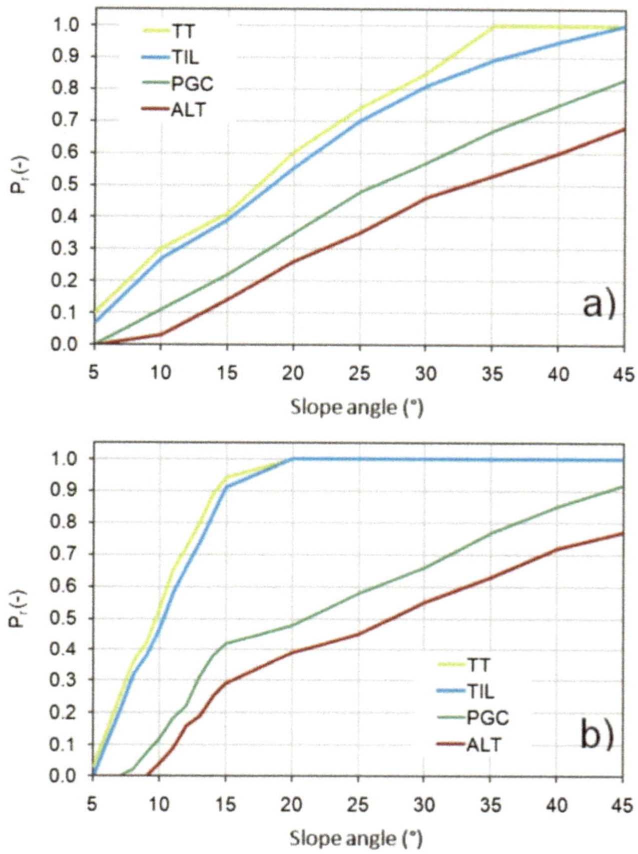

Fig. 6 Failure probability (P_f) as function of steepness for different land uses, considering **a** clayey silts/clayey-sandy silts soils or **b** clays with silts/silty clays

Discussions and Conclusions

The agronomical practices in vineyards, in particular the management of the inter-rows, influence soil properties and grapevine root development. In the context of the Oltrepò Pavese, representing viticulture in sloping landscapes of the northern Italian Apennines, the soil hydraulic conductivity is the most influencing parameter by the inter-row management. The macroporosity allows to increase the superficial (first 0.2 m of soil) hydraulic conductivity of inter-rows without tillage. In the remaining part of the soil, a high density of roots may represent an obstacle for hydraulic conductivity and the related water fluxes, as shown for vineyards with alternating tillage and permanent grass cover. Anthropic factors related to grapevine cultivation and harvesting (e.g. deep ploughing, spread of fertilizers, harvesting, plant pruning activities) during the year tend to standardize the other soil physical, geotechnical and hydrological properties.

Vineyards with alternation management of the inter-rows or with permanent grass cover promote a significant increase in root density and root reinforcement than other types of management. Thus, the failure probability decreases passing from vineyards with tilled inter-rows to the ones with grass covering or alternating inter-rows.

In the typical conditions of shallow landslides triggering, slopes with medium steepness (10° for clayey soils, 17–18° for silty soils) are unstable if inter-rows of vineyards are tilled. In the same conditions, permanent grass cover or alternation in the inter-rows promote the stability of slopes in a wider range of steepness (>21–25° for vineyards with permanent grass cover in the inter rows, 28–33° for vineyards with alternation in the inter rows). The slope stability analyses leading to these results were conducted with a 1D approach, which could be improved through a 2D approach in order to model better the soil fluxes and the variations in root reinforcement across a hillslope.

These results provide important indications for land use planning at catchment and regional scales able to reduce the proneness towards shallow landsliding, maintaining or increasing soil conservation.

Acknowledgements The field surveys and the analyses carried on the tested vineyards of this study were supported by the project "Oltrepò BioDiverso", funded by Fondazione Cariplo in the frame of AttivAree Program. We thank Salvatore Bianco, Daniela Gallo and Luca Lucchelli for help in the field surveys. We are also very grateful to Marco Tumiati for the help in the realization of laboratory analyses.

References

Bischetti GB, Chiaradia EA, Epis T, Morlotti E (2009) Root cohesion of forest species in the Italian Alps. Plant Soil 324:71–89

Blahut J, Glade T, Sterlacchini S (2014) Debris flows risk analysis and direct loss estimation: the case study of Valtellina di Tirano Italy. J Mountain Sci 11:288–307

Bordoni M, Meisina C, Valentino R, Bittelli M, Chersich S (2015) Site-specific to local-scale shallow landslides triggering zones assessment using TRIGRS. Nat Hazards Earth Syst Sci 15:1025–1050

Bordoni M, Vercesi A, Maerker M, Ganimede C, Reguzzi MC, Capelli E, Wei X, Mazzoni E, Simoni S, Gagnarli E, Meisina C (2019) Effects of vineyard soil management on the characteristics of soils and roots in the lower Oltrepò Apennines (Lombardy, Italy). Sci Total Environ 693:133390

Chiaradia EA, Vergani C, Bischetti GB (2016) Evaluation of the effects of three European forest types on slope stability by field and probabilistic analyses and their implications for forest management. For Ecol Manage 370:114–129

Cohen D, Schwarz M (2017) Tree-root control of shallow landslides. Earth Surface Dyn 5:451–477

Fonte N, Masciocco L (2009) A simplified physically-based approach for the assessment of hazard related to shallow landslides in Southern Piedmont (Italy). Geografia Fisica E Dinamica Quaternaria 32:193–202

Lu N, Godt JW (2008) Infinite slope stability under steady unsaturated seepage conditions. Water Resour Res 44:W11404

Organisation Internationale de la Vigne et du Vin (OIV) (2017) State of the Vitiviniculture World Market—April 2017. https://www.oiv.int/en/technical-standards-anddocuments/statistical-analysis/state-of-vitiviniculture.

Prosdocimi M, Cerdà A, Tarolli P (2016) Soil water erosion on Mediterranean vineyards: a review. CATENA 141:1–21

Rodrigo-Comino J (2018) Five decades of soil erosion research in "terroir". The State-of-the-Art. Earth Sci Rev 179:436–447

Smart DR, Schwass E, Lakso A, Morano L (2006) Grapevine rooting patterns: a comprehensive analysis and a review. Am J Enol Vitic 57:89–104

Wu TH (2012) Root reinforcement of soil: review of analytical models, test results and application to design. Can Geotech J 50:259–274

Zhang S, Zhao L, Delgado-Tellez R, Bao H (2018) A physics-based probabilistic forecasting model for rainfall-induced shallow landslides at regional scale. Nat Hazards Earth Syst Sci 18:969–982

2020 Kyoto Japan

Landslide Susceptibility in Two Secondary Rivers of La Ciénega Watershed, Nevado de Toluca Volcano, Mexico.

Sandra García Reyes, Gabriel Legorreta Paulin, Rutilio Castro Miguel, and Fernando Aceves Quesada

Abstract

Unstable areas along first-order tributary rivers and meander bends developed in volcanic poorly consolidated materials such as lahars, pyroclastic flows, and pumice fall deposits are common in Mexico. The present research is based on studies of the stream system of La Ciénega watershed on the eastern flank of Nevado de Toluca volcano, Mexico. The watershed is prone to landslides due to its climatic, topographic, geomorphologic, and geologic conditions that predispose the study area to episodic landslides and debris flows. Landslide volcaniclastic sediments are dragged by the streams and torrents during the rainy season and create a hazardous situation for people living along the stream system. Our work is focused on two secondary rivers located ian the southern portion of La Ciénega watershed. In both tributaries, a detailed landslide inventory and a geomorphological map were carried out to determine the landslide susceptibility by landforms. The results show that debris slides are the most frequent processes along the two secondary rivers, and three landforms out of fourteen have the highest landslide susceptibility. In these landforms, factors such as steep slopes, geological faults, and hillslope morphology influence the abundance and distribution of landslides.

Keywords

GIS • Landslide inventory map • Landslide susceptibility • Landforms • Nevado de toluca volcano

Introduction

In Mexico, landslides are very common geomorphological processes that occur in volcanic regions with stratovolcanoes and monogenetic fields. In spite of the importance of assessing such processes, there are few landslide inventory maps, landslide geo-data sets, and there is no practical and standardized landslide mapping method using Geographic Information Systems (GIS). This is the case of Nevado de Toluca volcano, the fourth highest mountain in Mexico (4680 m a.s.l). The stream system of the La Ciénega watershed on the eastern flank of Nevado de Toluca volcano has been selected as a case study area. The study area extends from the rim of the volcano to the foothills at an altitude of 2850 m a.s.l. The watershed has a great potential to produce landslides and debris flows because of its large area of weakened rocks at high altitudes and high seasonal rainfall. These conditions create a hazardous situation for people living along the stream system. For example, in 1940 rainfalls triggered landslides on the up-stream of La Ciénega watershed. The landslide sediments increased the destructive power of a debris flow that caused loss of life and property at the town of Santa Cruz (Peña 2006; Toscana and Valdez 2013; Aceves et al. 2014). The town belongs to the municipality of Tenango del Valle, State of Mexico. As a result of this event the inhabitants rebuilt the village that is currently called Santa

S. G. Reyes · F. A. Quesada
Facultad de Filosofía Y Letras, Universidad Nacional Autónoma de México, Ciudad Universitaria, 04510 Ciudad de México, México
e-mail: garcigeograf@gmail.com

F. A. Quesada
e-mail: acevesquesada.fernando@gmail.com

G. L. Paulin (✉)
Instituto de Geografía, Universidad Nacional Autónoma de México, Circuito Exterior, Ciudad Universitaria, 04510 Coyoacán, Ciudad de México, México
e-mail: legorretag@hotmail.com

R. C. Miguel
Posgrado en Geografía, Universidad Nacional Autónoma de México, Ciudad Universitaria, 04510 Ciudad de México, México
e-mail: rutilio.cm@gmail.com

F. Guzzetti et al. (eds.), *Understanding and Reducing Landslide Disaster Risk*,
ICL Contribution to Landslide Disaster Risk Reduction,
https://doi.org/10.1007/978-3-030-60227-7_17

Cruz Pueblo Nuevo. However, this relocation was made without adequate planning. The town was rebuilt in the lower part of the watershed on an alluvial fan, which is prone to new and future landslides and debris flows (Aceves et al. 2013). Despite its importance, there is a lack of landslide inventories that aid the assessment of landslide susceptibility in the watershed. To address this deficiency, this paper provides standardized methods in a GIS system for evaluating landslide susceptibility per geomorphological landform units. The present research is focused on two secondary rivers located in the southern portion of the watershed. These secondary rivers are referred as La Cieneguilla 1 (ASC1) and La Cieneguilla 2 (ASC2). In both secondary rivers, a detailed landslide inventory and geomorphologic cartography are carried out to determine the landslide susceptibility per landform. Each landform unit is derived by aerial photo interpretation, geological fieldwork, and morphometric parameters. For each landform unit, the landslide area rate (LAR) and the landslide frequency rate (LFR) are used to determine the landslide susceptibility. This method follows and adapts the Washington State, Department of Natural Resources, Forest Practices Division (WSDNRFPD), Landslide Hazard Protocol (LHZP) (WSDNRFPD 2006). The analysis divides the watershed into 23 geomorphologic landform units that are assigned slope susceptibility ratings from low to high. The ASC1 and ASC2 secondary rivers erode and dissect 14 out of the 23 landforms. Among the fourteen landform units, three of them have the highest landslide susceptibility.

Background

Worldwide, determining the landslide distribution, and landslide susceptibility has been addressed for multiple scopes and scales (WSDNRFPD 2006; Hervás and Bobrowsky 2009; Blahut et al. 2010).

In Mexico, numerous GIS-based applications have been used to assess landslide susceptibility (Pérez 2007; Oliva et al. 2012). In Nevado de Toluca volcano, investigations of different physical aspects have been made. For example, geomorphological studies (Aceves 2006, 2014; Espinosa et al. 2014), volcanic hazards (Capra et al. 2008; Aceves et al. 2006), river flooding and social aspects (Peña 2006; Tosacana and Valdez 2013) were conducted. In La Ciénega watershed, assessment of the relationship between land use change and landslides has been established (Álvarez 2015).

In spite of this, there is no practical and standardized landslide susceptibility mapping method for landslides that occur continually along the stream systems in the watershed.

Study Area

The study area is a small endorheic watershed of 28 km^2. La Ciénaga watershed is located on the eastern flank of Nevado de Toluca volcano (NTV). NTV in central Mexico is the fourth highest elevation in Mexico with an approximate altitude of 4680 m a.s.l. (Macías 2005). Eruptions produced a complex sequence of pyroclastic deposits that have affected the area at least 18 times during the last 100,000 years. Thirteen eruptions and the destruction of at least three domes have all occurred in the last 42,000 years as well as three sector collapses in the last 100,000 years (Aceves et al. 2006; Aceves 2007; Caballero and Capra 2011). NTV rises 2100 m above the upper basin of the Lerma River, and 3100 m above Ixtapan de la Sal and Tonatico. To the east, the volcanic piedmont continues with active fans composed of pyroclastic materials and alluvial sediment. The ravines are 100–300 m deep. They start in an old glacial cirque and continue through a long dendritic drainage. The slope is 20–35° in the elevated portions while in the piedmont it is 6–12° and in the plain it is 2–6°. Pyroclastic flow deposits are widely spread around the volcano, filling the stream valleys. In La Ciénaga watershed volcanic deposits have an average thickness of 5 m; the maximum distance reached by these deposits is 12 km from the crater towards the east. The block and ash flows form massive units interstratified with surge horizons covered by lahars (Aceves et al. 2013). These deposits are related to dome collapses. The secondary rivers La Cieneguilla 1 (ASC1) and La Cieneguilla 2 (ASC2) that are studied in this work are located in the central and southern part of La Ciénega watershed (Fig. 1).

The mean annual precipitation is high (1200–1100 mm/yr at > 4000 m a.s.l, with temperature between −2 and 12 °C, and 1100–800 mm/yr at elevations between 2650 and 4000 m a.s.l. (García-Enriqueta 2004), the bulk of which falls as rain during seasonal storms in summer.

In response to the volcanic lithology, the secondary rivers ASC1 and ASC2 have a dendritic pattern with subparallel features. The La Ciénega main river is a Strahler's fourth order stream, while ASC1 and ASC2 are of third order.

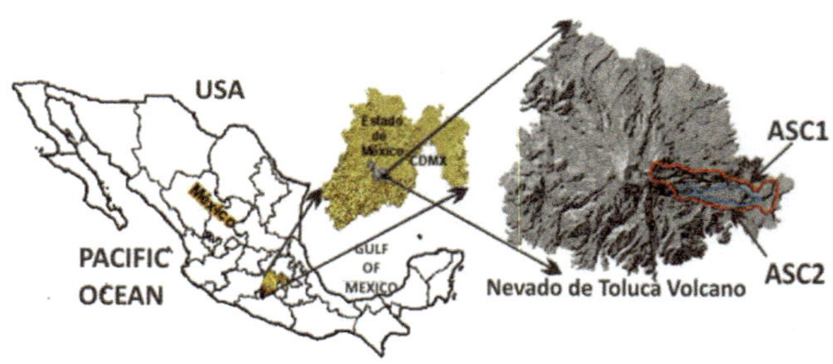

Fig. 1 Study area

ASC1 and ASC2 rivers erode hillslopes of highly weathered pyroclastic flows, pumice fall deposits, blocks and ash flow deposits, and lahars deposits (Capra et al. 2008; Aceves et al. 2014). The unstable areas frequently create debris flows and debris slicing affecting the human settlements. On June 24, 1940, a large debris flow partially destroyed the town of Santa Cruz Pueblo Nuevo. Today, the town is settled in the alluvial fan of the old debris flows deposit. It is a highly probable that the town will be affected again by landslides and debris flows.

Method

The methodology encompasses four main steps of analysis:

Step 1: Gathering the basic information and create new thematic GIS information layers. Bibliographic information and previous studies in geology, geomorphology, and geography were collected. Also, topographic maps at scale 1: 50,000 (INEGI 2013), a geological map at scale 1: 250,000 (SGM 2002), SPOT-6 satellite images with 1.5 m resolution, GoogleEarth images 2015, and digital elevation model (resolution 10 m) were obtained.

Step 2: Building a historic landslide inventory using GIS, photo interpretation, and fieldwork. This study develops a detailed landslide inventory based on fieldwork and interpretation of digital aerial photographs and satellite images (Chuvieco 1996; Verstapen 1977). Some landslides were compiled from the database of Álvarez 2015. For the landslide mapping and classification, we follow the landslide hazard zonation protocol (2006) of Washington State, Department of Natural Resources, Forest Practices Division, Cruden and Varnes (1996), and Wieczorek (1984). Landslides were mapped as the following types: shallow landslides, debris flow, debris slides, deep-seated landslides, meanders' incision, and rockfalls.

Step 3: Building the geomorphologic map using GIS, maps compilation (Capra et al. 2008; Torres 2011; Aceves et al. 2014), photo interpretation, and fieldwork geology. The geomorphologic map is derived from the interpretation of the altimetric map, the slope map, the dissection density map, the vertical erosion map, and the morphology map. The values obtained from these maps facilitated the delimitation of landforms. The classification of landform units is defined by following the LHZP of the WSDNRFPD (2006), Lugo (1988), based on Chemekov (1972) and Bashenina et al. (1975).

Step 4: Elaboration of the landslide susceptibility map applying the LHZP of the WSDNR (WSDNRFPD 2006). For the assignment of the landslide susceptibility to each geomorphologic landform, we determined the LFR and LAR. These quantitative rates are derived from values that correspond to the total number and area (in ha) of landslides within each landform. They are normalized for the total period of time spanned by aerial photographs and/or fieldwork (in our case 34 years) used to elaborate the landslide inventory. After the quantitative rates were determined, the LFR and LAR values are entered in a matrix to determine the landslide susceptibility for each landform unit. As a result, each landform unit is assigned a susceptibility rating of Low, Moderate, and High. Figure 2 shows the procedure for rating each landform unit.

Results

During the assessment of La Ciénega watershed, a representative sample of 179 mass-wasting features was inventoried (Fig. 3). They affect 0.04% of the watershed. The landslide inventory along the secondary rivers (ASC1 and ASC2) identified 6 types of landslide: 155 were debris slides that represent 86.6% of the total gravitational processes in the study area. This type of landslide is the predominant mass-wasting features in both streams. They were followed by 8 deep-seated landslides (4.5%), 5 shallow landslides (2.8%), 5 debris flows (2.8%), 4 meanders' incision (2.2%), and 2 rockfalls (1.1%).

During the assessment of the watershed, 23 homogeneous landform units were identified (Fig. 4) as well as the landslide susceptibility associated with each one (Fig. 5). Among the 23 landforms, only 14 of them are eroded by the ASC1 and ASC2 rivers. 59 and 84% of the ASC1 and ASC2 streams respectively, eroded landforms with moderate and high landslide susceptibility. Among the fourteen landform units, three of them have the highest landslide susceptibility. The lower hillslopes, debris avalanche hillslopes covered with pyroclasts, and hillslopes covered with pumice and fall deposits landform units were the most susceptible to landslides for both streams.

The results show that factors such as steep slopes, geological faults, and hillslope morphology that characterized each landform unit influence the abundance and distribution of landslides.

The risk associated with debris flow is increased by the coalescence of up-stream landslides in the ASC1 and ASC2 streams. This creates a hazardous situation for 1634 people living in the town of Santa Cruz Pueblo Nuevo.

Conclusion

Landslide susceptibility is difficult to model due to the continuous changes in the topography by landslides, local land use, and environmental conditions. The main aim of this study was to model and assess landslides susceptibility in terms of landform units.

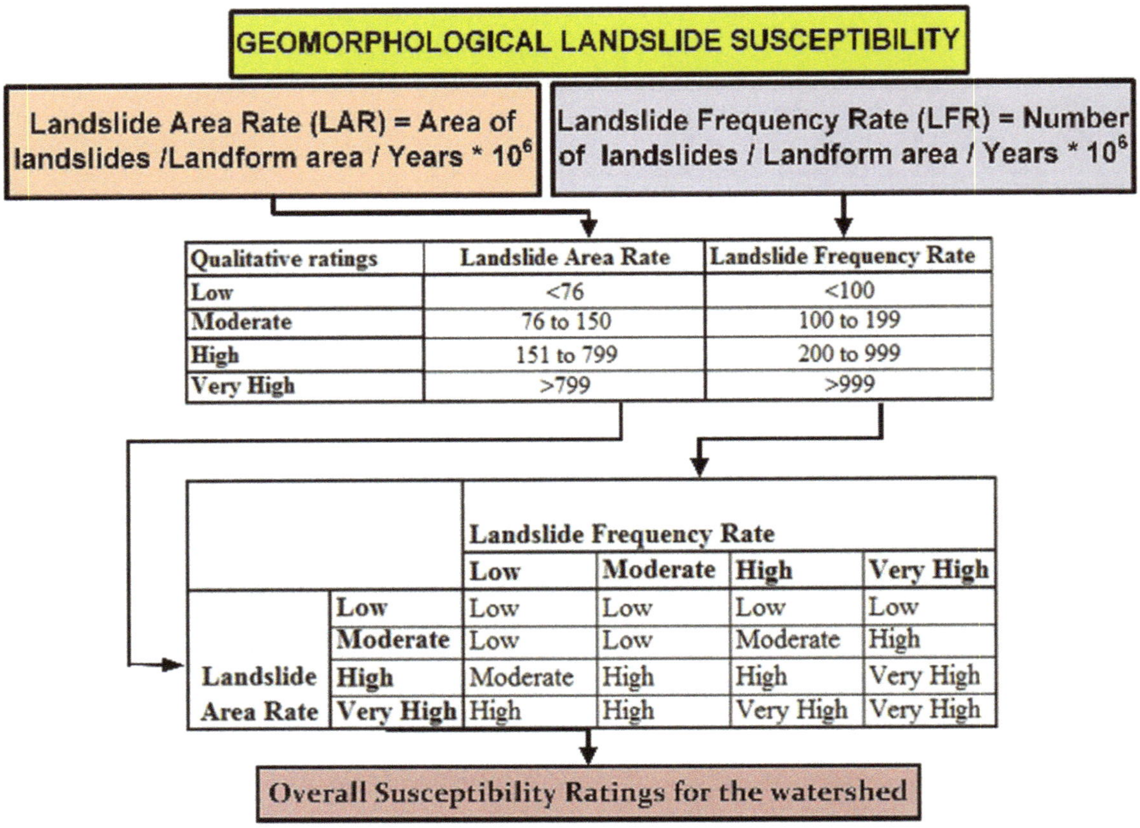

Fig. 2 General procedure to assign landslide susceptibility

Fig. 3 Landslide inventory

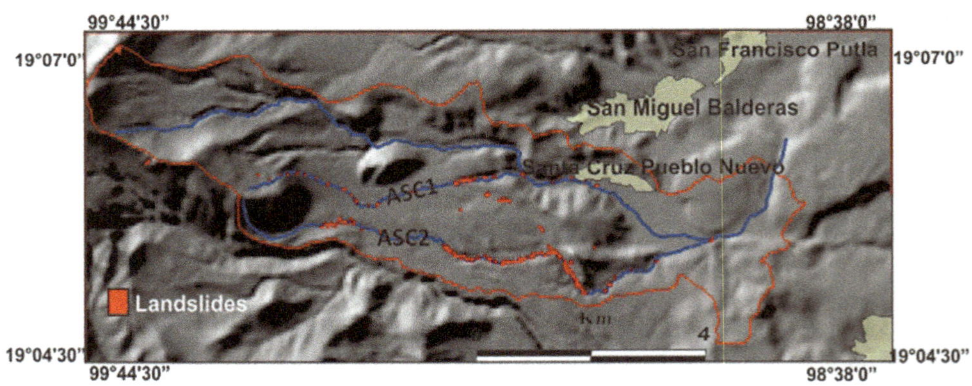

The approach uses aerial photographs, fieldwork surveys, and geomorphological evaluation to map landslides and establish landforms units. Then, quantitative rates are calculated and entered into a semi-quantitative matrix to determine the landslide susceptibility for each landform.

The results indicate that the landform units and their landslide susceptibility may be useful in the quantification, assessment, and modeling of landslides and debris flows in the future. The assessment of landslide susceptibility in homogeneous units will potentially allow to compare landslide intensity and sediment production delivered to lower lands among units.

This study is a first step towards a more comprehensive research about landslides susceptibility assessment in one of the highest volcanoes in Mexico.

The methodology applied here is an alternative procedure for the construction of landslide susceptibility maps in areas with a scarce and of low spatial resolution information.

By directly addressing the landslide characterization per landform units, local authorities in Mexico such as the civil protection agencies of the state of Mexico and other governmental organizations as well as the inhabitants will benefit to mitigate landslide hazards.

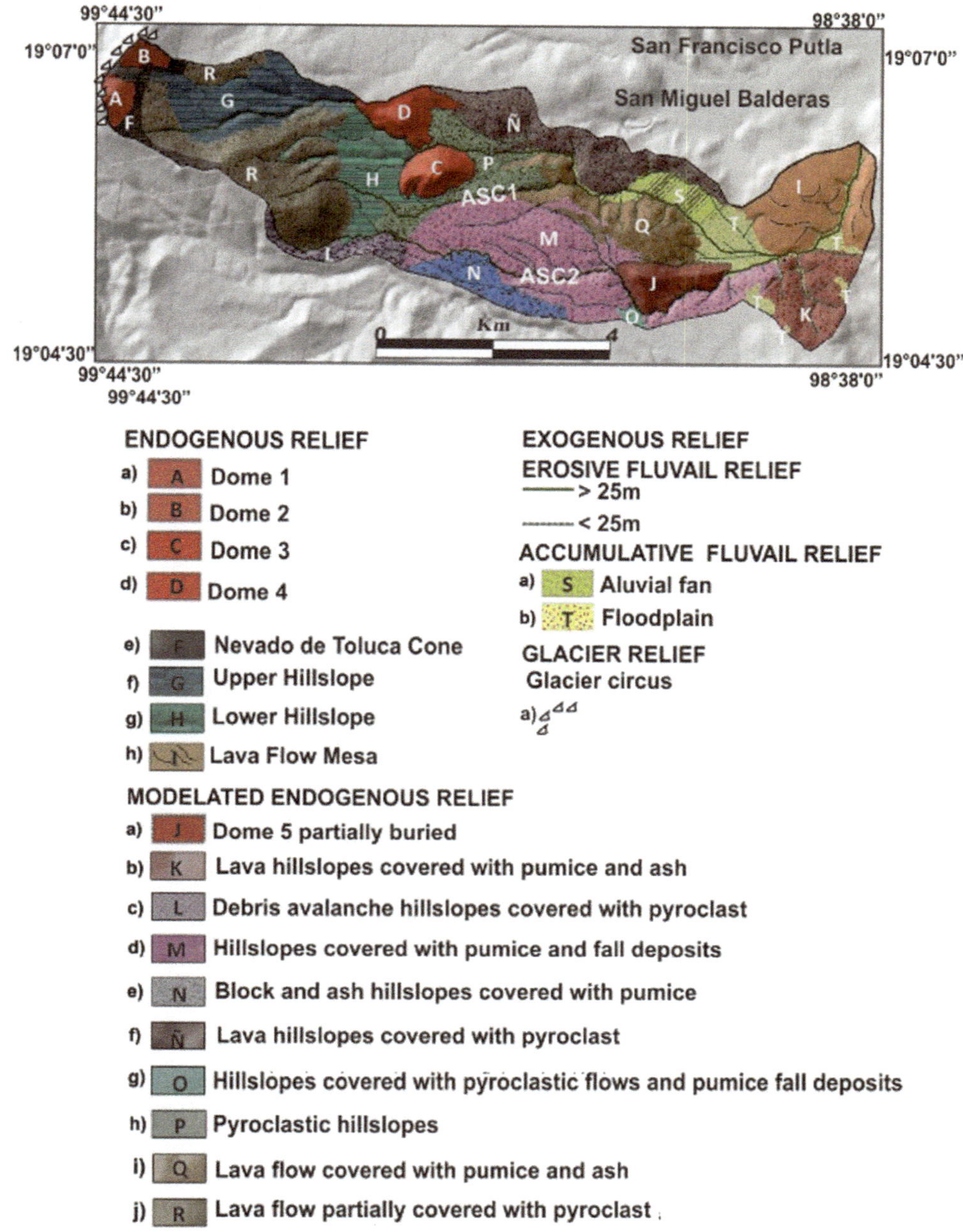

Fig. 4 Geomorphologic landforms

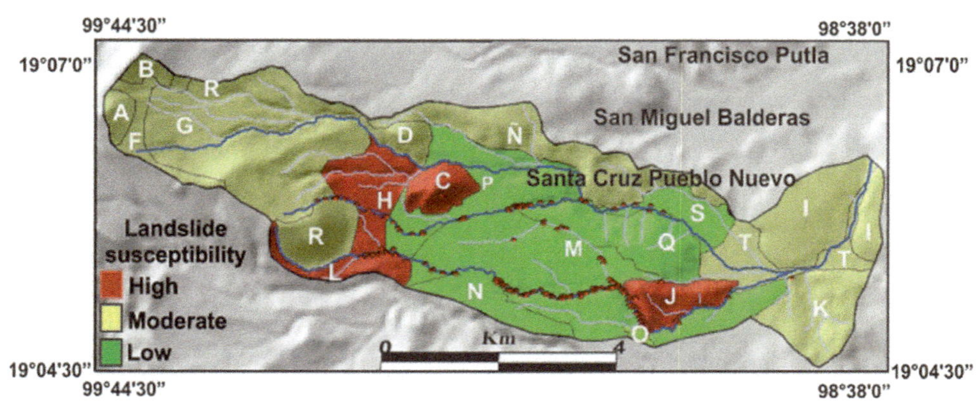

Fig. 5 Landslide susceptibility in the Río la Ciénega watershed

Acknowledgements The authors thank authorities from the Department of Physical Geography from the Institute of Geography, UNAM for their approval and help. This research was supported by the Programa de Apoyos para la Superación del Personal Académico de la UNAM (PASPA) de la Dirección General de Asuntos del Personal Académico (DGAPA), UNAM.

References

Aceves QJF (2007) Mapas de riesgo volcánico basados en Sistemas de Información Geográfica: volcán Nevado de Toluca. PhD thesis. UNAM. Ciudad de México, México

Aceves QJF, López BJ, Martin PA (2006) Determinación de peligros volcánicos aplicando técnicas de evaluación multicriterio y SIG en el área del Nevado de Toluca, centro de México. Revista Mexicana De Ciencias Geológicas 23(2):113–124

Aceves QJF, Luna VMM, Legorreta PG (2013) Gravitational processes in the eastern flank of the Nevado de Toluca México. In Landslide Science and Practice. Springer, Berlin, Heidelberg, pp. 211–219

Aceves QJF, Legorreta PG, Álvarez RY (2014) Cartografía geomorfológica para el inventario de procesos gravitacionales en la cuenca endorreica del arroyo La Ciénega, flanco oriental del volcán Nevado de Toluca. Boletín De La Sociedad Geológica Mexicana 66(2):329–342

Álvarez YR (2015) Relación entre el cambio de cobertura—uso de suelo, y los deslizamientos, año 1983 y 2014, en la Cuenca la Ciénega, volcán Nevado de Toluca. BS thesis. Facultad de Ciencias. UNAM. Ciudad de México, México

Bashenina NV, Gellert JE, Joly F, Klimaszewski E (1975) Leyenda unificada para cartas geomorfológicas de detalle, en La cartografía geomorfológica en escalas grandes, Ed. MGU, Moscú, pp 18–68

Blahut J, Horton P, Sterlacchini S, Jaboyedoff M (2010) Debris flow hazard modelling on medium scale: Valtellina di Tirano, Italy. Nat Hazards Earth Syst Sci 10(11):2379–2390

Caballero L, Capra L (2011) Textural analysis of particles from El Zaguán debris avalanche deposit, Nevado de Toluca volcano, Mexico: Evidence of flow behavior during emplacement. J Volcanol Geoth Res 200(1):75–82

Capra L, Norini G, Groppelli G, Macías JL, Arce JL (2008) Volcanic hazard zonation of the Nevado de Toluca volcano, México. J Volcanol Geoth Res 176(4):469–484

Chemekov YF (1972) Manual de investigaciones geomorfológicas, Ed. Niedra, Leningrado (en ruso)

Chuvieco SE (1996) Fundamentos de teledetección espacial. Rialp, Madrid, p 224p

Cruden DM, Varnes DJ (1996) Landslide types and processes. In: Turner AK y Schuster RL (eds) Landslides; Investigation and mitigation, transportation research board; special report 247. National Academy Press; Washington D.C., pp. 36–75

Espinosa RL, Balderas MÁ, Cabadas HV (2014) Caracterización geomorfológica del área natural protegida Nevado de Toluca: complejo de volcanes Nevado de Toluca y San Antonio. Ciencia UAT 9(1):6–14. ISSN 2007-7521

García E (2004) Modificaciones al sistema de clasificación climática de Koppen. Instituto de Geografía. Serie de Libros núm. 6. Universidad Nacional Autónoma de México, P 246

Hervás J, Bobrowsky P (2009) Mapping: inventories, susceptibility, hazard and risk. In: Sassa K, Canuti P (eds) Landslides—disaster risk reduction. Springer, Berlin, pp 321–349. ISBN 978-3-540-69966-8

Lugo HJI (1988) Elementos de geomorfología aplicada: métodos cartográficos. Instituto de geografia, UNAM. (ISBN 9683605605), p 128

Macías JL (2005) Geología e historia eruptiva de algunos de los grandes volcanes activos de México. Boletín De La Sociedad Geológica Mexicana Volumen Conmemorativo Del Centenario Temas Selectos De La Geología Mexicana 57(3):379–424

Instituto Nacional de Estadística y Geografía (INEGI) (2013) Continuo de Elevación Mexicano 3.0 (CEM 3.0). https://www.inegi.org.mx. Last accessed: Feb 1 2019

Oliva AOG, Navarro AR, Salgado RM, Nicieza CG, Fernández MIÁ (2012) Urban development and human activity as factors in terrain instability in Tijuana. Eng Fail Anal 19(1):51–62

Peña VE (2006) Análisis de la Vulnerabilidad Social e Inundaciones en la Cuenca de La Ciénega, parte alta de la cuenca del Río Lerma. MS thesis Universidad Nacional Autónoma de México. Ciudad de México, México

Pérez GR (2007) Análisis de la vulnerabilidad por los deslizamientos en masa, caso: Tlacuitlapa, Guerrero. Boletín De La Sociedad Geológica Mexicana 59(2):171–181

Servicio Geológico Mexicano (SGM) (2002) Carta Geológica Minera E14–2 Ciudad de México. https://www.sgm.gob.mx. Last accessed: March 9 2020

Torres R O, (2011) Volcanismo efusivo en el área del Nevado de Toluca: Distribución y génesis de magmas. MS thesis. UNAM. Ciudad de México, México

Toscana AA, Valdez VP (2013) Representaciones Sociales del desastre de 1940 en Santa Cruz Pueblo Nuevo, Estado de México. Investigaciones Geográficas. Boletín del Instituto de Geografía. 2014(83):88–101

Verstappen H (1977) The use of aerial photographs in geomorphological mapping. In ITC Text Book VII-5, Enschede, The Netherlands

Washington State Department of Natural Resources (DNR), Forest Practices Division (2006) Landslide Hazard Zonation Protocol (LHZP), Mapping Protocol, version 2.1. https://www.dnr.wa.gov/programs-and-services/forest-practices. Last accessed: Feb 14 2015

Wieczorek GF (1984) Preparing a detailed landslide inventory map for hazard evaluation and reduction. Bull Eng Geol 21(3):337–342

An Ordinal Scale Weighting Approach for Susceptibility Mapping Around Tehri Dam, Uttarakhand, India

Naorem Sarju Singh, Sharad Kumar Gupta, Chandra Shekhar Dubey, and Dericks P. Shukla

Abstract

The landslides are natural hazards, which cause damage to both property and life every year, especially in the Himalayas. Detailed studies of landslide susceptible areas are instrumental in getting fast and safe mitigation actions and doing future planning for any construction work. The study area lies under the Lesser Himalaya in the Tehri-Garhwal region of Uttarakhand, India. This area consists of weak and unstable lithology, highly fragile rocks due to the complicated tectonic settings; consequently, landslide movement is a common phenomenon in the area. We have used several landslide controlling factors such as slope, lithology, thrust buffers, relative relief, land use land cover, lineaments, and stream buffer in order to generate a susceptibility map. We have prepared these parameters from geological (structural and lithological) maps, Landsat TM, and ASTER GDEM data and field investigation data. We have integrated the data based on the ordinal scale weightage rating technique to generate the landslide susceptibility index (LSI) values. The LSI frequency distribution is divided into five zones (i.e., very low, low, moderate, high, and very high susceptibility) based on the geometric interval as well as the standard deviation to enhance the classes with minimal frequency. These zones account for 3.30%, 20.88%, 47.99%, 41.13%, and 1.83% of total area respectively. Furthermore, the final susceptibility map is validated using the field data of landslide occurrences, which depicts that more than 50% of landslides occur in very high and high zones. These zones lie in the north-eastern side of the Tehri reservoir, which is traversed by North Almora Thrust (NAT), while just 16% of landslides have fallen in low and very low susceptible zones.

Keywords

Landslide susceptibility zonation (LSZ) • Ordinal weighting scheme • Garhwal himalaya • Landslides

Introduction

A landslide is the movement of rock, debris, or earth along a slope with a wide range of movements such as falling, sliding, and flowing under the influence of gravity. They are one of the widespread and destructive hazards in the Himalayan region causing immense damage to life and property. Landslide study has global consideration owing to the ever-increasing demand for urbanization in the hilly region along with increasing awareness of its socioeconomic impacts (Aleotti and Chowdhury 1999). Among the natural disasters that have been occurred globally from 1990 to 2005, landslides constituted 4.89% (Gupta and Shukla 2018). This landslide movement may persist in the future also due to an increase in regional precipitation as an effect of changing climatic conditions in landslide-prone areas and increasing unplanned urbanization and development with continuous deforestation (Schuster 1996). Many devastating landslides have occurred in the past, especially in the Uttarakhand Himalayas. Some of the devastating landslides include Ukhimath, Uttarakhand landslide of 1998 killing 109 people (Kanungo et al. 2009); Malpa landslide of 1998 killing 209 people and wiping the whole Malpa village (Juyal 2002); Phata landslide of 2001 (Naithani et al. 2002);

N. S. Singh · C. S. Dubey
Department of Geology, Delhi, 110007, India
e-mail: sarjunaoria@gmail.com

C. S. Dubey
e-mail: csdubey@gmail.com

S. K. Gupta (✉) · D. P. Shukla
School of Engineering, Indian Institute of Technology Mandi, Mandi, 175005, India
e-mail: sharadgupta27@gmail.com

D. P. Shukla
e-mail: dericks@iitmandi.ac.in

F. Guzzetti et al. (eds.), *Understanding and Reducing Landslide Disaster Risk*,
ICL Contribution to Landslide Disaster Risk Reduction,
https://doi.org/10.1007/978-3-030-60227-7_18

Budha Kedar landslide of 2002 (Rautela and Pande 2005), and Uttarkashi landslide of 2003 (Vinod Kumar et al. 2008). Several significant past landslides activities are given in Table 1 which caused large-scale human tragedies, resource damage, and associated environmental-social hazards.

The prediction of landslide incident is very challenging in space and time. Hence, an area could be categorized into near-homogeneous zones concerning degrees of possible hazard due to mass movements (Varnes 1984). This paper describes the development of a methodology designed for the preparation of a landslide susceptibility zonation (LSZ) map. The preparation of LSZ maps has been carried out for the past four decades with the help of several qualitative and quantitative studies. With the advent of remote sensing and GIS technologies, a variety of spatial and non-spatial data can be collected, manipulated, and analyzed (Saha et al. 2005; Shukla et al. 2016; Gupta et al. 2018). In an area, mitigation of landslides can be successful only when comprehensive knowledge is acquired from probable frequency, properties, and degree of the landslides. The LSZ map may be the basis for any landslide disaster mitigation and can be helpful to planners and policymakers. The study aims to identify the landslide affected area distribution pattern using landslide influencing spatial features (such as slope, land cover, geology, and landform). Moreover, the LSZ maps are prepared using ordinal scale weightage rating technique.

Study Area

The study area as sh'own in Fig. 1, is bounded by 78° 3′ 20.62″–78° 57′ 1.73″ E longitude and 30° 2′ 47.29″–30° 47′ 9.33″ N latitude and covering an area of about 7087 km^2. The study area falls under the Survey of India (1950) topo sheets number 53 J/6, J/7, J/10, and J/11 at 1:50,000 scale. In the Himalayan terrain, susceptibility to landslide is high due to variable slope and relative relief, existing crustal movements, complex geological setting, heavy precipitation as well as increasing human interference in the ecosystem.

The area under investigation has a rugged topography and is tectonically very active. The study area falls under the seismic zone V as per the Bureau of Indian Standards (BIS 1893, Part 1:2002) and has witnessed two significant earthquakes in the past [Uttarkashi (M = 6.1, 1991) and Chamoli (M = 6.7, 1999)]. Most recently, in June 2013, the cloudburst event followed by numerous landslides and flooding killed thousands of people and affected several families in the whole Uttarakhand (Dubey et al. 2013; Martha et al. 2015). The study area consists of weak and unstable lithology, highly fragile rocks due to the setting of the litho-units due to continent–continent collision; consequently, the landslide is a common phenomenon in the area.

Table 1 Major landslides in Uttarakhand between 1998 and 2013

S. No	Years	Area	Live and property losses	References
1	1998	Ukhimath	Over 60 persons killed and numerous houses were damaged	Kanungo et al. (2009)
2	1998	Malpa village	The whole Malpa village in Uttaranchal wiped out by Malpa landslide and killed 210 people in the year 1998	Juyal (2002)
3	2001	Phata-Byung Village	Death of 27 persons and serious loss of livestock. Several houses were damaged	Naithani et al. (2002)
3	2002	Bhudha Kedar	28 persons killed together with 99 cattle	Rautela and Pande (2005)
4	2003	Uttarkashi	In this landslide due to timely evacuation, no life was lost; however, heavy losses to the infrastructure and livelihood were reported	Vinod Kumar et al. (2008)
5	2012	Ukhimath	51 people and a significant loss of property	Martha and Vinod Kumar (2013)
6	2013	Kedarnath	According to the Uttarakhand state govt., an estimated 6074 people have died due to landslides and floods, with maximum deaths reported at the Kedarnath town	Martha et al. (2015)

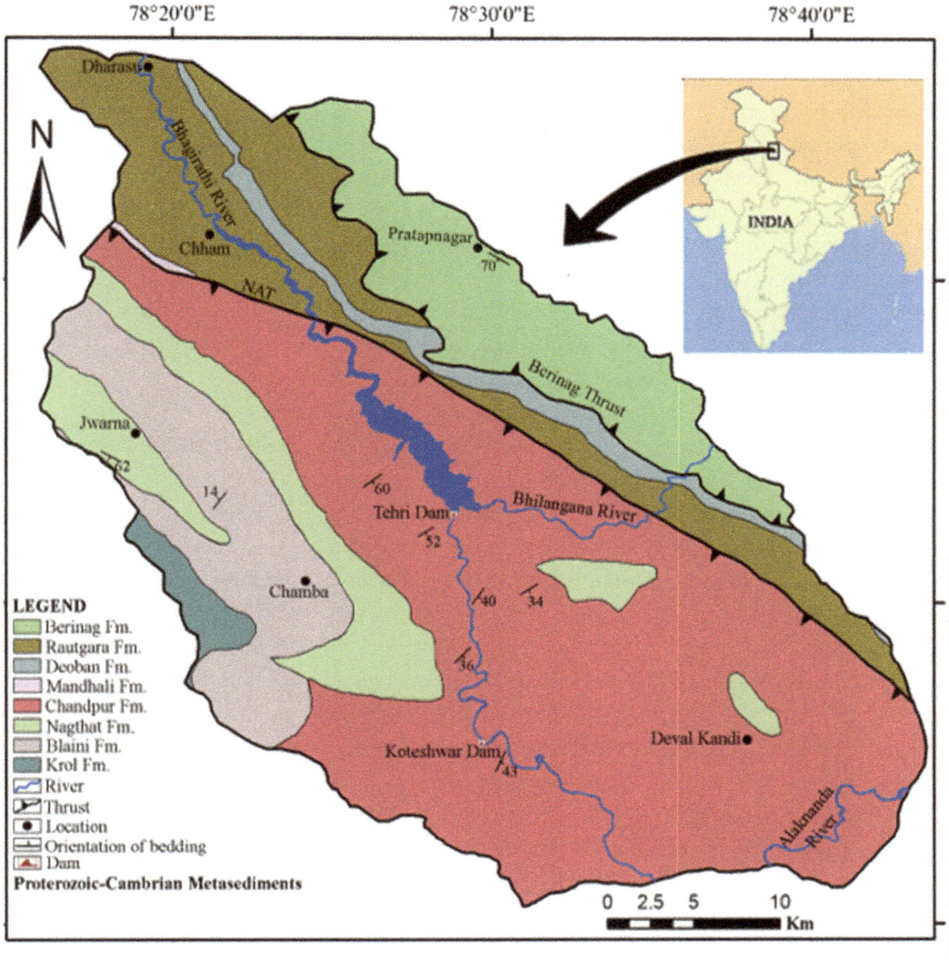

Fig. 1 Study area map (Geology of the area is shown on the map)

Geological Setting

The study area lies in the Lesser Himalayan region of in the Tehri-Garhwal districts of Uttarakhand. The Lesser Himalayan Sequence (LHS) comprises of mafic volcanic rocks, quartz arenites of the Berinag formation, phyllites, siltstones of the Chandpur formation, quartzite, schistose quartzite with intercalations of slates of the Nagthat formation, conglomerate/diamictite of the Blaini formation, dolomite of the Deoban formation, quartzite, slate, phyllite of the Rautgara formation, and Schist, Limestone of the Mandali, Krol formation.

The Blaini formation is a very distinctive sequence of conglomerate, siltstone, greywacke, shale of grey, olive green, and black colors and impersistent lenticular beds of pinkish limestone/dolomite associated with pinkish shale and sandstone. It unconformably overlies the grayish-green and purplish quartz arenite of the Mandhali formation (Saklani 1993). The Nagthat formation represented by medium to coarse-grained, pink, purple, and buff-grey, thickly bedded (10–100 m) massive quartzite, and schistose quartzite with intercalations of slates. It unconformably lies over the Blaini formation. Nagthat quartzite also occurs in the form of a NW–SE trending synclinal window around New Tehri. It shows well developed inclined to overturned anticlines plunging towards the north near Nail. This quartzite is very much ferruginous. Cross bedding and ripple, marks are noticed in Nagthat quartzite near New Tehri.

Especially in the Tehri area, the dominant rock types observed are phyllites, quartzites, and quartzitic phyllites of the Chandpur formation overlain by recent colluvial and alluvial materials (Gupta and Anbalagan 1997). The Chandpur formation is constituted of phyllites. The phyllite comprises of olive green to gray sericite-chlorite-quartz abundant slaty member and grayish-green sericite-chlorite abundant schistose member. Quartz vein is intruded into phyllite rock and is finely interbedded by siltstones and basic volcanics at many places. Saklani and Nainwal 1989, considered the Chandpur Formation as Dharmandal Group and defined two basal members, Gwar Phyllite and Jakhnidhar Schist in the Bhagirathi Valley near Tehri. Near the Tehri area, the Chandpur formation is restricted towards the north by the distinct NAT, trending approximately northwest-southeast and dipping towards the southwest. The rocks in the intra-thrust zone are highly fractured in nature.

The Berinag Formation has medium to coarse-grained massive white quartzite, with a few intercalated beds of sericite schist, and generally sericitic quartz arenite of white, pale purple and green color with metamorphosed amygdaloidal vesicular basalts and tuffites (Valdiya 1997). Overlying Berinag formation is the Deoban formation that represents the second carbonate facies of the Garhwal Group conformably overlies the Rautgara Formation. The study area includes the part of the tectonically active and structurally complex zone of the Lesser Himalaya. This zone has several thrusts and fault zones, many nappes and klippen, (Valdiya 1980). The tectonic contact between Chandpur phyllite and Berinag formation is named as North Almora Thrust (NAT). It can be traced for 70 km from Dangchunra (southeast of Chandrabadni) to the northwest of Nagal with a NW–SE trend and southwesterly dips. A highly subdued topographical expression with distinct erosional characteristics, developed in Chandpur Phyllites as against the bold topographical expression of Berinag Quartzite is indicative of the North Almora Thrust (NAT) on satellite imagery (Valdiya 1980).The presence of debris avalanches, active and stabilized uplifted terraces, deep gorges are the signature of recent movement along the North Almora Thrust (NAT). The Berinag thrust is a branch of North Almora thrust, designated as Pratapnagar thrust (Saklani 1971). It spreads over about 35 km in the area from Kandi-Banali Calc Zone (limestone-slate-quartzite) along with the Pratapnagar thrust in tectonic contact with Pratapnagar quartzite. A relative absence of surface drainage in the limestone member and subdued topographical expression of the slaty member of the Kotga-Banali Calc Zone helps in distinguishing it from the Pratapnagar quartzites on satellite images. Gneisses and quartzites commonly occur at the base of the hanging wall, whereas dolomites and schist of the Deoban Formation typically occur directly below the thrust. The thrust zones are highly crushed and show unstable slopes along the strike. Several minor irregularly oriented faults are also present.

Data Used and Methodology

Preparation of LSZ requires various types of data, such as satellite images, digital elevation models, field data. Details of different types of datasets used in the study are described below.

i. Topo sheet maps collect from Survey of India at a scale of 1:50,000.
ii. Geological (lithological and structural) maps (compiled after Célérier et al. 2009 and Valdiya 1980).
iii. Satellite sensor data. LANDSAT TM multispectral digital data of the year 2010 in three bands (blue, green, red) with 30 m spatial resolution and ASTER global digital elevation model (DEM) with 30 m spatial resolution.
iv. Fieldwork data, including clarifications on landslides, geology, structure, and Land-use/Land-cover. The landslide inventory has been prepared from the field data, Geological Survey of India (GSI) reports, and satellite images.

Selection of Parameters for LSZ Map

The LSZ is used to understand the relation of the susceptibility or risk caused due to slope failure processes, which requires the initial identification and detailed understanding of several terrain features that manage the slope stability. The methodology derived from the present studies takes into consideration some well-known factors like lithology, fault/lineament, slope, drainage, land-use/land cover, and incidence of landslides in the area (Table 2). The essential parameters of each layer have been selected for analysis as an input to the weighted linear model described in later sections.

i. Slope angle layer—This layer has been selected from the digital slope raster map with six classes.
ii. Lithology layer—This layer has been prepared from the geology map with eight lithological rock types.
iii. Buffer map of the Fault/lineament layer—We have prepared this layer from the lineaments map of the study area.
iv. Relative relief layer—This layer has been selected from the digital raster with five relief classes.
v. Buffer map of Drainage Density layer—This layer has been prepared from a digital vector map of streams.
vi. Land-use layer—We have prepared this layer from satellite images. It consists of eight classes.
vii. We have prepared the landslide inventory raster layer from landslide locations consisting of 1240 landslides.

Slope and Relative Relief

The slope angle is one of the essential parameters for preparing LSZ, which indicates the slope at a specific site. The shear stress in soil or other unconsolidated material generally rises as the angle of slope increases. Hence, chances of occurrence of landslides are more predominant in

Table 2 Different data layers and landslide susceptibility weightage–rating system estimated in this study

S. No	Data layers	Classification		Rating value	Weightage
1	Slope angle	A	>60°	9	9
		B	46–60°	8	
		C	36–45°	7	
		D	26–35°	5	
		E	16–25°	3	
		F	<15°	1	
2	Lithology/Litho-Stratigraphic unit	A	Rautgara and Chandpur Formation	9	8
		B	Berinag and Nagthat Formation	7	
		C	Mandali and Krol Formation	5	
		D	Deoban Formation	3	
		E	Blaini Formation	1	
3	Thrust (buffer)	A	<500 m	9	7
		B	500–1000 m	6	
		C	>1000 m	3	
4	Relative relief	A	>2000 m	9	6
		B	1601–1800 m	7	
		C	1201–1600 m	5	
		D	801–1200 m	3	
		E	<800 m	1	
5	Land-use/land-cover	A	Barren	9	5
		B	Sparse vegetation (with pine trees)	8	
		C	Open area	5	
		D	Agriculture	4	
		E	Dense (mixed) forest	3	
		F	Settlement	2	
		G	Water body	1	
6	Photo-lineament (buffer)	A	<500 m	7	4
		B	500–1000 m	5	
		C	>1000 m	3	
7	Drainage (buffer)	A	Influence of drainage (<50 m)	3	2
		B	Rest of the area	1	

the steep slope regions than in moderate and low slope regions—slope map generated by using Digital Elevation Model (DEM) in a GIS environment. We have classified the slope areas into six different slope facets (in degrees), i.e., 0–15°, 16°–25°, 26°–35°, 36°–45°, 40°–45°, 46°–60° and above 60°, as shown in Fig. 2. The rating values are given according to the steepness of the slope (Table 2).

The relative relief is defined as the variation in minimum and maximum altitude within a defined grid (e.g., in a 3 × 3 window). The study area has a high relative or local relief. The lower relief value shows the mature topography, whereas higher relief values indicate the presence of faults and a rapid rise in elevation. We have divided the relative relief into five classes in terms of elevation ranging from more than 2000 m,

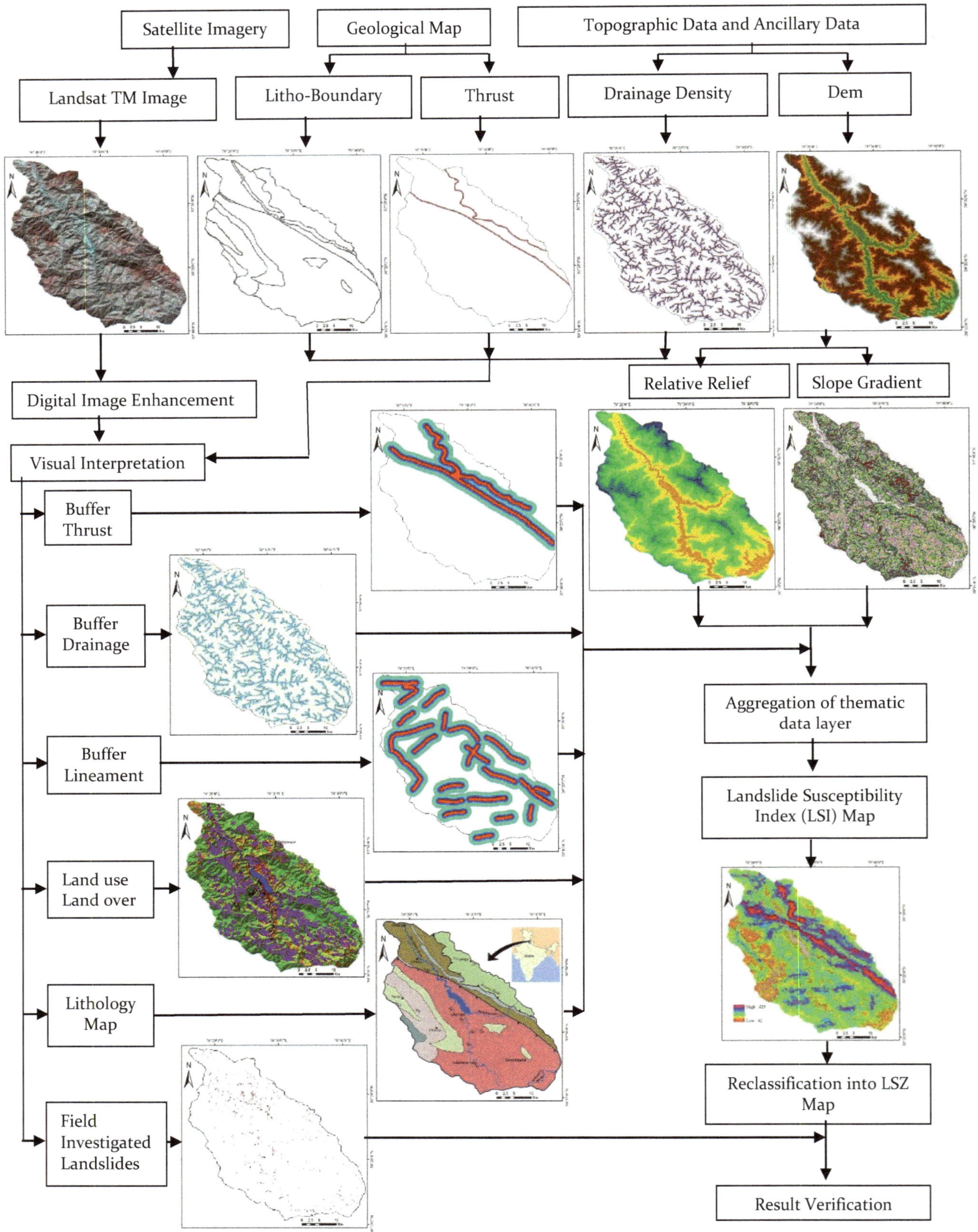

Fig. 2 Methodology and GIS analysis for integrating various thematic data layers and preparing LSZ map

1600–2000 m, 1200–1600 m, 800–1200 m, and less than 800 m from mean sea level, respectively, as shown in Fig. 2. Landslides are more susceptible to higher elevated areas as compared to the lower elevated areas. Hence higher ratings are given to high relative relief and a low rating for lower relief class (Table 2).

Structure and Lithology

There are two primary thrusts, which pass through the area, i.e., North Almora Thrust (NAT) and Berinag Thrust. These thrusts are distinct by shear zones and demarcated from the regional geological map of the study area. Furthermore, there are other faults and photo-lineaments, which are identified and digitized on Landsat TM satellite image. The buffer zones of each thrust and photo-lineament are generated by applying distance function and classified accordingly. The distance from these structural features is given less rating as compared to near buffer zones because the probability of occurrence of landslides may decrease with increasing distance from these features.

Land Use and Land Cover

The land use/land cover (LULC) represents the utilization of land parcel. Hence, we have identified eight prominent classes in the study area, i.e., settlement, agriculture, dense vegetation, moderate vegetation, open land, sparse land, water bodies, and barren land. We have used Landsat TM bands, 2 (green), 3 (red), and 4 (near-infrared) for the preparation of the LULC map. After that, we have prepared the LULC map using the maximum likelihood classification (MLC) algorithm (Foody et al. 1992; Saha et al. 2005), a supervised classification technique. An approach of hybrid classification was adopted to overcome the hill shadow problems. We have used the ground truth locations for evaluating the quality of LULC.

Drainage

The drainages in the area were digitized from Survey of India topo sheets 53 J/6, J/7, J/10, and J/11 on the scale of 1:50,000. The streams create specific patterns within the drainage basin depending on the slope surface, climatic condition, and underlying rock structure. We have prepared drainage buffers, where the area was falling within 50 m of streams. The sides of rivers get a higher rating while the rest of the area gets less rating (Table 2).

Landslide Distribution

The landslides were demarcated manually on satellite images based on the contrast between the landslide and its surroundings. Landslide shows high reflectance due to the non-availability of vegetation and commonly circular to elliptical. We have mapped several landslides in the field. However, most of the landslides were debris slides. In some instances, rotational and complex types of failures were also present.

We have also identified a few rockfalls of small size in the field. Some of the main contributing factors for landslides, as observed in the field, are excessive water flow along the drainage channels, particularly during the rains, and steep slopes with weathered disintegrated rocks in the proximity of faults/structural lineaments. The landslide inventory map contains 1240 landslides showing the areas occupied by sliding activity.

Methodology

Weightage and Rating of the Data Layers

The preparation of LSZ involves the preparation of several thematic layers such as slope, lithology and geological structure, relative relief, land use/land cover, photo lineaments, drainages. All the parameters of each class were carefully analyzed to create their relation to landslide susceptibility. The steps followed for LSZ preparation with the help of remote sensing and GIS technique is also shown in Fig. 2. The weightage value depends on the relative importance of causative parameters. All the factors are given weights from 0 to 9 in order of importance based on ranking provided by the experts. The weightage and rating values have been re-grouped using weighted overlay and error method, and the weight rating values have been re-adjusted. The slope angle, lithology, and thrust/faults are the main parameters, which cause direct landslide activity. In the study area, most of the slopes are within the range of 35–60°, which may be very prone to landslide due to the reason that colluvial deposits form at approximately 35° slope.

A slope steeper than 60° with barren rock exposures is more susceptible to rockfall than landslides. An increase in the slope angle represents the increase in the probabilities of failure.

Therefore, the slope is one of the most crucial parameters for stability consideration. Hence the most significant weight is provided to the slope. The geological formations play a significant role in landslide activity. The rocks of the Rautgara and Chandpur Formation are weathered and low-grade metamorphic rock, highly jointed, and form steep slopes, and hence receive the highest rating. Another direct effect of landslide activity is thrust/fault as they indicate an area of sheared rocks. The landslide hazard decreases with an increase in the distance from the buffer (thrust). Hence, we have provided the rating in decreasing order of distance from the thrust. The photo-lineaments are small structures as compare to thrusts. Accordingly, we have given a lesser rating to each buffer class as compared to the thrust buffers. In the study area, the different classes of LULC are considered, such as dense vegetation, moderate vegetation, sparse land, open land, barren land, water bodies, settlement, and agriculture. Here highest rating is assigned to the barren areas, as they are most prone to landslide.

The stability of slope tends to increase when vegetation density increases; therefore, relative ratings are assigned. The distance to drainages plays a vital role because of the toe erosion of slopes due to the river action, which may induce landslides. Hence, the highest rating has been given to the buffer of 50 m from the rivers. We have classified the values of relative relief into five classes and provided ratings according to the increasing order of relative relief. The weighted linear combination method is applied to make LSZ maps based on the relative importance of many causative factors described in the previous sections. Then we multiply the thematic maps (landslide causative factors) with the corresponding weights and then summed up to produce the landslide susceptibility index (LSI) for each cell (Gupta et al. 1999). The Landslide Susceptibility Index is given by

$$LSI = \sum weightage * datalayer(attributes)$$

The cumulative score obtained after summation was rearranged into five different classes.

Results and Discussions

Landslide Susceptibility Zonation

After combining all the controlling factors and corresponding weights using the mathematical expression previously presented, the LSI lies within the range of 42 and 423. We have computed the frequency chart (or histogram as shown in Fig. 3) of LSI values for classification in 5 zones to prepare the LSZ map. We have used the following LSI threshold values, i.e., 110, 154, 184, and 239, based on the geometric interval as well as standard deviation. These values define the range for five different zones, such as very high, high, moderate, low, and very low susceptibility.

This study is an effort to make a thorough LSZ map of the study area. We have validated the LSZ map using the areal distribution of the existing landslides, i.e., by overlaying the

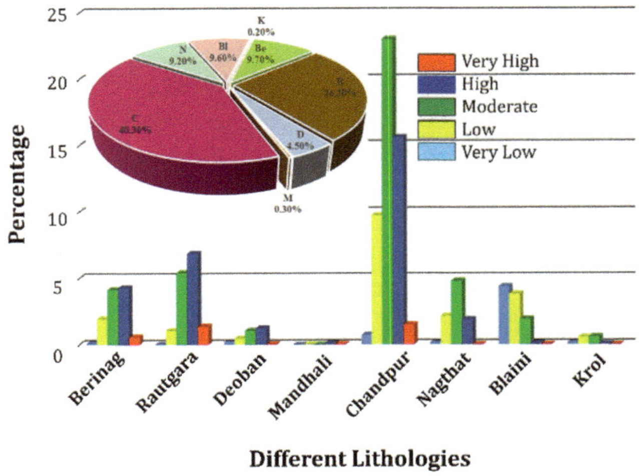

Fig. 3 Histogram chart showing comparative distribution of various susceptibility zones in different lithologies. Pie chart shows the area percentage of different lithologies. Where K-Krol, Be-Berinag, R-Rautgara, D-Deoban, M-Mandhali, C-Chandpur, N-Nagthat and Bl-Blaini

landslide locations on LSZ. A comparison of area vs. landslide incidences for each susceptibility zonation class has been shown in Fig. 4. The study area can be divided into following LULC classes with the area in each class as vegetation (24%), moderate vegetation (19%), sparse land (15%), open land (10%), agriculture (23%), settlement (4%), water bodies (4%) and barren surfaces (1%). The settlements/built-up area occupies about 4% of the total area, whereas agricultural activities occupy over 23% of the total area. These statistics imply that land use in the study area under direct human occupation has very high intensity, mainly in the valley floor region.

The area under very high susceptibility zone is highly unstable and under continuous threat from landslides, predominantly during and after intense rainfall. In the study

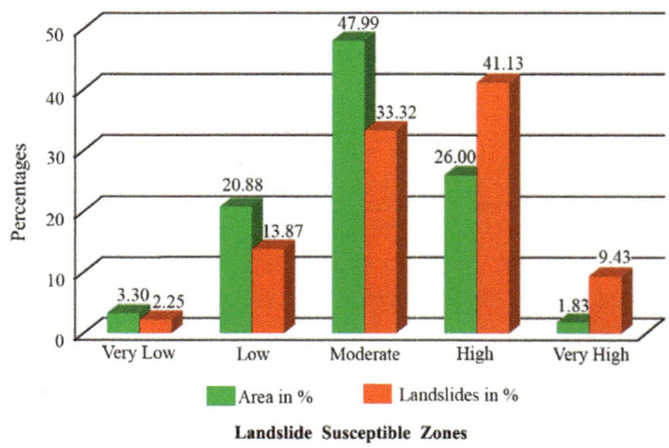

Fig. 4 Histogram chart showing relative distribution of various susceptibility zones and landslide percentage occurrence within each zone

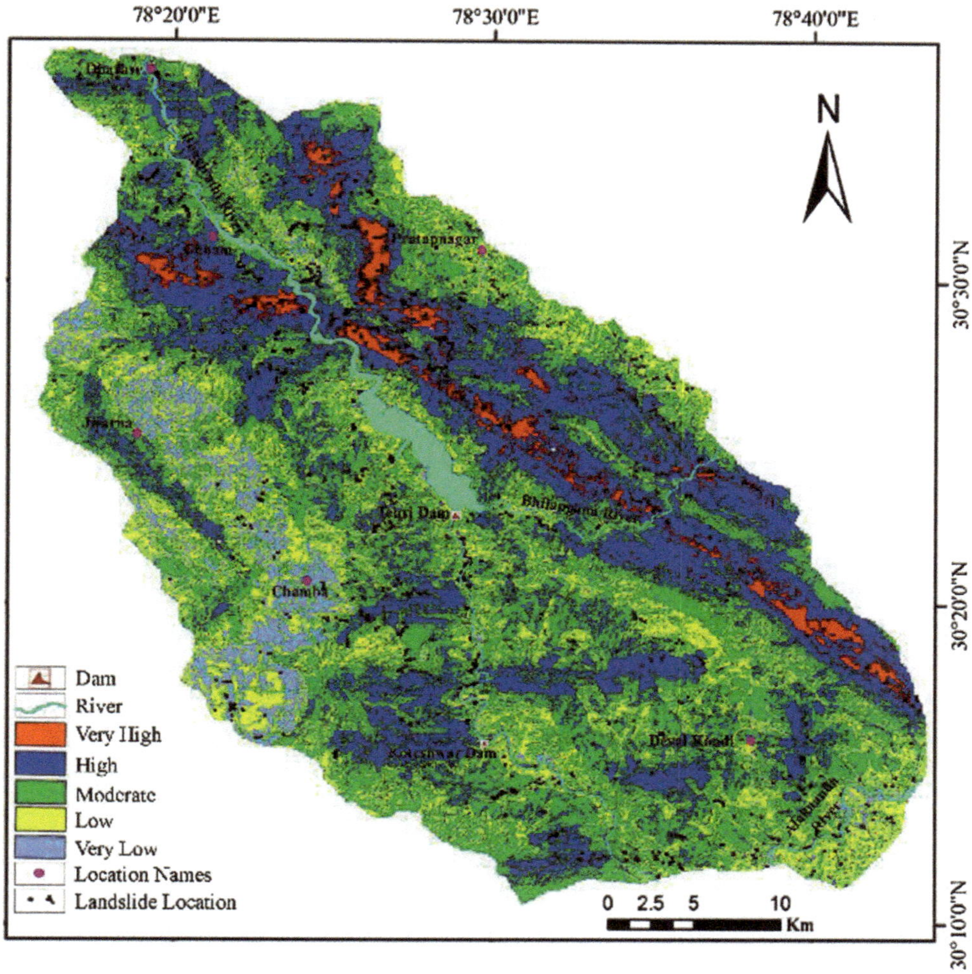

Fig. 5 Landslide Susceptibility Zonation Map with Landslide locations

area, very high susceptibility zones are found at or near the fault and tectonically weak zones. The landslide frequency of a very high zone contains 9.43%, whereas the total area percent is only 1.93%. Meanwhile, as this area is highly susceptible to landslides, no human settlement or any other structural project should be allowed in this zone. This zone may restore itself with natural vegetation to get natural stability in due course of time. The high susceptibility zone lies around the very high susceptibility zone, and in many settlement and developmental areas. This zone covers a region of a steep slope that is prone to landslide when disturbed, so, geologically this zone is unstable. Any slope failure may be produced mainly after heavy rainfall.

In this zone, persistent erosion exists, which is due to streams and river action. Moreover, the soft rock types and loose overlying burden enhance the problem. This zone occupies 26.00% of the total area and discovers 41.13% frequency of the landslide occurrence. Implementation of structural projects and human settlements should be discouraged within this zone. The moderate susceptibility zone is generally considered as stable if its present status is preserved. This area covers moderately small and hard rocks, moderate slope, moderate or dense forest. The moderate susceptibility zone covers the maximum percentage of the area within the study area, and many human settlements lie under this zone. Even though this area may have contained steep slope (more than 45°), the orientation of the bedrock and lack of overlying loose sediments and human effort make them less hazardous.

Slope instability can be triggered by heavy rainfall and seismic activity, so it is suggested not to interrupt the natural drainage system along with changes in slope surface should be avoided. These zones cover about 47.99% of the total area and discover 32.33% of the landslide occurrence. Any constructional activity should be well planned to preserve its existing environment. LSZ map with landslide locations is shown in Fig. 5.

The low susceptibility zone is generally more stable and suitable for human settlement and other developmental schemes because no evidence of instability is detected, and mass movement may not occur unless there are significant changes in controlling factors. This area comprises the combination of several controlling factors, which is unlikely to affect the slope stability adversely. In this zone, slope angles are generally low, with dense vegetation and flatlands. This zone occupies 20.88% of the total area and contains 13.87% of the landslide occurrence.

The very low zone usually comprises valley fill and flatlands. This zone is presumed to be free from existing and future landslide susceptibility. The 'very low susceptibility zone' covers about 3.30% of the total area and includes 2.25% of landslide occurrence.

Conclusions

Landslide is one of the most frequently occurring natural disasters and considered a significant problem encountered while implementing the developmental activities in the mountainous regions. Thus, the preparation of LSZ is very important for developmental planning in hilly regions. The identification of landslide susceptible zones using geospatial techniques is widely applied and appreciated. Hence in this work, the LSZ map for the Tehri area of Uttarakhand is prepared using an ordinal scale relative weightage technique. The LSZ map shows that most landslides occur in high and very high susceptible zones, which cover roughly 28% of the study area. The areas that fall in high and very high susceptibility zones are mostly located on the north-eastern side of the reservoir, and the North Almora Thrust (NAT) passes from there in the NW–SE direction. Thus, the presence of tectonically active NAT thrust seems to be the reason for making this area prone to landslide movements. Hence any construction activity in the vicinity of NAT needs to be carried out very diligently to take proper preventive and mitigation measures.

References

Aleotti P, Chowdhury R (1999) Landslide hazard assessment: summary review and new perspectives. Bull Eng Geol Environ 58:21–44

Célérier J, Harrison TM, Webb AAG, Yin A (2009) The Kumaun and Garwhal Lesser Himalaya, India: part 1. Structure and stratigraphy. Bull Geol Soc Am 121:1262–1280

Dubey CS, Shukla DP, Ningreichon AS, Usham AL (2013) Orographic control of the Kedarnath disaster. Curr Sci 105:1474–1476

Foody GM, Campbell NA, Trodd NM, Wood TF (1992) Derivation and applications of probabilistic measures of class membership from the maximum-likelihood classification. Photogramm Eng Remote Sens 58:1335–1341

Gupta P, Anbalagan R (1997) Slope stability of Tehri Dam Reservoir Area, India, using landslide hazard zonation (LHZ) mapping. Q J Eng Geol Hydrogeol 30:27–36

Gupta RP, Saha AK, Arora MK, Kumar A (1999) Landslide Hazard Zonation in a part of the Bhagirathi Valley Garhwal Himalyas, using integrated remote sensing-GIS. Himal Geol 20:71–85

Gupta SK, Shukla DP (2018) Application of drone for landslide mapping, dimension estimation and its 3D reconstruction. J Indian Soc Remote Sens 46:903–914

Gupta SK, Shukla DP, Thakur M (2018) Selection of weightages for causative factors used in preparation of landslide susceptibility zonation (LSZ). Geomatics, Nat Hazards Risk 9:471–487

Juyal GP (2002) Landslides in the Uttaranchal Himalaya and their rehabilitation by bio-engineering measures. In: Geodynamics and environment management of Himalaya, HNB Garhwal University, Srinagar Garhwal, India, pp 182–190

Kanungo D, Arora M, Sarkar S, Gupta R (2009) Landslide Susceptibility Zonation (LSZ) mapping–a review. J South Asia Disaster Stud 2:81–105

Martha TR, Vinod Kumar K (2013) September 2012 landslide events in Okhimath, India—an assessment of landslide consequences using very high-resolution satellite data. Landslides 10:469–479

Martha TR, Roy P, Govindharaj KB et al (2015) Landslides triggered by the June 2013 extreme rainfall event in parts of Uttarakhand state, India. Landslides 12:135–146

Naithani AK, Kumar D, Prasad C (2002) The catastrophic landslide of 16 July 2001 in Phata Byung area, Rudraprayag District, Garhwal Himalaya, India. Curr Sci 82(8):921–923

Rautela P, Pande RK (2005) Traditional inputs in disaster management: the case of Amparav, North India. Int J Environ Stud 62:505–515

Saha AK, Gupta RP, Sarkar I et al (2005) An approach for GIS-based statistical landslide susceptibility zonation-with a case study in the Himalayas. Landslides 2:61–69

Saklani PS (1971) Structure and tectonics of the Pratapnagar area, Garhwal Himalaya. Himal Geol 1:75–91

Saklani PS (1993) Geology of the lower Himalaya: (Garhwal). In: International books and periodicals supply service

Saklani PS, Nainwal DC (1989) Microstructural strain path in Pratapnagar quartzite, Yamuna valley, Garhwal Lesser Himalaya. Himal Mt Build 45–62

Schuster RL (1996) Socioeconomic significance of landslides. Landslides: investigation and mitigation. Transportation Research Board Special Report, 247. National Academy Press, Washington, DC, pp 12–35

Shukla DP, Gupta S, Dubey CS, Thakur M (2016) Geo-spatial technology for landslide Hazard zonation and prediction. In: Marghany M (ed) Environmental applications of remote sensing. InTech, Rijeka, Coratia, pp 281–308

Valdiya KS (1962) An outline of the stratigraphy and structure of the southern part of Pithoragarh District, Uttar Pradesh. J Geol Soc India 3:27–48

Valdiya KS (1980) Geology of kumaun lesser Himalaya. Wadia Institute of Himalayan Geology, Dehradun

Valdiya KS (1997) Himalaya, the northern frontier of east Gondwanaland. Gondwana Res 1:3–9

Varnes DJ (1984) Landslide hazard zonation: a review of principles and practice. In: International association of engineering geology. Commission on landslides and other mass movements on slopes. UNESCO.

Vinod Kumar K, Lakhera RC, Martha TR et al (2008) Analysis of the 2003 Varunawat landslide, Uttarkashi, India using Earth observation data. Environ Geol 55:789–799

Potential Analysis of Deep-Seated Landslides Caused by Typhoon Morakot Using Slope Unit

2020 Kyoto Japan

Meei-Ling Lin, Jian-Fang Wang, Yen-Cheng Chen, and Te-Wei Chen

Abstract

Taiwan is prone to landslide hazard due to fragile geological conditions, frequent earthquakes, and heavy rainfall in the monsoon seasons. The severe rainfall often caused deep-seated landslides. This study utilized the slope unit and discriminant analysis to assess the deep-seated landslide potential in four sub-basins in the Namasia area, Kaoshiung City. The deep-seated landslides were mapped using LiDAR data and aerial photos of the study area. Delineation method of the slope unit was developed, and the resulting slope units consisted well with boundaries of deep-seated landslides. Typically, one deep-seated landslide enclosed one to few slope units. The assessment model was constructed using discriminant analysis on two sub-basins, C and D, with slope angle, aspect, height, and area as causal factors. The model was validated and provided good accuracy for the fitting and validation data. The model was then adopted to assess the landslide potential and moderate results were obtained. Differences in property distributions among sub-basins and choice of factors could cause some discrepancy in assessment results. The potential maps of the study were developed accordingly, and good to moderate results were obtained.

Keywords

Deep-seated landslide • Potential analysis • Slope-unit • Discriminant function

Introduction

Taiwan situates in the convergence zone of tectonic plates which results into fragile geological conditions and frequent earthquakes, and severe monsoon rainfall often caused significant landslide hazard. Typhoon Morakot struck Taiwan carrying more than 3000 mm accumulated rainfall, which induced severe landslides in southern Taiwan in 2009. A significant number of deep-seated landslides were triggered, which lead to massive landslide mass being transported and affected the downstream area. The deep-seated landslides involved large amount of mass movement and often caused sever and secondary hazard in the watershed. Thus, the regional analysis of deep-seated landslide potential is very important for regional hazard mitigation policy.

In this study, deep-seated landslides caused by Typhoon Morakot in the Namashia area, Kaoshiung City, southern Taiwan, were mapped. The regional potential analysis was conducted using the mapped deep-seated landslides in four sub-basins. The slope unit was used as the basic unit in the analysis for its ability of better descriptions of the characteristics of the deep-seated landslide. A discriminant analysis was adopted for construction of the potential assessment model, and the causal factors were determined. The model was then applied to separate set of data for validation of the potential assessment model. Regional potential map of the study area was then constructed accordingly.

M.-L. Lin (✉) · J.-F. Wang · Y.-C. Chen · T.-W. Chen
Department of Civil Engineering, National Taiwan University, Taipei, 106, Taiwan
e-mail: linml@ntu.edu.tw

J.-F. Wang
e-mail: r01521133@ntu.edu.tw

Y.-C. Chen
e-mail: nonametochoice@gmail.com

T.-W. Chen
e-mail: acura0308@gmail.com

F. Guzzetti et al. (eds.), *Understanding and Reducing Landslide Disaster Risk*,
ICL Contribution to Landslide Disaster Risk Reduction,
https://doi.org/10.1007/978-3-030-60227-7_19

Study Area and Mapping of the Deep-Seated Landslides

Study Area

The study area locates in the Namashia area, Kaoshiung City, south Taiwan, with Chishan River flowing through from northeast to the southwest direction as shown in Fig. 1. The SPOT6 satellite image taken in 2013 after Typhoon Morakot was used in Fig. 1. The geological map of the Namashia area is as shown in Fig. 2. The geological properties of Namashia area were characterized as the Western Foothill Geological Zone in Taiwan. The geological formations are mainly composed of un-metamorphic rock formations of the Miocene with sandstone, siltstone, shale, and sandstone/shale interlayer formations. Four sub-basins were selected as the study area as shown in Figs. 1 and 2. The geological formations in the four sub-basins mainly include Changchihkeng Formation, which is a sandstone/shale interlayer, and a small portion of Hunghuatzu Formation, which is mainly silty sandstone. Distribution of geological formations is relatively uniform of the four sub-basins so that the effects of geological properties could be neglected in the subsequent potential analysis.

Mapping of the Deep-Seated Landslides

The deep-seated landslide was considered as large-scale landslide by various researchers; e.g. Chigira (2007) proposed that the landslide involving mass volume larger than 100,000 m^3 be considered as the large-scale landslide. Further discussions extended the criteria to include landslides with landslide area larger than 10 hector and/or depth of sliding surface of more than 10 m (NCDR 2012). In this study, the deep-seated landslide was considered following the criteria of NCDR (2012). Mapping of the deep-seated landslides was conducted using Digital Terrain Model (DTM) derived from Light Detecting and Ranging data to identify the lineation, which was then compared to the satellite images and aerial photos after Typhoon Morakot. The preliminarily mapped landslides were validated by examining the results using the geomorphological characteristics of the deep-seated landslide proposed by Varnes (1978) (Lin et al. 2014). The resulting deep-seated landslides were as shown in Fig. 3.

The landslides thus mapped would include all the potential deep-seated landslides satisfying designated characteristics. Lin et al. (2009) proposed to classify these landslides into four types; i.e. type A—which is the current landslide event, type B—which is the prior landslide event, type C—which involves scar with some geomorphologic evidences, and type D—which involves some geologic/geomorphologic properties as listed in Table 1. The type A, type B, and type C deep-seated landslides of the Namashia area distributed extensively as shown in Fig. 3. Due to the uncertain natures of the type C and type D landslides, only the type A and type B landslides were adopted in this study.

Delineation of Slope Unit

Various delineation units have been used for statistical type analysis when assessing landslide potential. Different statistic unit suggests different topographical characteristics in the analysis during derivation of the analysis factors. The objective of selecting a delineation unit was to be able to represent the characteristics of the slope and landslide properties properly. The slope unit was used widely and being suggested as having good ability to represent the properties and continuity of topography (Guzzetti et al. 1999; Erener et al. 2012). However, adaptation of the slope unit for shallow landslide potential analysis often leads to over-estimation of high potential area due to the small area involved with the shallow landslide. The deep-seated landslides usually associate with large landslide areas, and thus the slope unit was considered appropriate for representing topographic characteristics of the landslides in this study.

Giles and Franklin (1998) proposed to use the break of slope to delineate the slope unit. Xie (2007) proposed to delineate slope unit based on watershed, and was used widely. This delineation method started with finding the watershed based on generating the first order river system using DTM in the Geographic Information System (GIS), and then reversed the DTM of the watershed. By overlapping the DTM and reverse DTM of the watershed, the jointed crest and valley lines would delineate the slope unit. The procedures were straight forward and easy to apply in GIS, and thus was used widely. However, the slope unit thus mapped often includes fragmented unit or units not conforming to the topographic characteristics, which requires manual edition referring to slope aspect, slope angle, and stream line. An individual delineation method was developed by Lin and Lin (2012) by overlaying shaded maps derived at different shading angles and the resulting slope units suggested a better capability of displaying detail features of topography. But the resulting slope unit still requires manual edition for minor irregularities. To improve the shaded map delineation method, the watershed was generated first and then overlaid with shaded maps at different shading angles to delineate the slope unit. The resulting slope unit appears to provide good descriptions of topographic features, and does not require much edition. The

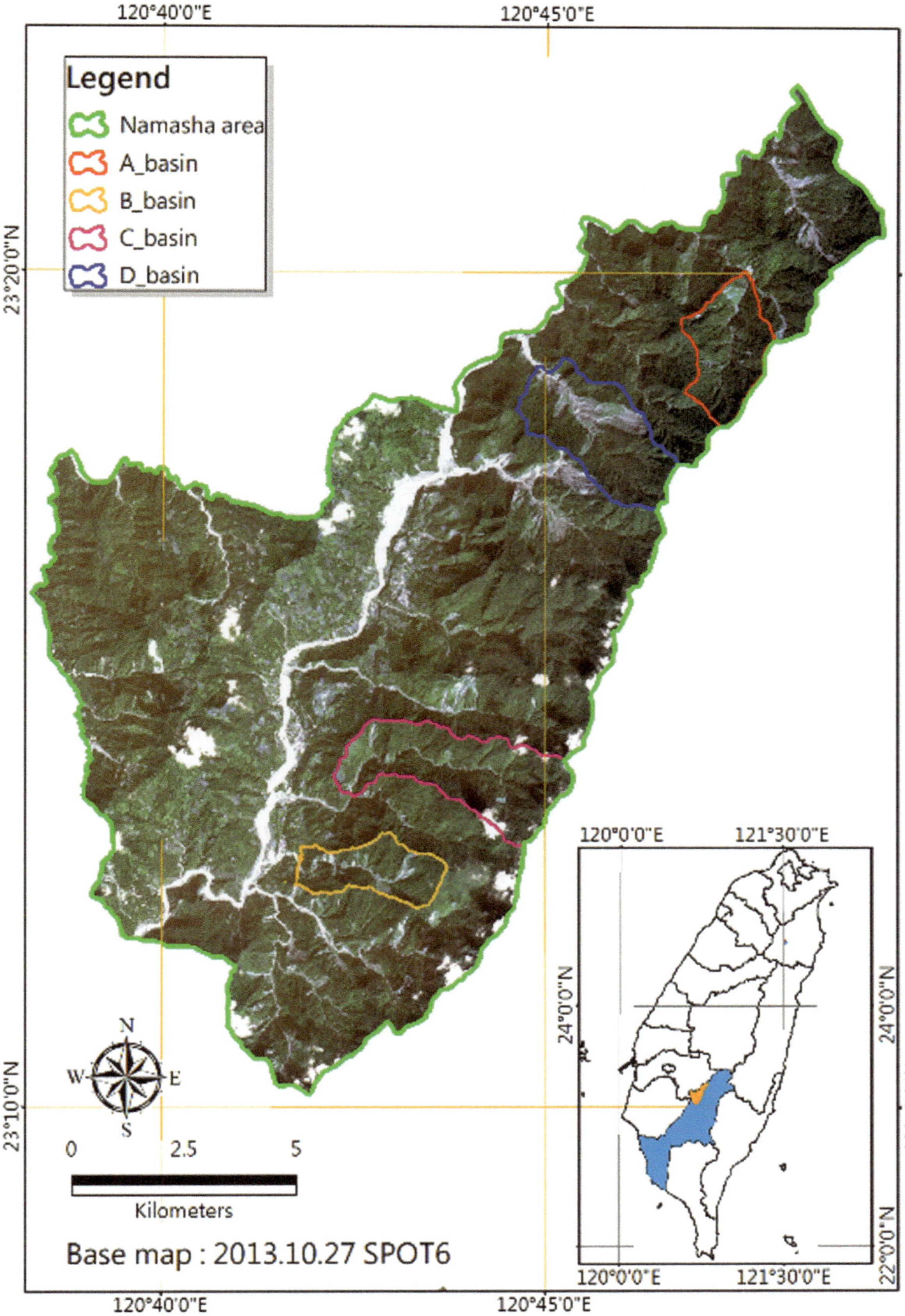

Fig. 1 The study area of four sub-basins in Namashia area, Kaohsiung City, south Taiwan

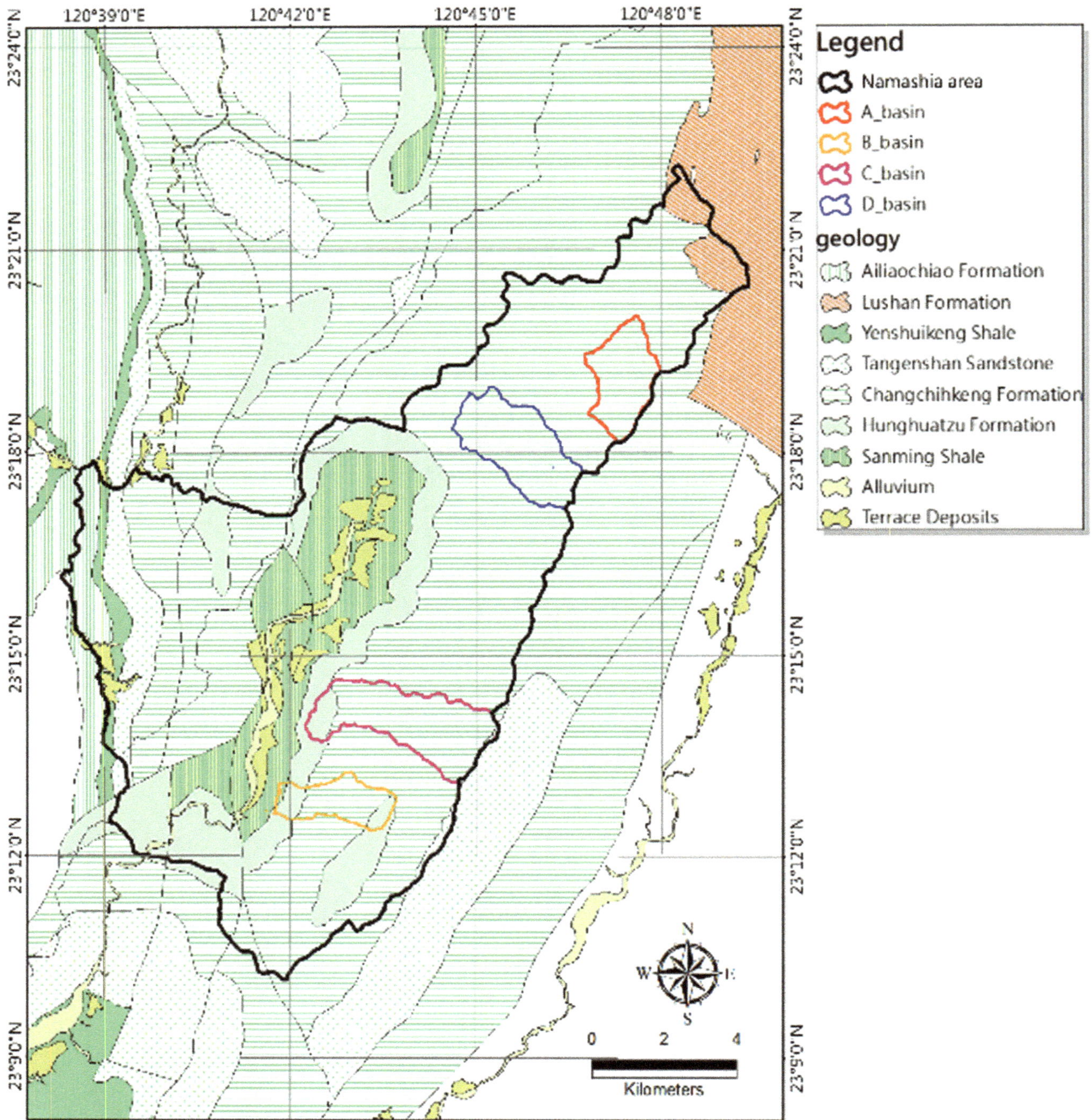

Fig. 2 The geological map of Namashia study area with four sub-basins. (modified from CGS 2013)

resulting slope unit on the shaded map of the four sub-basins are as shown in Fig. 4. From the shaded map in Fig. 4, it was found that the delineated slope units have a good presentation of the slope features.

The sub-basins A and D with delineated slope units were overlaid with the mapped deep-seated landslides as shown in Fig. 5. Figure 6 shows the sub-basins B and C with delineated slope units and mapped deep-seated landslides. It can be observed that the boundaries of most of the mapped deep-seated landslides conform relatively well with the delineated slope units with typically one to few slope units enclosed in each deep-seated landslide. These slope units were considered as landslides units, and the rest of the slope units were considered as non-landslide slope units. The resulting number of landslide and non-landslide slope units of the four sub-basins are listed in Table 2.

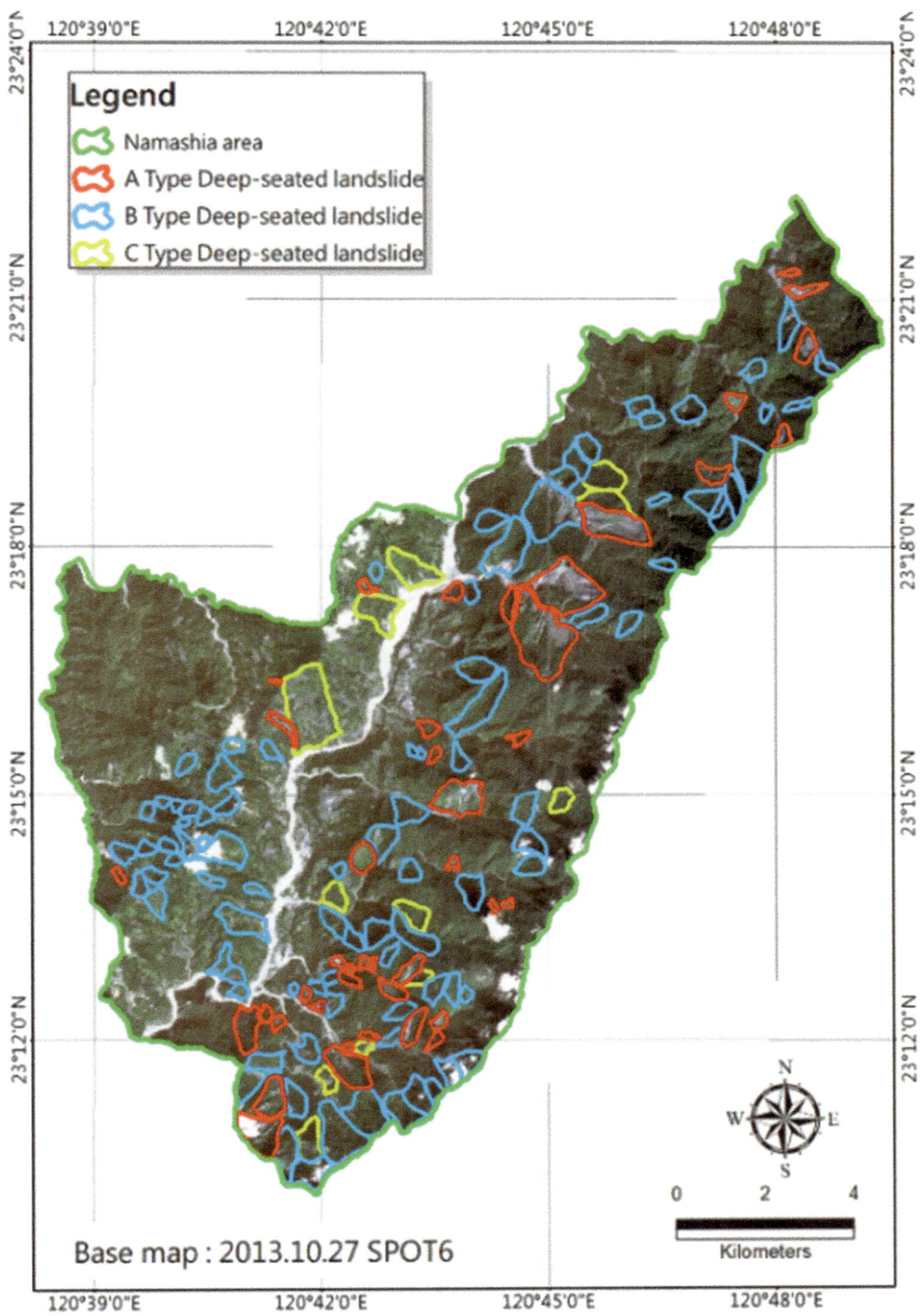

Fig. 3 The distribution of the mapped deep-seated landslides in Namashia area

Discriminant Analysis for Potential Assessment

Causal Factors

To conduct the statistical type of analysis, the causal factors related to topographic features of the deep-seated landslide were selected. Preliminary selection of factors included height, mean elevation, the maximum elevation, elevation of toe, slope angle, area, length to width ratio, and aspect of slope unit. Due to the nature of a categorical type factor, the aspect factor was represented using landslide area ratio based on eight different aspect ranges. All factors were derived from DTM of the study area for both the landslide and non-landslide slope units. In the discriminant analysis, a linear relation of multiple factors was established to discriminate between two different groups of parameters, i.e. the landslide slope unit and the non-landslide slope unit. In order to conduct the discriminant analysis, factors of the

Table 1 Classifications of the deep-seated landslides (Lin et al. 2014)

Type	Classification	Landslide occurrence	Properties
A	Current event	Landslide	With clear geomorphologic evidences and significant scarps
B	Prior event	Landslide	With geomorphologic evidences but no significant scarp
C	With geomorphologic evidences	Uncertain	With some geomorphologic evidences
D	Without geomorphologic evidences	No slide yet	With some geological and/or geomorphologic properties

slope units should satisfy normal distribution condition. More than 30 data were needed for both groups to satisfy this requirement; thus, the sub-basins C and D were combined for construction of the assessment model.

In addition to the normal distribution requirement, each factor was considered as a valid factor only when satisfying independency and significance checks. The differences in distribution property between the landslide and non-landslide groups were examined to ensure the significance of a factor. Then the T test was conducted to find the variances and a significance level of 0.1 was taken in this study. For the independency of the factors, the Pearson correlation coefficients were used. The factor was considered as independent when the absolute value of correlation coefficients with respect to one another factors being smaller than 0.2. The causal factors passing through the significance and independency checks were area, slope angle, height, and aspect of slope unit following the order of significance level.

Discriminant Analysis

The discriminant analysis is conducted using a discriminant function, which is a linear relation of multiple causal factors with coefficients showing the weighting of each factor. The linear function can be written as:

$$y = a_0 + a_1x_1 + a_2x_2 + a_3x_3 + \cdots + a_ix_i \quad (1)$$

where, y is the discriminant score, x_i is the causal factor, and a_i is the coefficient of the factor and a constant a_0. The higher the score, the more likely the landslide will occur. The causal factors used in the analysis were area, slope angle, height, and aspect of slope unit. To eliminate the effects of different units and ranges of value among different factors, all causal factors were normalized by subtracting its mean value and then divided by the standard deviation. Thus, the value will be -1 for mean value subtracted by one standard deviation, and 1 for mean value adding one standard deviation with a mean value of 0. A large portion of the values will distribute in the range from -1 to 1. The coefficient of each factor suggests the weighting of the factor and trends of score variation with the factor. Slope units of sub-basins C and D was used for constructing the discriminant function. The Bernoulli's random sampling method was used, and about 60% of the landslide group slope unit and about same number of non-landslide group slope unit were sampled for fitting of the discriminant function. The resulting discriminant function was achieved by adding one factor at a time to evaluate its effectiveness and eventually reached the final discriminant function. The resulting coefficient of each factor is listed in Table 3 in a decreasing order of significance as suggested by its value. It can be observed that all the coefficients are positive, which suggests that the discriminant score increases positively with all the factors. In Table 3, the slope angle appears to be the most significant factor followed by aspect, area, and lastly the height. Noted that the order of significance of the factors in Table 3 is different from that estimated during the significance check.

The rest 40% of the landslide slope units and about the same number of non-landslide slope units were then adopted for validation of the discriminant function established. Results of both the fitting and validation groups were listed in Table 4. The average accuracies were 88.6% for fitting group and 71.9% for validation group. Thus, the discriminant function appeared to be valid, and provided good accuracy for the randomly selected data set of sub-basins C and D.

Results of Potential Analysis of the Deep-Seated Landslide

The discriminant function was adopted to assess landslide potential of sub-basins A through D, and the overall accuracy was listed in Table 5. From Table 5, the overall accuracy was about 62% for sub-basin A, B, and C, and 73% for sub-basin D. The resulting accuracy appeared to be good for sub-basin D, and moderately good for all other sub-basins. It was noted that sub-basin D had fewer number of slope units, and most of the slope units had been incorporated during

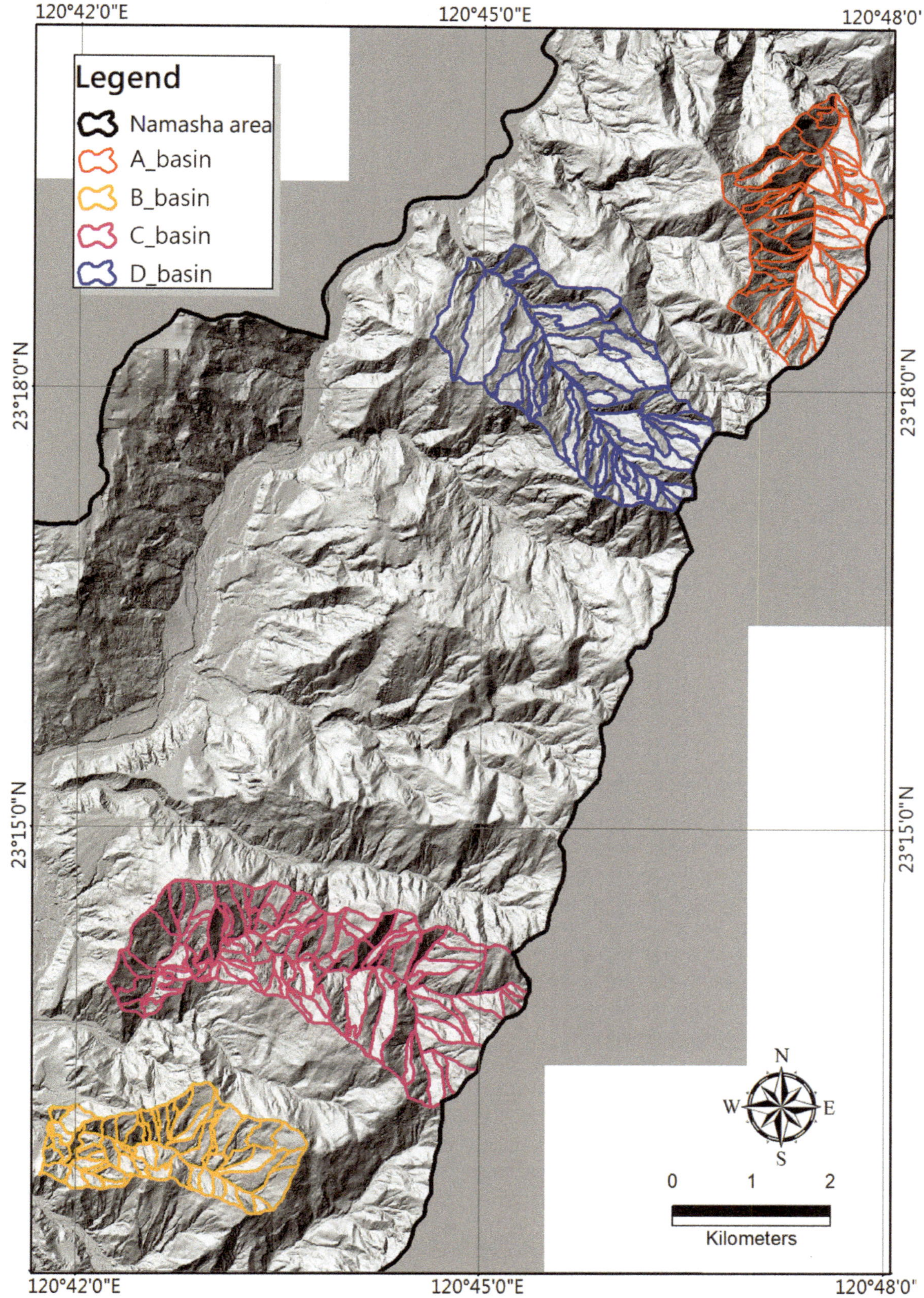

Fig. 4 The delineated slope unit results of four sub-basins

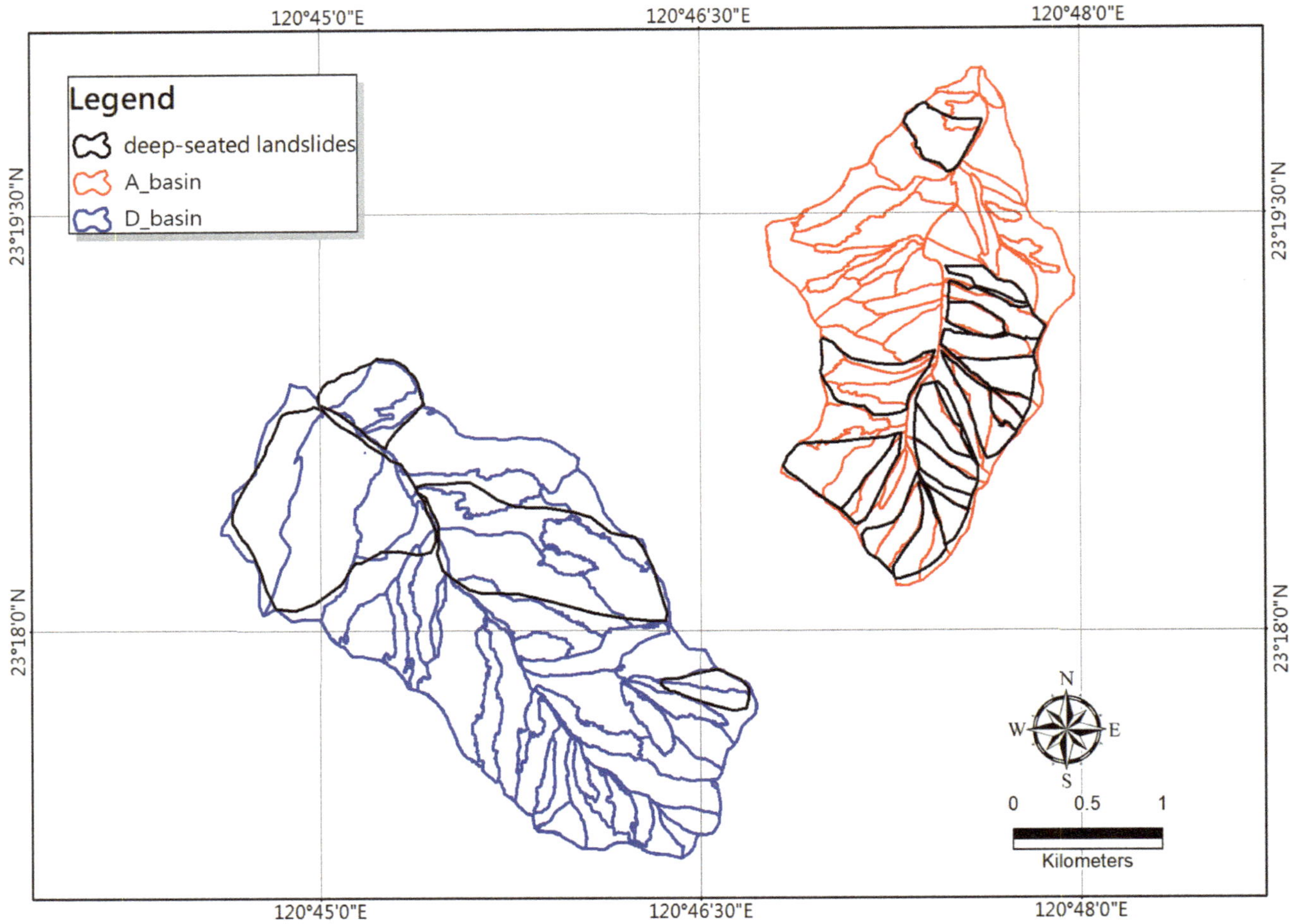

Fig. 5 The sub-basins A and D with delineated slope units and deep-seated landslides

fitting and validation of the model. Thus a more consistent accuracy with the validation result can be expected. The number of non-landslide units were higher than the number of landslide units as shown in Table 1, which could be the cause of lower accuracy. However, the overall accuracies were relatively consistent for sub-basins A, B, and C, which suggested that the discriminant function had a steady performance in estimation of landslide potential.

The discriminant ability might be improved by including more effective factors. Further examination of distribution properties of factors might be helpful. In addition, the geological factor was not considered in this study, which might also affect the results. Utilizing the discriminant analysis implies a linear relation of the factors, thus the possible non-linear relation of factors is not accounted for.

The potential map could be developed using the calculated discriminant scores of the study area. The valid discriminant score was −0.5, and for slope units with scores smaller than −0.5 were considered as with low potential. The slope units with scores larger than −0.5 and smaller than the mean value of landslide group distribution were considered as with moderate potential, and slope units with scores larger than mean value of landslide group distribution were considered as with high potential. The potential map of sub-basins A and D is shown Fig. 7, and that of sub-basins C and D is shown in Fig. 8. It appeared that most of the deep-seated landslides fell in the moderate potential class. The distribution of landslides consisted with the potential map better for the sub-basins C and D, which were used for construction of assessment model. There might be regional property differences of sub-basins A and B not been accounted for, therefore showing a larger discrepancy of the landslide distribution and estimated potential. Further study suggests that the differences in property distribution can attribute to the discrepancy.

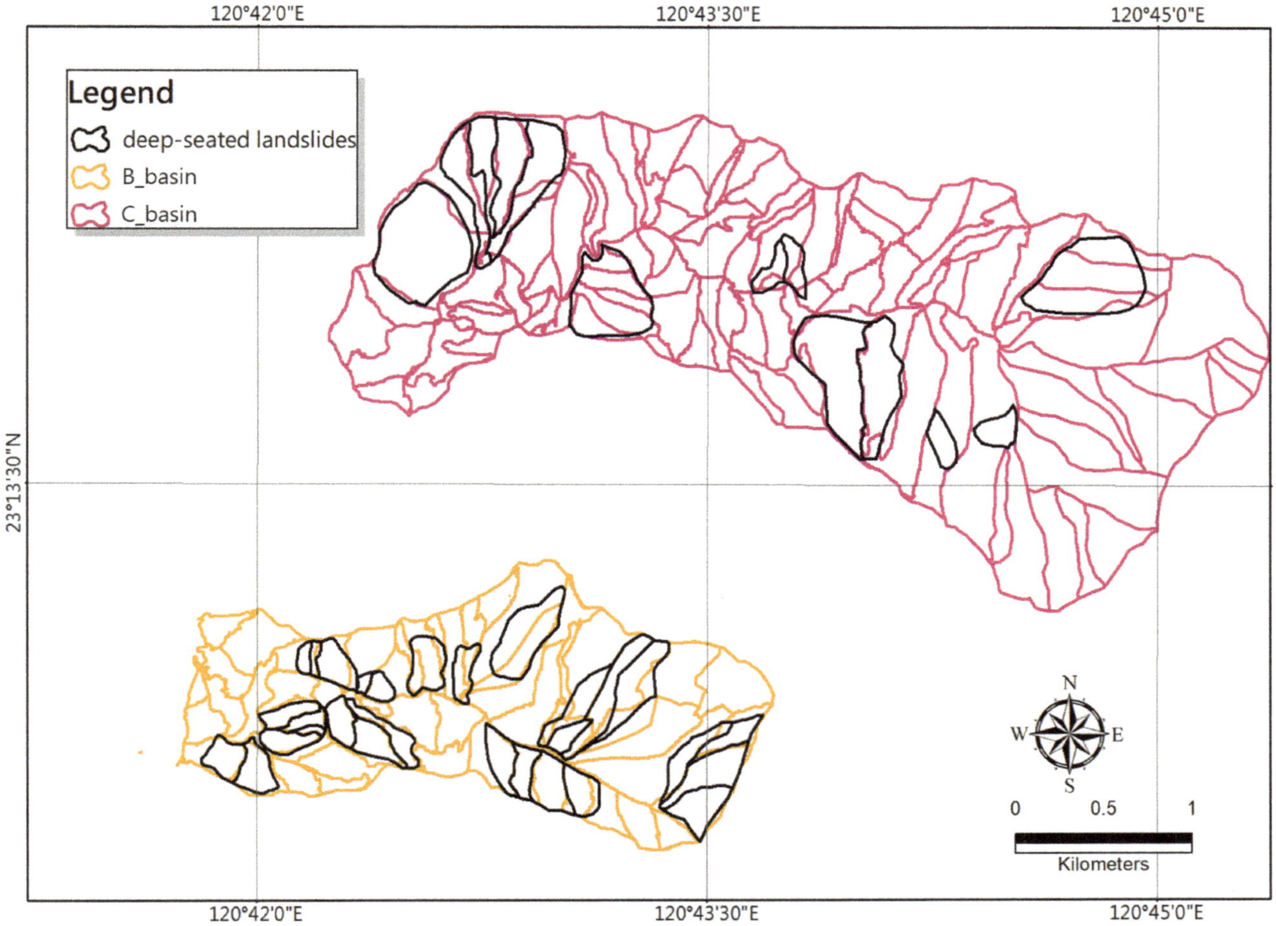

Fig. 6 The sub-basins B and C with delineated slope units and deep-seated landslides

Table 2 Number of landslide and non-landslide slope units in the study area

Study area	Sub-basin A	Sub-basin B	Sub-basin C	Sub-basin D
Total number of slope unit	56	55	104	48
Non-Landslide slope unit	33	30	88	34
Landslide slope unit	23	25	16	14

Table 3 Coefficients of the discriminant function

Factor	Slope angle	Aspect	Area	Height
Coefficient	1.003	0.611	0.447	0.271

*The constant in the discriminant function is −0.272

Table 4 Results of fitting and validation of the discriminant analysis on sub-basins C and D

Study group		Non-landslide	Landslide
Fitting result	Non-landslide	15 (83.3%)	1 (5.9%)
	Landslide	3 (16.7%)	16 (94.1%)
Validation result	Non-landslide	14 (73.7%)	4 (30.8%)
	Landslide	5 (26.3%)	9 (69.2%)

*Percentage of each group is presented in the parentheses

Table 5 Overall accuracy of potential assessment for sub-basins A, B, C, and D

Study area	Sub-basin A	Sub-basin B	Sub-basin C	Sub-basin D
Accuracy, %	60.6	61.8	62.5	72.9

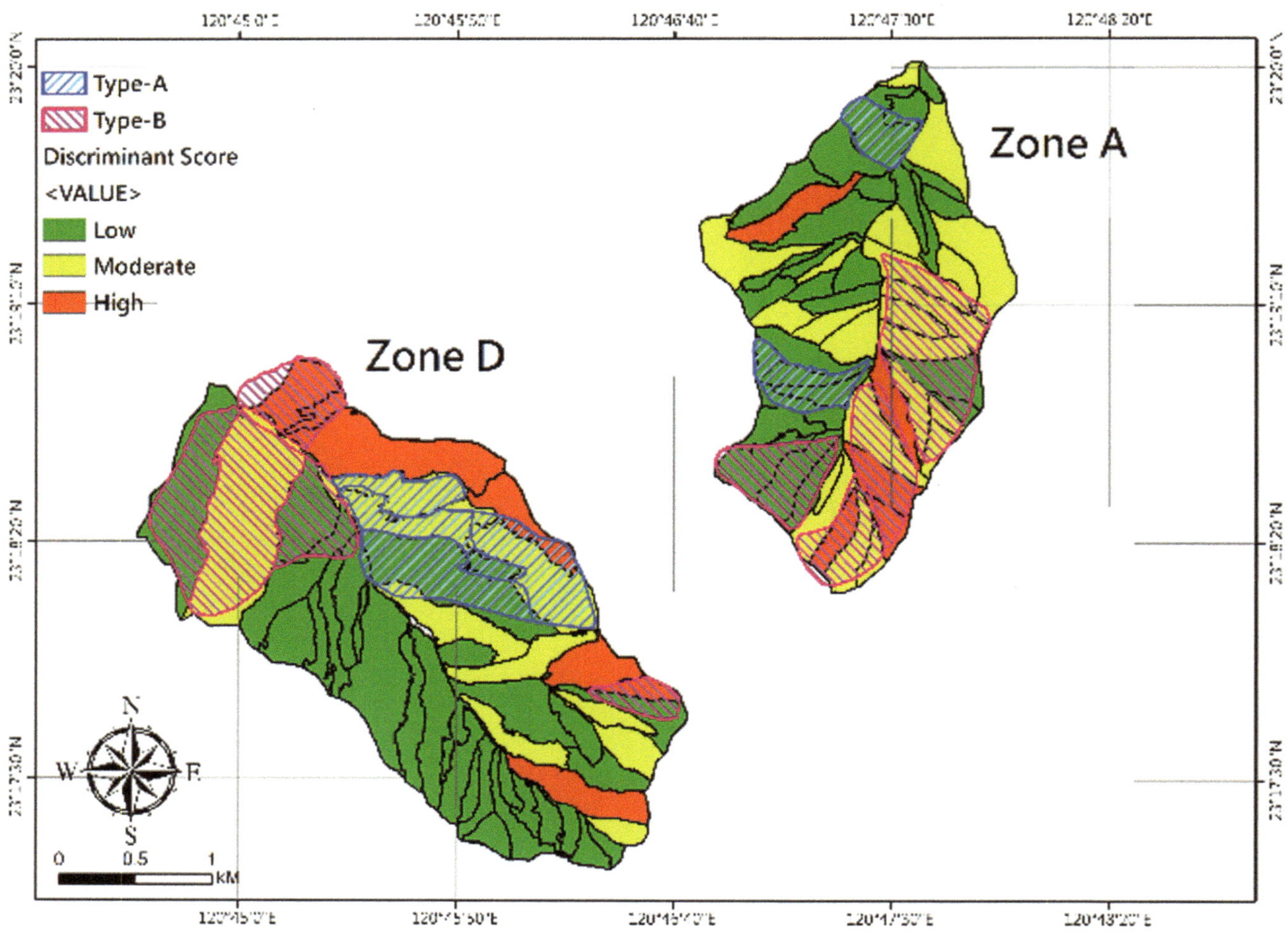

Fig. 7 The potential map of sub-basins A and D

Conclusion

The slope unit delineation method was developed and used for discriminant analysis to assess the potential of deep-seated landslide of the study area. The delineation method was straight forward, and the derived slope units consisted well with the deep-seated landslides. One deep-seated landslide typically enclosed one to few slope units. Discriminant analysis was conducted using the slope units of sub-basins C and D, and the slope angle, aspect, area, and height factors were adopted following an order of decreasing significance. The model provided good accuracies for both fitting and validation groups. Good to moderate results of potential assessment were obtained for the four sub-basins, and the potential maps were generated using discriminant scores. Most of the deep-seated landslides fell in the moderate potential class in this study and could be used as references for hazard mitigation. The assessment method developed in this study was sturdy and provided moderate to good results. Further improvement of the method might be to incorporate more effective factors, taking into account the geological factor, and consideration of the differences in property distribution among different sub-basins.

Acknowledgements Part of this research was supported by the National Science Council, project number: NSC102-2119-M-002-019.

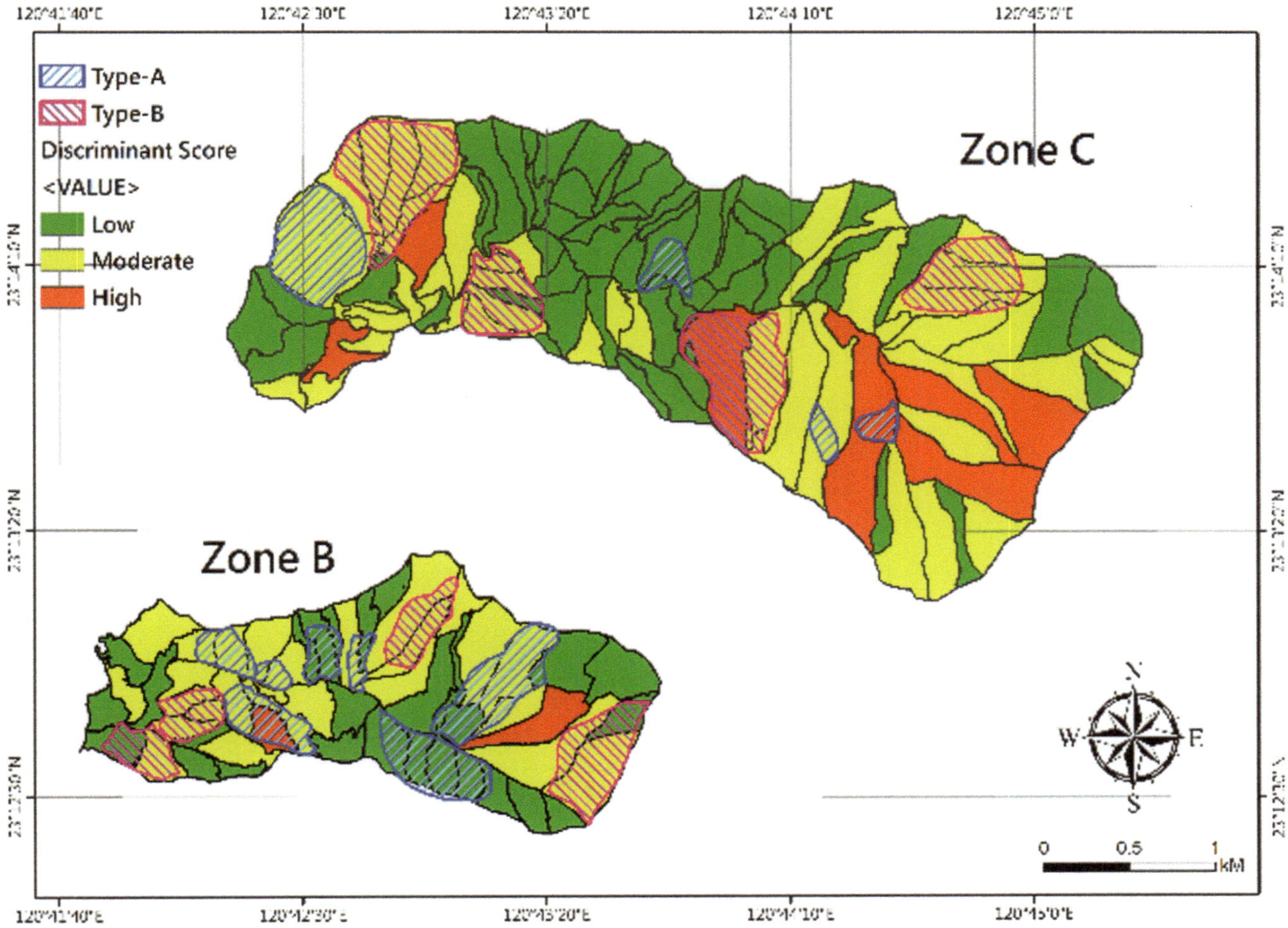

Fig. 8 The potential map of sub-basins B and C

References

Central Geological Survey (2013) Geological investigation & database construction for upstream of flood prone area

Chigira M (2007) Hokai no basho: Daikibo hokai no hassei basho yosoku, Kinmiraisha, Nagoya, p 256. (in Japanese)

Erener A, Du¨zgu¨n HSB (2012) Landslide susceptibility assessment: What are the effects of mapping unit and mapping method? Environ Earth Sci 66:859–877

Giles PT, Franklin SE (1998) An automated approach to the classification of the slope units using digital data. Geomorphology 21:251–264

Guzzetti F, Carrara A, Cardinali M, Reichenbach P (1999) Landslide hazard evaluation: a review of current techniques and their application in a multi-scale study, central Italy". Geomorphology 31:181–216

Lin CW, Lin ML, Chang JB, Wu MJ, Wang TT, Chen TC (2009) Analysis of geological sensitive properties of hazardous area by typhoon Morakot, I, Report, Central Geological Survey, Taipei, Taiwan, p 262. (in Chinese)

Lin ML, Chen TW, Lin CW, Ho DJ, Cheng KP, Yin HY, Chen MC (2014) Detecting large-scale landslides using lidar data and aerial photos in the Namasha-Liuoguey area. Taiwan, Remote Sens 6:42–63

Lin SC, Lin ML (2012) Delineating slope unit using shading method for landslide potential analysis, Report (unpublished), p 20

National Center of Technology for Disaster Reduction (NCDR) 2012 Scale assessment procedures for large-scale landslide, Report, NCDR

Varnes DJ (1978) Slope movement types and processes. In: Landslides, analysis and control. Special Report 176, Transportation Research Board, Washington, pp 11–33

Xie M (2007) Spatial three-dimensional landslide susceptibility mapping tool and its applications. Earth Sci Front 14(6):73–84

Landslide Susceptibility Assessment Using Binary Logistic Regression in Northern Philippines

Dymphna Nolasco-Javier and Lalit Kumar

Abstract

In a landslide prone area in mountainous northern Philippines, landslide susceptibility using binary logistic regression was investigated. Landslide data were randomly divided into training and validation sets using 80% and 20% proportions, respectively. A detailed logistic regression procedure was applied and presented herein. Nine landslide conditioning factors were used. Based on the coefficients obtained, the most influential factors were NDVI followed by land use/land cover, slope aspect, lithology and slope angle. Distance to lineament, distance to road, plan and profile curvature showed no influence in the model generated. Training and validation accuracy were good, amounting to 91% and 86%, respectively. Using training data, 80%, 11%, 5%, 2% and 2% of the landslides were associated with the very high, high, moderate, low and very low susceptibility classes, respectively. Using validation data, the proportions were 82%, 10%, 5%, 3% and 0%, respectively. The strong influence of NDVI affirms its major role in modelling landslide susceptibility. It supports the strong potential of revegetation of precarious slopes in complementing ongoing structural slope stabilization and rehabilitation measures.

Keywords

Landslide susceptibility • Binary logistic regression • Landslide • Philippines

D. Nolasco-Javier (✉)
College of Science, University of the Philippines Baguio, Governor Pack Road, 2600 Baguio City, Philippines
e-mail: dnjavier@up.edu.ph

L. Kumar
School of Environmental and Rural Science, University of New England, Armidale, NSW 2351, Australia
e-mail: lkumar@une.edu.au

Introduction

The northern mountainous Philippines ranks among the most landslide prone in the world (Kirschbaum et al. 2015; Le et al. 2017; Petley 2012). The geologic setting, the seasonal high rainfall and human activities such as unabated deforestation and rapid urbanization are among the recognized causes (Nakasu 2011; Nolasco-Javier and Kumar 2018; Peñaflor et al. 1987; Yumul et al. 2013). Fatalities and damages from rainfall-induced landslides (RIL) have been serious especially in the areas where scant resources need to be allocated to more pressing concerns such as poverty and health. In October 2009, Typhoon Parma caused more than 200 fatalities and more than PHP 90 million (USD 2 million) damages were reported by disaster coordinating agencies in Benguet province alone (National Disaster Coordinating Council 2009). The need to identify landslide hazards and mitigate the annual occurrence of RIL is thus urgent.

Many studies have shown the applicability of the binary logistic regression (BLR) in landslide susceptibility assessment (LSA). BLR is free of data distribution issues (i.e. predictor variables do not need to be normally distributed) and can handle a variety of datasets, such as continuous, categorical, and binary, common types of instability factor data used in landslide studies (e.g. Lee 2005; Lee and Sambath 2006; Nefeslioglu et al. 2008; Yesilnacar and Topal 2005). Due to the flexibility and the non-linear nature of the logistic regression model, this method has the prospect of being able to deal with a wider range of data. The non-parametric nature of BLR also allows analysis of instability factors that are nonsymmetrical or having skewed distributions, typical of those found in the natural environment (Nandi and Shakoor 2010). Moreover, Brenning (2005) stated that overfitting is not a problem for logistic regression.

This study focused on a 40 km^2 area in the municipality of Tublay, province of Benguet in northern Philippines (Fig. 1). The approximate boundaries are 16.45°–16.52° north latitude and 120.6°–120.7° east longitude. The highest

F. Guzzetti et al. (eds.), *Understanding and Reducing Landslide Disaster Risk*, ICL Contribution to Landslide Disaster Risk Reduction, https://doi.org/10.1007/978-3-030-60227-7_20

24-h rainfall for the Philippines (>1000 mm) has been recorded in a nearby synoptic weather station of Baguio City, 20 km south of the study site. Tublay is gradually accommodating the rapid urbanization from adjacent centers, chiefly Baguio City and La Trinidad, the former a recognized economic, educational and tourism center in northern Luzon (Delias and Daly 2016). Located on a major trade route, Tublay is a hub of agriculture and artisanal mining activity (Municipality of Tublay Benguet Housing and Land Use Regulatory Board 2014). It is located in a seismically active, geologically young mountain range in the headwater regions of three drainage basins. It is underlain by deeply weathered, highly fractured, faulted, and intrinsically weak pyroclastic, intrusive and sedimentary rocks. The geomorphology and geologic setting have been detailed in published literature (Balce et al. 1980; David et al. 1999; Peña 1998; Yumul et al. 2008). These predisposing geologic features, together with high rainfall and human activities have placed the area at high risk of landslides and related hazards. The occurrence of RIL triggered by tropical cyclones and the southwest monsoon has caused significant damage to life and property, agriculture and infrastructure. Municipal reports indicate that the cost of damage can at times exceed the municipality's annual budget (A.Lawangen, pers.comm). Fifteen deaths were reported after typhoon Parma, mostly from RIL.

Data and Methods

Generation of Landslide Inventory and Derivation of Factors

A landslide inventory map was generated using satellite images, field survey data and local knowledge after the extreme rainfall event spawned by Typhoon Parma in October 2009. The selected landslides were randomly divided into 80% training data and 20% validation data.

The landslide conditioning factors were generated from the available data sources: 10 m × 10 m SAR-based DEM, 1:50,000 scanned geological and soils maps, and high resolution satellite image (HRSI). The DEM was provided by the Nationwide Disaster Risk and Exposure Assessment for Mitigation (DREAM) Program of the Department of Science and Technology and the University of the Philippines Training Center for Applied Geodesy and Photogrammetry. It was obtained through Radarsat-2 for the period February 2012–September 2013. The absolute vertical accuracy LE90 is <10 m. The twelve initially selected factors are among those recognized to influence shear stresses, slope resistance to instability, subsurface flow velocity, runoff rate and water retention (Budimir et al. 2015; Jaafari et al. 2014; Pourghasemi et al. 2013). Six landslide conditioning factors were derived from the DEM using ArcMap Spatial Analyst: (1) elevation, (2) slope angle, (3) slope aspect, (4) plan curvature, (5) profile curvature and (6) distance to drainage. Three factors were derived from scanned maps: (1) lithology, (2) soil type, and (3) distance to fault/lineament. These were sourced from 1:50,000 geologic and soil maps (Mines and Geosciences Bureau 1995; Mines and Geosciences Bureau Lands Geology Division 1996) and the Bureau of Soils and Water Management, respectively. The soil map was downloaded from the website www.philgis.org, a locally initiated GIS data clearing house. From the HRSI, the following were derived: (1) land use/land cover, (2) distance to road, and (3) NDVI (normalized difference vegetation index). Selected factors were verified in the field.

Implementation of BLR

To apply BLR, the following were implemented in SPSS v.22: (1) test for training data multicollinearity (i.e. those with tolerance (tol) < 0.1 and variance inflation factor (VIF) > 2 are indicative of multicollinearity), and conditioning index, eliminating factor variables, if any, that show redundancy with others, (2) implement BLR on training data using the forward step likelihood ratio (fstep) method; which starts with a model without explanatory variables and at each step the variable with the largest score statistic whose significance value is less than the specified value (0.05, for 95% confidence interval) is added to the model until the remaining variables all have significance values larger than 0.05; (3) selection of significant factors, using the p-value significance test; estimated coefficients having a significance value (Sig.) of less than 0.05 were regarded as significantly different from zero and thus designated as influential predictor variables; and (4) apply coefficients and BLR equations on training data.

Prior to application of BLR, the absence of serious multicollinearity must be ensured, i.e. factors are not linear functions of each other. As the degree of multicollinearity increases, the coefficient estimates become unstable and the associated standard errors can be inflated (Kleinbaum et al. 2010; Midi et al. 2010). The multicollinearity indicator tolerance (tol) is calculated to estimate the percent of variance in the factor that is unaccounted for by the other factors. Another indicator is the variance inflation factor (VIF), which is the inverse of tol (1/tol). The square root of VIF estimates how much larger the standard error is, compared with what it would be if that variable were uncorrelated with the other explanatory variables. In addition to tol and VIF, the conditioning indices in the collinearity diagnostics matrix were also examined to evaluate multicollinearity.

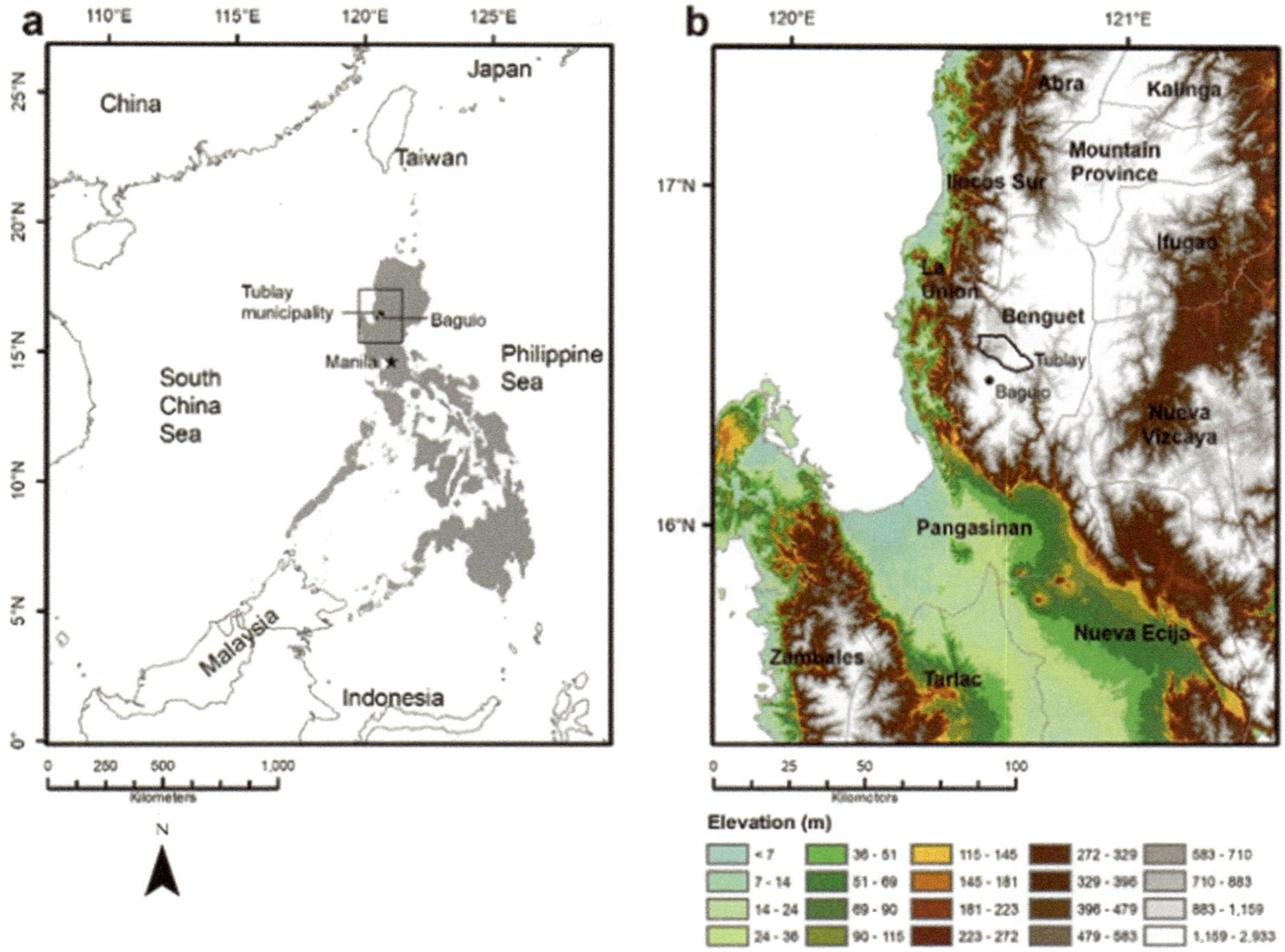

Fig. 1 Location of study area. **a** Index map, **b** Elevation map and neighboring provinces

In BLR, the dependent variable is dichotomous and is assigned a value of 1 or 0 to denote landslide presence or absence, respectively. The following equations are applied:

$$\text{P(landslide)} = \text{Y} = 1/(1 + e^{-z}) \tag{1}$$

$$z = B_o + B_1X_1 + B_2X_2 + \ldots + B_nX_n \tag{2}$$

$$\text{Logit P} = \ln(P/(1-P)) = z = B_o + B_1X_1 + \cdots + B_nX_n \tag{3}$$

where P is the estimated probability of landslide occurrence, ranging between 0 and 1, P/(1 − P) is called the odds ratio, B_o is the intercept and B_1, B_2, …, B_n are the regression coefficients that estimate the contribution of the independent variable or factors X_1, X_2, …, X_n to the variations of Y, the probability of landslides. Values equal or close to zero represents no or low likelihood of a landslide. Values equal or close to one signifies definite or high likelihood of a landslide. The coefficients are calculated using a maximum likelihood approach. A positive coefficient indicates that the factor increases the probability of a landslide assuming that the other variables in the model are held constant. If a coefficient is negative, the odds of the event occurring decreases. A coefficient of zero does not change the odds one way or the other. LR analysis relates the probability of landslide occurrence (having values from 0 to 1) to z (where $-1 < z < 0$ for higher odds of non-occurrence and $0 < z < 1$ for higher odds of occurrence). The advantage of using the fstep method in BLR is that the algorithm "chooses" factors based on their significant contribution to model improvement.

Model Evaluation

The Nagelkerke R square estimates the proportion of variance in the outcome variable that is accounted for by the factors or predictor variables entered into the BLR analysis.

A threshold of above 0.3 is regarded as an acceptable model (Kleinbaum et al. 2010).

One measure of model sufficiency is obtained by estimating the number of landslide cells within five classes of susceptibility: very high, high, moderate, low and very low. The quantile classification method was adopted. Acceptable model sufficiency is observed when the number of landslide points or cells is nil or low in the low classes, increases from low to high classes, and is highest in the high-end classes (Tsangaratos et al. 2016). Model sufficiency estimates were performed in Excel.

The training accuracy and validation accuracy were derived by generating the receiver operating characteristic and estimating the area under the curve of the receiver operating characteristic (AUROC) in SPSS v.22. The advantage of ROCs is the generation of a threshold-independent measure of predictive ability, i.e. it is unnecessary to define a threshold of modelled landslide probability below which susceptibility is recognised (Heckmann et al. 2014).

Results and Discussion

Landslide Inventory

A total of 305 rainfall-induced landslides were identified, with area ranging from 20 to 70,215 square meters, covering 2.8% (1.13 km^2) of the 40 km^2 study area. Training and validation data consisted of 244 and 61 landslide points, respectively.

Multicollinearity Test

From an initial set of 12 predictor variables, we omitted those with high VIF (>2.0) and low Tol (<0.1) values and those that, based on the SPSS collinearity diagnostics output, yielded high variance proportions (i.e. >30) in dimensions with high condition indices (Abu-Bader 2011; Field 2005; Kleinbaum et al. 2010). The omitted variables were soil type, elevation and profile curvature. Thus the nine variables considered in BLR analysis were the categorical variables slope aspect, lithology, LULC, plan curvature and the continuous variables slope angle, distance to drainage, distance to fault/lineament, distance to road and NDVI. Using these variables, the lowest tol value is 0.5 while the highest VIF and condition index yielded 1.7 and 29, respectively. Hence, the retained variables are regarded as adequate.

BLR Model

Based on the Nagelkerke R square, which is above 30%, the factors selected in the BLR analysis contribute considerably in the prediction of landslide occurrence.

The various coefficients are shown in Table 1. The most influential predictor variables are NDVI > land use/land cover > slope aspect > lithology > slope angle. The factors plan curvature, distance to drainage, distance to fault/lineament and distance to road were not selected by the fstep algorithm. The high magnitude negative coefficient of the NDVI factor confirm that with decreasing NDVI—or diminishing vegetation cover-landslide probability increases. The inverse relationship of NDVI and landslide occurrence in the study site agrees with several previous studies (Pradhan 2010; Pradhan and Lee 2010). The highest BLR coefficient associated with NDVI (−9.027) and their corresponding Wald statistic confirm that it is the most influential factor. Being a measure of surface vegetation density, NDVI can reveal the vegetated areas where slopes are likely to be stable as well as barren ground where slopes are potentially more landslide prone.

A limited number of papers have recognized that NDVI as strongly influential in landslides (Ahmed 2015; Budimir et al. 2015; Lee and Pradhan 2007). The high negative coefficient is quite unique when compared with previous BLR studies where the most influential predictor variables recognized are those, among others, indicative of geomorphology (e.g. elevation, slope angle, river proximity), geology (proximity to fault) and human influence (e.g. road proximity, distance to residential residences) (Ayalew and Yamagishi 2005; Mathew et al. 2007; Tsangaratos and Ilia 2016). The low NDVI areas coincide with landslide areas, several of which are reactivated. Several are proximal to mining sites, such as the central region of the study area. Some are at higher elevations, such as in the eastern and southeastern part of the area. In the study site, the presence of comparatively dense vegetation areas does not offer full protection against landslides because most of these high NDVI areas are dominated by shallow rooted vegetation, like pine, alnus and other fruit trees. Most trees—regarded as secondary growth—are likely to yet develop deep and extensive root systems—before their landslide mitigating potential may become evident. The results support the strong potential of revegetation of precarious slopes in complementing ongoing slope stabilization and rehabilitation measures, as is already being implemented by local government officials (A. Lawangen, pers. comm. 2014). The influence of

Table 1 Factors and coefficients of BLR equation. (*Note* Reference category is indicated by dash '-')

Landslide conditioning factor	Coef
Slope angle	0.062
NDVI	−9.027
Slope aspect	
North	–
Northeast	0.880
East	1.395
Southeast	2.059
South	1.781
Southwest	1.895
West	1.009
Northwest	−0.530
Land use/land cover	
Dense vegetation	−0.291
Sparse vegetation	–
Bare ground	−0.332
Drainage	−2.407
Built up	−3.879
Lithology	
Balatoc dacite	−1.800
Black mountain quartz diorite porphyry	−0.468
Cordillera central diorite complex	0.434
Zigzag Formation	−1.906
Pugo formation	–

NDVI appears understated in the literature as only a select number use it in LSA.

The other significant predictor variables are land use/land cover, where landslides were observed in sparsely vegetated and bare ground. Another is slope aspect, known to affect evaporation and moisture retention. The influential directions are southwest, south and southeast. The other is lithology, most landslides being observed in the Central Cordillera diorite complex. As in many published literature, slope angle also plays a role.

Landslide Susceptibility

High to very high landslide susceptibility zones are evident in the center, south and east. A low susceptibility band is evident in the west, where slopes were observed to be gentler (Fig. 2).

Model Performance

High to very high landslide susceptibility zones are evident in the center, south and east. A low susceptibility band is evident in the west and north, where slopes were observed to be gentler (Fig. 2).

Using training data, 80%, 11%, 5%, 2% and 2% of the landslide points were associated with the very high, high, moderate, low and very low susceptibility classes, respectively. Using validation data, the proportions were 82%, 10%, 5%, 3% and 0%, respectively (Fig. 3). The decreasing trend in landslides in the selected categories affirms a good model outcome. The high to very high susceptibility areas in the training and validation data covered areas of 0.82 km^2 and 0.21 km^2, respectively.

Training and validation accuracy, amounting to 91% and 86%, respectively, also confirms favourable performance.

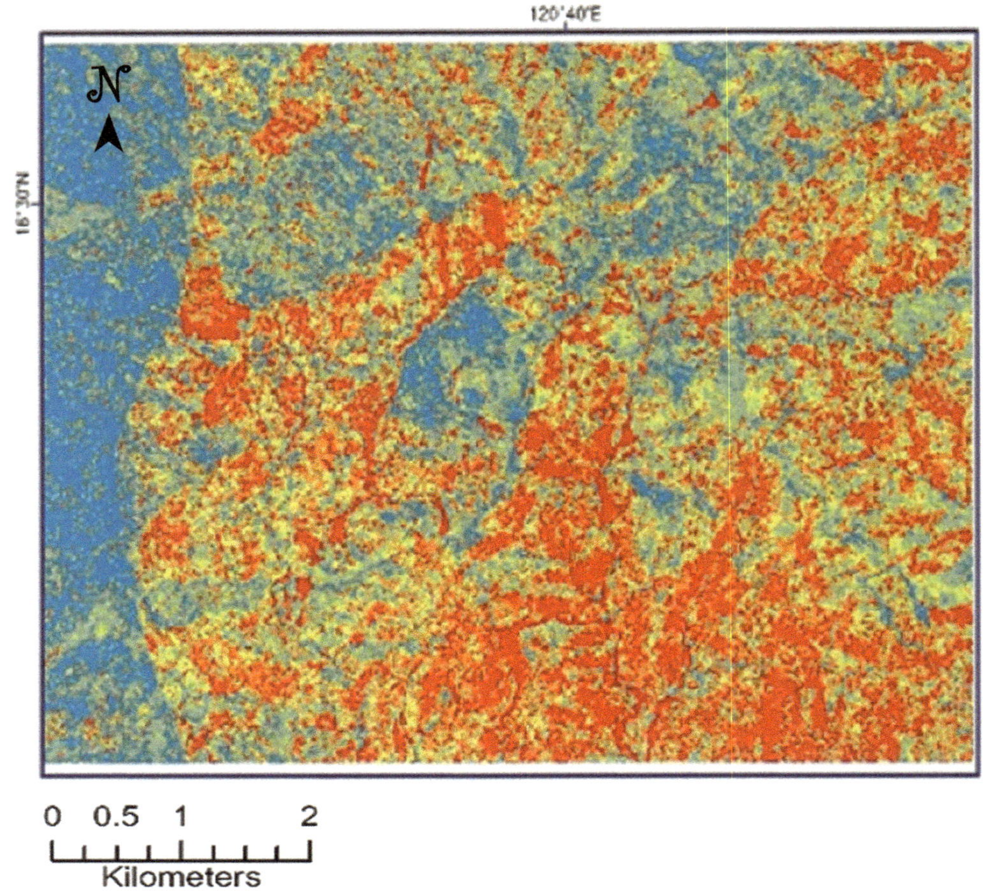

Fig. 2 Landslide susceptibility map. Very high–red, high-orange, moderate–yellow, low-green, very low–blue

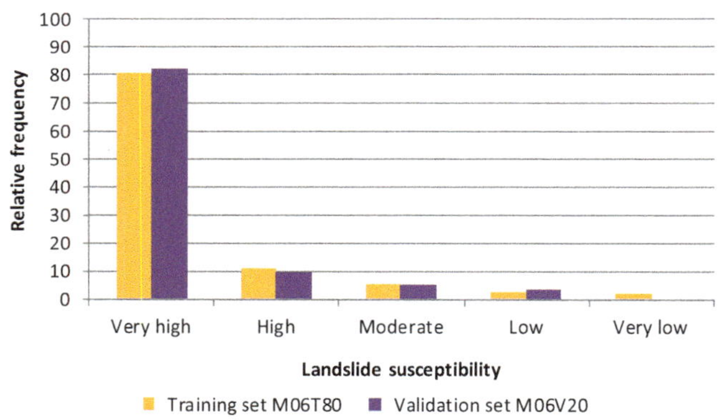

Fig. 3 BLR Model sufficiency for study site

Conclusion

Three hundred and five landslide points were randomly divided into training and validation data using 80%-20% proportion. The binary logistic regression was applied in order to generate a sound landslide susceptibility map of the area. Of nine landslide conditioning variables used in modelling, only five were regarded as major contributors to predicting landslide occurrence. In decreasing order, these are NDVI > land use/land cover > slope aspect > lithology > slope angle. The strong influence of NDVI affirms its importance in modelling landslide susceptibility by the BLR method. It supports the strong potential of revegetation in slope stabilization efforts.

The BLR method performed well in modelling landslide susceptibility in the site. Ninety-two per cent of the validation data coincided with the high and very high landslide susceptibility categories. Training and validation accuracy were 91% and 86%, respectively. The method utilized here holds promise in modelling similar landslide prone areas in the region.

Acknowledgements The authors would like to thank the University of New England and the University of the Philippines Baguio for supporting this research which was part of the dissertation of DNJ at UNE. We are grateful also to the government officials and residents of the municipality of Tublay for the fieldwork support extended.

References

Abu-Bader SH (2011) Advanced and multivariate statistical methods for social science research. Lyceum Books Inc., Chicago, Illinois, USA

Ahmed B (2015) Landslide susceptibility modelling applying user-defined weighting and data-driven statistical techniques in Cox's Bazar Municipality, Bangladesh. Nat Hazards 79(3):1707–1737

Ayalew L, Yamagishi H (2005) The application of GIS-based logistic regression for landslide susceptibility mapping in the Kakuda-Yahiko Mountains Central Japan. Geomorphology 65(1):15–31

Balce GR, Encina RY, Momongan A, Lara E (1980) Geology of the Baguio district and its implication on the tectonic development of the Luzon Central Cordellera. Geol Paleont Southeast Asia 21:265–287

Brenning A (2005) Spatial prediction models for landslide hazards: review, comparison and evaluation. Nat Hazards Earth Syst Sci 5 (6):853–862

Budimir M, Atkinson P, Lewis H (2015) A systematic review of landslide probability mapping using logistic regression. Landslides 12(3):419–436

David S Jr, Stephan J, Popoff M (1999) The Baguio Basin: Investigation of a transtensional relay basin using digital elevation models and drainage network analysis. J Geol Soc Philipp 54:35–47

Delias D, Daly P (2016) The challenges of disaster risk reduction in rapidly expanding urban environments: Baguio City, Philippines since the 1990 Luzon Earthquake. Rebuilding Asia following natural disasters: approaches to reconstruction in the Asia-Pacific Region, 57

Field A (2005) Discovering statistics using SPSS, 2nd edn. Sage Publications Ltd, London UK

Heckmann T, Gegg K, Gegg A, Becht M (2014) Sample size matters: investigating the effect of sample size on a logistic regression susceptibility model for debris flows. Nat Hazards and Earth Syst Sci 14(2):259

Jaafari A, Najafi A, Pourghasemi H, Rezaeian J, Sattarian A (2014) GIS-based frequency ratio and index of entropy models for landslide susceptibility assessment in the Caspian forest, northern Iran. Int J Environ Sci Technol 11(4):909–926

Kirschbaum D, Stanley T, Zhou Y (2015) Spatial and temporal analysis of a global landslide catalog. Geomorphology 249:4–15

Kleinbaum DG, Klein M, Pryor E (2010) Logistic regression: a self-learning text. Springer, New York, USA

Le L, Lin Q, Wang Y (2017) Landslide susceptibility mapping on a global scale using the method of logistic regression. Nat Hazards Earth Syst Sci 17(8):1411

Lee S (2005) Application of logistic regression model and its validation for landslide susceptibility mapping using GIS and remote sensing data. Int J Remote Sens 26(7):1477–1491

Lee S, Pradhan B (2007) Landslide hazard mapping at Selangor, Malaysia using frequency ratio and logistic regression models. Landslides 4(1):33–41

Lee S, Sambath T (2006) Landslide susceptibility mapping in the Damrei Romel area, Cambodia using frequency ratio and logistic regression models. Environ Geol 50(6):847–855

Mathew J, Jha V, Rawat G (2007) Application of binary logistic regression analysis and its validation for landslide susceptibility mapping in part of Garhwal Himalaya, India. Int J Remote Sens 28 (10):2257–2275

Midi H, Sarkar S, Rana S (2010) Collinearity diagnostics of binary logistic regression model. J Interdiscip Math 13(3):253–267

Mines and Geosciences Bureau (1995) Geologic map of the Baguio quadrangle

Mines and Geosciences Bureau Lands Geology Division (1996) Geologic map of Atok quadrangle

Municipality of Tublay Benguet Housing and Land Use Regulatory Board (2014) Comprehensive land use plan Tublay, Benguet Planning Period 2014–2023

Nakasu T (2011) The exacerbation of human suffering and disaster response caused by tropical storm ondoy and typhoon pepeng disaster-cases of NCR and Baguio City

Nandi A, Shakoor A (2010) A GIS-based landslide susceptibility evaluation using bivariate and multivariate statistical analyses. Eng Geol 110(1–2):11–20

National Disaster Coordinating Council (2009) Situation report no. 40 on Tropical Storm Ondoy (Ketsana) and Typhoon Pepeng (Parma)

Nefeslioglu HA, Gokceoglu C, Sonmez H (2008) An assessment on the use of logistic regression and artificial neural networks with different sampling strategies for the preparation of landslide susceptibility maps. Eng Geol 97(3–4):171–191

Nolasco-Javier D, Kumar L (2018) Deriving the rainfall threshold for shallow landslide early warning during tropical cyclones: a case study in northern Philippines. Nat Hazards 90(2):921–941

Peña RE (1998) Further notes on the stratigraphy of the Baguio district. J Geol Soc Philipp 53(3–4):141–157

Peñaflor ES, Calalo AC, Taule SL (1987) Preliminary report on the detailed geological-geochemical survey Barangay Ambassador. Tublay, Benguet

Petley D (2012) Global patterns of loss of life from landslides. Geology 40(10):927–930

Pourghasemi H, Pradhan B, Gokceoglu C, Moezzi KD (2013) A comparative assessment of prediction capabilities of Dempster-Shafer and weights-of-evidence models in landslide susceptibility mapping using GIS. Geomatics Nat Hazards Risk 4 (2):93–118

Pradhan B (2010) Landslide susceptibility mapping of a catchment area using frequency ratio, fuzzy logic and multivariate logistic regression approaches. J Indian Soc Remote Sens 38(2):301–320

Pradhan B, Lee S (2010) Delineation of landslide hazard areas on Penang Island, Malaysia, by using frequency ratio, logistic regression, and artificial neural network models. Environ Earth Sci 60 (5):1037–1054

Tsangaratos P, Ilia I (2016) Comparison of a logistic regression and Naïve Bayes classifier in landslide susceptibility assessments: the influence of models complexity and training dataset size. CATENA 145:164–179

Tsangaratos P, Ilia I, Hong H, Chen W, Xu C (2016) Applying information theory and GIS-based quantitative methods to produce landslide susceptibility maps in Nancheng County, China. Landslides, pp 1–21

Yesilnacar E, Topal T (2005) Landslide susceptibility mapping: a comparison of logistic regression and neural networks methods in a medium scale study, Hendek region (Turkey). Eng Geol 79(3–4):251–266

Yumul GP, Dimalanta CB, Tam TA, Ramos EGL (2008) Baguio mineral district: an oceanic arc witness to the geological evolution of northern Luzon, Philippines. Island Arc 17(4):432–442

Yumul GP Jr, Dimalanta CB, Servando NT, Cruz NA (2013) Abnormal weather events in 2009, increased precipitation and disastrous impacts in the Philippines. Climatic Change 118(3–4):715–727

Landslide Hazard Mapping of Penang Island Malaysia Based on Multilayer Perceptron Approach

2020 Kyoto Japan

Ilyas A. Huqqani, Lea Tien Tay, and Junita Mohamad-Saleh

Abstract

Landslide is one of the natural disasters in Malaysia. It causes property damages, infrastructure destruction, injuries and causalities. Landslide hazard mapping is one of the efforts to identify the landslide prone areas with the purpose of reducing the risk of landslide hazards. In this paper, landslide hazard map of the study area, Penang Island Malaysia, is produced using artificial neural network model. Penang Island dataset is collected and its data samples are used to train the artificial neural networks. This study deals with the hidden layer of ANNs. The number of hidden neurons in hidden layer is one of the important parameters of the neural network. Although the hidden layer is not interacted with the external environment but it has tremendous influence on the final output. The different number of hidden neurons of artificial neural networks applied on landslide data produce landslide hazard maps with distinct accuracies and computation time. Finally, Receiver of Characteristics curve is applied on whole Penang Island dataset to validate the accuracy and effectiveness of trained artificial neural model.

Keywords

Landslide hazard map • Artificial neural network • Multilayer perceptron • Receiver of characteristics

I. A. Huqqani · L. T. Tay (✉) · J. Mohamad-Saleh
School of Electrical and Electronic Engineering, Universiti Sains Malaysia, 14300 Nibong Tebal, Penang, Malaysia
e-mail: tay@usm.my

I. A. Huqqani
e-mail: ilyashuqqani@gmail.com

J. Mohamad-Saleh
e-mail: jms@usm.my

Introduction

Landslides occur in the form of serious hazards and evolution of landscapes in many areas around the world (Guzzetti etc. 2012). Numerous causalities, damages to buildings, properties etc. are caused by landslide occurrence. It also results in the indirect cost for society, loss of productivity, reduction of real estate value, loss of tax revenue and other induced economic effects (Scaioni etc. 2014). Landslide is a geo-hazard phenomenon which involves movement of debris or soil down a slope or a mass of rock considered (Cruden 1991). Research works have been carried out to study various landslide hazardous areas of different regions in the world. The main emphasizes of studies are inventory of landslides, impact of geological, topographical, hydrological and anthropogenic factors on the occurrences of landslides and the role of several triggering factors like precipitation and other climatic changes. Landslide occurrences depend on several causative factors that are divided into various categories like geomorphology, geology, metrological soil, land cover and hydrologic conditions (Varnes 1984; Hutchinson 1995).

Landslides occur frequently in Malaysia due to heavy rainfall throughout the year. Most of the landslides happened in annual monsoon i.e. Southwest monsoon (May–September) and Northeast monsoon (November–March). Many damages have been done due to landslides in Malaysia from 2000 to 2009 (Pradan and Lee 2010; Murakami etc. 2014). A lot of research and methods have been proposed to reduce and minimize the damages caused by landslides (Quoc etc. 2018; Ya'acob etc. 2019). This is accomplished by predicting the risky areas and act as a warning so that proper measures can be taken (Pradan and Lee 2010; Lee and Talib 2005; Tay etc. 2014). It is not possible to predict a landslide occurrence in time and space, therefore a region of landslide is considered and is classified into different categories of potential hazards (Varnes 1984).

F. Guzzetti et al. (eds.), *Understanding and Reducing Landslide Disaster Risk*, ICL Contribution to Landslide Disaster Risk Reduction,
https://doi.org/10.1007/978-3-030-60227-7_21

For evaluating landslide hazards, Geographical Information System (GIS) and various evaluation techniques have been applied. Advanced computational tools such as artificial neural network (ANN) and decision tree have been applied successfully in generating landslide hazard maps (Alkhasawneh etc. 2013a, b, 2014a, b). ANN is used in many applications of natural sciences such as speech recognition, human face recognition, satellite image classification, shape and texture recognition (Kawabata and Bandibas 2009). The main advantage of ANN is that it can handle data at any measurement scale ranging from nominal, ordinal to linear and ratio, and any form of data distribution. In addition, it also deals with qualitative variables that are widely used in integrated analysis of spatial data from multiple sources for prediction and classification (Kawabata and Bandibas 2009).

The purpose of this study is to generate the landslide hazard map of the Penang Island, Malaysia using an ANN model with different number of hidden neurons. The assessment is carried out using accuracy with confusion matrix and Receiver of Characteristics (ROC) curve.

Study Area

Landsides frequently occurred in many regions of Malaysia. One of the most affected area is Penang Island, Malaysia, and it is chosen as the study area in this paper. Penang is located on the North West of Malaysia Peninsula. It is surrounded with the state of Kedah (North and East), the state of Perak (South) and the Strait of Malacca and Sumatra, Indonesia (West). Penang consists of Island as well as a coastal strip on the mainland which is also named as Province Wellesley. In this research, the study area of landslide hazard analysis is the island of Penang. The total area of Penang Island is 285 km^2. It is positioned at $5^\circ 15'$ N–$5^\circ 30'$ N latitudes and $100^\circ 10'$ E–$100^\circ 20'$ E longitudes. The Penang Island consists of plants, forest, greenland, swamp and urban areas. The range of slope varies from 0° to 87° and elevation of the terrain above the sea level is 0 to 820 m. The temperature of Penang Island ranges from 29 to 32 °C. The average amount of precipitation lies between 2254 and 2903 mm annually. The major types of vegetation in Penang Island are forest and fruit plantation. This island is also affected by fault lines. These lines exist in the center of the island from North to South. Figure 1 shows the map of Penang Island.

Fig. 1 Penang Island (Google map)

Data Set

Remote sensing method and Geographical Information System (GIS) are used to collect data on geological database of Penang Island. Numerous images of Penang Island are acquired from several governments sectors in Malaysia such as Department of Meteorological Malaysia, Department of Irrigation and Drainage (DID), Department of Agriculture (DOA), Department of Minerals and Geo-Science (JMG) etc. The occurrence points of landslide in Penang Island are also accumulated and converted into spatial datasets. The dataset with resolution of 5 m is used in this study. The hilly topography of the island is in the central of the island and considered as the most occurred landslide area. The landslides mainly involve shallow rotational debris slides, debris flows and rock falls.

The relationship of causative factors with slope failure and identification of instability should be done to know the main cause of landsliding (Guzzetti etc. 1999). The datasets of landslide causative parameters are gathered and transformed into spatial dataset. There are 12 landslide causative factors used to train the ANN model to classify the hazard level of landslide. By using topographic database, a digital elevation model (DEM) with a resolution of 5 m is developed. Elevation is obtained from DEM and then further used to calculate the slope gradient angle, slope aspect and slope curvature. The road map and drainage map are used to find the distance from road and drainage respectively. The buffer zones for line features should be prepared as 5 m (Van Westen etc. 2003; Van Westen 1993). The distances from fault lines are divided into 100 m segregations. The soil texture of the island is also acquired, and it consists of six types of soil texture. Geology map covers six types of different granites of Penang Island. The natural triggering factors of landslide in this island is precipitation. Due to heavy rainfall, the soil dampens, debris and rocks are washed away. The inverse weight distance interpolation method is used to produce the precipitation map because of limited rain measurement stations in the island. The landslide causative factors maps are shown in Fig. 2.

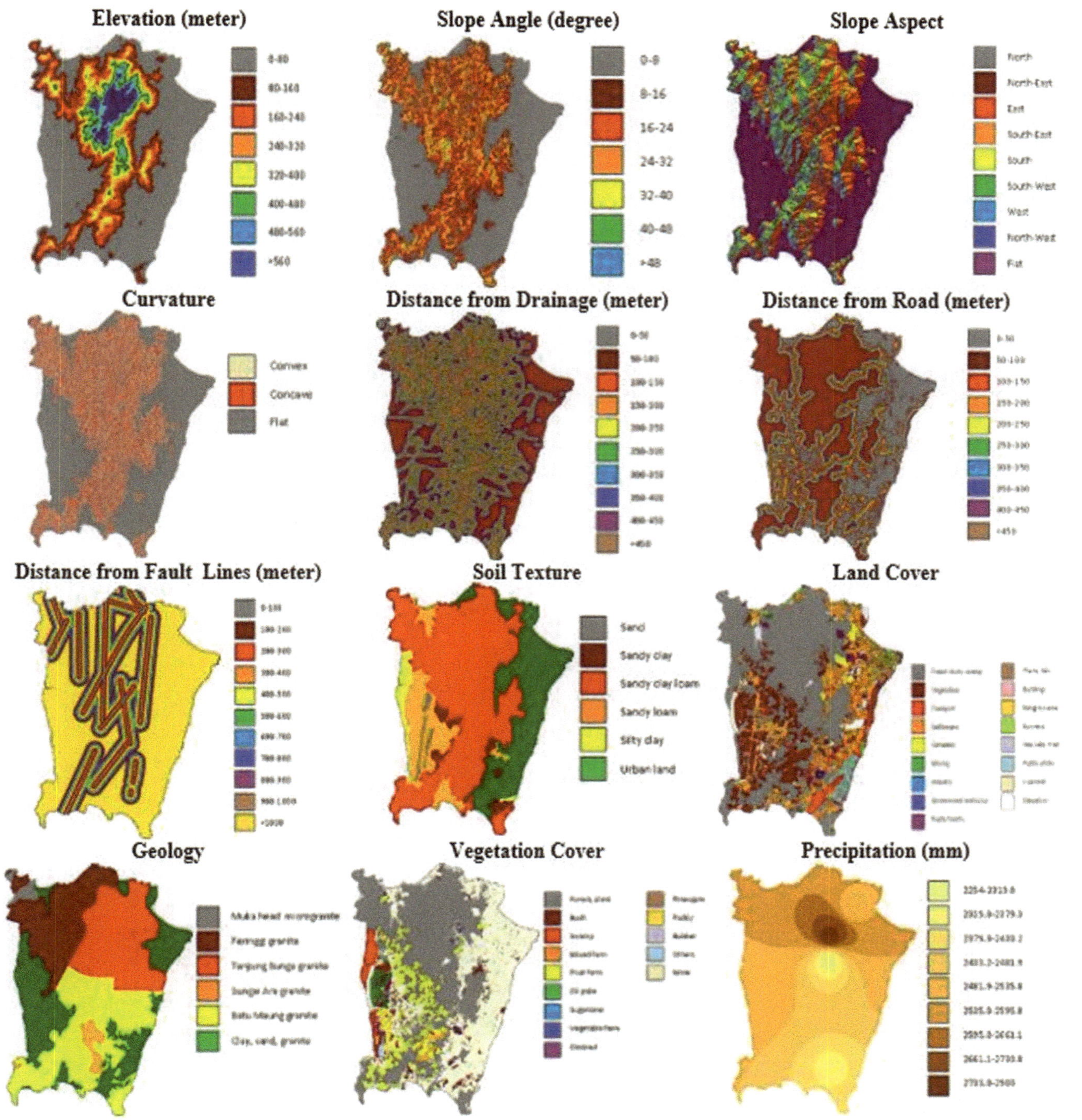

Fig. 2 The maps of landslide causative factors

Methodology

An ANN is important and powerful computational model which is inspired by the biological neural networks. ANN has tendency to adapt the changes in hidden layer that propagates the information in training phase. It is used in many applications due to its strong capability of nonlinear mapping, high accuracy for learning and good robustness (Sheela and Deepa 2013). ANN is divided in two categories, i.e. feed forward and feedback. The most popular back propagation model is the feedforward and it is used in the learning of the neural network especially the multiplayer perceptron (MLP). MLP is one of the widely used tool for classification and prediction problems. The structure of MLP is based on three adjacent layers: input, hidden and output layers (Haykin 1998) as illustrate in Fig. 3. Each layer consists of independent processing elements called neurons. These neurons are connected to other layer's neurons through the weight. These weights are altered during learning phase and MLP develops a mapping function between the inputs and outputs (James and Garrett 1994). The backpropagation learning algorithm is used to adjust the weights between the neurons.

In learning phase, the external environment feeds the data to input neurons denoted by $x_1, x_2, \ldots, x_N$ and forward them to hidden neurons through weights denoted by $w_{11}, w_{21}, \ldots, w_{ij}$. The results obtained from hidden neurons are mapped onto the threshold function of each neuron and final result is then produced. These values act as input values to all neurons in output layer. The final output values are obtained at each output neuron denoted by $y_1, y_2, \ldots, y_k$. The error value is calculated by taking difference between actual input and output from the MLP. The overall training phase can be terminated when an acceptable small error is achieved. After completing learning phase, MLP can produce output solutions of any set of input data (Paola and Schowengerdt 1995).

The landslide occurrence is denoted by value 1 and landslide non-occurrence is denoted by value 0. The total landslides occurrences in Penang Island is 68,786. For learning phase of MLP, this data is divided into two datasets, i.e. training dataset and validation dataset. Figure 4 describes the complete flow of landslide data distribution. The simulation and implementation of landslide dataset are performed on MATLAB® by using built-in function of Neural Network Toolbox. The whole Penang Island data is then applied on the trained MLP to produce the landslide hazard map. Besides, the accuracies of confusion matrix and area under the curve (AUC) of Receiver of Characteristics (ROC) are computed for the study area, Penang Island.

Result and Discussions

In this study, the accuracies and average computation time of training and validation data achieved by MLP during the learning phase are given in Table 1. It is observed that as number of hidden neurons increases, the accuracies of training and validation are also increased. However, it is observed that the accuracy increment reduces as number of hidden neurons increases. The average computational time increases as the number of hidden neurons increases due to rise of computation required.

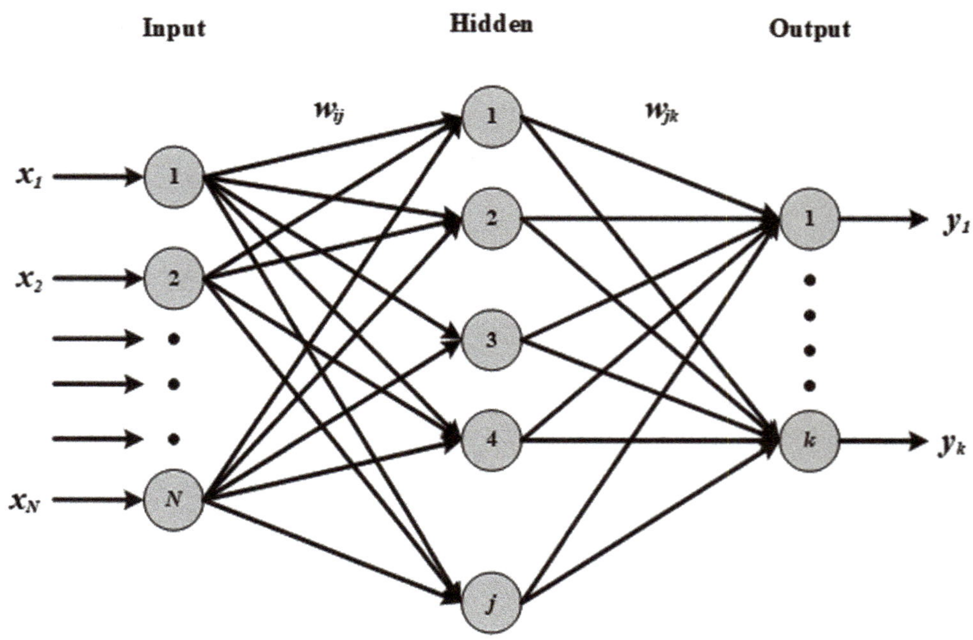

Fig. 3 Structure of multilayer perceptron (MLP)

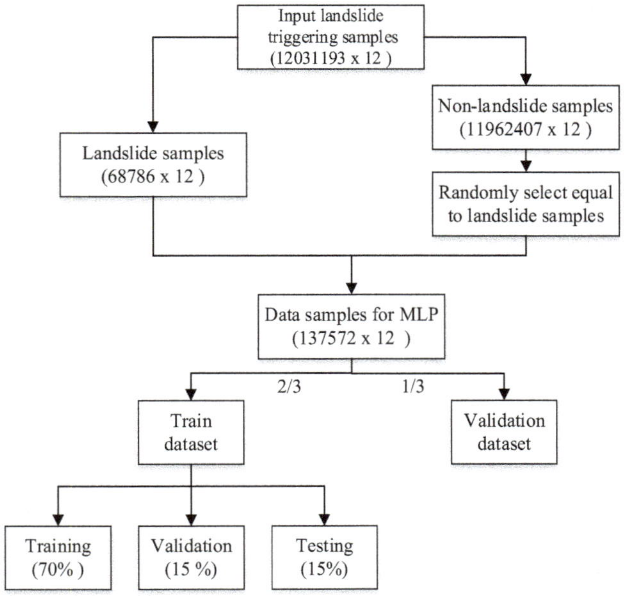

Fig. 4 Data distribution of MLP learning

Table 1 Accuracies of training and validation and average computational time of MLP during learning phase

Hidden neurons	Accuracy		Avg. time (sec)
	Training data	Validation data	
	0.8132	0.8097	11.02
20	0.8562	0.8310	35.66
30	0.8875	0.8560	74.46
40	0.8952	0.8740	112.31
50	0.9052	0.8800	148.89
60	0.9156	0.8960	238.35
70	0.9200	0.9020	288.15
80	0.9257	0.9079	418.20
90	0.9270	0.9122	537.12
100	0.9316	0.9159	892.74

The accuracy of landslide hazard mapping is verified by using Receiver Operating Characteristic (ROC) and confusion matrix method. There are two assumptions taken to validate the maps. First, the landslides are linked to spatial data like topography height and slope gradient. Second, the triggering factors such as rainfall affects the future landslide (Chung and Fabbri 1999). The first step in verification by ROC method is to arrange the landslide hazard index (LHI) from ANN in a descending sequence. These indexes are then segregated into one hundred (100) subclasses on y-axis with 1% separations on x-axis (Pradhan and Lee

Table 2 Overall accuracies of whole Penang Island

Hidden neurons	Accuracy	
	Confusion matrix	ROC
10	0.6744	0.8538
20	0.7425	0.8830
30	0.7813	0.9003
40	0.7985	0.9065
50	0.8131	0.9116
60	0.8256	0.9178
70	0.8403	0.9232
80	0.8423	**0.9266**
90	0.8393	0.9249
100	**0.8507**	0.9244

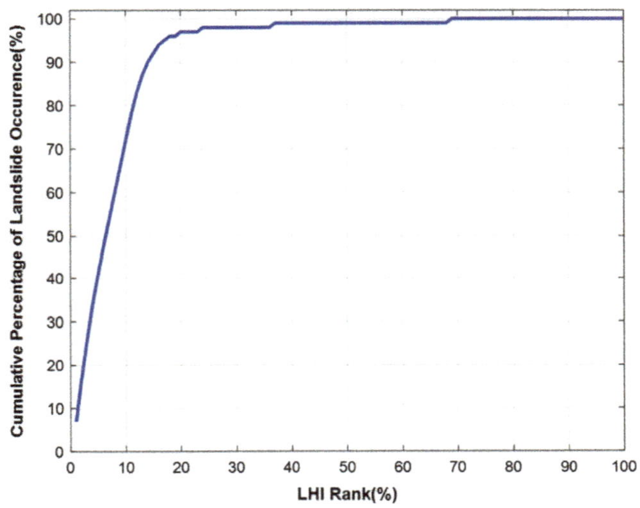

Fig. 5 ROC curves of Whole Penang Island

2010). The curve illustrates the accuracy of the approaches and factors predict the upcoming landslides. The accuracy of the approach is obtained by area under the curve (Chung and Fabbri 1999; Begueria 2006). Table 2 depicts the accuracies of confusion matrix and ROC of landslide hazard map of Penang Island using ANN. Accuracy of landslide hazard maps of Penang Island produced by ANN range from 67.44 to 85.07% assessed using confusion matrix when the number of hidden neurons increases from 10 to 100. The best accuracy of the confusion matrix is 85.07% with 100 hidden neurons. AUC of ROC increases from 85.38 to 92.66% when the number of hidden neurons increases from 10 to 80. It then drops slightly to 92.49% and 92.44% for hidden neuron of 90 and 100, respectively. The best accuracy based on ROC curve shows accuracy of 92.66% with 80 hidden neurons and its ROC curves is shown in Fig. 5.

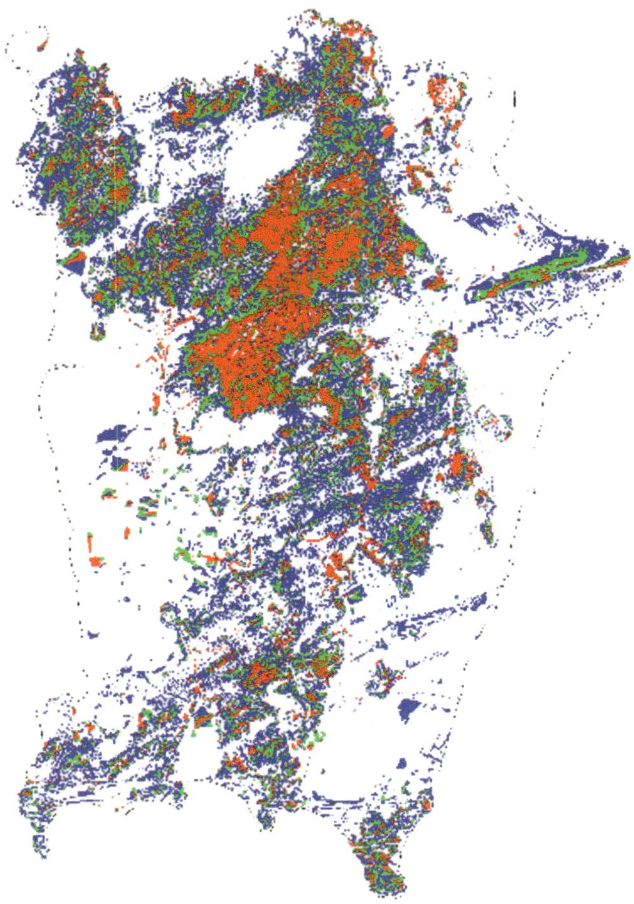

Fig. 6 Landslide hazard maps produced using ANN with 80 hidden neurons

Landslide hazard maps are produced by LHI obtained from trained ANN of whole Penang Island dataset. LHI is divided in four major categories. First category is the 'highly hazardous area' which is most prone to landslide area and 10% values of the maximum of LHI (90–100%). The second category is named as 'hazardous area'. The value of this category is of following 10% of LHI (80–90%). Third category of 'Moderately hazardous area' corresponds next 20% of LHI (60–80%) and finally 'Non-hazardous area' category is for the remaining 60% of LHI (0–60%).

Landslide hazard map of Penang Island generated using ANN technique is shown in Fig. 6. In the figure, red color indicates for highly/seriously hazardous area, green color for hazardous areas, blue color for moderately hazardous areas and white color for non-hazardous areas.

Conclusions

Landslide is one of the most hazardous disasters occurred around the globe. The assessment of landslide hazards and their associated risks are studied by worldwide research institutions and governments. Recently, a lot of construction and development has carried out in the hilly regions due to the limited flat terrain such as in Penang Island. Therefore, landslide hazard map is important to alert organization and people about the occurrence of landslides. In this study, performance of landslide hazard mapping of Penang Island using artificial neural network approach is assessed. The ROC of the generated landslide hazard map shows good accuracy of 85.38–92.66% for different number of hidden neurons. ANN proves to perform better compared to the conventional probabilistic methods and thus it has huge potential in producing more accurate hazard maps which are very useful in the planning of future urban developments.

Acknowledgements The author would like to thank Malaysia Education Ministry/Kementerian Pendidikan Malaysia (KPM) for providing the financial support under research grant (FRGS—Geran Penyelidikan Fundamental 203/PELECT/6071390) in this project.

References

Alkhasawneh MS, Ngah U, Tay LT, Mat Isa NA, and Al-Batah M (2013a) Determination of Important topographic factors for landslide mapping analysis using MLP network. Sci World J 2013:415023

Alkhasawneh MS, Ngah U, Tay LT, and Mat Isa NA (2013b) Determination of importance for comprehensive topographic factors on landslide hazard mapping using artificial neural network. Environ Earth Sci 72

Alkhasawneh MS, Ngah U, Tay LT, Mat Isa NA, Al-Batah M (2014) Modeling and Testing Landslide Hazard Using Decision Tree," J. Appl. Math., vol. 2014

Alkhasawneh MS, Tay LT, Ngah U, Al-Batah M, and Mat Isa NA (2014) Intelligent landslide system based on discriminant analysis and cascade-forward back-propagation network. Arab J Sci Eng 39

Beguería S (2006) Validation and evaluation of predictive models in Hazard assessment and risk management. Nat Hazards 37(3):315–329

Chung CJF, Fabbri A (1999) Probabilistic prediction models for landslide hazard mapping. Photogramm Eng Remote Sens 65:1389–1399

Cruden DM (1991) A simple definition of a landslide. Bull Int Assoc Eng Geol—Bull l'Assoc Int. Géologie l'Ingénieur 43(1):27–29

Guzzetti F, Mondini AC, Cardinali M, Fiorucci F, Santangelo M, Chang KT (2012) Landslide inventory maps: new tools for an old problem. Earth-Sci Rev 112(1):42–66

Guzzetti F, Carrara A, Cardinali M, Reichenbach P (1999) Landslide hazard evaluation: a review of current techniques and their application in a multi-scale study, Central Italy. Geomorphology 31(1):181–216

Haykin S (1998) Neural networks: a comprehensive foundation, 2nd edn. Upper Saddle River. Prentice Hall PTR, NJ, USA

Hutchinson JN (1995) Landslide hazard assessment. Keynote paper. In: Bell DH (ed) Landslides, in 6th international symposium on landslides, pp 1805–1841

James J, Garrett H (1994) Where and why artificial neural networks are applicable in civil engineering. J Comput Civ Eng 8(2):129–130

Kawabata D, Bandibas J (2009) Landslide susceptibility mapping using geological data, a DEM from ASTER images and an artificial neural network (ANN). Geomorphology 113(1):97–109

Lee S, Talib JA (2005) Probabilistic landslide susceptibility and factor effect analysis. Environ Geol 47:982–990

Murakami S et al (2014) Landslides Hazard map in Malay peninsula by using historical landslide database and related information. J Civ Eng Res 4(3A):54–58

Paola JD, Schowengerdt RA (1995) A review and analysis of backpropagation neural networks for classification of remotely-sensed multi-spectral imagery. Int J Remote Sens 16(16):3033–3058

Pradhan B, Lee S (2010) Delineation of landslide hazard areas on Penang Island, Malaysia, by using frequency ratio, logistic regression, and artificial neural network models. Environ Earth Sci 60 (5):1037–1054

Quoc A, Tran DT, Dinh C, Tien BD (2018) Flexible configuration of wireless sensor network for monitoring of rainfall-induced landslide, Indones. J Electr Eng Comput Sci 12:1030–1036

Scaioni M, Longoni L, Melillo V, Papini M (2014) Remote sensing for landslide investigations: an overview of recent achievements and perspectives. Remote Sens 6(10):9600–9652

Sheela K, Deepa SN (2013) Review on methods to fix number of hidden neurons in neural networks. Math Probl Eng 2013

Tay LT, Alkhasawneh MS, Lateh H, Hossain MK (2014) Landslide hazard mapping of Penang Island using poisson distribution with dominant factors. J Civ Eng Res 4:72–77

Van Westen CJ, Rengers N, Soeters R (2003) Use of geomorphological information in indirect landslide susceptibility assessment. Nat Hazards 30(3):399–419

Van Westen CJ (1993) Application of geographic information systems to landslide hazard zonation.

Varnes DJ (1984) Landslide hazard zonation: a review of principles and practice, 3:63. Natural Hazards.

Ya'acob N, Tajudin N, Azize A (2019) Rainfall-landslide early warning system (RLEWS) using TRMM precipitation estimates, Indones. J Electr Eng Comput Sci 13:1259–1266

Landslide Susceptibility Mapping Based on the Deep Belief Network: A Case Study in Sichuan Province, China

2020 Kyoto Japan

Wei-Dong Wang, Zhuolei He, Zheng Han, and Yange Li

Abstract

Landslides dataset of Sichuan Province in China, containing 1551 historical individual landslides, is a result of two teams' effort in the past few years, based on which landslide susceptibility can be mapped. Considering complex internal relations among the triggering factors, logistic regression (LR) and shallow neural networks, such as back propagation neural network (BPNN), are often limited. In this paper, we make a straightforward development that the deep belief network (DBN) based on deep learning technology is introduced to map the regional landslide susceptibility. Seven key factors with respect to geomorphology, geology, and hydrology are considered, and a DBN model containing three pre-trained layers of Restricted Boltzmann Machines (RBM) by stochastic gradient descent (SGD) method is configured to obtain the landslide susceptibility. In the receive operator characteristic (ROC) analysis, comparing DBN with LR and BPNN shows that DBN has a better prediction precision, with lower false alarm rate and fake alarm rate. This research will contribute to a better-performance model for regional-scale landslide susceptibility mapping, in particular at the area where triggering factors show complex relation and relative independence.

Keywords

Landslide • Susceptibility mapping • Deep learning • Deep belief network • Sichuan area

Introduction

Landslides are the most widespread geological hazard worldwide (Alexander 2004). China for instance, landslides in the past years were accounted to almost 76% of the annual geological disasters (Ministry of Land and Resources 2016). This kind of landscape forming process often transports large volume of deposits, disrupting traffic, blocking rivers, burying villages, and therefore posing serious risks to humans and local society (Wang et al. 2012; Han et al. 2015a).

Given the severe risk arisen by landslide hazards, the issue of landslide prevention and mitigation has received considerable attention. A key aspect regards to landslide susceptibility (LS) mapping, depending on which rational countermeasures against landslide can be planned. LS is defined as the likelihood of a landslide occurring in an area on the basis of a comprehensive analysis of the relevant geological and topographic conditions (Brabb 1984; Reichenbach et al. 2018). Existing research has recognized groups of triggering factors of landslide. Remarkable studies have been done by Kanungo et al. (2011), Raja et al. (2017), Chen et al. (2017), Shirzadi et al. (2017), in which factors found to be influencing, i.e., topography, geological setting, rainfall condition, etc., have been explored. Based on the above studies, index systems for LS mapping are now well established. Extensive studies on this aspect laid a solid foundation for the purpose of LS mapping.

Assessment model is another key issue for LS mapping. Studies over the past decades have provided a variety of alternative models, which in general can be divided into two categories, i.e., qualitative models and quantitative models.

W.-D. Wang · Z. He · Z. Han (✉) · Y. Li
School of Civil Engineering, Central South University, Shaoshan South Road 22, Changsha, 410075, China
e-mail: zheng_han@csu.edu.cn

W.-D. Wang
e-mail: 701139@csu.edu.cn

Z. He
e-mail: babyba@vip.qq.com

Y. Li
e-mail: liyange@csu.edu.cn

F. Guzzetti et al. (eds.), *Understanding and Reducing Landslide Disaster Risk*,
ICL Contribution to Landslide Disaster Risk Reduction,
https://doi.org/10.1007/978-3-030-60227-7_22

Qualitative model is knowledge-driven that rather based on expert experience. Remarkable studies include geomorphological analysis methods (Listo and Carvalho 2012), the combination or overlay of index maps (Avtar et al. 2011) and analytic hierarchy processes (Abedini and Tulabi 2018), etc. These models depend on historical experience and expert expertise to a great extent, therefore, are often limited owing to uncertainty and subjectivity (Polykretis and Chalkias 2018).

The past decades have seen the rapid development of Geographic Information System (GIS) technology in many fields, inspired by which up-to-date studies use quantitative methods for LS mapping. Quantitative methods are a means of data-driven solution based on numerical expressions, such as logistic regression (LR) analysis (Meten et al. 2015; Sahana and Sajjad 2017; Aditian et al. 2018) and frequency ratio (Ramesh and Anbazhagan 2015). Recently, with the development of machine learning, there is a growing body of literature that apply artificial neural networks (Arora et al. 2004; Polykretis et al. 2015; Polykretis and Chalkias 2018), decision tree analysis method (Park and Lee 2014; Wu et al. 2014; Palamakumbure et al. 2015; Mao et al. 2017), neuro-fuzzy system method (Pradhan et al. 2010; Sezer et al. 2011; Sdao et al. 2013), support vector machine method (Ballabio and Sterlacchini 2012; San 2014; Wu et al. 2014; Su et al. 2015), etc., for the purpose of LS mapping. These shallow neural networks aim at the optimized weights of the triggering indexes. In view of this, quantitative models are more deterministic and objective comparing to the qualitative models.

In most of the quantitative models, the acquisition of knowledge about the landslide probability requires dealing with some challenging issues. Recent work by Polykretis and Chalkias (2018) recognized the difficulties in LS mapping due to the complex nonlinear relations among triggering factors, lack of relevant data, and the integration of the dynamic changes in the environment. However, to date most of the classification and regression learning methods, such as back propagation neural network (BPNN), are shallow structure algorithms. The key features of data are not detected completely and the nonlinear relationship between different features is not fully reflected in current models (Chen et al. 2017). In order to address the issue concerning the complex influence of triggering factors, a much-debated question has arisen that whether a neural network with deeper hidden layers or more neurons is better for LS mapping. Recent evidence suggests that deep learning technology performs better to deal with the above issues. The essence of deep learning is to learn more useful features by building machine learning models with many hidden layers and massive training data. Consequently, it benefits to improve the accuracy of classification or prediction.

In this study, we report an application using the deep belief network (DBN), a typical model based on deep learning technology, for LS mapping. The training and testing data are the landslide dataset in Sichuan Province, China, which contains information of 1551 landslide in total. Seven key factors with respect to geomorphology, geology, and hydrology are prepared, and two comparison methods, i.e., LR and BPNN reported in our previous studies are selected for examining the effectivity and advantages of DBN method.

Description of Landslide Dataset

Overview of the Study Area

The training and testing data in this paper is based on the landslide dataset in Sichuan Province, China. Sichuan is located in southwestern China. The area covers 486,000 km^2 area, which is one of the four major basins in China. It has a complex topography that mainly composed of plains, hills, mountains and plateaus, accounting for 8.2%, 10.3%, 74.2% and 7.3% of the total area, respectively.

Sichuan Basin, with a topographic terrain generally higher in the northwest and lower in the southeast, is adjacent to Tibet Plateau in the west, the mountainous regions of Hunan and Hubei in the east, Qinling Mountains and Loess Plateau in the north, and Yunnan-Guizhou Plateau in the south. Sichuan Basin covers an area of 260,000 km^2 with a variety of landforms, including mountainous regions on the edge and lowlands at the bottom (Zhang et al. 2019). The area is located at the seismic distribution zones, including Minjiang Fault Zone, Longmenshan Fault Zone, Jinshajiang Fault Zone and Daliangshan Fault Zone (Liu et al. 2018). The tectonic process in this area is extensive, providing complex geological conditions for triggering landslides, making it vulnerable to geological disasters. The strongest earthquake in Sichuan is the Wenchuan earthquake with a magnitude of 7.9 on the Richter scale in 2008, which triggered plenty of coseismic giant landslides, resulting in huge loss of life and property safety (Liu et al. 2020). The study region has abundant water resource due to the existence of dense river networks in a low altitudinal basin surrounded by mountains and hills, with annual precipitation ranging from 900 to 1200 mm (Xu et al. 2018). In addition, climate variation is obvious due to the difference between eastern and western landforms. The Eastern Plain is a typical subtropical monsoon climate type with drier and colder winters and hot and humid summers; the Western Plateau is a plateau monsoon climate, with lower annual temperature, more sunny days and longer sunshine duration. Because of the action of the terrains, vertical climate is obvious in the

Western Plateau. Overall, various climates along subtropical to permafrost all occur in the vertical direction in Sichuan Basin (Shao et al. 2012). Owing to the complex geomorphology, complicated geological setting, and abundant hydrological condition, Sichuan suffers from many landslide disasters every year (Han et al. 2018a).

Historical Landslide Dataset

LS mapping in Sichuan area has long arisen wide attention. A spatial dataset that represents former landslides provides critical information for quantitative modeling. It is beneficial for exploring the relationship between the triggering factors and ultimately predict the incidence of each individual landslide.

In the past few years, a historical landslide dataset in Sichuan area has been gradually developed through two teams' persistent effort. The basic data was collected by the team from the Chinese Academy of Sciences, while the team from Central South University digitized and visualized these basic data into a landslide distribution map under ArcGIS environment. To date, the landslide dataset has been expanded to 1551 individual landslide (as shown in Fig. 1), key information that required by LS mapping, such as spatial locations, landslide areas, etc., are embedded using geographical registration function in ArcGIS environment.

The historical individual landslides in Fig. 1 have been pre-transformed from surface elements to point elements in ArcGIS environment to highlight the landslide distribution. It is apparent from Fig. 1 that the historical landslides are generally concentrated in the south, central and eastern parts in Sichuan area.

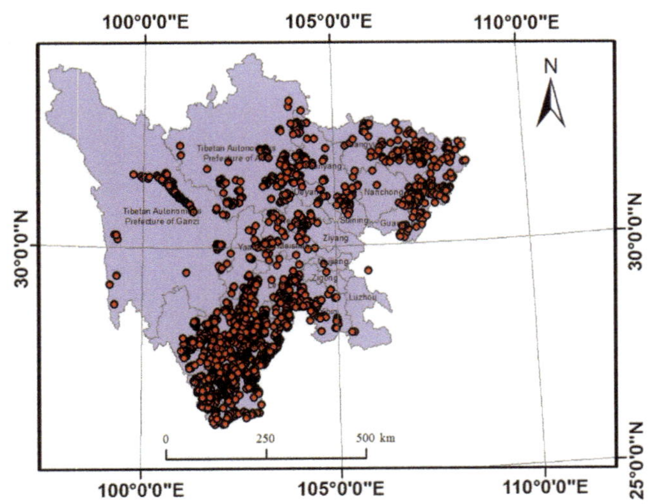

Fig. 1 Historical landslide dataset in Sichuan area contains key information of 1551 individual landslides

Digital Data of Triggering Factors

Considering the geomorphological, geological, and hydrological conditions in Sichuan Area, our previous study (Wang et al. 2009) has recognized key triggering factors that can be categorized into inoculation factors and inducing factors. These recognized factors are clearly supported by the following studies (Kanungo et al. 2011; Raja et al. 2017; Chen et al. 2017; Shirzadi et al. 2017). The factors refer to the natural conditions that influence the formation and occurrence of landslides in the study area, including altitude, topography, lithology, slope, tectonic line and distance to drainage network. In this paper, we follow the previous studies, the above seven key triggering factors are chosen.

The data source of the seven triggering factor layers are collected from the open database. The altitude data was available in the geospatial cloud digital elevation model (DEM) data of the Chinese Academy of Sciences that 30 m in resolution. The topography, tectonic lines and drainage networks were accessed in the database of the National Basic Geographic Information System, China. The lithology data was digitalized from the geological map of Sichuan Province. The average annual rainfall data was derived from the meteorological data of the China Meteorological Administration. The slope data was processed by the DEM under ArcGIS environment. These factor layers and their influence are illustrated below.

(1) **Altitude**. Altitude distribution is shown in Fig. 2a. A greater the altitude often results in a higher potential energy and greater stress in the slope. Beside of that, altitude intensively controls runoff direction and rate of drainage density (Hosseinzadeh et al. 2009).
(2) **Topography**. Topography exhibited in Fig. 2b represents the relief of earth surface in Sichuan area. It provides basic condition for landslide formation. Different types of topography have significant differences in the impact of landslides.
(3) **Lithology**. Regional lithology layer in Fig. 2c has different mechanical properties and deformation characteristics, resulting in different slope performance (Carrara et al. 1991).
(4) **Slope gradient**. Slope gradient layer is revealed in Fig. 2d. It closely relates to the altitude layer. Slope gradient has a direct impact on the shape and internal stress distribution of the slope, majorly affecting the slope stability. An abrupt slope with a greater slope gradient is prone to triggering landslides (Han et al. 2015b; Youssef et al. 2016).
(5) **Distance to tectonic lines**. Distance of a slope to the tectonic lines is mapped in Fig. 2e. Regions with tectonic lines across often represent the geological weak

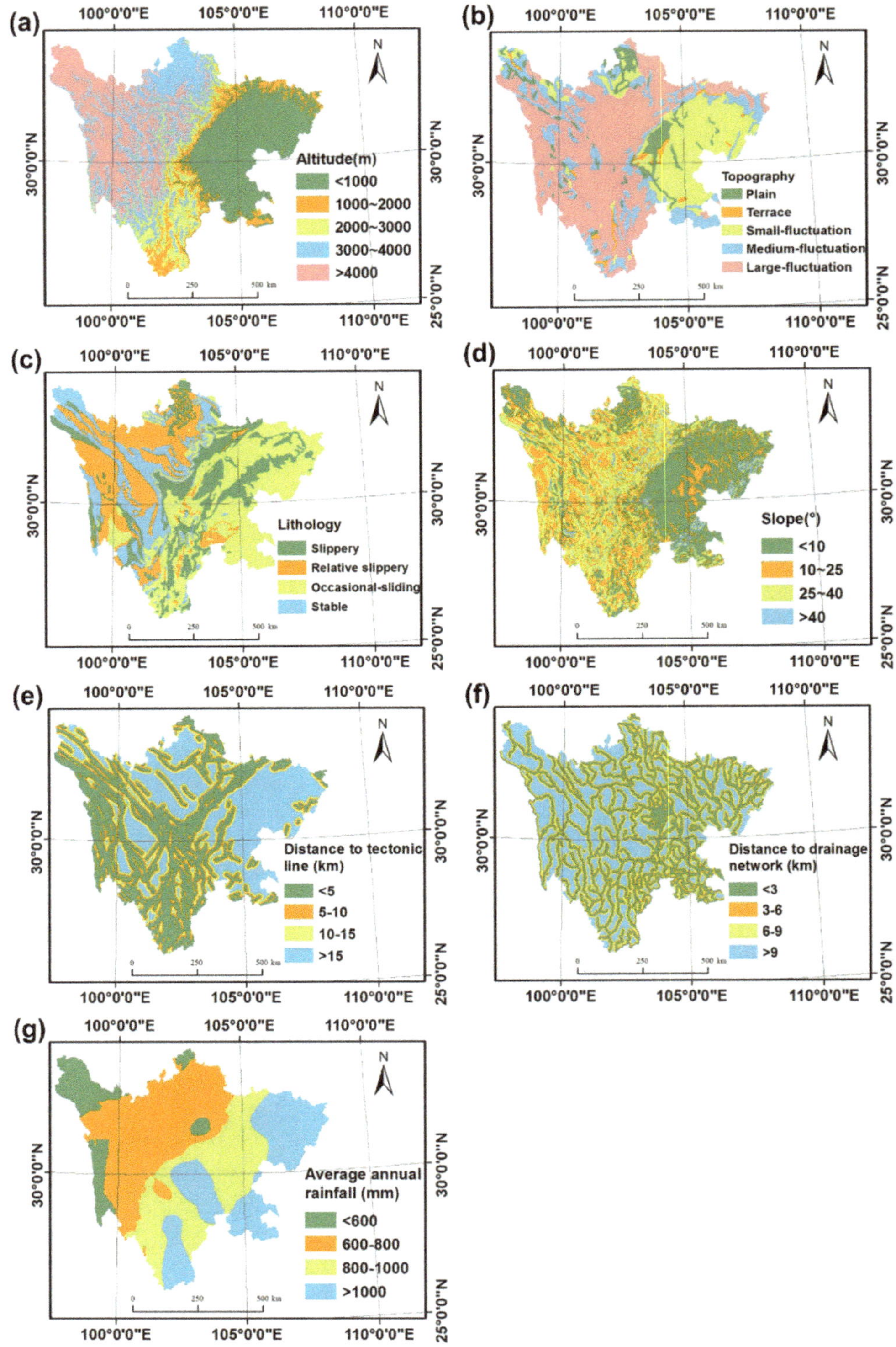

Fig. 2 Digital layers of seven key triggering factors, **a** altitude, **b** topography, **c** lithology, **d** slope angle, **e** distance to tectonic line, **f** distance to drainage network, **g** average annual rainfall

and earthquake-prone areas. Under the action of long-term crustal activity, the shear strength decreases, providing geological dynamic conditions for geological hazards (Abedini and Tulabi 2018).

(6) **Distance to drainage network**. The drainage network that composed of rivers and streams contribute greatly to the occurrence of landslides (Fig. 2f). This can be explained by the fact that as drainage network poses significant deposit erosion to the slope and saturates the underwater part (Demir et al. 2013; Han et al. 2015c).

(7) **Average annual rainfall**. Rainfall strength that represented by the average annual rainfall is shown in Fig. 2g. Rainfall strength is a common cause of landslides and related geological disasters (Cheng et al.

2018; Cogan et al. 2018; Han et al. 2018b). Rainfall intensity and duration play a critical role in landslides occurrence that depends on climate conditions, topography, slope structure and permeability (Lydia and Daniel 2002).

Methodology

Deep Belief Network (DBN)

The concept of deep learning originates from the study of artificial neural networks. It combines low-level features to form a more abstract high-level representation (attribute class or feature) to discover distributed features of data (Dalal et al. 2006). DBN is an efficient unsupervised learning algorithm in deep learning. To date, it has been widely used in image classification, speech recognition and other fields on the basis of Python (Ghahabi and Hernando 2014; Rioux and Giguere 2014; Zhu et al. 2014).

A significant advantage of DBN in LS mapping is that it learns essential genetic features from the features that may be critical to landslide occurrence. The seven triggering factors in this paper are often used by researchers, but the complex interaction and internal relation between the factors is difficult to interpret and quantify. Traditional network training methods such as BPNN and Radial Basis Function (RBF) have high dependence on the original features of factors, lacking the ability to reveal the combined effects of factors. In this context, DBN is supposed as a better solution to this problem. In view of this, our study attempted to apply DBN for LS mapping in Sichuan Province.

DBN is a two-way deep network and is a probabilistic generative model that is composed of Restricted Boltzmann Machines (RBM). RBM consists of input data layer (visual layer υ) and hidden layer h. No connection exists between the neurons in each layer (Fig. 3).

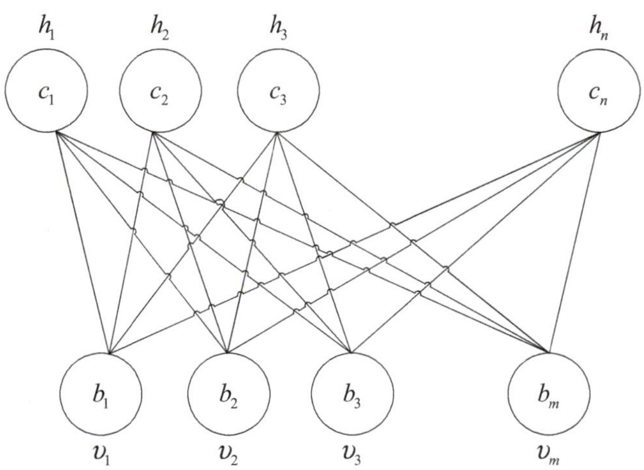

Fig. 3 Schematic illustration of RBM network

In RBM, a weight W between any two connected neurons denotes the connection strength. Each neuron has a bias coefficient b (for the visual neuron) and c (for the hidden neuron) to represent its own weight. The energy of RBM can be expressed as

$$E(\upsilon, h) = -\sum_{i=1}^{N_\upsilon} b_i \upsilon_i - \sum_{j=1}^{N_h} c_j h_j - \sum_{i,j=1}^{N_\upsilon, N_h} W_{ij} \upsilon_i h_j \quad (1)$$

where N_υ and N_h denote the number of visual neurons and hidden neurons, respectively. The probability P of activation of hidden neurons h_j is:

$$P(h_j|\upsilon) = \sigma\left(b_j + \sum_i W_{ij} x_i\right) \quad (2)$$

Because DBN is a two-way connection, the neurons in the visual layer are activated by neurons in the hidden layer:

$$P(\upsilon_j|h) = \sigma\left(c_j + \sum_j W_{ij} h_j\right) \quad (3)$$

where σ is Sigmoid function. Because there is independence between neurons in the same layer, the probability density satisfies the following independence:

$$P(h|\upsilon) = \prod_{j=1}^{N_h} P(h_j|\upsilon) \quad (4)$$

$$P(\upsilon|h) = \prod_{i=1}^{N_\upsilon} P(\upsilon_i|h) \quad (5)$$

Training process of RBM can be described by the following steps.

(1) Given a sample data x, which will be assigned to the visual layer υ_1, and then the probability of activation of each neuron $P(h_1|\upsilon_1)$ in the hidden layer is calculated by Eq. (2).
(2) Gibbs sampling method is used to extract a sample from the probability distribution of the calculation:

$$h_1 \sim P(h_1|\upsilon_1) \quad (6)$$

(3) h_1 is used to reconstruct the visual layer, that is, the hidden layer is used to retrieve the visual layer, and Eq. (3) is used to calculate the activation probability $P(\upsilon_2|h_1)$ of each neuron in the visual layer.
(4) Similarly, Gibbs sampling is used to extract a sample from the calculated probability distribution:

$$v_2 \sim P(v_2|h_1) \tag{7}$$

(5) The probability of activation of each neuron in the hidden layer is calculated again by v_2, and the probability distribution is obtained.

(6) Update weight using an adjusting coefficient λ:

$$W \leftarrow W + \lambda(P(h_1|v_1)v_1 - P(h_2|v_2)v_2) \tag{8}$$

$$b \leftarrow b + \lambda(v_1 - v_2) \tag{9}$$

$$c \leftarrow c + \lambda(h_1 - h_2) \tag{10}$$

A DBN is structured by concatenating several RBMs. The hidden layer of the previous RBM is the visual layer of the next RBM, and the output of the previous RBM is the input of the next RBM. During the process, the previous RBM requires an adequateness training before training the current RBM until the last layer (Fig. 4).

Deep Belief Network (DBN)

(1) **Logistic regression (LR)**

LR model is a generalized linear regression analysis model that commonly used in classification problems. In the process of LS mapping, LR analyzes the linear relationship among the triggering factors, classifies and predicts the landslide incidence of each point. The possibility of landslide can be expressed by

$$P = \frac{1}{1 + e^{-z}} \tag{11}$$

where P indicates the possibility of landslides in the range of 0–1. The linear relationship between triggering factors can be represented by

$$z = B_0 + B_1 I_{1j} + \cdots + B_n I_{nj} \tag{12}$$

where i is the number of triggering factors, j is the ordinal number of the factor categories, B_0 is a constant term, B_i is the influence coefficient of each triggering factor, and I_{ij} represents the weight of triggering factors.

Each triggering factor contains several secondary categories. The area of all secondary categories A_{ij} and the corresponding historical landslide area a_{ij} were obtained with the statistical function of ArcGIS. Then the area proportion of the categories R_{ij} and the area proportion of the corresponding historical landslide area r_{ij} can be calculated:

$$R_{ij} = A_{ij} / \sum_{j=1}^{m} A_{ij} (i = 1, 2, \ldots n; j = 1, 2, \ldots m) \tag{13}$$

$$r_{ij} = a_{ij} / \sum_{j=1}^{m} a_{ij} (i = 1, 2, \ldots n; j = 1, 2, \ldots m) \tag{14}$$

where m is the number of factor categories. Finally, the dimensionless relative weight of each category I_{ij} is calculated according to the landslide density of each category T_{ij}:

$$T_{ij} = r_{ij} / R_{ij} (i = 1, 2, \ldots n; j = 1, 2, \ldots m) \tag{15}$$

$$I_{ij} = T_{ij} / \sum_{j=1}^{m} T_{ij} (i = 1, 2, \ldots n; j = 1, 2, \ldots m) \tag{16}$$

The weight of each category is recorded in Appendix Table 1. Subsequently the historical landslide dataset and all

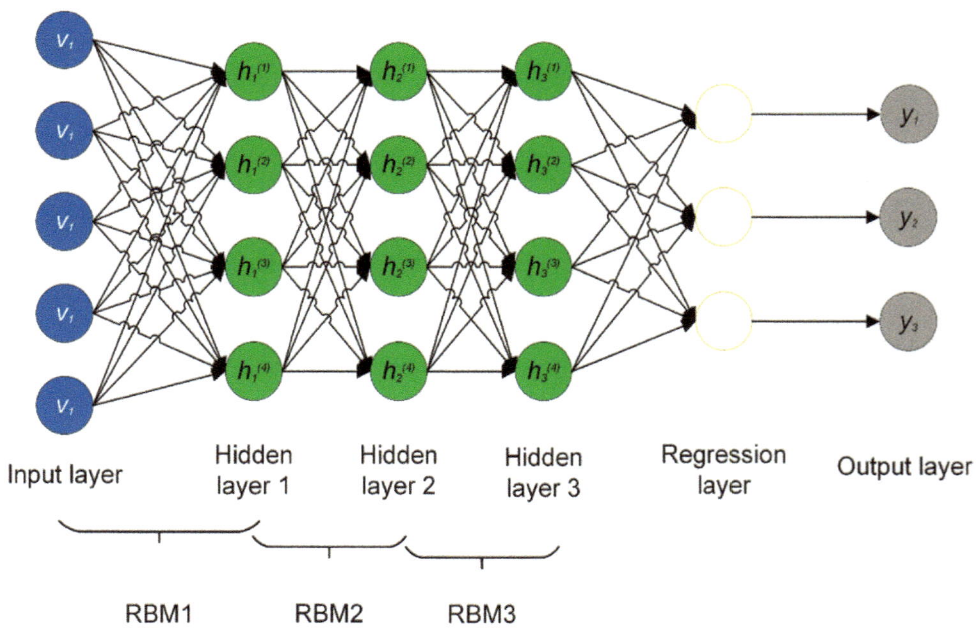

Fig. 4 Schematic illustration of the structure of DBN network

Table 1 Indexes in ROC analysis

Predicted susceptibility	Historical data		Sum
	Observed	No observed	
Risky unit	TP	FP	TP+FP
Stable unit	FN	TN	FN+TN
Sum	TP+FN	FP+TN	

triggering factor layers are superimposed in 6391 surface units under the ArcGIS platform.

(2) **BP neural network (BPNN)**

BPNN is the most widely used and representative model of artificial neural network. It is a kind of multi-layer feed-forward network trained by error back-propagation algorithm. BPNN has the characteristics of large-scale parallel processing ability, self-adaptability, self-learning ability, strong fault tolerance, robustness and distributed information storage. Artificial neural network toolbox of MATLAB environment is chosen for modeling analysis.

Results

Results by DBN

In our study, a typical multi-layer DBN is selected, containing three RBMs and an output layer. Considering that the information represented by RBMs and output layer is probability, we choose sigmoid function as the activation function of all layers. The entire training process is divided into two steps. First, the RBM is pre-trained. The numbers of neurons in three RBMs are configured as 100, 50 and 20, respectively. The training method was stochastic gradient descent (SGD), with the learning rate of 1.0, 100 updates per batch, and mean squared error (MSE) as the performance function.

The DBN is subsequently trained after pre-train of RBMs. Seven nodes in the input layer represent the above seven key triggering factors. The final one node in the last layer represents the Landslide susceptibility level of the unit. Adaptive Moment Estimation (ADAM) is selected as the optimization function and MSE as the loss function. The learning rate is 0.01 and the number of iterations is 1000. We randomly select 1241 historical landslides, counting to 70% of the dataset as to the training dataset, while the remaining 310 landslide data as the test set. Finally, we attain an 83% accuracy of the entire network.

Based on the well-trained DBN, the LS of Sichuan Province is mapped by combining the weight of categories of each unit and the historical landslides. The attained susceptibility result ranges from 0 to 0.9905. According to the natural breakpoint method, the generated LS map was classified into four levels, i.e., low (0–0.2055), moderate (0.2055–0.5093), high (0.5093–0.7603), and very high (0.7603–0.9905) susceptibility. Results by DBN are shown Fig. 5a.

Results by LR

In our LR analysis, the individual historical landslide is considered as dependent variable. The existence marker is set to be 1, and the non-existence marker is 0. The data is exported to SPSS, and the value of B_i and z in Eq. (12) are obtained below.

$$\begin{aligned} z = & -5.273 + 1.444 I_{\text{Altitude}} + 2.285 I_{\text{Topography}} \\ & + 3.705 I_{\text{Lithology}} + 0.867 I_{\text{Slope}} \\ & + 2.068 I_{\text{Tectonic\_line}} + 3.151 I_{\text{Drainage}} + 3.569 I_{\text{rainfall}} \end{aligned} \tag{17}$$

Combining with the weight of triggering factors at each point, the distribution range of P in Sichuan area is deduced between 0.0195 and 0.7991 based on Eqs. (11) and (17). The LS map is then classified into four levels based on the classification method of natural breakpoints, i.e., low (0.0195–0.1606), moderate (0.1606–0.3186), high (0.3186–0.5089) and very high (0.5089–0.7991), respectively. Results by LR are shown Fig. 5b.

Results by BPNN

A three-layer BPNN is used to train the input data. The transfer functions of hidden layer and output layer are Tansig and Logsig, respectively, and the training function is Traingdx, the learning function of threshold and weight is Learngd, and the performance function is MSE. The input layer contains seven nodes that consistent to the triggering factors, while the output layer has 1 output node, indicating the landslide susceptibility of the unit. The number of hidden layer nodes is determined according to the reference formula $n_1 = \sqrt{l+k} + a$, where l is the number of input layer nodes, k is the number of output layer nodes, and a is the constant value in the [1, 10] interval. Therefore, the number of hidden layer nodes in this case is set to be 7, within the suggested range [5, 14]. The same with the DBN training process, 70% of the data is randomly selected as the training samples of the model, and the remaining 30% as the validation samples of the model.

According to the natural breakpoint method under ArcGIS environment, the susceptibility was classified into four levels, i.e., low (0.1072–0.4194), moderate (0.4194–0.5845), high (0.5845–0.7685) and very high (0.7685–0.9950). Results by BPNN are shown Fig. 5c.

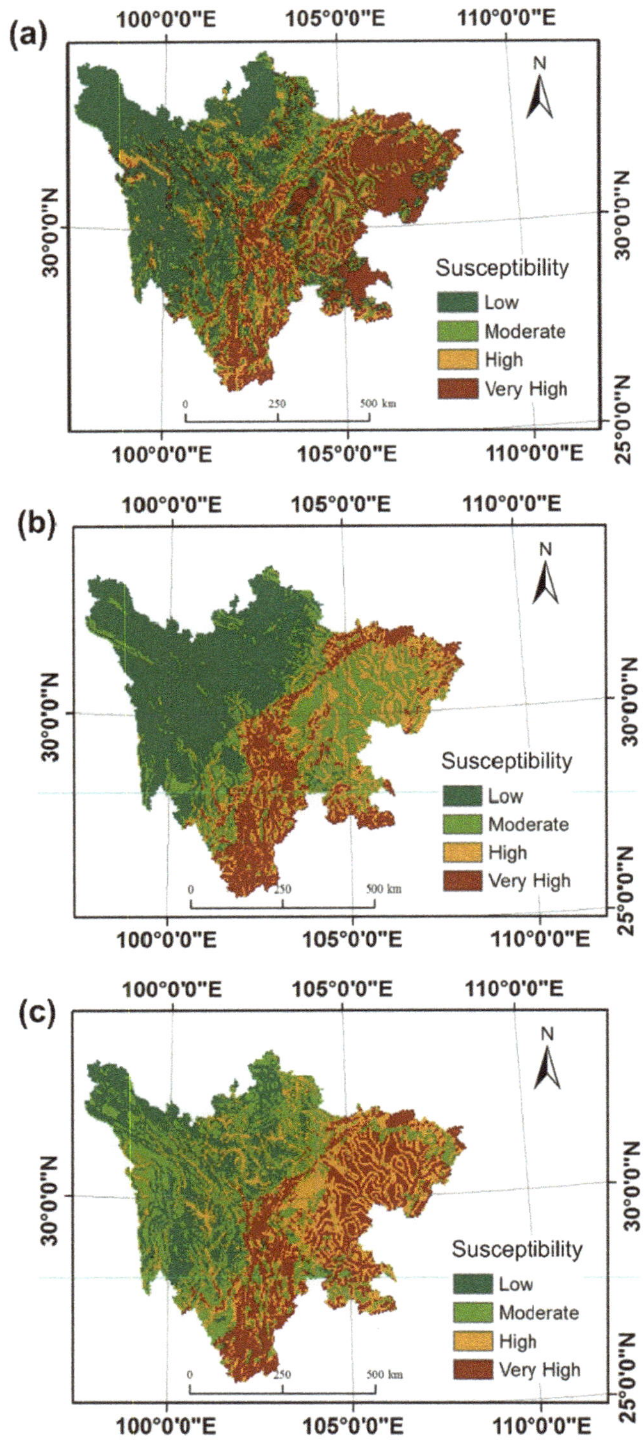

Fig. 5 Landslide susceptibility (LS) maps produced by the different models. **a** DBN, **b** LR, **c** BPNN

Discussions

Results Comparison

Figure 5 is quite revealing in several ways. First, LS mapping results by three models have a similar feature that the low and moderate susceptibility areas are mainly concentrated in the west, by contrast, the high and very high susceptibility areas distribute in the south. Second, three models predict different proportions of 4 susceptibility levels. It seems that LR model somewhat underestimates the landslide susceptibility in Sichuan area. To highlight the differences between DBN and BPNN, LR, we compared the results by different models in Fig. 6. It is apparent that the differences between DBN and BPNN are no such significant, but differences between DBN and LR are majorly concentrated in the northwestern and eastern area, where DBN made a conservative susceptibility prediction. In LR model, there is no highly sensitive area in the northwest and only a few areas in the East are high-risk areas, while in the DBN model, some areas in the northwest and most areas in the East are divided into landslide prone areas. It can be explained that each factor is linearly superposed in LR model, while in DBN and BPNN model, it is treated as non-linear relationship by using more powerful data processing ability.

We explore the differences of the three groups of results in Figs. 7 and 8, where the predicted proportion of the area covered by a susceptibility level (Fig. 7) and percentage of historical landslides located in different susceptibility areas (Fig. 8) are shown in detail. Overall, what stands out in two figures is that DBN predicts a conservative susceptibility among three models, that the area covered by "very high-" and "high-" susceptibility counts to 27.57 and 18.57% of the total area. Meanwhile, 52.72% of total historical landslides that have been confirmed are classified as "very high" susceptibility, better than the results of 42.10% by BPNN and 43.06% by LR. Similar finding could be indicated if we count the historical landslides in "high" susceptibility in addition, that DBN predicts totally 81.04% of "very high-" and "high-" susceptibility landslides that have been confirmed, best among all the three models.

Interestingly, BPNN is observed to control better the false alarm, with only a minority of historical landslides classified as "low" susceptibility. However, in return, BPNN classifies

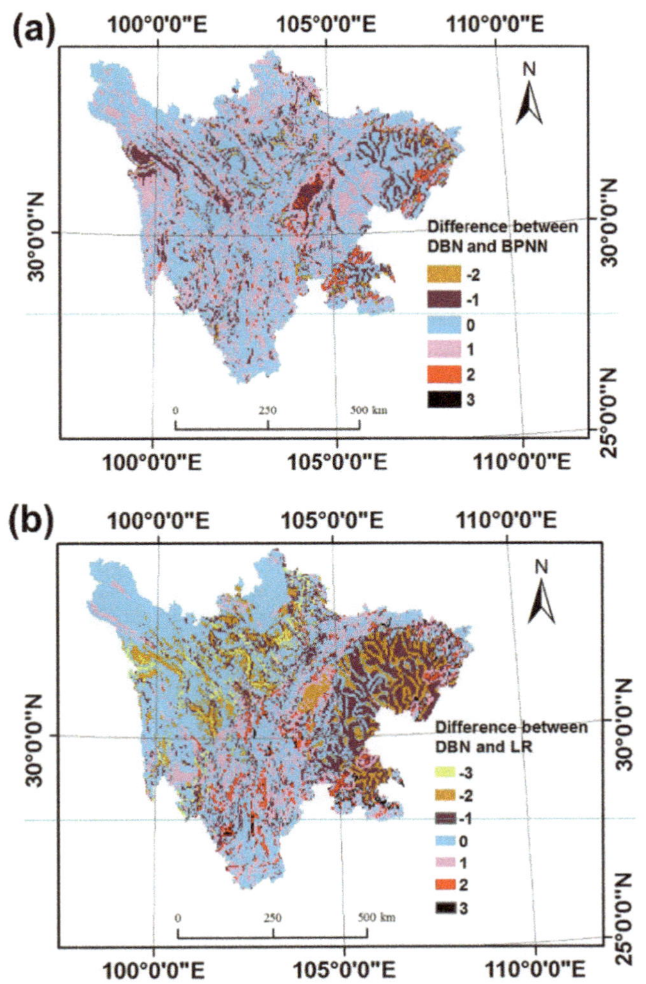

Fig. 6 Spatial distribution of the susceptibility differences between DBN and the compared models. **a** Difference between DBN and BPNN, **b** difference between DBN and LR

more historical landslides as "moderate" susceptibility among three model. In this sense, DBN is found to have a better effect limiting the false alarm, because totally 18.96%

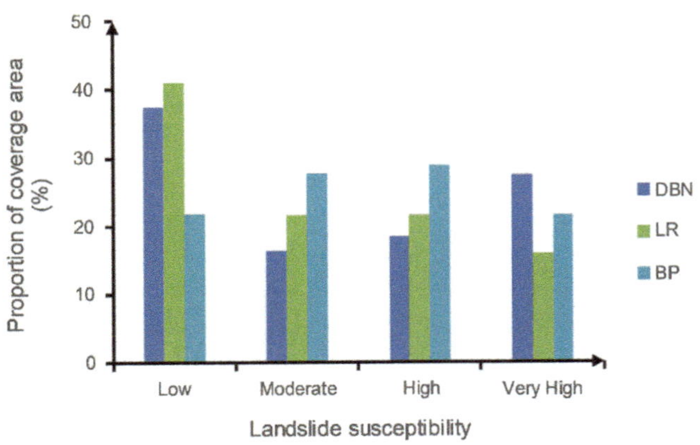

Fig. 7 Proportion of coverage area of each landslide susceptibility level using different models

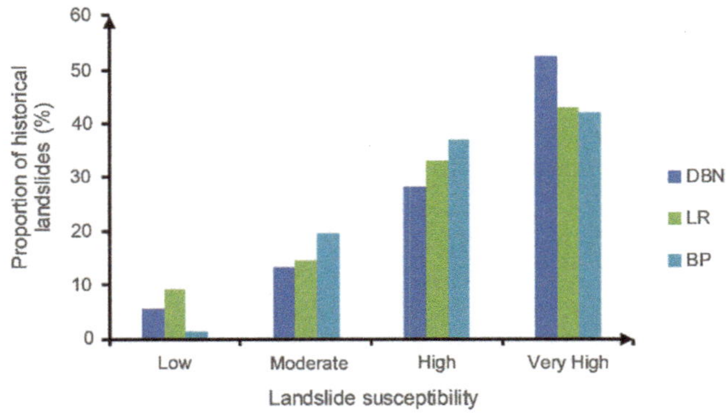

Fig. 8 Percentage of historical landslides located in different susceptibility areas

historical landslides are falsely classified as "low-" and "moderate-" susceptibility, comparing to 20.93% by BPNN.

Accuracy Evaluation Using ROC Curves

To further discuss the advantages of the DBN, we use Receiver operating characteristic (ROC) curves to evaluate the accuracy of the LS mapping results by three models. ROC is a statistical decision theory-based curve. It was initially applied to the evaluation of radar signal acceptance (Arian and Peter 1998). With the promotion of the scope of application, it has also been well developed in precision evaluation (Kwiatkowski et al. 1999; Park et al. 2004; Nguyen and Nguyen 2019). In the generated LS map, we suppose the units in "low-" and "moderate-" susceptibilities are stable, by contrast, in "high-" and "very high-" susceptibilities are risky. Each unit is then compared to the historical landslide data. As Table 1 shows, we evaluate the susceptibility result of each element according to the following criterions. True positive (TP) indicated historical landslide units those are correctly identified as risky. False positive (FP) indicated the stable units those are incorrectly identified as risky. True negative (TN) means the units without historical landslides observed are correctly identified as stable. False negative (FN) is that historical landslide units are incorrectly identified as stable.

ROC curves analysis combines the sensitivity and specificity of the model to evaluate the reliability of the model with two indexes,

$$\text{False positive rate} = 1 - \frac{TP}{TP + FN} \tag{18}$$

$$\text{True positive rate} = \frac{TN}{FP + TN} \tag{19}$$

The area under the curve (AUC) is the final evaluation criterion of the model precision. The ROC curves of DBN,

LR, and BPNN are shown in Fig. 9, with AUC values of 0.899, 0.782, and 0.867, respectively. From this data, DBN appears to have the best prediction precision among the three models. ROC analysis in our study broadly supports the work of many current studies (Polykretis and Chalkias 2018; Chen et al. 2017) indicating the advantage of a deep neural network to deal with complex features of factors.

Sensitivity to the Triggering Factors

To investigate the impact of triggering factors using different models, a further statistical test is performed to explore the proportion of the predicted landslide susceptibility to the areas of each level of different triggering factors (Fig. 10). Indicated by this figure, both DBN and BPNN highlight a positive correlation between the factors of the average annual rainfall and distance to drainage network with landslide occurrence, because the predicted "high-" and "very high-" susceptive landslides are concentrated in the area with a greater annual rainfall intensity and a shorter distance to drainage network. This finding is consistent with the conclusion by previous studies (Demir et al. 2013; Han et al. 2015c; Cheng et al. 2018; Cogan et al. 2018; Han et al. 2018b).

The factor of altitude also played an essential role as implied in Fig. 10. Results by LR and BPNN indicate that the area that 1000–2000 m in altitude covers most "very high-" susceptive landslides, while DBN prefers the area that <1000 m. Although this finding is statistically opposed to the conclusion by Hosseinzadeh et al. (2009) who demonstrated that the area with a greater altitude is prone to landslides. This discrepancy can be explained by the landslide dataset in our study, because landslide information in such high altitude is difficult to investigate.

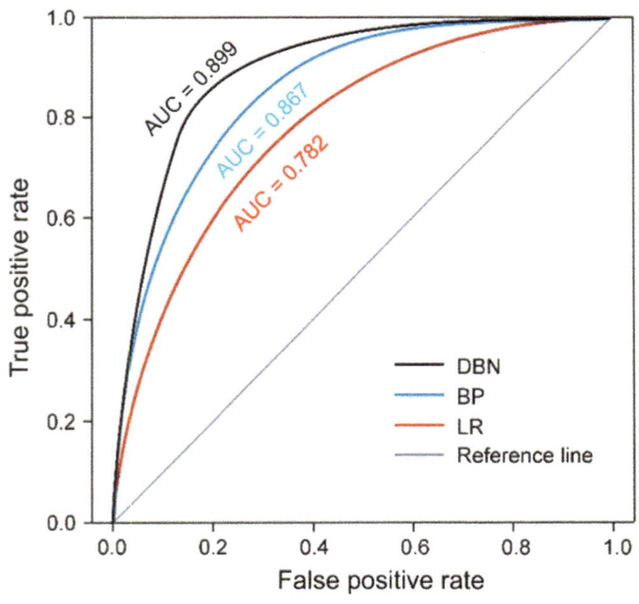

Fig. 9 ROC curves for the three models using the landslide dataset to train the network

What is striking about the figure is that in all the three models, the factor of slope gradient has no such statistically significant influence on the landslide occurrence comparing to the others, because landslides distribute almost uniform in different levels of this factor. It is somewhat contrary to expectations and conclusions of the studies focusing on the mechanism of individual landslide (Abedini and Tulabi 2018). One reasonable explanation is that susceptibility mapping in our study only present the spatial probability of landslide, rather than temporal probability. Many landslides occurred long, the following measured regional DEM data only reflect the landforms shaped by these old landslides. A further study with more focus on the temporal variations of the triggering factor is therefore suggested.

Together these results in Fig. 10 provide a deep understanding into the responses of the three methods with respect to the triggering factors. In Fig. 11, the results by DBN are used as the benchmark, differences to LR and BPNN are plotted in different colors. Overall, results by DBN and BPNN show a quite similar distribution feature, with little differences regarding the factors of altitude, distance to drainage network, and average annual rainfall. Differences between DBN and BPNN majorly exist regarding the remaining four factors. From Fig. 11, results by LR appear to have significant differences to DBN and BPNN, in particular the factor of topography, because LR predicts a majority of highly susceptible landslides in terrace topography. This finding, while preliminary, may help to understand the responses of each model with respect to the triggering factors.

Conclusion

In this paper, we report on the application of deep belief network (DBN), a kind of deep learning technology for the landslide susceptibility mapping in Sichuan Province of China. Seven key factors suggested in previous studies, with respect to geomorphology, geology, and hydrology are considered, and a DBN model containing three pre-trained layers of RBMs is configured. We randomly select 70% of the historical landslide dataset in Sichuan region as to the training dataset, attaining 83% accuracy based on the test of the remaining 30% data. The susceptibility mapping result indicates that landslides are much prone in the southwest and southeast of Sichuan.

To validate the performance of the DBN in landslide susceptibility mapping, we use the BPNN and LR to make a comparison. Statistical analysis on the attained landslide

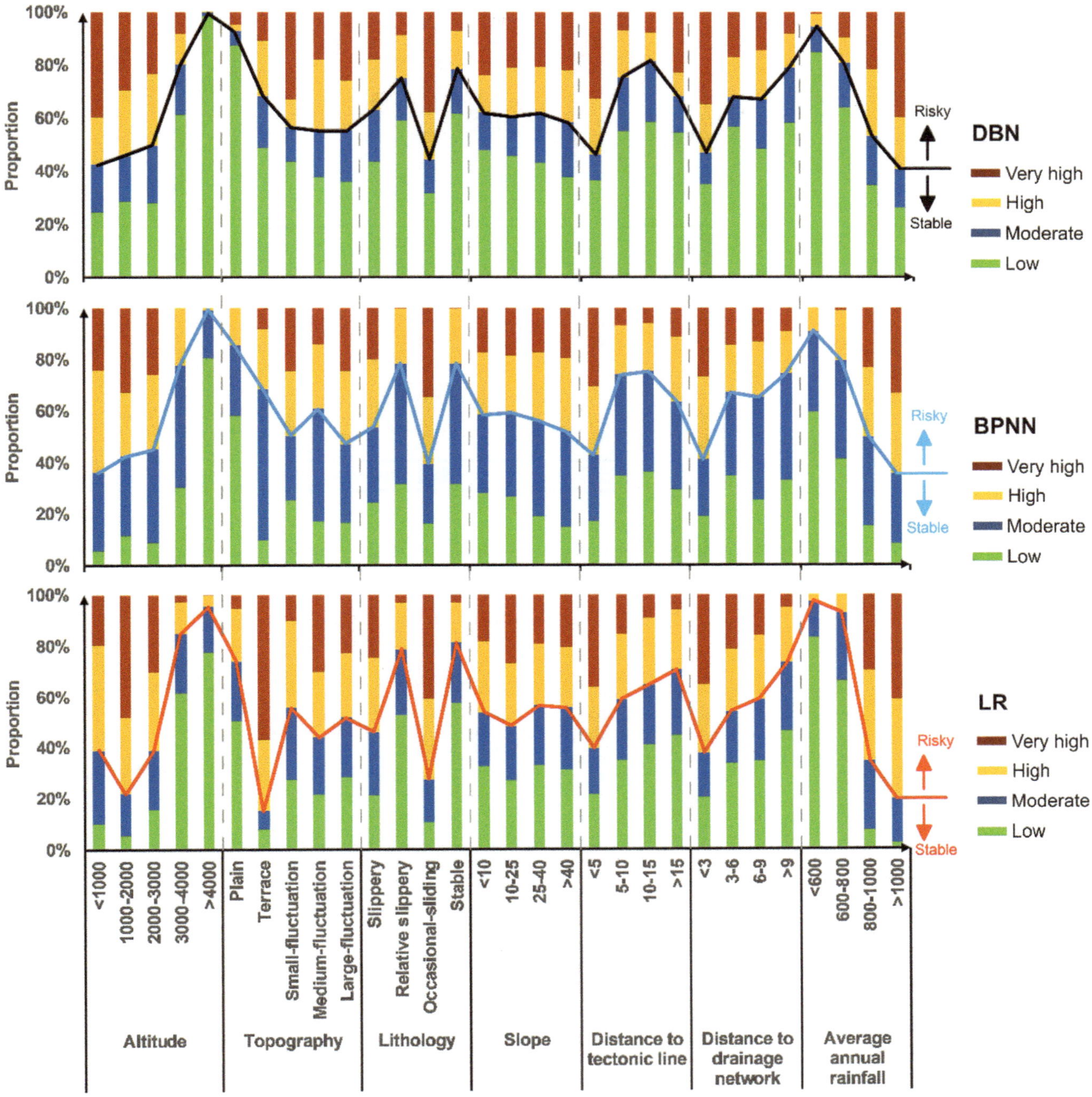

Fig. 10 Proportion of landslide susceptibility to the areas of each level of different triggering factors

susceptibility demonstrates that LR underestimates the landslide susceptibility in Sichuan area, while DBN and BPNN generate a quite similar result. Further ROC curve test on the three models show that DBN has a better performance to control the false alarm and fake alarm. In sensitivity analysis on the triggering factors, it is found that the factors of altitude, distance to drainage network, and average annual rainfall play a key role in landslide susceptibility mapping.

However, the generalizability of these results is subject to certain limitations. For instance, important factors in common wisdom, such as slope gradient, show no such statistically significant influence on the landslide. Considerably more work will need to be done to determine the temporal relativity of factor data and landslide susceptibility. In the future, the DBN model will be applied to the regions with frequent occurrence of soil landslides and rock landslides respectively to find the specific influence of slope factors in

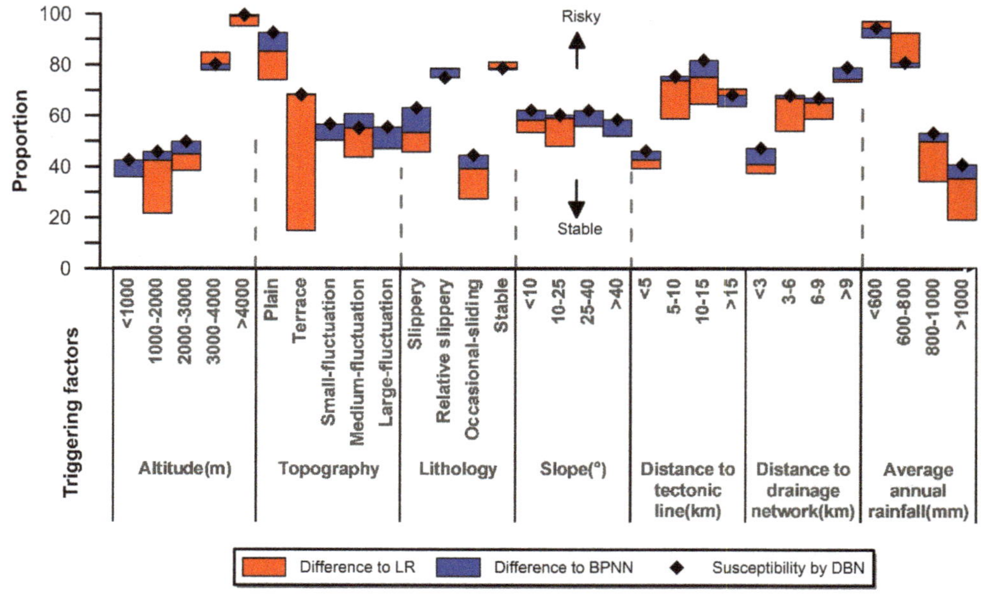

Fig. 11 Overall comparison of landslide susceptibility to the areas of each level of different triggering factors

different landslides. Meanwhile, DBN will be used for landslide susceptibility mapping in more areas, especially in small areas, to verify the applicability of deep learning model in landslide mapping.

Acknowledgements This study was financially supported by the National Natural Science Foundation of China (Grant No. 51478483, W. Wang; Grant No. 41702310, Z. Han); the National Key R&D Program of China (Grant No. 2018YFC1505401, Z. Han); and the Natural Science Foundation of Hunan (Grant No. 2018JJ3644, Z. Han). These financial supports are gratefully acknowledged.

References

Abedini M, Tulabi S (2018) Assessing LNRF, FR, and AHP models in landslide susceptibility mapping index: a comparative study of Nojian watershed in Lorestan province. Iran Environ Earth Sci 77 (11):405

Aditian A, Kubota T, Shinohara Y (2018) Comparison of gis-based landslide susceptibility models using frequency ratio, logistic regression, and artificial neural network in a tertiary region of Ambon, Indonesia. Geomorphology 318:101–111

Alexander DE (2004) Vulnerability to landslides. In: Glade T et al (eds) Landslide Hazard and Risk. Wiley, Hoboken, pp 175–198

Arian R, Peter M (1998) Receiver operating characteristic (ROC) analysis: Basic principles and applications in radiology. Eur J Radiol 27 (2):88–94

Arora MK, Das Gupta AS, Gupta RP (2004) An artificial neural network approach for landslide hazard zonation in the Bhagirathi (Ganga) Valley, Himalayas. Int J Rem Sens 25(3):559–572

Avtar R, Singh CK, Singh G, Verma RL, Mukherjee M, Sawada H (2011) Landslide susceptibility zonation study using remote sensing and GIS technology in the Ken-Betwa River Link area, India. Bull Eng Geol Env 70(4):595–606

Ballabio C, Sterlacchini S (2012) Support vector machines for landslide susceptibility mapping: the Staffora river basin case study, Italy. Mathem Geosci 44(1):47–70

Brabb EE (1984) Innovative approaches to landslide hazard mapping. In: Proceedings of the 4th international symposium on landslides, Toronto 1, pp 307–324

Carrara A, Cardinali M, Detti R, Guzzetti F, Pasqui V, Reichenbach P (1991) GIS techniques and statistical models in evaluating landslide hazard. Earth Surf Proc Land 16:427–445

Chen W, Xie XS, Wang JL et al (2017) A comparative study of logistic model tree, random forest, and classification and regression tree models for spatial prediction of landslide susceptibility. CATENA 151:147–160

Cheng DQ, Cui YF, Su FH, Jia Y, Choi CE (2018) The characteristics of the Mocoa compound disaster event, Colombia. Landslides 15:1223–1232

Cogan J, Gratchev I, Wang GH (2018) Rainfall-induced shallow landslides caused by ex-Tropical Cyclone Debbie, 31st March 2017. Landslides 15(6):1215–1221

Dalal N, Triggs B, Schmid C (2006) Human detection using oriented histograms of flow and appearance. Springer, Berlin Heidelberg, pp 428–441

Demir G, Aytekin M, Akgün A, Ikizler SB, Tatar O (2013) A comparison of landslide susceptibility mapping of the eastern part of the North Anatolian Fault Zone (Turkey) by likelihood-frequency ratio and analytic hierarchy process methods. Nat Hazards 65:1481–1506

Ghahabi O, Hernando J (2014) Deep belief networks for i-vector based speaker recognition. In: Proceedings of 2014 international conference on acoustics, speech and signal processing. Florence: IEEE 2014, pp 1700–1704

Han Z, Chen GQ, Li YG et al (2015a) Numerical simulation of debris-flow behavior incorporating a dynamic method for estimating the entrainment. Eng Geol 190:52–64

Han Z, Li YG, Du YF, Wang WD, Chen GQ (2018a) Noncontact detection of earthquake-induced landslides by an enhanced image binarization method incorporating with Monte-Carlo simulation. Geomat Nat Hazards Risks 10(1):219–241

Han Z, Wang WD, Li YG, Huang JL, Chen GQ et al (2018b) An integrated method for rapid estimation of the valley incision by debris flows. Eng Geol 232:34–45

Han Z, Chen GQ, Li YG et al (2015b) Exploring the velocity distribution of debris flows: an iteration algorithm based approach for complex cross-sections. Geomorphology 241:72–82

Han Z, Chen GQ, Li YG et al (2015c) Assessing entrainment of bed material in a debris-flow event: a theoretical approach incorporating Monte Carlo method. Earth Surf Proc Land 41:1877–1890

Hosseinzadeh M, Servati M, Mansouri A, Mirbagheri B, Khezri S (2009) Landslide hazard zonation using logistic regression, the way Sanandaj-Dehgolan, Iran. J Geol 11:27–37

Kanungo DP, Sarkar S, Sharma S (2011) Combining neural network with fuzzy, certainty factor and likelihood ratio concepts for spatial prediction of landslides. Nat Hazards 59(3):1491

Kwiatkowski F, Gachon F, Charrier S et al (1999) Decision-making strategy in biology—ROC analysis. Bull Cancer 86(9):787–789

Listo FDLR, Carvalho VB (2012) Mapping of risk and susceptibility of shallow-landslide in the city of Sao Paulo, Brazil. Geomorphology 169–170:30–44

Liu G, He J, Li R et al (2018) Topological and dynamic complexity of rock masses based on GIS and complex networks. Physica A 512:1240–1248

Liu XW, Chen Q, Zhao JJ et al (2020) The spatial response pattern of coseismic landslides induced by the 2008 Wenchuan earthquake to the surface deformation and Coulomb stress change revealed from InSAR observations. Int J Appl Earth Observ Geoinf 87

Lydia EE, Daniel B (2002) Land slide hazard and risk zonation mapping in the Rio Grande basin, central Andes of Mendoza, Argentina. Mt Res Dev 22(2):177–185

Mao Y, Zhang M, Sun P, Wang G (2017) Landslide susceptibility assessment using uncertain decision tree model in loess areas. Environ Earth Sci 76(22):752

Meten M, Bhandary NP, Yatabe R (2015) GIS-based frequency ratio and logistic regression modelling for landslide susceptibility mapping of Debre Sina area in central Ethiopia. J Mt Sci 12(6):1355–1372

Ministry of Land and Resources (2016) National geological disaster bulletin. Technical Guidance Center for Geological Disaster Emergency Response of Ministry of Land and Resources

Nguyen HT, Nguyen LT (2019) ROC curve analysis for classification of road defects. Brain: Broad Res Artif Intell Neurosci 10(2):65–73

Palamakumbure D, Flentje P, Stirling D (2015) Consideration of optimal pixel resolution in deriving landslide susceptibility zoning within the Sydney Basin, New South Wales, Australia. Comput Geosci 82:13–22

Park I, Lee S (2014) Spatial prediction of landslide susceptibility using a decision tree approach: a case study of the Pyeongchang area, Korea. Int J Remote Sens. 35(16):6089–6112

Park SH, Goo JM, Jo CH (2004) Receiver operating characteristic (ROC) curve: practical review for radiologists. Korean J Radiol 5 (1):11–18

Polykretis C, Chalkias C (2018) Comparison and evaluation of landslide susceptibility maps obtained from weight of evidence, logistic regression, and artificial neural network models. Nat Hazards 93(3):1–26

Polykretis C, Ferentinou M, Chalkias C (2015) A comparative study of landslide susceptibility mapping using landslide susceptibility index and artificial neural networks in the Krios River and Krathis River catchments (northern Peloponnesus, Greece). Bull Eng Geol Env 74 (1):27–45

Pradhan B, Sezer EA, Gokceoglu C, Buchroithner MF (2010) Landslide susceptibility mapping by neuro-fuzzy approach in a landslide prone area (Cameron Highland, Malasia). IEEE Trans Geosci Remote Sens 48(12):4164–4177

Raja NB, Cicek I, Turkoglu N, Aydin O (2017) Landslide susceptibility mapping of the Sera River Basin using logistic regression model. Nat Hazards 85(3):1323–1346

Ramesh V, Anbazhagan S (2015) Landslide susceptibility mapping along Kolli hills Ghat road section(India) using frequency ratio, relative effect and fuzzy logic models. Environ Earth Sci 73:8009–8021

Reichenbach P, Rossi M, Malamud BD et al (2018) A review of statistically-based landslide susceptibility models. Earth Sci Rev 180:60–91

Rioux L, Giguere P (2014) Sign language fingerspelling classification from depth and color images using a deep belief network. In: Proceedings of 2014 Canadian conference on computer and robot vision. Quebec: IEEE 2014, pp 92–97

Sahana M, Sajjad H (2017) Evaluating effectiveness of frequency ratio, fuzzy logic and logistic regression models in assessing landslide susceptibility: a case from Rudraprayag district, India. J Mount Sci 14(11):2150–2172

San BT (2014) An evaluation of SVM using polygon-based random sampling Inlandslide susceptibility mapping: the Candir catchment area (western Antalya, Turkey). Int J Appl Earth Obs Geoinf 26:399–412

Sdao F, Lioi DS, Pascale S, Caniani D, Mancini M (2013) Landslide susceptibility assessment by using a neuro-fuzzy model: a case study in the Rupestrian heritage rich area of Matera. Nat Hazards Earth Syst Sci 13:395–407

Sezer EA, Pradhan B, Gokceoglu C (2011) Manifestation of an adaptive neuro-fuzzy model on landslide susceptibility mapping: Klang valley, Malaysia. Expert Syst Appl 38(7):8208–8219

Shao JA, Li YB, Ni JP (2012) The characteristics of temperature variability with terrain, latitude and longitude in Sichuan-Chongqing region. J Geog Sci 22(2):223–244

Shirzadi A, Bui DT, Pham BT et al (2017) Shallow landslide susceptibility assessment using a novel hybrid intelligence approach. Environ Earth Sci 76(2):60

Su C, Wang L, Wang X, Huang Z, Zhang X (2015) Mapping of rainfall-induced landslide susceptibility in Wencheng, China, using support vector machine. Nat Hazards 76:1759–1779

Wang WD, Guo J, Fang LG, Chang XS (2012) A subjective and objective integrated weighting method for landslides susceptibility mapping based on GIS. Environ Earth Sci 65(6):1705–1714

Wang WD, Xie CM, Du XG (2009) Landslides susceptibility mapping based on geographical information system, GuiZhou, south-west China. Environ Geol 58(1):33–43

Wu X, Ren F, Niu R (2014) Landslide susceptibility assessment using object mapping units, decision tree, and support vector machine models in the Three Gorges of China. Environ Earth Sci 71:4724–4738

Xu C, Wu W, Ge Q (2018) Impact assessment of climate change on rice yields using the ORYZA model in the Sichuan Basin, china. Int J Climatol (D18)

Youssef AM, Pourghasemi HR, El-Haddad BA, Dhahry BK (2016) Landslide susceptibility maps using different probabilistic and bivariate statistical models and comparison of their performance at Wadi Itwad Basin, Asir Region, Saudi Arabia. Bull Eng Geol Environ 75(1):63–87

Zhang L, Guo XM, Zhao TL et al (2019) A modelling study of the terrain effects on haze pollution in the Sichuan Basin. Atmos Environ 196:77–85

Zhu CC, Yin JP, Li Q (2014) A stock decision support system based on DBNs. J Comput Inf Syst 10(2):883–893

A Comparative Study of Deep Learning and Conventional Neural Network for Evaluating Landslide Susceptibility Using Landslide Initiation Zones

2020 Kyoto Japan

Jie Dou, Ali P. Yunus, Abdelaziz Merghadi, Xie-kang Wang, and Hiromitsu Yamagishi

Abstract

Considerable effort has been advanced to comprehend where seismically triggered landslides may occur in the forthcoming because they pose an extraordinarily and disastrous risk in tectonically active mountainous areas. The objectives of this research are to investigate and compare two advanced artificial intelligence models (AI), i.e., artificial neural network (ANN) and deep learning (DL) technique to evaluate the susceptible zones using landslide initiation zone polygons (i.e., scarp areas). For this, a comprehensive landslide inventory map comprising the representative of the landslide scarp is constructed using high resolution aerial photographs and Lidar digital elevation models (Lidar DEM) for the 2018 Hokkaido earthquake-affected sites. Next, 11 causative factors were prepared, including seismic, topographic, and hydrological factors. Our results show that DL has better predictive performance than the traditional ANN obtained model. Also, the importance of factor ranks indicates that topography has played the leading role in the landslide occurrences. The deep learning model shows a promising way for rapid response in the field of landslide hazard mitigation.

Keywords

Deep learning • Scarp • Landslide susceptibility • Artificial intelligence • Lidar DEM

Introduction

In tectonically active terrains, seismicity causes numerous landslides to occur, such as the 1999 Chichi earthquake (Taiwan), 2004 Chuetsu earthquake (Japan), 2008 Wenchuan earthquake (China), 2015 Gorkha Earthquake (Nepal), 2017 Kaikoura earthquake (New Zealand) and 2018 Hokkaido Iburi earthquake (Japan) (Dou et al. 2015b, 2019b; Valagussa et al. 2019; Yunus et al. 2020). Seismically triggered-landslides pose exorbitant and frequently fatal threats to people and infrastructure. Extensive efforts have been devoted in the past to discerning where such potential landslides may occur in the future. Yet some local efforts have succeeded, though no consistent agreement on a universally accepted methodology is presented for forecasting the probability and spatial extent of seismically triggered-landslides.

In the past, several researchers constructed the landslide susceptibility maps (LSM) using the historical inventories of either landslide centroid points or landslide boundary polygons. To our knowledge, there are seldom studies that manage to uses landslide scarp polygons to build the landslide inventory maps accurately (Dou et al. 2019a, d; Pham et al. 2019). In this study, we purposely use two advanced

J. Dou (✉)
Department of Civil and Environmental Engineering, Nagaoka University of Technology, 1603-1 Kami-Tomioka, Nagaoka, Niigata, 940-2188, Japan
e-mail: douj888@gmail.com

A. P. Yunus
State Key Laboratory of Geo-Hazard Prevention and Geo-Environment Protection, Chengdu University of Technology, Chengdu, China
e-mail: yunusp@csis.u-tokyo.ac.jp

A. Merghadi
Research Laboratory of Sedimentary Environment, Mineral and Water Resources of Eastern Algeria, Larbi Tébessi University, Tebessa, Algeria
e-mail: abdelaziz.merghadi@univ-tebessa.dz

X. Wang
State Key Laboratory of Hydraulics and Mountain River Engineering, College of Water Resource and Hydropower, Sichuan University, Chengdu, 610065, People's Republic of China
e-mail: wangxiekang@scu.edu.cn

H. Yamagishi
Hokkaido Research Center of Geology, Hokkaido, Japan
e-mail: hiromitsuyamagishi88@gmail.com

F. Guzzetti et al. (eds.), *Understanding and Reducing Landslide Disaster Risk*, ICL Contribution to Landslide Disaster Risk Reduction,
https://doi.org/10.1007/978-3-030-60227-7_23

artificial intelligence (AI) models, i.e., deep learning (DL) and neural network to predict the landslide-prone areas using the scarp polygons as representative data. Here, the scarp indicates the area that initiated the coseismic landslides, and are potentially unstable because of the increased value of slope angle (>45o in most cases). Any amount of instability in the scarp region leads to the crown collapses in the future.

To understand the coseismic landslide controls and back predicts the accuracies of susceptibility models, we selected the ground motion from the earthquake-triggered event (e.g., peak ground acceleration and velocity), combined with extensively available landslide conditioning factors such as topographic variables, lithologic and hydrological predictors of the test site.

Case Study

The case study area is situated in the Iburi-Tobu region of southern Hokkaido, Japan, covering a part of the Yubari mountain range (Fig. 1). The study area is characterized by gently rolling mountains with elevations ranging between 100 and 800 m above sea level (Ito et al. 2020). On September 6, 2018, at 3:08 a.m., Japan standard time (JST), a powerful earthquake with a moment magnitude Mw 6.6 struck this region. The earthquake caused severe damage to the slopes, killed 41 people, and 691 people were injured. The surficial geology in this area is encompassed of deposits originated from multiple volcanic events. The pumice layers with several meters thick are distributed throughout the affected area, which was later exposed with a layer of denser soil.

Materials

The overall research workflow can be summarized in Fig. 2 that includes three key processes: (1) data collection by field and remote sensing methods; (2) construction of two AI models;(3) producing LSM maps and accuracy evaluation.

Construction of Coseismic Landslide Inventory

Construction of an accurate landslide inventory maps is essential for forecasting landslide-prone areas (Dou et al. 2019c; Chang et al. 2019; Li et al. 2020a). Thus, the initial step in the landslide assessment is to produce a comprehensive landslide inventory. The earthquake affected slopes were interpreted for landslide scarps by means of the 0.2 m aerial photographs from Geospatial Information Authority of Japan (GSI). We used 2 m Lidar DEM from Hokkaido Government and Chibouzu (CBZ) for eliminating the effect shadow in the aerial photographs. Landside scarps represent the source of disposition, which is the actual zone of initiation; hence, in this study, we interpreted landslide scarp instead of whole landslide boundary for susceptibility assessment in Fig. 3. A total of 10120 landslides were interpreted from aerial photographs and LiDAR DEM, creating one of the comprehensive and detailed landslide inventories for the 2018 Hokkaido earthquake event. Most landslides were shallow failures (Yamagishi and Yamazaki 2018). The comparison of the probability density distribution of landslides in Hokkaido and the other recent historical earthquake in Japan are shown in Fig. 4. Compared to 1994 Kobe, and 2011 Tohoku earthquakes, Hokkaido earthquake triggered numerous large landslides. Landslides triggered during Niigata and Hokkaido earthquakes are mostly of similar sizes with the exception that the former has some large rockslides.

Causative Factors

The spatial location of seismically triggered-landslides is in need of some physical characteristics of the region where they happen. Studies recommend that the seismic, topographic, and lithology factors have a strong influence on the landslide spatial distribution, among other causative factors (Keefer and Larsen 2007; Chang et al. 2019). Causative factors to test in the models are selected based on the relationships shown in previous studies, as well as the availability of datasets that can be used as proxies for each of the studied variables (Yamagishi and Yamazaki 2018).

Based on these guidelines, we included the following predictor variables: ground motion produced by the earthquake (distance to epicentre, peak ground acceleration and velocity); topographic elevation, slope, aspect and curvature; material strength in the form of lithology layer, hydrologic factors such as distance to streams and their density; and topographic wetness. Previously, many studies show that rivers undercut the slopes and are one of the potential areas of failures (Panikkar and Subramanyan 1996). Furthermore, streams of high density are areas of considerable erosion and also high soil moisture contents (Li et al. 2020b). We would like to check whether any correlation exists between these potential zones of failures with earthquake-induced landslides.

Methods

Artificial Neural Network

Artificial neural networks (ANN) are black-box models defined (Hecht-Nielsen 1989) as a computing system made

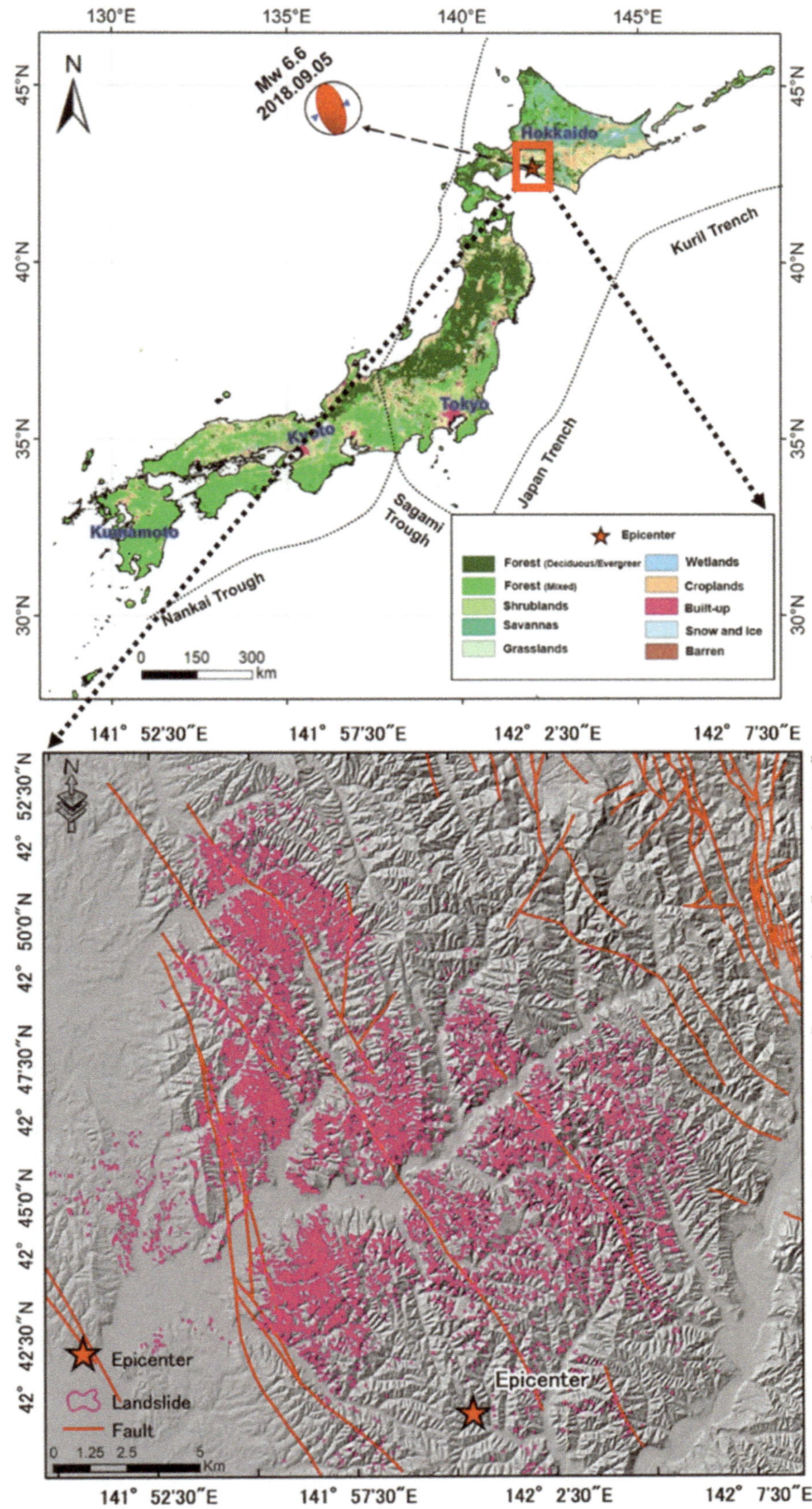

Fig. 1 The location of the case study

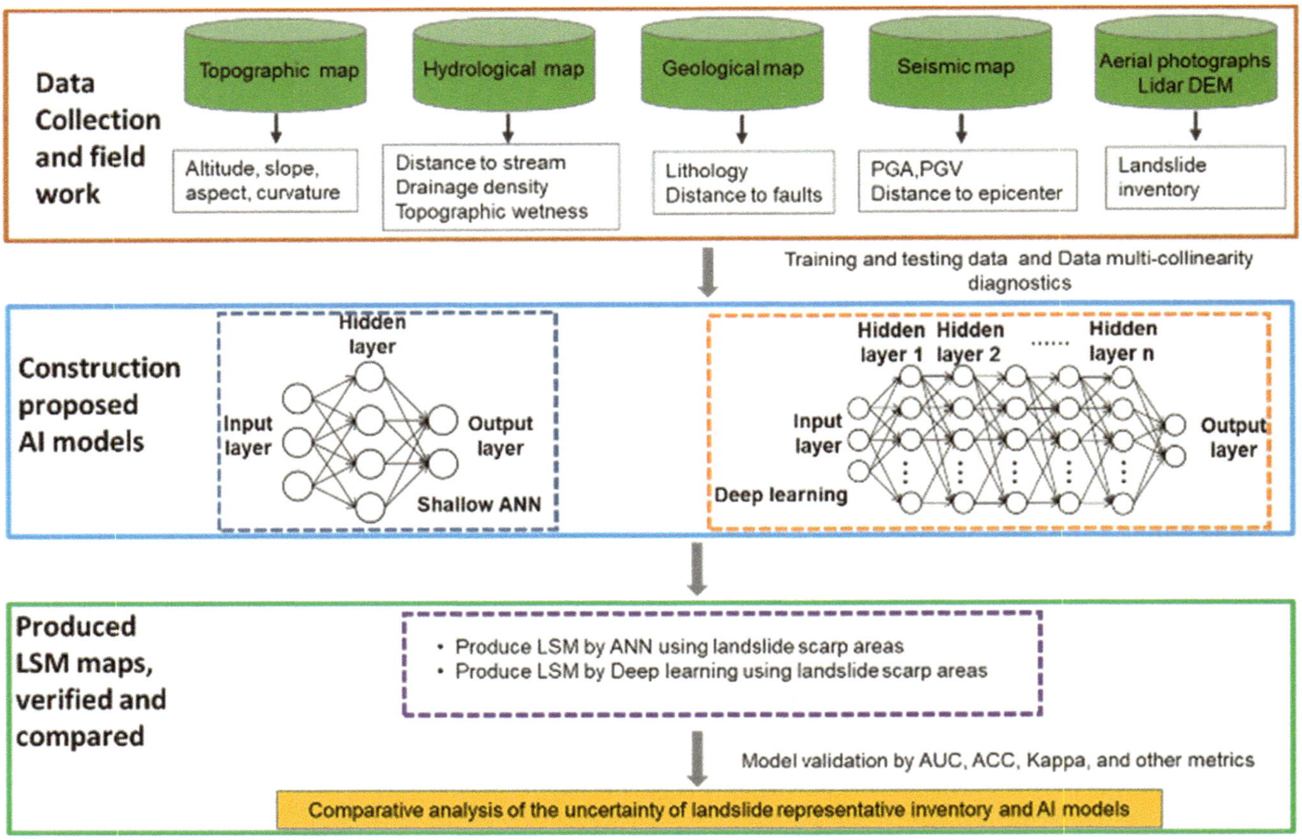

Fig. 2 The research workflow used for this case study

Fig. 3 Landslide representative map **a** graphical sketch of landslide scarp, **b** interpret using the aerial photographs from GSI

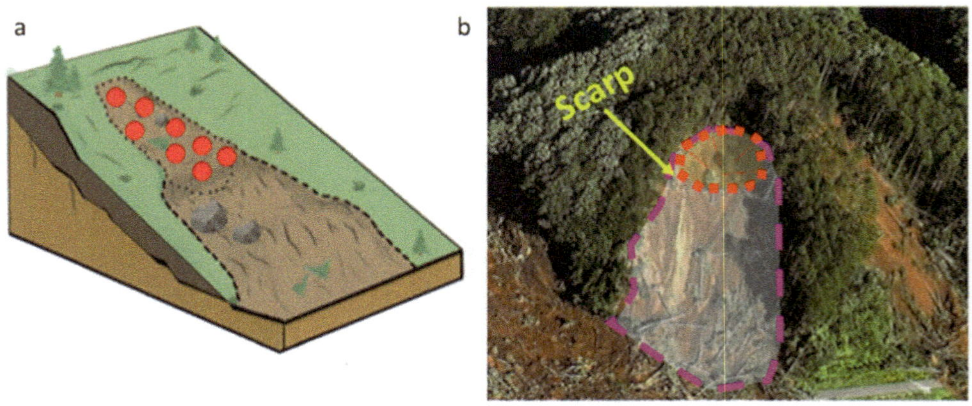

up of a number of simple, highly interconnected processing elements, which process information by their dynamic state response to external inputs.

Most ANN models are composed of simple and highly interrelated processing units (neurons) that are in permanent connection with each other. Generally, neurons are located in different layers, and ANNs are characterized on the basis of the number of layers and the training procedures. Connections between processing units are physically represented by weights, and each neuron has a rule for summing the input weights and a rule for calculating an output value. More than one layer of neurons can be included in the perceptron in order to cope with non-linearly separable problems, and a multilayer perceptron (MLP) can be obtained.

Two-stage processes are involved in using MLP for classification problems: the training stage in which the internal weights are adjusted and the classification stage. Typically, the back-propagation algorithm trains the network until the desired minimal error is achieved between the

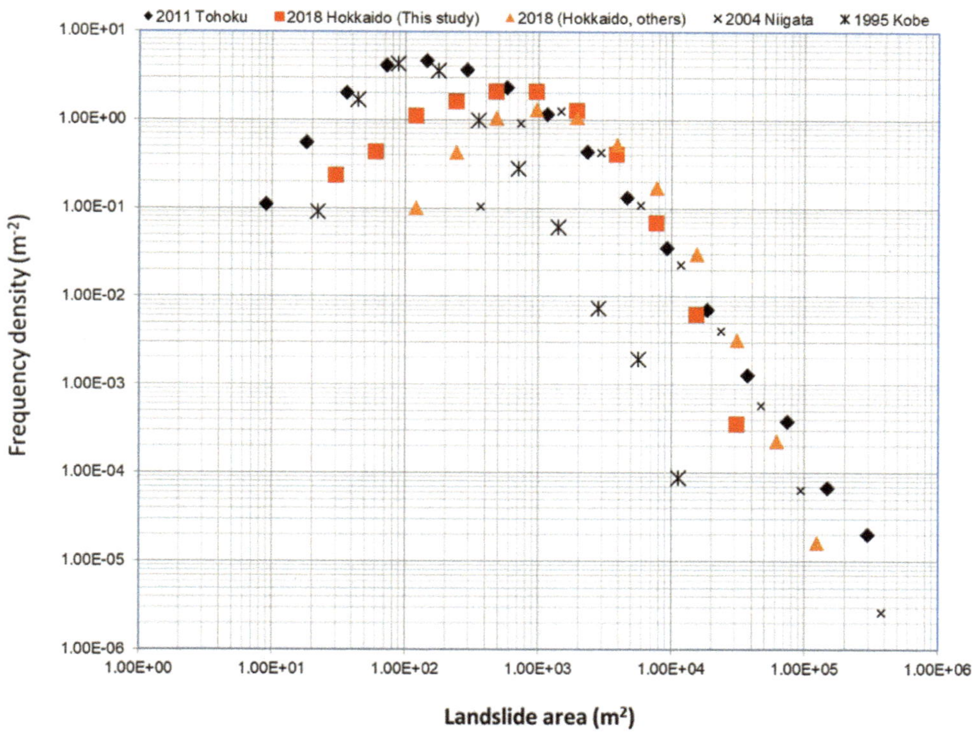

Fig. 4 Probability density functions of landslides in the Hokkaido earthquake and other major earthquakes in Japan

desired and actual output values of the network. Once the training is complete, the network is used as a feed-forward structure to produce a classification for the entire data (Paola and Schowengerdt 1995).

Deep Learning

Deep learning (DL) is non-probabilistic, generative, supervised deep models with greedy layer-wise training much like the ANN training (Bengio 2009). In addition to the supply of good initialization points, the DL comes with other attractive properties. First, the learning algorithm makes effective use of unlabelled data. Second, it can be interpreted as a probabilistic generative model. Third, the over-fitting problem, which is often observed in the models with millions of parameters such as ANN, and the under-fitting problem, which often occurs in ANN, can be effectively alleviated by the generative pre-training step. It has been noted that DL models are capable of detecting complex data patterns much more efficiently than traditional ANN algorithms such as object detection or instance segmentation with high precision. Theoretically speaking, DL is just layers of ANN on top of each other stacked on a certain layout that assure detecting as much data pattern as possible.

Using hidden layers with many neurons in a DL significantly improves the modelling power and creates many jointly optimal configurations. Even if parameter learning is trapped into a local optimum, the resulting DL can still perform quite well since the chance of having a poor local optimum is lower than when a small number of neurons are used in the network.

Collinearity Analysis

Generally, a large number of causative factors are used in hazard assessment models for obtaining the highest accurate products. Nevertheless, the overabundance of data undoubtedly causes an overfitting prognostic model and undermines the statistical significance of an independent variable. Consequently, three statistical tests have been executed for diagnosing data multicollinearity to avoiding impacting the models' accuracy, namely, Pearson Correlation matrix (PCM), Variance Inflation Factors (VIF), and entropy-based descriptor, Information Gain (IG).

Performance Evaluation

Although there exist several statistical metrics to evaluate the performance of machine learning models, in this study, we applied accuracy (classification rate-Acc), kappa, and area under the receiver operatic characteristic curve (AUC) metrics. The accuracy is computed using for possibility indices, including true positive (TP), true negative (TN), false positive (FP), and false-negative (FN). The TP and TN are the number of landslide pixels that correctly classified as landslide and non-landside pixels. However, FP and FN are a number of landslide pixels that incorrectly classified as landslide and non-landside pixels.

Results and Discussion

Rank Importance of Factors

The output results of the PCM (Fig. 5) shows that certain variables are negatively or positively interrelated with each other. Nonetheless, they didn't surpass the acceptable threshold of 0.7 for the relationship except for PGV and Altitude. Although parameters such as PGV and Altitude are having a good correlation between them (0.72), but this may be accidental as there is no clear-cut relationship that exists between peak-ground velocity and altitude, thus can be neglected. It can be confirmed from additional exercises such as VIF analysis (Table 1), that shows the implemented models are safe to use since the VIF is less than the critical threshold of 5 that indicated the no existence of multi-collinearity (Dou et al. 2020) (Figs. 6 and 7; Table 2).

According to Table 1, the obtained IG values for conditioning factors are greater than 0.1, except for distance to streams. Explicitly, parameters such as Epicentre, Altitude, Slopes, stream density, aspects, curvature, and TWI have very high IG contribution to the executed models.

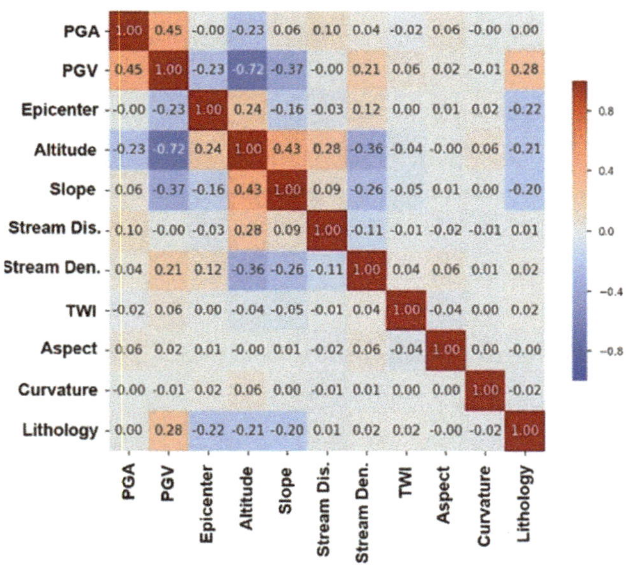

Fig. 5 The output results of the PCM for landslide modelling

Table 1 The multi-collinearity test for landslide conditioning factors using IG and VIF indexes

Variable	IG	VIF
PGA	0.203	1.453
PGV	0.149	2.969
Epicentre	0.612	1.341
Altitude	0.633	3.015
Slope	0.646	1.535
Stream distance	0.012	1.206
Stream density	0.58	1.236
TWI	0.559	1.008
Aspect	0.643	1.01
Curvature	0.663	1.013
Lithology	0.142	1.166

Earthquake-triggered LSM

After the assessment of the rank factors in Sect. 4.1, it is found that all the factors have influenced the triggering of landslides. Next, the landslide susceptibility maps are produced by ANN and DL using the initiation of polygon boundaries (scarp areas). The achieved LSM maps are shown in Fig. 6. The susceptibility index is classified into five levels using natural break classification, i.e., Very low (0–0.05), low (00.05–0.3), moderate (0.3–0.55), high (0.55–0.75), and very high (0.75–1). From the visual interpretation aspect, their results are somewhat similar. On the left bottom corner, most areas are predicted more susceptible zones using ANN than DL; however, the DL model illustrates a visually pleasing map with the transition from very low to very high landslide susceptibilities. ANN-based maps were decent and showed the more sparse distribution of moderate landslide susceptibility class. These results can be explained by the better model fit (i.e., generalization) of scarp dataset models.

Model Validation

The original landslide dataset are randomly divided into two groups: 70% of total landslides for training and left data for testing data. The performance results (Table 2 and Fig. 7) show that both models produced very good results (AUC > 0.859, Acc > 0.89, and kappa > 0.617). Generally, DL outperforms the traditional ANN

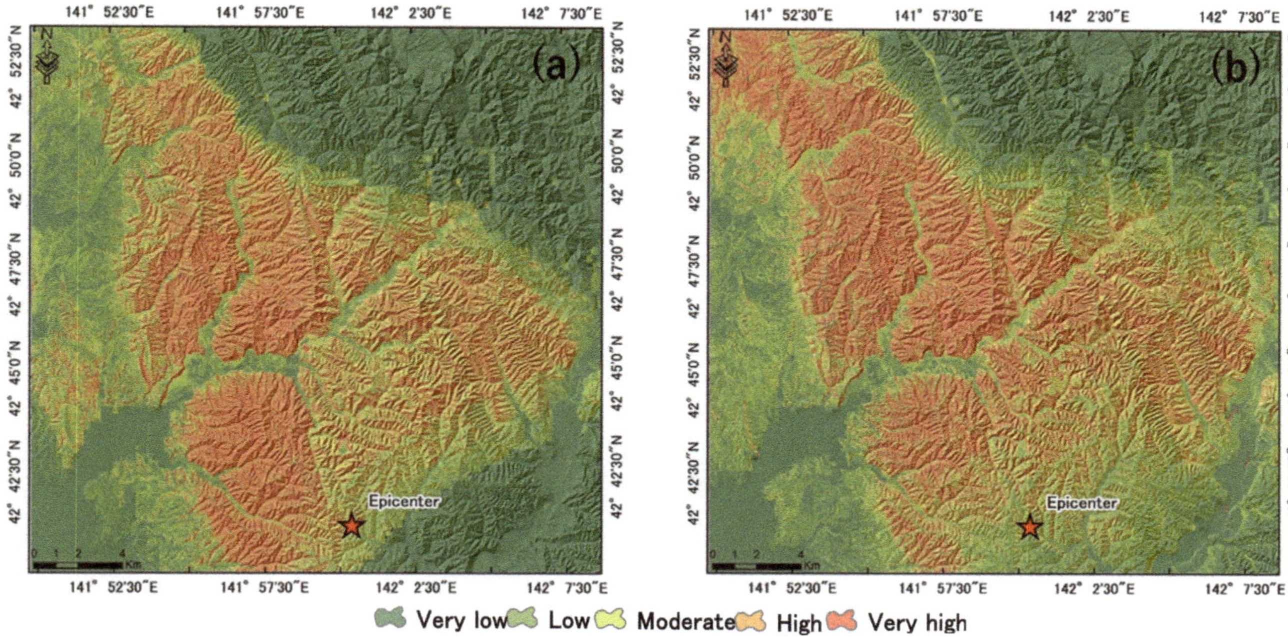

Fig. 6 Landslide susceptibility maps produced by ANN, **a** and DL, **b** models using the scarp areas

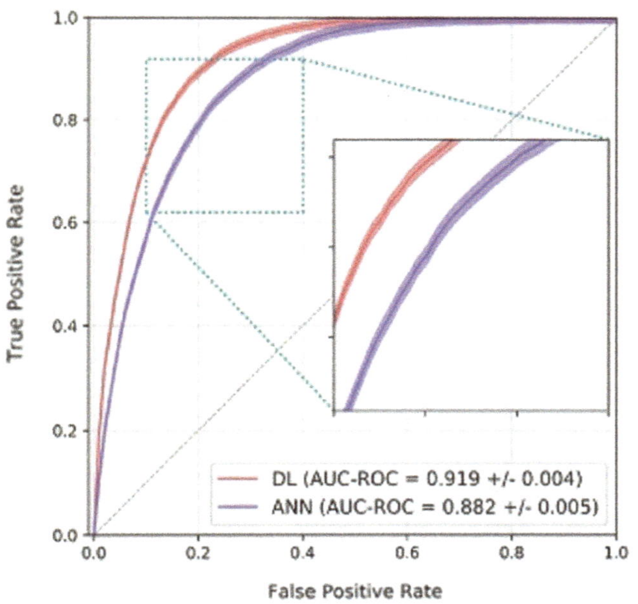

Fig. 7 AUC of Roc curves of the models: ANN and DL, respectively

Discussion

The quality of landslide inventory maps plays very important in the landslide assessment in the mountain areas (Chang et al. 2019). In our research, we prepared a different data source to advance the quality of landslide inventory maps. It exists some problems in interpreting landslides in the shadow regions, and densely forested areas using aerial photographs, yet, the high-resolution LiDAR elevation models improves the visualization and interpretation of topographic features (Görüm 2019), providing accurate landslide inventory mapping. The benefits of DL with a gradient descent method contain fast and simple execution and comparatively rapid conjunction as compared to approaches, like Genetic Algorithms, ANN (Dou et al. 2015a, c). Also, in this study, we used the scarp representation, i.e., the initiation zones and unstable areas for potential failures for improving the accuracy of models. Because the area is covered by thick layers of Quaternary

Table 2 The performance results of DL and ANN models, respectively

Variable		Metric						
		AUC	ACC	Kappa	TP	TN	FP	FN
D	Mean	0.919	0.847	0.693	2099.33	2324	288.33	513
L								
	Std.	0.004	0.006	0.012	36.935	8.485	19.754	15.895
A		0.882	0.809	0.617	1988	2236.33	376	624.333
N	Mean							
N	Std.	0.005	0.005	0.011	44.967	19.754	8.485	21.638

pumice and ash layers, which are volatile for failures during snowmelt seasons or by heavy rainfall, continuous monitoring is essential for mitigating any potential risk.

The causative factors to the models have attracted substantial conditions in the field of landslide hazard mitigation (Tien Bui et al. 2019; Li et al. 2020b; Dou et al. 2020). The result of the assessment of input causative factors importance related to landsliding using IG is shown in Table 1. The results suggest that topographical factors such as curvature, aspect, slope, and altitude are directly correlated with the occurrence of coseismic landslides in the study area. While curvature reflects the topographic amplification factor, the southeast (SE) aspect is found parallel to the strike of the seismogenic source assessed from the moment tensor solutions offered by the United States Geological Survey (USGS).

Concluding Remarks

A comprehensive inventory of over 10,000 landslides observed in the aftershock September 6, 2018, Mw6.6 Hokkaido earthquake has been interpreted. Landslide inventory maps were created using high resolution aerial photographs and Lidar DEM. Two artificial intelligence models, namely, deep learning, and ANN, were used to execute the LSM maps based on scarp inventory for disaster hazard mitigation. The AUC values of models are over 0.859 with excellent results. We found that the proposed DL method has better performance (AUC = 0.919) than ANN (AUC = 0.882). Hence, our findings can deliver a reasonable inventory criterion for the landslide assessment. The results suggest that topographic amplification of the terrain leads to enormous landslides. Our study can be useful for developing appropriate hazard management practices.

Acknowledgements This work has been supported by the JSPS Program, the National Natural Science Foundation of China (Grant No. 51639007), and the opening fund from State Key Laboratory of Hydraulics and Mountain River Engineering, Sichuan University (Grant No. SKHL1903) and we thank Hokkaido Government for providing useful 2m_DEM Lidar data.

References

Bengio Y (2009) Learning deep architectures for AI. Found Trends Mach Learn 2:1–27. https://doi.org/10.1561/2200000006

Chang K-T, Merghadi A, Yunus AP et al (2019) Evaluating scale effects of topographic variables in landslide susceptibility models using GIS-based machine learning techniques. Sci Rep 9:12296. https://doi.org/10.1038/s41598-019-48773-2

Dou J, Chang KT, Chen S et al (2015a) Automatic case-based reasoning approach for landslide detection: Integration of object-oriented image analysis and a genetic algorithm. Remote Sens 7:4318–4342. https://doi.org/10.3390/rs70404318

Dou J, Paudel U, Oguchi T et al (2015b) Shallow and deep-seated landslide differentiation using support vector machines: a case study of the Chuetsu area, Japan. Terr Atmos Ocean Sci 26:227. https://doi.org/10.3319/tao.2014.12.02.07(eosi)

Dou J, Yamagishi H, Pourghasemi HR et al (2015c) An integrated artificial neural network model for the landslide susceptibility assessment of Osado Island, Japan. Nat Hazards 78:1749–1776. https://doi.org/10.1007/s11069-015-1799-2

Dou J, Yunus AP, Bui DT et al (2019a) Improved landslide assessment using support vector machine with bagging, boosting, and stacking ensemble machine learning framework in a mountainous watershed, Japan. Landslides 17:641–658. https://doi.org/10.1007/s10346-019-01286-5

Dou J, Yunus AP, Merghadi A et al (2020) Different sampling strategies for predicting landslide susceptibilities are deemed less consequential with deep learning. Sci Total Environ 720:137320. https://doi.org/10.1016/j.scitotenv.2020.137320

Dou J, Yunus AP, Tien Bui D et al (2019b) Evaluating GIS-Based Multiple Statistical Models and Data Mining for Earthquake and Rainfall-Induced Landslide Susceptibility Using the LiDAR DEM. Remote Sens 11:638. https://doi.org/10.3390/rs11060638

Dou J, Yunus AP, Tien Bui D et al (2019c) Assessment of advanced random forest and decision tree algorithms for modeling rainfall-induced landslide susceptibility in the Izu-Oshima Volcanic Island, Japan. Sci Total Environ 662:332–346. https://doi.org/10.1016/j.scitotenv.2019.01.221

Dou J, Yunus AP, Xu Y et al (2019d) Torrential rainfall-triggered shallow landslide characteristics and susceptibility assessment using ensemble data-driven models in the Dongjiang Reservoir Watershed, China. Nat Hazards 97:579–609. https://doi.org/10.1007/s11069-019-03659-4

Görüm T (2019) Landslide recognition and mapping in a mixed forest environment from airborne LiDAR data. Eng Geol 258:105155

Hecht-Nielsen R (1989) Theory of the backpropagation neural network. IEEE, pp 593–605

Ito Y, Yamazaki S, Kurahashi T (2020) Geological features of landslides caused by the 2018 Hokkaido Eastern Iburi Earthquake in Japan. Geol Soc London, Spec Publ SP501–2019–122. https://doi.org/10.1144/sp501-2019-122

Keefer DK, Larsen MC (2007) GEOLOGY: Assessing Landslide Hazards. Science (80-) 316:1136–1138. https://doi.org/10.1126/science.1143308

Li H, Xu Y, Zhou J et al (2020a) Preliminary analyses of a catastrophic landslide occurred on July 23, 2019, in Guizhou Province, China. Landslides 17:719–724. https://doi.org/10.1007/s10346-019-01334-0

Li Y, Liu X, Han Z, Dou J (2020b) Spatial proximity-based geographically weighted regression model for landslide susceptibility assessment: a case study of Qingchuan area, China. Appl Sci 10:1107. https://doi.org/10.3390/app10031107

Panikkar SV, Subramanyan V (1996) A geomorphic evaluation of the landslides around Dehradun and Mussoorie, Uttar Pradesh, India. Geomorphology 15:169–181. https://doi.org/10.1016/0169-555X(95)00121-K

Paola JD, Schowengerdt RA (1995) A review and analysis of backpropagation neural networks for classification of remotely-sensed multi-spectral imagery. Int J Remote Sens 16:3033–3058

Pham BT, Prakash I, Dou J et al (2019) A novel hybrid approach of landslide susceptibility modelling using rotation forest ensemble and different base classifiers. Geocarto Int 0:1–25. https://doi.org/10.1080/10106049.2018.1559885

Tien Bui D, Shirzadi A, Shahabi H et al (2019) New ensemble models for shallow landslide susceptibility modeling in a semi-arid watershed. Forests 10:743. https://doi.org/10.3390/f10090743

Valagussa A, Marc O, Frattini P, Crosta GB (2019) Seismic and geological controls on earthquake-induced landslide size. Earth Planet Sci Lett 506:268–281. https://doi.org/10.1016/j.epsl.2018.11.005

Yamagishi H, Yamazaki F (2018) Landslides by the 2018 Hokkaido Iburi-Tobu Earthquake on September 6. Landslides 15:2521–2524. https://doi.org/10.1007/s10346-018-1092-z

Yunus AP, Fan X, Tang X et al (2020) Decadal vegetation succession from MODIS reveals the spatio-temporal evolution of post-seismic landsliding after the 2008 Wenchuan earthquake. Remote Sens Environ 236:111476. https://doi.org/10.1016/j.rse.2019.111476

Landslide Susceptibility Assessment by Ensemble-Based Machine Learning Models

Mariano Di Napoli, Giuseppe Bausilio, Andrea Cevasco, Pierluigi Confuorto, Andrea Mandarino, and Domenico Calcaterra

Abstract

Slope failures are among the most hazardous natural disasters, causing severe damage to public and private properties. Casualties owing to landslides have been growing in many areas of the world, especially since the increase of climate changes and precipitations. To this, decision-makers need trustworthy information that may be employed to decide the spatial solution plans to protect people. Statistical landslide susceptibility mapping is facing a constant evolution, especially since the introduction of Machine Learning algorithms (ML). A new methodology is here presented, based on the ensemble of Artificial Neural Network, Generalized Boosting Model and Maximum Entropy ML algorithms. Such an approach has been used in Cinque Terre National Park (Northern Italy), severely affected over the years by landslides, following precipitation events, causing extensive damage in a World Heritage Site. Nine predisposing factors were selected and assessed according to the knowledge of the territory, including slope angle, aspect angle, planform curvature, profile curvature, distance to roads, distance to streams, agricultural terraces state of activity, land use and geological information, whilst a database made of ca. 400 landslides was used as input. Four different Ensemble techniques were applied, after the averaging of 150 stand-alone methods, each one providing validation scores such as ROC/AUC curve. Therefore, the results obtained through Ensemble modeling showed improved values, confirming the reliability and the suitability of the proposed approach for decision-makers in land management at local and regional scales.

Keywords

Landslide • Machine learning • Cinque terre • Susceptibility • Spatial distribution model • Ensemble model

Introduction

Identification and mitigation of landslide risks are extremely complex tasks for decision-makers, being the spatio-temporal prediction of these phenomena still characterized by large uncertainties. For this reason, the assessment of landslide susceptibility is fundamental to facilitate their understanding. Landslide susceptibility is defined as the likelihood of landslide occurrence in an area based on the local terrain and environmental condition (Brabb 1984). No standard susceptibility procedure has yet been established, therefore, numerous algorithms and methodologies have been tested for such purposes. Among the new statistical approaches, Machine Learning algorithms (ML) are gaining increasing consideration. However, such methodologies still convey uncertainties and many authors have proposed methods to minimize it, relying on combining the predictions yielded by multiple algorithms, as testified by the works of Umar et al. (2014), Kim et al. (2018), Bueechi et al. (2019).

Anyhow, Ensemble Modeling (EM) provides a concrete contribution to minimize uncertainty and to refine and increase the prediction accuracy, the key parameter to consider when working on Landslide Susceptibility Mapping (LSM). In this study, three different MLAs were used to assess the landslide susceptibility of the Cinque Terre National Park territory (Northern Italy, eastern Liguria Region) (Fig. 1).

M. Di Napoli · A. Cevasco · A. Mandarino
Department of Earth, Environmental and Life Sciences, University of Genova, Corso Europa 26, Genoa, Italy

G. Bausilio · P. Confuorto · D. Calcaterra (✉)
Department of Earth Sciences, Environment and Resources, Federico II University of Napoli, Via Cinthia 21, Naples, Italy
e-mail: domenico.calcaterra@unina.it

F. Guzzetti et al. (eds.), *Understanding and Reducing Landslide Disaster Risk*, ICL Contribution to Landslide Disaster Risk Reduction, https://doi.org/10.1007/978-3-030-60227-7_24

Fig. 1 Setting of the study area, in red the Cinque Terre National Park. In the top-right inset, Italy and the location of the Park

Fig. 2 **a**, **b** Landslides located in the National Park area, **c** example of agricultural terraces in the Cinque Terre (photo M. Di Napoli)

Study Area

Geological and Geomorphological Setting

The Cinque Terre is a wonderful naturalistic scenario, named in 1997 as a UNESCO World Heritage Site and recognized as a National Park since 1999. This area is a typical example of anthropic landscape, characterized by well-known century-old agricultural terraces retained by thousands of km of dry-stone walls. Geologically, the Cinque Terre is part of an NW-SE oriented segment of the Northern Apennine, a orogenic chain formed during the Tertiary (Abbate 1969; Terranova et al. 2006). It is made up of a nappe sequence that includes five overlapping tectonic units, from top to bottom: Gottero Unit, Ottone Unit, Marra Unit, Canetolo Unit and Tuscan Nappe (Raso et al. 2019a).

Local morphology is strongly influenced by the geo-structural and lithological configuration: most of the slopes are SE to SW oriented, with high slope gradient values (above 60°) for more than 74% of the total area (ca. 38 km^2). The hydraulic network is deeply influenced by the tectonic activity: the profiles of the main streams are short and very steep, with an anti-Apennine direction, overlapped to an approximately N-S oriented faulting system. These creeks show, for the most part, a torrential behaviour, characterized by high rates of solid transport mainly due to the erosional processes affecting the surrounding slopes, especially during extraordinary rainfall events.

The landscape of the Cinque Terre has been almost totally modified by human activities during the last hundred years through agricultural terraces building (Terranova 1984; Brandolini 2017), contributing to the worldwide notoriety of the Cinque Terre National Park, and, at the same time, having also a fundamental role for slope stability (Fig. 2).

New spaces were gained by local farmers to cultivate mainly vineyards and olive groves, due to the lack of fertile terrains, reworking and retaining shallow eluvial-colluvial soil covers by constructing dry stone-wall, over the most suitable bedrock (Brandolini 2017). The agricultural terraced slope at Cinque Terre extends from the shoreline, just above the cliff edge or from the toes of coastal landslides, to an average altitude of 400–450 m a.s.l., up to 500 m in some cases (Terranova 1984; Terranova et al. 2006). According to Terranova (1984), only 19% of the total terraced areas (about 60% of the territory of the Cinque Terre) were still cultivated at that time.

This issue is a testimony of the social changes that took place in the second half of the 1950s and had an impact on the land use at the Cinque Terre (Schilirò et al. 2018). The lack of maintenance of dry-stone walls and the clogging of drainage channels due to farmland abandonment are considered among the main causes behind the increase in landslide susceptibility of terraced slopes within the Cinque Terre (Cevasco et al. 2014; Brandolini et al. 2018).

Data and Methods

Landslide Inventory

Recurring landslides have historically plagued the Cinque Terre area, mainly in response to extreme rainfall events

(Cevasco et al. 2015). Raso et al. (2019a, 2019b) redacted a thorough landslide inventory map of the Cinque Terre, where a total of 459 landslides, having areal extent higher than 100 m^2, were identified and classified according to Cruden and Varnes (1996) and Hungr et al. (2001) schemes. A large number of debris slides (22.9%) can be associated with the vulnerability of dry-stone terraces, while rockfalls (17.6%) are concentrated along the coast. Debris flows (11.1%) are also diffused, especially in the western sector of Cinque Terre (Monterosso and Vernazza hamlets).

The landslide inventory used in this work was obtained considering only those landslide characterized by rapid kinematics and attesting in the shallower levels of the ground. Thus, four hundred landslides have been implemented for the generation of the LSM.

Predisposing Factors

Nine predisposing factors (PFs) were chosen. PFs encompass slope and aspect angle, planform and profile curvature, distance to roads and streams, agricultural terraces state of activity, land use and geolithological map (Cevasco et al. 2014; Schilirò et al. 2018). It must be stated that the analysis is focused on the environmental variables, thus triggering factors were not taken into consideration (e.g. intense rainfall or earthquakes).

Modeling Procedures

Three different MLAs, implemented in the framework of the "biomod2" package (Thuiller et al. 2016), developed in R environment (R Development Core Team 2019), were applied. Specifically, Artificial Neural Network (ANN) (Zurada 1992), Generalized Boosting Model (GBM) (Schapire 1989) and Maximum Entropy (MaxEnt) (Phillips and Dudík 2008) algorithms were chosen, for their previous good performance (Elith et al. 2006).

A k-fold cross-validation approach was used to implement and evaluate the models obtained with the different ML algorithms (Araújo and New 2007). Cross-validation is one of the most widely accepted approaches for testing the predictive accuracy of Species Distribution Models (SDMs). A definite percentage of the data is kept for calibration (i.e. training data) while the residue is used to test (i.e. testing data) the prediction of the model. The whole approach is hence repeated several times for every single model and the average predictive accuracy is finally reported (Araújo and New 2007; Thuiller et al. 2009). Thus, each SDM algorithm was performed 50 times, and in every run, input data were subdivided in different combination of training (80%) and testing data (20%).

Ensemble Modeling

The Ensemble Modeling approach (EM) was firstly introduced by Burnham and Anderson (2002), who made the average of different regression models. Through ensembling, uncertainty and bias of both variables and models are reduced. The first step (Fig. 3) allowed to recognize landslide occurrence data as the response variable and the predisposing factors as predictors. In the second step, the Variance Inflation Factor (VIF) was measured, and a value of 0.7 was set as a threshold, to reduce the collinearity among predictors. The VIF detects multicollinearities that cannot always be easily identified through a simple correlation. In the next step, the above mentioned SDMs were run to model the study area. Furthermore, during this phase, the importance of each environmental variable was evaluated (Fig. 3).

To assess the results reliability of each SDM, the ROC method (Receiver Operating Characteristic) was chosen. Prediction accuracy is considered to be similar to random for

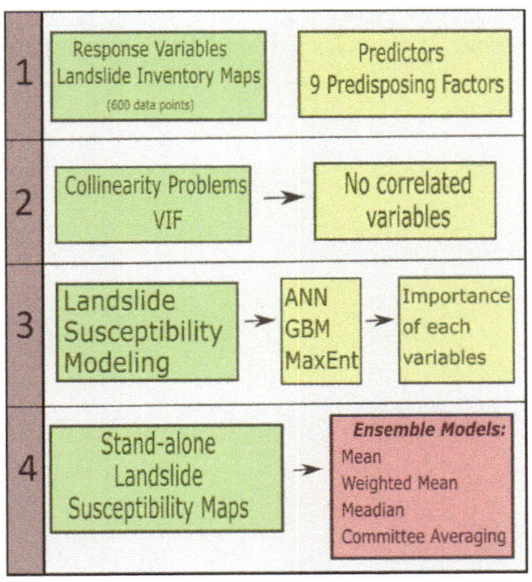

Fig. 3 Flow chart of the proposed approach for Landslide Susceptibility Modeling

ROC/AUC values lower than 0.5, poor for values in the range 0.5–0.7, fair in the range 0.7–0.9, and excellent for values higher than 0.9 (Swets 1988; Fressard et al. 2014).

Lastly, the outputs of the stand-alone methods were combined by only applying four ensemble techniques, namely (Thuiller et al. 2016): mean of probabilities (PM), median of probabilities (PME), binary mean or committee averaging (CA) and weighted mean of probabilities (PMW).

Results and Conclusions

Ensemble Forecasting

Four different Ensemble methods were selected, depending on the method used to aggregate the probability values: mean, weighted mean, median and binary mean. The usage of EM has ensured an evident improvement of the evaluation scores and therefore of the model reliability, compared to stand-alone methods (Table 1). Here, The Weighted Mean method (PMW) was exclusively adopted because of its higher validation performance (AUC = 0.9), which makes it more effective from a previsional standpoint. All the subsequent analyses are carried out by using the above mentioned methodology (Fig. 4).

To verify the performance of the models used, statistical analyses were performed by intersecting the inventoried landslides with the weighted mean map. The distribution of landslides in the various susceptibility ranges is shown in Fig. 5: it is characterized by an increasing trend, with the highest number of inventoried landslides present in the high susceptibility class (ca. 36%). On the contrary, the areal distribution shows a decreasing trend, with extensive portions of territory with low susceptibility values (ca. 38%) and a limited areal extension characterized by higher values.

Table 1 AUC scores for the stand-alone and ensemble ML algorithm used

Stand-alone methods	AUC
GBM	0.78
ANN	0.74
MaxEnt	0.8
Ensemble methods	**AUC**
Mean	0.89
CA	0.86
Median	0.87
Wmean	0.9

Through the application of the biomod2 library, is possible to obtain the estimation of the variable importance for each model. Through this peculiar function, landslide dynamics in the study area and the role of representative PFs can be assessed.

The sum of all predictors, for any model, is not equal to 1, because every variable must be considered individually. Aspect, slope and terraces state of activity variables exhibited a higher score thant those in every other model, with values of 0.24, 0.22 and 0.38, respectively. The other PFs showed moderate levels of importance, such as land use (0.15), planform curvature (0.18) and geological factors (0.10).

Discussion and Conclusions

With the aim of a proper landslide risk mitigation, accurate landslide susceptibility mapping dealing with locally-based data is essential. For this reason, statistic-probabilistic methods represent an optimal synthesis. However, one of the constraints of statistical methods is represented by the uncertainty associated with every methodology. One of the main objectives is the minimization of such effects, also through a multiple data computation, as done in this work, implementing EM. The combination of several models demonstrates higher soundness and reliability, if compared to the application of single models. To remark this statement, evaluation scores present higher values when EM has been applied, as well as most of landslides inventoried in the Cinque Terre area are located into higher susceptibility classes.

Environmental variables chosen as PFs were selected according to the knowledge of the territory and many of them were collected through accurate field surveys. Such an assumption is fundamental because many aspects here presented are strictly related to the peculiar geomorphological setting and of the anthropic activity of the Cinque Terre area. Indeed, the selection of landslide PFs is not based on universally recognized guidelines, but it was determined according to the type of landslides and the slope dynamics of the Cinque Terre area. Geomorphological variables, such as

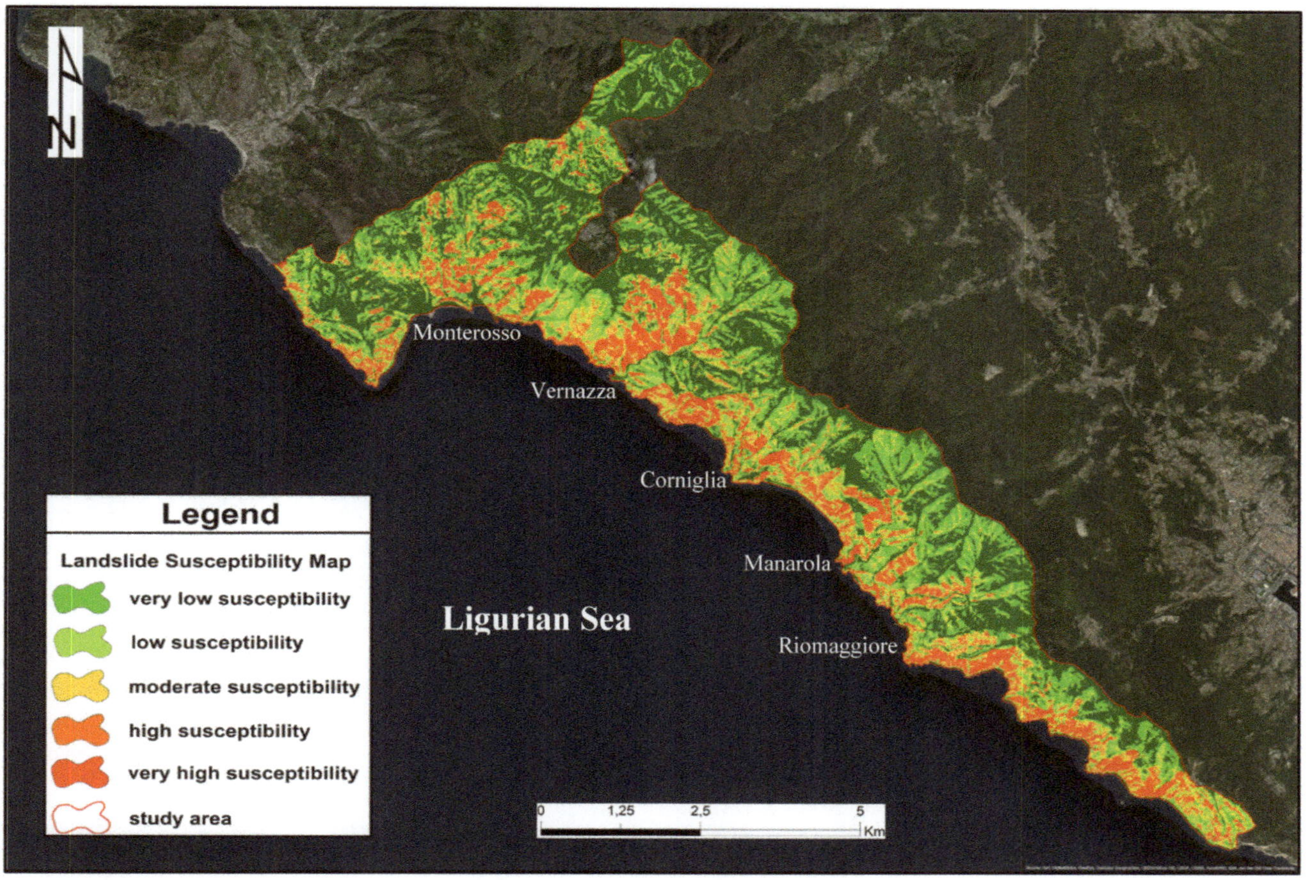

Fig. 4 Landslide susceptibility map obtained with weighted mean method (PMW)

slope and aspect, are very significant, influencing the debris cover deposition, the hydrological circulation, evapotranspiration processes and thus erosional processes interacting with the peculiar land use. Among the local variables, an important role is played by human activities such as terraces state of activity and land use. These factors have been already highlighted in previous works (Cevasco et al. 2013, 2014; Brandolini et al. 2018), revealing that terraces abandoned for a short time showed the highest landslide susceptibility degree, as well as slope failures affecting cultivated zones were characterized by a lower magnitude than those occurred on abandoned terraced slopes. However, a further refinement of the input data, such as constant update and new-data implementation is necessary to obtain susceptibility models which may consider all potential factors leading to landslide triggering.

Finally, further improvements and newer awareness about ML algorithms for landslide susceptibility mapping are necessary, to obtain more accurate estimations and reduce uncertainty. A further improvement toward more precise forecasting, both in time and space, of landslide occurrence may be based on the integration of landslide spatial prediction and triggering factors such as rainfalls, aiming at the full support for civil protection agencies and for an optimal set-up of warning systems. This may potentially represent a fundamental tool for the safety of both people living in the area and tourists that every year visit the Cinque Terre National Park.

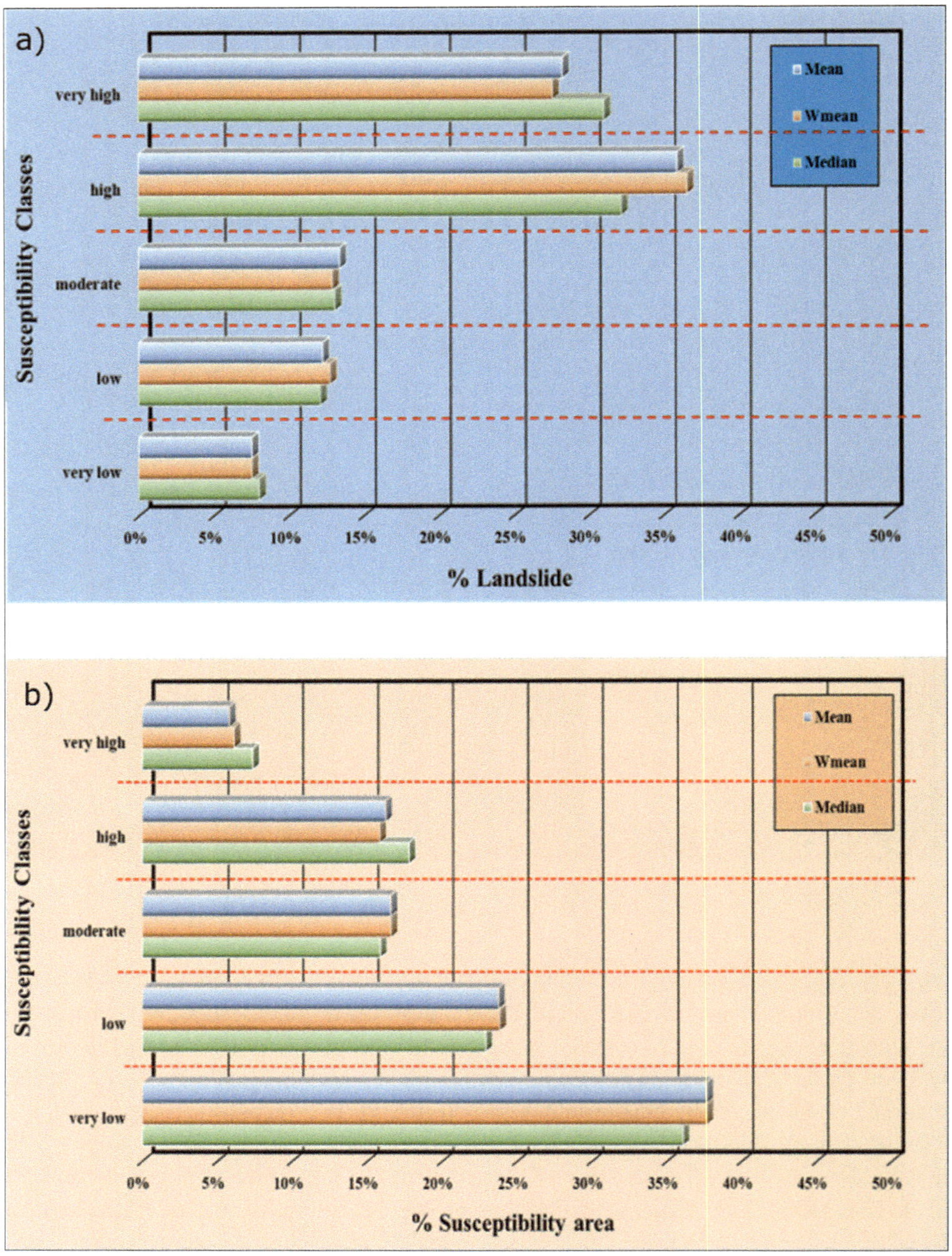

Fig. 5 **a** Number of landslides in the different susceptibility classes, **b** extension area of the different susceptibility classes

Acknowledgements Research funded by Cinque Terre National Park, which is greatly acknowledged. An anonymous reviewer is thanked for providing helpful comments on an earlier draft of the manuscript.

References

Abbate E (1969) Geologia delle Cinque Terre e dell'entroterra di Levanto (Liguria Orientale). Memorie della società geologica Italiana, vol 8. Arti Grafiche Pacini Mariotti, Pisa

Araújo MB, New M (2007) Ensemble forecasting of species distributions. Trends Ecol Evol (Amst) 22:42–47. https://doi.org/10.1016/j.tree.2006.09.010

Brabb EE (1984) Innovative approaches to landslide hazard and risk mapping. In: International landslide symposium proceedings Toronto Canada, vol 1, pp 17–22

Brandolini P (2017) The outstanding terraced landscape of the Cinque Terre coastal slopes (Eastern Liguria). In: Landscapes and landforms of Italy, Springer, Cham, pp 235–244

Brandolini P, Cevasco A, Capolongo D, Pepe G, Lovergine F, Del Monte M (2018) Response of terraced slopes to a very intense rainfall event and relationships with land abandonment: a case study from Cinque Terre (Italy). Land Degrad Dev 29:630–642. https://doi.org/10.1002/ldr.2672

Bueechi E, Klimeš J, Frey H, Huggel C, Strozzi T, Cochachin A (2019) Regional-scale landslide susceptibility modelling in the Cordillera Blanca, Peru—a comparison of different approaches. Landslides 16:395–407. https://doi.org/10.1007/s10346-018-1090-1

Burnham KP, Anderson DR (2002) Model selection and inference: a practical information-theoretic approach (2nd edn). Springer-Verlag, New York. http://dx.doi.org/10.1007/b97636

Cevasco A, Pepe G, Brandolini P (2014) The influences of geological and land use settings on shallow landslides triggered by an intense rainfall event in a coastal terraced environment. Bull Eng Geol Environ 73:859–875. https://doi.org/10.1007/s10064-013-0544-x

Cevasco A, Diodato N, Revellino P, Fiorillo F, Grelle G, Guadagno FM (2015) Storminess and geo-hydrological events affecting small coastal basins in a terraced Mediterranean environment. Sci Total Environ 532:208–219. https://doi.org/10.1016/j.scitotenv.2015.06.017

Cruden DM, Varnes DJ (1996) Landslides: investigation and mitigation. Chapter 3-Landslide types and processes. Transportation research board special report (247)

Elith J, Graham CH, Anderson RP, Dudík M, Ferrier S, Guisan A, Hijmans RJ, Huettmann F, Leathwick JR, Lehmann A, Li J, Lohmann LG, Loiselle BA, Manion G, Moritz C, Nakamura M, Nakazawa Y, Overton J, McC M, Townsend Peterson A, Phillips SJ, Richardson K, Scachetti-Pereira R, Schapire RE, Soberón J, Williams S, Wisz MS, Zimmermann NE (2006) Novel methods improve prediction of species' distributions from occurrence data. Ecography 29(2):129–151

Fressard M, Thiery Y, Maquaire O (2014) Which data for quantitative landslide susceptibility mapping at operational scale? Case study of the Pays d'Auge plateau hillslopes (Normandy, France). Nat Hazards Earth Syst Sci 14:569–588. https://doi.org/10.5194/nhess-14-569-2014

Hungr O, Evans SG, Hutchinson IN (2001) A review of the classification of landslides of the flow type. Environ Eng Geosci 7(3):221–238

Kim HG, Lee DK, Park C, Ahn Y, Kil S-H, Sung S, Biging GS (2018) Estimating landslide susceptibility areas considering the uncertainty inherent in modeling methods. Stoch Environ Res Risk Assess 32:2987–3019. https://doi.org/10.1007/s00477-018-1609-y

Phillips SJ, Dudík M (2008) Modeling of species distributions with Maxent: new extensions and a comprehensive evaluation. Ecography 0:080328142746259-??? doi: 10.1111/j.0906-7590.2007.5203.x

R Core Team (2019) R: a language and environment for statistical computing [Internet]. Vienna, Austria: R Foundation for Statistical Computing; Available from https://www.R-project.org/

Raso E, Mandarino A, Pepe G, Di Martire D, Cevasco A, Calcaterra D, Firpo M (2019a) Landslide inventory of the Cinque Terre National Park, Italy. In: IAEG/AEG Annual Meeting Proceedings, San Francisco California 2018. Springer, Cham, vol 1, pp 201–205

Raso E, Cevasco A, Di Martire D, Pepe G, Scarpellini P, Calcaterra D, Firpo M (2019b) Landslide-inventory of the Cinque Terre National Park (Italy) and quantitative interaction with the trail network. J Maps 15:818–830. https://doi.org/10.1080/17445647.2019.1657511

Schapire RE (1989) The strength of weak learnability. In: 30th Annual Symposium on Foundations of Computer Science, pp 28–33

Schilirò L, Cevasco A, Esposito C, Scarascia Mugnozza G (2018) Shallow landslide initiation on terraced slopes: inferences from a physically-based approach. Geomat Nat Hazards Risk 9:295–324. https://doi.org/10.1080/19475705.2018.1430066

Swets JA (1988) Measuring the accuracy of diagnostic systems. Science 240:1285–1293. https://doi.org/10.1126/science.3287615

Terranova R (1984) Aspetti geomorfologici e geologico-ambientali delle Cinque Terre: rapporti con le opere umane. Studi e ricerche di Geografia 7:38–90 (In Italian)

Terranova R, Zanzucchi G, Bernini M, Brandolini P, Campobasso S, Faccini F, Zanzucchi F (2006) Geologia, geomorfologia e vini del parco Nazionale delle Cinque Terre (Liguria, Italia). Bollettino della Società Geologica Italiana Spec 6:115–128 (In Italian)

Thuiller W, Lafourcade B, Engler R, Araújo MB (2009) BIOMOD—a platform for ensemble forecasting of species distributions. Ecography 32:369–373. https://doi.org/10.1111/j.1600-0587.2008.05742.x

Thuiller W, Georges D, Engler R, Breiner F (2016) biomod2: Ensemble platform for species distribution modeling. R package version 3.3–7

Umar Z, Pradhan B, Ahmad A, Jebur MN, Tehrany MS (2014) Earthquake induced landslide susceptibility mapping using an integrated ensemble frequency ratio and logistic regression models in West Sumatera Province, Indonesia. CATENA 118:124–135. https://doi.org/10.1016/j.catena.2014.02.005

Zurada JM (1992) Introduction to artificial neural systems, vol 8. West publishing Company, St. Paul

Application of Machine Learning Algorithms and Their Ensemble for Landslide Susceptibility Mapping

2020 Kyoto Japan

Bahareh Kalantar, Naonori Ueda, Vahideh Saeidi, and Parisa Ahmadi

Abstract

Landslide hazard assessment followed by susceptibility map is the primary purposes of predicting the landslides and reducing the risk of landslide occurrences. Such information which can be obtained through detailed analyses of landslide susceptible area is a significant step in developing adequate models in landslide prone zones. In this research, three methods of machine learning, namely Generalized Linear Model (GLM), Boosted Regression Trees (BRT), Support Vector Machine (SVM), and their ensemble model were applied in a clipped region of Sajadrood, Iran. The evaluation of spatial correlations between 14 landslide conditioning factors and classifying their importance have been done, as well to identify the most important cause of landslide in the area. Eventually, accuracy assessment has been carried out using the area under the curve (AUC), correlation coefficient (COR), and total sum of squares (TSS). The results of AUC showed that GLM produced the highest prediction accuracy, with the value of 0.86, followed by SVM (0.84), ensemble model (0.84) and BRT (0.83). Consequently, Stream Power Index (SPI) was the least important factor for landslide prediction. The results also showed that curvatures and distance to road had the most significant effect on the occurrence of landslide in Sajarood.

Keywords

Landslide susceptibility mapping • GLM • BRT • SVM • Conditioning factors • Ensemble

B. Kalantar (✉) · N. Ueda
RIKEN Center for Advanced Intelligence Project, Goal-Oriented Technology Research Group, Disaster Resilience Science Team, Tokyo, 103-0027, Japan
e-mail: Bahareh.kalantar@riken.jp

N. Ueda
e-mail: naonori.ueda@riken.jp

V. Saeidi
Department of Mapping and Surveying, Darya Tarsim Consulting Engineers Co. Ltd, 1457843993 Tehran, Iran
e-mail: saeidi@daryatarsim.com

P. Ahmadi
Institute of Graduate Studies Building, Institute of Ocean and Earth Science (IOES), University of Malaya, 50603 Kuala Lumpur, Malaysia
e-mail: parisa@um.edu.my

Introduction

Soil and rock mass movement or landslide is a type of natural hazard, which commonly happens due to natural phenomena and human activities around the world (Al-Najjar et al. 2019). This failure endangers the environment, economy, lives, residences, transportations and infrastructures (Gupta et al. 2018). Susceptibility mapping is a valuable and suitable method to display the probability of the risks in certain areas (Kalantar et al. 2019). Influential components have remarkable effects to landslide prone areas; for instance, mining, heavy rainfall, volcanic activities, earthquakes and iceberg melting can trigger the landslides (Mousavi et al. 2011). Accordingly, study and modelling the landslide conditioning factors and their relationship to provide landslide susceptibility map (LSM) might lead to understand the nature of many landslide events for better mitigating, preventing and early alarming (Al-Najjar et al. 2019; Gupta et al. 2018).

Recently, machine learning methods have been widely used in the field of natural disaster prediction modelling including landslide and their superiority to other qualitative and probabilistic methods have been proven (Gupta, et al. 2018). Several studies have also been conducted on the comparison of different machine learning techniques, including support vector machine (SVM) and decision tree (DT) in modelling processes, which, depending on the risk under studies and the research conditions, they have yielded

F. Guzzetti et al. (eds.), *Understanding and Reducing Landslide Disaster Risk*, ICL Contribution to Landslide Disaster Risk Reduction,
https://doi.org/10.1007/978-3-030-60227-7_25

different results. In some cases, DT depicted a better result (Singh et al. 2009), while in others, SVM (Marjanović et al. 2011; Wu et al. 2014). In this regards, Ada and San (2018) applied SVM and random forest (RF) for landslide susceptibility mapping using a sampling strategy named two-level random sampling (2LRS). The results showed the highest performance was recorded using both methods. Goetz et al (2015) applied logistic regression (LR), generalized additive models (GAM), weights of evidence (WOE), SVM, RF, and bootstrap aggregated classification trees (bundling) with penalized discriminant analysis (BPLDA) in spatial prediction and modelling of landslides which led to acceptable accuracy and robust prediction results. Some scholars also presented an ensemble technique like AdaBoost, MultiBoost, Bagging, and Rotation Forest, which is combination of different machine learning methods for producing hybrid models to overcome complexity of the model (Hong et al. 2018).

Through an in-depth investigation in literature, it appears that the involvement of various combinations of compelling factors and methods is highly recommended to gain more precise susceptibility maps. This is because diverse factor combination and models have particular impact on hazard mapping accuracies. For instance, adding or omitting a factor in the combination process could lead to different accuracy and inference. Hence, in this study, our aim is to examine the effectiveness of 14 conditioning factors to extract the most determinant ones, which have a higher probability of causing landslides. As there is no standard guideline to rank these factors (Roy & Saha 2019), we evaluate the most dominant factors by the literatures through the model processing. We also examine sensitivity and impact of various supervised machine learning modelling namely Generalized Linear Model (GLM), SVM, Boosted Regression Trees (BRT) and their ensemble model to the group of the dataset. Finally, their efficiency and validation have been evaluated against area under the curve (AUC), correlation coefficient (COR), and total sum of squares (TSS).

Study Area and Data Preparation

Geological Characteristic of the Region

The study was conducted in Sajadrood catchment located in Mazandaran province, Iran between latitudes of 36°9′N, 36° 10′N and longitudes of 52°30′E, 52°40′E (Fig. 1). The study area is about 119 km^2 mostly covered by agricultural, forestry and paddy land use. In terms of weather, the average temperature fluctuated annually from −3 °C in February to 38 °C in August, and the average annual rainfall is about 680 mm.

Landslide Inventory Map

A landslide inventory map was developed using geomorphological surveys and satellite images as well as field surveying. Furthermore, 227 historical landslides were located in the region and 70% of the inventories were randomly utilized for training the algorithms and remains (30%) for the testing and validation (Al-Najjar et al. 2019).

Landslide Conditioning Factors

Landslide is affected by a combination of geographical and environmental factors, each of which has a different effect on landslide phenomena (Hong et al. 2018). A topographic map with the scale of 1: 25,000 was used to generate digital elevation model (DEM) with 10 meters spatial resolution as a base and consequently for other DEM derivatives, as well. In this study, 14 conditioning factors were provided. The selected factors are namely Altitude, Aspect, Curvature, Plan curvature, Profile curvature, Slope, SPI, Topographic Roughness Index (TRI), Topographic Wetness Index (TWI), land use/cover (LULC), lithology, Distance to fault, Distance to road and Distance to stream. Below, we briefly mention about the landslide causal factors.

Topography Factors

Topography factors or altitude, slope degree, slope aspect, curvature, profile curvature, and plan curvature, were extracted from the 10 m DEM by spatial analyst tool and were classified (Kalantar et al. 2019) using ArcGIS. Altitude has effect on geomorphology and geology of the region, hence, it considers as one of the landslide causal factor (Kalantar et al. 2019). The altitudes of the study area were categorized into 5 using the natural break classification. The altitude classes range from 74 meters (representing in the northern part) to 1500 meters height in the southern area.

Slope has considerable influence on land stability and weathering and soil strength (Calvello & Ciurleo, 2018). Based on quantile scheme (Khosravi et al. 2018) the slope degree was classified into 5 classes; from angles less than 8.4° to maximum angel of 48°.

Aspect represents the orientation and direction of the ground slope; therefore, it influences soil characteristics of the study area (Al-Najjar et al. 2019). Here, the aspect map was divided into nine classes as (1) flat, (2) north, (3) northeast, (4) east, (5) southeast, (6) south, (7) southwest, (8) west, and (9) northwest.

Curvature, profile curvature, and plan curvature define the surface shape and soil run off so they might influence the speed of soil run off and slide, and flow orientation on the

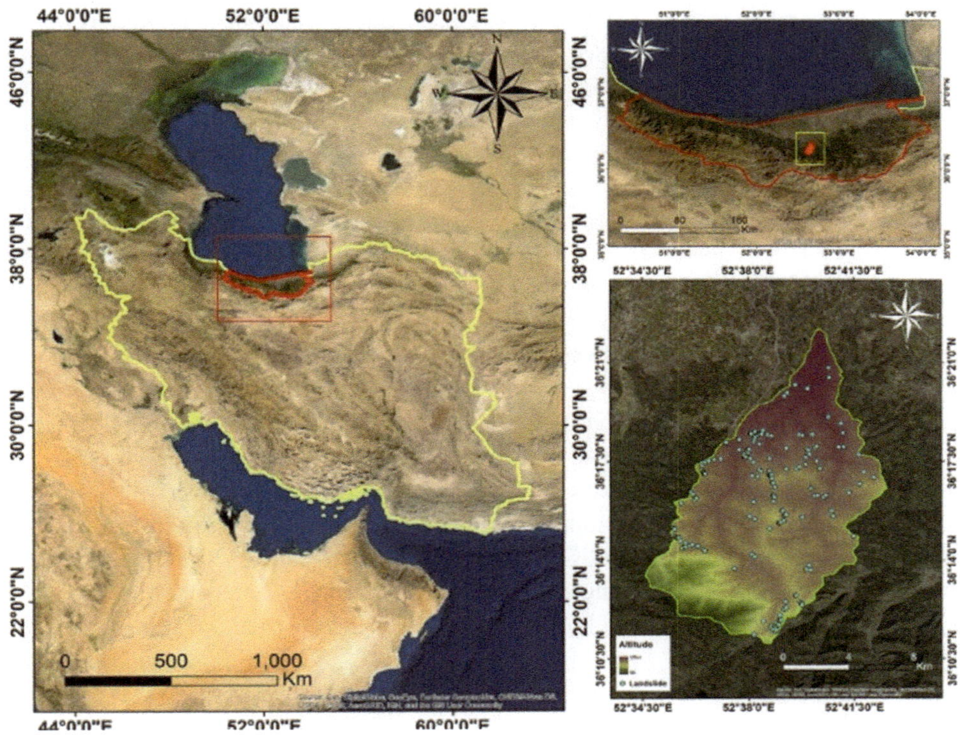

Fig. 1 Study area and Landslide inventory map

surface (Al-Najjar et al. 2019). Curvature is a definition for deviation from a straight line whereas, Profile curvature is parallel to the direction of maximum slope measuring the flow acceleration/deceleration, and plan curvature is perpendicular to the direction of maximum slope, representing convergence/divergence of flow across the surface (Saleem et al. 2019). These factors were classified into three classes namely concave, flat and convex using ArcGIS 10.2.

Hydrological Factors

SPI and TWI represent the hydrological characteristics of the soil and surface (Al-Najjar et al. 2019). TWI (Eq. 1) depends on slope and DEM and evaluates soil moisture and the tendency of water to accumulate and turn to runoff and also the position where water converges (Saleem et al. 2019). SPI (Eq. 2) is an index for stream power which measures the erosion power and intensity of slope surface runoff using DEM (Moore et al. 1991). The higher SPI increases the erosion risk, as well (Al-Najjar et al. 2019).

The following formulas implemented to extract hydrological factors:

$$TWI = \log_e\left(\frac{A_s}{\tan\beta}\right), \tag{1}$$

$$SPI = A_s \tan\beta \tag{2}$$

where β (in radians) is the gradient of the slope, A_s area of the catchment in m^2. Another hydrological factor in landslide modelling is distance to river. Furthermore, stream adjacency triggers landslide and erosion and distance to stream directly changes the soil moisture (Al-Najjar et al. 2019).

Topographic Roughness Index (TRI)

TRI is one of the morphometric indices to define heterogeneity of terrain surface (Saleem et al. 2019). TRI reflects the terrain roughness (Eq. 3) and indicates the local height variation in neighboring pixels in the terrain (Al-Najjar et al. 2019). It is calculated from DEM by Eq. 3.

$$TRI = Abs\sqrt{\left(\overset{2}{\max} + \overset{2}{\min}\right)} \tag{3}$$

where max and min are the highest and lowest intensity of pixel values, subsequently, in nine rectangular altitude neighborhoods.

Lithology and Land Use/Cover (LULC)

Lithology or soil/rock characteristics and land use/cover maps are listed as the important cause of landslide hazard towards soil strength, human activities and erosion (Al-Najjar et al. 2019; Saleem et al. 2019). Land use map of the study area was produced by National Geographic Organization (NGO) from Landsat Enhanced Thematic

Mapper (2017) image with accuracy of 90%. Land use map contained six land use categories, namely agriculture, orchards, paddy field, dense forest, residential land, and harvested forest. Additionally, lithology maps were provided by Geological Survey and Mineral Exploration of Iran. The dominant lithological units of the study area were sandstone, silty marl, mudstone, limy marl and marly limestone according to lithology map (Mousavi et al., 2011).

Distance to Fault and Distance to Road

Additionally, distance to faults and road are considered as landslide conditioning factors. Road construction and fault adjacency might increase the risk of soil mass slide (Al-Najjar et al. 2019). These factors were generated from the topographic map of the region using Euclidean distance function in ArcGIS software (Hong et al. 2018; Golkarian et al. 2018).

Methodology

Three algorithms namely GLM, BRT and SVM and their ensemble have been used in order to predict, cluster and classify the data for the possibility of future landslides occurrence. The models were run using R programming. Figure 2 shows the flowchart of the methodology implemented in the current study.

SVM is a non-probabilistic binary linear classifier algorithm which is widely applied in data mining. The SVM model is based on the concept of decision planes that define decision boundaries. A decision plane is one that separates a set of objects having different class memberships. The strength and accuracy of BRT is due to the combination of non-parametrical regression method. Boosting machine learning algorithm, also known as Stochastic Gradient Boosting (SGB) and Gradient Boosted Machine (GBM). In the model process, at each level misclassification or residual error of past steps has been modified by re-weighting and allocating new regression trees, the quantity of trees is determined by inward cross-validation as well as the number of tree nodes, rate and fraction set through try and error (Colin et al. 2017). The use of linear regression was reported to fit the data, assume that the distribution of data is normal but we know that this usually does not happen in the real situation, thus to overcome this problem GLM or logistic regression is applied. GLM has a statistical basis and model a function of the dependent variable as a linear combination of the dependent variable. This method has presented satisfied results in landslide modelling prediction (Conoscenti et al. 2014). Additionally, the metrics used for classifier evaluation are the area AUC, COR, and TSS. The landslide inventory was randomly utilized for training (70%) the algorithms and testing and validation (30%).

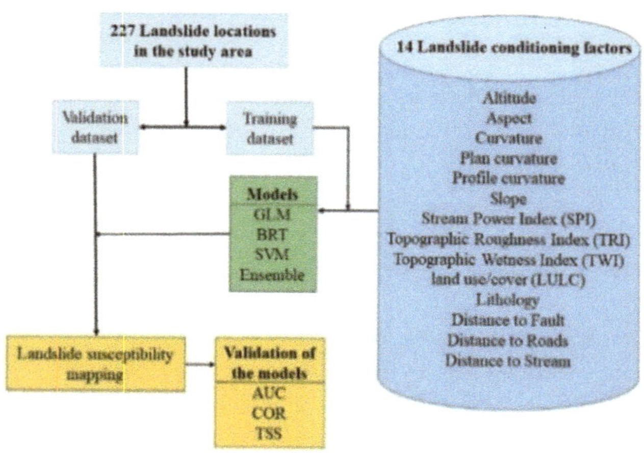

Fig. 2 Flowchart of the methodology used in this study

Results

The training dataset was used to train GLM, BRT, SVM, and ensemble models for LSM and the performance of each model was analysed by the testing dataset. The accuracy and performance of the results are presented in Table 1 against the area AUC, COR, and TSS. Acquired AUC values for all models passed the standard value and varied from 0.84 to 0.86. Moreover, COR values were scored from minimum of 0.59 (SVM) to maximum of 0.68 (GLM). In addition, maximum TSS value was obtained by GLM model (0.7) whereas the minimum TSS belonged to SVM algorithm (0.56).

The variable importance was calculated using CroTest and AUCtest for all three models as shown in Table 2. The results showed that curvatures and distance to road was the most important factors by GLM. However, SVM and BRT showed that distance to road was the most important factor.

Figure 3 illustrated the LSM produced by the three aforementioned models and their ensemble into five classes of probability namely, very low (<0.2), low (0.2–0.4),

Table 1 Accuracy assessment

Methods	AUC	COR	TSS
GLM	0.86	0.698	0.86
SVM	0.84	0.59	1.07
BRT	0.83	0.6	1.06
Ensemble	0.84	0.62	0.99

Table 2 Variable Importance for GLM, SVM, and BRT using CroTest and AUCtest

Methods	GLM		SVM		BRT	
Variables	CroTest	AUCtest	Crotest	AUCtest	Crotest	AUCtest
Altitude	0.0074	0.0036	0.0517	0.047	0	0
Aspect	0.0001	0	0.0309	0.0178	0	0
Curvature	0.8747	0.731	0.0202	0.0064	0	0
Dis. to Stream	0.0029	0.0032	0.0497	0.045	0	0
Plan curvature	1	0.8892	0.0349	0.02	0	0
Profile Curvature	1	0.8434	0.018	0.012	0	0
Dis. to road	0.8779	0.6964	0.5304	0.391	0.9801	0.7032
Slope	0.0503	0.0436	0.0273	0.025	0	0
SPI	0.0003	0.0006	0.0039	0.0032	0	0
TRI	0.0057	0.005	0.0228	0.0254	0	0
TWI	0.0196	0.0156	0.0955	0.0348	0	0.0374
Landuse	0.0202	0.0166	0.06	0.0218	0	0
Dis. to faults	0.0113	0.0062	0.0333	0.0476	0	0
Lithology	0.0088	0.0072	0.0766	0.0474	0	0.0156

moderate (0.4–0.6), high (0.6–0.8), and very high (>0.8) susceptible areas. Between the four maps, the outcome of BRT model looks rather different from others as it seems to miss "low", "moderate", and "high" classes in classification.

Discussion

The precision results of applying three machine learning algorithms and their ensembled model on 14 factors were promising. Statistically, ranking the importance of explanatory predictors and data modeling are a challenging subject due to different variables characteristics such as spatial resolution and scale, diverse data pre-processing, site situation and landslide types and sizes (Pawluszek et al. 2018; Tarolli, 2018). Hence, various algrithms and complex modeling might lead to different results and ranking (Goetz et al. 2015).

According to AUC test for aforementioned machine learning methods, although the accuracies were very similar, GLM exhibited higher accuracy of 0.86, comparing to SVM, BRT and ensemble model. The least precision was obtained by BRT (AUC = 0.83). The ensemble model resulted better than BRT and similar to SVM, hence the improvement was proved. Among all 14 conditioning factors modelling by GLM, the plan curvature, profile curvature, curvature and distance to the road had the highest AUCtest value of 0.8892, 0.8434, 0.7310 and 0.6964 respectively, which means that those factors made the most significant effect on the landslide prediction mapping. In terms of applying SVM and BRT models, the highest AUCtest factor evaluation belonged to the distance to road. Besides, using CroTest, distance to road was labelled as the most effective factor by all machine learning algorithm. Conversely, aspect, SPI, and distance to stream did not have a significant impact on the landslide occurance by applying GLM model. Accordingly, this could be interpreted that the hydrological factors were not influential for landslide hazards in this area. However, from different point of view and visual examination of susceptibility maps (Fig. 3) it was revealed that the landslide prone areas were partially clustered around the streams in the valleys, as well. Therefore, for profound and comprehensive investigation regarding the causative factors in a study area, the numerical and visual examinations seem complementary. Moreover, SPI factor was marked by SVM as the least important factor, as well. However, BRT did not gently ranked all the 14 factors and all the factors were mostly listed as lowest imporance. This might contribute to the odd appearance on the map resulted from BRT modeling and the discrete classification by BRT was extremely clear. The produced map by BRT was mainly covered by very low and very high probability classes whereas, the presence of all 5 probaility classes in other maps is obvious. Although each of the landslide explanatory variables made different contributions to the models but from our research it was obvious that plan curvature, profile curvature, distance to the road and curvature were depicted as the most influencing factors. Our findings were in agreement with (Nguyen et al. 2019; Pradhan et al. 2017; Tsangaratos & Ilia 2016) as they emphasized on the importance of distance to road and interpreted its relationship with inappropriate road construction, excavation and changing the land cover, in their study areas. Moreover, the curvatures were determined as significant variables for erosion and landslides by related

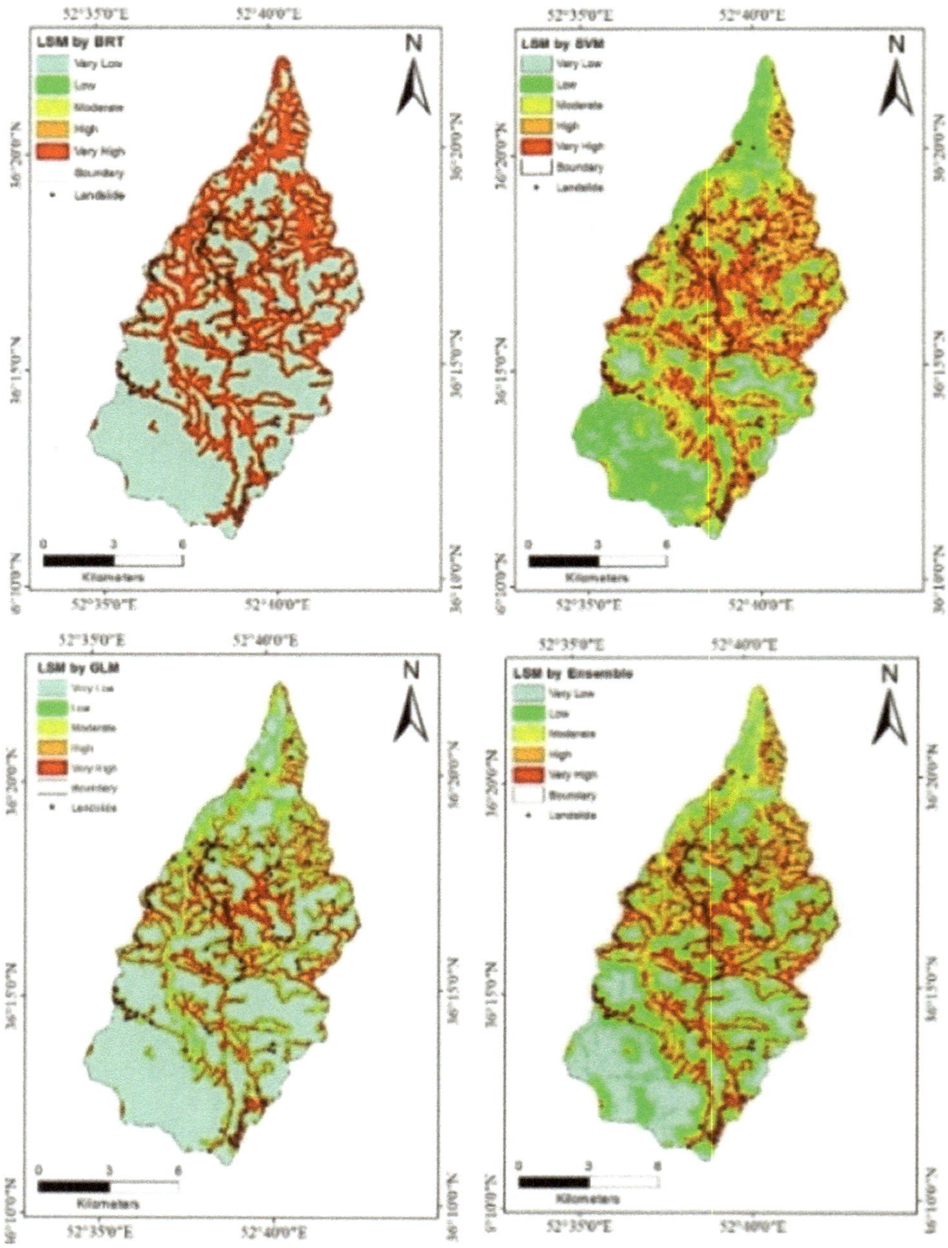

Fig. 3 Overlaying of the landslide inventory map and the final susceptibility maps

researcher (Lee et al. 2017) and it is consistent with our result. However, the curvatures were not highly remarkable factors in (Märker et al. 2016; Fan et al. 2017)'s studies, for landslide events which it demonstrates the site's dependency of the explanatory variables. Considerably, the curvature factors' correlations need more investigation due to their selection as significant factors, as they might exhibit more comparability. Since those three methods agreed on the importance of distance to road in LSM, it could be concluded that human activities (such as road construction) along with the natural surface shapes (curvatures) in the study area trigger the landslides.

Conclusion

In this research 14 conditioning factors were explored by three machine learning GLM, BRT, SVM models and their ensemble to produce LSM. From the modelling point of view, BRT model was not as successful as GLM and SVM to evaluate the variables and create smooth susceptibility map. The outcome proved the nature of BRT model is best fit for two-condition classification as existence/non-existence (Mousavi et al. 2017) rather than producing probability map in 5 classes. Additionally, GLM, SVM and ensemble models showed encouraging results and ability to handle various factors and data for this specific purpose. The outcome suggest that some mitigation programs are needed during road construction due to the surface shape and curvatures in the study area.

References

Ada M, San BT (2018) Comparison of machine-learning techniques for landslide susceptibility mapping using two-level random sampling (2LRS) in Alakir catchment area, Antalya, Turkey. Nat Hazards 90 (1):237–263

Al-Najjar HAH, Kalantar B, Pradhan B, Saeidi V (2019, October) Conditioning factor determination for mapping and prediction of landslide susceptibility using machine learning algorithms. In Earth Resources and Environmental Remote Sensing/GIS Applications X, vol 11156, p. 111560 K. International Society for Optics and Photonics

Calvello, M. and Ciurleo, M, (2018). Optimal use of thematic maps for landslide susceptibility assessment by means of statistical analyses: case study of shallow landslides in fine-grained soils. In: Landslides and engineered slopes. Experience, theory and practice. CRC Press, pp 537–544

Colin B, Clifford S, Wu PP, Rathmanner S, Mengersen K (2017) Using boosted regression trees and remotely sensed data to drive decision-making. Open J Statist 7(5):859–875

Conoscenti C, Angileri S, Cappadonia C, Rotigliano E, Agnesi V, Märker M (2014) Gully erosion susceptibility assessment by means of GIS-based logistic regression: a case of Sicily (Italy). Geomorphology 204:399–411

Fan W, Wei XS, Cao YB, Zheng B (2017) Landslide susceptibility assessment using the certainty factor and analytic hierarchy process. J Mt Sci 14(5):906–925

Goetz JN, Brenning A, Petschko H, Leopold P (2015) Evaluating machine learning and statistical prediction techniques for landslide susceptibility modeling. Comput Geosci 81:1–11

Gupta SK, Shukla DP, Thakur M (2018) Selection of weightages for causative factors used in preparation of landslide susceptibility zonation, vol 5705

Hong H, Liu J, Bui DT, Pradhan B, Acharya TD, Pham BT, Zhu AX, Chen W, Ahmad BB (2018) Landslide susceptibility mapping using J48 decision tree with AdaBoost, bagging and rotation forest ensembles in the Guangchang area (China). CATENA 163:399–413

Kalantar B, Ueda N, Al-Najjar HAH, Gibril MBA, Lay US, Motevalli A (2019) An evaluation of landslide susceptibility mapping using remote sensing data and machine learning algorithms in Iran. ISPRS Annals of Photogrammetry, Remote Sensing and Spatial Information Sciences, pp 503–511

Khosravi K, Panahi M, Tien Bui D (2018) Spatial prediction of groundwater spring potential mapping based on an adaptive neuro-fuzzy inference system and metaheuristic optimization. Hydrol Earth Syst Sci 22(9):4771–4792

Lee S, Lee MJ, Jung HS (2017) Data mining approaches for landslide susceptibility mapping in Umyeonsan, Seoul, South Korea. Appl Sci 7(7):683

Märker M, Hochschild V, Maca V, Vilímek V (2016) Stochastic assessment of landslides and debris flows in the Jemma basin, Blue Nile, Central Ethiopia. Geografia Fisica e Dinamica Quaternaria 39 (1):51–58

Moore ID, Grayson RB, Ladson AR (1991) Digital terrain modelling: a review of hydrological, geomorphological, and biological applications. Hydrol Process 5(1):3–30

Mousavi SM, Golkarian A, Naghibi SA, Kalantar B (2017) GIS-based groundwater spring potential mapping using data mining boosted regression tree and probabilistic frequency ratio models in Iran, vol 3, no. March, pp 91–115

Mousavi SZ, Kavian A, Soleimani K, Mousavi SR, Shirzadi A (2011) GIS-based spatial prediction of landslide susceptibility using logistic regression model. Geomatics, Natural Hazards and Risk 2 (1):33–50

Nguyen PT, Tuyen TT, Shirzadi A, Pham BT, Shahabi H, Omidvar E, Vu TB (2019) Development of a novel hybrid intelligence approach for landslide spatial prediction. Appl Sci 9(14):2824

Pawluszek K, Borkowski A, Tarolli P (2018) Sensitivity analysis of automatic landslide mapping: numerical experiments towards the best solution. Landslides 15(9):1851–65

Pradhan B, Seeni MI, Kalantar B (2017) Performance evaluation and sensitivity analysis of expert-based, statistical, machine learning, and hybrid models for producing landslide susceptibility maps. In: Laser scanning applications in landslide assessment, Springer, Cham, pp 193–232

Roy J, Saha S (2019) Landslide susceptibility mapping using knowledge driven statistical models in Darjeeling District, West Bengal, India, Geoenvironmental Disasters, vol 6, no 1

Saleem N, Enamul Huq M, Twumasi NYD, Javed A, Sajjad A (2019) Parameters derived from and/or used with digital elevation models (DEMs) for landslide susceptibility mapping and landslide risk assessment: a review. ISPRS Int J Geo-Inf 8:12

Singh Y, Kaur A, Malhotra R (2009) Comparative analysis of regression and machine learning methods for predicting fault proneness models. Int J Comput Appl Tech 35(2/3/4):183–193

Tsangaratos P, Ilia I (2016) Comparison of a logistic regression and Naïve Bayes classifier in landslide susceptibility assessments: the influence of models complexity and training dataset. CATENA 145:164–179

Overcoming Data Scarcity Related Issues for Landslide Susceptibility Modeling with Machine Learning

Anika Braun, Katrin Dohmen, Hans-Balder Havenith, and Tomas Fernandez-Steeger

Abstract

Landslide susceptibility maps can be a useful tool to support holistic urban planning in mountainous environments. Data-driven methods for landslide susceptibility modeling work well even in data scarce areas, and there is an increasing relevance of machine learning methods that help analyze efficiently large and complex datasets. In this contribution we present some of our study examples to show how data quality, quantity, complexity, and preparation can have major effects on the outcomes of landslide susceptibility modeling. The aforementioned aspects are too often neglected in spite of their relevance, both in data scarce, but also data rich areas. We also use these examples to discuss the way we evaluate landslide susceptibility models, as the spatial performance of landslide susceptibility maps often differs from the mathematical performance. We finally discuss the necessity of standards for input data, modeling results and result communication to improve the usability of landslide susceptibility models in urban planning.

Keywords

Landslide susceptibility • Data quality • Machine learning

Introduction

With current trends of increasing intensity of climatic events and rapid urbanization in many areas of the world, it is becoming more important to systematically consider geological hazards in urban and land-use planning and build more resilient cities (UNISDR 2015). Landslide susceptibility mapping, which is defined as the analysis of the spatial probability of landslide occurrence, is a valuable tool for implementing slope stability in urban and land-use planning in inhabited mountainous areas (Fell et al. 2008). Data-driven, or statistically based methods work very well in data scarce areas, because although they require certain amounts of data, the required data can be relatively simple and is easily accessible as compared for instance to physically based methods. Usually, a landslide inventory, morphological and hydrological information derived from a digital elevation model (DEM), and some geological information are the minimum of data used in landslide susceptibility modeling (Reichenbach et al. 2018). Increasing availability of satellite data and methods for automatic landslide detection, as well as increasing availability of elevation data, make it possible to generate landslide susceptibility models even in data scarce areas.

The goal of data-driven or statistically based landslide susceptibility modeling is to quantify the spatial relationship between occurred landslides and related factors and to ultimately identify locations likely to be affected by future landslides by using only information about the factors. This can be achieved with simple bivariate statistical or multivariate methods, while in the last decade machine learning methods have increasingly proven to be useful for this task, as they can be employed to yield high accuracies while being able to handle large datasets. They are particularly relevant considering the increasing collection and accessibility of data. One main idea of data-driven methods is to overcome subjectivity based on decisions of the operator (van Westen et al. 2006). Landslide susceptibility can furthermore help to

A. Braun (✉) · K. Dohmen · T. Fernandez-Steeger
Engineering Geology Department, Institute of Applied Geosciences, Technische Universität Berlin, Ernst-Reuter-Platz 1, 10587 Berlin, Germany
e-mail: anika.braun@tu-berlin.de

H.-B. Havenith
Department of Geology, Georisks and Environment, University of Liège, Liège, Belgium

F. Guzzetti et al. (eds.), *Understanding and Reducing Landslide Disaster Risk*, ICL Contribution to Landslide Disaster Risk Reduction, https://doi.org/10.1007/978-3-030-60227-7_26

develop process understanding for landslides in a regional or even wider scale.

There is a huge amount of publications on the topic of statistically based landslide susceptibility modeling, as Reichenbach et al. (2018) demonstrated in their extensive review of 565 articles published between 1983 and 2016. As many of these authors showed, the methods for data-driven landslide susceptibility modeling work very well. They yield impressive performances measured by skill scores such as the area under the receiver operator characteristics (AUROC) curve. However, apart from tweaking models to reach better performances, two important topics are rarely ever discussed: data quality and usefulness of results.

As the name already suggests, in data-driven modeling we are replacing the expert reasoning with data, not algorithms. No matter how complex or advanced an algorithm is, it can only discover effects that are in the data. This point is often neglected in the literature, and until now, there exists no standard or unified approach for assessing the required quality, consistency, quantity, and preparation of input data for landslide susceptibility modeling.

The other point is the usefulness of resulting models. Evaluation criteria for landslide susceptibility models should depend on the goals of a study (Hearn and Hart 2019; Teimouri and Kornejady 2019). This point also implies the issue of the spatial plausibility of the resulting model and a reality check by local experts or through field verification, which could even help to develop new process understanding. To sum it up, we need better, standardized tools to define goals and assess the usefulness of landslide susceptibility models.

In this contribution we want to discuss the aspects data quality, consistency, quantity, and pre-processing of landslide inventories and input factor datasets, respectively, as well as result usefulness in landslide susceptibility modeling and their implications for data scarce areas using some examples from our research with machine learning tools on different scales.

Data and Methodology

Study Areas

In the following sections we want to use examples from three of our studies related to landslide susceptibility mapping with machine learning methods to discuss the above-mentioned aspects. The studies differ in the size of the study area and their environment, as well as in their specific goals and challenges. Two rather small study areas are the urban area of Tegucigalpa, capital city of Honduras (353 km^2, see Table 1), and the rural area of Ningnan County in southwestern China (726 km^2). The third study with a very large extent covers a good part of the Kyrgyz and Tajik Tien Shan Mountains (115,000 km^2).

The goal of the Tegucigalpa study was to generate a susceptibility map for urban planning with the challenge related to the availability of input data. The goal of the Ningnan study was to model landslide susceptibility based on parameters expressing rock quality and weathering, with a challenge regarding the landslide inventory quality. The goal of the Tien Shan study was to try to grasp the big picture, while the challenge was to handle a large dataset with uncertainties regarding the consistency of data for a large area covering two different countries.

Input Data

For all studies referred to in this contribution the freely available ALOS 30 m World DEM has been used (JAXA 2015–2019) as the basis to derive primary and secondary terrain and hydrological parameters with ArcGIS and SAGA GIS. In terms of data quality of the DEM all our studies are comparable.

The landslide inventory of the Ningnan study was mapped during field campaigns in 2015 and 2017. It will be discussed in more detail in the following section. The Tegucigalpa landslide inventory was generated in 2013 in collaboration with the Japan International Cooperation Agency (JICA) and the National Autonomous University of Honduras (UNAH), based on stereoscopic aerial image interpretation and field surveys (Braun et al. 2019). The Tien Shan landslide inventory has been created by Havenith et al. (2015) using Google Earth imagery.

For Ningnan a geological map was available that was reclassified regarding the geotechnical properties of the lithologies. For Tegucigalpa, different geological maps were available. For the Tien Shan only a classification into soft or hard rock was available ready-to-use.

All data was prepared on a pixel basis matching the ALOS 30 m DEM that was resampled to 30 m cells in UTM projection.

Modeling

For all three study areas landslide susceptibility was analyzed with the IBM SPSS Modeler. First, the data was transferred from a GIS into a table format and imported to the modeller. In a first step, the datasets were explored regarding their completeness and quality, distributions and inter-correlations of variables. In order to optimize the mathematical representation of the data, the variables were

Table 1 Key figures about the study areas presented in the examples in this contribution

Location	Area (km^2)	Landslides (%)	Variables (n)
Tegucigalpa	353	7	19
Ningnan	726	15	8
Tien Shan	115,000	0.8	25

transformed, scaled and recoded if necessary. More details about this workflow can be found in Braun et al. (2019).

In all studies, different classifiers, mainly Artificial Neural Networks (ANN) and Decision Trees (DT), were explored for their capabilities to model landslide occurrence in the different study areas using separate training, test, and sometimes validation subsets of the data using different compositions of the parameter sets. The models are described in Braun et al. (2019). The modeling results were evaluated with different skill scores, such as the total percentage of correct classifications, the percentage of correctly classified landslides (hit rate), false positives (false alarms), and false negatives (misses). Moreover, the results, namely the raw propensities (confidence that a cell is a landslide) and the binary classification result were plotted back in a GIS to evaluate the spatial quality, plausibility, and usefulness of the resulting models.

Landslide Inventories

Inventories of past landslides are the most crucial input for landslide susceptibility modeling, and also a great source of uncertainty and bias. Until recently, landslide inventories have mainly been created by experts, e.g. in field reconnaissance and through aerial or satellite image interpretation. Depending on experience, method, data quality, goal, scale, landslide type and landslide discretization (e.g. point or polygon) the results can vary greatly, e.g. in terms of accuracy and completeness, making it difficult to compare studies.

However, with the increasing availability of satellite data, techniques for automatic landslide detection, such as InSAR (Schlögel et al. 2015) or optical object recognition (Behling et al. 2014), are advancing. Another way to make landslide inventories more comparable is the use of slope units to discretize landslide occurrence (Alvioli et al. 2016; Schlögel et al. 2018).

In this section, we want to show how we used slope units to fix a landslide inventory with high uncertainties, and we want to discuss the problem of underrepresentation of landslides in datasets and how it can be solved using sampling or balancing techniques.

Improving a High Uncertainty Landslide Inventory with Slope Units

The landslide inventory available for the Ningnan study had spatial uncertainties. Landslides were mapped during an extensive field campaign, but mainly based on communications from local villagers. The inventory was updated in another field campaign and using Google Earth, but the spatial discretization of most landslides remained difficult. The reports of the villagers were however believed to be reliable, and thus a slope unit approach was implemented to spatially discretize slopes where failures have occurred in the past.

A slope unit is a region of space delimited between ridges and valleys under the constraint of homogenous slope aspect distribution (Carrara et al. 1991), which corresponds to either the left or right side of a sub-basin of any order into which a watershed is subdivided. Slope units were delineated in the study area by further subdividing watersheds computed in ArcGIS according to the main slope aspect and three geotechnical classes. Each pixel in a slope unit containing a landslide location was classified as a landslide pixel. The landslide susceptibility analysis was then carried out on a pixel basis. The advantage was that by assigning an event to an entire slope unit, a landslide to non-landslide pixel ratio of 15/85 could be achieved, which made it possible to model without any further balancing or sampling steps (see explanation below).

Underrepresentation of Landslide Cases

A very common problem in modeling with real-world datasets is the underrepresentation of the target class, making it hard for machine learning algorithms to capture it. In general, ratios of the target class of 20/80 to 30/70 are considered ideal (Pyle 1999).

There are different techniques to enhance the representation of the target class, such as balancing and under-sampling. In balancing techniques, duplicates of target datasets are generated for the training of the models. In under-sampling, a fraction of the non-target class is sampled to reach the desired distribution. Balancing is useful in small study areas to maintain a maximal sample size. For large areas, under-sampling is ideal, as it helps reduce the computational effort by decreasing the total sample size.

When working with these techniques, care has to be taken to maintain the original distributions of the variables within the dataset. Also, with a higher representation of the target class in the training data, models might tend to overestimate its occurrence and produce increased numbers of false alarms.

We have worked with balancing in the Tegucigalpa study (Braun et al. 2019) and with under-sampling in the Tien Shan study (Dohmen 2019). Exploring different ratios of landslides to non-landslides revealed that a 30/70 ratio was ideal to increase the hit rates of models but still keep the false alarms reasonably low. Figure 1 shows for the Tien Shan and the Tegucigalpa studies how the hit rate (hr) of ANN increased significantly with an increasing ratio of landslides. At the same time, the number of false positives (fp) increased, which can be useful to a certain degree depending on the goal of the study (Teimouri and Kornejady 2019). For DT on the other hand we found they responded less to balancing, it even promoted over fitting (Braun et al. 2019).

Input Factor Sets—Sometimes More Is More

When it comes to the composition of the input dataset, we have made the experience that more is sometimes really more. In science we like to follow the principle of parsimony and strive for the simplest solution, which means in landslide susceptibility modeling to create a model with a minimum of influence factors. This makes a lot of sense for more basic bivariate statistical methods for landslide susceptibility mapping, as it also helps improve the interpretability by experts and end-users. In data mining however, the main idea is to use as much data as possible to maximize the chances that patterns can be discovered. Some algorithms, such as ANN, depending on their complexity actually require a certain number of input variables and samples, although there is no general rule as to how much data is enough.

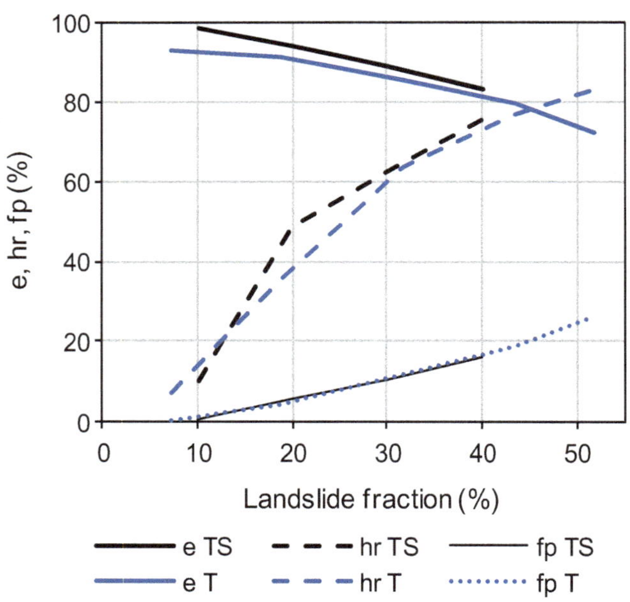

Fig. 1 Effect of balancing. Hit rates (hr), efficiency (e), and false positives (fp) for test run of ANN models trained with different landslide to non-landslide ratios for the Tien Shan (TS, black) and Tegucigalpa (T, blue)

Maximizing Information

In the Tien Shan study, we explored the response of ANN and DT models to the complexity of the set of input variables in a very large dataset. We used different combinations of input parameter sets, with a very basic set of parameters that is usually used in bivariate statistical modeling (elevation, slope, aspect, landforms, distance to rivers and faults, lithology), a complex set of 25 parameters (morphological, hydrological, climatic, geological, seismic, and anthropogenic factors), and a set of the 10 most important parameters as identified in the latter model. Then, to examine the effect of the continuous vs. nominal nature of the data, sets with only nominally coded data and with only continuous data were tested, respectively.

Interestingly, as it can be seen in Fig. 2, while the performance in terms of hit rate (hr) of the ANN increased with increasing complexity of the input parameter set, the DT already reached their optimal performance with 10 parameters. Two interesting aspects could be concluded from this. For the ANN more data really meant a better performance, even though most of the added data was derived from the same DEM. Moreover, it could be concluded that the DT model is an interesting choice when a simpler model is anticipated. Then again, the DT models produced significant artefacts wherever a variable was split into different branches of the DT. In the spatial validation, it also turned out that the ANN containing both, continuous and nominally discretized parameters, had a tendency to form spatial artefacts as well, while the artefacts did not occur when only continuous parameters were considered (Fig. 4). This underlines how skill scores alone do not enable a meaningful evaluation of landslide susceptibility models.

The Role of Lithological Information—Can Less Be Ok?

One parameter that is not always available in some areas, or not in a consistent form, but which is used in most

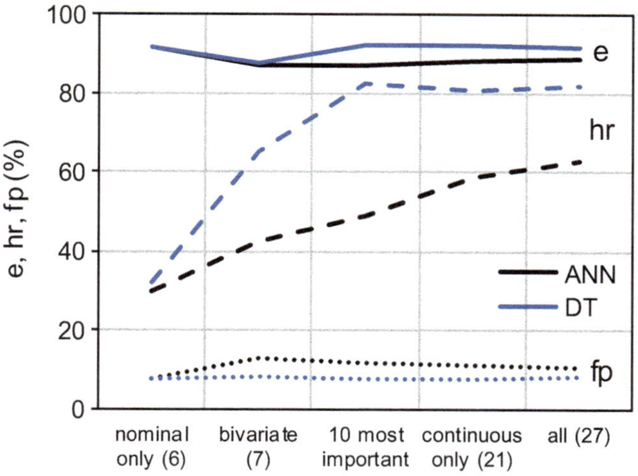

Fig. 2 Effect of dataset complexity. Hit rates (hr), efficiency (e), and false positives (fp) for validation run of ANN and DT models trained with different input parameter sets in the Tien Shan

publications (Reichenbach et al. 2018), is lithology. For Tegucigalpa, different geological maps were available, but they were not consistent. With some "expert interpolation" we tried to fix the geological information into a consistent map. In order to find out the effect of this highly uncertain information we generated ANN and DT with and without lithology as an input. We found that while the ANN performed a little poorer in terms of hit rate without the information about the lithology, the DT actually performed better without the lithology information (Fig. 3).

One explanation is that the nominal nature of the lithology variable is unfavourable for the DT because the dataset can only be split between classes, giving only few options

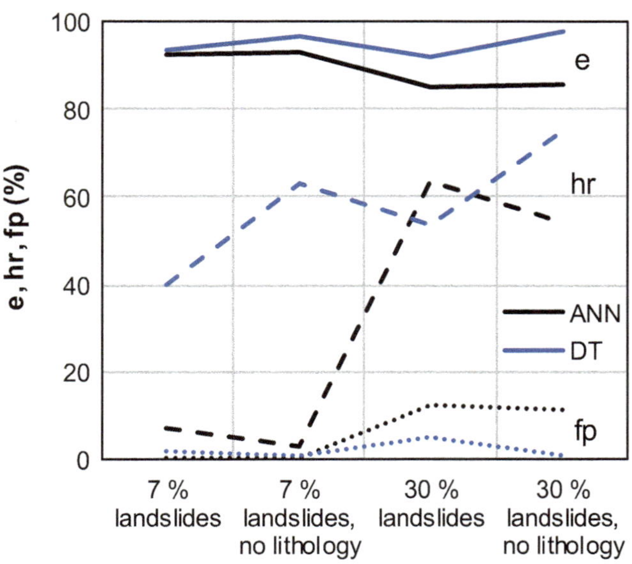

Fig. 3 Hit rates (hr), efficiency (e), and false positives (fp) for validation run of ANN and DT models trained with different input parameter sets in Tegucigalpa

for the tree to split with this variable. It should be noted that in the Tegucigalpa study ANN and DT developed quite different skills. While the DT developed very high accuracies with a tendency to over-fitting, the ANN produced more false positives, which formed however consistent patterns that seemed to be more useful for zoning. In the spatial context, the ANN trained with lithological information showed a more concise and coherent pattern than the one trained without it. Thus, in this case, in spite of being noisy, information on lithology could improve the model.

In the large study area of the Tien Shan on the contrary, as already explained above, the ANN had a better, artefact-free, spatial performance when no nominal parameters, such as the lithology class, were included (Fig. 4). However, the spatial evaluation of the entire, very large area showed that their accuracy differs strongly in different areas, which might be related to local effects. The analysis of this large area showed in the end that some landslides can be explained with such a general model, while others cannot. It will be interesting to explore this in future research to distinguish local from global effects.

Conclusions

In this contribution we presented three studies of landslide susceptibility mapping with machine learning methods to discuss the importance of different aspects regarding data quality, quantity, complexity, and preparation for data-driven landslide susceptibility modeling. We showed how we used slope units to fix a messy landslide inventory. We also showed how balancing, dataset complexity, and scale can have significant effects on both, the mathematical and spatial performance of different types of machine learning models of landslide susceptibility. The message of this is how important the data itself is, but also that there are techniques to overcome data quality related issues.

Perspectives

One point we want to discuss is the usefulness of landslide susceptibility models. There is a clear lack of studies regarding the implementation of landslide susceptibility maps in urban planning (Hearn and Heart 2019). In spite of the great potential that has been demonstrated in hundreds of publications, it remains difficult to implement machine learning methods for landslide susceptibility assessment into planning due to the lack of transparency for the end-user. Decision makers are most likely not impressed by high AUROC numbers that they are unable to interpret. In order to build confidence in landslide susceptibility maps as a tool for

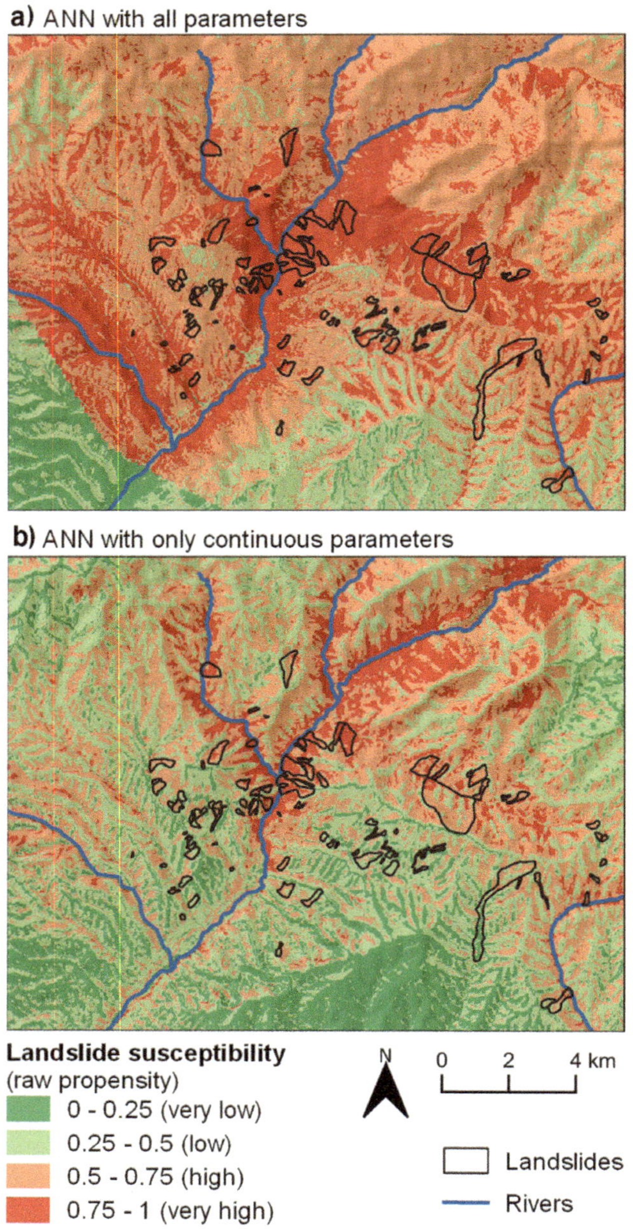

Fig. 4 Landslide susceptibility maps for the Maily Say area in the Tien Shan study

holistic urban planning, we need to define universal standards for input data, methodology, and evaluation criteria.

We also need to discuss more the way we communicate landslide susceptibility to make it more accessible for target audiences. The final medium of communication is the resulting map, which should look plausible and reliable to the end-user. Maps like the ones in Fig. 4 may be accurate but seem rather difficult to communicate. Slope units could be a tool that should be explored more for risk communication, as has been done in Italy for decades. It is easier to communicate that a whole slope is susceptible, rather than explain to the citizen why his house is on a red pixel, while his neighbours house is on a green pixel.

With the increasing challenges we are facing with the ongoing climate change and increasing urbanization we need to work on the way we do landslide susceptibility modeling to create a more useful tool for planning resilient cities. The data and methods are there.

Acknowledgements Many people contributed to the presented case studies, namely Luqing Zhang, Xueliang Wang, Zhenhua Han, and Jian Zhou, Elias Leonardo Garcia Urquia, Rigoberto Moncada Lopez, and Hiromitsu Yamagishi. Part of this work was funded by the Natural Science Foundation of China (Grant No. 41402285) and the Chinese Academy of Sciences President's International Fellowship Initiative (Grant No. 2016PZ032).

References

Alvioli M, Marchesini I, Reichenbach P, Rossi M, Ardizzone F, Fiorucci F, Guzzetti F (2016) Automatic delineation of geomorphological slope units with r.slopeunits v1.0 and their optimization for landslide susceptibility modeling. Geoscientific Model Dev 9 (11):3975–3991

Behling R, Roessner S, Kaufmann H, Kleinschmit B (2014) Automated spatiotemporal landslide mapping over large areas using rapideye time series data. Remote Sens 6(9):8026–8055

Braun A, Garcia Urquia EL, Lopez RM, Yamagishi H (2019) Landslide susceptibility mapping in Tegucigalpa, Honduras, using data mining methods. In IAEG/AEG Annual Meeting Proceedings, San Francisco, California, vol 1, pp 207–215 (2018)

Carrara A, Cardinali M, Detti R, Guzzetti F, Pasqui V, Reichenbach P (1991) GIS techniques and statistical models in evaluating landslide hazard. Earth Surf Proc Land 16(5):427–445

Dohmen K (2019) Landslide factors and susceptibility analysis using data mining methods for large study areas: a case study from the Tien Shan Mountains, Central Asia. Master thesis, Technische Universität Berlin, Berlin, Germany

Fell R, Corominas J, Bonnard C, Cascini L, Leroi E, Savage W (2008) Guidelines for landslide susceptibility, hazard and risk zoning for land-use planning. Eng Geol 102(3–4):99–111

Havenith HB, Torgoev A, Schlögel R, Braun A, Torgoev I, Ischuk A (2015) Tien Shan geohazards database: landslide susceptibility analysis. Geomorphology 249:32–43

Hearn GJ, Hart AB (2019) Landslide susceptibility mapping: a practitioner's view. Bull Eng Geol Env 78(8):5811–5826

JAXA (Japan Aerospace Exploration Agency) ALOS Global Digital Surface Model 'ALOS World 3D—30 m' (AW3D30 DSM Ver.1.0, 2.0 and 2.1), data available from the JAXA web interface (2016). http://www.eorc.jaxa.jp/ALOS/en/aw3d30/data/index.htm

Pyle D (1999) Data Preparation for Data Mining. Morgan Kaufmann, San Francisco. https://books.google.de/books?hl=de&lr=&id=hhdVr9F-JfAC&oi=fnd&pg=PA6&dq=Data+Preparation+for+Data+Mining&ots=6h9RbOGw5w&sig=elC6MNiVplWM14O5nx1ajw6drj0#v=onepage&q&f=false

Reichenbach P, Rossi M, Malamud B, Mihir M, Guzzetti F (2018) A review of statistically-based landslide susceptibility models. Earth Sci Rev 180:60–91

Schlögel R, Marchesini I, Alvioli M, Reichenbach P, Rossi M, Malet J (2018) Optimizing landslide susceptibility zonation: effects of DEM spatial resolution and slope unit delineation on logistic regression models. Geomorphology 301:10–20

Schlögel R, Doubre C, Malet J, Masson F (2015) Landslide deformation monitoring with ALOS/PALSAR imagery: a D-InSAR geomorphological interpretation method. Geomorphology 231: 314–330

Teimouri M, Kornejady A (2019) The dilemma of determining the superiority of data mining models: optimal sampling balance and end users' perspectives matter. Bull Eng Geol Environ

UNISDR (United Nations International Strategy for Disaster Reduction) (2015) Sendai framework for disaster risk reduction 2015–2030. http://www.wcdrr.org/uploads/Sendai_Framework_for_Disaster_Risk_Reduction_2015-2030.pdf. Accessed 28 Jan 2020

Van Westen CJ, Van Asch TW, Soeters R (2006) Landslide hazard and risk zonation—why is it still so difficult? Bull Eng Geol Env 65 (2):167–184

Practical Accounting for Uncertainties in Data-Driven Landslide Susceptibility Models. Examples from the Lanzhou Case Study

2020 Kyoto Japan

Jewgenij Torizin, Michael Fuchs, Dirk Kuhn, Dirk Balzer, and Lichao Wang

Abstract

Modeling of a complex environment is inevitably associated with uncertainties arising from the model design or data errors. In the uncertainty assessment, the bias (related to accuracy) and the random error (related to precision) are distinguished. Recent reviews of case studies, which used data-driven methods for landslide susceptibility assessment (LSA), indicate a general lack of appropriate evaluation of uncertainties. In this paper, we discuss practical techniques to account for uncertainties in LSA, relying majorly on the examples from the project "Landslide Hazard and Risk Assessment for Lanzhou" (LHARA).

Keywords

Uncertainty • Random subsampling • Landslides • Data-driven • Susceptibility

J. Torizin (✉) · M. Fuchs · D. Kuhn · D. Balzer
Federal Institute for Geosciences and Natural Resources (BGR), Hannover, 30655, Germany
e-mail: jewgenij.torizin@bgr.de

M. Fuchs
e-mail: michael.fuchs@bgr.de

D. Kuhn
e-mail: dirk.kuhn@bgr.de

D. Balzer
e-mail: dirk.balzer@bgr.de

L. Wang
China Institute of Geo-Environment Monitoring (CIGEM), Beijing, 100081, China
e-mail: wanglc@cigem.cn

Introduction

Modeling of a complex environment is always associated with uncertainties. The uncertainties can derive from the simplification of nature and hence the model framework or errors in the input data. The uncertainties may be epistemic, and thus, reducible by increasing knowledge. However, they can also be aleatory as an inherent property of a random nature.

In the past, many authors discussed concepts for uncertainty assessment in environmental modeling and related issues (e.g., Jolma and Norton 2005; Refsgaard et al. 2007). Nevertheless, Reichenbach et al. (2018) show while reviewing more than 565 case studies dealing with data-driven statistical landslide susceptibility assessment (LSA) that only a tiny fraction of screened studies (approx. 3%) evaluated the model uncertainty.

Most studies concerned with uncertainties in LSA focused on the discussion of uncertainties in landslide inventories, including location, mass movement type and, more general, completeness of the inventory (e.g., Zêzere 2002; Guzzetti et al. 2005; Galli et al. 2008; Guzzetti et al. 2012; Steger et al. 2017). A lesser number of studies explicitly considered and tried to account for different effects governed by the quality and resolution of predisposing factors (e.g., Qin et al. 2013; Fressard et al. 2014; Segoni et al. 2020). Only a few analyze the effect of sample size (e.g., Petschko et al. 2014).

In this study, we focus on some bias and sampling error issues, provide some examples from the Lanzhou case study (Torizin et al. 2018), and discuss some practical techniques on how to account for such uncertainties in data-driven analyses.

F. Guzzetti et al. (eds.), *Understanding and Reducing Landslide Disaster Risk*, ICL Contribution to Landslide Disaster Risk Reduction,
https://doi.org/10.1007/978-3-030-60227-7_27

Study Area

Lanzhou, the capital of the Gansu Province, NW China, is in the focus of the sino-german research project Landslide Hazard and Risk Assessment for Lanzhou (LHARA) (Tian et al. 2017). The study area covers approximately 500 km^2 and encompasses the eastern part of Lanzhou city and its surrounding loess mountains (Fig. 1) (Torizin et al. 2018). The loess thickness range from an average of 30 m to more than 300 m (Derbyshire 2001; Zheng et al. 2014). Economic growth in the past two decades increased the demand for suitable building ground. It triggered building land reclamation activity affecting the original geo-morphological conditions in the Lanzhou area (e.g., Li et al. 2014).

Landslides frequently occur in the Lanzhou area. Spatial plans for newly developed areas account for these phenomena based on hazard assessment introducing massive slope protection measures such as piles and retaining walls. The allover anthropogenic activity, such as greening, irrigation, and reshaping of the slopes, introduces a complex dynamic that also affects the landslide controlling factors. Fuchs et al. (2019) quantified the areas reshaped by land reclamation providing valuable datasets to catch the dynamic in the geomorphological factors (Fig. 1). Torizin et al. (2018) compiled a multi-temporal landslide inventory for the area based on previous research (Derbyshire et al. 2000) and high-resolution imagery from Google Earth. In total, they could identify about 920 landslide events. Utilizing multi-temporal landslide and controlling factor datasets, Torizin et al. (2018) tested the applicability of a data-driven LSA approach in the dynamic environment. As a result of this, they emphasized the limitations, highlighted increasing requirements on data quality and study design, but also the inevitability of detailed uncertainty analyses.

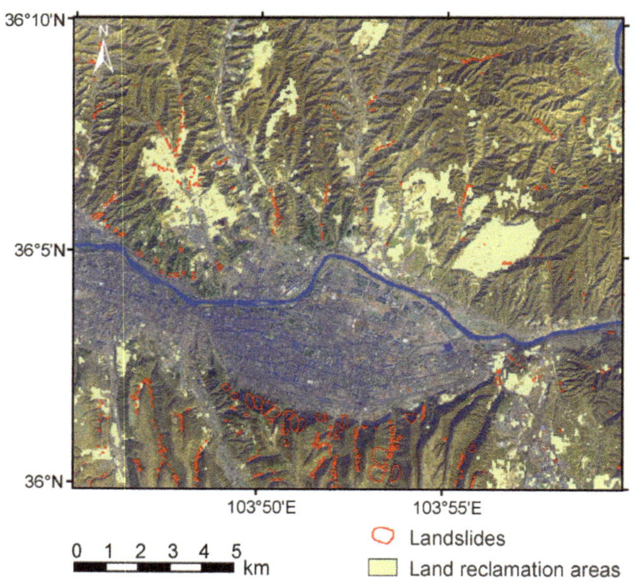

Fig. 1 Study area centers on the eastern outskirts of Lanzhou City (Brovey sharped Landsat 8 Imaginary [bands 8, 5, 2], from March 13th, 2014) (modified after Torizin et al. 2018)

Bias-Related Uncertainty in Data-Driven LSA

The systematic error, or merely the deviation of the modeled or measured result from the truth (accuracy), is usually called bias. All LSA studies include various sources of bias related to data (measured) or experience of the researcher (subjective). For example, landslides inventories derived from aerial photographs can have biases due to the high subjectivity in the visual interpretation process and lacking standards (e.g., Guzzetti et al. 2012). Further, landslide inventories may be incomplete and thus not representative for a certain areal extent (e.g., Steger et al. 2017). Also, thematic layers considered as explanatory factors exhibit bias due to the resolution, generalization of information, georeferencing errors, or misclassification, which were only seldom considered in LSA yet. Both options are equally worse: to have an accurately labeled dataset on landslide observations paired with bad thematic layers, or to have accurate thematic maps and inaccurate inventory. In both cases, a data-driven model will learn false associations, which, if not recognized, ultimately leads to crude misinterpretation of the results.

Bias in the data becomes evident only if we can compare a particular dataset with a reliable reference dataset. In practice, it may become extremely challenging to find an appropriate reference for regional analyses.

Digital elevation models (DEM) are the primary source of topographic variables such as slope, aspect, curvature, morphometric features, and many others, considered in the LSA as controlling factors for the landslide occurrence. Today, a palette of DEMs with a ground resolution between 1-90 m ground resolution is available from open and commercial space satellite missions. DEMs provide a vertical and horizontal accuracy error in terms of a statistical root mean square error (RMSE) computed on limited ground control points. For extended regions with sharp relief, the local errors may, however, be substantially larger. Thus, DEMs with comparable ground resolution and error metrics may differently describe local conditions. Working with a single DEM dataset, those measurement biases usually cannot be revealed by statistical analysis.

Therefore, to sensitize for measurement bias, we briefly compare two DEMs and, in particular, their derivatives. One DEM is based on the ZY-3, a Chinese optical sensor from 2012. The second DEM is the Elevation 10, a product of Airbus DS based on TerraSAR-X and acquired in summer 2016. Both DEMs have a ground resolution of 10 m.

To compare the data, we removed all areas identified or suspected as anthropogenically altered in the period between the acquisition dates of both DEMs based on the study from Fuchs et al. (2019). After, we compared the masked DEMs using the pixel-by-pixel approach. Figure 2 visualizes the comparison of the datasets. The joint plot in Fig. 2 depicts the marginal distributions of the slope gradient values along the x and y axes. The main body of the plot represents the scatter plot and the corresponding bivariate kernel density delineating the relationship between each pixel pair of gradient values. The marginal distributions indicate that the Elevation10 model exhibits more substantial portions of flat areas. The scatterplot shows significant deviances for pairwise comparison. Linear fit with R^2 of 0.31 and the kernel density pattern show a weak trend pointing to a general correlation of the datasets but with substantial local deviances in values.

To estimate the effects of these differences, we computed two simple models with both DEMs utilizing the identical inventory datasets. As a classifier, we used the well-known weight-of-evidence (WofE) method (e.g., Teerarungsigul et al. 2015; Torizin 2016) and compared the outcomes using the receiver operating characteristics (ROC) curve. The ROC curve is a popular tool in binary classification problems to estimate the goodness of the classifier (e.g., Fawcett 2006). It depicts the relation of the True Positive Rate (Sensitivity) against False Positive Rate (1-Specificity) of the model. The quantitative measure to compare ROCs is the *area under curve* (AUC) index. Here it is notably to remark that the ROC curve is a statistical metric and not an appropriate index to make any conclusions about the reliability of the model. High AUC indices must not result in better predictive models (e.g., Rossi et al. 2010). Also, models exhibiting equal AUC must not be equal in prediction patterns (e.g., Vakhshoori and Zare 2018).

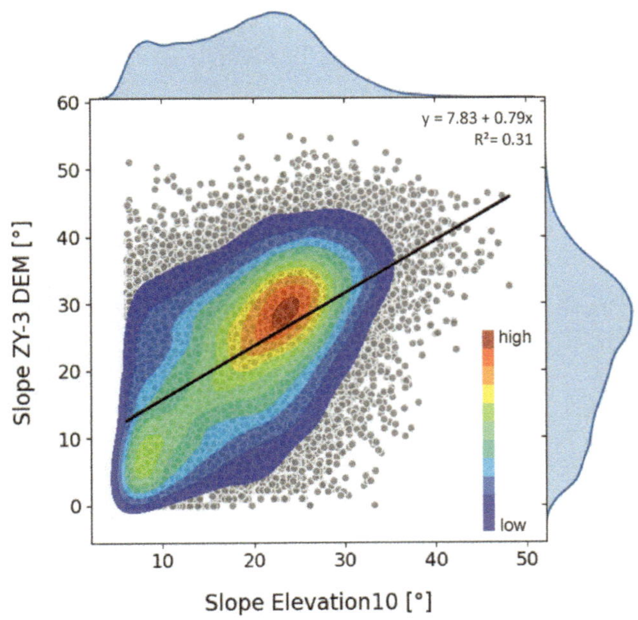

Fig. 2 Joint plot for slope gradient derived from ZY-3 DEM, and Elevation10

Further, for comparison, we use only landslides in areas not affected by reshaping activities between 2012 and 2016. The inventory is consisting of 629 landslide polygons. In the first step, it was randomly subset in 100 samples, each of which contains 70% of the inventory. After the polygons in the subsets were converted into points using the polygon centroid. In doing so, we obtained corresponding subsets with point and polygon representations of the events. Note that we did not want to discuss explicitly, which type of data is better to use; instead, we wanted to see how sensitive the data types are to the different spatial distributions in the thematic layers. In the next step, we reclassed the slope gradient into 5-degree classes for both DEMs.

The analysis ran with all 100 subsamples. Figure 3 summarises the results of the study. Plots in Fig. 3a, and Fig. 3b depicts metrics for the point data type, plots in Fig. 3c, d for polygons. The general relative distribution of the weights varies depending on the geometry type of the datasets. Compared to the polygons, the point geometries are more sensitive to differences in the distribution of the slope classes, especially for steeper areas. Further, we used the computed weights to rank the classes applying a color code to emphasize the obtained spatial patterns. The results in Fig. 4 indicate that the same model learned four different relative patterns governed by inherited differences from the primary data distributions in DEMs and different geometry types of the data.

Taking into account models with additional variables, and in particular, applying more complex classifiers such as machine learning (ML) algorithms, which are more sensitive to the bias in the data than the used WofE, the differences in the final models would even increase.

Uncertainty Related to Sampling Error

Sampling error (SE) is the inevitable uncertainty in statistical inference. Different from the bias, which relates to the accuracy of the model, the SE relates to precision. Drawing a sample of the dataset, we introduce a SE that we have to account for in further steps of the analysis. For supervised classification, it is a common practice in data science to split the event datasets into training, test, and, sometimes, validation parts (summarized as cross-validation methods). For ML methods, such as logistic regression or artificial neural networks, there is also a need to generate non-event samples. Generally, simple cross-validation does not account for SE and can exhibit substantial variance. Therefore, the

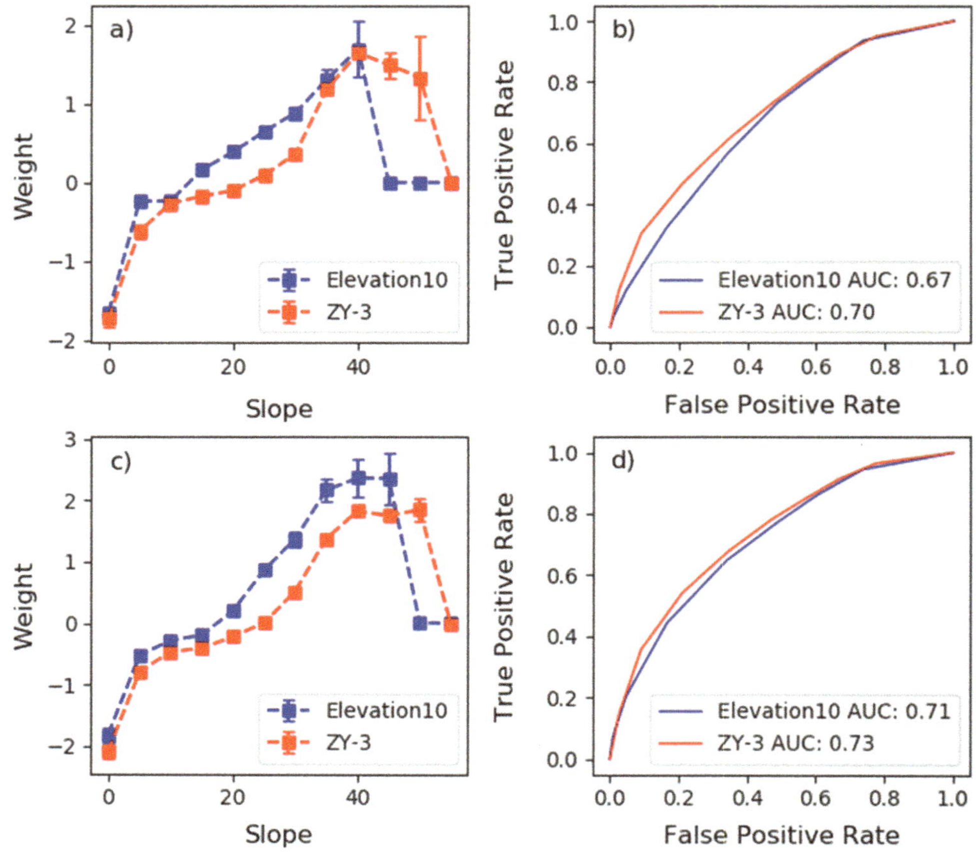

Fig. 3 Weight distribution and corresponding ROC curves for the slope gradient derived from Elevation10, and ZY-3 DEM and directly trained with identical samples: **a**, **b** depict weights and ROCs for point data; **c**, **d** depict weights and ROCs for polygon data

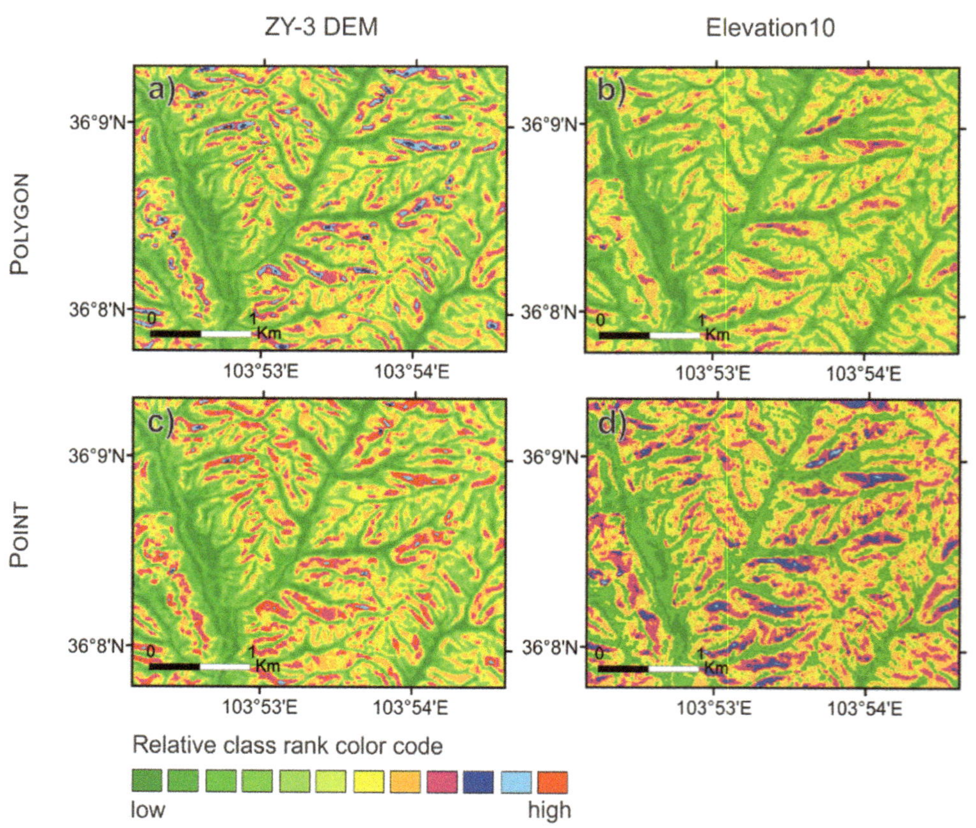

Fig. 4 Relative class ranking patterns for ZY-3 DEM and Elevation10: **a**, **b** for polygon data type, **c**, **d** for point data

iterative cross-validation (e.g., Fabbri and Chung 2019), also known as random subsampling or Monte Carlo (MC) cross-validation (e.g., Torizin et al. 2017) represents a well-applicable technique for control of SE and its potential effects in the model. The procedure is straightforward, and therefore, feasible to implement in all types of analyses. The disadvantage, of course, is higher computational expenses due to numerous repetitions.

The majority of studies on LSA have been using simple cross-validation evaluating the models with ROC curves, and thus, do not account for SE. The ROC curve for training inventory (success) and the ROC curve for test inventory (prediction) usually show differences frequently interpreted as model bias. However, due to different sizes of training and test datasets, it can also be related to the SE. The importance of understanding the source of uncertainty is crucial for correct reasoning, e.g., when comparing models.

For the Lanzhou inventory, we can show that the sample size can have very significant effects. Therefore, we introduced a small sensitivity study. In the first step, we computed a model with factors slope gradient and lithology without subsampling utilizing the entire training dataset (polygon data type). After, the same training dataset was randomly subsampled, drawing 100 samples with 1%, 10%, 20%, 40%, and 80% of the training dataset, respectively. We evaluated the model with all drawn subsamples. The results emphasize that the sample size can have an enormous impact on model evaluation (Fig. 5). Note that though the model was aware of all subsampled events, with a small sample, it may appear performing bad or excellent by chance. Further, the spatial distribution of the events, their geometry type, the complexity of the classifier, and trained joint distribution influence the magnitude of the sample size effect.

The significance of the obtained results becomes clear if we try to evaluate the model with a new dataset with a specific size, which is not known to the model. For example, for a dataset of a size corresponding to 10% of the training dataset and for which we assume that it follows the same joint distribution, we can expect the model performance within the 95% confidence interval of the SE as being accurate. In other words, we cannot falsify a model with a small test sample, if the results are within the possible SE range.

The depicted precision uncertainty in the model has direct consequences on the next step of zonation (e.g., Fabbri and Chung 2019). To emphasize this, we created a model stack of 100 models consisting of slope gradient and lithology and generated with random subsamples (70% of training inventory each). We classified all models in the stack based on the same approach into five susceptibility classes: *Very high*, *High*, *Moderate*, *Low*, and *Very low*. The thresholds for the classes were selected based on the ROC curve taking the classes encompassing 50%, 30%, 15%, 3%, and less 2% of landslide areas, respectively (Fig. 6b). After that, we used the stack of reclassified models for computing the probability of an arbitrary raster cell to be in a specific landslide susceptibility class (Fig. 6d). Note that in Fig. 6d, the maximum probability for the class *Low* is about 0.89, pointing to a high uncertainty for this class definition. The entire class *Low* is within a possible class range of the neighbor classes *Moderate* and *Very Low* (Fig. 6b). Based on the probabilities, we were able to generate different maps. Figure 6c shows the maximum likelihood map of the susceptibility classes. Other options could be a conservative map depicting maximum class values and an optimistic map depicting minimum class values. Further, the analysis of the class probability map allows depicting class areas within a specified confidence interval.

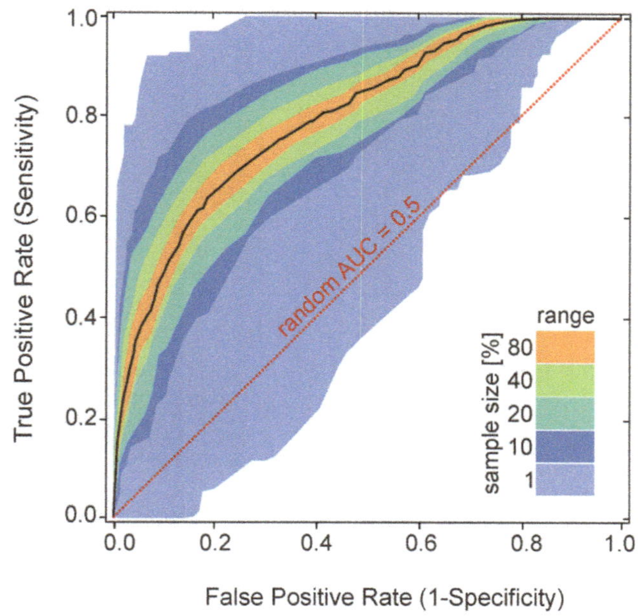

Fig. 5 The effect of the sample size on model performance evaluated with ROC based on random samples without replacement from the training dataset. The shaded areas represent the range for the respective sample size

Discussion and Conclusion

Based on the examples from the Lanzhou case study, we show resampling techniques to assess uncertainties related to bias and SE. The examples are not exhaustive but provide a general idea of how to approach the types of uncertainty.

The uncertainties related to the bias in the factor data cannot be revealed statistically in the data-driven analysis

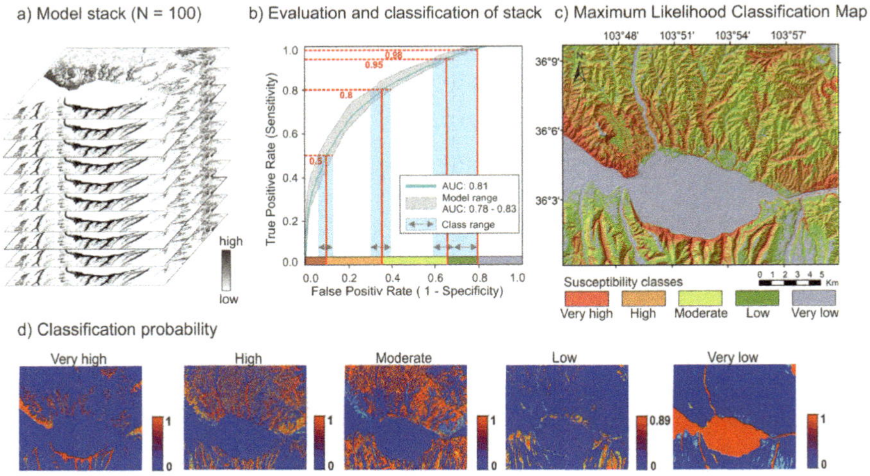

Fig. 6 Example of SE propagation and assessment based on MC cross-validation: **a** a model stack generated with 100 subsamples, **b** model stack evaluation using a ROC curve with subsequent classification into five susceptibility classes, **c** an example of the zonation map derived from the stack based on the maximum likelihood classification of the class probabilities, **d** probability of a raster pixel to be a member of one of the susceptibility zones

since all analysis steps inherently suffer from the initial data uncertainty. Comparing different datasets and alternative models could identify biases. It remains, however, still tough to correct those from such comparisons because then we will need to decide which of the datasets or models is the most reliable reference. Having more than two datasets could justify such decisions moving from subjective to more systemic approaches by, e.g., eliminating layers or models with the largest deviances to the mean values or introducing stochastic approaches to account for these uncertainties.

In LSA, we usually try to model complex environmental conditions with comparably sparse observational data. The uncertainties related to the SE in sparse datasets can be very dramatic. Given ideally accurate datasets, the variance due to sampling becomes the most severe source of uncertainty. Therefore, accounting for SE must become a standard procedure in LSA.

In most cases, we do not have very accurate factor layers and sparse observational datasets, and thus, will have both types of uncertainty. Understanding the nature of both will allow estimating the so-called bias-variance tradeoff, and thus select the most appropriate model to characterize the landslide susceptibility patterns. Further, reporting the uncertainties results will increase the general reliability of the modeling procedure. Also, the depiction of uncertainties will establish better limits for usage of the model, identifying areas for further investigations.

Acknowledgements We conducted this work in the framework of a scientific-technical cooperation project between the Federal Institute for Geosciences and Natural Resources (BGR) and the China Geological Survey (CGS) co-funded by the German Ministry of the Economic Affairs and Energy (BMWi) and Ministry of Natural Resources of the People's Republik of China.

References

Derbyshire E (2001) Geological Hazards in loess terrain, with particular reference to the loess regions of China. Earth Sci Rev 54:231–260

Derbyshire E, Dijkstra TA, Unwin D et al (2000) Slope instability distribution mapping. In: Derbyshire E, Meng X, Dijkstra TA (eds) Landslides in the thick loess terrain of North-West China, Wiley, pp 230–242

Fabbri A, Chung C-J (2019) Landslide susceptibility prediction maps: from blind-testing to uncertainty of class membership: a review of past and present developments. In: Pourghasemi HR, Rossi M (eds) Natural hazards GIS-based spatial modeling using data mining techniques. Springer Nature, Switzerland (ISBN 978-3-319-73382-8)

Fuchs M, Torizin J, Wang L, Tong B, Balzer D, Chen L, Li A, Wan L (2019) Identification and temporally-spatial quantification of geomorphic relevant changes by construction projects in loess landscapes: case study Lanzhou City, NW China. Big Earth Data 3(4):395–410

Fawcett T (2006) An introduction to ROC analysis. Pattern Recogn Lett 27(8):861–874

Fressard M, Thiery Y, Maquaire O (2014) Which data for quantitative landslide susceptibility mapping at an operational scale? Case study of the Pays d'Auge plateau hillslopes (Normandy, France). Nat Hazard Earth Sys 14:569–588

Galli M, Ardizzone F, Cardinali M et al (2008) Comparing landslide inventory maps. Geomorphology 94:268–289

Guzzetti F, Mondini AC, Cardinali M et al (2012) Landslide inventory maps: new tools for an old problem. Earth Sci Rev 112:42–66

Guzzetti F, Reichenbach P, Cardinali M, Galli M, Ardizzone F (2005) Probabilistic landslide hazard assessment at the basin scale. Geomorphology 72:272–299

Jolma A, Norton J (2005) Methods of uncertainty treatment in environmental models. Environ Modell Software. 20:979–980

Li P, Qian H, Wu J (2014) Environment: accelerate research on land creation. Nature 510:29–31

Petschko H, Brenning A, Bell R, Goetz J, Glade T (2014) Assessing the quality of landslide susceptibility maps—case study Lower Austria. Nat Hazard Earth Sys. 14:95–118

Qin C-Z, Bao L-L, Zhu A-X, Wang R-X, Hu X-M (2013) Uncertainty due to DEM error in landslide susceptibility mapping. Int J Geogr Inf Sci 27:1364–1380

Refsgaard JC, Van der Sluijs JP, Højberg AL, Vanrolleghem PA (2007) Uncertainty in the environmental modeling process—a framework and guidance. Environ Modell Softw 22(11):1543–1556

Reichenbach P, Rossi M, Malamud BD, Mihir M, Guzzetti F (2018) A review of statistically-based landslide susceptibility models. Earth Sci Rev 180:60–91

Rossi PH, Guzzetti F, Reichenbach P, Mondini AC, Perruccacci S (2010) Optimal landslide susceptibility zonation based on multiple forecasts. Geomorphology 114(3):129–142

Segoni S, Pappafico G, Luti T, Catani F (2020) Landslide susceptibility assessment in complex geological settings: sensitivity to geological information and insights on its parameterization. Landslides. Online. https://doi.org/10.1007/s10346-019-01340-2

Steger S, Brenning A, Bell R, Glade T (2017) The influence of systematically incomplete shallow landslide inventories on statistical susceptibility model and suggestions for improvement. Landslides. 14:1767–1781

Teerarungsigul S, Torizin J, Fuchs M et al (2015) An integrative approach for regional landslide susceptibility assessment using weight of evidence method: a case study of Yom River Basin, Phrae Province, Northern Thailand. Landslides. 13(5):1151–1165

Tian T, Balzer D, Wang L, Torizin J, Wan L, Li X, Chen L, Li A, Kuhn D, Fuchs M, Lege T, Tong B (2017) Landslide hazard and risk assessment Lanzhou, province gansu, China - Project introduction and outlook. In: Mikoš M, Tiwari B, Yin Y, Sassa K. (eds) Advancing culture of living with landslides, pp. 1027–1033. WLF 2017. Springer, Cham

Torizin J (2016) Elimination of informational redundancy in the weight of evidence method: an application to landslide susceptibility assessment. Stoch Env Res Risk A. 30(2):635–651

Torizin J, Fuchs M, Awan AA et al (2017) Statistical landslide susceptibility assessment of the Mansehra and Thorgar districts, Khyber Pakhtunkhwa Province, Pakistan. Nat Hazards 89(2):757–784

Torizin J, Wang L, Fuchs M, Tong B, Balzer D, Wan L, Kuhn D, Li A, Chen L (2018) Statistical landslide susceptibility assessment in a dynamic environment: a case study for Lanzhou City, Gansu Province, NW China. J Mt Sci 15(6):1299–1318

Vakhshoori V, Zare M (2018) Is the ROC curve a reliable tool to compare the validity of landslide susceptibility maps? Geomat Nat Haz Risk. 9(1):249–266

Zêzere JL (2002) Landslide susceptibility assessment considering landslide typology. A case study in the area north of Lisbon (Portugal). Nat Hazard Earth Sys 2:73–82

Zheng RQ, Meng XM, Wasowski J, Dijkstra T, Bovenga F, Xue YT, Wang SY (2014) Ground stability detection using PS-InSAR in Lanzhou, China. Q J Eng Geol Hydroge 47:307–321

Assessment of Shallow Landslides Susceptibility Using SHALSTAB and SINMAP at Serra Do Mar, Brazil

Victor Carvalho Cabral and Fábio Augusto Gomes Vieira Reis

Abstract

The Serra do Mar mountain range in Brazil is highly susceptible to shallow landslides initiation, due to its relief and high rainfall index. The aim of this study is the application and comparative performance analysis of SHALSTAB and SINMAP in the landslide susceptibility assessment of the Perequê watershed at the Serra do Mar escarpments. The application of physically based models is an objective method used in landslide susceptibility to predict slope failure under different geotechnical scenarios, supporting hazard assessment studies. Model calibration is based on the landslide scars inventory of the 1985 and 1994 events, while topographic parameters are DEM sourced and geotechnical parameters obtained from soil samples. Using performance classifiers based on a contingency table to assess model performance, the results indicate that SHALSTAB is the best-fit model at the chosen scale due to higher accuracy and higher concentration of scars in unstable areas (>62%). SINMAP was less accurate, also exhibiting higher false positive results and lower density of landslides scars in unstable areas. For future studies, the compartmentalization of the study area according to geology is suggested as an improvement in the representativeness of modeling results.

Keywords

Shallow landslides • SHALSTAB • SINMAP • Performance assessment • Serra do mar

V. C. Cabral (✉) · F. A. G. V. Reis
Applied Geology Department, São Paulo State University (UNESP), Av. 24A 1515, Rio Claro, 13506-752, Brazil
e-mail: victor.carvalho@unesp.br

V. C. Cabral
Center for Applied Geosciences, Eberhard Karls Universität Tübingen, Hölderlinstr. 12, Tübingen, Germany

F. Guzzetti et al. (eds.), *Understanding and Reducing Landslide Disaster Risk*,
ICL Contribution to Landslide Disaster Risk Reduction,
https://doi.org/10.1007/978-3-030-60227-7_28

Introduction

Gravitational mass movements are part of the natural evolution of the landform and can represent great hazards to life and infrastructures in mountain regions. Climate change, increasing deforestation and urbanization contribute to the intensification of the frequency and hazard of these phenomena (Dietrich et al. 1998; Goetz et al. 2011).

Shallow landslides are the predominant type of mass movement at Serra do Mar (IPT 1986; Tatizana et al. 1987) and the characterization of landslide prone areas is essential to mitigate associated hazards—though it is not an easy task. Physically-based models stand out as an objective method that describes shallow landslides susceptibility by coupling hydrological and slope stability models, supporting urban planning and hazard assessment studies (Bel et al. 2016).

"Stability Index Mapping"—SINMAP (Pack et al. 1998) and "Shallow Landslide Stability Model"—SHALSTAB (Montgomery and Dietrich 1994; Dietrich et al. 1998) are two of the most commonly used models and most landslide susceptibility studies are performed at slope or small catchment scales (e.g. Goetz et al. 2011; Michel et al. 2014; Nery and Vieira 2014; Alvioli and Baum 2016).

This study assesses the landslide susceptibility of the Perequê watershed in Cubatão (São Paulo State, Brazil), using SHALSTAB and SINMAP at a 1:25,000 scale, and compares the performance of the models to identify the most representative of the chosen area. SHALSTAB and SINMAP are selected due to their spatially explicit examination of hydrological changes and geomorphic processes, ease of use and application. The definition of the best performing model in the chosen scale can aid preliminary studies in mountain regions that exhibit similar geology and climate settings.

Study Area

Figure 1 shows the location of the Perequê watershed and the 1:50,000 geological map of the study region. The 28.3 km^2 watershed is located at the Cubatão escarpments (Serra do Mar), characterized by steep slopes and high rainfall annual rates (up to 4000 mm) (IPT 1986).

Several major landslide and debris-flow events have been recorded in Perequê. The January 23–24, 1985, and February 6–7, 1994, landslide events are chosen to assess model results, due to their large magnitude and the availability of stereoscopic aerial photographs post-event. 480 shallow landslide scars were mapped in the 1985 event and 442 in the 1994 event (Fig. 1).

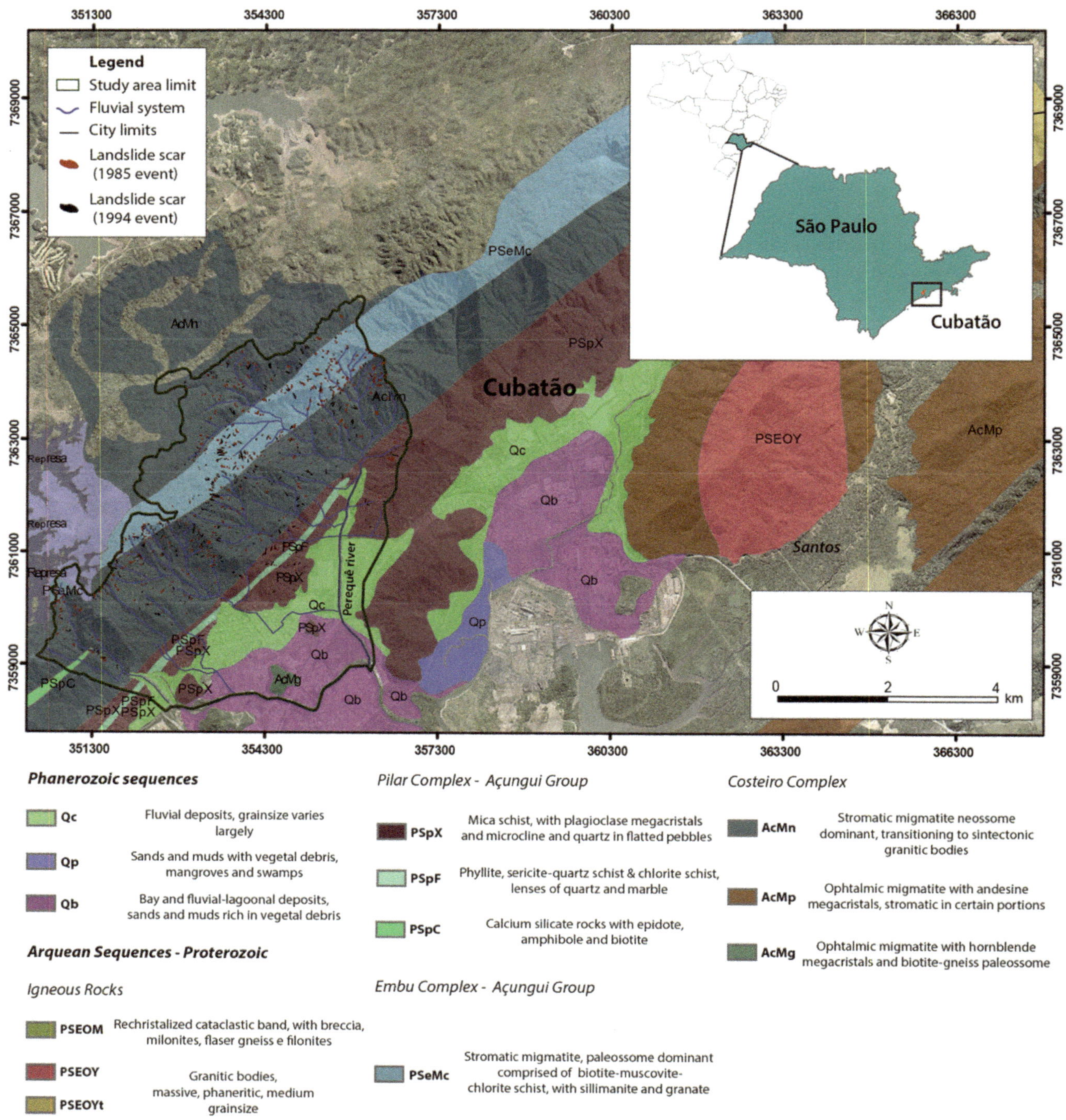

Fig. 1 Location of the study area and the geological map of the region, based on IPT (1986). The landslide scars of the 1985 and 1994 event are indicated in red and black, respectively

The geology of the hillslopes is mainly comprised of metamorphic (orthogneiss) rocks (Fig. 1) that, combined with the physiographic characteristics, results in thin residual soil with average depth of 1 to 2 m (Wolle and Carvalho 1994; Vieira et al. 2015). Perequê's soil exhibits high sand concentration, with cohesion ranging from 1 to 4 kPa and saturated hydraulic conductivity (**K$_s$**) from 10^{-6} m/s to 10^{-5} m/s (Wolle and Carvalho 1994).

Materials and Methods

Landform and Rainfall Database

The digital elevation model (DEM) sources the topographic parameters required for landslide modeling. A 5 m resolution DEM is adopted, based on pre-event 1:10,000 topographic map provided by the Geographic and Cartographic Institute of the State of São Paulo—IGC.

Landslide scars were identified using stereoscopic aerial photographs, also provided by IGC at a 1:25,000 scale. The identification criteria were: lack of vegetation, characteristic morphology and drainage conditions of hillslopes.

According to the Department of Water and Energy of the São Paulo State (DAEE), in 1985 the 24 h accumulated rainfall reached 265 mm, with peak rainfall of 84 mm/h, and in 1994 the 24 h accumulated reached 214 mm, with peak rainfall of 60 mm/h.

Input Geotechnical Parameters

The input parameters are based on four (4) soil samples collected in the study area (Fig. 1) by the Institute of Technological Research (IPT), presented in Wolle and Carvalho (1994) (Table 1). The average value of the parameters is used in both SINMAP and SHALTAB modeling. Since SINMAP is a stochastic model, the standard deviation (sd) is used to establish the lowest and highest limits in the input parameters range.

The dimensionless cohesion (**C'**) used in SINMAP (Eq. 1) is based on soil cohesion (**C$_s$**), specific weight (**ρs**) and depth (**z**) from Wolle and Carvalho (1994), and the standard gravity (**g**) of 10 m/s^2. Root cohesion (**C$_r$**) is challenging to estimate and is excluded from calculation (set as 0), as suggested by Meisina and Scarabelli (2007).

$$C' = \frac{C_s + C_r}{z * \rho_s * g} \tag{1}$$

Physically-Based Models

SHALSTAB and SINMAP are based on the coupling of infinite-slope and hydrology models in FS calculation (Montgomery and Dietrich 1994; Pack et al. 1998). The infinite-slope model is based on the *Mohr-Coulomb* Law: in the moment of slope failure, shear forces (**τ**) are superior to shear-resistant forces, such as soil cohesion (**c**) and internal friction angles (**φ**), due to Normal Stress (**σ**) in the rupture surface (Eq. 2). **μ** is soil's pore-pressure that opposes the Normal.

$$\tau = c + (\sigma - \mu) * \tan \tag{2}$$

The most common hydrological concept used in slope stability modeling is the steady-state subsurface flow, described in TOPOG (O'Loughlin 1986). This concept assumes a uniform recharge that simulates the spatial variation of the water level during rainfall. According to O'Loughlin (1986), Wetness (**W**) is given by the ratio between precipitation (**q**) and soil's transmissivity (**T**). Equation 3 shows the final formulation of the steady-state hydrological model, where **a** is drained area, **b** is contour length element, **θ** is mean slope (approximately 33° in the study region), **h** is water level and **z** is soil depth. Equation 4 shows soil's **T** calculation, where **K$_s$** is the saturated hydraulic conductivity.

$$W = \frac{q * a}{b * T * \sin\theta} = \frac{h}{z} \tag{3}$$

$$T = K_s * z * \cos\theta \tag{4}$$

SINMAP, based on the FS, calculates the probability of shallow landslide occurrence (Pack et al. 1998) and classifies the area into six stability classes (Table 2). A more detailed description of the model and its governing equations can be found on Pack et al. (1998).

Table 1 Geotechnical parameters (Wolle and Carvalho 1994)

Samples	Soil depth (m)	Specific Weight (Kg/m^3)	Cohesion (kPa)	Internal friction angle (°)
1	1	1710	1	34
2	2	1950	4	39
3	1	1820	1	36
Average	1.5	1873	2.4	37
sd	0.5	77	1.4	2

Table 2 SINMAP stability classes

Stability index (Factor of Safety)	Stability classes
>1.5	Stable
1.5–1.25	Quasi-stable
1.25–1.0	Moderately stable
1.0–0.5	Lower threshold
0.5–0	Upper threshold
<0	Unstable/defended

Table 3 SHALSTAB stability classes

Stability index	Stability classes
Stable	Unconditionally stable and saturated
>−2.2	Unconditionally stable and non-saturated
−2.5 to −2.2	Stable and non-saturated
−2.8 to −2.5	Unstable and non-saturated
−3.1 to −2.8	Unstable and saturated
<−3.1	Unconditionally unstable and non-saturated
Unconditionally unstable	Unconditionally unstable and saturated

SHALSTAB assumes that the local topography is the main controlling factor in shallow landslide occurrence and the susceptibility is calculated based on the ratio of effective rainfall to soil transmissivity (**q/T**) (Montgomery and Dietrich 1994). SHALSTAB classifies the area into seven stability classes (Table 3) and a more detailed description of its governing equations is presented in Montgomery and Dietrich (1994).

Performance Assessment

A contingency table is applied in performance assessment (Fawcett 2006). The table compares model results to the landslide scar maps, classifying them into four classes (Fig. 2). These four classes are the foundation for the performance classifiers shown in Fig. 3. An ideal susceptibility map maximizes the coincidence between known and predicted landslides and minimizes unstable area outside known and predicted landslides. Additionally, two field campaigns were made in the region to confirm modeling results.

Results and Discussions

Shalstab

Table 4 shows the modeling results. 26.91% of the study region is classified as potentially unstable and 63.09% of the 1985 landslide scars (299) are within areas that indicate instability; for the 1994 event, 70.23% of the landslide scars (309) are in these areas. These results stand out due to the high density of landslides scars in Low(q/T) grid cells, as shown in Table 4. The stability map of the study region is shown in Fig. 4.

The performance classifiers shown in Table 5 can further aid analyzing modeling results, indicating that SHALSTAB was accurate (0.73), with a high true positive index (0.6 and 0.7) and fairly low false positive results (0.26). The model, therefore, successfully delineates unstable areas (low false positive index) at the same time that encompasses areas where instabilities are proven to have occurred (high true positive index). The satisfactory performance of SHALSTAB shows that it is adequate describing landslide susceptibility at the chosen scale.

Moreover, areas classified as 'chronically unstable' are commonly associated to regions that are unstable even without precipitation or run-off (Montgomery and Dietrich 1994). As investigated during fieldwork, lowest log(q/T) areas (6.8% of the total area) often exhibit soil and tree cover, although bedrock outcrops are also observed locally (Fig. 6a). These 'chronically unstable' areas exhibit steep slopes (>30°) with thin soil cover (0–1 m depth) and, in some portions, landslide scars (Fig. 6a) are observed without a recent record of expressive rainfall. Figure 4 also shows

Positive (P)	Negative (N)	
True Positive (TP)	False Positive (FP)	**True Positive:** Landslide scars in unstable areas **False Positive:** Unstable areas, no landslide scars
False Negative (FN)	True Negative (VN)	**True negative:** Stable areas, no landslide scars **False negative:** Landslide scars in stable areas

Fig. 2 Contingency table used in model performance assessment, based on Fawcett (2006)

CLASSIFIERS	PARAMETERS
Accuracy	$\frac{TP + TN}{P + N}$
Precision	$\frac{TP}{TP + FP}$
True Positive Rate (TPR)	$\frac{TP}{P}$
False Positive Rate (FPR)	$\frac{FP}{N}$

Fig. 3 Performance classifiers based on the contingency table. Retrieved from Fawcett (2006)

Table 4 Statistical analysis of SHALSTAB modeling results. Area in Km^2 and density in unit/km^2. #= number of landslide scars

Stability class	Area	# 1985	Dens.	# 1994	Dens.
Stable	14,7	48	3,25	59	4,00
−2,2	3,20	65	20,34	31	9,70
−2.5/−2.2	2,74	62	22,66	41	14,98
−2.8/−2.5	2,42	61	25,18	62	25,59
−3.1/−2.8	1,50	47	31,39	54	36,07
−3.1	1,78	60	33,71	63	35,39
Unstable	1,92	131	68,37	130	67,85

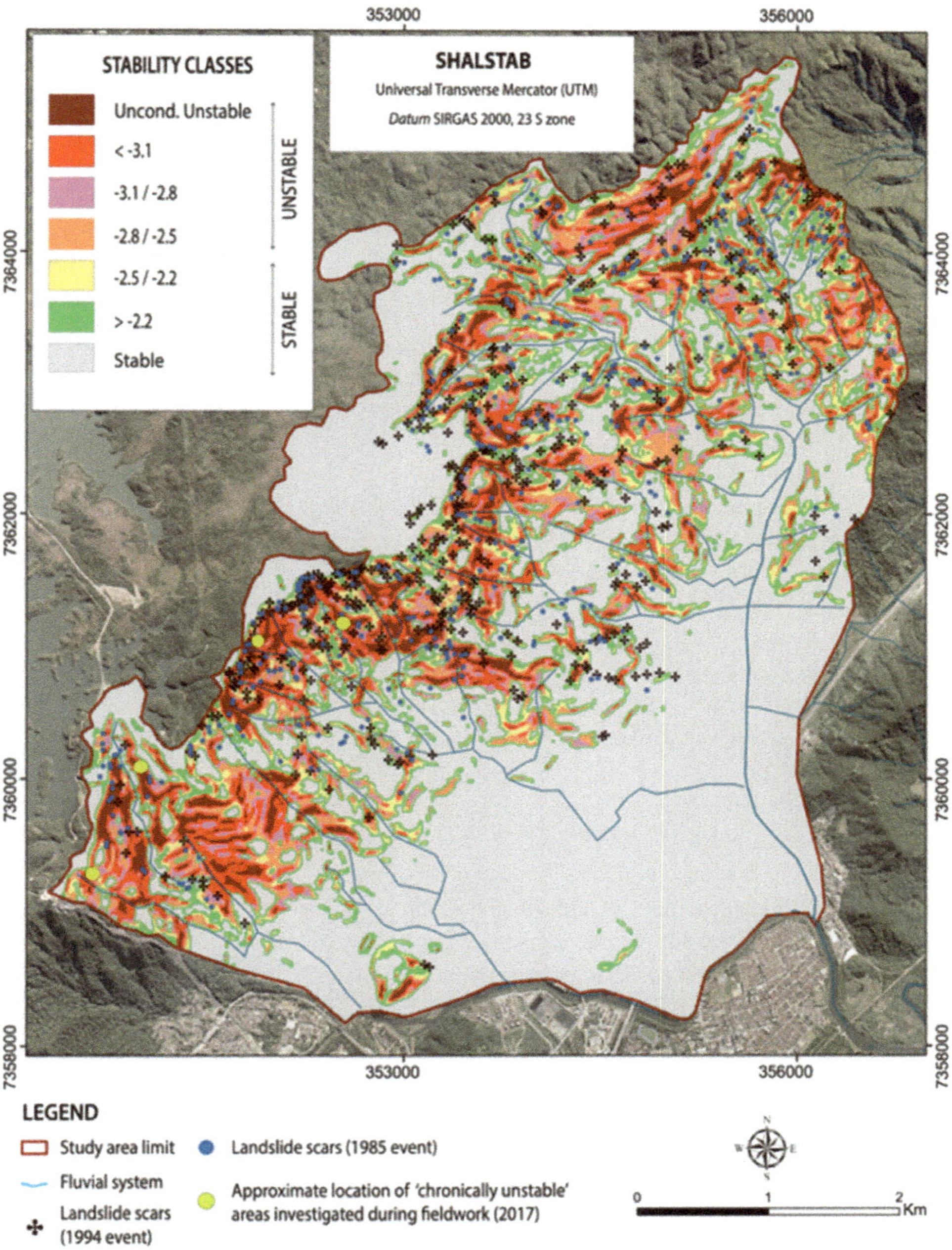

Fig. 4 SHALSTAB stability map. Scale: 1:25,000

Table 5 Performance classifiers of the modeling results using SHALSTAB

Year	1985	1994
Accuracy	0733	0733
Precision	0020	0012
FPR	0266	0267
TPR	0631	0702
TPR/FPR	2371	2630

the approximate location of areas where recent landslide scars and/outcropping bedrock were observed during field campaigns, indicating that the classification of these areas in the lowest stability class is potentially adequate.

Several studies in the Serra do Mar escarpments have attested SHALSTAB efficiency in landslide susceptibility studies in slope and catchment scales (1:5000 to 1:10,000) (Michel et al. 2014; Sbroglia et al. 2018). SHALSTAB, therefore, also successfully represents landslide susceptibility in regional scale (1:25,000).

SINMAP

SINMAP modeling results are shown in Table 6 and Stability map in Fig. 5. A single calibration region was used due to the relative homogeneity of the region's geology and consequent geotechnical characteristics, as well as due to the limited sampling location.

Potentially unstable areas (FS < 1) classified by the model comprise 31.8% of the area (9 km^2), encompassing 245 (1985) and 208 (1994) landslide scars (56.9 and 61.75%, respectively). This results in medium accuracy (0.68 and 0.53) and high true positive results (0.62 and 0.8) when model performance is analyzed (Table 7).

Potentially stable classes (FS > 1) include 152 (1985) and 157 (1994) landslide scars (38.3 and 43%, respectively), representing 68.2% of the study region (19.3 km^2). While SINMAP exhibits high true positive results, the medium/high false positive results and medium accuracy indicate that the model often fails to correctly classify potentially unstable (accuracy) at the same time that overestimates unstable areas (high false positive rate). These results suggest that SINMAP is not completely successful identifying potentially unstable areas in the study region.

Furthermore, as observed during fieldwork, some areas classified as potentially stable in SINMAP exhibit very steep slopes and thin soils, suggesting that they can potentially fail if the sufficient rainfall strikes (Fig. 5). Studies from Nery and Vieira (2014) and Cardozo et al. (2019), however, indicate that SINMAP is adequate in describing landslide susceptibility at Serra do Mar. These studies are in their majority focused on small catchments, using lower density of landslide scars in performance assessment, as well as differing in the methodology used to assess the results (Fig. 6).

Comparative Performance Analysis

The comparative performance analysis is mainly based on the performance classifiers, shown in Tables 5 and 7. Model performance assessment based on real landslide events ensures the reliability of the results, especially when large magnitude events are employed.

SHALSTAB is more accurate predicting sites of landslide initiation, due to higher accuracy and lower false positive results, indicating it did not classify large portions of the area as unstable to correctly depict landslide susceptibility. SINMAP, on the other hand, fails to correctly identify unstable areas with landslide scars (low accuracy) while overestimating these areas.

The satisfying performance of SHALSTAB is also highlighted when the rate between true positive and false positive results (TPR/FPR) is considered (Table 5). This rate analyzes the ratio between the total amount of unstable areas that coincide with the mapped landslide scars (true positive) and the amount of unstable areas without landslide scars (false positive)—the higher the number, the better the performance. SHALSTAB is, therefore, suggested as more adequate to represent landslide susceptibility based on the conditions used in this study.

Studies from Meisina and Scarabelli (2007) and Michel et al. (2014) have similarly compared the performance of SHALSTAB and SINMAP in rainfall triggered events, concluding that SHALSTAB exhibits more adequate also when applied in slope and small catchment scales (1:10,000 and 1:5,000). The adoption of a single set of parameters for a broad region might generalize geotechnical features, which can represent a setback in studies that require greater detail.

Table 6 Statistical analysis of the modeling results using SINMAP. Area in Km^2 and density in unit/km^2.# = number of landslide scars

Stability Class	Area	# 1985	Dens.	# 1994	Dens.
Stable	12,5	35	2,8	47	3,8
Mod Sta	2,4	27	11,2	26	10,8
Quasi Sta	4,4	90	20,3	84	18,9
Lower	4,1	96	23,2	80	19,4
Upper	2,6	81	30,8	79	30,0
Defended	2,2	68	30,4	49	21,9

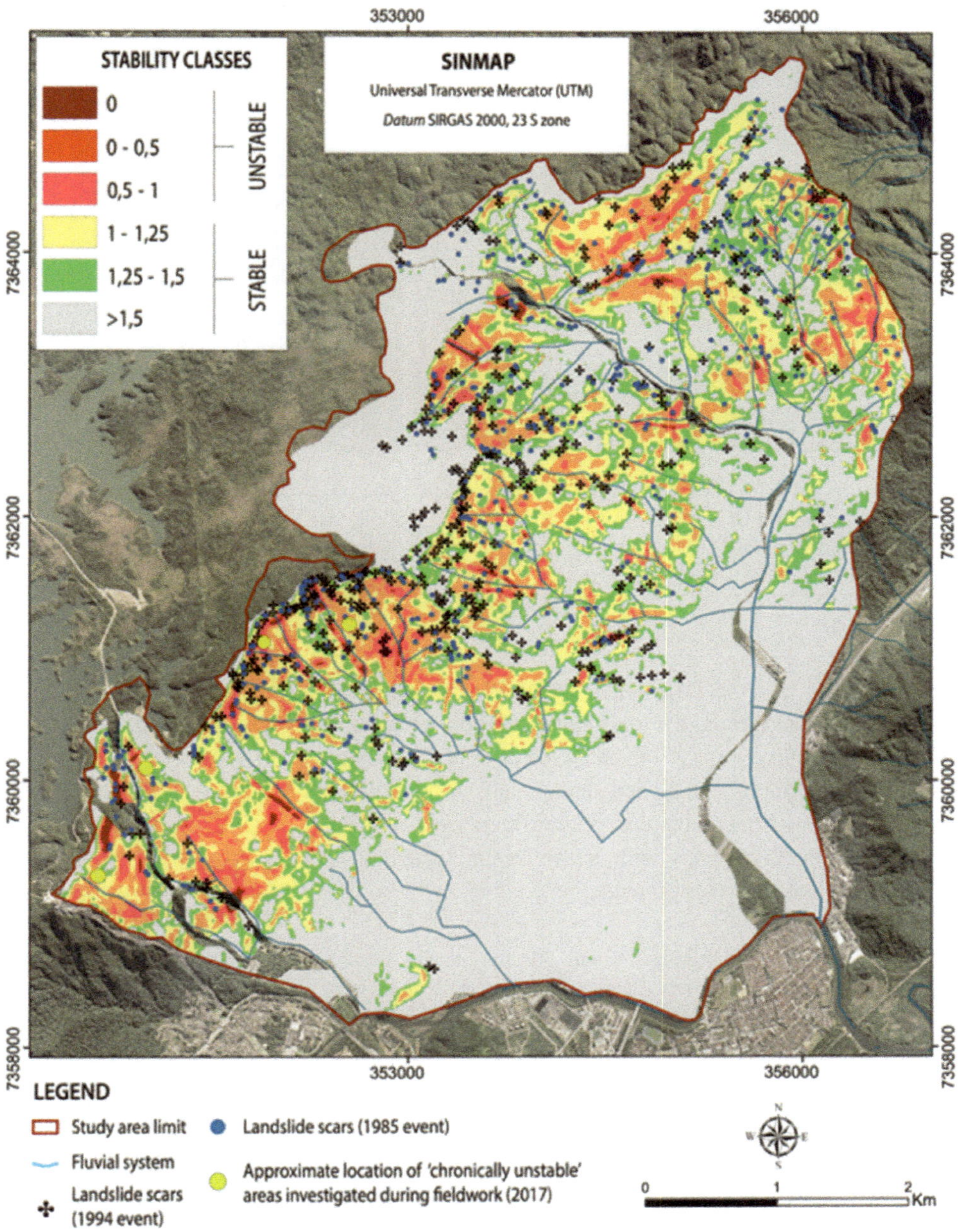

Fig. 5 SINMAP stability map. Scale: 1:25,000

Table 7 Performance classifiers of the modeling results using SINMAP

Year	1985	1994
Accuracy	0.682	0527
Precision	0.017	0008
FPR	0.317	0474
TPR	0.617	0800
TPR/FPR	1.945	1688

Fig. 6 **a** Bedrock exposed area (Coordinates: 362,258 m E; 7,364,004 m N), **b** shallow landslide scar in steep slope (>40°). Soil depth in steeper slopes ranges from 0 to 1 m (Coordinates: 352,780 m E; 7,360,811 m)

Conclusion

By comparing SHALSTAB and SINMAP's performance, SHALSTAB emerges as the one that best represents landslide susceptibility in the chosen scale at the Perequê watershed. To improve the representativeness of modeling results, the compartmentalization of the watershed according to geology units, coupled with geotechnical sampling in each compartment, is recommended.

Acknowledgements The authors would like to thank Brazil's National Council for Scientific and Technological Development (CNPq) and the "Fundação de Apoio à Pesquisa, Ensino e Extensão" (FUNEP) for the financial support.

References

Alvioli M, Baum R (2016) Parallelization of the TRIGRS model for rainfall-induced landslides using the message-passing interface. Environ Modell Softw 81:122–135

Bel C, Liébault F, Navratil O, Eckert N, Bellot H, Fontaine F, Laigle D (2016) Rainfall control of debris-flow triggering in the Réal Torrent, Southern French Alps. Geomorphology 291:17–32

Cardozo CP, Lopes ESS, Monteiro AMV (2019) Calibration of physically based slope-stability models: a case study in Nova Friburgo (Rio de janeiro, Brazil). Geociências 38(2):535–548

Dietrich WE, Asua RR, Trso M (1998) A validation study of the shallow slope stability model, SHALSTAB, in forested lands of Northern California. Department of Geology and Geophysics University of California, Berkeley

Fawcett T (2006) An introduction to ROC analysis. Pattern Recogn Lett 27(8):861–874

Goetz J, Guthrie R, Brenning A (2011) Integrating physical and empirical landslide susceptibility models using generalized additive models. Geomorphology 129(3–4):376–386

IPT—Instituto de Pesquisas Tecnológicas do Estado de São Paulo (1986) Programa Serra do Mar—levantamentos básicos nas folhas de Santos e Riacho Grande, Estado de São Paulo. Anexo A—Estudos geológicos e geomorfológicos. São Paulo: Rel 23.394 (2): 120 p

Meisina C, Scarabelli S (2007) A comparative analysis of terrain stability models for predicting shallow landslides in colluvial soils. Geomorphology 87:207–223

Michel GP, Kobiyama M, Goerl RF (2014) Comparative analysis of SHALSTAB and SINMAP for landslide susceptibility mapping in the Cunha River basin, southern Brazil. J Soils Sediments 14 (7):1266–1277

Montgomery D, Dietrich W (1994) A physically based model for the topographic control on shallow landsliding. Water Resour Res 30 (4):1153–1171

Nery TD, Vieira BC (2014) Susceptibility to shallow landslides in a drainage basin in the Serra do Mar, São Paulo, Brazil, predicted using the SINMAP mathematical model. Bull Eng Geol Environ 74 (2):369–378

O'Loughlin E (1986) Prediction of surface saturation zones in natural catchments by topographic analysis. Water Resour Res 22(5):794–804

Pack RT, Tarboton DG, Goodwin CN (1998) The SINMAP approach to terrain stability mapping. In: 8th Congress of the international association of engineering geology, Vancouver, British Columbia, Canada, 25 p

Sbroglia RM et al (2018) Mapping susceptible landslide areas using geotechnical homogeneous zones with different DEM resolutions in Ribeirão Baú basin, Ilhota/SC/Brazil. Landslides 15(10):2093–2106

Tatizana C et al (1987) Análise de correlação entre chuvas e escorregamentos na Serra do Mar, município de Cubatão. In: Congresso Brasileiro De Geologia De Engenharia. São Paulo, São Paulo

Vieira BC, Ferreira FSG, Villaça MC (2015) Propriedades físicas e hidrológicas dos solos e os escorregamentos rasos na Serra do Mar paulista. Revista Ra'e Ga - O espaço geográfico em análise 34:269–287

Wolle CM, Carvalho CS (1994) Taludes Naturais. In: Falconi FF, Junior AN (Org.), Solos do Litoral de São Paulo. ABMS, São Paulo, pp 180–203

WLF 2020 Kyoto Japan

Regional Slope Stability Analysis in Landslide Hazard Assessment Context, North Macedonia Example

Miloš Marjanović, Biljana Abolmasov, Igor Peshevski, James Reeves, and Irena Georgievska

Abstract

This paper is representing a successful application and comparison of heuristic and deterministic landslide hazard assessment modelling. The advantages of deterministic model, which quantify hazard more plainly and more transparently, are herein emphasized, as they are usually better accepted by potential end users, i.e., decision makers. However, applying deterministic model on large scale is always challenging due to data shortage and uncertainty. Presented example appears to be applicable in road management, i.e. assessment of landslide hazard exposure of the road network. The case study involved Polog region in North Macedonia, modeled for two types of landslides by two different models. The first included shallow translational sliding mechanism and implementation of SINMAP model, while the second included flow mechanism and implementation of RAMMS model. Both models resulted in concurrent map products, suitable for further use in road network decision making. The latter was identified particularly useful when it can be back analyzed on the basis of recent real flow example with sufficient documentation, such as examples from Polog region in 2015 when massive failures occurred following rainstorm and flooding. In comparison to conventional heuristic map which was created in previous research for the same area, the new maps were more difficult to parameterize, with sufficient certainty, so back analysis is a very useful convenience of this particular case study. In conclusion, regional scale deterministic landslide assessment is desirable tool for standard applications in planing and decision making, but it is also recommendable to use it in combination with expert-driven heuristic outputs.

Keywords

Landslide hazard • SINMAP • Regional scale • North Macedonia

Introduction

Road network could be well considered a backbone of a modern society. Communication, trade and travel are still chiefly conducted along roads, while cities, factories and other facilities are commonly grown along their paths. This only causes further densification and expansion of roads, widening of their profiles, increasing their capacity and transport speed.

However, the areas where such expansion commonly takes place is perpetually confined by terrain physiognomy, its geological setting and other environmental conditions and phenomena. Often taken for granted, these natural factors keep reminding us how our societal backbone is fragile, and how quickly our society can be paralyzed. Adding the rapidly changing climate into the picture further worsens the situation. In effect, we seem to witness climate or climate-related extremes, such as floods, landslides, wildfires, etc., more frequently and more intensively than before.

There are numerous reminders of severe natural disasters affecting road networks worldwide, but developing regions are especially sensitive (Haque et al. 2016). Once struck, poor societies resort to quick but temporary repairments,

M. Marjanović · B. Abolmasov (✉)
Faculty of Mining and Geology, University of Belgrade, Đušina 7, 11000 Belgrade, Serbia
e-mail: biljana.abolmasov@rgf.bg.ac.rs

M. Marjanović
e-mail: milos.marjanovic@rgf.bg.ac.rs

I. Peshevski
Faculty of Civil Engineering, Ss. Cyril and Methodius University, Bul. Partizanski odredi 24, Skopje, 1000, Macedonia
e-mail: pesevski@gf.ukim.edu.mk

J. Reeves · I. Georgievska
IMC Worldwide Ltd, Road Redhill, London, R1 1LG, Surrey, UK
e-mail: james.reeves@imcworldwide.com

I. Georgievska
e-mail: irena.georgievska@imcworldwide.com

F. Guzzetti et al. (eds.), *Understanding and Reducing Landslide Disaster Risk*, ICL Contribution to Landslide Disaster Risk Reduction,
https://doi.org/10.1007/978-3-030-60227-7_29

with little or no investigation, innovation and insightful investment, which further accumulates their loss over time, as the phenomenon repeats over and over, each time requiring higher unaffordable costs (Mutter 2015). North Macedonia in the Balkan region in the South Europe (Fig. 1) is just one example of developing country, which has suffered severe calamities in the past couple of decades, including a devastating rainstorm episode in 2015 in its northern parts, leaving 6 casualties behind, 85,000 affected and € 30 million worth of damage. Floods, torrents and landslides that followed, demonstrated shear natural force and interrupted traffic in several major cities in the north, for days. However, the message was received and understood, which is why local and state governments in North Macedonia, under auspices of the World Bank and other global developing initiatives, put their effort and resources in research of natural phenomena and their impact instead of jumping to cheap solutions for coping with acute disaster aftermath. By 2018–19, a comprehensive project of the Public Enterprise for State Roads (PESR) in North Macedonia was conducted and involved preparation of a more resilient road network design. In short, the project started with localizing the hotspots of landslide and flood hazard and estimating of the exposure of the road network over time. The actual choice of suiting preventive or remedial, short- or long-term solution is allocated based on its cost–benefit context and estimated priority, which is a systematic and rational alternative to hasty and bad investment.

In this paper, particular aspects of understanding and modelling landslides as one of the most notorious adversary of road network will be exampled for the Polog Region in North Macedonia. This aspect has previously been elaborated (Peshevski et al. 2019), so the Polog case study will partly be used as a tool to compare results of different landslide assessment techniques, but primarily as a tool to reflect landslide influence on the particular road network.

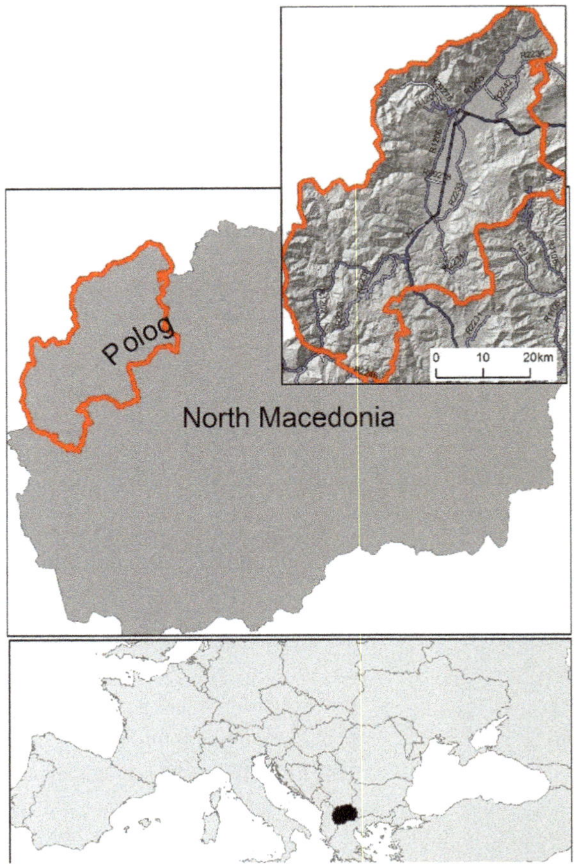

Fig. 1 Location of the study area

Materials and Methods

Inputs

Common landslide assessment practice zones areas prone to landslide hazard and uses several approaches to that end, including heuristic (arbitrary or expert-based multi-criteria framework), statistic (data-based multi-criteria) and deterministic (physical, based on limit equilibrium, i.e. Factor of Safety concept) modelling. The choice is scale-dependent, wherein downscaling entails more complex models, i.e. advanced statistic and deterministic, while heuristic work well in regional scale. The choice is always a balance between available data, their level of detail and model complexity (Marjanović 2014). In this paper, the regional scale is targeted, and both heuristic (commonly used) and deterministic models are compared. The former was previously studied (Peshevski et al. 2019) and will be briefly overviewed herein, while the focus remains on the latter.

Common practice further implies using various input data, usually called conditioning factors/parameters, which include various geological, morphologic, and environmental data relevant to establish the spatial connection between landslide occurrence and conditioning factors. Deterministic modelling is slightly different, as it requires very specific, geotechnical properties as input parameters, which are seldomly available at large scales. The connection between landslide occurrence and susceptibility zone is therein established by calculation based on the balance of driving and resisting forces acting upon slopes (Factor of Safety concept—F_S). The forces are approximated by index properties of the soil material and geometry of the hosting slopes.

Verification is provided by using historic landslide data, i.e., previous occurrences. High hazard zones are supposed to coincide with locations of previous occurrences, and the accuracy of the model can be quantified in this context.

Finally, the road network itself is the targeted element at risk. The influence of high hazard zones can be overlapped and quantified along the road network by allocating particular level of hazard for each road section. Resulting low—high hazard exposure map of road network is an excellent tool to aid decision makers in planning and investment choices.

Deterministic Models

Stability Index MAPping—SINMAP model (Pack et al. 2001) based on Stability Index concept (*SI*) is one of the most commonly used in large scale deterministic landslide hazard tools. *SI* couples a simple transient hydrologic model for simulating rainfall triggering with an even simpler infinite slope stability model, which simulates stress balance of the soil. The former is described by the precipitation recharge *R*, soil transmissivity (water conductivity times the depth) *T*, and contributing upslope area *a*. The latter, stability model, is described by topographic slope angle θ, material bulk density γ, dimensionless cohesion C_{soil} ($c/\gamma h$) and friction angle φ. A simple formula (Eq. 1) summarizes the balance of all encountered resisting and driving forces, i.e. available soil strength versus its stress state. *SI* only further calculates the probability of failure, i.e. considers $F_S < 1$ with 0–1 probability to fail, while all areas with $F_S > 1$ it considers stable. Naturally, it is somewhat difficult to localize and regionalize all required parameters, especially *T* and mechanical parameters *c* and φ, which should be determined for all soil/rock units present in the area of interest. Such approximated model is recommendable only for shallow translational landsliding mechanism, common for thick weathering crust (which is the case in the subject area of interest) in rock formations prone to weathering, such as flysch, low crystalline schists, weak clastites, etc.

$$F_s = \frac{C_{soil} + \left(1 - \min\left(\frac{aR}{T\sin\theta}, 1\right)\frac{\gamma_w}{\gamma_{soil}}\right)\cos\theta tg\varphi_{soil}}{\sin\theta}; \quad (1)$$

(dimensionless)

Another type of mechanism that is common for high yield precipitation and thick weathering crust is debris/earth flow. It cannot be simulated by F_S/*SI* concept, as it requires fluid dynamic analysis. Instead a RApid Mass Movements Simulation—RAMMS model can be applied (Hussin et al. 2012). RAMMS is based on fluid friction model. It is best used when its parameters are back analysed from the existing occurrences (which was the case with Polog in 2015), from which flow runout and height, source areas and other features are extracted. Assuming the accurate geometry is available from elevation data, the critical part is to estimate the bulk density, friction angle (determines the flow behaviour in deceleration) and turbulence factor ξ (determines the fluidisation degree in acceleration). Velocity v is calculated from the topographic constraints, whereas simulated frictional resistance *S* (Eq. 2) is back-analysed against the ground truth (real conditions) until the runout becomes realistic.

$$S = h\gamma_{soil}\cos\theta tg\varphi_{soil} + \frac{\gamma_{soil}v^2}{\xi}; \ (\mathrm{Pa}) \quad (2)$$

Study Area

The Polog Region General Characteristics

The Polog Region is located in the NW part of the North Macedonia covering cca. 2420 km^2, with a wide altitude range (350–2700 m a.s.l.). The wide Vardar river valley crosses through the middle, while Mt. Shar Planina and Korab to the W-NW and Mt. Suva Gora to the E-SE dominate the landscape. Main cities are Tetovo and Gostivar, but there are also many villages on the steep slopes. There is a well-developed state and local road network, railway, and hydro-energetic system Mavrovo, consisting of 130 km of water distribution channels accompanied by 167 km of service roads (Fig. 1). Hence, considerable linear infrastructure is potentially exposed to landslides, torrents and floods, which are all recorded in the past and have even greater potential to develop in the future, given the climate change context.

Geological setting is complex, as the area belongs to an intensively tectonized zone, stressed, sheared and flexed many times in its geological history. Palaeozoic, Mesozoic, Pliocene and Quaternary formations are represented. Igneous rocks include many varieties: granodiorite; rhyolite; serpentinite; gabbro; diabase etc., because volcanic and magmatic activity was intense in the Pliocene age. The Palaeozoic formations are represented by thick complexes of metamorphic rocks, mostly phyllitic schists, meta-conglomerates, metasandstones, quartzites, quartz-chlorite schists and marbles. Most landslides reported in the area lie at the contact of the weak schist and their overbearing weathering crust. Major structures are aligned with general orientation of mountain ranges and river valleys striking NE-SW. This geologically young-forming area is seismically active and reflects the stress distribution from subducting African plate. MSK seismic intensities range from 7 to 9 with return periods of 100 and 500 years.

The region is also very dynamic in terms of landsliding. They are mainly shallow and hosted in the thick schistose regolith. Earlier research (Peshevski et al. 2019) introduced a sound inventory with 1178 occurrences, 13% of which are categorised as active, while remaining are either passive or suspended. About 19% are categorized as deep-seated, while remaining are mostly shallow, up to 5 m deep. They mainly affect traffic infrastructure (28%), while 8% affect populated areas. The inventory is partly available for preview at: https://pmm.nasa.gov/landslides/index.html.

The road network of Polog region is developed by preference of natural topography routes, around river valleys and ridges, and includes various road categories, ranging from highways and motorways (labelled A) through categorized lower-order roads (labelled P1-2) to low-order uncategorized local roads. Total length of the categorized network is 498 km, split in 274 sections. Each section was considered a separate hazard exposure unit, with their lengths ranging from 4 to 26 km.

Previous Landslide Assessment of the Polog Region

The area of interest was previously studied (Peshevski et al. 2019) in similar respect, with similar objectives, but using heuristic approach, i.e., arbitrary scoring method. Therein, a set of conditioning factors, including lithologic units, slope inclination, seismic intensity, average annual rainfall, and land use, was prepared for subsequent scoring. Each factor was split into 2–6 classes and each class was scored on a scale 0–3. The final scoring and summing of factors ware preceded by sensitivity analysis, so that preference weight was given to one of the factors, while other factors receive equal importance. The sensitivity procedure was repeated (giving preference to a different factor each time) until the appropriate accuracy is achieved. The most representative model WLT-TLSR 500 reached about 75% accuracy (Fig. 2), which is considered satisfactory for regional scale landslide modelling and comparable with other quantitative methods (Peshevski et al. 2019).

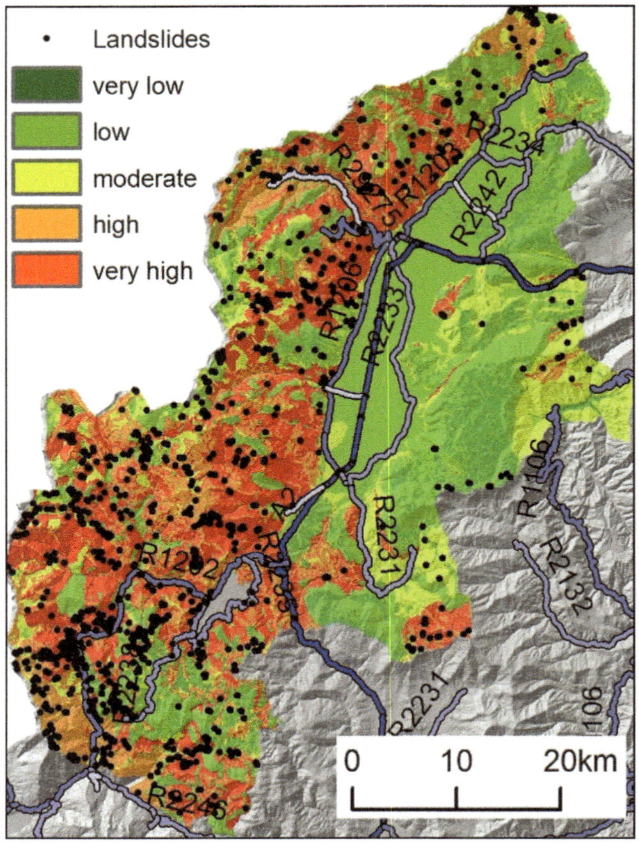

Fig. 2 Heuristic model of landslide susceptibility in Polog

Results

Translational Shallow Landslide Model

Factors used in the previous Polog research, lithology in particular, were useful for extracting relevant geotechnical parameters per each unit (Table 1). The rainfall input is the one that allowed introducing temporal component, which is necessary in hazard assessment as an upgrade towards landslide susceptibility assessment. Rainfall data were collected from National Hydrometeorological Service (https://uhmr.gov.mk/), and the extreme daily precipitation from the baseline period (1981–2010). This part of the tool allows for simulating even higher precipitations, for longer periods, or for future rainfall projections, based on climate change scenarios. Topographic layer, which provides surface geometry and slope angle, was used from the previous Polog research—5 m Digital Terrain Model. Landslide inventory was also used from previous Polog research and includes 950 shallow landslide occurrences (out of 1172 total landslides), which is suitable for the SINMAP modelling approach.

After several iterations, and fine adjustments the following results were achieved and verified (Figs. 3 and 4). The range of F_S values suggests that most of the landslide occurrences (about 80%) lie below $F_S < 1$.

In terms of landslide hazard all areas indicated by $F_S < 1$ can be interpreted as hazardous for annual rainfall extremes (because this type of precipitation was used as input), i.e. for annual return period. The result is comparable to previous Polog model based on heuristic approach. Finally, the map was overlain with the road network layer, enabling further exposure analyses.

Flow Landslide Model

The same DTM from previous Polog research was used for RAMMS model, but its 5 m resolution was too detailed for computational capacity of conventional hardware. For this reason, the northern section of the Polog region was

Table 1 Adopted geotechnical parameters

Unit	c^a	γ^b	φ^c	R/T^d
Solid rock	30–50	27	30–40	1500
Quaternary sediments and weathered rock	0.1–2	18	20–25	100
Clay and flysch	1–10	22	20–30	300
High-crystalline complex	10–30	24	25–30	500
Claystone-shale-schist	0.5–1	20	20–25	1000
Low-crystalline schist	1–2	19	15–20	600

[a]Dimensionless cohesion (cohesion divided by bulk density per soil depth)
[b]Bulk density (kN/m^3)
[c]Shear friction angle (°)
[d]Recharge (mm) versus Transmissivity (water conductivity times the depth mm^3/day)

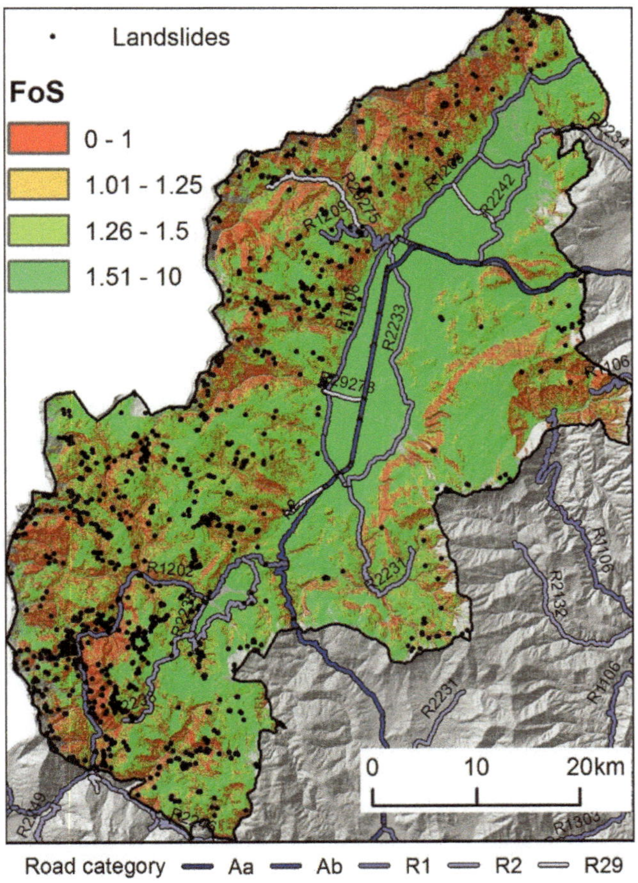

Fig. 3 SINMAP shallow landslide mechanism in Polog

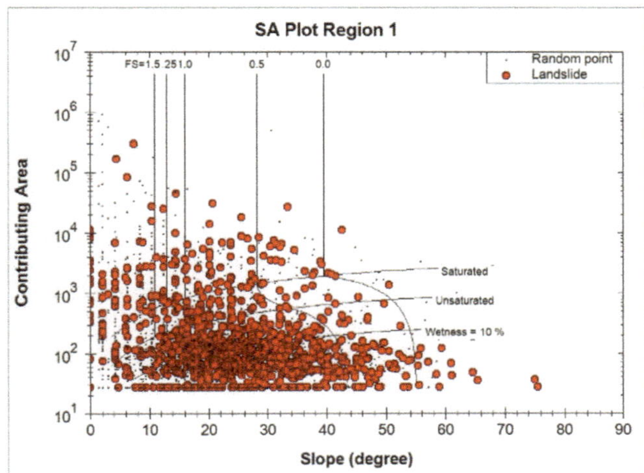

Fig. 4 SINMAP verification plot for Unit 1

subsampled for flow modelling, i.e. for the potential hazards induced by flow mechanism. This model was used as an indirect hazard assessment tool, since there was no temporal component involved. However, the hazard can also be assessed by using other available outputs such as flow height, velocity or kinetic energy. The model also required identification of release areas and estimation of friction parameters within. In addition to parameters from Table 1, turbulence factor was also fitted, to meet evidence from back analysed cases of flows in the Polog region in 2015. Release sources were located using simple heuristic model in GIS environment by conditioning specific slope range and lithology throughout the subsampled area of interest. The identified zones were then visually filtered, so that only upper hollow parts of the slopes are manually selected. In this sense, the tool is more appropriate for back analysis of previous occurrences, but it can be used for simulating future scenarios (Fig. 5), locating places where deposit depth might overcome the culvert capacity, where velocity/kinetic energy might become destructive etc. The flow model shows that there is no major threat for the state road network, but local roads might be easily affected, which was the case in 2015 in a series of torrents and debris flows that were running down the mountain slopes and gullies (Fig. 6).

Conclusions

This example showed a successful application of deterministic modelling along landslide prone parts of the road network in North Macedonia. The primary focus was on advantages of applying regional deterministic modelling and

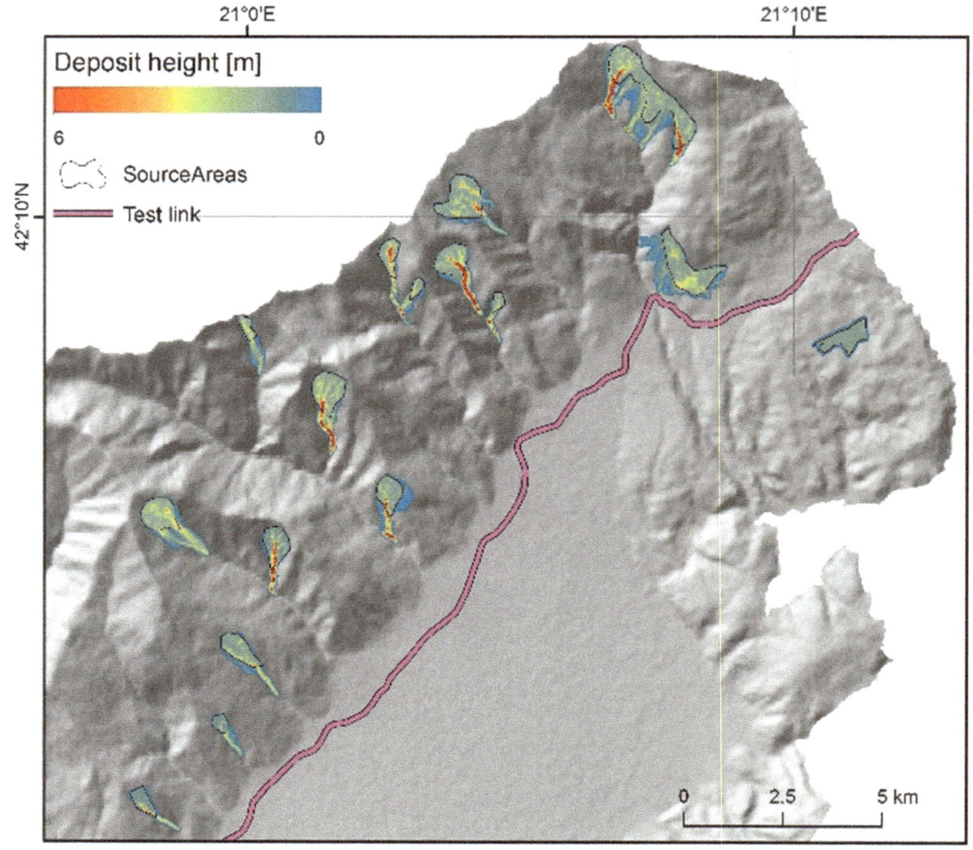

Fig. 5 RAMMS model of debris flow in the subsampled area

Fig. 6 Flow in the SE facing slopes in the Polog region in 2015. Note large boulders and debris deposits adjacent to the stream. The culvert was interrupted at the time. *Photo* Marjanović (2019)

aspects of its contribution in respect to a more common application of heuristic procedure at regional scales. Firstly, it is safe to conclude that deterministic models produce concurrent results and fit very well with the historical data, even better than heuristic approach. Another advantage is that deterministic is easier to communicate to decision makers, especially those with engineering background who understand the concepts of safety factor, energy, velocity or other measurable parameters better than more abstract concepts of landslide susceptibility and hazard zoning. Such outputs allow for a better design and investment plans, including determining the remedial, protective or preventive measure type and dimensioning.

Acknowledgements This paper was supported by the project of Ministry of Education of Republic of Serbia TR36009. The entire research was performed under the framework of World Bank project P148023.

References

Haque U, Blum P, da Silva FP, Andersen P, Pilz J, Chalov SR, Malet J-P, Jemec Auflič M, Andres N, Poyiadji E, Lamas PC, Zhang W, Peshevski I, Pétursson HG, Kurt T, Dobrev N, García-Davalillo JC, Halkia M, Ferri S, Gaprindashvili G, Engström J, Keellings D (2016) Fatal landslides in Europe. Landslides 13(6):1545–1554

Hussin HY, Quan Luna B, van Westen CJ, Christen M, Malet J-P, van Asch ThWJ (2012) Parameterization of a numerical 2-D debris flow model with entrainment: a case study of the Faucon catchment, Southern French Alps. Nat Hazards Earth Syst Sci 12:3075–3090

Marjanović M (2014) Conventional and machine learning methods for landslide assessment in GIS. Palacký University, Department of Geoinformatics, Olomouc, 204 p. ISBN 978-80-244-4169-6

Mutter JC (2015) The disaster profiteers: how natural disasters make the rich richer and the poor even poorer. St. Martin's Press, New York, 288 p. ISBN 978-1137278982

Pack RT, Tarboton DG, Goodwin CN (2001) Assessing terrain stability in a GIS using SINMAP. In: 15th annual GIS conference, GIS 2001, Vancouver, 19–22 Feb 2001, pp 1–9

Peshevski I, Jovanovski M, Abolmasov B, Papic J, Đurić U, Marjanović M, Haque U, Nedelkovska N (2019) Preliminary regional landslide susceptibility assessment using limited data. Geol Croat 72(1):81–92

Applying the Newmark Model in the Assessment of Earthquake Triggered Landslides During the 2017 Ms 7.0 Jiuzhaigou Earthquake, China

Xiaoli Chen, Xinjian Shan, Mingming Wang, Chunguo Liu, and Nana Han

Abstract

The 8 August 2017 Jiuzhaigou, China, earthquake (Ms 7.0) occurred within Jiuzhaigou County, northern Aba Prefecture, Sichuan province, China. The earthquake generated 4834 coseismic landslides with individual areas >7.8 m^2 over a 600 km^2 region. Both the quantity of landslides and the areas that were affected by landslides are similar to those typical for earthquakes of similar magnitude in the eastern Tibetan Plateau, suggesting that the regional geologic structure significantly effects seismic attenuation. Instead of correlating geological and topographic factors with the co-seismic landslide distribution pattern, this study focuses on analyzing the seismic landslide susceptibility, which comes from a calculation of critical acceleration values using a simplified Newmark block model analysis. Results show that seismic landslide susceptibility plays an important role in the co-seismic landslide pattern. Based on the seismic landslide susceptibility, it becomes feasible to provide a quick evaluation of earthquake-triggered landslides when combined with a peak ground acceleration map for an event. Moreover, the correlation between the characteristics of seismic landslide susceptibility and previously observed landslides suggest that an unmapped extension of the Huya fault was the source fault of the Jiuzhaigou earthquake.

Keywords

Coseismic landslides • Seismic landslide susceptibility • Newmark method • 2017 Ms 7.0 Jiuzhaigou earthquake

Introduction

On 8 August 2017, a Ms 7.0 earthquake (33.20°N, 103.82° E) occurred within Jiuzhaigou County, of northern Aba Prefecture, Sichuan province, China at a focal depth of 20 km. The complex geologic and geographic conditions in southwest China have historically caused serious earthquake-triggered landslides resulting in significant damage to both human lives and property (Fig. 1) (Yi et al. 2002; Chen et al. 2012; Qi et al. 2010; Xu and Xiao (2013); Yin et al. 2009). Since the 2008 Ms 8.0 Wenchuan earthquake, several strong earthquakes have occurred along the eastern and southeastern margins of the Tibetan Plateau, including the 2013 Ms 7.0 Lushan earthquake, the 2014 Ms 6.5 Ludian earthquake, and the 2017 Ms 7.0 Jiuzhaigou earthquake (Yin et al. 2009; Dai et al. 2010; Qi et al. 2010; Gorum et al. 2011; Chen et al. 2015; Dai et al. 2017; Su et al. 2017). These earthquakes occurred on lesser-known active faults in the eastern and southeastern Tibetan Plateau, demonstrating the likely occurrence of future high magnitude earthquakes in the region. An understanding of the factors controlling the distribution pattern of coseismic landslides in the area as well as a reliable method for evaluating post-seismic landslides is in need.

Previous studies of earthquake-triggered landslides in southwestern China show that local geological conditions and historical earthquakes have an important influence on the resulting landslide distribution (Chen et al. 2012). Generally, areas affected by landslides have similar outlines and

X. Chen (✉) · X. Shan
Institute of Geology, China Earthquake Administration, Beijing, 100029, China
e-mail: chenxl@ies.ac.cn

X. Shan
e-mail: xjshan@ies.ac.cn

M. Wang
Sichuan Earthquake Administration, Chengdu, 610041, China
e-mail: dannywang9088@163.com

C. Liu
China Earthquake Networks Center, Beijing, 100045, China
e-mail: liuchguo@126.com

N. Han
Seismological Bureau of Shanghai, Shanghai, 200062, China
e-mail: hannanasky@163.com

F. Guzzetti et al. (eds.), *Understanding and Reducing Landslide Disaster Risk*,
ICL Contribution to Landslide Disaster Risk Reduction,
https://doi.org/10.1007/978-3-030-60227-7_30

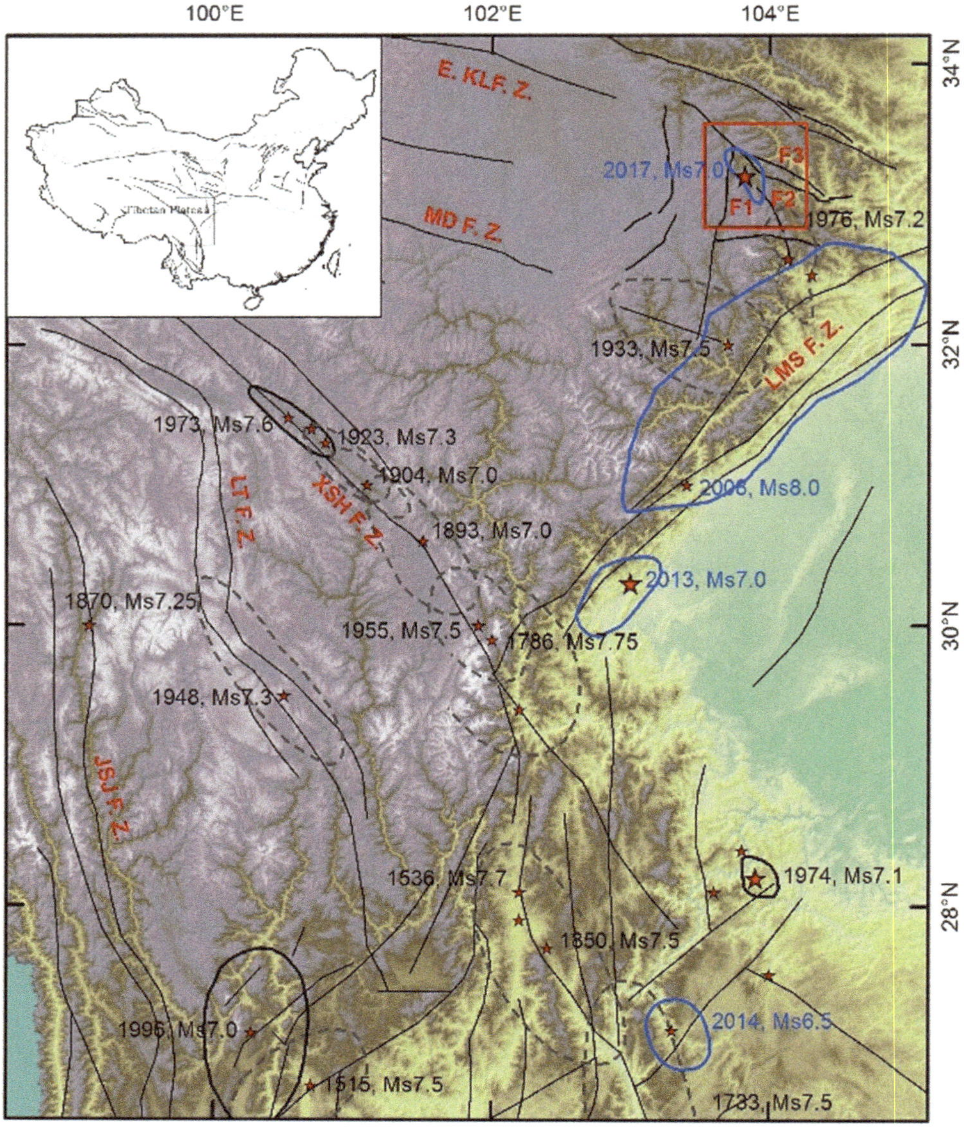

Fig. 1 Map showing earthquake trigged landslides in the southwestern China. The red square is this study area. Solid black lines represent the areas with detailed earthquake induced landslides records, dash lines show landslide data inferred from the historical documents, which using seismic intensity line of VII for instead, blue lines indicate landslides from recent earthquake events. Red stars represent earthquake epicenters and circles represent areas of associated landslides. *Abbreviations* F1: the Mingjiang fault; F2: the Huya fault; F3: the Tazang fault

sizes (Fig. 1). However, the historical events-based method can only coarsely provide the extent where an area would be affected by landslides. Although it is useful for the first stage in preliminary landslide assessment, the knowledge is not sufficient to meet the requirements for an evaluation of coseismic landslides.

In this study, instead of correlating geologic and topographic factors with the co-seismic landslide distribution pattern, which was conducted by Fan et al. (2018) and Tian et al. (2019), here we relate the observed landslide distribution to existing seismic landslide susceptibility models. Results of this study contribute to the development and improvement of landslide modeling, which we hope will allow for a faster evaluation of earthquake-triggered landslides after future events in the Tibetan Plateau.

Landslides Triggered by the Ms 7.0 Jiuzhaigou Earthquake

The Ms 7.0 Jiuzhaigou earthquake occurred within the eastern Kunlun fault zone, where the fault system splays into a horsetail structure composed of the Huya fault (F2 in Fig. 1), Tazang fault (F3 in Fig. 1), and the Minjiang fault (F1 in Fig. 1) (Xu et al. 2017). Focal-mechanism solutions for the Jiuzhaigou event show a steeply dipping left-lateral strike-slip fault within the horsetail structure, though there is no obvious ground surface rupture or fault mapped. The lack of evidence for surface rupture raises the question of which fault within the Kunlun fault zone produced the Jiuzhaigou earthquake, and the southern branch of the Tazang fault or an extension of the Huya fault may be the most probable causative faults (https://www.eq-igl.ac.cn/upload/images/2017/8/991724632).

After the event, using 0.5 m-resolution Geoeye-1 post-seismic images (shot on August 14, 2017) and pre-seismic Google Earth (GE) images, the team of Dr. Xu interpreted 4834 coseismic landslides by the quake (personal communication), which is the most exhaustive inventory reported so far (Fig. 2). The planar areas of these landslides are several to hundred thousand m^2, with the smallest 7.8 m^2 and maximum 236,338 m^2, covering 9.64 km^2. Post-earthquake field investigations show that shaking triggered slope failures along steep river banks near the epicenter and along steep road cuts within the Jiuzhaigou parkland. The size of the failures ranged from a few cubic meters rock fall to large rock avalanches estimated to be a million cubic meters. Most of the landslides are distributed along the northern extension of the Huya fault in NW direction and as much as 10 km from the mapped fault zone, which is consistent with the characteristics of landslides caused by earthquakes on strike slip faults.

Methods and Data

According to Newmark (1965), landslide potential is simply modeled as a rigid block on an inclined plane with a known critical acceleration (a_c) necessary to overcome the shear resistance at its base. The cumulative permanent displacement of the block is modeled relative to the base of the block as it is subjected to the effects of an earthquake's acceleration, and is used to predict the slope behavior during a shaking event. This model first proposed that the seismic stability of slopes should be evaluated in terms of cumulative permanent displacement rather than a traditional minimum factor of safety (F_S).

The critical acceleration, which is a simple function between the static factor of safety and slope geometry (Eq. 1), represents a measure of intrinsic slope properties that are independent of any ground-shaking scenario. It is a connection between static and dynamic slope stability analysis and portrays seismic landslide susceptibility (Chang et al. 1984; Jibson et al. 2000).

$$a_c = (F_s - 1)g \times \sin a \quad (1)$$

And, F_S is the static factor of safety and can be expressed as:

$$F_S = \frac{c'}{\gamma t \sin \alpha} + \frac{\tan \varphi'}{\tan \alpha} - \frac{m\gamma_w \tan \varphi'}{\gamma \tan \alpha} \quad (2)$$

Here, the variables in Eq. (2) are related to the slope material characteristics and slope geometry. Table 1 shows the variables in Eq. (2) and their descriptions.

As indicated above, the simplified rigid-block model applied here requires only inputs of local topographic slope and material strength. For this study, the slope angle is estimated using a slope map derived from 30 m Shuttle Radar Topography Mission (SRTM) elevation data (https://srtm.csi.cgiar.org/SELECTION/inputCoord.asp) (Fig. 2a). Meanwhile, geologic units from 1:200,000-scale geologic maps (CGS 2001) are used to estimate material properties for the region. The study region has exposures of the strata from the Devonian to Quaternary periods, while there is a lack of Jurassic to Cretaceous sequences (Fig. 2b). Almost all bedrocks appearing in this region are weathered and deformed due to intensely tectonic movements.

Though precise material strength parameters play an important role in a physical-based model for calculating slope stability, it is impractical and beyond the scope of this study to test the parameters across such a large region (Jibson et al. 2000). Therefore, we use a set of representative shear-strength values for each of the geologic units. First, the rocks in this study region are classified into four types, which including a hard rock group (type I), a moderately hard rock group (type II), a soft rock group (type III) and a second soft rock group (type IV) General Administration of Quality Supervision, Inspection and Quarantine of the People's Republic of China, Ministry of Construction of the People's Republic of China (1995). Then, rock material shear-strength parameters are assigned based on the "Standard for engineering classification of rock masses" (GB50218-94, China) and some additional relevant references (i.e., Jibson et al. 2000; Dreyfus et al. 2013; Shinoda and Miyata 2017; Chen et al. 2014). Table 2 shows material strength values for the rock types in the study region.

Results

Using the topographic data and geologic maps, we calculated and mapped critical accelerations for slopes throughout the study area using the Newmark method and evaluated the co-seismic landslide potential by analyzing the difference between the seismic landslide susceptibility of a particular

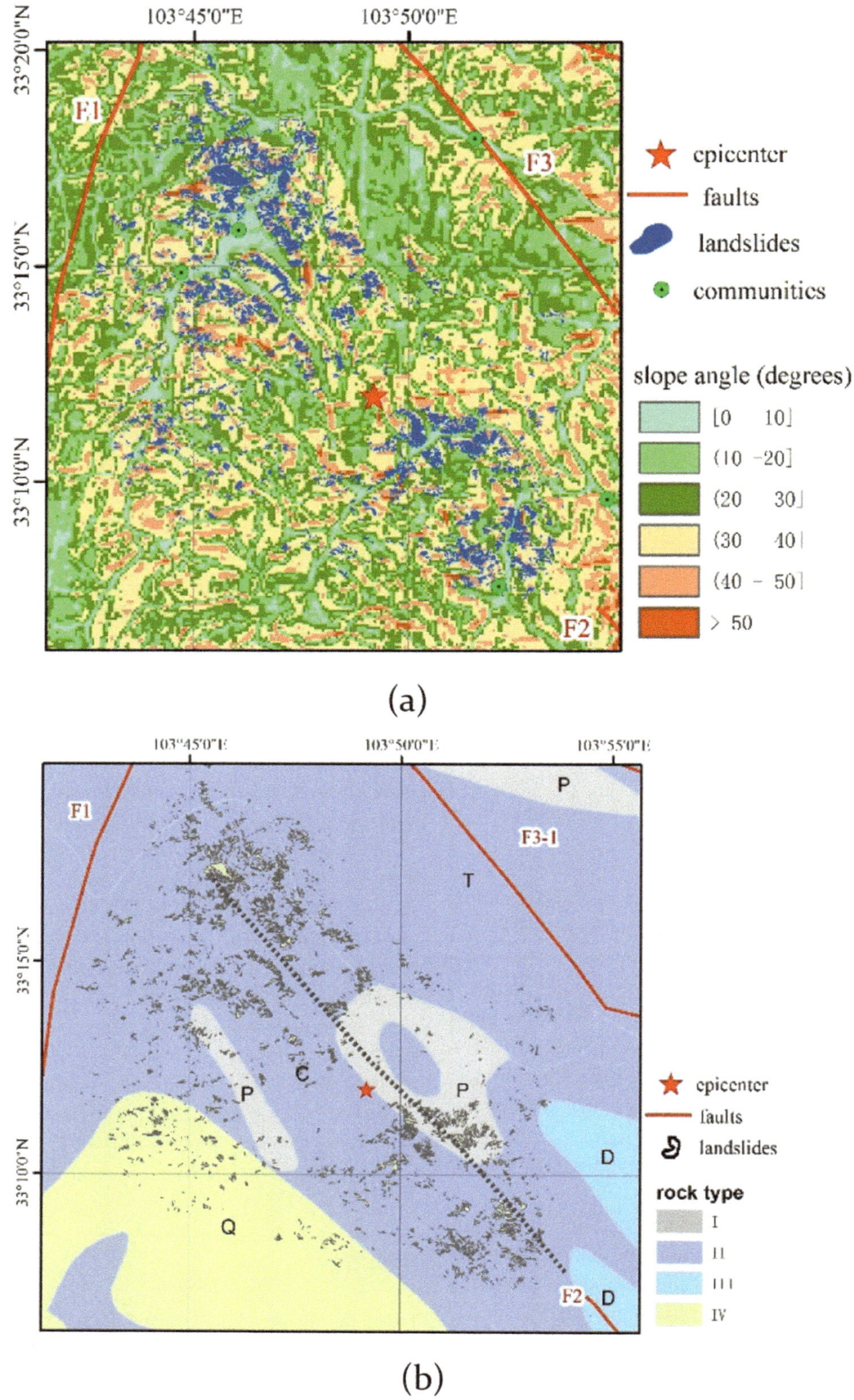

Fig. 2 Map showing the distribution of slope (**a**) and rock types (**b**) in the Jiuzhaigou earthquake region. Geology: Q: Quaternary; N: Neogene; E: Paleogene; T: Triassic; P: Permian; C: Carboniferous; D: Devonian; F1: the Mingjiang fault; F2: the Huya fault; F3: the Tazang fault. The black dashed line represents the extension of the Huya fault

Table 1 Variables in Eqs. (1) and (2)

Variable	Description
a_c	Critical acceleration in terms of g
g	Acceleration of earth's gravity
F_S	The static factor of safety
α	The angle from the horizontal direction
φ'	The effective frictional angle
c'	The effective cohesion
γ	Material unit weight
γ_w	Water unit weight
t	The slope-normal failure slab thickness
m	The proportion of the slab thickness saturated

Table 2 Rock parameters in the study area

Rock type	c′ (MPa)	φ' (°)	γ (kN/m^3)
I	0.035	40	27.0
II	0.027	35	25.0
III	0.020	20	22.0
IV	0.015	10	15.0

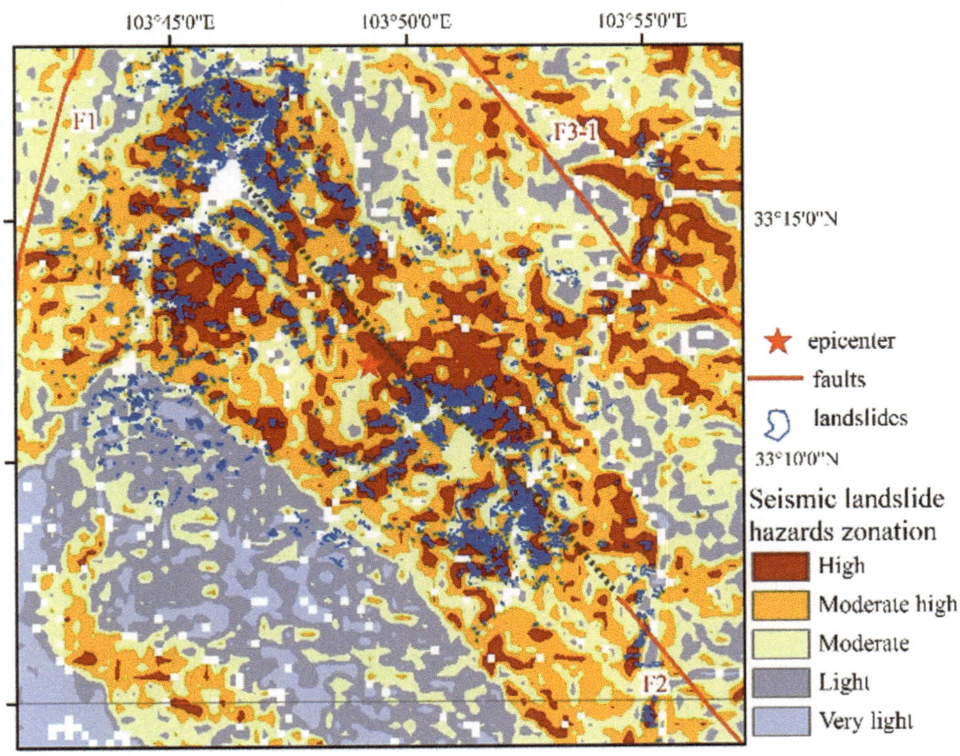

Fig. 3 Evaluation of the co-seismic landslides hazards zonation. The black dashed line represents an extension of the Huya fault. F1: the Mingjiang fault; F2: the Huya fault; F3: the Tazang fault

slope and the peak ground acceleration at that site following the Ms 7.0 Jiuzhaigou earthquake. Figure 3 shows the results of our analysis using Newmark block analysis to determine coseismic landslide hazards, which reflecting the result of influencing from the shaking event on pre-earthquake seismic landslide susceptibility.

Potential hazard zones are divided into 5 levels: high, moderately high, moderate, light, and very light hazards. Figure 3 shows a higher concentration of landslide prone areas along the Tazang fault (F3-1, F3-2) and the northern extension of the Huya fault (F2). We observe a strong correlation between this model and the location of actual co-seismic landslides: proportions of landslides identified in different hazard classes from the very high level to the very light level are 62%, 14%, 15%, 6% and 0%, respectively, demonstrating the strength of the model.

Discussion

Similar to the 2013 Ms 7.0 Lushan earthquake and the 2014 Ms 6.5 Ludian earthquake, the 2017 Ms 7.0 Jiuzhaigou earthquake did not produce an obvious surface rupture. Additionally, the earthquake epicenter was located between the Minjiang fault, Tazang fault, and an extension of the Huya fault. The absence of a surface rupture and the fact that the event is located in a densely faulted region makes it difficult to determine which fault actually produced the earthquake. However, existing research has shown that the fault type has a notable effect on the distribution of co-seismic landslides, helping to determine the fault responsible for the Ms 7.0 Jiuzhaigou earthquake. Co-seismic landslides tend to concentrate on the hanging wall of a thrust fault, whereas those associated with a strike-slip fault occur within a narrow zone along the sides of the fault (Harp and Jibson 1996; Jibson et al. 2004).

As seen in Fig. 3, the landslides are closely clustered near the northern extension of the Huya fault. By combining that observation with the seismic landslide susceptibility data, we favor the northern Huya fault as the earthquake source. The co-seismic landslides plotted in Fig. 3 are distributed along the extension of the Huya fault and within 10 km of the fault, an area that has been classified as having high seismic landslide susceptibility. This observation is consistent with the characteristic landslide patterns expected from strike-slip faults (Jibson et al. 2004; Xu et al. 2013). If the Tazang fault were the source of the Jiuzhaigou earthquake, we would expect a concentration of landslides in the areas classified at high susceptibility near this fault. Because this is not the case, we have confidence in assuming the Huya fault is the source of the Jiuzhaigou earthquake.

Conclusions

1. The range of the landslides induced by the Ms 7.0 Jiuzhaigou earthquake is consistent with what is normal for earthquakes of similar magnitude in the eastern Tibetan Plateau, suggesting that the regional geologic structure here significantly effects seismic acceleration attenuation.
2. Modeled seismic landslide susceptibility controls the co-seismic landslide distributions, most of the landslides occurred in the areas with high seismic landslide susceptibility. Despite some drawbacks, the simplified Newmark analysis is a feasible and practical way to quickly evaluate earthquake-triggered landslides.
3. The correlation between the characteristics of seismic landslide susceptibility and the locations of observed landslides suggests that the extension of the Huya fault is the source fault of the Jiuzhaigou earthquake.

Acknowledgements We thank Xu Chong for his landslide inventory provided. This work was supported by the National Key Research and Development Program of China (Project No. 2017YFC1501004) and the Basic Scientific Fund of the Institute of Geology, China Earthquake Administration (Grant No. IGCEA1604).

References

Chang CJ, Chen WF, Yan JTP (1984) Seismic displacements in slopes by limit analysis. Journal of Geotechnical Engineering (CX) (7):850–874

China Geological Survey (CGS) (2001) Regional geological map of Sichuan Province (1:200, 000), Geological Press

Chen XL, Zhou Q, Ran HL, Dong RS (2012) Earthquake-triggered landslides in southwest China. Nat Hazards Earth Syst Sci 12:351–363

Chen XL, Zhou Q, Liu CG (2015) Distribution pattern of co-seismic landslides triggered by the 2014 Ludian, Yunnan, China Mw 6.1 earthquake: special controlling conditions of local topography. Landslides 12(6):1159–1168

Chen XL, Liu CG, Yu L, Lin CX (2014) Critical acceleration as a criterion in seismic landslide susceptibility assessment. Geomorphology 217:15–22

Dai FC, Xu C, Yao X, Xu L, Tu XB, Gong QM (2010) Spatial distribution of landslides triggered by the 2008 Ms 8.0 Wenchuan earthquake, China. J Asian Earth Sci 40:883–895

Dai LX, Xu Q, Fan XM, Chang M, Yang Q, Yang F, Ren J (2017) A preliminary study on spatial distribution patterns of landslides triggered by Jiuzhaigou earthquake in Sichuan on August 8th, 2017 and their susceptibility assessment. J Eng Geol 25(4):1151–1164 (in Chinese)

Dreyfus D, Rathje EM, Jibson RW (2013) The influence of different simplified sliding-block models and input parameters on regional predictions of seismic landslides triggered by the Northridge earthquake. Eng Geol 163:41–54

Fan XM, Scaringi G, Xu Q, Zhan WW, Dai LX, Li YS, Pei XG, Yang Q, Huang RQ (2018) Coseismic landslides triggered by the 8th August 2017 Ms 7.0 Jiuzhaigou earthquake (Sichuan, China): factors controlling their spatial distribution and implications for the seismogenic blind fault identification. Landslides. https://doi.org/10.1007/s10346-018-0960-x

General Administration of Quality Supervision, Inspection and Quarantine of the People's Republic of China, Ministry of Construction of the People's Republic of China (1995) Standard for engineering classification of rock masses. Standards Press of China, Beijing (in Chinese)

Gorum T, Fan XM, van Westen CJ, Huang RQ, Xu Q, Tang C, Wang G (2011) Distribution pattern of earthquake-induced landslides triggered by the 12 May 2008 Wenchuan earthquake. Geomorphology 133(3):152–167

Harp EL, Jibson RW (1996) Landslides triggered by the 1994 Northridge, California earthquake. Bull Seismol Soc Am 86(1B): s319–s332

Jibson RW, Harp EL, Michael JA (2000) A method for producing digital probabilistic seismic landslide. Eng Geol 58:271–289

Jibson RW, Harp EL, Schulz W, Keefer DK (2004) Landslides triggered by the 2002 M-7.9 Denali fault, Alaska, earthquake and the inferred nature of the strong shaking. Earthq Spectra 20:669–691

Newmark NM (1965) Effects of earthquakes on dams and embankments. Geotechnique 15:139–160

Qi SW, Xu Q, Lan HX, Zhang B, Liu JY (2010) Spatial distribution analysis of landslides triggered by 2008.5.12 Wenchuan earthquake, China. Eng Geol 116:95–108

Shinoda M, Miyata Y (2017) Regional landslide susceptibility following the Mid-NIIGATA prefecture earthquake in 2004 with NEWMARK'S sliding block analysis. Landslides. https://doi.org/10.1007/s10346-017-0833-8

Su LJ, Xu XQ, Genge XY, Liang SQ (2017) An integrated approach for investigating hydrogeological characteristics of a debris landslide in the Wenchuan earthquake area. Eng Geol 219:52–63

Tian YY, Xu C, Ma SY, Xu XW, Wang SY, Zhang H (2019) Inventory and spatial distribution of landslides triggered by the 8th August 2017 Mw 6.5 Jiuzhaigou earthquake, China. J Earth Sci 30(1):206–217

Xu C, Xiao JZ (2013) Spatial analysis of landslides triggered by the 2013 Ms 7.0 Lushan earthquake: a case study of a typical rectangle area in the northeast of Taping town. Seismol Geol 35(2):436–451 (in Chinese)

Xu C, Xu X, Yu G (2013) Landslides triggered by slipping-fault-generated earthquake on a plateau: an example of the 14 April 2010, Ms 7.1, Yushu, China earthquake. Landslides 10 (4):421–431

Xu XW, Chen GH, Wang QX, Chen LC, Ren ZK, Xu C, Wei ZY, Lu RQ, Tan XB, Dong SP, Shi F (2017) Discussion on seismogenic structure of Jiuzhaigou earthquake and its implication for current strain state in the southern Qinghai Tibet Plateau. Chin J Geophys 60(10):4018–4026 (in Chinese)

Yi GX, Wen XZ, Xu XW (2002) Study on recurrence behaviors of strong earthquakes for several entireties of active fault zones in Sichuan-Yunnan region. Earthq Res China 18(3):267–276 (in Chinese)

Yin YP, Wang FW, Sun P (2009) Landslide hazards triggered by the 2008 Wenchuan earthquake, Sichuan, China. Landslides 6(2):139–152

Evaluation of Secondary Landslide Susceptibility for the Rescue Activity Using LiDAR UAV Data

Shoji Doshida

Abstract

On April 11th, 2018, a large-scale landslide occurred in Yabakei Town, Nakatsu City, Oita Prefecture, Japan (Yabakei landslide). The characteristic of the Yabakei landslide is that occurred without any special causes such as rainfall and earthquakes. In this research, we analyzed the LiDAR UAV data (Nakanihon Air Service Co., LTD.) measured immediately after the Yabakei landslide, and considered the evaluation of secondary landslide susceptibility during the rescue activities. And I considered the cause of the time lag landslide which occurred without any special causes based on the features of Yabakei landslide. I thought that the main cause was groundwater.

Keywords

Secondary landslide susceptibility • The time lag landslide • Rescue activity • LiDAR UAV

Introduction

At the landslide area, such as a catastrophic landslide, a slope failure and a debris flow, I have to consider the danger of secondary landslide in order to carry out rescue activity. However, it was very difficult to acquire sufficient information for carrying out rescue activity at the moment of the first response, and I had to carry out the old rescue activity for depending on a few information.

At 3:48 on April 11th, 2018, a large-scale landslide occurred in Yabakei Town, Nakatsu City, Oita Prefecture, Japan (Yabakei landslide) (Figs. 1 and 2). The Yabakei landslide killed 6 persons. The Yabakei landslide occurred suddenly without any special causes such as rainfall and earthquakes (such a landslide I call "the time lag landslide") (Doshida and Araiba 2019).

In this research, I analysed the LiDAR UAV data (Nakanihon Air Service Co., LTD.) measured immediately after the Yabakei landslide, and considered the evaluation of secondary landslide susceptibility during rescue activities. And I considered the cause of the time lag landslide based on the features of Yabakei landslide.

The Topographical Data (DEMs) and Geological Units

In this research, I use two kinds of topographic data (DEM: Digital Elevation Model). One is 5 m mesh data of GSI (Geographical survey Institute, Japan) and before the disaster (Fig. 3 left). Another is LiDAR UAV data (Light Detection and Ranging, Unmanned Aerial Vehicle) [Nakanihon Air Service Co., LTD. supply] and after the disaster (Fig. 3 right). The LiDAR UAV data is measured on April 13th, two days after the disaster, and the resolution is 0.1 m mesh. Both of the DEMs are DTM (Digital Terrain Model), so the influence of vegetation is removed.

I compared the topographic data before and after the disaster, and analysed calculating the difference of elevations. And based on the difference, I classified the topographic feature of the landslide, and considered the evaluation of secondary landslide susceptibility in each area. I am clarifying the features of the Yabakei landslide by analysing DEMs too.

The Geological and Geomorphological Setting of the Yabakei Landslide

The geological setting around the Yabakei area consists of non-alkaline pyroclastic flow volcanic rocks (Early Pleistocene) in the upper part and non-alkaline mafic volcanic

S. Doshida (✉)
National Research Institute of Fire and Disaster, Chofu, 182-8508, Tokyo, Japan
e-mail: sdoshida@fri.go.jp

F. Guzzetti et al. (eds.), *Understanding and Reducing Landslide Disaster Risk*, ICL Contribution to Landslide Disaster Risk Reduction,
https://doi.org/10.1007/978-3-030-60227-7_31

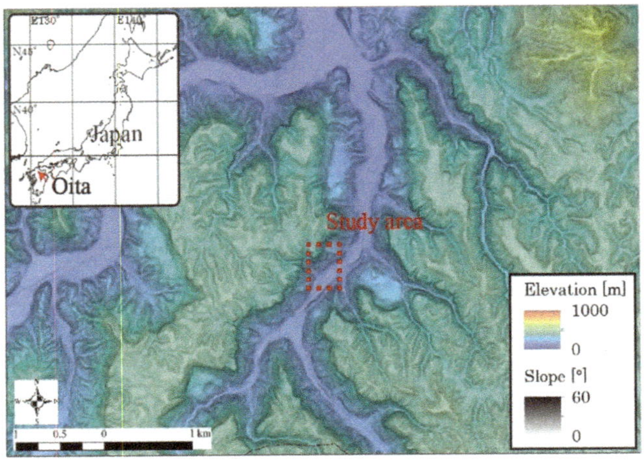

Fig. 1 Index map of the Yabakei landslide

Fig. 2 Photographs of the Yabakei landslide. The photo above was taken on the ground, and the photo below taken by the UAV

rocks (Late Miocene to Pliocene) in the lower part (Geological Survey of Japan 2009). In the Yabakei landslide, tuff-breccia and the debris are mainly distributed.

The geomorphological setting around the Yabakei area is a plateau (Fig. 1). The end of a plateau is a steep slope and changes to a gradual slope. Most of these gradual slopes are landslide topography. The Yabakei landslide also occurred in landslide topography.

The Yabakei landslide looks such as a rockslide (Cruden and Varnes 1996), but the main landslide (Fig. 4A) consisted of debris (secondary sediments of basement rock), so the main Yabakei landslide is a debris slide. The sub east landslide (Fig. 4B) consisted of basement rock (tuff-breccia), the sub east landslide is a rockslide. The sub west landslide (Fig. 4C) consisted of the secondary sediments of the main landslide (Fig. 4A), the sub west landslide is a debris slide.

The Results and Discussions

The Difference of the Elevations

Figure 4 shows the difference of elevations before and after the Yabakei landslide. The base maps are the slope figure and 1 m contour map of LiDAR UAV. The cold colors show the negative change and the warm colors shows the positive change. In consideration of the accuracy of the difference, the colors are not displayed for location less than 1 m greater than −1 m. The red stars in Fig. 4 indicate the spring location on April 13th and 28th. By comparison, I am able to estimate the depth of the landslide (maximum depth about 20 m) and the volume of the landslide (about 6 million cubic meter).

Figure 5 shows the profiles of the Yabakei landslide before and after the disaster. Vertical distance of the Yabakei landslide is about 112 m, and Horizontal distance is about 220 m. The equivalent coefficient of friction is about 27.0°. The relatively gradual equivalent coefficient of friction can assume that the water is related to the landslide movement. This is consistent with the results of a local hearing that heard the sound of a flowing waterfall immediately after the disaster.

As a result of field survey, I considered that the slip surface was around the spring point, because the area below the spring point was not moving, and the layer of soft clay was exposed at the spring point. The presence of springs and a layer of clay indicate the presence of contrast of permeability and an aquifer in the upper part of the slope. But there was almost no rain before this landslide. I determined that the direct cause of this landslide was groundwater.

From direction of the fallen tree, it was assumed that the Yabakei landslide was a slump type landslide. It was checked by following investigation that these are right.

Topographic Features Classification

Figure 6 shows the topographic features classification based on LiDAR UAV data and the difference of the DEMs. I have classified the landslide into 11 areas. Details of each area

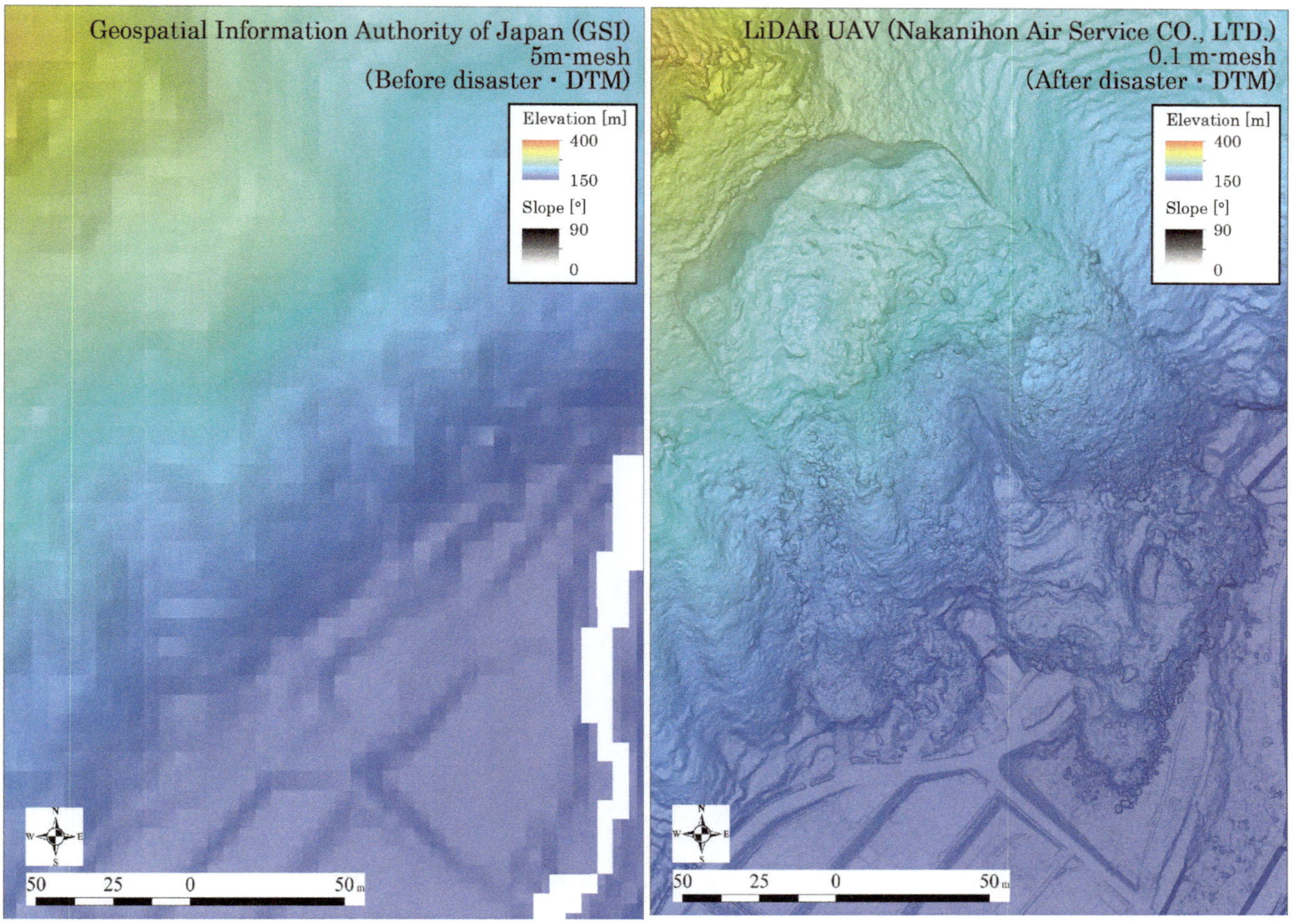

Fig. 3 Topographical maps of the Yabakei landslide. The left map is GSI 5 m mesh DEM before the disaster and the right map is 0.1 m mesh DEM measured by LiDAR UAV (Nakanihon Air Service Co., LTD.) after the disaster

shown in Table 1. As a result of calculating the amount of deposit in the landslide polygon, the amount of deposition increased only by 1.67%, so the consistency of the amount of deposition before and after the disaster was obtained.

The topographic features classification shows that Area 1–3 are negative change, Area 4–10 are positive change, and Area 11 is fixed. In the Area 2, there is no deposit of an upside landslide (Area 1). So, it is considered that the Area 2 landslide occurred after the Area 1 landslide. The slip surface of Area 2 was formed by the crack of the basement rock (tuff-breccia). Area 3 is a landslide, but looks like a fixed by the volume change. I assume that the deposit from Area 1 landslide has slipped down as it is in Area 3. From above, The Area 4, 5 is the part where the deposit of the Area A landslide remained in the slope. Therefore, it is surmised that the risk of a re-slide is high.

Evaluation of Secondary Landslide Susceptibility for Rescue Activity

In the Yabakei landslide, the rescue activity was carried out in the deposition part, Area 6–10. In the case of the rescue activity at Area 7, it is necessary to watch out for the Area 2 and Area 4. In Area 2, the re-slide of the basement rock around the spring needed to be warned. In Area 4, the risk of falling rocks is higher than in other areas because the deposit of Area 1 remains. On the other hand, I can judge that even if the cliff of the main landslide (Area 1), there is almost no effect on the rescue activity site. By comparing the topographic date before and after the disaster and the topographic classification, I can clarify what to watch out for on the rescue site.

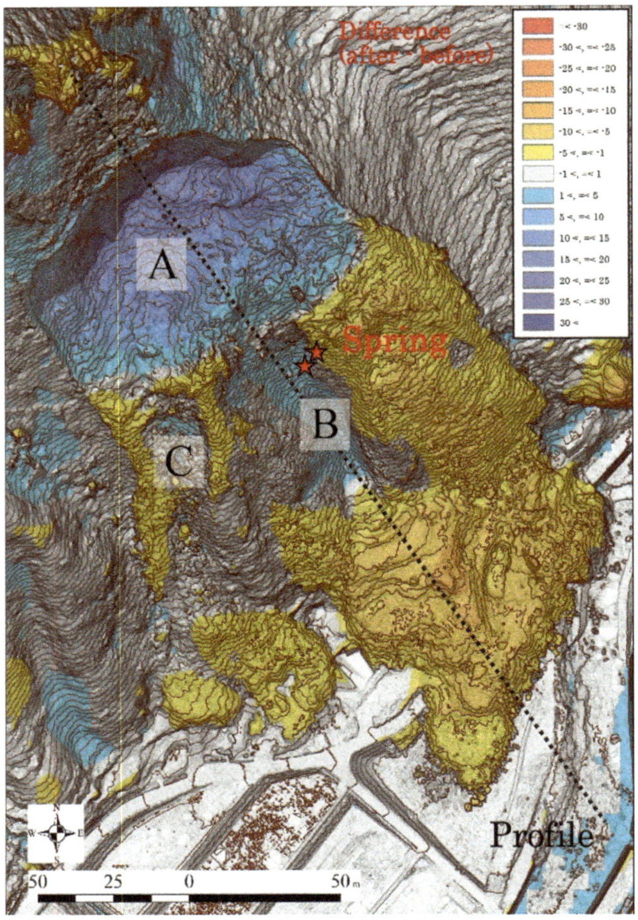

Fig. 4 The difference of the DEMs before and after the disaster. The cold colors show the negative change and the warm colors shows the positive change

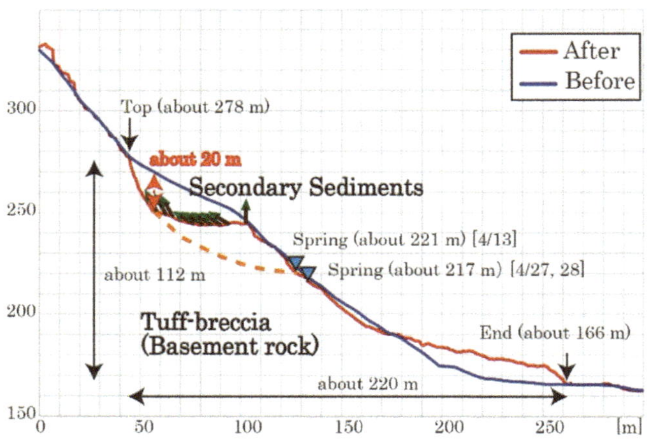

Fig. 5 The profiles of the Yabakei landslide before and after the disaster

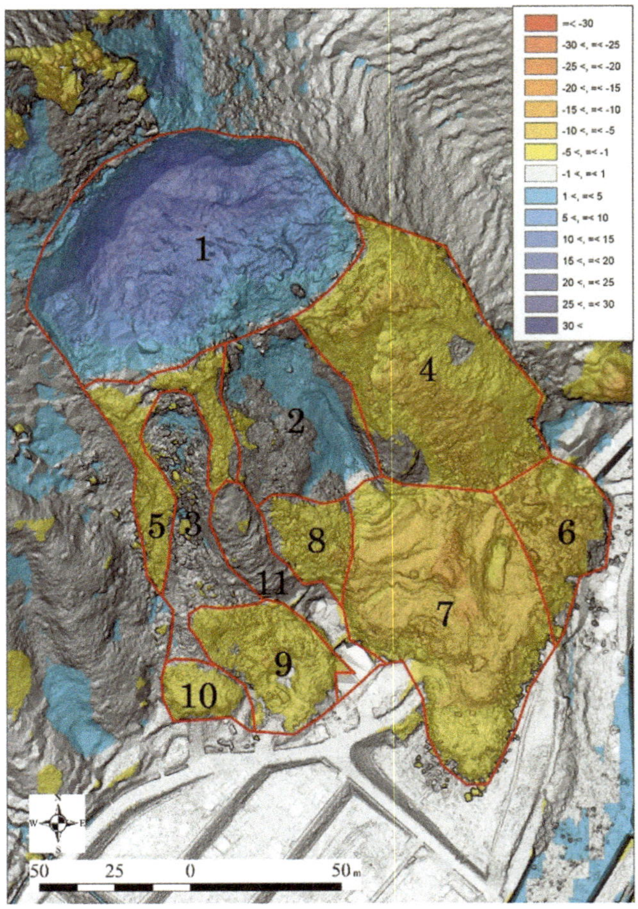

Fig. 6 Topographic features classification of the Yabakei landslide

The Reasons of the Time Lag Landslide

I considered the cause of the time lag landslide based on the features of Yabakei landslide by analysing DEMs and filed survey.

I assume that the main cause of the time lag of the Yabakei landslide is groundwater (Fig. 7). The Yabakei landslide have a deep slip surface. The Yabakei landslide bodies consisted of secondary sediments mainly and the slip surface formed the boundary of water permeability. And the Yabakei landslides are distributed in the landslide topography. It is thought that the time lag landslide occurred owing to the features.

In the future, I need to think about where is dangerous, what to observe, and how long to observed the time lag landslide for the rescue activity. In order to do, it is necessary to investigate the mechanism of various time lag landslides in detail.

Table 1 The negative and positive change of the topographic classification area

Id	Area (m^2)	Name	Ave. of dif. (m)	Volume (m^3)
1	6417.6	Main landslide	8.9	57,373.4
2	2239.4	Sub east landslide	0.9	1993.1
3	1447.5	Sub west landslide	−0.0	−43.4
4	4377.8	Flow section—east	−3.9	−17,073.3
5	1302.9	Flow section—west	−1.6	−2123.7
6	1252.9	Main sediment part—east	−5.5	−6903.7
7	4618.6	Main sediment part—center	−6.0	−27,573.2
8	733.3	Main sediment part—west	−2.6	−1928.7
9	1405.3	Sub sediment part—east	−2.4	−3330.6
10	502.3	Sub sediment part—west	−2.0	−1019.6
11	1037.0	Fixed area	−0.4	−362.9

+: negative change, −: positive change

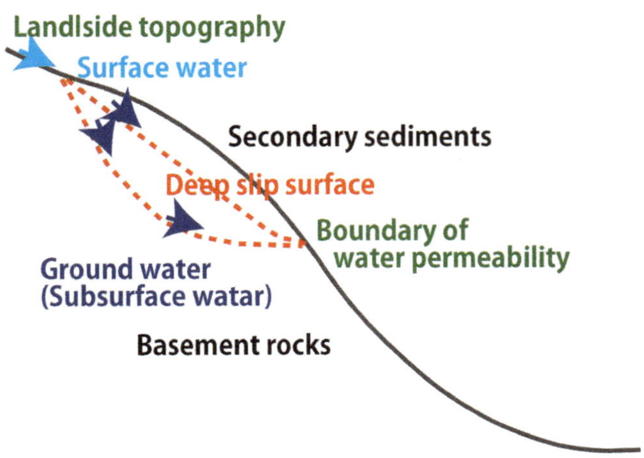

Fig. 7 The schematic image of the time lag landslide

Conclusions

I analysed the LiDAR UAV data measured immediately after the Yabakei landslide, and considered the evaluation of secondary landslide susceptibility during rescue activities. By comparing the topographic date before and after the disaster and performing topographic features classification using detailed DEM data after the disaster, I can clarify what to watch out for on the rescue site.

I considered the cause of the time lag landslide based on the features of Yabakei landslide by analysing DEMs and filed survey. I assume that the main cause of the time lag of the Yabakei landslide is groundwater.

Acknowledgements LiDAR UAV data used in this research provided by Nakanihon Air Service Co., LTD.

References

Cruden DM, Varnes D (1996) Landslide type and processes. Special report 247. Transportation Research Board, National Research Council. National Academy Press, Washington, pp 36–75

Doshida S, Araiba K (2019) Geomorphological features of the time lag landslide. In: 2019 AGU fall meeting, NH33D-0942

Geological Survey of Japan, AIST (ed) (2009) Seamless digital geological map of Japan 1: 200,000. 5 June, 2019 version. Research Information Database DB084. Geological Survey of Japan, National Institute of Advanced Industrial Science and Technology

Methodology for Landslides Assessment Causing River Channel Obstructions and the Consequent Water Shortage in Rural Communities

Johnny Alexander Vega and César Augusto Hidalgo

Abstract

Landslides are considered one of the natural hazards that cause the most significant losses worldwide. In countries like Colombia, landslides events cause the highest number of deaths and economic losses and are related to river flooding caused by landslides in basins. This paper presents a methodology to assess the associated risk with landslides in water supply basins. The hazard is assessed considering probabilistic methods that include the effects of rainfall and earthquakes. Furthermore, this study assesses the probability that a sliding mass reaches riverbeds forming a natural landslide deposit known as landslide dam (LD), calculating the probability of obstructions in its channel. Besides, damage degree (DD) or vulnerability is assessed using damage curves based on the obstruction height of the stream channel. This methodology is based on probability methods, such as the first order second moment method (FOSM) and the point estimate method (PEM). As study case, this methodology was applied in the La Liboriana River basin, in the southwest of Colombia, where morphodynamic and hydrometeorological conditions have generated several natural disasters that have left dead, injured and damage to the infrastructure. The model results show a high coincidence of affected areas with landslides according to the inventory of events in the study zone, with areas of high probability of failure predicted by the proposed model, indicating its coherence to identify areas to be studied with more detail.

Keywords

Landslide • Landslide dam • River channel obstruction • Water shortage

Introduction

Each year, landslides cause several casualties and they are responsible for enormous infrastructure damages and both direct and indirect costs worldwide. Because of the high impact levels associated with landslides, a great research interest has been generated worldwide regarding to understand the physical and economic aspects related to landslides (Vega and Hidalgo 2017).

Landslides are a complex problem that should be considered in case of mountainous terrains, where the financial costs may include costs associated with the relocation of communities, rebuilding of structures, and restoration of water quality in water sources (Dragicevic et al. 2015).

The interference between landslide and river courses is a topic of great relevance, because to date many catastrophic events have occurred in the world as a consequence of breaching of dams produced by landslides (Dal Sasso et al. 2013). A landslide dam (LD) is part of a natural landslide deposit that blocks a river (partial or totally) and causes damming of the river and water shortage downstream for several purposes. This phenomenon is a complex topic, due to the numerous variables involving both hillslope and river dynamics at the same time. LD can be formed by rapid deposition of landslide, debris flow, or rockfall materials (Zhang et al. 2016). Intense rainfall and earthquake are the most common LD trigging factor (Xuan et al. 2016).

Once the river is blocked, the water level may rise quickly, and the LD may fail by variety of process including overtopping, abrupt collapse or progressive failure, within a short period after the formation of the dam.

J. A. Vega (✉) · C. A. Hidalgo
Civil Engineering Program, School of Engineering, Universidad de Medellin, Medellin, Colombia
e-mail: javega@udem.edu.co

C. A. Hidalgo
e-mail: chidalgo@udem.edu.co

J. A. Vega · C. A. Hidalgo
Civil Engineering Program, Faculty of Engineering, Universidad de Medellin, Medellin, Colombia

F. Guzzetti et al. (eds.), *Understanding and Reducing Landslide Disaster Risk*,
ICL Contribution to Landslide Disaster Risk Reduction,
https://doi.org/10.1007/978-3-030-60227-7_32

If the landslide dam fails, the discharge of a huge volume of both water and sediment can results in a devastating floods that may greatly threaten the lives and property of populations downstream of the event (Chen et al. 2019). Over the last decades, the damages caused by LD disaster have been increased significantly due to the effect of climate change, and the risks posed by a LD are rather high.

In mountainous areas of tropical countries like Colombia, mass movements and floods are the natural disasters that denote the most severe risks (UNGRD 2020; DESINVENTAR 2019). These natural disasters are often caused by the diverse geographic and physiographic features, triggered by both natural as well as anthropogenic factors. Alike, blockage of a main stream by tributary debris flow events is a natural phenomenon of local river evolution between the confluences of the streams.

Majority of the little Colombian villages as well as rural areas receive fresh water from small superficial waterways, known as "creeks," given their topographic features. These sources are vital both for human consumption and for agricultural, mining, and industrial activities in areas such as the Colombian Andean zone. However, if a significant amount of solid material from near slopes can be rapidly delivered to the water streams by the inflow of a tributary debris flow, the volume of sliding mass can be deposited in the main stream allowing the formation of a landslide dam with the consequent water shortage.

Southeastern Antioquia (Colombia) was chosen as study area, because is frequently affected by mass movements, causing numerous casualties, wounded and affected victims, and hefty monetary losses; many of these mass movements are associated with the supply river basins such as the La Liboriana River (case study).

La Liboriana River basin, is a tropical and mountainous terrain in northwestern Colombian Andes, where on May 18th, 2015, more than 40 landslides and an associated flash flood and debris flow afterwards killed more than 100 inhabitants and caused infrastructure damage with significant economic losses.

The probability that a sliding mass of a landslide could block a water stream can be assessed quantitatively using empirical or physically-based approaches. With reference to the LD creation, several authors proposed some geomorphic indexes to forecast landslide dam behaviour which take in account mainly geomorphic variables characterizing both the landslide and the river channel (Dal Sasso et al. 2013).

This paper presents a methodology to estimate the landslide hazard in the basins supplying fresh water. This study assesses the probability that a sliding mass reaches riverbeds and the probability of obstructions in its channels, considering whether the deposition height of the sliding mass is higher than the water depth of the main stream. Besides, the damage degree of stream channel is assessed using damage curves based on the obstruction height of the river channel. The consequent flood downstream of the obstruction site due to a potential LD failure is not considered neither analyzed in this work.

Methodology

To evaluate the landslide hazard, the first-order second-moment (FOSM) method is applied to determine the failure probability (Vega and Hidalgo 2016; Hidalgo et al. 2018). The hazard estimation model was implemented using the EPADYM software, which was developed in Python language, working under the open source MapWinGIS application interface, which imports data from the geographic information system.

This software was developed by the Civil Engineering Research Group (GICI) at the University of Medellin to calculate the reliability index, failure probability, and security factor of a slope under seismic and static conditions. This methodology, presented graphically in Fig. 1, calculates the slope failure probability (*TPF*) according to the total probability theorem using the following equation:

$$TPF = P_{fs} \times P_s + P_{fns} \times (1 - P_s) \quad (1)$$

where P_{fs} is the slope failure probability due to the action of an earthquake under soil saturation conditions, P_{fns} is the failure probability under unsaturated conditions, P_s is the marginal probability of the saturated soil, and $(1 - P_s)$ represents the marginal probability of unsaturated soil. The slope failure probabilities in saturated and unsaturated conditions are calculated independently. The accumulated effect of rainfall and occurrence of landslides are likely to be related via the failure thresholds or physical-based numerical modelling to estimate the saturation probability (Hidalgo et al. 2018).

The material detached from unstable areas can obstruct the water stream exposed to the landslide. The height h_T of the obstruction (landslide dam height) depends on the volume of removed material, the distance that the material must travel, the slope of the terrain, and the shape of the geometry of the stream channel. The greater the h_T, the greater will be the intensity of the effect.

The effect on the stream due to obstruction can be considered in several ways:

1. interruption of the aqueducts and other downstream supplies;
2. impact on the aquatic ecosystem; and

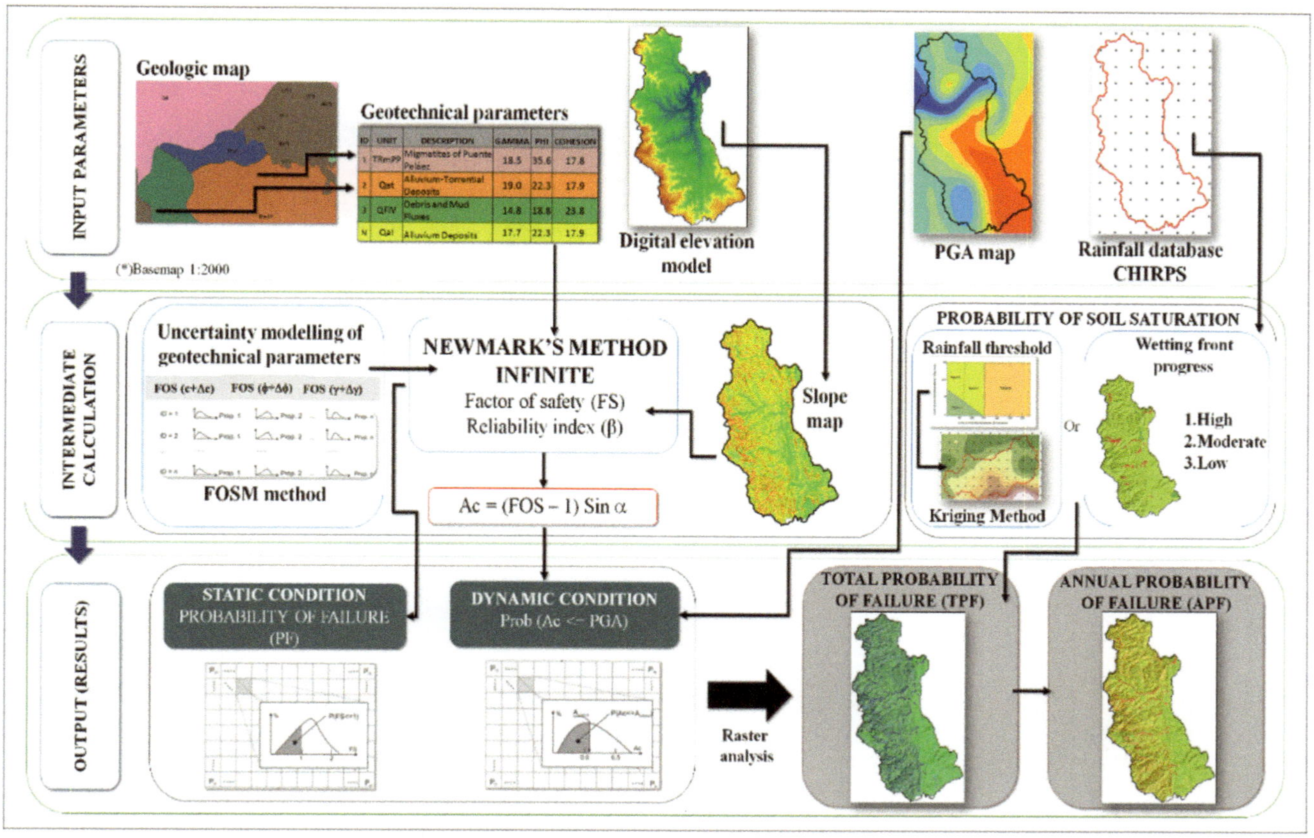

Fig. 1 Methodology for landslide hazard assessment in terms of total probability of failure (*TPF*) and annual probability of failure (*APF*)

3. possible generation of avalanches because of the generated reservoir.

The obstruction may be complete, as depicted in Fig. 2, or may affect only a part of the creek, where the level of damage will change. The obstruction degree of a creek will depend on various factors such as the amount and type of sliding material and the shape of the creek channel (height, width, and inclination of the slopes).

Figure 2 shows the methodology that has been adopted for calculating total landslide deposition height (obstruction height or LD height) in specific site of the stream in this study. For this purpose, a method based on slope gradient (Z), as a parameter that reflects the potential and kinetic energies involved with the movement of soil mass has been used, which conveniently serves as an indicator of the landslide intensity.

Once the value of Z has been calculated for each of the cells in the potential slip path from the D8 flow direction model, the probability that the material of each of these cells will reach the riverbed (P_R) is calculated as a probabilistic estimate value using the punctual estimates method (PEM).

Then, the deposition height of each cell (h_i) of the slip potential trajectory can be calculate for the final calculation of the total deposition height of the slide acting on the creek (h_T), as depicted in Fig. 2. Thus, the cumulative effect of each cell of the potential slip trajectory is calculated via a probabilistic approach.

The damage degree (*DD*) of the stream channel is taken as a random variable that follows a logistic distribution generating a "damage curves" characterized by the average height values h_T. These curves describe the probability of exceeding a predetermined limit state of exposure of the creeks under the hazard of a landslide, given a damage degree measurement of the same, as illustrated in Fig. 3.

Study Case

For the development and explanation of the proposed methodology, the La Liboriana river basin was taken as a case study (Fig. 4) because is well documented for the process of validation of the results obtained with the proposed model, because on May 18th, 2015, more than 40 landslides and an associated flash flood and debris flow were triggered by intense rainfall. Even, La Liboriana stream has caused large torrential avenues, which in turn have been linked to the previous occurrence of rainfall-induced landslides.

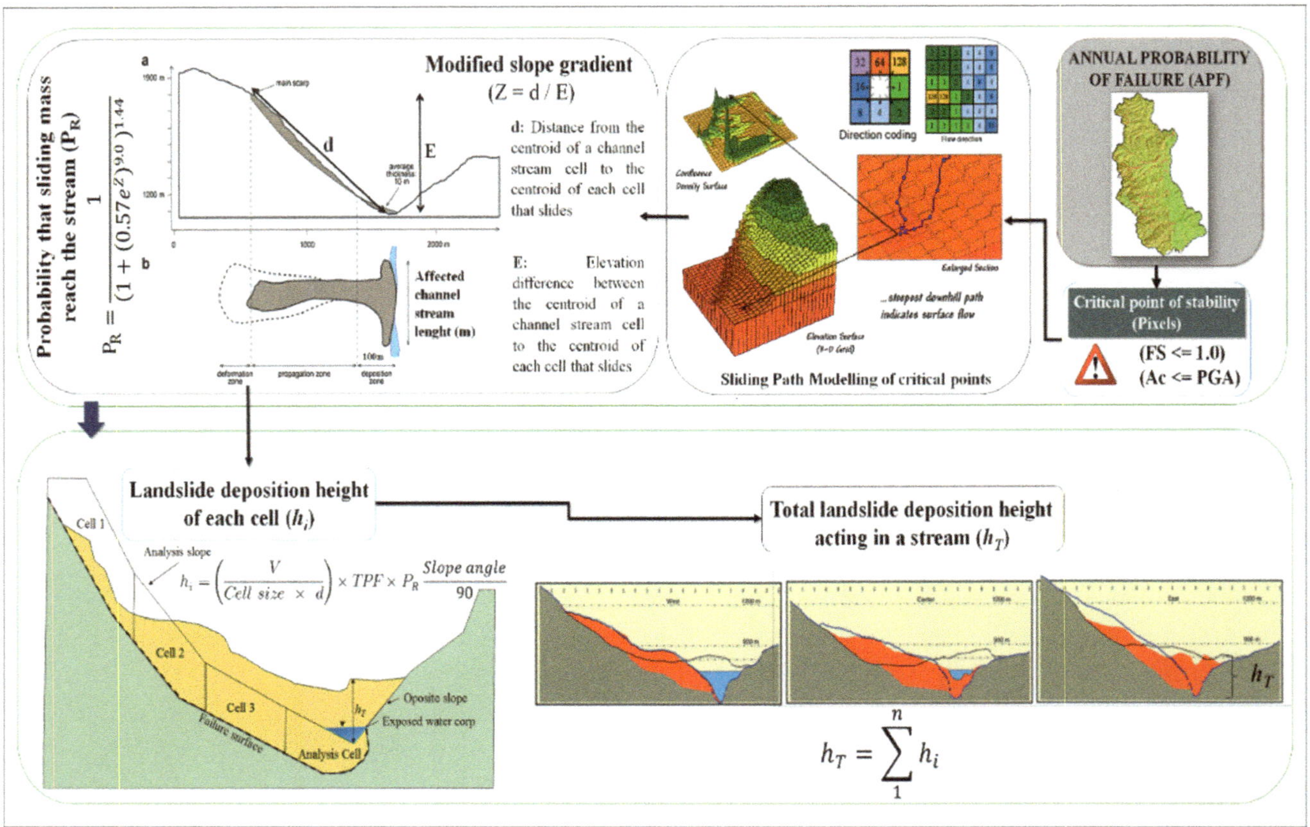

Fig. 2 Methodology for deposition height (landslide dam height) assessment acting in a stream

La Liboriana river basin joins El Barroso river basin, and both drain to Cauca River. Two main populated areas extend along the lower zone of the La Liboriana river valley, La Margarita village and Salgar urban zone, which have a population of almost 9000 inhabitants.

The geomorphology of the basin exhibits in the upper part a mountain region with a rugged morphology, narrow valleys, and very steep forested hillslopes. Vegetation and land use vary considerably within the basin.

In the upper zone of La Liboriana basin, there are crop fields and tropical deep forests. In the middle and lower zones grasslands (pastures) and coffee plantations. The basin exhibits grazing areas and urban development near the river banks. This basin has geological (predominately by a Cretaceous sedimentary rock formation and an intrusive Miocene body, forming well-developed saprolite and residual soils), and climatic conditions that make it particularly susceptible to landslides, especially in the top of the basin.

Results and Discussion

The total probability of failure (*TPF*) was calculated using the soil saturation probability (P_s) obtained based on the rainfall threshold and wet front progression methodologies for a horizontal earthquake acceleration of 0.2 g with a coefficient of variation of 50% to consider probable amplification according to the Colombian building code NSR-10 (AIS 2010).

The failure surface at an average depth of 2 m, with a coefficient of variation of 25%, was considered to simulate a shallow landslide, which corresponds to the characteristic type of mass movement in tropical mountainous terrain such as the study area, which are usually induced by high intensity or long duration rainfalls mainly. The rainfall used for the evaluation of the probability of soil saturation and the annual probability of failure (*APF*) has a return period of 20 years.

Critical zones ("Factor of safety" $FS < 1.1$ and $TPF > 0.50$), are presented in greater percentage in areas of steep slopes (>30°) in areas of high mountain forest vegetation, in a combination of areas of pastures and crops with steep slopes.

Regarding the annual probability of failure (*APF*), it was found that the study basin presents a high hazard according to the classification established by Chowdhury et al. (2010), with values ranging between 2.3 and 2.8%.

Analysing the study area, the damage degree of streams channels in an area taking an elongated area of 100 m on each side of the stream, an average value of 0.052 (damage

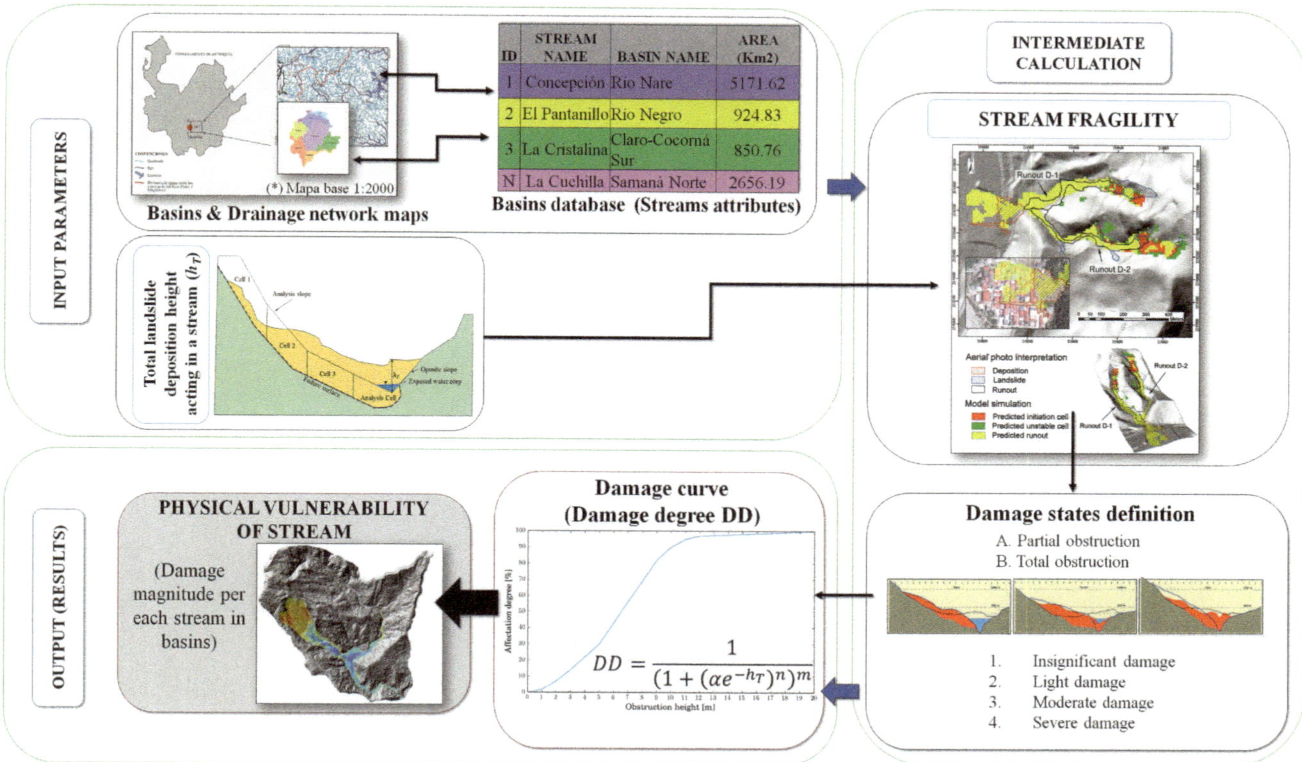

Fig. 3 Methodology for damage degree (*DD*) assessment in a stream channel

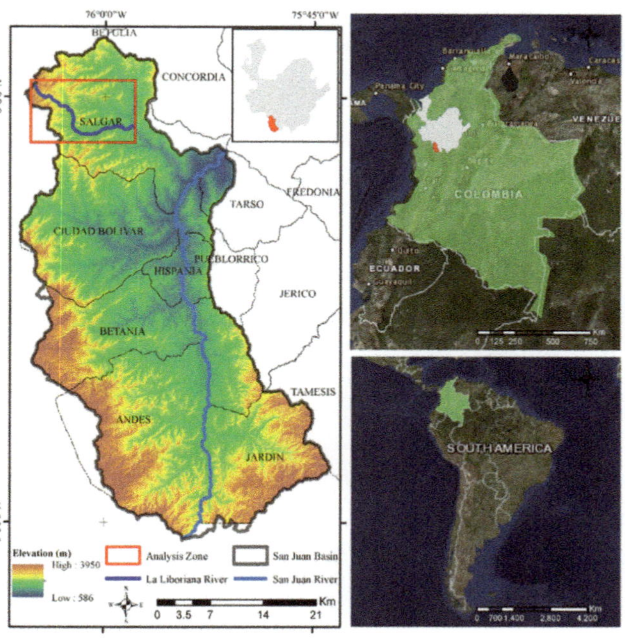

Fig. 4 Location of study area

degree of 5.2%) is obtained. As an example, Fig. 5 shows the damage results obtained for the La Liboriana River.

To assess the prediction capacity of the proposed landslide hazard assessment model, La Liboriana River basin data was used. This basin is well documented following the tragedy that occurred in May 2015, where dozens of landslides were triggered by rainfall, with its consequent torrential avenue and flow of sludge and debris, causing tens of deaths and a considerable economic loss. For the validation process, a landslide inventory was used in the La Liboriana basin raised and documented in Ruiz-Vásquez and Aristizábal (2018), which was generated from a multi-temporal analysis of satellite images and aerial photographs, obtaining an extension of landslides in the area that covers approximately 0.6 km^2 corresponding to 1% of the basin.

In order to establish a validation process around the predictive capacity of the proposed model, a ROC (Receiver Operating Characteristic) analysis was carried out. This method can be applied for the validation of physically based models, by using a threshold value in the stability indicator

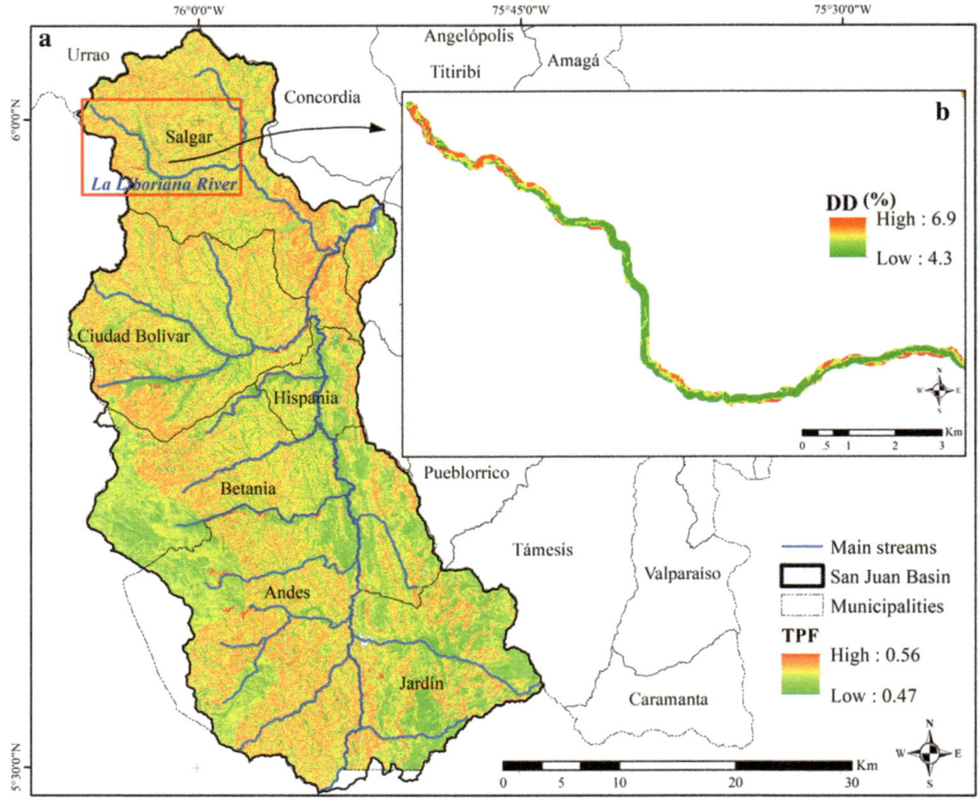

Fig. 5 Results of total probability of failure (**a**) and the damage degree of La Liboriana river (**b**)

used (in this context it corresponds to the static saturated safety *FS*), whose threshold value to indicate a condition of terrain instability corresponds to a value less than 1.1 (SGC 2015). According to the adopted FS threshold value, a hit rate (*TPR*) of 54%, and false alarm rate (*FPR*) of 39% were obtained, which places this point in the central left part of the ROC graph. These results can be considered "conservative".

According to the obtained results with the proposed model, there is a 74% coincidence between instability zones (FS < 1.1), with areas corresponding to the landslide inventory polygons used as reference. This indicates a good approach in the evaluation of this complex phenomenon. Likewise, in the 17% of the remaining polygons in the inventory, the average values of FS were between 1.1 and 1.25, which represent conditions of potential instability. The upper limit corresponds to the threshold established by the Colombian construction code NSR-10 (AIS 2010) for slopes in normal groundwater conditions for the construction stage. Likewise, this range of safety factors is a high risk indicator for risk studies due to landslides (SGC 2015).

Conclusions

A methodology for hazard assessment of landslide triggered by earthquakes and rains in water supply basins is presented. The proposed hazard assessment model has shown the ability to identify unstable areas, using secondary information. Based on this, the landslide hazard in the La Liboriana river basin was evaluated and the areas of greatest hazard were identified. These areas were compared with the areas affected by mass movements reported in disaster databases, evidencing a high coincidence of zones affected, with areas of high probability of failure, which indicates that the model is robust to identify areas to be studied in greater detail.

The damage degree assessment shows that, in general, the area of La Liboriana River has a potential damage between low and medium. However, it should be taken into account that the cumulative effect of events with different return periods has not been considered. It is also important to determine damage curves from records of mass movements in different places and times.

The methodology developed based on the failure threshold yields results comparable to those obtained by the wetting front methodology, which represents an economy in computation time and in the acquisition of soil data. Using cost analysis, the investment amount required to reduce the risk to acceptable levels with respect to the exposed infrastructure in the river basin can be estimated.

Although the proposed model has exhibited robustness in identifying unstable areas, the data quality should be improved to ensure that the results can be used for decision-making while planning rural areas and supply basins.

Acknowledgments The information for the development of this study was obtained from the project "Programa Integral Red Agua (Piragua)' carried out between the University of Medellin and Corantioquia Corporation. The development was implemented in EPADYM software, a innovation process product of University of Medellin. The study is framed within the guidelines and products of the research program "Vulnerability, resilience and risk of communities and supplying basins affected by landslides and avalanches" code 1118-852-71251, project "Functions for vulnerability assessment due to water shortages by landslides and avalanches: micro-basins of southwest Antioquia", contract 80740- 492-2020 held between Fiduprevisora and the University of Medellin, with resources from the National Financing Fund for science, technology and innovation, "Francisco José de Caldas".

References

AIS (2010) Colombian code for earthquake-resistant construction (NSR-10). Association of Earthquake Engineering, Colombia (in Spanish)

Chen KT, Chen XQ, Hu GS, Kuo YS, Huang YR, Shieh CL (2019) Dimensionless assessment method of landslide dam formation caused by tributary debris flow events. Geofluids 2019. Article ID 7083058. https://doi.org/10.1155/2019/7083058

Chowdhury R, Flentje P, Bhattacharya G (2010) Geotechnical slope analysis. Taylor & Francis, London

Dal Sasso SF, Sole A, Pascale S, Sdao F, Bateman Pinzón A, Medina V (2013) Assessment methodology for the prediction of landslide dam hazard. Nat Hazards Earth Syst Sci Discuss 1:5663–5694. https://doi.org/10.5194/nhessd-1-5663-2013

DESINVENTAR (2019) Disaster effects inventory system. National Historical Inventory of Disasters, Colombia. Website: https://online.desinventar.org/ (in Spanish)

Dragicevic S, Lai T, Balram S (2015) GIS-based multicriteria evaluation with multiscale analysis to characterize urban landslide susceptibility in data-scarce environments. Habitat Int 45:114–125

Hidalgo CA, Vega JA, Parra M (2018) Effect of the rainfall infiltration processes on the landslide hazard assessment of unsaturated soils in tropical mountainous regions. In: Hromadka II TVV, Rao P (eds) Engineering and mathematical topics in rainfall. IntechOpen, London, pp 163–185. https://doi.org/10.5772/intechopen.70821

Ruiz-Vásquez D, Aristizábal E (2018) Landslide susceptibility assessment in mountainous and tropical scarce-data regions using remote sensing data: a case study in the Colombian Andes. In: EGU general assembly 2018. Geophysical research abstracts, vol 20, EGU2018-3408

SGC (2015) Methodological guide for studies of mass movements hazard, vulnerability and risk. Servicio Geológico Colombiano, Bogotá-Colombia, 179 p (in Spanish)

UNGRD (2020) Digital library from national unit for disaster risk management-UNGRD. Website: https://portal.gestiondelriesgo.gov.co/ (in Spanish)

Vega JA, Hidalgo CA (2016) Quantitative risk assessment of landslides triggered by earthquakes and rainfall based on direct costs of urban buildings. Geomorphology 273:217–235. https://doi.org/10.1016/j.geomorph.2016.07.032

Vega JA, Hidalgo CA (2017) Risk assessment of earthquake-induced landslides in urban zones. In: Advancing culture of living with landslide—advances in landslide science, vol 2. https://link.springer.com/chapter/10.1007/978-3-319-53498-5_108

Xuan KD, Minseok K, Thao Nguyenc HP, Kwansue J (2016) Analysis of landslide dam failure caused by overtopping. In: 12th international conference on hydroinformatics. Procedia Eng 154:990–994. https://doi.org/10.1016/j.proeng.2016.07.587

Zhang LM, Peng M, Chang D, Xu Y (2016) Dam failure mechanisms and risk assessment. Wiley, Singapore. ISBN 9781118558546

Part IV

Landslide Hazard Assessment and Zonation—Temporal and Size Modelling

Landslide Size Distribution Characteristics of Cretaceous and Eocene Flysch Assemblages in the Western Black Sea Region of Turkey

Aykut Akgun, Tolga Gorum, and Hakan A. Nefeslioglu

Abstract

The main purpose of the study is to determine the general characteristics of the landslide sizes observed in Cretaceous and Eocene aged flysch assemblages at the Western Black Sea region of Turkey by using magnitude and frequency relations. For this purpose, the magnitude and frequency relations were investigated by considering power-law scaling characteristics of the geological formations. The probability distributions were also examined by considering Double Pareto and Inverse Gamma distribution models. According to the power-law relations, the rollover effects were observed at 0.047 and 0.048 km^2, and the fractal dimensions of the distributions were obtained as −1.97 and −1.41 for Cretaceous and Eocene flysch assemblages, respectively. Considering the probability distributions, the best-fits were acquired from the models Double Pareto with three parameters estimated by maximum likelihood estimation for Cretaceous flysch and Kernel density estimation for Eocene flysch. When we compared these results with the results of a study carried out in the same flysch but in another sub-catchment, it is concluded that rollover effects and fractal dimensions may not be generalized, that means, the parameters may differ site to site depending on not only spatial resolution but also morphological, climatic, and anthropogenic features of the region in concern, and conversely, landslide size distributions fit Double Pareto distribution models in general.

Keywords

Cretaceous flysch • Eocene flysch • Landslide size distributions • Western Black Sea region of Turkey

Introduction

Landslides are natural phenomena that frequently encountered both in the World and in Turkey. Due to major reasons such as precipitation and earthquake, landslides have a severe damage potential on to humans. Considering this point, an appropriate hazard management requirement arises. The first and crucial step for an effective landslide hazard management is a detailed landslide mapping procedure. It is very well-known that landslide susceptibility, hazard and risk mapping are the indispensable works for a hazard management effort. Knowing the number, area, volume and frequency-size distribution of landslides is essential to determine landslide susceptibility and hazard (Guzzetti et al. 2005, 2009). For this reason, precisely quantification of distributions by accurate and reliable methods is essential (Guzzetti 2006).

In the last four decades, the characterization of landslide size distribution has been investigating by many researchers for different purposes such as comprehension of landslide dynamics, quantification of erosion caused by landslides and the evaluation of landslide event magnitude for natural hazard assessments (Fujii 1969; Pelletier et al. 1997; Guthrie and Evans 2004; Malamud et al. 2004; Korup 2005; Frattini and Crosta 2013; Regmi et al. 2014; Tanyas et al. 2019; Gorum 2019a).

When we make a close inspection of these studies, some points on which a general agreement exits, can be inferred: (a) the frequency distribution exhibits power-law scaling for

A. Akgun (✉)
Geological Engineering Department, Karadeniz Technical University, Trabzon, Turkey
e-mail: aykutakgun@gmail.com

T. Gorum
Eurasia Institute of Earth Sciences, Istanbul Technical University, Istanbul, Turkey
e-mail: tgorum@itu.edu.tr

H. A. Nefeslioglu
Department of Geological Engineering, Hacettepe University, Ankara, Turkey
e-mail: hanefeslioglu@gmail.com

F. Guzzetti et al. (eds.), *Understanding and Reducing Landslide Disaster Risk*, ICL Contribution to Landslide Disaster Risk Reduction,
https://doi.org/10.1007/978-3-030-60227-7_33

landslides larger than a size threshold, (b) below this threshold the distribution shows a deflection and deviates from power-law. As stated by Malamud et al. (2004), the deflection aforementioned here occurs right below the modal peak of distribution and this is called to rollover. As stated by Frattini and Crosta (2013) and Tanyas et al. (2019), the meaning of the deflection and the power-law scaling is not explained.

In the literature, the reason of power-law scaling was interpreted to be a stochastic process of rupture propagation (Stark and Guzzetti 2009), heterogeneity of slopes and fracture system (Katz and Aharonov 2006) and self-organized critically (Bak et al. 1988; Hergarten and Neugebauer 1998). Frattini and Crosta (2013) stated that the reason of the deflection from power law below a size threshold had been explained to be the result of undersampling (Stark and Hovius 2001), the effect of the soil moisture distribution for shallow landslides (Pelletier et al. 1997), the consequences of the transition from friction-controlled to cohesion-controlled strength when decreasing the area and depth of the landslide (Malamud et al. 2004; Stark and Guzzetti 2009), and the consequence of slope length constraint on the downslope propagation of long-runout landslides (Guthrie and Evans 2004).

In this study, the primary purpose is: to determine the general characteristics of the landslides observed in Cretaceous and Eocene aged flysch assemblages at the Western Black Sea region of Turkey by using magnitude and frequency relations. To carry out the purpose, the landslide inventory map produced by Duman et al. (2005a) was used. The magnitude and frequency relations were investigated by considering power-law scaling characteristics of the geological formations. The probability distributions were also examined by considering Double Pareto and Inverse Gamma distribution models. At the end, the results obtained for the flysch assembles were compared and discussed with respect to the landslide size distribution characteristics.

Characteristics of the Study Area

The study area has average altitudes of 800 m and is limited by the Blacksea coast and North Anatolian Fault which is one of the most active and large strike-slip fault system in the world, from the north and the south, respectively (Fig. 1a). The landscape of the study area is mountainous, and structural features such as the attitude of bedding planes and rock strength differences control the morphology of the hillslopes. The region has a maritime temperate climate, with a mean annual precipitation of 1100 mm, and the mean annual precipitation in the study areas (1020 mm) is considerably close to the regional averages (Gorum 2019a). The study focuses on the upper catchment of the northward-draining Filyos and Bartın rivers which are the largest river systems in the Western Blacksea region.

The study area covers a 716 km^2 area between the Istanbul-Zonguldak zone and located in the Western Pontides. The Pontide mountain belt constitutes the margins of the study area and is an east-west-trending orogenic belt representing a coalesced tectonic entity in the northern section of Turkey (Channell et al. 1996). The lower section of the Istanbul-Zonguldak zone consists of Precambrian gneiss, metagranite, and amphibolite Pan-African basement rock series (Okay and Tüysüz 1999). The series on the

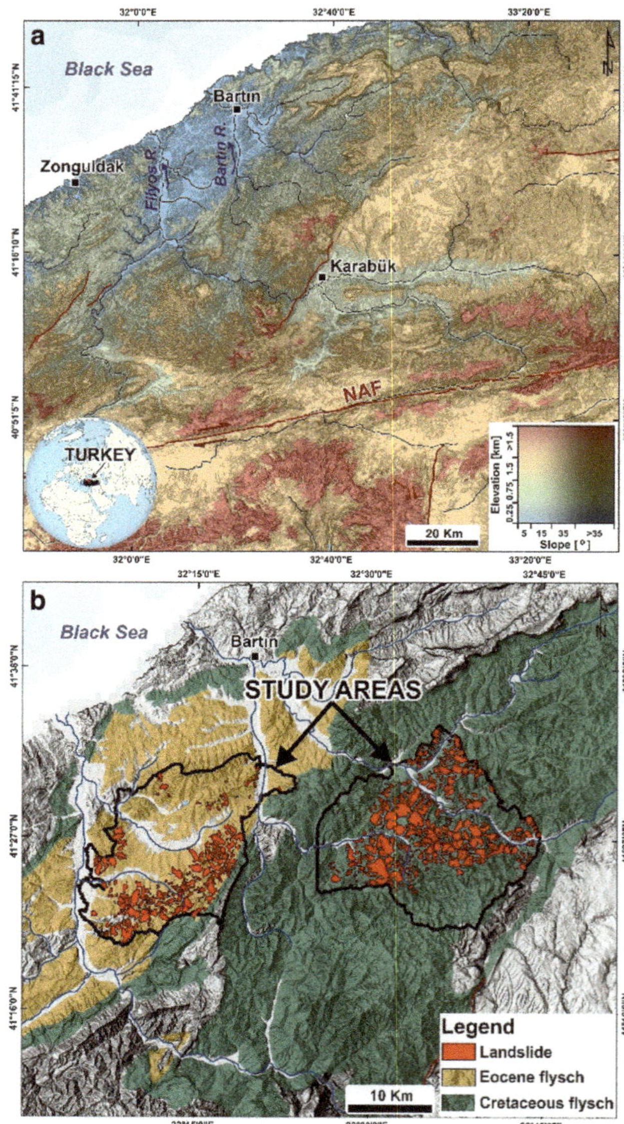

Fig. 1 Physiographic setting and location map of the study area. **a** The physiographic map of the Western Pontides and major river (blue lines) network of the area. Red lines are active faults (Emre et al. 2013). **b** Landslide distribution in Eocene and Cretaceous flysch assemblages. Thick black lines delimit the boundary of study areas. Landslides were digitized from https://yerbilimleri.mta.gov.tr, which are based on Duman et al. (2005b)

Pan-African basement rocks starts with the Ordovician-Carboniferous clastic and carbonate sedimentary sequence which is not affected by metamorphism (Okay et al. 1994). The series from lower sections to the top continues with a Triassic terrestrial unit containing river and flood plain sediments, Jurassic clastic units consisting of sandstone, mudstone, and siliciclastic turbidites, and an Upper Jurassic-Lower Cretaceous homogenous platform carbonate sequence (Gorur et al. 1997; Okay and Tüysüz 1999; Duman et al. 2005a; Turer et al. 2008).

Eocene Flysch

In general, Eocene flysch formation characterized by low topographic relief values within the rugged topography of the region (Fig. 1b). The average height of the unit varies between 20 and 970 m above sea level (a.s.l.), and the average height of the unit is around 320 m (a.s.l.). Sedimentologically, the Eocene flysch unit begins with the (1) lowest turbidite facies alternates with pebble, sandstone, and shale and continues with (2) lytic tuff, andesitic tuff breccias and volcanoclastic sandstone and siltstones. In the middle part of this flysch series (3), there is a series characterized by crystal-vitric tuffs and constitutes relatively fine grain-sized compared to the lower series. (4) The upper sections of the Eocene flysch series consist of shallow marine deposits consisting of Middle Eocene sandstone and claystone alternations (Duman et al. 2005b).

The landslides occurred in the Eocene flysch unit are generally of slide type. During the heavy and prolonged rainy periods, the flow type landslides were also recorded in Eocene flysch units (Can et al. 2005). The shallow to intermediate deep-seated landslide occurrences in the Eocene flysch unit is broadly individual, rotational and complex slides, and they are commonly observed within the contact zone of fine-grained sandstone and weak-strength clay-silt stone successive levels (Duman et al. 2005b). On the other hand, deep-seated slides are generally occurred at the clayey levels of thick-bedded sandstones and in areas where the topographic slope is on averagely higher than 12°.

Cretaceous Flysch

The Cretaceous flysch series starts volcanoclastic rocks at the bottom of the series and overlain by sandstone and shale sequential units in upper levels. The formation consists of pelagic clayey limestone, biomicrite limestone with shale, sandstone, and conglomerate intercalations towards the middle sections of the series (Altıner et al. 1991).

Landslides that occur in Cretaceous flysch are generally in the slide and flow type (Duman et al. 2005b). These landslides were specially formed in the specific facies of the Cretaceous flysch series. Landslides are dense in the thin bedding and low strength levels of mudstone, claystone and siltstone levels corresponding to lower sections of the flysch unit. The rock units of the Cretaceous flysch series, which are affected by various orogenic phases, are extremely deformed and jointed. In this respect, rotational slides are intensely observed in this flysch series (Nefeslioglu et al. 2012). Shallow landslides are relatively frequent in this unit in contrast to the Eocene flysch, and abundant in the steep valley side slopes and the depositional parts of the large-scale slope failures (Gorum 2019b).

Landslide Size Distributions

The landslides of both sub-catchments defined on the flysch assembles were digitized from the Geoscience Map Viewer and Drawing Editor web site of the General Directorate of Mineral Research and Exploration (Duman et al. 2011). Accordingly, 436 landslides were identified on Cretaceous flysch while 245 landslides were digitized on Eocene flysch. The type of mass movement was defined as deep-seated (depth > 5 m) slides (Duman et al. 2005a). The parameter landslide size was evaluated as the landslide area. The minimum and maximum landslide area values were obtained as follows: 0.009 and 4.491 km^2 on Cretaceous flysch, and 0.006 and 3.085 km^2 on Eocene flysch, respectively. Landslide size and frequency relations for flysch assemblages were explored by considering power-law relations. The rollover determinations were also performed. According to the power-law relations, the rollover effects were observed at 0.047 and 0.048 km^2, and the fractal dimensions of the distributions were obtained as −1.97 and −1.41 for Cretaceous and Eocene flysch assemblages, respectively (Fig. 2).

The probability distributions of landslide sizes observed in the flysch assemblages were also investigated. For the purpose, an R script, namely LStats, was written by Rossi and Malamud (2014) was employed. Double Pareto and Inverse Gamma distribution models were examined by using Kernel density and maximum likelihood estimators. Accordingly, the best-fits were acquired from the models Double Pareto with three parameters estimated by maximum likelihood estimation for Cretaceous flysch and Kernel density estimation for Eocene flysch (Fig. 3). According to

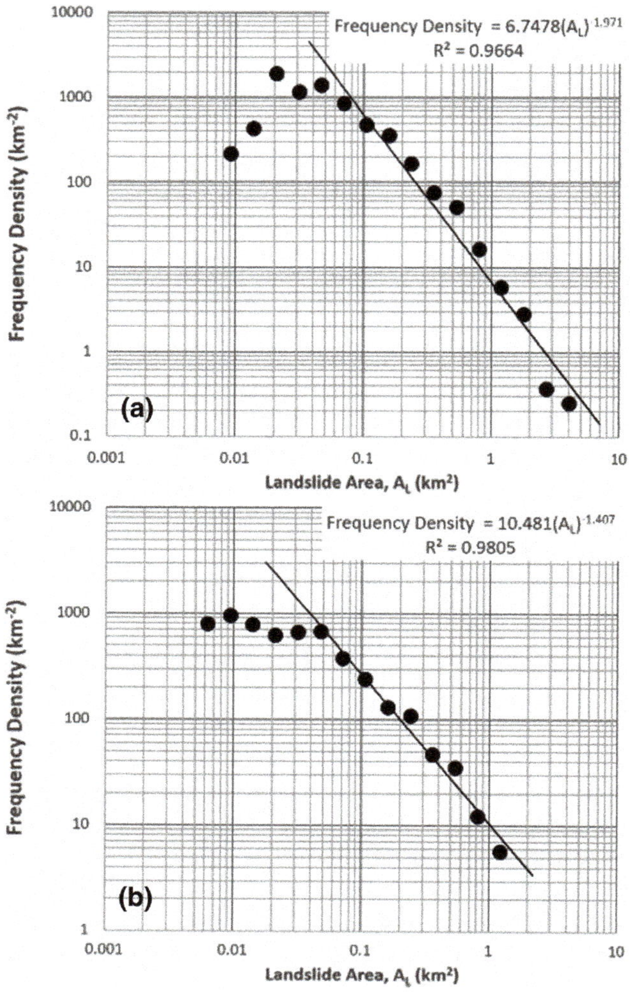

Fig. 2 The power-law relations between landslide size and frequencies in Cretaceous (a) and Eocene flysch (b) assemblages in the sub-catchments

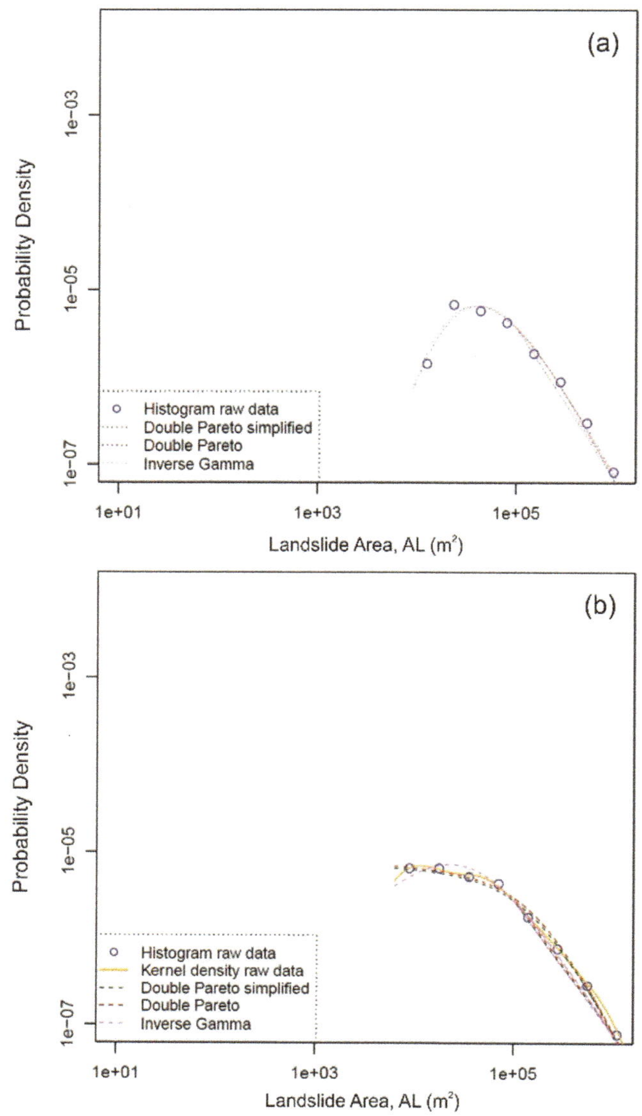

Fig. 3 Landslide size probability distributions in Cretaceous flysch, estimated by maximum likelihood estimation (a) and Eocene flysch, estimated by Kernel density estimation (b)

the one-sampled Kolmogorov-Smirnov tests, *P*-values of the estimations were calculated as 0.32 and 0.468, respectively. The distribution parameters α, β, and t were obtained as follows: 1.28, 7.5, and 45,871.72 for Cretaceous flysch, and 1.84, 3.13, and 308,517 for Eocene flysch.

Discussion and Conclusions

Size distribution characteristics of landslides are important outputs that may help to determine possible landslides having specific morphological characteristics in a concentration area. To understand these morphological characteristics may also help in a hazard management framework. Of course, not only consideration of size characteristics of landslides but also lithological effects should be taken into account in such an assessment.

In this study, size characteristics of landslides occurred in the same type lithology, and having different stratigraphic ages were investigated by considering the power-law relations and probability distributions.

When a summarize is done for the values obtained, it is clear that the rollover values indicate very close relations between the lithological assemblages. Additionally, fractal dimensions are highly close to each other, which means the landslide size characteristics are similar.

In another research performed on Eocene flysch unit close to our sub-catchments, the rollover effect and fractal dimension were calculated as 0.018 km^2 and −2.3 (Nefeslioglu and Gorum 2020). This result may indicate that rollover effects and fractal dimensions may not be generalized, that means, the parameters may differ site to site depending on not only spatial resolution but also morphological, climatic, and anthropogenic features of the region in concern. On the other hand, in the same research conducted within Eocene flysch, the best fit for the landslide size probability distributions was obtained for Double Pareto distribution model. This result points out that landslide size distributions fit Double Pareto distribution models in general.

References

Altıner D, Koçyiğit A, Farinacci A, Nicosia U, Conti MA (1991) Jurassic, Lower Cretaceous stratigraphy and paleogeographic evolution of the southern part of north-western Anatolia. Geol Romana 28:13–80

Bak P, Tang C, Wiesenfeld K (1988) Self-organized criticality. Phys Rev A38:364–374

Can T, Nefeslioglu HA, Gokceoglu C, Sonmez H, Duman TY (2005) Susceptibility assessments of shallow earthflows triggered by heavy rainfall at three catchments by logistic regression analyses. Geomorphology 72(1–4):250–271

Channell JET, Tüysüz O, Bektas O, Sengör AMC (1996) Jurassic-Cretaceous paleomagnetism and paleogeography of the Pontides (Turkey). Tectonics 15(1):201–212

Duman TY, Can T, Emre Ö, Keçer M, Doğan A, Ateş Ş, Durmaz S (2005a) Landslide inventory of northwestern Anatolia, Turkey. Eng Geol 77(1–2):99–114

Duman TY, Emre O, Can T, Nefeslioglu HA, Kecer M, Dogan A, Durmaz S, Ates S (2005b) Landslide inventory map of Turkey-1/500 000 scaled Zonguldak section. MTA special publication series, 24 p [in Turkish]

Duman TY, Çan T, Emre Ö (2011) 1/1.500.000 Türkiye Heyelan Envanteri Haritası, Maden Tetkik ve Arama Genel Müdürlüğü Özel Yayınlar Serisi-27. Ankara. ISBN 978-605-4075-85-3

Emre O, Duman TY, Ozalp S, Elmaci H, Olgun S, Saroglu F (2013) Active fault map of Turkey with and explanatory text. Special publication series 30. General Directorate of Mineral Research and Exploration, Ankara

Frattini P, Crosta G (2013) The role of material properties and landscape morphology on landslide size distributions. Earth Planet Sci Lett 361:310–319

Fujii Y (1969) Frequency distribution of the magnitudes of landslides caused by heavy rainfall. J Seismol Soc Jpn 22:244–247

Gorum T (2019a) Tectonic, topographic and rock-type influences on large landslides at the northern margin of the Anatolian Plateau. Landslides 16(2):333–346. https://doi.org/10.1007/s10346-018-1097-7

Gorum T (2019b) Landslide recognition and mapping in a mixed forest environment from airborne LiDAR data. Eng Geol 258:105–155. https://doi.org/10.1016/j.enggeo.2019.105155

Gorur N, Monod O, Okay AI, Sengor AMC, Tuysuz O, Yigitbas E, Sakınc M, Akkok R (1997) Palaeogeographic and tectonic position of the carboniferous rocks of the western Pontides (Turkey) in frame of the Variscan belt. Bull Soc Geol Fr 168:197–205

Guthrie RH, Evans SG (2004) Analysis of landslide frequencies and characteristics in a natural system, coastal British Columbia. Earth Surf Process Landf 29:1321–1339

Guzzetti F (2006) Landslide hazard and risk assessment. PhD thesis, Bonn University

Guzzetti F, Reichenbach P, Cardinali M, Galli M, Ardizzone F (2005) Probabilistic landslide hazard assessment at the basin scale. Geomorphology 72(1):272–299

Guzzetti F, Ardizzone F, Cardinali M, Rossi M, Valigi D (2009) Landslide volumes and landslide mobilization rates in Umbria, central Italy. Earth Planet Sci Lett 279:222–229

Hergarten S, Neugebauer NJ (1998) Self-organized criticality in a landslide model. Geophys Res Lett 25:801–804

Katz O, Aharonov E (2006) Landslides in vibrating sandbox: what controls types of slope failure and frequency magnitude relations? Earth Planet Sci Lett 247(3–4):280–294

Korup O (2005) Distribution of landslides in southwest New Zealand. Landslides 2:43–51

Malamud BD, Turcotte DL, Guzzetti F, Reichenbach P (2004) Landslide inventories and their statistical properties. Earth Surf Process Landf 29:687–711

Nefeslioglu HA, Gorum T (2020) On the use of landslide hazard maps to determine mitigation priorities in a dam reservoir and its protection area. Land Use Policy 91:104363

Nefeslioglu HA, San T, Gokceoglu C, Duman TY (2012) An assessment on the use of Terra ASTER L3A data in landslide susceptibility mapping. Int J Appl Earth Obs Geoinf 14(1):40–60

Okay AI, Tüysüz O (1999) Tethyan sutures of northern Turkey. Geol Soc Lond Spec Publ 156(1):475–515. https://doi.org/10.1144/GSL.SP.1999.156.01.22

Okay AI, Sengör AMC, Görür N (1994) Kinematic history of the opening of the Black Sea and its effect on the surrounding regions. Geology 22(3):267–270

Pelletier JD, Malamud BD, Blodgett T, Turcotte DL (1997) Scale-invariance of soil moisture variability and its implications for the frequency-size distribution of landslides. Eng Geol 48:255–268

Regmi NR, Giardino JR, Vitek JD (2014) Characteristics of landslides in western Colorado, USA. Landslides 11:589–603

Rossi M, Malamud BD (2014) Prototype software for determination of landslide statistics from inventory maps (LStats). Landslide modelling and tools for vulnerability assessment preparedness and recovery management (LAMPRE). In: Seventh framework programme collaborative project, WP 5. Research and tools for triggered event landslide mapping, 19 p

Stark CP, Guzzetti F (2009) Landslide rupture and the probability distribution of mobilized debris volumes. J Geophys Res 114 (F00A02):16

Stark CP, Hovius N (2001) The characterization of landslide size distributions. Geophys Res Lett 28(6):1091–1094

Tanyas H, van Westen CJ, Allstadt KE, Jibson RW (2019) Factors controlling landslide frequency-area distributions. Earth Surf Process Landf 44:900–917

Turer D, Nefeslioglu HA, Zorlu K, Gokceoglu C (2008) Assessment of geo-environmental problems of the Zonguldak province (NW Turkey). Environ Geol 55(5):1001–1014

A Statistical Exploratory Analysis of Inventoried Slide-Type Movements for South Tyrol (Italy)

Stefan Steger, Volkmar Mair, Christian Kofler, Massimiliano Pittore, Marc Zebisch, and Stefan Schneiderbauer

Abstract

Landslides of the slide-type movement represent common damaging phenomena in the Italian province of South Tyrol. Up to January 2019, the landslide inventory of the province lists 1928 accurately mapped landslides that required intervention by e.g. the local road service or the provincial geological survey. Thus, this landslide data set mainly includes events that caused damage. The aim of this contribution was to investigate and critically interpret statistical associations between the inventoried slide-type movements and a variety of spatial environmental variables. The assessment of conditional frequencies and the discriminatory power of single variables revealed conditions that are typically present at landslide mapping locations, e.g. topography, land cover, rock types, and proximity to infrastructure. A critical interpretation of the statistical results highlighted the need to consider the landslide data origin (i.e. background information) in order to avoid misleading statements and wrong inferences. The findings of the here presented work show that the availability of detailed landslide information does not always ensure that valid process-related conclusions can be drawn from subsequent statistical analyses (e.g. identification of important landslide controls). Despite considerable methodical advancements in the field of statistical data analysis and machine learning, we conclude that the principle '*correlation does not necessarily imply (geomorphic) causation*' remains of particular relevance when exploiting available landslide information.

Keywords

Landslide inventory • South Tyrol • Exploratory data analysis • IFFI • Sampling bias • Susceptibility

S. Steger (✉) · C. Kofler · M. Pittore · M. Zebisch · S. Schneiderbauer
Institute for Earth Observation, Eurac Research, Viale Druso 1, 39100 Bolzano, Italy
e-mail: stefan.steger@eurac.edu

C. Kofler
e-mail: christian.kofler@eurac.edu; christian.kofler1@natec.unibz.it

M. Pittore
e-mail: massimiliano.pittore@eurac.edu

M. Zebisch
e-mail: marc.zebisch@eurac.edu

S. Schneiderbauer
e-mail: stefan.schneiderbauer@eurac.edu; schneiderbauer@ehs.unu.edu

V. Mair
Office for Geology and Building Materials Testing, Autonomous Province of Bolzano-South Tyrol, 39053 Cardano, Italy
e-mail: volkmar.mair@provinz.bz.it

C. Kofler
Faculty of Science and Technology, Free University of Bozen-Bolzano, 39100 Bozen-Bolzano, Italy

S. Schneiderbauer
GLOMOS Program, United Nations University, Institute for Environment and Human Security, Bonn, Germany

Introduction

The Italian province of South Tyrol is an Alpine area frequently affected by natural hazard phenomena, such as floods, snow avalanches and different landslide types. Landslides of the slide-type movement (cf. Cruden and Varnes 1996; Hungr et al. 2014) cause damage on infrastructure, agricultural fields and private properties every year. The most extensive landslide inventory for South Tyrol is managed by the provincial geological survey and contains 1928 positionally accurate points that relate to the scarp location of slide-type movements.

This contribution aims to investigate statistical associations between inventoried shallow slide-type movements and frequently applied spatial environmental variables. Particular

F. Guzzetti et al. (eds.), *Understanding and Reducing Landslide Disaster Risk*,
ICL Contribution to Landslide Disaster Risk Reduction,
https://doi.org/10.1007/978-3-030-60227-7_34

emphasis is placed on a critical interpretation of the results by taking information on the underlying landslide data collection explicitly into account. Potential implications for the statistical analysis of already available inventory data are discussed.

Study Area and Data

South Tyrol

South Tyrol extends over 7400 km^2 and is situated at the northernmost part of Italy (Fig. 1). The area is characterized by a substantial morphological, geological and climatic heterogeneity (e.g. annual precipitation from <500 mm up to >1000, altitude from 200 to 3900 m a.s.l) (Stingl and Mair 2005; Piacentini et al. 2012).

In the study area, landsliding is controlled by an interplay of manifold (de)stabilizing variables. Besides predisposing factors related to lithology, the properties of weathered material and topography, also slowly changing preparatory factors (e.g. vegetation, prolonged rain and snow melt) and potential triggers (e.g. heavy rainfall, intensive snow melt, storms) are known to affect the location and frequency of slope instability. Human activities are as well an important landslide influencing factor, particularly due to land use practices and construction works (Tasser et al. 2003; Corsini et al. 2005; Stingl and Mair 2005; Borgatti and Soldati 2010; Piacentini et al. 2012).

Landslide Data and Environmental Factors

The available landslide inventory is based on the Italian Landslide Inventory project (IFFI: Inventario dei Fenomeni Franosi in Italia) which was launched in 1999 (Trigila et al. 2010). Up to January 2019, the South Tyrolean IFFI data consists of 7573 registered events, among them 1928 shallow landslides of the slide-type movement. Positionally accurate points (i.e. GPS-based mapping in the field) at the landslide scarp positions represent these slide-type movements (Fig. 1). Deep-seated landslides were not considered, because of the necessity to analyse deep-seated phenomena separately from their shallow counterparts and a low sample size of deep-seated movements. Due to the underlying provincial landslide mapping strategy only events that required an intervention by the geological survey for the purpose of documentation, mitigation and/or prevention are registered within this data set. Therefore, the inventory mainly relates to events that caused damage.

Information on spatial environmental variables were accessed via the provincial Geodata platform (Geokatalog 2019). Topographic information was derived from a 2.5 m × 2.5 m Light Detection and Ranging (LiDAR) digital terrain model (DTM) which was resampled to 10 m spatial resolution in order to achieve an adequate generalization (e.g. smoothing of noise) and to avoid a close description of post-failure topography (Steger et al. 2020). Spatial information on infrastructure (i.e. roads, pathways, buildings footprints), perennial water courses, geology (i.e.

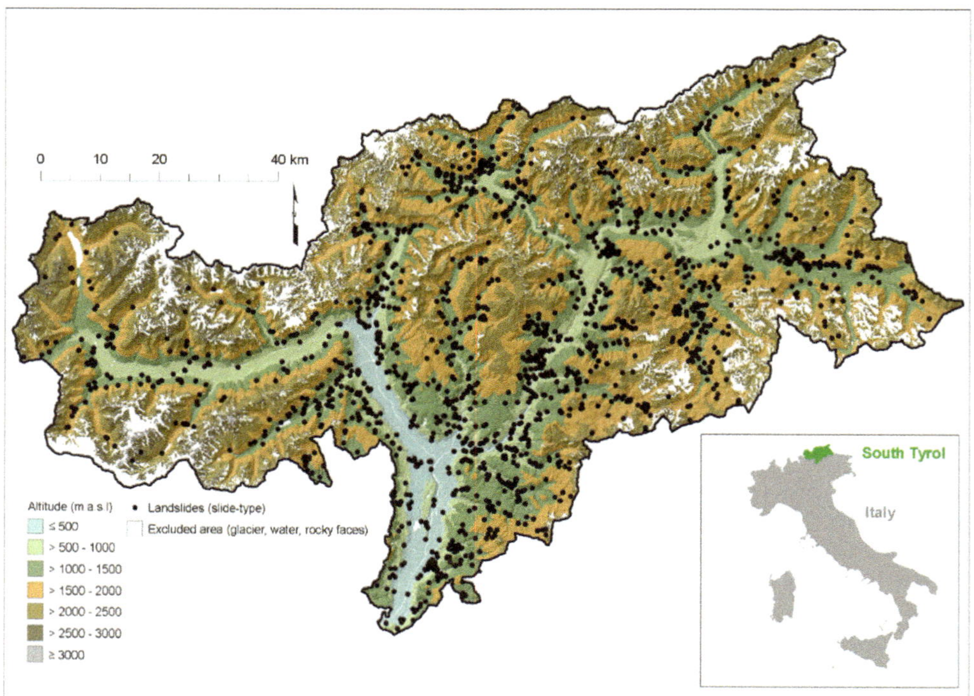

Fig. 1 Overview of the study area and spatial distribution of inventoried slide-type movements (one point per landslide scarp). Areas shown in white (i.e. glaciers, water bodies, rocky faces) were a-priori excluded from the analysis

'*Geologische Übersichtskarte Südtirol*') and land use/cover ('*Realnutzungskarte*' version 2015; 1:10,000) were also extracted from the open Geodata platform (Geokatalog 2019).

Methods

Data Preparation

A binary variable that consist of (i) landslide presences (i.e. inventory data) and (ii) randomly sampled landslide absence locations builds a common basis to investigate landslide controls or landslide susceptibility (Brenning et al. 2015; Schmaltz et al. 2017; Reichenbach et al. 2018). For this study, 1928 absence locations (1:1 ratio presences to absences) were randomly distributed across the study site while glaciers, water bodies, rock faces and locations in close proximity to inventoried landslides (i.e. 50 m buffer) were a-priori disregarded (Fig. 1). Trigonometric calculations based on a slope angle map were used to control sampling intensity of absences in order to account for discrepancies between the dimension of an area in planar view and the actual surface area (i.e. the steeper the terrain, the higher the sampling probability of a cell). This procedure was assumed to be of utility to counterbalance bias related to the presence of "trivial" terrain (cf. Steger and Glade 2017) and an overrepresentation of landslides in steep terrain.

SAGA GIS (Conrad et al. 2015) and the R package RSAGA (Brenning 2008) were used to calculate the commonly utilized DTM derivatives slope angle, slope aspect, convergence index and the upslope contributing catchment area. The latter variable was transformed logarithmically to decrease skewness (Brenning et al. 2015). GRASS GIS (Lennert 2017) was utilized to derive 10 initial geomorphological phonotypes (cf. Jasiewicz and Stepinski 2013) which were heuristically merged to derive six distinct landforms. Euclidean distances were calculated to obtain 10 m raster layers that represent the distance to rivers (i.e. perennial rivers and streams), roads (i.e. asphalt roads, highways), paths (i.e. unpaved roads and trails), transport networks (i.e. streets and paths merged) and to buildings. Nearest neighbour resampling was applied to rasterize land use/cover information and the geological map (Geokatalog 2019).

Exploratory Data Analysis

Relationships between landslide presences/absences and spatial variables were elaborated by means of conditional frequency plots (c.f. Brenning 2008; Steger et al. 2016). The plots depict how the observed landslide presence to absence ratio varies across variable values. Due to the balanced number of observation in the initial sample (1:1 presence to absence ratio), a conditional frequency of 0.5 can be interpreted as equal-ratio-threshold that separates '*relative high landslide densities*' (>0.5) from '*low densities*' (<0.5).

Insights into the efficiency of single variables to '*explain*' the distribution of landslide presences and absences were gained by evaluating the individual discriminatory power using the Area Under the Receiver Operating Characteristic curve (AUROC) in analogy to Steger et al. (2020). In summary, predictions based on a Generalized Additive Model (Wood 2006) were used to evaluate the ability of single explanatory variables to separate landslides presences from absences (AUROC of 0.5: random separation, 1: perfect separation).

Results and Discussion

Conditional frequencies showed that medium inclined terrain (Fig. 2a), south and east exposed sides (Fig. 2b), footslopes (Fig. 2c), areas below 1600 m a.s.l (Fig. 2d), and areas close to rivers (Fig. 2f) were associated with comparably high conditional landslide frequencies. In contrast, landslide scarps were more sparsely registered for flat areas (Fig. 2a, c) and very steep terrain (Fig. 2a), for north facing slopes (Fig. 2b), for higher slope positions (i.e. ridges, spurs) and high altitudes (Fig. 2c, d). Low landslide frequencies were also observed for extraordinarily convex and concave shaped terrain (Fig. 2e), for locations distant from rivers (Fig. 2f) and for areas that exhibit a very high contributing area (Fig. 3a). The plots that depict the distance to infrastructure (Fig. 3b–e) provided clear evidence that landslides were frequently registered for areas close to asphalt roads (Fig. 3b), pathways (Fig. 3c) and buildings (Fig. 3e).

Considering that the underlying landslide data represents mainly events documented by the geological survey for intervention, mitigation and/or prevention, we inferred that the observed associations have to be interpreted with caution. For instance, the observed high landslide frequencies close to infrastructure and at lower slope positions might mainly be attributed to the provincial landslide mapping strategy, which concentrates on events that led to interventions and neglects areas far from potentially exposed assets (e.g. roads, major infrastructures, buildings).

In this context, we deduced that the observed varying conditional landslide frequencies among rock types (Fig. 3f) and land cover units (Fig. 4a) do not only reflect slope stability effects, but can partly be ascribed to the underlying landslide mapping design. We expect an overrepresentation of inventoried landslide occurrence within the units (e.g.

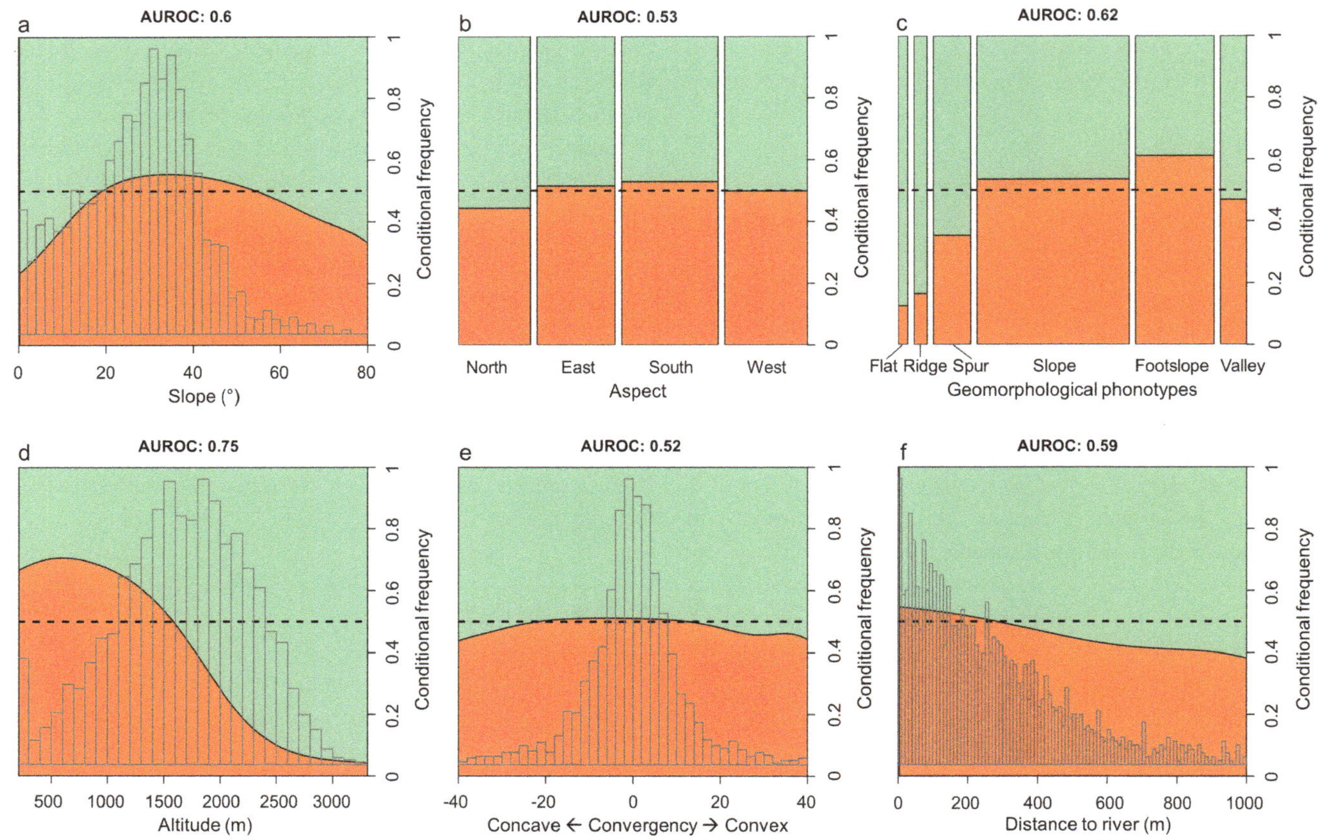

Fig. 2 Exploratory analysis for the variables slope angle (**a**), aspect (**b**), geomorphic phonotypes (**c**), altitude (**d**), convergency index (**e**) and distance to river (**f**). The conditional frequency illustrates the ratio of landslide presences (orange area) to absences (green area) for the respective variable. The histograms (grey bars) relate the distribution of continuous variables (**a**, **d**, **e**, **f**) and the width of the bars depicts data density for categorical variables (**b**, **c**). The AUROC shows the discriminatory power for single variables (AUROC 0.5: random separation, 1: perfect separation)

geological classes, land cover types) that spatially coincide with densely built hillsides (i.e. denser road network, many buildings).

Comparisons of the statistical associations among environmental variables (e.g. slope angle vs. land cover in Fig. 4c, d) underlined that medium inclined terrain was simultaneously associated with both a higher portion of forested area (Fig. 4d) and a high landslide density (Fig. 2a). Thus, the observed slightly above average landslide density in forests (Fig. 4b) does not necessarily highlight that forested areas are slightly less stable than their non-forested equals. We would rather conclude that in South Tyrol forests are likely located on slope inclinations that favour slope instability, while also the dense network of forest roads might play a role on why forest were associated with comparably high landslide mapping densities. It can be assumed that the observed low landslide density for steep forested areas above ~1600 m a.s.l. and at a greater distance from agricultural land and infrastructure/settlements strongly reflects the underlying data collection process. The fact that farmers consciously avoid using unstable zones for agricultural purposes also influences the results.

Ultimately, such interpretations highlight the need for a joint consideration of multiple variables when assessing and interpreting statistical relationships (e.g. multiple-variable models) and the necessity to take the underlying data background explicitly into account.

General guidelines for the AUROC interpretation (cf. Hosmer and Lemeshow 2000) revealed that altitude was the only topographic factor with an acceptable ability to separate landslide presences from absence observations (AUROC > 0.7, Fig. 2d). Acceptable to excellent AUROCs were

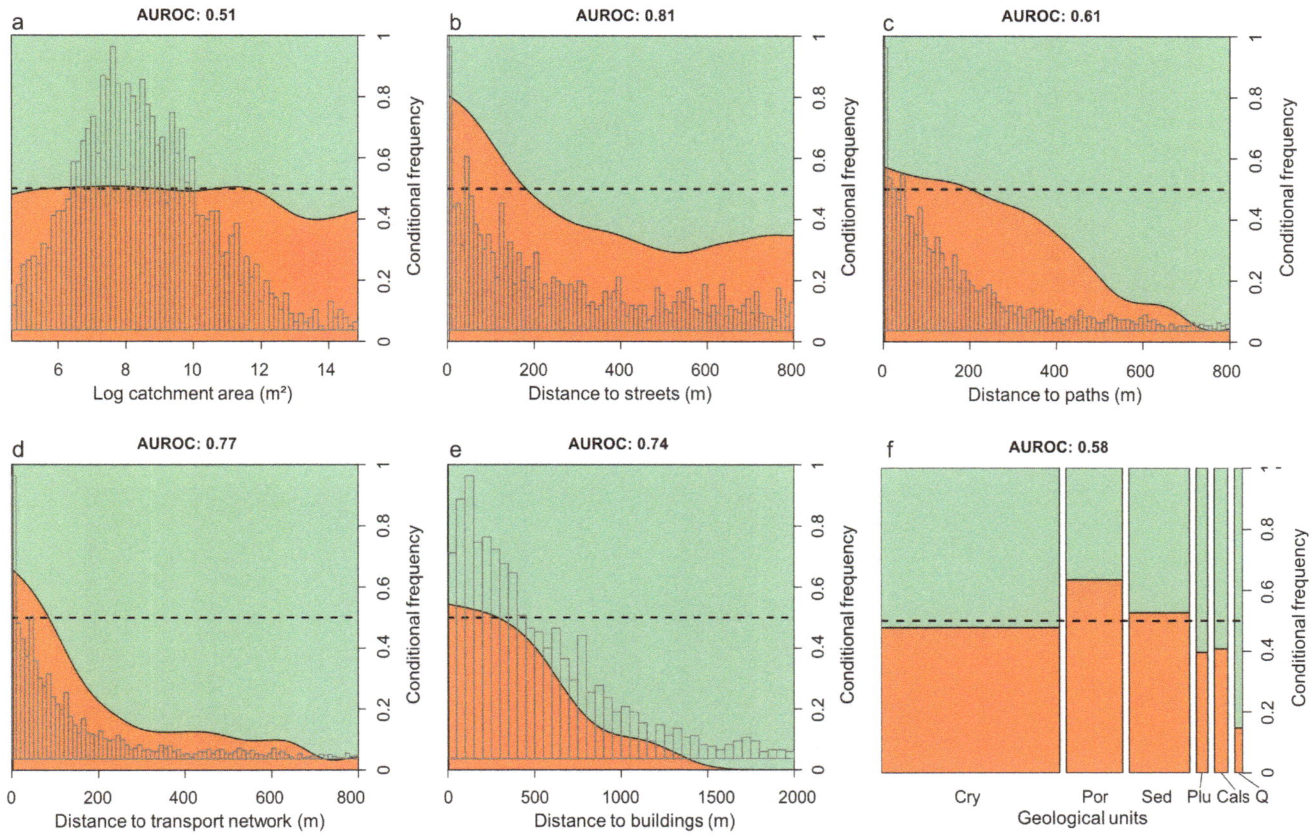

Fig. 3 Exploratory analysis for the variables upslope contributing catchment area (**a**), distance to streets (**b**), distance to pathways (**c**), distance to transport network (streets and pathways; **d**), distance to buildings (**e**) and rock types (**f**). The conditional frequency illustrates the ratio of landslide presences (orange area) to absences (green area) for the respective variable. The histograms (grey bars) relate the distribution of continuous variables (**a**–**e**) and the width of the bars in f depicts data density for geological units (Cry: Crystalline basement, Por: Porphyry, Sed: Sedimentary rocks, Plu: Plutonite, Cals: Calc-schists with ophiolites, Q: Quaternary). The AUROC shows the discriminatory power for single variables (AUROC 0.5: random separation, 1: perfect separation)

observed for the variables that depict the proximity to important infrastructure: distance to buildings (0.74, Fig. 3e) and distance to roads (0.81, Fig. 3b).

These numbers do not provide evidence that undercutting or overloading of hillslopes due to infrastructure construction are important landslide predisposing factor for South Tyrol. The obtained AUROCs reflect the provincial mapping strategy that explicitly focuses on important infrastructure and settlements. We conclude that statistically '*important*' or '*significant*' associations between landsliding and environmental factors have to be interpreted in the light of the underlying data characteristics and by questioning their geomorphological plausibility (Steger et al. 2016).

Conclusion and Outlook

This contribution highlights how erroneous conclusions are likely to follow from statistical analyses when the background information associated with available landslide data is ignored. A look at published literature indicates that recalling the well-known principle '*correlation does not imply (geomorphic) causation*' seems to be of particular relevance in the field of statistically-based landslide analysis (e.g. susceptibility modelling), also because (i) the vast majority of available landslide inventories exhibit a spatially heterogeneous completeness and because (ii) analysts still

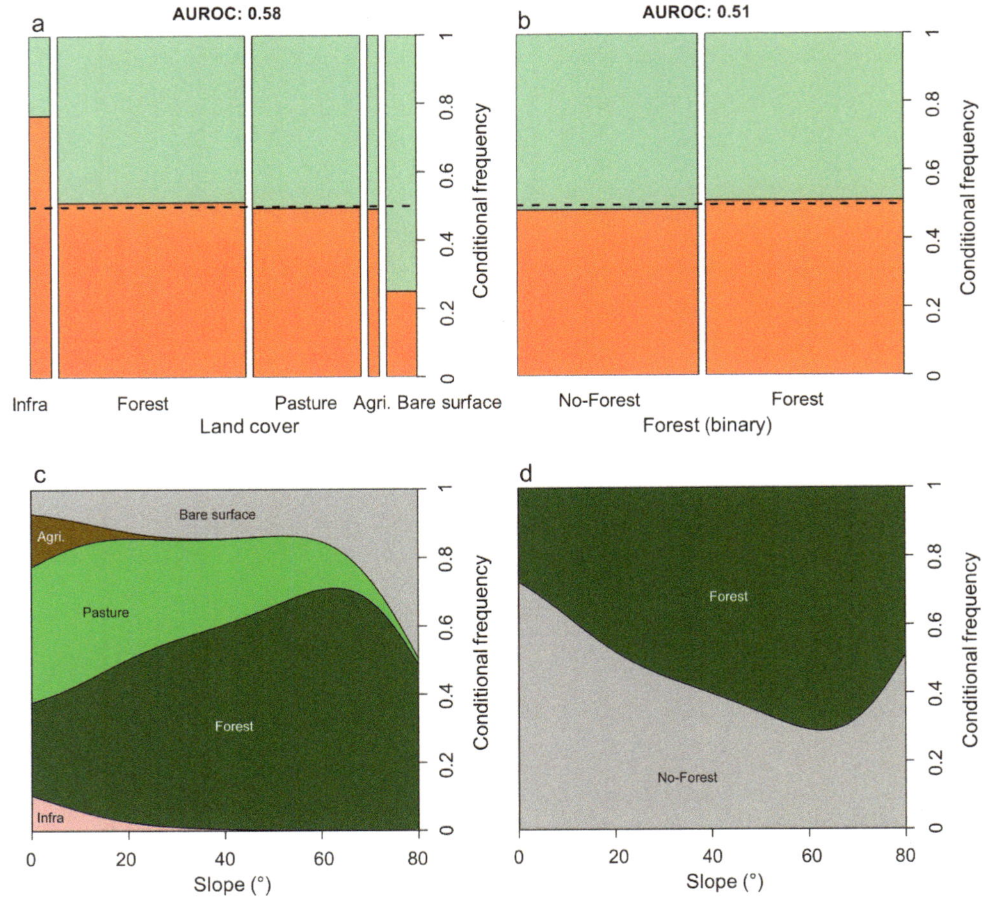

Fig. 4 Exploratory analysis of the variables land cover (**a**) and forest/no-forest (**b**) and interrelations among land cover and slope angle (**c**, **d**). The conditional frequency in **a** and **b** illustrates the ratio of landslide presences (orange area) to absences (green area) for the respective class while the width of the bars depicts data density. The AUROC relates to the discriminatory power of the variables (AUROC 0.5: random separation, 1: perfect separation). Conditional frequencies in **c** and **d** show the fraction of land cover units conditional on slope angle

need to assign meaning and geomorphological plausibility to the obtained numbers (Steger et al. 2016, 2017).

Based on the results, we will focus on the co-development (i.e. modellers and landslide data provider) of non-linear multiple-variable spatial landslide models that explicitly account for the underlying data origin. The modelling results will also be presented at the World Landslide Forum 5.

Acknowledgements The authors thank the Autonomous Province of South Tyrol for providing spatial input data.

References

Borgatti L, Soldati M (2010) Landslides as a geomorphological proxy for climate change: a record from the Dolomites (northern Italy). Geomorphology 120:56–64

Brenning A (2008) Statistical geocomputing combining R and SAGA: the example of landslide susceptibility analysis with generalized additive models. Hamburger Beiträge Phys Geogr Landschaftsökol 19:410

Brenning A, Schwinn M, Ruiz-Páez AP, Muenchow J (2015) Landslide susceptibility near highways is increased by 1 order of magnitude in the Andes of southern Ecuador, Loja province. Nat Hazards Earth Syst Sci 15:45–57

Conrad O, Bechtel B, Bock M et al (2015) System for automated geoscientific analyses (SAGA) v. 2.1.4. Geosci Model Dev 8:1991–2007

Corsini A, Pasuto A, Soldati M, Zannoni A (2005) Field monitoring of the Corvara landslide (Dolomites, Italy) and its relevance for hazard assessment. Geomorphology 66:149–165

Cruden DM, Varnes DJ (1996) Landslide types and processes landslides: investigation and mitigation. TRB special report, 247. National Academy Press, Washington, pp 36–75

Geokatalog (2019) Open geodatabase of the Autonomous Province of South Tyrol. Accessible via https://geokatalog.buergernetz.bz.it/geokatalog/. Accessed 20 Feb 2019

Hosmer DW, Lemeshow S (2000) Applied logistic regression. Wiley, New York

Hungr O, Leroueil S, Picarelli L (2014) The Varnes classification of landslide types, an update. Landslides 11:167–194

Jasiewicz J, Stepinski TF (2013) Geomorphons—a pattern recognition approach to classification and mapping of landforms. Geomorphology 182:147–156

Lennert M (2017) GRASS development team. Geographic resources analysis support system (GRASS) software, version 7

Piacentini D, Troiani F, Soldati M et al (2012) Statistical analysis for assessing shallow-landslide susceptibility in South Tyrol (south-eastern Alps, Italy). Geomorphology 151–152:196–206

Reichenbach P, Rossi M, Malamud BD et al (2018) A review of statistically-based landslide susceptibility models. Earth Sci Rev 180:60–91

Schmaltz EM, Steger S, Glade T (2017) The influence of forest cover on landslide occurrence explored with spatio-temporal information. Geomorphology 290:250–264

Steger S, Glade T (2017) The challenge of "trivial areas" in statistical landslide susceptibility modelling. In: Proceedings of the 4th world landslide forum, 29 May–2 June, Ljubljana. Springer, Ljubljana

Steger S, Brenning A, Bell R et al (2016) Exploring discrepancies between quantitative validation results and the geomorphic plausibility of statistical landslide susceptibility maps. Geomorphology 262:8–23

Steger S, Brenning A, Bell R, Glade T (2017) The influence of systematically incomplete shallow landslide inventories on statistical susceptibility models and suggestions for improvements. Landslides 14:1767–1781

Steger S, Schmaltz E, Glade T (2020) The (f)utility to account for pre-failure topography in data-driven landslide susceptibility modelling. Geomorphology 354:107041

Stingl V, Mair V (2005) Einführung in die Geologie Südtirols: [aus Anlass des 32. Internationalen Geologischen Kongresses im Sommer 2004 in Florenz]. Autonome Provinz Bozen-Südtirol. Amt Geol Baustoffprüfung

Tasser E, Mader M, Tappeiner U (2003) Effects of land use in alpine grasslands on the probability of landslides. Basic Appl Ecol 4:271–280

Trigila A, Iadanza C, Spizzichino D (2010) Quality assessment of the Italian Landslide Inventory using GIS processing. Landslides 7:455–470

Wood SN (2006) Generalized additive models: an introduction with R. Chapman and Hall/CRC, Taylor and Francis Group, Boca Raton

2020 Kyoto Japan

Assessing Landslide Volume for Landform Hazard Zoning Purposes

Gabriel Legorreta Paulin, Lilia Arana-Salinas, Rutilio Castro Miguel, Jean-François Yves Pierre Parrot, and Trevor A. Contreras

Abstract

We model the relationship between surface area and volume of landslides caused by rain, together with a geomorphologic analysis of 15 landforms to characterize slope instability. The use of this method allows a better understanding of the landslide susceptibility along homogeneous units. The analysis is supported by Geographic Information Systems (GIS) to create a comprehensive method for landslide volume estimation for each landform. This approach is applied to the Río Chiquito-Barranca del Muerto watershed on the south flank of Pico de Orizaba volcano, Mexico. The watershed is prone to gravitational processes because highly weathered volcanic and sedimentary deposits that are affected by extreme seasonal precipitation and deforestation. In the area, more 600 landslides have been mapped and grouped into the landform units. Representative landslides in the watershed were measured in detail with differential GPS and a drone to establish an empirical relationship between landslide area and volume. This relationship expressed as a power law is used to estimate the potential contribution of material delivered from each volcanic and sedimentary landform.

Keywords

GIS • Landslide inventory map • Landslide volume • Drone • Power law

Introduction

In mountainous volcanic terrains, landslides form a major natural hazard that cause economic and human losses, especially during the volcanic dormant activity when gravitational processes occur. This is the case for the central region of Mexico, where stratovolcanoes, domes, monogenetic volcanos, lava flows, and ash fields are very common. In this central region, the Trans Mexican Volcanic Belt (TMVB) physiographic province is an active Neogene volcanic chain that extends 1000 km from west to east across central-southern Mexico, from the Pacific Ocean to the Gulf of Mexico (Ferrari et al. 2012). The physiographic province has at least 14 stratovolcanoes with great potential to produce landslides and volcanic debris flows owing to their earthquakes and volcanic eruptions history, heavy rainfall during the wet season, and steeps slopes covered with weathered and poorly consolidated pyroclastic deposits (CENAPRED (Centro Nacional de Prevención de Desastres) 2017). This study refers in more detail to the eastern portion of the TMVB, where the highest dormant stratovolcano in Mexico, Pico de Orizaba (5675 m a.s.l.) rises. The Río Chiquito-Barranca del Muerto watershed on the south flank of Pico de Orizaba is the study area. In the watershed, geomorphologic, geologic, and land use conditions

G. L. Paulin (✉)
Instituto de Geografía, Universidad Nacional Autónoma de México, Circuito Exterior, Ciudad Universitaria, 04510 Coyoacán, Ciudad de México, México
e-mail: legorretag@hotmail.com

L. Arana-Salinas
Universidad Autónoma de la Ciudad de México, Av. La Corona 320, Col. Loma la Palma, Del. Gustavo A. Madero, 07160 Ciudad de México, México
e-mail: lilia.ar@gmail.com

R. C. Miguel
Posgrado en Geografía, Universidad Nacional Autónoma de México, Ciudad Universitaria, 04510 Ciudad de México, México
e-mail: rutilio.cm@gmail.com

J.-F.Y. P. Parrot
Laboratorio de Análisis Geoespacial, Universidad Nacional Autónoma de México, Circuito Exterior, Ciudad Universitaria, 04510 Coyoacán, Ciudad de México, México
e-mail: parrot@igg.unam.mx

T. A. Contreras
Washington State Department of Natural Resources, Washington Geological Survey, Olympia, WA, USA
e-mail: Trevor.Contreras@dnr.wa.gov

F. Guzzetti et al. (eds.), *Understanding and Reducing Landslide Disaster Risk*,
ICL Contribution to Landslide Disaster Risk Reduction,
https://doi.org/10.1007/978-3-030-60227-7_35

predispose the area to landslides. As a result of the remobilized sediment from landslides toward the inner gorges, episodic debris flows occur along the stream system during the wet season. For example, the tropical storm Ernesto on June 5, 2003, brought heavy rainfall (260 mm of precipitation in a single day), which caused debris runoff through the streams of the watershed. The debris caused destruction and loss of lives at the town of Balastrera (a settlement at the SW base of Pico de Orizaba) (Rodríguez et al. 2011). In 2011, a new landslide and debris flow event destroyed the retention-walls and gabions that the government had installed along the river to prevent flooding and to decrease the destructive power of debris flows. The hazard associated with landslides that constantly affect the stream system with accumulation of rocks, soil, and debris is increased by the expansion of human settlements and economic activity along the Pico de Orizaba flanks. This creates a dangerous situation for inhabitants of important cities such as Córdoba, Orizaba, Rio Blanco, Nogales and Ciudad Mendoza on the southern flank of the volcano. Despite the importance of assessing such processes, there are few landslide inventories and seldom any modeling and evaluation of materials added to the streams by landslides.

Worldwide homogeneous areas, called landform units are used to characterize and understand landslide susceptibility and the volume of material displaced by a landslide in a watershed that is rather complex from the geomorphological standpoint (Hungr et al. 2008; Broeckx et al. 2016). Sediment production by landslides has been addressed for multiple scopes and scales by the use of heuristic, statistical, or deterministic approaches embedded in GIS applications (Guzzetti et al. 2009; Chen et al. 2014; Broeckx et al. 2016; Vanmaercke et al. 2017). To estimate landslide volumes, the physical form of the landslide is obtained by using detailed Digital Elevation Models (DEMs) through the use of imaging equipment (digital cameras, multispectral sensors, Light Detecting and Ranging (LIDAR) instruments and Synthetic Aperture Radars (SAR)) carried by drones (Rango and Laliberte 2010; UNEP 2013) and differential GPS (Travelletti and Malet 2012; Young 2015). To determine the volume of landslide materials in large affected areas, empirical relationships have been established by using a power-law function that links geometric measurements of landslide area to landslide volume (Kaldaron-Asael et al. 2008; Guzzetti et al. 2009; Wenkey et al. 2011).

To address this deficiency, this paper provides standardized methods in a GIS system for evaluating landslide susceptibility per geomorphologic landform and the sediment production by landslides; this will assist governmental authorities in hazard mitigation and landscape planning in Mexico. This study divides the watershed into 15 geomorphologic landforms. Each landform is derived by using GIS, photo interpretation, geological fieldwork, hypsometry, energy relief, vertical erosion, and slope. For each landform, the total landslide volume is estimated. The landslide volume is obtained by detailed geometric values (area and volume) of representative landslides measured during fieldwork using a drone and a differential GPS. These geometric values are then used to establish an empirical relationship that takes the form of a power law with a scaling exponent of $\alpha = 1.1263$. The empirical relationship is used to estimate the potential total volume of material delivered from all landslides (607) in the watershed and per landform. The study shows that landslides in the watershed have the potential to deliver $\sim 1{,}547{,}657$ m^3 of sediments.

Study Area

The main goal of the models is to provide a systematic methodology for modeling and mapping volume of shallow and deep-seated landslides in unstable areas. A case study of landslide volume calculation is conducted in the Rio Chiquito-Barranca del Muerto located between Puebla and Veracruz states, on the south flank of Pico de Orizaba volcano, in Mexico (Fig. 1). The area covers approximately 105 km^2. The study area is prone to landslides due to its natural conditions which are exacerbated by deforestation and agricultural activities. Andesitic and dacitic Tertiary and Quaternary volcanic rocks and deposits form the main lithology in the study area follows by Cretaceous limestone and shales (almost 30% of the total watershed area comprising sedimentary landforms).

Rainfall in the study area is concentrated mainly in the boreal summer (from May to November) and it is mostly due to its location north of the Intertropical Convergence Zone, the influence of westerly winds, and the formation of low-pressure systems in the Pacific region. The maximum annual rainfall averages ~ 1100 mm, mostly concentrated in the mountain area at >4000 m a.s.l. The minimum precipitation is recorded in the lower parts of the watershed where annual rainfall is ~ 927 mm/yr. The contribution of rainfall from hurricanes is important because some of these events affect the hillslopes of Pico de Orizaba. Deforestation and modification of the slopes in favor of anthropic activities have left the main river and its tributaries with a buffer of only 50–100 m of forest (Alanis-Anaya 2017).

The study area was selected to modeling landslide volumes delivered to stream systems. The study area has hazardous deep-seated and shallow landslides that continuously occur (Fig. 2).

Landslide occurrence was ascertained through an inventory of 607 landslides (Fig. 3); landslides that were substantially reforested and estimated to have been covered by

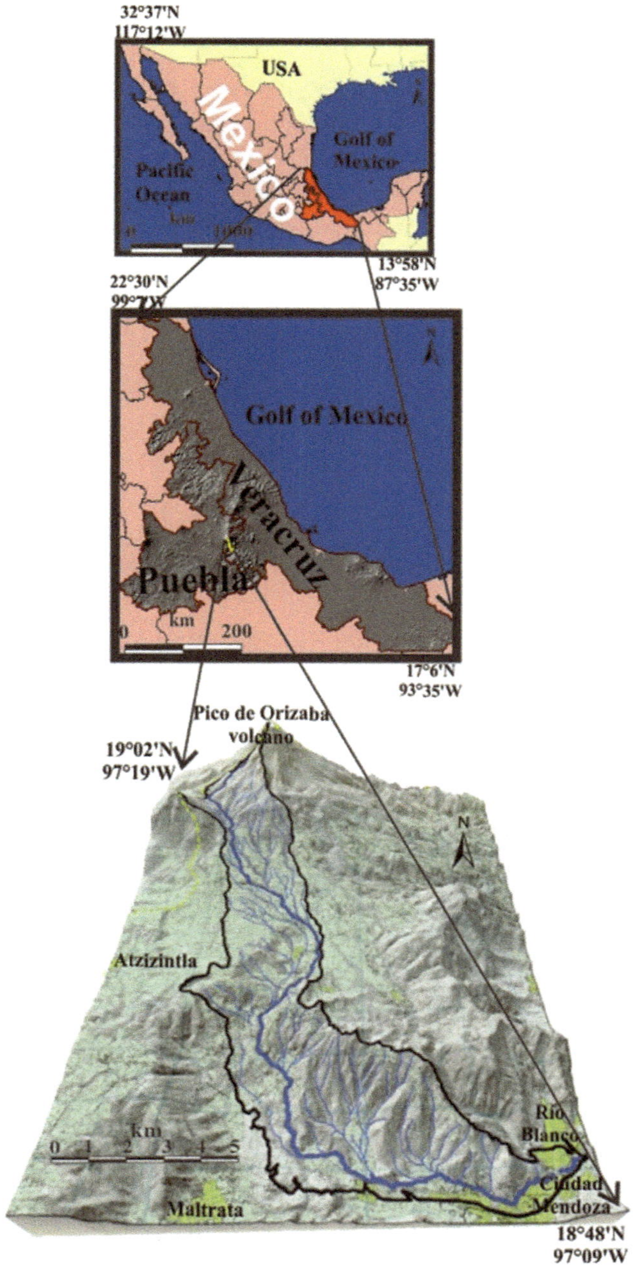

Fig. 1 Study area

vegetation for more than 17 years were not included in the inventory or in the evaluation of volume. Also, landslides whose area was less than 0.2 ha (minimum mapping size) were not used.

Method

Landslide occurrence was ascertained through an inventory of 607 landslides (Fig. 3); landslides that were substantially reforested and estimated to have been covered by vegetation for more than 17 years were not included in the inventory or in the evaluation of volume. Also, landslides whose area was less than 0.2 ha (minimum mapping size) were not used. This landslide inventory map uses the Landslide Hazard Zonation Protocol (LHZP) of the Washington State Department of Natural Resources, Forest Practices Division (2006).

Fig. 2 **a** and **b** Deep-seated landslide and a debris flow on volcanic terrains; **c** and **d** deep-seated landslide and a debris slide on sedimentary terrains

For mapping landforms, this research used 15 landforms previously defined by Alanis-Anaya (2017). The landforms defined in this study are: relict of volcanic collapse, historical lava flows (1545, 1566 and 1613), lava flows (Upper Pleistocene), lava flow hillsides from Sierra Negra (Pleistocene), pyroclastic hillslopes, pyroclastic ramp, undifferentiated pyroclastic ramp, volcanic plateau, basaltic hillslopes, folded mountains (Cretaceous limestone), folded hillslopes (Cretaceous limestone), piedmont (3–6°), piedmont (<3°), fluvioglacial ramp, and fluvial valleys. We use geomorphological criterion that include the land use, hypsometric map, the energy relief map, and the vertical erosion map to describe the landforms.

For the landslide volume contribution analysis, the fluvial valley landform is not considered because this landform is contained in the remaining 14 landforms. Instead, the area of fluvial valleys that affects each landform is given.

Sixty-one landslides ($\sim 10\%$) were verified in the field to measure their area and volume with tape, stadia rod, laser range finder, differential GPS and a drone. Both shallow landslides (including debris flows, and debris slides) and deep-seated landslides were used; 51 were shallow and 10 were deep-seated. DEMs at 5 m resolution were derived. These DEMs were used to calculate the landslide volume in

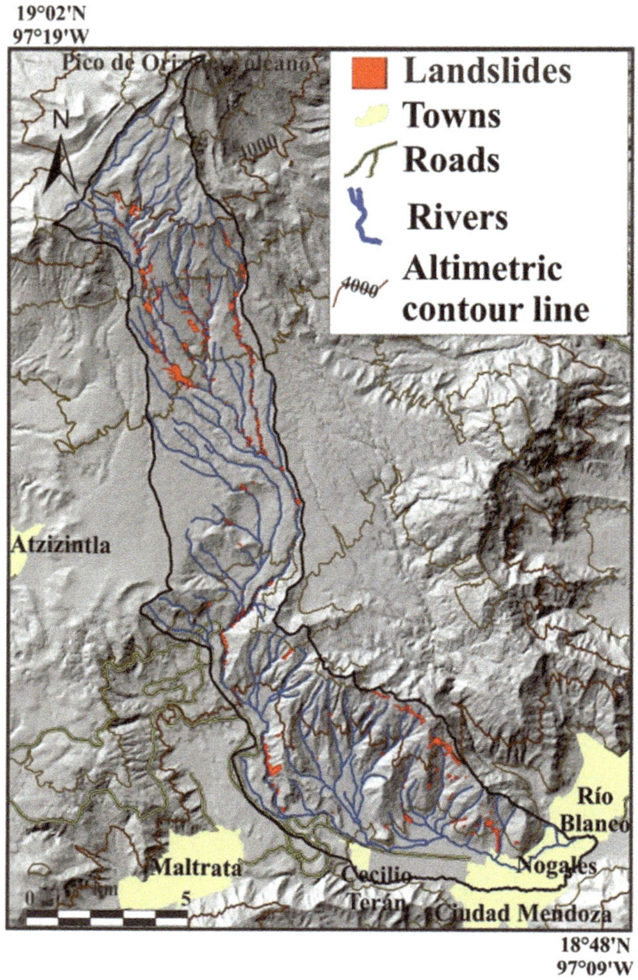

Fig. 3 Landslide inventory

ArcMap. In ArcMap, the current sliding and pre-sliding surfaces were created. The pre-sliding surface was created by means of interpolation (ESRI 2001). Nodes from polygons that represent the landslide border were used to extract the elevation data from the current sliding surface. The nodes were interpolated in ArcMap to create the pre-sliding surface. The landslide volume was calculated by subtracting the real current landslide surface from the pre-sliding landslide surface and multiplying by pixel size. The geometrical measurements were used to develop an empirical relationship between area and volume (Kaldaron-Asael et al. 2008; Guzzetti et al. 2009; Wenkey et al. 2011) that took the form of a power law $V = 0.70242506 * A^{1.1263}$. With the landslide inventory map and related geo-dataset, the planimetric area of the remaining 546 landslides was obtained. The planimetric area was the input for the empirical relationship area-volume in the power law. Finally, we added the calculated volumes of all landslides to estimate the potential total volume that landslides can contribute per landform to the stream system.

Results

The landslide inventory identified 6 types of landslide: 295 are shallow landslides (SL), 126 debris flows (DF), 85 debris slides (DS), 77 rockfalls (RF), 19 deep-seated landslides (DSL), and 5 earthflows (EF). In the watershed, the abundance and types of landslides are determined by geomorphological conditions in conjunction with land use change: 405 (66.7%) are in volcanic terrain (mostly in pyroclastic deposits where the land use has changed from forest activities to agriculture) and the rest (33.3%) in sedimentary terrain (mostly in folded and faulted Cretaceous limestones covered by broad-leaf and coniferous forest). Among the landslide types, SL predominate (48.6%), followed by DF (20.8%), DS (14%), RF (12.7%), DSL (3.1%), and EF (0.8%).

The landforms and their contribution of materials are as follows (Fig. 4):

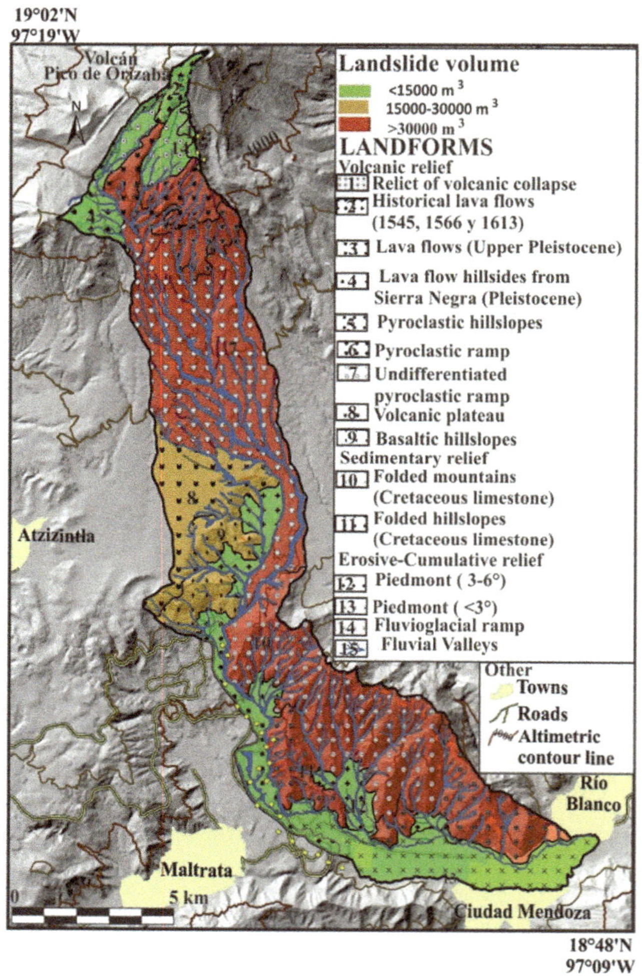

Fig. 4 Lanforms and landslide volumes

1. Relict of volcanic collapse. This is a very eroded, fractured, and weathered andesitic collapse structure of the ancestral Pico de Orizaba volcano (first constructional stage, 0.25 Ma). The structure lies in the northeastern part of the watershed and covers 0.36% of its area. The landform is not affected by fluvial valleys. The altitudinal range is 4260–4740 m a.s.l., slopes are between 3° and 70°, relief energy between 293 and 414 m, and fluvial dissection density between 0.5 and 1.4 km/km^2. The landslide susceptibility is low, but it has the potential for RF. No visible landslides are mapped in this landform.
2. Historical lava flows (1545, 1566 and 1613). This landform is formed by 3 dacitic lava flows associated with the end of the effusive period of the present Pico de Orizaba volcano. The lava flows lie to the southwest of the crater and in the northwestern end of the watershed, covering 1.46% of its area. Of its 0.38 km^2, 3.43% is affected by fluvial valleys. The altitudinal range is 4040–5600 m a.s.l., slopes are between 9° and 65°, relief energy between 246 and 492 m, and fluvial dissection density between 0.3 and 1.7 km/km^2. The landslide susceptibility is low, but it has the potential for rock falls. No visible landslides are mapped in this landform.
3. Lava flows (Upper Pleistocene). This landform is formed by 3 andesitic and dacitic lava flows associated with the second constructional stage of the Pico de Orizaba volcano (0.09 ± 0.053 Ma). The lava flows lie to the south of the crater and cover 4.44% of the watershed. Of its 4.68 km^2, 11.38% is affected by fluvial valleys. The altitudinal range is 3460–4080 m a.s.l., slopes are between 5° and 55°, relief energy between 248 and 357 m, and fluvial dissection density between 0.9 and 3.6 km/km^2. The landslide susceptibility is high and includes 2.14% of the mapped landslides, which contribute 2.90% of the watershed sediments.
4. Lava flow hillsides from Sierra Negra (Pleistocene). This landform comprises andesitic and dacitic lava flows from the Sierra Negra volcano, which were emplaced before those of the ancestral Pico de Orizaba volcano. The lava flow hillsides lie in the northwestern part of the watershed and cover 1.68% of its area. Of the 1.77 km^2, 11.42% is affected by fluvial valleys. The altitudinal range is 3720–4540 m a.s.l., slopes are between 3° and 65°, relief energy between 248 and 357 m, and fluvial dissection density between 0.4 and 3 km/km^2. The landslide susceptibility is moderate and includes 0.34% of the mapped landslides, which contribute 0.42% of the watershed sediments.
5. Pyroclastic hillslopes. The landform is correlated with the eruptive sequence of Loma Grande, which occurred 8200 ± 120 yr B.P. It consists of ash and blocks that have been re-deposited along the highest regions of the watershed. The pyroclastic hillslopes lie in the northwestern part of the watershed and cover 3.78% of its area. Of their total area (3.99 km^2), 20.88% is affected by fluvial valleys. The altitudinal range is 1550–2180 m a.s.l., slopes are between 2° and 48°, relief energy between 135 and 219 m, and fluvial dissection density between 0.9 and 2.8 km/km^2. The landslide susceptibility is low, but it has the potential for SL and DSL. No visible landslides are mapped in this landform.
6. Pyroclastic ramp. This landform is formed by 3 basaltic and andesitic units (0.65 ± 0.71 Ma). The basal unit is an andesitic-basaltic lava that is overlaid by andesitic blocks, lava, and breccia covered by lahars and tephra. The upper unit is a hummocky andesitic lava flow covered by pyroclastic deposits. The landform lies in the northern part of the watershed and covers 3.11% of its area. Of its 3.27 km^2, 14.06% is affected by fluvial valleys. The altitudinal range is 3480–4080 m a.s.l., slopes are between 4° and 50°, relief energy between 272 and 342 m, and fluvial dissection density between 1 and 3.6 km/km^2. The landslide susceptibility is high and includes 2.47% of the mapped landslides, which contribute 2.40% of the watershed sediments.
7. Undifferentiated pyroclastic ramp. This landform is formed by different deposits (pyroclastic, scoria, and pumice flows) originating from the successive phases of formation of the Pico de Orizaba volcano. This landform lies on all flanks of the volcano and covers 24.49% of the watershed. Of its area (21.85 km^2), 18.28% is affected by fluvial valleys. The altitudinal range is 2400–3820 m a.s.l., slopes are between 1° and 42°, relief energy between 103 and 338 m, and fluvial dissection density between 0.8 and 3.3 km/km^2. The landslide susceptibility is very high and includes 58.48% of the mapped landslides, which contribute 52.89% of the watershed sediments.
8. Volcanic plateau. This landform is formed by a sequence of pyroclastic flows with andesitic-dacitic blocks and ashes dated between the years 1860 and 1910. The landform lies to the south of the crater, close to the town of Texmalaquilla, and covers 7.65% of the watershed. Of its 8.07 km^2, 10.86% is affected by fluvial valleys. The altitudinal range is 2580–3000 m a.s.l., slopes are <45°, relief energy between 96 and 252 m, and fluvial dissection density between 0.5 and 2.9 km/km^2. The landslide susceptibility is low and includes 1.81% of the mapped landslides, which contribute 1.27% of the watershed sediments.
9. Basaltic hillslopes. This landform is formed by a sequence of Pliocene and Early Pleistocene basaltic lava flows associated with small cinder cones. It lies in the central portion of the watershed and covers 4.65%

of the area. Of its 4.90 km^2, 10.82% is affected by fluvial valleys. The altitudinal range is 2280–2800 m a.s.l., slopes are between 5° and 70°, relief energy between 149 and 277 m, and fluvial dissection density between 1.5 and 2.8 km/km^2. The landslide susceptibility is low and includes 1.32% of the mapped landslides, which contribute 0.45% of the watershed sediments.

10. Folded mountains (Cretaceous limestone). This landform is formed by a sequence of limestones and shales that were formed from the Upper Cretaceous reefs. It lies in the southwestern part of the watershed and covers 18.50% of the area. Of its 19.50 km^2, 13.93% is affected by fluvial valleys. The altitudinal range is 1400–2680 m a.s.l., slopes are between 3° and 72°, relief energy between 139 and 405 m, and fluvial dissection density between 0.9 and 4 km/km^2. The landslide susceptibility is very high and includes 18.45% of the mapped landslides, which contribute 13.10% of the watershed sediments.

11. Folded hillslopes (Cretaceous limestone). This landform is formed by a sequence of limestones that were formed from the Lower Cretaceous reefs. It lies in the southeastern part of the watershed and covers 10.62% of the area. Of its 11.19 km^2, 10.97% is affected by fluvial valleys. The altitudinal range is 1300–1900 m a.s.l., slopes are between 6° and 55°, relief energy between 144 and 361 m, and fluvial dissection density between 0.9 and 3.5 km/km^2. The landslide susceptibility is very high and includes 14.99% of the mapped landslides, which contribute 26.57% of the watershed sediments.

12. Piedmont (3–6°). This is formed by a sequence of reworked deposits along the base of the highest regions of the watershed and covers 6.32% of the area. Of its 6.66 km^2, 23.04% is affected by fluvial valleys. The altitudinal range is 1400–2580 m a.s.l., slopes are between 3° and 6°, relief energy between 118 and 353 m, and fluvial dissection density between 1.6 and 3.8 km/km^2. The landslide susceptibility is low. No visible landslides are mapped in this landform.

13. Piedmont (<3°). This is formed by a sequence of reworked sedimentary and volcanic deposits along the base of the highest regions of the watershed and covers 9.56% of the area. Of its 10.07 km^2, 6.15% is affected by fluvial valleys. The altitudinal range is 1280–1410 m a.s.l., slopes are <3°, relief energy between 118 and 353 m, and fluvial dissection density between 0.3 and 2.6 km/km^2. The landslide susceptibility is low. No visible landslides are mapped in this landform.

14. Fluvioglacial ramp. This is formed by a sequence of moraine and reworked fluvial deposits associated with the actual glacial dynamics that take place on the summit of the volcano. It lies in the southern part of the Pico de Orizaba cone and covers 3.38% of the area. Of its 3.57 km^2, 13.26% is affected by fluvial valleys. The altitudinal range is 3980–5675 m a.s.l., slopes between 11° and 40°, relief energy between 246 and 426 m, and fluvial dissection density between 0.4 and 3.2 km/km^2. The landslide susceptibility is low. No visible landslides are mapped in this landform. Thirteen percent of its area is affected by fluvial valleys.

15. Fluvial valley. This is distributed throughout the entire watershed and is found in each of the other 14 landforms. It is characterized by intermittent rivers with a dendritic pattern that form fluvial deposits. This landform covers an area of 14.78 km^2. The altitudinal range is 1280–4833 m a.s.l., slopes are between 30° and 70°, relief energy between 106 and 4441 m, and fluvial dissection density between 0.5 and 3.9 km/km^2. The landslide susceptibility is very high and includes 58.5% of all mapped landslides, which contribute 48.64% of the watershed sediments.

A power law of $V = 0.70242506 * A^{1.1263}$ was determined by using 61 landslides. Then, the planimetric area of each remaining landslides (546) in the inventory map was used in the power law to calculate volume. Finally, we added the calculated volume of all landslides to estimate the potential total volume for each landform. The volume was classified into three ranges: low less than 15,000 m^3, average 15,000–30,000 m^3 and high greater than 30,000 m^3 (Fig. 4). The results show that landslides have a potential contribution of $\sim$1,547,657 m^3 of material to the basin. Between the relief forms, a volcanic unit and two sedimentary relief forms contribute the greatest amount of sediments per square kilometer.

Discussion and Conclusion

The study shows that the landforms differ in their patterns of gravitational processes as a result of their geology, land use, and geomorphometric properties. More than 58% of the mapped landslides are on poorly consolidated ash fall, scoria, pomes, and other pyroclastic deposits that form the undifferentiated pyroclastic ramp landform. This landform, originally covered by forest, has fertile soils that have favored a change to large agricultural areas; however, unsuitable treatment of the remaining forest has triggered the high incidence of gravitational processes and erosion. This landform contributes ~54% of the total volume of sediments in the watershed; of this contribution, almost 75% is derived from landslides along the fluvial valley that affect this landform. The other two landforms (Cretaceous mountains and hillslopes) with very high susceptibility, number of landslides, and sediment contribution are in sedimentary areas. In these landforms, the folded and faulted sedimentary rocks are prone to landslides and increase the volumes involved. The Cretaceous mountains and hillslopes account for 18.45% and 14.99% of landslides, and these contribute 13.10% and 26.57%, respectively, of the sediments in the watershed.

The assessment of landslide volume delivered to the stream provided an important step for planning disaster management in the study area. However, note that the power law and the volume calculation is performed only for existing landslides and does not include the potential landslide areas, so the calculated volume of sediments is well below the total potential volume that landslides can bring in extraordinary events.

Acknowledgements The authors thank authorities from the Washington State Department of Natural Resources (DNR) Washington Geological Survey, and the Department of Physical Geography from the Institute of Geography, UNAM for their approval and help. This research was supported by he Programa de Apoyos para la Superación del Personal Académico de la UNAM (PASPA) de la Dirección General de Asuntos del Personal Académico (DGAPA), UNAM.

References

Alanis-Anaya RM (2017) Influencia del cambio de uso de suelo en la inestabilidad de la ladera en la subcuenca del Río Chiquito-Barranca del Muerto, volcán Pico de Orizaba. Thesis, Facultad de Filosofía y Letras, UNAM

Broeckx J, Vanmaercke M, Bălteanu D, Chendeş V, Sima M, Enciu P, Poesen J (2016) Linking landslide susceptibility to sediment yield at regional scale: application to Romania. Geomorphology 268:222–232

CENAPRED (Centro Nacional de Prevención de Desastres) (2017) Elaboración de escenarios de peligro por inundación y deslizamiento de laderas, mediante información obtenida por Vehículos aéreos no tripulados-drones (VANT). Informe trimestral del ejercicio de recursos y el informe técnico del avance de proyecto. H00-DG/0540/017. Ciudad de México, p 13

Chen Q, Cheng H, Yang Y, Liu G, Liu L (2014) Quantification of mass wasting volume associated with the giant landside Daguangbao induced by the 2008 Wenchuan earthquake from persistent scatterer InSAR. Remote Sens Environ 152:125–135

Environmental Systems Research Institute (, ESRI,) (2001) ArcGistm spatial analyst: advanced GIS spatial analysis using raster and vector data. ESRI, California

Ferrari L, Orozco-Esquivel T, Manea V, Manea M (2012) The dynamic history of the Trans-Mexican Volcanic Belt and the Mexico subduction zone. Tectonophysics 522:122–149

Guzzetti F, Ardizzone F, Cardinali M, Rossi M (2009) Landslide volumes and landslide mobilization rates in Umbria, central Italy. Earth Planet Sci Lett 279:22–229

Hungr O, McDougall S, Wise M, Cullen M (2008) Magnitude-frequency relationships of debris flows and debris avalanches in relation to slope relief. Geomorphology 96(3–4):355–365

Kaldaron-Asael B, Katz O, Aharonov E, Marco S (2008) Modeling the relation between area and volume of landslides. Report for Steering Comittee for Earthquakes, Jerusalem, Israel

Rango A, Laliberte A (2010) Impact of flight regulations on effective use of unmanned aircraft systems for natural resources applications. J Appl Remote Sens 4:1

Rodríguez S, Mora I, Murrieta J, Morales BWV (2011) Peligros geológicos más frecuentes en el estado de Veracruz. Universidad Veracruzana Xalapa, Veracruz

Travelletti J, Malet JP (2012) Characterization of the 3D geometry of flow-like landslides: a methodology based on the integration of heterogeneous multi-source data. Eng Geol 128:30–48

Vanmaercke M, Ardizzone F, Rossi M, Guzzetti F (2017) Exploring the effects of seismicity on landslides and catchment sediment yield: an Italian case study. Geomorphology 278:171–183

Washington State Department of Natural Resources, Forest Practices Division (2006) Landslide hazard zonation (LHZ) mapping protocol, version 2.1. Accessed at: https://www.dnr.wa.gov/Publications/fp_lhz_protocol_v2_1_final.pdf

Wenkey D, Jen CH, Böse M, Lin JC (2011) Assessment of sediment delivery from successive erosion on stream-coupled hillslopes via a time series of topographic surveys in the central high mountain range of Taiwan. Quatern Int. https://doi.org/10.1016/j.quaint.2011.02.018

Young AP (2015) Recent deep-seated coastal landsliding at San Onofre State Beach, California. Geomorphology 228:200–212

2020 Kyoto Japan

Empirical Relationships to Estimate the Probability of Runout Exceedance for Various Landslide Types

Marc-André Brideau, Saskia de Vilder, Chris Massey, Andrew Mitchell, Scott McDougall, and Jordan Aaron

Abstract

Estimating the potential runout distance of a landslide and it associated impacted area is an important component of landslide hazard and risk analysis. This can be done using empirical; or physics-based runout analysis methods. The current study uses the empirical method by compiling previously published international examples of landslide runout and unpublished New Zealand landslides triggered by recent earthquakes and rainstorm events. The results are presented in a series of ΔH/L versus volume plots for different landslide types, substrate (glacier versus unglaciated), and triggers (earthquake versus rainfall). The plots contain superimposed probability of runout exceedance lines. The results from this study provide landslide researchers and practitioners simple tools to conduct a forward-looking empirical-probabilistic runout analysis for debris avalanches, debris flows, and rock avalanches. The methodology presented is appropriate for a regional-level assessment or scoping-level site-specific assessment to identify areas warranting more detailed work. Applications and limitations of empirical-probabilistic runout methods are also discussed.

M.-A. Brideau (✉) · S. de Vilder · C. Massey
GNS Science, 1 Fairway Drive, Lower Hutt, 5010, New Zealand
e-mail: m.brideau@gns.cri.nz

S. de Vilder
e-mail: s.devilder@gns.cri.nz

C. Massey
e-mail: c.massey@gns.cri.nz

A. Mitchell · S. McDougall
Department of Earth, University of British Columbia, Ocean and Atmospheric Sciences, Vancouver, V6T 1Z4, Canada
e-mail: amitchell@eoas.ubc.ca

S. McDougall
e-mail: smcdouga@eoas.ubc.ca

J. Aaron
Department of Earth Sciences, ETH Zürich, 8092 Zürich, Switzerland
e-mail: jordan.aaron@erdw.ethz.ch

Keywords

Runout • Travel distance • Probability of runout exceedance

Introduction

The distance that landslide debris travels down a slope from a source area is usually referred to as the landslide travel distance or runout distance (e.g. McDougall 2017). Understanding the potential runout distance of a landslide and its associated impacted area is an important component of landslide hazard and risk analysis. The impacted area of a landslide is defined as the area of ground that travelling debris may cover, which comprises the source zone, transport zone, and the deposition zone (Fig. 1). Numerous techniques exist to estimate the potential landslide debris runout and inundation area. These typically fall into two categories: (1) empirical; or (2) physics-based methods. Empirically-based methods rely on the debris inundation area of past landslides of a given type, to estimate the anticipated debris inundation area of future landslides of a similar type. They are relatively simple to apply which makes them appropriate for regional-scale studies or initial analysis for site-specific assessments. Because empirically-based methods record the actual behaviour of landslides, they are not limited by our knowledge of the initial and boundary conditions or material properties (Iverson 2014).

Physics-based methods include a range of techniques that consider the site-specific topography—usually in 2D or 3D—landslide volume, and equivalent rheological parameters of the failed landslide mass (e.g. Hungr and McDougall 2009). Such techniques need to be calibrated by back analysing a suite of data from representative landslides of varying types, from which detailed information about the landslide can be obtained and used in the calibration. Iverson

F. Guzzetti et al. (eds.), *Understanding and Reducing Landslide Disaster Risk*,
ICL Contribution to Landslide Disaster Risk Reduction,
https://doi.org/10.1007/978-3-030-60227-7_36

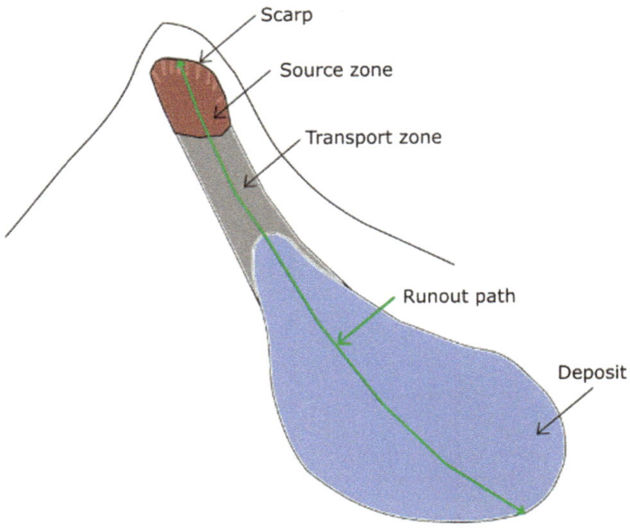

Fig. 1 Schematic representation of the source area and debris inundation area (source, transport, and deposit zones) of a landslide. Drawing modified from Mitchell et al. (2019)

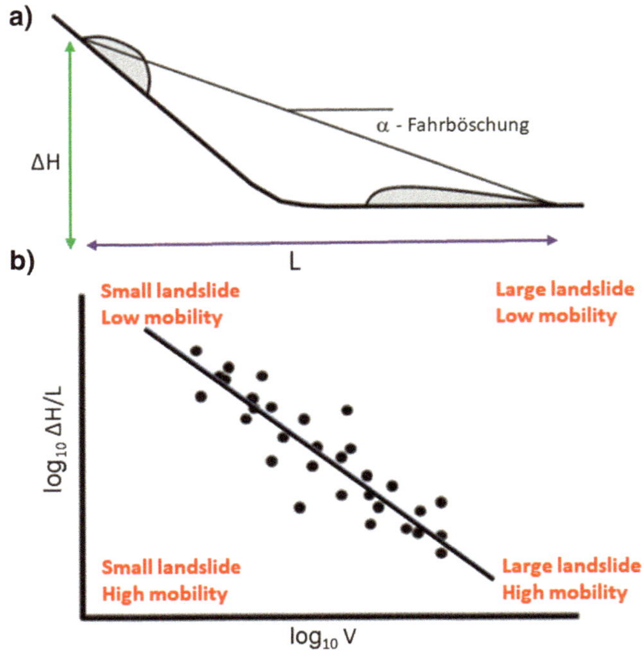

Fig. 2 **a** Schematic representation of the fall height (ΔH) and runout distance (L) parameters for a landslide and **b** the conceptual relationship between ΔH/L and the landslide volume (V) (modified from McDougall 2017)

(2014) noted that there is often not enough information available to fully constrain a physics-based debris-flow model, and as such empirical methods (i.e. back analyses of historical events) continue to play a dominant role in estimating the hazard associated with debris flows.

The present study compiles published back-analyses of international debris flows, debris avalanches, and rock avalanches and augment it with New Zealand landslides triggered by the 1929 Murchison Earthquake, 2010–2011 Christchurch Earthquake sequence, 2016 Kaikōura Earthquake, prolonged July 2017 rainfall in Wellington and the 2018 ex-tropical cyclone Gita rainfall in Kaikōura. The results from this study are presented as a series of ΔH/L versus volume plots for different landslide types, glacier versus unglaciated environment, and triggers. The plots contain superimposed probability of runout exceedance lines. Applications and limitations of empirical-probabilistic runout methods are also discussed.

Methodology

Heim (1932) proposed that the distance a landslide travels is proportional to its fall height. The tangent of the ratio of the fall height (ΔH) to horizontal runout distance (L) between the crest of the source zone and toe of the deposit, known as the "Fahrböschung" (Fig. 2a), has subsequently been correlated with landslide volume (V) (Fig. 2b).

Empirical relationships between the Fahrböschung and landslide volume have been derived for different landslide types (e.g. Corominas 1996). In this study, the landslide trigger (earthquake or rain) was used as a proxy for the initial water content of the landslide source material. This is to account that source materials with a higher water content (i.e. rain triggered) typically result in longer runout distances than source material that have lower water content (i.e. earthquake triggered). Measurement of the slope crest and deposit toe can be compiled for many different landslide types and trigger conditions to generate empirical relationships similar to the one in Fig. 2b. From these compilations, a best fit linear regression (in log–log space) between ΔH/L and V was derived for each landslide type (debris avalanche, rock avalanche, and debris flow) and trigger (seismic/dry and rainfall/wet).

Empirically-based relationships between ΔH/L and V represent the compilation of landslides with different site conditions (e.g. source geology, local topography, grain-size of failed material, water content of source material and erodible material along travel path, vegetation, buildings, the presence (or not) of mitigation measures). The variability in the landslide debris runout and inundation data also allows for the probability of runout exceedance, or limits of confidence to be defined for the prediction of runout for each landslide volume (e.g. Li 1983). For example, the best fit (orange) line in Fig. 3, represents an exceedance probability of 50%, i.e. half of the landslides of the given size and type will travel further than this line while the 10% line (yellow) represents a 10% chance that a similar landslide will travel further (McDougall 2017). Comparison between probability of exceedance function obtained by linear regression (this

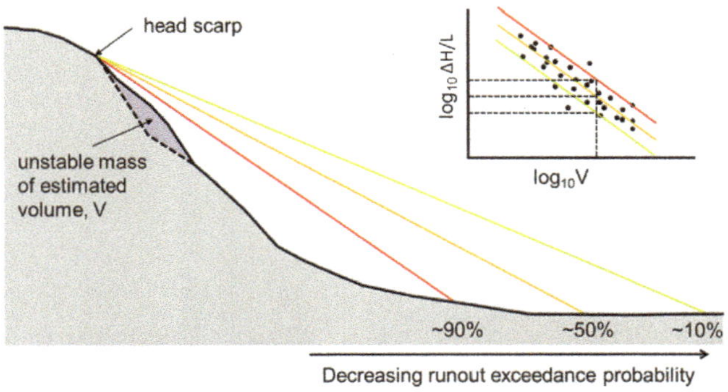

Fig. 3 Schematic representation of runout exceedance probability based on empirical relationships (modified from McDougall 2017)

study) and by survival function (Mitchell et al. 2019) were found to provide similar results.

Landslide Types Considered in This Study

Understanding what landslide type can occur at a given location is important since the different landslide types have typical behaviour (e.g. velocity of displacement, reoccurrence, travel distance) which affects how hazardous they can be. Based on the classification by Hungr et al. (2014), the rapidly moving landslide types assessed in this study are: (1) debris flows; (2) debris avalanches; and (3) rock avalanches.

Debris flows are defined by Hungr et al. (2014) as a "very rapid to extremely rapid surging flow of saturated debris in a steep channel. Strong entrainment of material and water from the flow path can occur along the runout path". The sediment concentration of debris flows exceeds 40% (typically 50–70%) and their size can be up to 1,000,000,000 m^3 (Iverson et al. 2011).

A debris avalanche is defined by Hungr et al. (2014) as a "very rapid to extremely rapid shallow flow of partially or fully saturated debris on a steep slope, without confinement in an established channel". In this study, the term debris avalanche was restricted to landslides with a volume less than 100,000 m^3 (above which landslides were considered to be a rock avalanche) and it excludes landslides from volcanic edifices (as they have been documented as having enhanced mobility; Griswold and Iverson 2014). Following Hungr et al. (2014), a debris flow refers exclusively to a channelised landslide, whereas a landslide consisting of flowing debris on an open slope is referred to as a debris avalanche.

A rock avalanche is defined by Hungr et al. (2014) as: "an extremely rapid (>5 m/s), massive, flowlike motion of fragmented rock from a large rock slide or rock fall. Large rock avalanches have a degree of mobility that exceeds the behaviour of a smaller-volume frictional flow of dry, angular, broken rock."

Results

Dry or Seismically-Triggered Debris Avalanche and Rock Avalanche

The dry debris avalanche and rock avalanche empirical relationship is intended to represent debris and rock slope failures that are associated with earthquake-triggered landslides (e.g. the New Zealand 2010–2011 Christchurch and 2016 Kaikōura events) or with large volume (>100,000 m^3) rock avalanches. In this project, we assume landslides >100,000 m^3 are rock avalanches. The rationale for considering the large volume rock avalanches as dry is that they do not represent water saturated masses in the same way to rainfall-triggered debris avalanches or debris flows. Figure 4 presents a compilation a previously published international rock avalanche case studies and mostly unpublished New Zealand examples. The bi-linear behaviour (in log–log space) of the relationship between ΔH/L ratio and volume had previously been noted by other researchers (e.g. Scheidegger 1973; Hermanns et al. 2012; Whittall 2019).

Rock Avalanche Travelling on Glacier

Rock avalanche travelling over glaciers can results in greater velocity and runout distance than similar landslides travelling over unglaciated terrain. Evans and Clague (1988, 1999) first presented ΔH/L versus V plots demonstrating the increased mobility of landslide debris as it travels over ice and incorporates snow, ice, meltwater, or wet sediment from the glacier. The compilation of rock avalanches travelling over glacier presented in Deline et al. (2014) is plotted in Fig. 5 and compared with the rock avalanches travelling over unglaciated terrain show in Fig. 4.

Wet or Rainfall-Triggered Debris Avalanches

Rainfall-triggered debris avalanches occur on open-slopes and have previously been referred to as open-slope debris flows. In this study the term debris flow is reserved for landslides travelling within a channelised path (Hungr et al. 2014). The range of source materials for the rainfall-triggered debris avalanches compiled (Fig. 6)

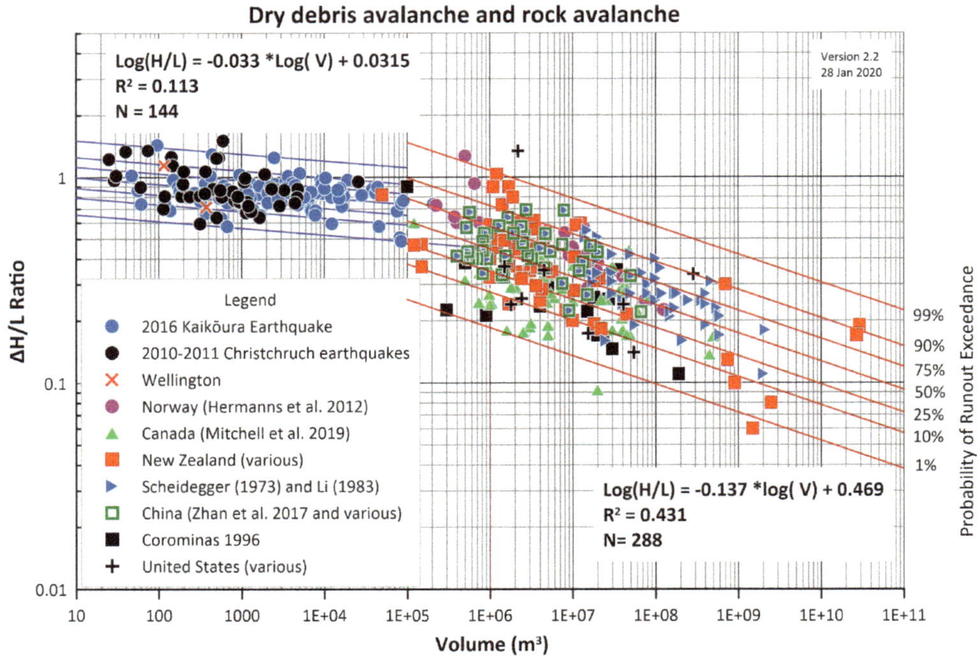

Fig. 4 Empirical relationship between ΔH/L ratio and volume of dry (i.e. earthquake-triggered) debris avalanches (<100,000 m^3) and rock avalanches (>100,000 m^3)

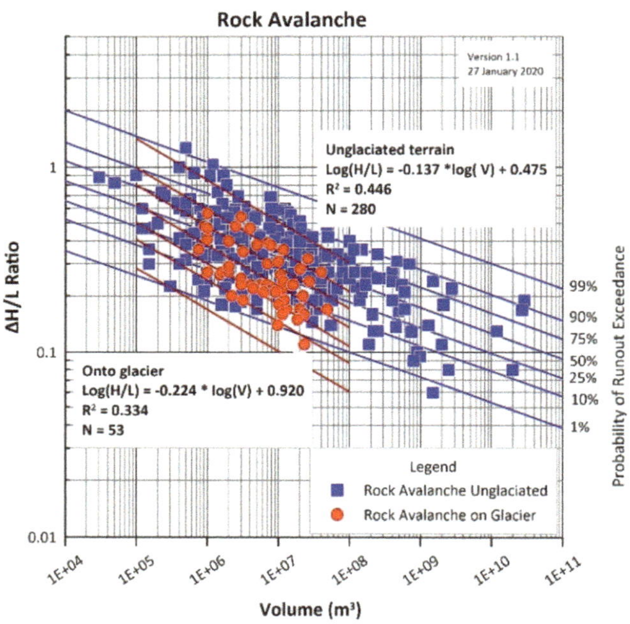

Fig. 5 Comparison of empirical relationship between ΔH/L ratio and volume for rock avalanches (>100,000 m^3) travelling over unglaciated and glaciated terrain

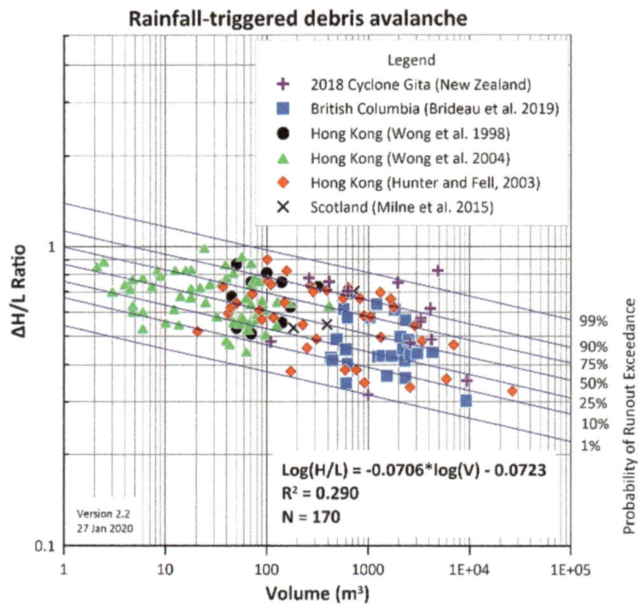

Fig. 6 Empirical relationship between ΔH/L ratio and volume for rainfall-triggered debris avalanches (open slope or non-channelised and sometimes previously referred to as open-slope debris flows)

include till (Milne et al. 2015; Brideau et al. 2019), weathered granite (Wong et al. 1998, 2004; Hunter and Fell 2003) and weathered greywacke bedrock/colluvium (Kaikōura). As rainfall-triggered debris avalanches interact with the surficial material and/or weathered bedrock along the travel path their volumes can vary, but typically they tend to be <100,000 m^3.

Wet or Rainfall-Triggered Debris Flows

The rainfall-triggered debris flows compiled in this study (Fig. 7) cover a wide range of geographical areas and source materials (till, colluvium, weathered granite, weathered greywacke). Only landslides with significant portion of their travel path being confined (e.g. gullies) were considered in this plot. Debris flows initiating from volcanic edifices (i.e.

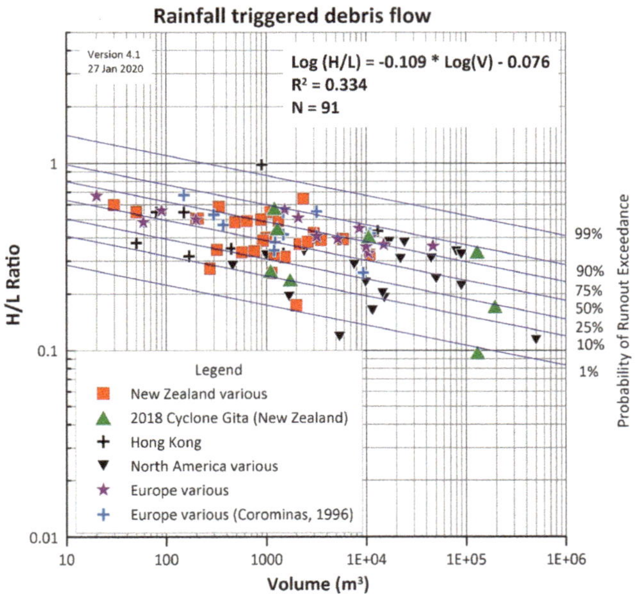

Fig. 7 Empirical relationship between ΔH/L ratio and volume for rainfall-triggered debris flows (confined or channelised)

lahars) were excluded from the current compilation as they have a documented high mobility (Griswold and Iverson 2014).

Discussion

The variability of the data in Figs. 4, 5, 6 and 7, can be attributed to a range of site-specific and analysis-specific conditions. Site-specific conditions might include vegetation (bare ground versus dense forest), topography (bends or obstacles), character of source material (texture, mineralogy, water content), and substrate (bedrock versus saturated alluvium) along the travel path. Analysis specific uncertainty are associated with different methods being used to estimate the landslide volume (visual estimate, change detection, area to volume relationship), and the age of the landslide event (i.e. fine-grained distal portion of the landslide deposit might be covered by vegetation and difficult to identify). Researchers may or may not consider distal fine-grained deposits to be part of the impacted area. In detailed landslide runout inventories where the measurements and characteristics are collected using a consistent methodology it is possible to divide the landslide runout inventory in subsets (e.g. Corominas 1996; Strom et al. 2019) or to use multiple linear regression analyses (e.g. Mitchell et al. 2019) to include the influence of travel path topography/confinement on the runout distance.

Figure 4 provides an example of how consistent methodology applied to characterising landslide volume and runout minimizes data variability by comparing 2010–2014 Christchurch Earthquake data and 2016 Kaikōura Earthquake debris avalanche data (collected by one research team) with the rest of the rock avalanche data (collected by numerous research teams).

A range of ΔH/L versus V plots with best fit linear regression have previously been published. Table 1 shows the variability between the calculated ΔH/L ratio using the different published empirical runout equations assuming a 10,000,000 m^3 rock avalanche and a ΔH of 500 m. This study and the work of Strom et al. (2019) are based on larger ($N > 200$) sample of case studies than the other works ($N < 100$).

It should be noted that the current study incorporates the published compilations of (Scheidegger 1973; Li 1983; Corominas 1996; Zhan et al. 2017) and as such likely represent an average of the site-specific conditions in each individual study. The calculated value from the linear regression proposed in this study is of 1500 m where the mean of all calculated values is 1600 m. A probability of runout exceedance value associated with a calculated runout value is one approach to capturing this variability and uncertainty. Additional work could look at using multi-linear regression where ΔH and the volume are both predictive variables for the calculated L value (e.g. Mitchell et al. 2019).

Limitation of Empirical-Probabilistic Approach

The empirical-probabilistic runout methods do not provide an estimate of the depth or velocity of the debris as it travels along the potential impacted area. Debris depth and velocity are important factors to estimate the destructiveness of the landslide and to design mitigation structures.

Empirically-based debris inundation model used in this study cannot incorporate the detailed site-specific conditions in the forward-looking estimates of the landslide impacted area.

Entrainment along the travel path is not explicitly modelled in an empirical-probabilistic runout method as the landslide volume is assumed to be constant from the start to end points. Multiple landslide volumes can be considered to assess the sensitivity of the runout result on potential material entrainment along the travel path.

Table 1 Comparison between selected empirical runout relationships of the calculated ΔH/L, Fahrböschung, and travel distance for an assumed potential 10,000,000 m^3 rock avalanche initiating 500 m above the valley floor

Source	Details	Assumed volume (Mm^3)	Calculated ΔH/L	Calculated Fahrböschung (°)	Assumed H (m)	Calculated L (m)
This study	>100,000 m^3	10	0.328	18	500	1500
Mitchell et al. 2019	Linear regression	10	0.257	14	500	1900
Mitchell et al. 2019	Multiple linear regression	10	N/A	N/A	500	1900
Strom et al. 2019	Frontally confined	10	0.354	19	500	1400
Strom et al. 2019	Laterally confined	10	0.330	18	500	1500
Strom et al. 2019	Unconfined	10	0.351	19	500	1400
Strom et al. 2019	All	10	0.347	19	500	1400
Legros 2002	Non-volcanic	10	0.319	18	500	1600
Corominas 1996	Translational slide—all	10	0.232	13	500	2200
Corominas 1996	Rock fall—all	10	0.280	16	500	1800
Li 1983	All	10	0.392	21	500	1300
Scheidegger 1973	All	10	0.337	19	500	1500

Based on the limitations outlined above, the plots presented in this study are appropriate for a regional-level assessment or preliminary site-specific assessment to identify areas warranting more detailed work. It should not be used to provide information to inform the design of landslide mitigation measures.

Conclusions

The results from this study provide landslide researchers and practitioners simple tools to conduct a forward-looking empirical-probabilistic runout analysis for debris avalanches, debris flows, and rock avalanches. The results from this study reveal that for the broadly recognized landslide runout behaviours:

- Confined landslides (debris flows) travel further than open slope landslides (debris avalanches).
- Rainfall-triggered landslides (wet debris avalanches and debris flows) have a greater mobility and travel further than earthquake-triggered landslides (dry debris avalanches) of similar volume,
- A change in mobility occurs in landslides at volumes between 100,000 and 1,000,000 m^3.
- Rock avalanches travelling over glaciers have a greater mobility than those travelling over non-glaciated terrain.

Acknowledgements The authors would like to acknowledge discussions over the years with M. Jakob, T. Davies, D. Stead, M. Jaboyedoff, R. Hermanns, J. Corominas, J. Whittall, M. Sturzenegger, and the late O. Hungr which have help shape the ideas presented in this paper.

References

Brideau M-A, Stead D, Millard T, Ward B (2019) Field characterisation and numerical modelling of debris avalanche runout on vancouver Island, British Columbia. Can Landslides 16:875–891

Corominas J (1996) The angle of reach as a mobility index for small and large landslides. Can Geotech J 33(2):260–271

Deline P, Hewitt K, Reznichenko N, Shugar D (2014) Rock avalanches onto glaciers. In: Davies T (ed.) Landslide hazards, risks, and disasters. Elsevier. pp 263–319

Evans SG, Clague JJ (1988) Catastrophic rock avalanches in glacial environments. In: Bonnard C (ed) Proceedings of the 5th international symposium on landslides, vol 2. Balkema, Rotterdam, pp 1153–1158

Evans SG, Clague JJ, (1999) Rock avalanches on glaciers in the coast and St. Elias mountains, British Columbia. In: Proceedings of the 13th annual vancouver geotechnical society symposium, vancouver, pp 115–123

Griswold JP, Iverson RM (2014) Mobility statistics and automated hazard mapping for debris flows and rock avalanches (ver. 1.1, April 2014): U.S. Geological survey scientific investigations report 2007–5276

Heim A (1932) Landslides and human lives—bergsturz und menschenleben. (trans: Skermer N). Vancouver (BC): BiTech Publishers. p 195

Hermanns RL, Oppikofer T, Anda E, Blikra LH, Bohme M, Bunkholt H, Crosta GB, Dahle H, Devoli G, Fisher L et al (2012) Recommended hazard and risk classification system for large unstable rock slopes in Norway. Nor Geol Unders Rep 2012:029

Hungr O, Leroueil S, Picarelli L (2014) The varnes classification of landslide types, an update. Landslides 11:167–194

Hungr O, McDougall S (2009) Two numerical models for landslide dynamic analysis. Comput Geosci 35:978–992

Hunter G, Fell R (2003) Travel distance angle for "rapid" landslides in constructed and natural soil slopes. Can Geotech J 40(6):1123–1141

Iverson RM (2014) Debris flows: behaviour and hazard assessment. Geol Today 30(1):15–20

Iverson RM, Reid ME, Logan M, LaHusen RG, Godt JW, Griswold JP (2011) Positive feedback and momentum growth during debris-flow entrainment of wet bed sediment. Nat Geosci 4(2):116–121

Legros F (2002) The mobility of long-runout landslides. Eng Geol 63:301–331

Li T (1983) A mathematical model for predicting the extent of a major rockfall. Z Geomorphol 27(4):473–482

McDougall S (2017) Landslide runout analysis—current practice and challenges. Can Geotech J 54(5):605–620

Milne FD, Brown MJ, Davies MCR, Cameron G (2015) Some key topographic and material controls on debris flows in Scotland. Q J Eng GeolHydrogeol 48(3–4):212–223

Mitchell A, McDougall S, Nolde N, Brideau MA, Whittall J, Aaron JB (2019) Rock avalanche runout prediction using stochastic analysis of a regional dataset. Landslides. https://doi.org/10.1007/s10346-019-01331-3

Scheidegger AE (1973) On the prediction of the reach and velocity of catastrophic landslides. Rock Mech 5(4):231–236

Strom A, Li L, Lan H (2019) Rock avalanche mobility: optimal characterization and effects of confinement. Landslides 16:1437–1452

Whittall J (2019) Runout estimates and risk-informed decision making for bench scale open pit slope failures. Can Geotech J. https://doi.org/10.1139/cgj-2018-0462

Wong HN, Ko FWY, Hui THH (2004) Assessment of landslide risk of natural hillsides in Hong Kong. Geotechnical Engineering Office. GEO Report No, Hong Kong, p 191

Wong HN, Lam KC, Ho KKS (1998) Diagnostic report on the November 1993 natural terrain landslides on Lantau Island. Geotechnical Engineering Office. GEO Report No, Hong Kong, p 69

Zhan W, Fan X, Huang R, Pei X, Xu Q, Li W (2017) Empirical prediction for travel distance of channelized rock avalanches in the Wenchuan earthquake area. Nat Hazards Earth Sys Sci 17:833–844

2020 Kyoto Japan

Rapid Sensitivity Analysis for Reducing Uncertainty in Landslide Hazard Assessments

Rex L. Baum

Abstract

One of the challenges in assessing temporal and spatial aspects of landslide hazard using process-based models is estimating model input parameters, especially in areas where limited measurements of soil and rock properties are available. In an effort to simplify and streamline parameter estimation, development of a simple, rapid approach to sensitivity analysis relies on field measurements of landslide characteristics, especially slope and depth. This method is demonstrated for a case study in Puerto Rico where widespread destruction resulted from tens of thousands of debris flows induced by Hurricanes Irma and María in Puerto Rico in 2017. The approach can be applied to estimation of shear strength as well as hydrologic parameters that control infiltration and flow of water in the subsurface and ultimately the timing of landslides resulting from heavy rainfall. Results narrow the possible range of cohesion and friction parameters as well as hydraulic conductivity and other soil water parameters by counting the fraction of field observations that can be explained by each combination of parameters. For cases studied in Puerto Rico, the method identified combinations of cohesion and friction values that explain more than 80–90% of observed landslide source areas.

Keywords

Hazard assessment • Debris flows • Puerto rico • Sensitivity analysis

Introduction

The tens of thousands of debris flows induced by Hurricanes Irma and María (2017) in Puerto Rico caused widespread destruction that raised awareness of the high hazard these rapidly moving landslides pose to people living throughout the mountainous parts of the main island (Bessette-Kirton et al. 2019a; Hughes et al. 2019). In the aftermath of Hurricane María, recent work by the U.S. Geological Survey is aimed at assessing potential for landslides and debris flows from future extreme rainfall, including hurricanes, as well as localized storms. One of the objectives of the work is to produce integrated maps of potential debris-flow initiation and inundation areas. Although much progress has been made in methods for assessing landslide susceptibility (e.g. Carrara et al. 1999; Chung and Fabri 2003; Lee et al. 2003; Canli et al. 2018) as well as debris-flow inundation (Reid et al. 2016; Aaron et al. 2017; Bessette-Kirton et al. 2019b), combining these two kinds of assessments into a single map for an area of hundreds of square kilometers remains challenging. One of the difficulties is making reasonable estimates of source area extent and depth. In addition, the diverse geology and rugged topography of Puerto Rico complicate the delineation of landslide and debris flow initiation locations, magnitude, and runout. This paper explores approaches for narrowing parameter uncertainty for models of landslide initiation to assess likely extent of potential source areas.

Study Area

Puerto Rico lies at the east end of the Greater Antilles (Fig. 1) and is characterized by rugged topography. The study areas lie in the east–west-trending Cordillera Central range, which spans most of the island. Faulted basement rocks, consisting mainly of oceanic crust, and volcaniclastic and intrusive rocks, underlie the range (Jolly et al. 1998).

R. L. Baum (✉)
Geologic Hazards Science Center, U.S. Geological Survey, M.S. 966, Box 25046 Denver, CO 80225-0046, USA
e-mail: baum@usgs.gov

F. Guzzetti et al. (eds.), *Understanding and Reducing Landslide Disaster Risk*, ICL Contribution to Landslide Disaster Risk Reduction,
https://doi.org/10.1007/978-3-030-60227-7_37

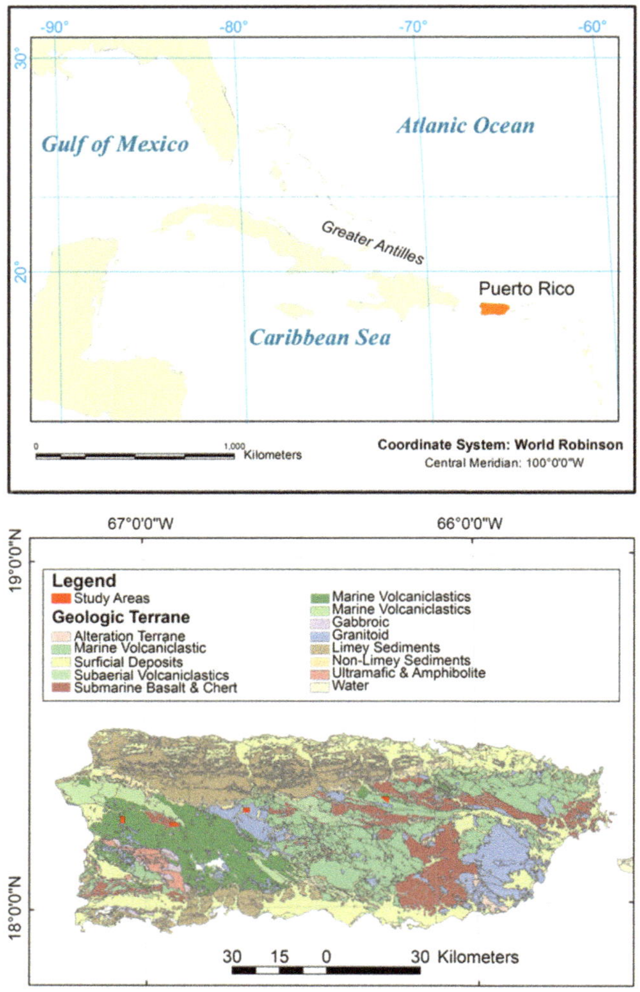

Fig. 1 Location map for Puerto Rico (above) and geologic map showing major terranes (below, Bawiec 1998)

Bawiec (1998) generalized the geology of Puerto Rico into 12 geologic terranes having similar rock types. Bessette-Kirton et al. (2019a, c) identified study areas in three geologic terranes (igneous intrusive, volcaniclastic, and submarine basalt and chert) where the highest densities of debris flows occurred. Soil mapping and databases published by the U.S. Department of Agriculture's Natural Resources Conservation Service (NRCS) indicate a range in the textures (particle-size distributions) and hydraulic properties of soils in the study areas (Soil Survey Staff 2018).

Materials and Methods

The work described in this short paper uses results from field studies, compilation of published data, and modelling studies. In recent field studies, observations and measurements of characteristics of debris-flow sources were collected in representative study areas where high debris-flow density was observed (Bessette-Kirton et al. 2017; Baum et al. 2018). Methods were developed for defining parameter distributions that could be used in a spatially distributed process-based model such as the Transient Rainfall Infiltration and Grid-Based Regional Slope-Stability Analysis (TRIGRS) program (Baum et al. 2010; Alvioli and Baum 2016). These methods included extraction of data from published geologic and soil mapping and modelling slope stability using a synthetic grid that includes a representative range of soil depth and slope angles.

Source Area Characterization

In recent field studies, observations and measurements of the size, shape, geologic materials, topographic setting, and other characteristics of debris-flow sources were collected (Baum et al. 2018, Bessette-Kirton et al. 2019c). GPS-enabled mobile devices recorded field observations and synchronized them to a cloud-based geodatabase. More than 100 shallow landslide source areas were examined and measured in the field during the weeks and months following the event. Two methods were used to obtain dimensions of landslide sources of the Hurricane María debris flows: (1) Direct measurements were made in the field using laser range finder, tape and clinometer. (2) Estimates of plan-view dimensions were made from detailed mapping based on post-María aerial photography (Bessette-Kirton et al. 2019c) matched to pre-event (U.S. Geological Survey 2017) airborne lidar.

Hydraulic and Soil Strength Parameter Compilation

Methods were developed for this study with the goal of defining parameter distributions that could be used in a spatially distributed process-based model such as TRIGRS (Baum et al. 2010; Alvioli and Baum 2016). Sampling and testing properties of materials in a statistically valid manner over a large geographic area is time consuming and costly. Therefore, considering the spatial variability of surficial deposits and weathered bedrock throughout the study area, ranges of parameters are needed that can result in similar outcomes for a set of modeled sequences of events that simulate rainfall conditions that either have or have not historically resulted in debris flows. For purposes of this study, model parameters were assigned to categories affecting either infiltration and pore pressure rise or slope stability. Parameters controlling the infiltration process vary with depth below the ground surface and include hydraulic properties of the soil: the hydraulic conductivity, K_s, saturated water content, θ_s, residual moisture content, θ_r, parameters describing the shape of the soil–water characteristic curve, including the height of capillary rise, α, as well as the rate at which water is supplied at the ground surface. Ground deformation leading to slope instability is linked to rainfall infiltration through induced pore pressure rise and a shared input parameter, depth to bedrock, where strength increases, and K_s typically decreases. Parameters controlling slope stability include depth, pore pressure, slope angle, density, and the soil strength and deformation properties. In previous studies (Godt et al. 2008; Zieher et al. 2017), incorporating the spatial distribution of soil types and properties into models of infiltration and slope stability has improved their accuracy and is likely to do so in the Puerto Rico study areas as well.

The TRIGRS program requires several input parameters to model infiltration and slope stability, as noted previously. The parameters can be assigned to zones or areas expected to have relatively homogeneous values based on geologic or soil mapping. The complex geologic history of Puerto Rico has resulted in a wide range of bedrock and soil types. Debris flows initiated in residual and colluvial soils formed on nearly every geologic unit in the Cordillera Central, but particularly in units dominated by igneous intrusive, submarine basalt and chert, and volcaniclastic rock types (Bawiec 1998; Bessette-Kirton et al. 2019a). Colluvial textural and geotechnical data were compiled from the literature (Jibson 1989; Simon et al. 1990; Larsen et al. 2004; Soil Survey Staff 2018). Following procedures similar to those outlined by Baum et al. (2019), soil units were grouped by their Unified Soil Classification System (USCS) classes and hydraulic conductivity, K_s (K_s ranges spanned 1 order of magnitude). Table 1 lists properties of dominant soil units in each target geologic terrane.

Ranges (soil classification, hydraulic conductivity) or means (θ_s-θ_r) of relevant parameters were summarized from the properties of dominant soil map units covering landslide-prone hillsides in study areas representing the three major terranes (Table 1). Properties not found in the USDA soil database (Soil Survey Staff 2018) were estimated based on texture and comparison with measured parameters. Inverse height of capillary rise, α (see Table 1), was estimated based on soil texture and descriptive statistics compiled by Carsel and Parish (1988). The ranges of soil strength parameters, cohesion, c', and angle of internal friction, ϕ', both for effective stress (Table 1), were estimated by comparing USCS texture class with average values tabulated at https://www.geotechdata.info/parameter/parameter.html (accessed 10/30/2018) and verified in textbooks (Hough 1969; Lambe and Whitman 1969; Terzaghi et al. 1996).

Infiltration and Slope Stability Model Initial and Boundary Conditions

The TRIGRS model assumes a steady background flux and initial water-table depth, d, to determine the infiltration initial condition. Infiltration boundary conditions are specified flux at the ground surface and an impermeable boundary, representing the low-permeability bedrock, at depth Z_{max}. The slope-stability initial condition is a factor of safety, FS, greater than 1 and the slope-stability boundary conditions are a stress-free ground surface and a sliding surface parallel to the ground surface at some depth, $Z \leq Z_{\max}$. A few measurements are available to constrain infiltration initial conditions (Bessette-Kirton et al. 2019a). The thin colluvial deposits drain rapidly so $d = Z_{\max}$ (initial water table at the impermeable basal boundary) is assumed.

Slope-Stability Modelling

To aid rapid interpretation of model results, the TRIGRS model was applied to synthetic grids in which each row of the slope grid has incrementally (0.5°) higher slope angle and each column of the soil depth grid has incrementally

Table 1 Estimated soil properties for colluvium in selected major bedrock types of Puerto Rico

Geologic terrane (Bawiec 1998)	USCS	Saturated hydraulic conductivity, K_s, $ms^{-1} \times 10^{-6}$	Inverse capillary rise, α, m^{-1}	Effective porosity, $\theta_s - \theta_r$	Angle of internal friction for effective stress, ϕ', degrees (°)	Cohesion for effective stress, c′, kPa
Volcaniclastic	OH	5.0–9.0	0.5–1.9	9.4–21.5	17°–35°	5–20
Submarine basalt and chert	CL, MH	0.9–7.0	0.5–3.6	22.2–38.0	27°–35°	5–20
Igneous intrusive	CL, SM	8.0–9.0	0.5–7.5	28.0–29.1	27°–41°	0–20

[Unified soil classification system (USCS) symbols, CL, low-plasticity clay; MH high-plasticity silts; SM, silty sand; OH, high-plasticity organic clay. Dual USCS symbols indicate a range of classes, rather than a borderline classification]

(0.1 m) greater depth (Fig. 2). This grid configuration made it possible to evaluate model performance throughout the range of observed source-area slopes and depths for a large number of parameter permutations using a small grid of 90 × 150 grid cells (Fig. 2). The model was first used (Stage 1) to find ranges of strength parameters that result in stability (as measured by the factor of safety, *FS*, with *FS* > 1) for dry soil conditions and instability (*FS* < 1) for positive pressure head, ψ, at various fractions *m*, of Z_{max}. Subsequently, the model was used to predict pore-pressure rise and *FS* for different combinations of hydraulic parameters and storm rainfall (Stage 2). The two model stages are intended to find likely combinations of parameters that can explain the observed pattern of debris-flow occurrence, including observed slope and depth combinations of source areas, for use in defining likely source areas for future debris flows.

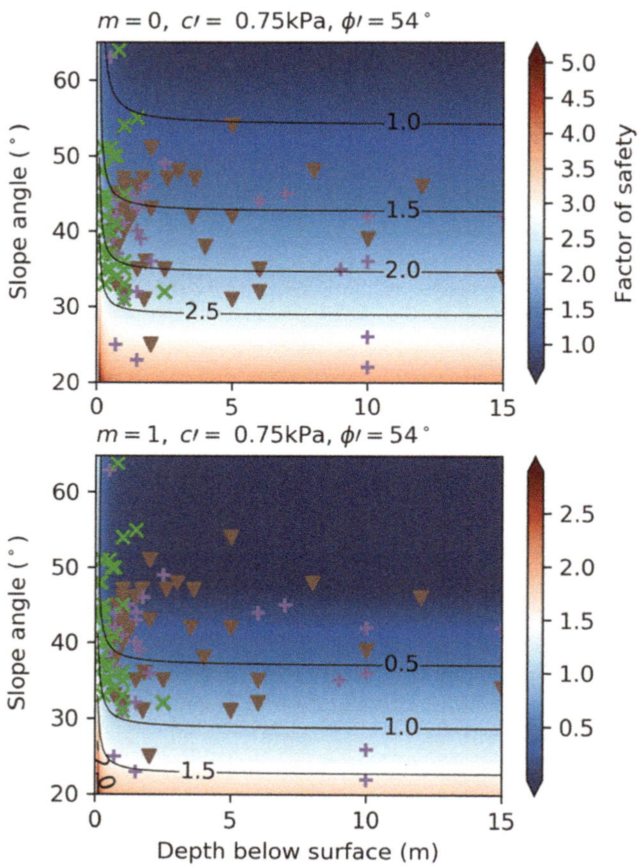

Fig. 2 Factor of safety results for selected combination of cohesion *c*′ and angle of internal friction, ϕ′, both for effective stress, and pressure head, where *m* is the ratio of pressure head to soil depth. Symbols denote observed slope angle and depth at debris-flow sources in various geologic terranes (Baum et al. 2018), green "x" granitoid intrusive; violet "+" submarine basalt; brown triangle, volcaniclastic. Factor of safety, *FS*, at slope and depth combinations observed at debris-flow sources indicates model success or failure. For the pair of *c*′ and ϕ′ values shown, *FS* > 1 for dry conditions (*m* = 0) at about 97% of sources and *FS* > 1 at 4% of sources for water table at the ground surface with flow parallel to the slope (*m* = 1). This indicates a success rate of about 93%

Results

Source-Area Characteristics

Source areas varied in size and shape among major bedrock types (Baum et al. 2018). Median source area length and width of the different terranes differ by a factor of 2, whereas lengths within a geologic terrane range over an order of magnitude from meters to tens of meters. Within study areas, depth ranges roughly over an order of magnitude from decimeters to meters (Fig. 2), and across the different terranes, median depth ranges over an order of magnitude from 0.5 to 5 m. Minimum slopes of source areas range from 26° to 37° and median slope of source areas ranges from 35° to 42°. Most source areas were fully evacuated, and shallow translational slides appear to be the most common type of movement prior to transforming to debris flows. Nevertheless, individual source area shapes are consistent with translational, rotational, or complex movement. Source areas exposed soil, saprolite, and bedrock (Baum et al. 2018). Soil matrix textures ranged from sand to clay; clast content increased with depth.

Parameter Ranges

Landslide source materials from most sites were highly weathered. In most areas underlain by intrusive igneous rocks, soils have weathered to sandy or silty clay (CH, CL). Landslide source materials from areas underlain by submarine basalt and chert are dominantly clay and silt soils (CH, MH, and CL), and may contain gravel-, cobble-, and boulder-sized rock fragments. In areas underlain by volcaniclastic sediments, the soils are likewise dominated by clayey and silty matrix and may contain gravel-, cobble-, and boulder-sized rock fragments. Table 1 lists soil classifications and estimated ranges of model parameters for modeling infiltration and slope stability in the study area. Values of K_s and θ_s-θ_r in Table 1 are reported ranges of soil map units (Soil Survey Staff 2018) composing each zone. The most common values of K_s range from 5.0×10^{-6} to 9.0×10^{-6} m/s. Ranges of α are $\bar{\alpha} \pm s_\alpha$ ($\bar{\alpha}$ is the mean and s_α is the standard deviation) tabulated by Carsel and Parish (1988) for representative soil textures. Cohesion listed in Table 1 is estimated soil cohesion. Measured values of K_s in the database fall within the ranges listed in Table 1.

Model Results

Model Stage 1 tested 1440 combinations of c' (0.0–10.0 kPa, in 0.25 kPa increments) and ϕ' (24°–60°, in 1° increments) for the synthetic terrain grid described previously. Results for a single combination of c′ and ϕ' are shown in Fig. 2. Comparing the numbers of correctly predicted depth-slope combinations observed at debris-flow source areas as explained in Fig. 2, indicates how successfully each pair of c'-ϕ' values predicts debris-flow source initiation. When aggregated for all c'-ϕ' pairs, the model predicted $FS > 1$ for dry conditions and $FS < 1$ for wet conditions for 70% or more of the observed source-area depths and slopes with about one fifth of c'-ϕ' pairs in a composite of the three terranes (Fig. 3). Calculating FS for intermediate pressure heads provided a simple test of the assumption that the water table was near the ground surface in landslides that were induced by Hurricane María. For pressure head equivalent to half the depth of landslide sources ($m = 0.5$), the best model prediction accounted for about 60% of observed source-depth combinations, a little over 80% for $m = 0.75$, and more than 80–90% for $m = 1$ (Fig. 3). Assuming that the soil requires up to 0.2 kPa of suction stress to maintain stability for dry conditions improves the $m = 0.5$ results to more than 70% and shifts the best performing ϕ' from 45°–47° to 38°–47°. Combinations that correctly predicted more than 80% of the observed values for the water table at ground-surface conditions clustered in slightly different parts of the c'-ϕ' space for each major bedrock type (Fig. 3).

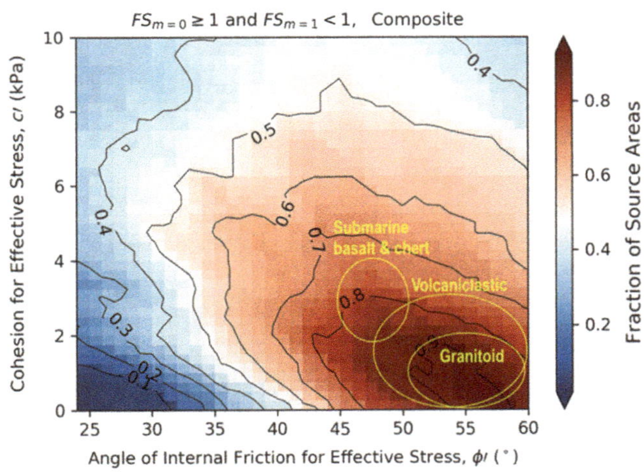

Fig. 3 Composite fraction of debris-flow sources predicted correctly as a function of cohesion, c', and angle of internal friction, ϕ', both for effective stress for the three major bedrock types in Puerto Rico. Yellow ovals show approximate ranges of most successful predictions for each type. Factor of safety (FS) for dry conditions is $FS_{m=0}$; factor of safety for water table at ground surface with slope-parallel flow is $FS_{m=1}$. Each grid cell represents the fraction $(N_{FSm=0} - N_{FSm=1})/N_t$, where $N_{FSm=0}$ is the number of source areas for $FS_{m=0} >= 1$, $N_{FSm=1}$ is the number of source areas for which $FS_{m=1} >= 1$, and N_t is the number of source areas in the bedrock type

Model Stage 2, though still preliminary, has narrowed the range of feasible hydraulic parameters from that listed in Table 1. Both c' and ϕ' were held constant ($c' = 0.75$ kPa and $\phi' = 54°$), a combination that correctly predicted more than 90 percent of observed debris-flow source depths and slopes (Fig. 2), while K_s and α varied. Values listed in Table 1 and the soil map databases (Soil Survey Staff 2018) for the bedrock formations having the greatest density of debris-flow sources (Bessette-Kirton et al. 2019a) guided selection of ranges for testing K_s (0.9×10^{-6}–8.0×10^{-5} m/s, in uneven increments) and α (1.5, 3.8, 7.5, and 11.2 m^{-1}). For the model, heavy rainfall of varying intensity associated with occurrence of the debris flows was represented by periods of steady rainfall of 27 mmh^{-1} over 15 h, 19 mmh^{-1} over 12 h, and 14 mmh^{-1} over 18 h to span the range of intensities recorded by U.S. Geological Survey gages near the study areas. The model was tested using combinations of K_s and α to explore their limits in predicting debris flow initiation for the storm. Model results showed that combined high values of K_s ($>7.0 \times 10^{-5}$ m/s) and α (>7.5 m^{-1}) drained too freely to produce sufficient rise in pressure head to cause instability, whereas reducing α to 3.8 m^{-1} produced instability on slopes as low as 28° in a narrow range centered on 0.5 m depth. Combinations of low values of K_s (0.9×10^{-6} m/s) and α (1.5 m^{-1}) produced considerable pore pressure rise and instability at depths <2 m but could not predict instability at depths of 3–5 m. Intermediate values, of K_s (9.0–28 $\times 10^{-6}$ m/s) and α (1.5–3.8 m^{-1}) produced the most successful predictions of slope instability of the observed debris-flow source depth-slope combinations (Fig. 4).

Discussion and Conclusion

Debris flows occurred over a broad range of elevation and geology. Differences between the characteristics of debris-flow sources in the different study areas seem consistent with weathering products of their respective lithologies (basalt, granodiorite, and volcaniclastic rocks) and depth of weathering. Many debris-flow sources were near roads and houses; however, sources on modified slopes were left in the input database to ensure that the results represent the majority of source areas. The observed variability indicates that successful modeling of potential debris-flow source areas for assessing future hazard must account for variable movement depths and mechanisms, source-material physical properties, topography, land disturbance related to agriculture as well as construction of roads and houses, drainage patterns, and rapid propagation of rainfall-induced pore pressure to 5–10 m depth.

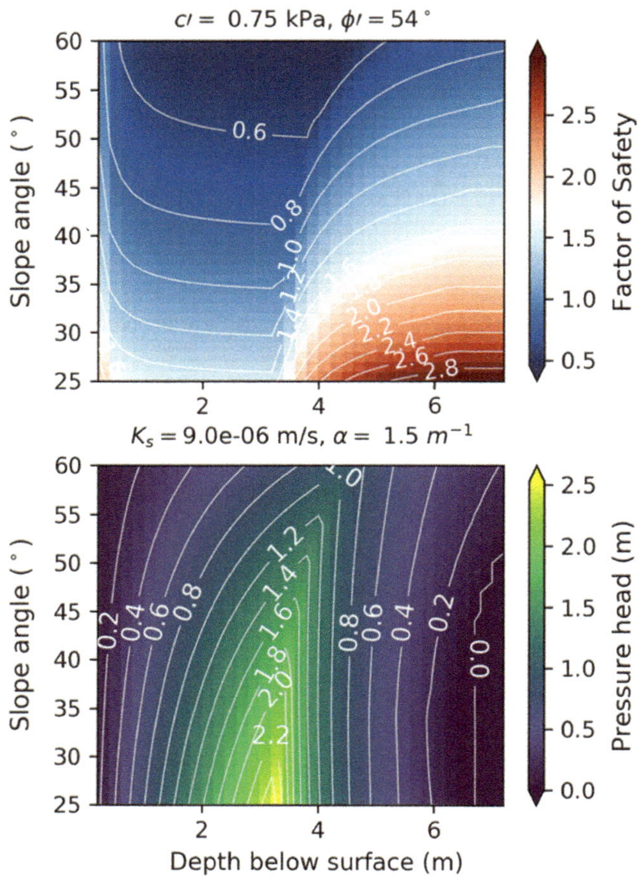

Fig. 4 Factor of safety (top) and pressure head (bottom) for slopes and depths typical of debris-flow sources in Puerto Rico. Cohesion, c', and angle of internal friction, ϕ', both for effective stress, are the same as for Fig. 2. Steady rainfall of 27 mmh^{-1} over 15 h was simulated in TRIGRS to obtain the pressure-head distribution, using saturated hydraulic conductivity, K_s of 9.0×10^{-6} and inverse capillary rise, α of 1.5 m^{-1}

Modeled friction angles (>45°) of the soil at the igneous intrusive sites seem unrealistically high and might result from a combination of several possibilities including that (1) suction stress contributes to their stability during dry conditions; (2) the field measurements underestimate the actual depths of the landslides; (3) the failure envelope of these soils is steeply curved at very low normal stress (consistent with lab tests, 40°–54°, Likos et al. 2010); or (4) a three-dimensional slope-stability model would more accurately represent the geometry and failure process at these source areas.

The proposed approach, which uses a synthetic grid, identified the range of parameters that explains the greatest number of debris-flow source areas. These results may be applied to map potential debris-flow source areas. This method is applicable to shallow landslide forecasting in other areas with similar initiation mechanisms (Baum et al. 2019). Results of this study show that this kind of sensitivity analysis can be used to inform parameter estimates for modelling.

Acknowledgements Adrian Lewis (USGS) compiled soil texture and geotechnical data for this study. This paper benefited from constructive reviews by Dennis Staley (USGS) and an anonymous reviewer. Any use of trade, firm, or product names is for descriptive purposes only and does not imply endorsement by the U.S. Government.

References

Aaron J, McDougall S, Moore JR, Coe JA, Hungr O (2017) The role of initial coherence and path materials in the dynamics of three rock avalanche case histories: geoenvironmental. Disasters 4(5):1–15. https://doi.org/10.1186/s40677-017-0070-4

Alvioli M, Baum RL (2016) Parallelization of the TRIGRS model for rainfall-induced landslides using the message passing interface. Environ Model Softw 81:122–135. https://doi.org/10.1016/j.envsoft.2016.04.002

Baum RL, Godt JW, Savage WZ (2010) Estimating the timing and location of shallow rainfall-induced landslides using a model for transient, unsaturated infiltration. J Geophys Res, Earth Surf. 115 (F03013). https://doi.org/10.1029/2009JF001321

Baum RL, Cerovski-Darriau C, Schulz WH, Bessette-Kirton E, Coe JA, Smith JB, Smoczyk GM (2018) Variability of hurricane María debris-flow source areas in Puerto rico—implications for hazard assessment, abstract NH14A-02 presented at 2018 fall meeting, AGU, Washington, DC, 10–14 Dec., https://abstractsearch.agu.org/meetings/2018/FM/NH14A-02.html

Baum RL, Scheevel CR, Jones ES (2019) Constraining parameter uncertainty in modeling debris-flow initiation during the September 2013 Colorado Front Range storm. In: Kean JW, Coe JA, Santi PM, Guillen BK (eds) Association of environmental and engineering geologists special publication vol. 28. pp 249–256. https://doi.org/10.25676/11124/173212

Bawiec WJ (1998) Geologic terranes of Puerto Rico, In: Bawiec WJ (ed) Geology, geochemistry, geophysics, mineral occurrences and mineral resource assessment for the commonwealth of Puerto Rico: U.S. Geological Survey Open-File Report 98–38. https://doi.org/10.3133/ofr9838

Bessette-Kirton EK, Coe JA, Godt JW, Kean JW, Rengers FK, Schulz WH, Baum RL, Jones ES, Staley DM (2017) Map data showing concentration of landslides caused by hurricane María in Puerto Rico: U.S. Geological Survey data release. https://doi.org/10.5066/F7JD4VRF

Bessette-Kirton EK, Cerovski-Darriau C, Schulz WH, Coe JA, Kean JW, Godt JW, Thomas MA, Hughes KS (2019a) Landslides triggered by Hurricane María: assessment of an extreme event in Puerto Rico. GSA Today 29:4–10

Bessette-Kirton EK, Kean JW, Coe JA, Rengers FK, Staley DM (2019b) An evaluation of debris-flow runout model accuracy and complexity in Montecito, California: Towards a framework for regional inundation-hazard forecasting. In: Kean JW, Coe JA, Santi PM, and Guillen BK (eds) Association of environmental and engineering geologists special publication vol 28. pp 257–264. https://mountainscholar.org/handle/11124/173051

Bessette-Kirton EK, Coe JA, Kelly MA, Cerovski-Darriau C, Schulz WH (2019c) Map data from landslides triggered by Hurricane María in four study areas of Puerto Rico: U.S. Geological Survey data release. https://doi.org/10.5066/P9OW4SLX

Canli E, Mergili M, Thiebes B, Glade T (2018) Probabilistic landslide ensemble prediction systems: lessons to be learned from hydrology: nat. Hazards Earth Syst Sci 18:2183–2202. https://doi.org/10.5194/nhess-18-2183-2018

Carrara A, Guzzetti F, Cardinali M, Reichenbach P (1999) Use of GIS technology in the prediction and monitoring of landslide hazard.

Nat Hazards 20:117–135. https://doi.org/10.1023/A:1008097111310

Carsel RF, Parish RS (1988) Developing joint probability distributions of soil water retention characteristics. Water Resour Res 24(5):769–775

Chung CF, Fabbri AG (2003) Validation of spatial prediction models for landslide hazard mapping. Nat Hazards 30:451–472. https://doi.org/10.1023/B:NHAZ.0000007172.62651.2b

Godt JW, Schulz WH, Baum RL, Savage WZ (2008) Modeling rainfall conditions for shallow landsliding in Seattle, Washington. In: Baum RL, Godt JW, Highland LM (eds) Geological society of America reviews in engineering geology vol. 20. pp 137–152. https://doi.org/10.1130/2008.4020(08)

Hough BK (1969) Basic soils engineering, 2nd edn. The Ronald Press Co, New York, p 634

Hughes KS, Bayouth-García D, Martínez-Milian GO, Schulz WH, Baum RL (2019) Map of slope-failure locations in Puerto Rico after Hurricane María: U.S. geological survey data release. https://doi.org/10.5066/P9BVMD74. Accessed 12 September 2019

Jibson RW (1989) Debris flows in southern Puerto Rico. Geol Soc Am Spec Pap 236:29–55

Jolly WT, Lidiak EG, Dickin AP, Wu Tsai-Way (1998) Geochemical diversity of Mesozoic island arc tectonic blocks in eastern Puerto Rico. In: Likiak EG, Larue DK (eds) Tectonics and geochemistry of the Northeastern caribbean: geological society of America special paper vol. 322. p 67–98. https://doi.org/10.1130/0-8137-2322-1.67

Larsen MC, Santiago M, Jibson R, Questell E (2004) Map showing susceptibility to rainfall-triggered landslides in the municipality of Ponce, Puerto Rico. U.S. geological survey scientific investigations map I-2818, 1:30,000, 1 sheet

Lambe TW, Whitman RV (1969) Soil mechanics. Wiley, New York, p 553

Lee S, Ryu J-H, Min K, Won J-S (2003) Landslide susceptibility analysis using GIS and artificial neural network. Earth Surf Process Landforms 28(12):1361–1376

Likos WJ, Wayllace A, Godt J, Lu N (2010) Modified direct shear apparatus for unsaturated sands at low suction and stress. Geotech Test J 33(4):286–298. https://doi.org/10.1520/GTJ102927

Reid ME, Coe JA, Brien DL (2016) Forecasting inundation from debris flows that grow volumetrically during travel, with application to the oregon coast range. USA Geomorphol 273:396–411. https://doi.org/10.1016/j.geomorph.2016.07.039

Simon A, Larsen MC, Hupp CR (1990) The role of soil processes in determining mechanisms of slope failure and hillslope development in a humid-tropical forest: eastern Puerto Rico. In: Kneuper PLK, McFadden LD (eds) Geomorphology, vol. 3. pp 263–286

Terzaghi K, Peck RB, Mesri G (1996) Soil mechanics in engineering practice, 3rd edn. Wiley, New York, p 549

Soil Survey Staff (2018) U.S. department of agriculture natural resources conservation service, 2018, soil survey geographic (SSURGO) database for Puerto Rico, all regions: https://websoilsurvey.sc.egov.usda.gov/App/WebSoilSurvey.aspx. Accessed 12 Sep 2019

U.S. Geological Survey (2017) 2015–2016 USGS Puerto Rico LiDAR: ftp://rockyftp.cr.usgs.gov/vdelivery/Datasets/Staged/Elevation/OPR/PR_PuertoRico_2015/. Accessed 12 Sep 2019

Zieher T, Rutzinger M, Schneider-Muntau B, Perzl F, Leidinger D, Formayer H, Geitner C (2017) Sensitivity analysis and calibration of a dynamic physically based slope stability model: Nat. Hazards Earth Syst Sci 17:971–992. https://doi.org/10.5194/nhess-17-971-2017

Applying Debris Flow Simulation for Detailed Hazard and Risk Mapping

2020 Kyoto Japan

Kana Nakatani, Yuji Hasegawa, and Yoshifumi Satofuka

Abstract

For houses situated in mountainous regions, evacuation routes during sediment disasters are limited. Thus, in the event of heavy rainfall, it may be difficult for inhabitants to evacuate from disaster areas. Therefore, it is important to understand the risk distribution before debris flow disasters in order to determine safe evacuation planning. Debris flow simulations are useful for the determination of hazard and risk mapping. When considering debris flow behavior and conducting debris flow simulations, there are two critical factors with respect to the debris flow flooding and deposition processes, namely, the scale of debris flow and landform conditions. The focus of this study was on various resolution digital elevation model (DEM) landform data in Japan. Simulations were performed for a residential area in Hiroshima, which was subject to debris flow events in 2014. Based on a comparison of the simulation results obtained using DEM data with a mesh resolution of 5 m, as sourced from the Geospatial Information Authority of Japan, and the results using high-resolution light detection and ranging (LiDAR) DEM data, the houses can be considered to describe the influence area and high risk region situated near the valley exit, in addition to locally dangerous areas and movement of debris flows on roads. To achieve a detailed hazard and risk mapping and determine safe evacuation routes and shelters, the application of high-resolution LiDAR DEM data and consideration of houses is critical.

Keywords

Debris flow • Simulation • Hazard and risk mapping • Digital elevation model (DEM)

Introduction

In Japan, there are numerous houses in the alluvial fan region near the mountain side, several of which are on the path of high-risk debris flows. In Hiroshima prefecture, which is located in the western side of Japan, many sediment disasters have occurred due to the weathered geological conditions as well as due to land-use, as many people are living in the mountainous areas. Furthermore, sediment disasters including 107 debris flows and 59 slope failures caused by torrential rains in Hiroshima City on August 20, 2014, resulted in severe damage to property and 74 fatalities. At Miiri observation station (Meteorological Agency), Asakita-ku Hiroshima City, 217.5 mm of rainfall was observed for three hours from AM 1:30, which represented the highest recorded value since observation began in 1976.

Debris flow simulations are critical for extensive hazard and risk mapping. When considering debris flow behavior and conducting debris flow simulations, there are two critical factors with respect to the debris flow flooding and deposition processes, namely, the scale of debris flow and landform conditions. The focus of this study was on the digital elevation model (DEM) landform data in Japan with different resolutions. A debris flow simulation was conducted in Hiroshima in 2014. The DEM settings that can most accurately describe the disaster case and fully realize hazard and risk mapping for safe evacuation planning before the disaster was considered in this study.

K. Nakatani (✉)
School of Agriculture, Kyoto University, Kitashirakawa, Oiwake-cho, Sakyo-ku, Kyoto, 6068502, Japan
e-mail: kana2151@kais.kyoto-u.ac.jp

Y. Hasegawa
School of Integrated Arts and Sciences, Hiroshima University, Kagami-yama 1-7-1, Higashi-Hiroshima, 7398521, Japan
e-mail: y-hasegawa@hiroshima-u.ac.jp

Y. Satofuka
Department of Civil Engineering, Ritsumeikan University, Noji-higashi 1-1-1, Kusatu, 5258577, Japan
e-mail: satofuka@se.ritsumei.ac.jp

F. Guzzetti et al. (eds.), *Understanding and Reducing Landslide Disaster Risk*,
ICL Contribution to Landslide Disaster Risk Reduction,
https://doi.org/10.1007/978-3-030-60227-7_38

Simulation Target and Methods

Debris Flow Occurred in 2014 Hiroshima

The simulation target was the Hiroshima sediment disaster on August 20, 2014 due to heavy rain (Nakatani et al. 2017). Most of the torrential debris flows that occurred were in places designated as high-risk debris flow regions by the local government before the 2014 disaster. In the mountainous housing area, evacuation routes are limited; thus, when heavy rainfall occurs, it may be difficult for inhabitants to move outside the designated area. It is therefore important to understand the risk distribution within the designated area before the disasters occur, in order to determine safe evacuation planning.

In this study, the target region was the Yagi Midorigaoka prefectural housing area in Asaminami-ku, Hiroshima City, which experienced severe damage due to debris flows. The torrent was designated as high-risk debris flow region after the 2014 disaster. The basin area was approximately 0.2 km^2 from the upstream of the valley exit. The target was located in a granite area and weathered geological conditions. Due to the debris flow, some of the deposited sediments were reported as accretionary wedge. Most of the downstream deposition sediment was distributed in 0.1–1.0 m diameter, and the velocity of the debris flow was not reported.

From the downstream of the valley exit, where prefectural houses are situated, there are roads in the north–south and east–west directions from the valley. From the upstream of the valley exit, new houses were constructed prior to the 2014 disaster. Moreover, many buildings were damaged including 3 buildings, situated near the valley exit, which were completely destroyed, and 27 people were killed.

Fig. 1 Apartments in prefecture after 2014 disaster (image captured by PASCO CO.)

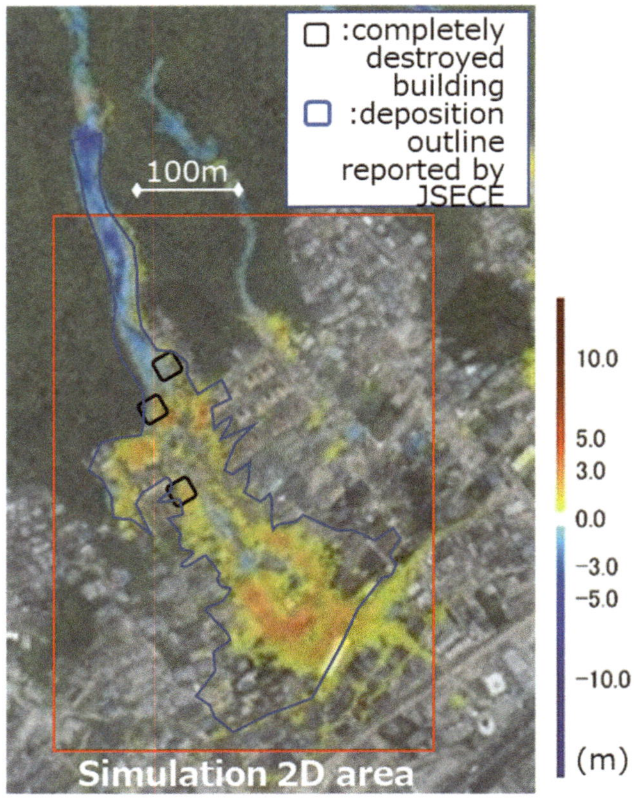

Fig. 2 Debris flow deposition in prefectural apartments (added to laser profile difference map before and after disaster, as obtained from Yoshino disaster reports, 2014)

Figure 1 presents an aerial image of the target area captured by PASCO CO., which shows the three buildings. Figure 2 presents the debris flow erosion/deposition distribution from a laser profile (LP) difference map before and after the disaster, along with the deposition outline in the residential area from the disaster survey conducted by the Japan Society of Erosion and Control Engineering (JSECE) (Kaibori et al. 2014; Yoshino 2014). Figure 2 also presents the two-dimensional (2D) simulation target area.

Debris Flow Simulation System

The authors used a geographic information system (GIS)-related Hyper KANAKO system (Nakatani et al. 2012) for the debris flow simulations. In the Hyper KANAKO system, the numerical simulation of the debris flow was based on the model presented by Takahashi et al. (2001), Takahashi (2007), considering the erosion/deposition due to equilibrium concentrations. These models also included equations of momentum, continuation, riverbed deformation, erosion/deposition, and riverbed shearing stress. Moreover, an integrated model (Wada et al. 2008) was developed and implemented. This model incorporated the influences on one-dimensional (1D) simulation areas such as the steep mountainous valley area, and 2D simulation areas, such as the alluvial fans. The fundamental 2D debris flow equations are shown below. The system applied the same equations used in 1D debris flow simulations, in addition to y-axis directional terms. The effect of sabo or erosion control dams in the 1D area was simulated based on a model developed by Satofuka and Mizuyama (2005).

The continuity equation for the total volume of the debris flow can be expressed as follows:

$$\frac{\partial h}{\partial t} + \frac{\partial uh}{\partial x} + \frac{\partial vh}{\partial y} = s_T \tag{1}$$

The continuity equation for determining the debris flow of particles can be expressed as follows:

$$\frac{\partial Ch}{\partial t} + \frac{\partial Chu}{\partial x} + \frac{\partial Chv}{\partial y} = s_T C* \tag{2}$$

The x-axis flow (main flow direction) is given by the following momentum equation:

$$\frac{\partial u}{\partial t} + u\frac{\partial u}{\partial x} + v\frac{\partial u}{\partial y} = -g\frac{\partial H}{\partial x} - \frac{\tau_x}{\rho h} \tag{3}$$

The y-axis flow (cross flow direction) can be expressed by a momentum equation, as follows:

$$\frac{\partial v}{\partial t} + u\frac{\partial v}{\partial x} + v\frac{\partial u}{\partial y} = -g\frac{\partial H}{\partial x} - \frac{\tau_y}{\rho h} \tag{4}$$

The equation for determining the change in the bed surface elevation can be expressed as follows:

$$\frac{\partial z}{\partial t} + s_T = 0 \tag{5}$$

In Eqs. 1–5, h is the flow depth(m), u is the x-axis flow velocity(m/s), v is the y-axis direction flow velocity(m/s), C is the volumetric sediment concentration in the debris flow, z is the bed elevation, t is the time(s), s_T is the erosion/deposition velocity, g is the acceleration due to gravity(m/s^2), H is the flow elevation H = h + z (m), ρ is the mass density of fluid phase in debris flow(kg/m^3), C* is the sediment concentration with respect to volume in the movable bed layer, and τ_x and τ_y are the riverbed shearing stresses in the x- and y-axis directions, respectively.

Landform Data Settings

In this study, in the DEM, the ground elevation was obtained, and the heights of the buildings and trees were excluded.

For the simulation conditions, the upstream torrent where debris flow occurred and the developed area were considered for the 1D simulation, and the downstream alluvial fan residential area was considered for the 2D simulation, as shown in Fig. 3. In Fig. 3, the area for the debris flow hazard, as designated by local government after the 2014 disaster, is shown.

Four sets of landform data conditions were employed, as shown in Table 1. For cases 1–2, the DEM data provided by the Geospatial Information Authority of Japan (GSI) have a mesh resolution of 5 m. The DEM data at mesh resolutions of 5 m and 10 m for all over Japan are available on the GSI website. The DEM data from GSI do not describe land-use

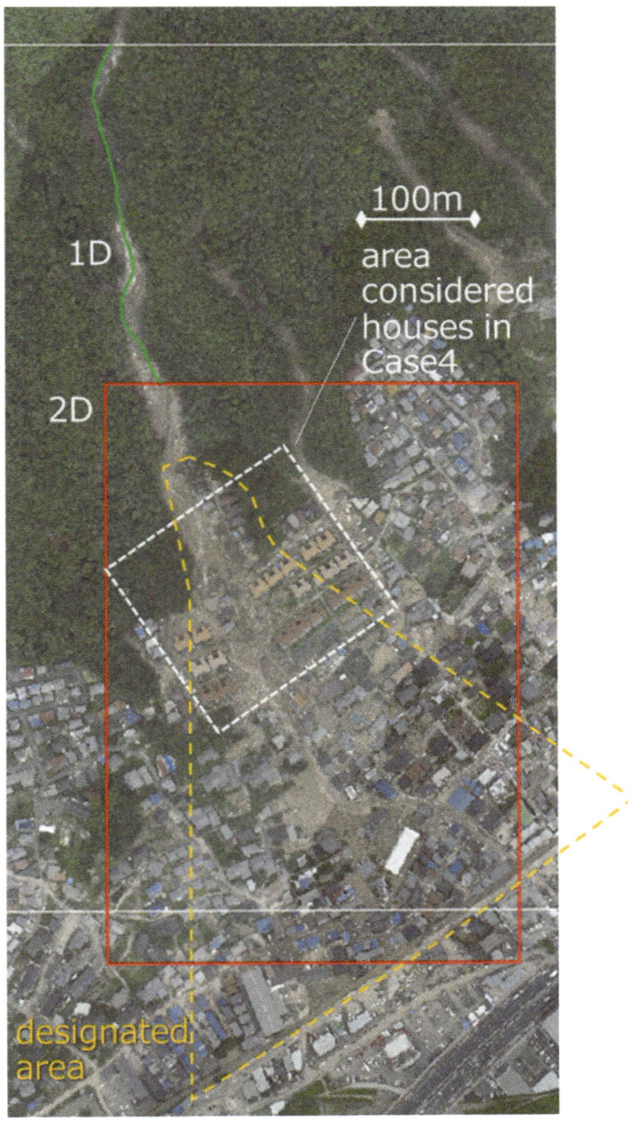

Fig. 3 Simulation target area of prefectural apartments (image obtained from Otagawa River Office, and the hazard area was designated after the 2014 disaster)

Table 1 Landform settings for simulations

Case	DEM data source, resolution	Considering houses height	2D sim. mesh size
1	GSI, 5 m	without	5 m
2	GSI, 5 m	consider	5 m
3	LiDAR, 1 m	without but house foundation considered	2 m
4	LiDAR, 1 m	consider in limited area shown in Fig. 3	2 m

such as housing in alluvial fan areas; however, users can obtain the contours of buildings from the GSI website. In Case 1, a DEM mesh resolution of 5 m was employed; and in Case 2, a building height of 6 m was employed, considering a two-story house in the ground elevation, using contour of buildings from GSI. The 2D simulation mesh resolution for Cases 1–2 was 5 m.

For Cases 3–4, DEM data with a mesh resolution of 1 m, as obtained from light detection and ranging (LiDAR) before the 2014 disaster, were employed. The data were provided by the Ministry of Land, Infrastructure and Transport, Japan (MLIT), Chugoku Regional Development Bureau, and Otagawa River Office. The high-resolution DEM data from LiDAR described the building foundation in the residential area.

For the 1D area, the LiDAR DEM data were used for all cases. Moreover, the same 1D landform data were employed to determine the effect of landform and land-use in the 2D area. In the 1-D area, the total length was 420 m from the upstream, and the average slope was approximately 18°. The river width was set as 10 m, and the interval of simulation points was set as 5 m.

Other Simulation Settings

Based on a previous study (Nakatani et al. 2017), the debris flow disaster in the Midorigaoka prefectural apartments area in 2014 was considered. Water was supplied from the 1D upstream end in the trapezoid-shape hydrograph with a peak discharge rate of 115 m^3/s in a continuous time of 350 s (peak continuous time of 250 s). Moreover, 33,000 m^3 of sediment included a void from disaster reports as the movable bed layer in the 1D area with uniform thickness. The case where debris flow and development occurred due to the erosion process in the 1D torrent area was considered.

A sediment diameter of 0.2 m from the disaster reports and field survey was applied. The erosion and deposition coefficients were set as 0.007 and 0.05, respectively, which are typical values for the debris flow simulations in Japan (Takahashi 2007). The other simulation parameters are shown in Table 2.

Simulation Results

Figures 4 and 5 present the 2D-area simulation results. Figure 4 presents the deposition thickness after the simulation, and Fig. 5 presents the maximum height (flow depth + deposition thickness) during the simulation. For Cases 1–2, the GSI DEM data with a mesh resolution of 5 m were employed. For Cases 3–4, the LiDAR DEM data with a mesh resolution of 1×1 m^2 were employed, which exhibited a similar trend with respect to deposition thickness. With focus on the details around areas of the houses and destroyed houses, there were differences between Cases 1 and 2 and between Cases 3 and 4. In Cases 2 and 4, with a comparison of the regions without houses, the deposition and maximum flow were spread around the houses due to the interruption of the straight movement of the flow by the houses. In particular, the results for Case 4 revealed that the debris flowed toward the roads in the residential area when compared with the other cases. In all the DEM data, roads were described, as they indicate the ground surface, and in the high-resolution DEM data, the road elevation could be described in detail. In residential areas, roads are generally situated between the houses. In the cases where houses were not considered, no significant difference was observed between the ground surface elevation at the roads and that at the houses. However, in Case 4, the heights of the houses were added to the ground elevation; thus, the roads exhibited relatively low elevations in the residential area, and the debris flow along the road was described.

A discussion on the deposition thickness distribution, as shown in Fig. 4, is presented below, in addition to a comparison with the disaster case. In the upstream of the 2D area where the first destroyed house (Area 1) was situated, local deposits larger than 300–500 cm^2 were observed in Cases 3 and 4, respectively. At the site where the second house was destroyed (Area 2), local deposits larger than 300 cm were observed in Case 4. Moreover, around Area 2, larger deposits than in other areas in Case 4 were observed; thus, the simulation results revealed that Area 2 was high-risk in comparison with the other areas. In Case 1, the areas with deposits larger than 100 cm were only at the upstream side

Table 2 Simulation parameters

Parameter	Value	Unit
Simulation time	600	s
Time step	0.01	s
Mass density of sediment	2650	kg/m^3
Mass density of fluid phase in debris flow	1000	kg/m^3
Concentration of movable bed	0.65	
Internal friction angle	35	deg
Acceleration of gravity	9.8	m/s^2
Manning's coefficient	0.03	s/m$^{1/3}$

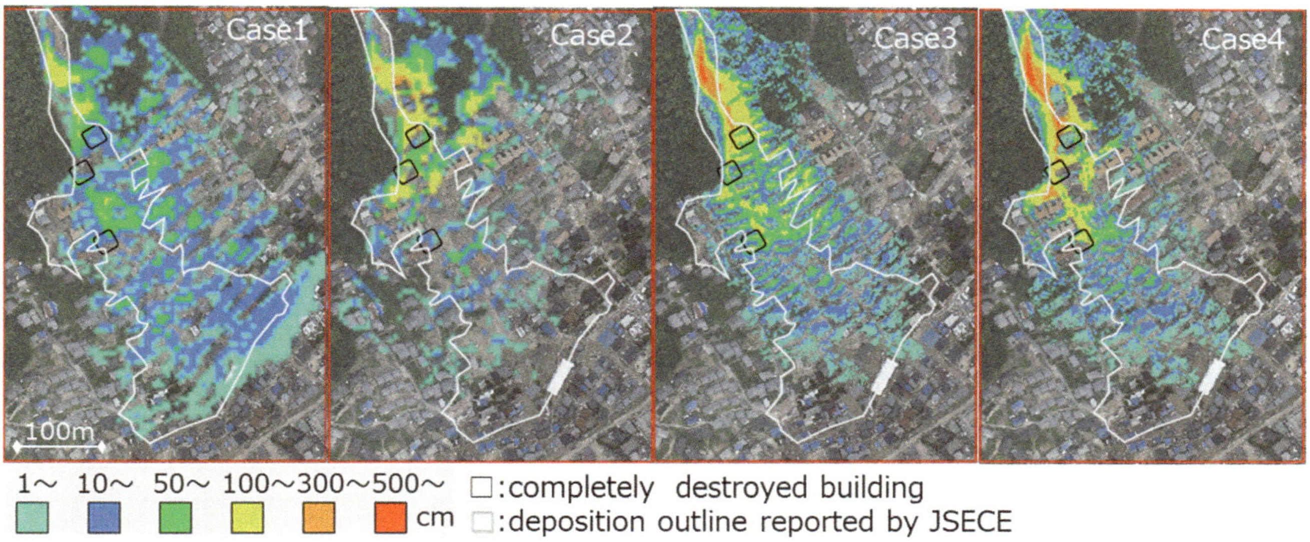

Fig. 4 Debris flow simulation results, and deposits at simulation end in prefectural apartments area

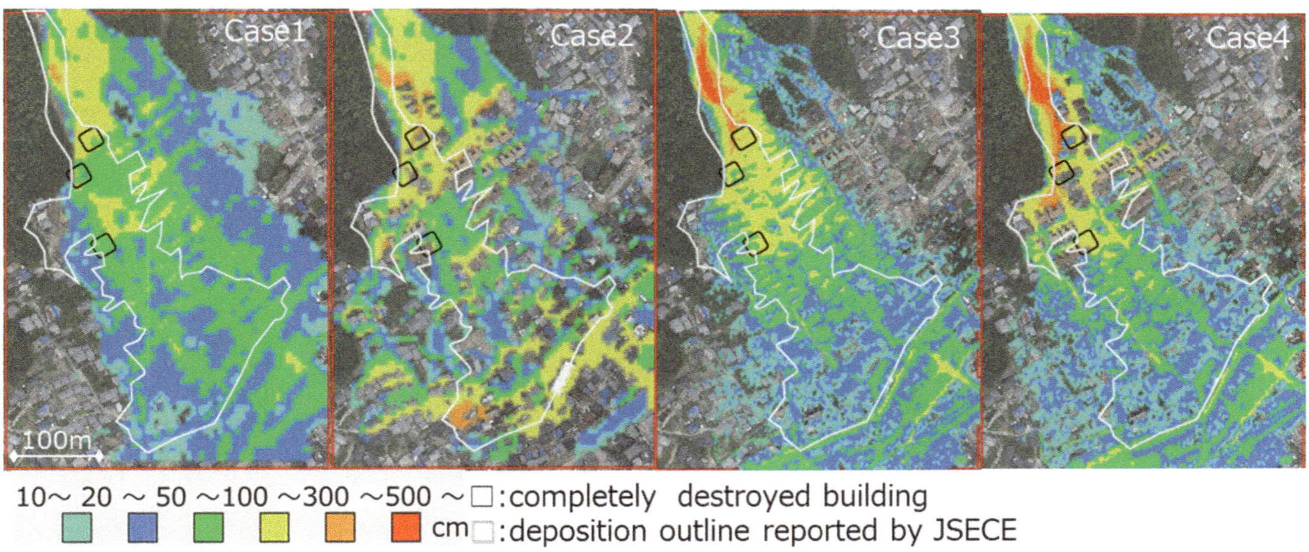

Fig. 5 Debris flow simulation results and maximum height (flow depth + deposition thickness) during the simulation at the simulation end in prefectural apartments area

of the 2D area, and no deposits larger than 100 cm were observed on the downstream side of Area 2. However, deposits larger than 10 cm were mostly spread toward the downstream side in all cases. In Case 2, Areas 1 and 2 of the destroyed buildings exhibited deposits larger than 100 cm; however, at Area 2 downstream, the deposits were small. In Cases 3 and 4, in the downstream of Area 2, the size of deposits decreased in accordance with the distance from the upstream side. Moreover, for the destroyed building at the furthest site downstream (Area 3), with respect to the outline of the deposits reported from JSECE, Cases 3 and 4 were in good agreement, and with respect to deposits larger than 10 cm, Case 4 was in good agreement. With respect to Areas 2 and 3, Case 4 exhibited deposits larger than 100 cm along the two houses. In the disaster case, debris flowed down toward the road and caused severe damage in Area 3. Therefore, the Case 4 simulation results accurately describe the disaster case.

Based on the simulation results, the high-risk area near the valley exit can be described in Cases 2–4. Case 2 was in better agreement with the case at the upstream side of the 2D area and around Area 1 than Case 1. However, with respect to local high-risk areas of destroyed buildings and the deposition distribution with the high-resolution LiDAR DEM data, Cases 3 and 4 yielded better results. Moreover, in Case 4, under the consideration of the house heights, the damage of Area 3 by debris flow toward the road was most consistent with the disaster conditions. Thus, when considering evacuation planning such as routes and shelters with a detailed risk distribution, Case 4 will be useful.

In the target area, based on the disaster case and Case 4 results, the damage was considerable around the road along the debris flow torrent. Around the road, deposition was larger and building damage occurred; thus, it was a high-risk area. The most downstream destroyed building was situated far from the valley exit; however, the debris flowed toward the road and caused severe damage. Thus, the road along the debris flow torrent was at high risk in the target area; thus, movement further away from the road and valley can ensure safety. Moreover, based on the disaster case and Case 4 results, east-side buildings (right-side in Fig. 4) did not exhibit damage or deposits. However, if the building is sufficiently strong, movement from the designated area is not required, and movement to the strong building on the east side area and upstairs can ensure safety.

Conclusions and Future Works

The focus of this study was on DEM landform data in Japan with different resolutions. A debris flow simulation of the disaster in Hiroshima in 2014 was conducted. We focused on deposition and maximum height distribution as an index that describes the hazard and risk mapping of debris flows. We also considered showing the high-risk area before the disaster occurs using simulations, because debris flows occur in a short time and it is challenging to evacuate in the event of heavy rainfall.

The DEM settings that can most accurately describe the disaster case and allow for the realization of a detailed hazard and risk mapping, in addition to safe evacuation planning, were determined. Compared with the simulation results using DEM data with a mesh resolution of 5 m, as provided by GSI, the results using high resolution LiDAR DEM data with a mesh resolution of 1 m under the consideration of house heights can accurately describe the influence area and high risk area close to the valley exit, in addition to the locally dangerous area and debris flow unto roads. For the realization of a detailed hazard and risk mapping, along with the determination of safe evacuation routes and shelters, the application of high-resolution LiDAR DEM data under the consideration of houses is effective.

However, only one debris flow scale case was considered in this study. The landform conditions and debris flow scale size have an influence on the debris flow behaviours. In future work, multiple debris flow scales will be considered, in addition to the landform DEM settings.

Acknowledgements We would like to give great thanks to the Ministry of Land, Infrastructure and Transport, Japan, Chugoku Regional Development Bureau, Otagawa River Office, for providing LP data. We also would like to great thanks to PASCO CO., Japan for providing us aerial photos. And in this study, we also applied DEM data provided by the Geospatial Information Authority of Japan (GSI).

References

Kaibori M, Ishikawa Y, Satofuka Y, Matsumura K, Nakatani K, Hasegawa Y, Matsumoto N, Takahara T, Fukutsuka K, Yoshino K, Nagano E et al (2014) Sediment related disasters induced by a heavy rainfall in Hiroshima-city on 20th August, 2014. J Jpn Soc Erosion Control Eng 67(4):49–59 ((in Japanese))

Nakatani K, Iwanami E, Horiuchi S, Satofuka Y, Mizuyama T (2012) Development of "Hyper KANAKO", a debris flow simulation system based on Laser profiler data, Proceedings of INTERPRAEVENT 2012, April 2012, Grenoble, France, vol 1. pp 269–280

Nakatani K, Kosugi M, Satofuka Y, Mizuyama T (2017) Influence of housing and roads on debris flow flooding and deposition in alluvial fan areas: case study on debris flows in Hiroshima, Japan, in August 2014. J Jpn Soc Erosion Control Eng 69(5):3–10 ((in Japanese with English abstract))

Satofuka Y, Mizuyama T (2005) Numerical simulation of a debris flow in a mountainous river with a sabo dam. J Jpn Soc Erosion Control Eng 58(1):14–19 ((in Japanese with English abstract))

Takahashi T, Nakagawa H, Satofuka Y, Kawaike K (2001) Flood and sediment disasters triggered by 1999 rainfall in Venezuela: a river restoration plan for an alluvial fan. J Nat Disaster Sci 23(2):65–82

Takahashi T (2007) Debris flow: mechanics, prediction and countermeasures. Taylor and Francis, Leiden, CRC Press, London, UK. (ISBN _978-0-415-43552-9) p 448

Wada T, Satofuka Y, Mizuyama T (2008) Integration of 1- and 2-dimensional models for debris flow simulation. J Jpn Soc Erosion Control Eng 61(2):36–40 ((in Japanese with English abstract))

Yoshino K (2014) Characteristics of rainfall triggers and sediment movement in 2014 Hiroshima disaster, reports on 2014 Hiroshima sediment disaster conducted by Japan society of erosion control engineering, JSECE publication, no. 74, pp 31–36 (in Japanese)

Debris-Flow Peak Discharge Calculation Model Based on Erosion Zoning

Xudong Hu, Kaiheng Hu, Jinbo Tang, Xiaopeng Zhang, Yanji Li, and Chaohua Wu

Abstract

Earthquakes trigger large numbers of landslides that provide abundant material for rainfall-induced debris flows and hence increase the frequency and magnitude of post-quake debris flows. A physically-based model incorporating a quantitative function of how much landslide deposit transforms into debris flows is developed to calculate debris-flow peak discharge by dividing a catchment into non-erosion, hydraulic erosion, and gravitational erosion zones. The model was implemented with a distributed hydrological computation program, and applied to Qipan catchment that was strongly influenced by the 2008 Wenchuan earthquake and where the debris flow peak discharge in 2013 was more than 10 times that before the earthquake. Remote sensing images and digital elevation model (DEM) data were used to obtain the spatial distribution of the earthquake-induced landslides. In situ tests were conducted to measure the hydrological parameters of the landslides. Comparisons of field survey and simulation results were carried out to verify the model validity. At 14 cross-sections, the peak discharge errors range from 3 to 20%. Moreover, the error of the inundation area is 2.1%, which suggests that the model is highly applicable to hazard mapping.

Keywords

Debris flow • Peak discharge • Earthquake-induced landslides • Distributed hydrological computation

X. Hu · K. Hu (✉) · J. Tang · X. Zhang · Y. Li · C. Wu
Institute of Mountain Hazards and Environment, Chinese Academy of Sciences, Chengdu, 610041, China
e-mail: khhu@imde.ac.cn

X. Hu
e-mail: hxd@ctgu.edu.cn

J. Tang
e-mail: jinbotang@imde.ac.cn

X. Zhang
e-mail: xpzhang_96@163.com

Y. Li
e-mail: 501830073@qq.com

C. Wu
e-mail: wuchaohua@imde.ac.cn

X. Hu
College of Civil Engineering and Architecture, China Three Gorges University, Hubei Yichang, 443002, China

X. Zhang · Y. Li · C. Wu
University of Chinese Academy of Sciences, Beijing, 100049, China

Introduction

Debris flows occur in regions with sufficient loose material, steep relief, and occasional or durative rainfall, which pose great threats to life and property (Jakob and Hungr 2005). The Wenchuan earthquake, which occurred on May 12, 2008 (Ms 8.0) triggered more than 56,000 landslides (Cui et al. 2011), providing abundant source materials for debris flows during rainy seasons (Tang et al. 2011; Dai et al. 2011). As a result, the frequency and magnitude of debris flows increased significantly in the earthquake-affected area (Cui et al. 2009). Similarly, the peak discharge of post-quake debris flows increased more greatly than pre-quake triggered by rainfalls with the same return period (Hu et al. 2010). Such augmentation is attributed to different soil erosions: gravitational erosion and hydraulic erosion. Gravitational erosion generally takes place together with hydraulic erosion, often produces extensive loose materials during or just after a heavy rainfall, such as avalanches, landslides or mudflows (Montgomery and Dietrich 1994; Tsai and Yang

F. Guzzetti et al. (eds.), *Understanding and Reducing Landslide Disaster Risk*, ICL Contribution to Landslide Disaster Risk Reduction,
https://doi.org/10.1007/978-3-030-60227-7_39

2006; Xu et al. 2015). Compared with gravitational erosion, the sediment amount caused by hydraulic erosion accounts for a small proportion to recharge debris flow (Jin et al. 2008). Therefore, the gravitational erosion of landslide deposits is very important when estimating the peak discharge in the earthquake-affected area, because more materials will be gravitationally eroded into debris flow in a zone with landslide deposits than a zone with bare soil or one covered with vegetation (Wichmann and Becht 2003).

The peak discharge is not only the key variable for the design of debris flow control engineering such as culverts, flumes, barriers, and check dams (Chen et al. 2009), but also the input important parameter in non-structural measures, like hazard zoning and risk assessment (Iverson et al. 1998; Rickenmann et al. 2006; Hungr and McDougall 2009). Indirect methods such as morphological investigation, matching method, comprehensive analysis, mathematical statistics, geo-analysis and hydrologic method (Kang 1985; Chen et al. 2007) are widely used to estimate the peak discharge of rainfall-induced debris flows. However, none of them takes the effect of gravitational erosion into account. The main objective of this paper is to develop a debris-flow peak discharge calculation model that is reliably applicable to engineering design and hazard mapping of debris flows in earthquake-affected areas. A catchment in the 2008 Wenchuan earthquake zone where a large-magnitude debris-flow event happened on July 11, 2013 is chosen as the case study of our model.

Debris-Flow Peak Discharge Calculation Model

Overland and channel flows can be both simulated by approximations of the Saint–Venant equations (Chaudhry 1993; Jain et al. 2004). The conservation of mass equation (Fig. 1b) is

$$h_{i,j}^{m+1} = h_{i,j}^{m} + \frac{\Delta t}{l^2}\left(Q_{in_{i,j}}^{\ m} - Q_{out_{i,j}}^{\ m}\right) \tag{1}$$

where i and j is the cell number (Fig. 1a); l is the cell width, m; Δt is the time interval, s; m is the time step; $Q_{in_{i,j}}^{\ m}$represents the total inflow of cell (i, j), m^3/s; $Q_{out_{i,j}}^{\ m}$ represents the total outflow of cell (i, j), m^3/s; and $h_{i,j}^{m+1}$, $h_{i,j}^{m}$ represents the depth of flow in cell (i, j) at $m + 1$ and m time, respectively, m.

A debris-flow catchment is divided into clear water or non-erosion zone (I), hydraulic erosion zone (II), and gravitational erosion zone (III) for estimating the discharge of debris flow. Each of these zones has a different contribution rate to recharge debris flows. Considering the difference of recharge flux for grid cells in the three zones, the total inflow in Eq. (1) can be written as

$$Q_{in_{i,j}}^{\ m} = \begin{cases} q_{in_{i,j}}^{\ m} & \text{I} \\ \alpha * q_{in_{i,j}}^{\ m} & \text{II} \\ \beta * q_{in_{i,j}}^{\ m} & \text{III} \end{cases} \tag{2}$$

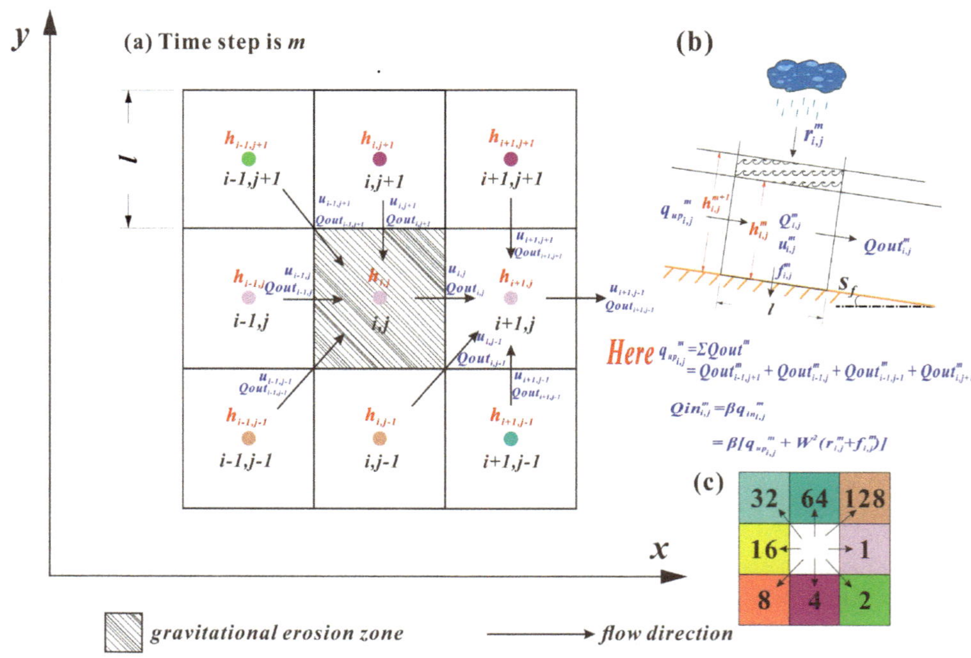

Fig. 1 Illustration of one-dimensional finite difference computation. **a** configuration of computation grid, **b** the conservation of mass, and **c** D8 algorithm

where $q_{in_{i,j}}^{m}$ represents the total water inflow in cell (i, j) at m time, m^3/s; α represents the augment coefficient of sediment caused by hydraulic erosion; β represents a simplified conversion coefficient between landslide deposits and debris flows, caused by gravitational erosion.

The model is completely implemented with GIS, and D8 algorithm (O'Callaghan and Mark 1984) is applied to determine flow direction of each cell. For the D8 algorithm, each cell only has one outflow direction, but could have multiple inflow directions (Fig. 1a).

Therefore, the total water inflow of cell (i, j) at m time is equal to the sum of rainfall intensity ($r_{i,j}^{m}$), infiltration rate ($f_{i,j}^{m}$), and upstream cells runoff ($q_{up_{i,j}^{m}}$) (Fig. 1b). The equation is

$$q_{in_{i,j}}^{m} = q_{up_{i,j}}^{m} + W^2\left(r_{i,j}^{m} - f_{i,j}^{m}\right) \tag{3}$$

where W is the cell width, m; $q_{up_{i,j}^{m}}$ represents the total discharge from upstream cells outflow, $q_{up_{i,j}^{m}} = \Sigma\, Q_{out}{}^{m}$, m^3/s.

The equation of total outflow in Eq. (1) is

$$Q_{out_{i,j}}^{m} = u_{i,j}^{m} A_{i,j}^{m} = u_{i,j}^{m} h_{i,j}^{m} W \tag{4}$$

where $u_{i,j}^{m}$ is the average velocity of flow in cell (i, j), m/s; $A_{i,j}^{m}$ is the cross-sectional area of cell (i, j), m^2.

It should be noted that the D8 algorithm has eight directions (Fig. 1c). The flow width of a cell is equal to the cell size if the flow direction is vertical or horizontal with the cell boundary, and to $\sqrt{2}$ times the cell size if the flow direction is diagonal. Hence, the symbol of W was used in Eq. (3) in lieu of l in Eq. (1), and W can be written as

$$W = \begin{cases} l & \textit{if direction is } 1, 4, 16, \textit{or } 64 \\ \sqrt{2} * l & \textit{if direction is } 2, 8, 32, \textit{or } 128 \end{cases} \tag{5}$$

The average velocity of flow can be described by one-parameter Manning resistance law

$$u_{i,j}^{m} = \frac{1}{n_{i,j}}\left(h_{i,j}^{m}\right)^{\frac{2}{3}} \times \left(s_{fi,j}^{m}\right)^{\frac{1}{2}} \tag{6}$$

where n is the Manning resistance coefficient and s_f is friction slope in flow direction, which were determined by field investigation or by hydraulic empirical method; h is the depth of flow (m).

In addition, the clear water or non-erosion zone, which is defined as the runoff area that only produces water for debris flow, is equal to the flow accumulation area. The hydraulic erosion zone is defined as the area that contains water and sediment caused by rainfall splash and runoff erosion. The augment coefficient of sediment α is used to represent the influence of runoff erosion on the peak discharge of debris flow, supposing

$$\alpha = (V_w + V_s)/V_w \tag{7}$$

where V_w is the volume of rainfall runoff, which is equal to the net rainfall depth multiplied by the watershed area; V_s is the sediment volume eroded by an individual rainfall, equal to the product of sediment yield (S_y) and the zone area. S_y can be calculated by MUSLE.

The gravitational erosion zone is defined as the area that not only provides water and sediment, but also supplies grains, gravels, and rocks from landslides. In a certain gravitational erosion zone, not all landslide material would participate in debris flows. So a landslide contribution rate of φ is introduced to estimating the value of β as follows

$$\beta = \frac{V_w + \phi V_l}{V_w} \tag{8}$$

$$V_l = h_{d_{i,j}^{x}} * l^2 \tag{9}$$

where V_l is the total volume of landslide; $h_{d_{i,j}^{x}}$ is the landslide depth of cell (i, j) in the gravitational zone.

From Wichmann and Becht's (2003) analysis, the cell's slope gradient and vertical distance to the channel network play a key role in landslides contribution rate to debris flow. Consequently, φ is defined as the function of cell's gradient (θ_g) and vertical distance to the channel network (d_v).

$$\varphi = f(\theta_g - d_v) \tag{10}$$

Moreover, each variable has the following relationship with φ.

$$\varphi \propto \theta_g \tag{11}$$

$$\varphi \propto 1/d_v \tag{12}$$

Suppose the cell (i, j) in the gravitational erosion zone that is closest to the channel network in vertical distance, has a 100% contribution rate. Based on the similarity principle for that zone, other cells in zone x are considered to calculate φ as follows

$$\varphi = \left(\frac{d_{v100\%}^{x}}{d_{v_{i,j}}^{x}} \frac{\theta_{g_{i,j}}^{x}}{\theta_{g100\%}^{x}}\right)^{\gamma} \tag{13}$$

where $\theta_{g_{100\%}^{x}}$, $d_{v_{100\%}^{x}}$ is the cell's gradient and vertical distance to the channel network of cell (i, j) in zone x with 100% contribution rate, respectively; $\theta_{g_{i,j}^{x}}$, $h_{d_{i,j}^{x}}$, $d_{v_{i,j}^{x}}$ is the cell's gradient, landslide depth and vertical distance to the channel network, respectively; and γ is the similarity coefficient to measure the topographic similarity between 100% contribution rate cell and other cells in landslide areas, which was determined by digital terrain analysis of DEM.

Once the debris-flow discharge is calculated at each cell along the streamline by the aforementioned model, the inundation area of the flow at each cross section and the two-dimensional distribution of flow depth along the main channel can be obtained by hydraulic method (Fig. 2).

Moreover, the Eq. (1) is stable if the Courant–Friedrichs–Lewy (CFL) criterion is satisfied. The criterion is

$$\frac{\Delta t}{W} \leq \frac{1}{u+\sqrt{gh}} \tag{14}$$

where g is the gravitational acceleration (m/s^2).

The initial condition for the kinematic wave equation is

$$h_{i,j}^{0} = Q_{out_{i,j}}^{0} = 0 \tag{15}$$

and the boundary condition is

$$Q_{up_{i,j}}^{0} = 0 \tag{16}$$

$$h_{outlet_{i,j}}^{m} = h_{outlet_{I,J}}^{m} \tag{17}$$

where $h_{outlet_{i,j}}^{m}$ is the flow depth at the outlet of the catchment; $h_{outlet_{I,J}}^{m}$ is the flow depth downstream of the outlet.

Model Application in Qipan Catchment

Study Area

Qipan catchment with the drainage area of 52.4 km^2 (31°26′39.19″N, 103°32′40.49‶E) is a tributary of Min River in China (Fig. 3). Its elevation ranges from 1320 to 4360 m above sea level (a.s.l.). The mainstream is 15.8 km in length and 170‰ in average gradient.

Parameter Settings

The total rainfall recorded by Weizhou meteorological station was 111.6 mm 4 days before the debris flow event on July 11, 2013 (Fig. 4). The triggering rainfall was approximately 6.4 mm/h at 3:00 AM on July 11. The event lasted for about 30 min (Zeng et al, 2014).

According to the remote sensing interpretation, up to 63 earthquake-induced landslides at various scales are distributed in Qipan (Fig. 5). Hence, there are 63 gravitational erosion zones. The non-erosion zones (I) include buildings, lake, forest land and grassland, the hydraulic erosion zones (II) include channel, terrace and bare land, and the gravitational erosion zones (III) include the landslides.

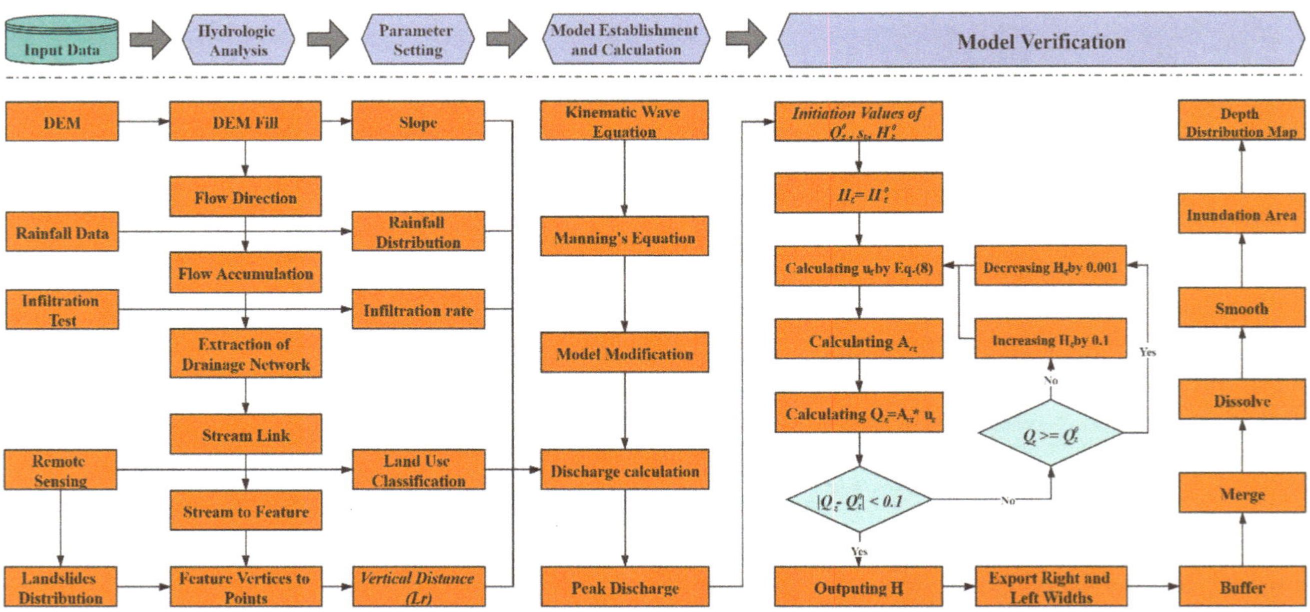

Fig. 2 Framework of the methodology and its verification method

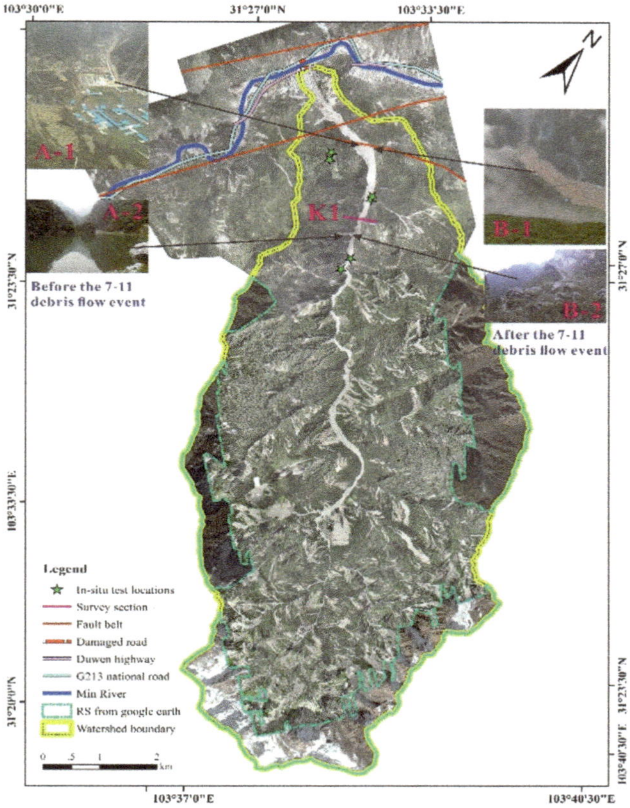

Fig. 3 Unmanned aerial vehicle (UAV) image of the study area (The resolution of image is 0.5 m × 0.5 m; Part of the RS was downloaded from Google Earth history images on December 31, 2013; the K1 section is the investigated cross section of the event after July 11, 2013.)

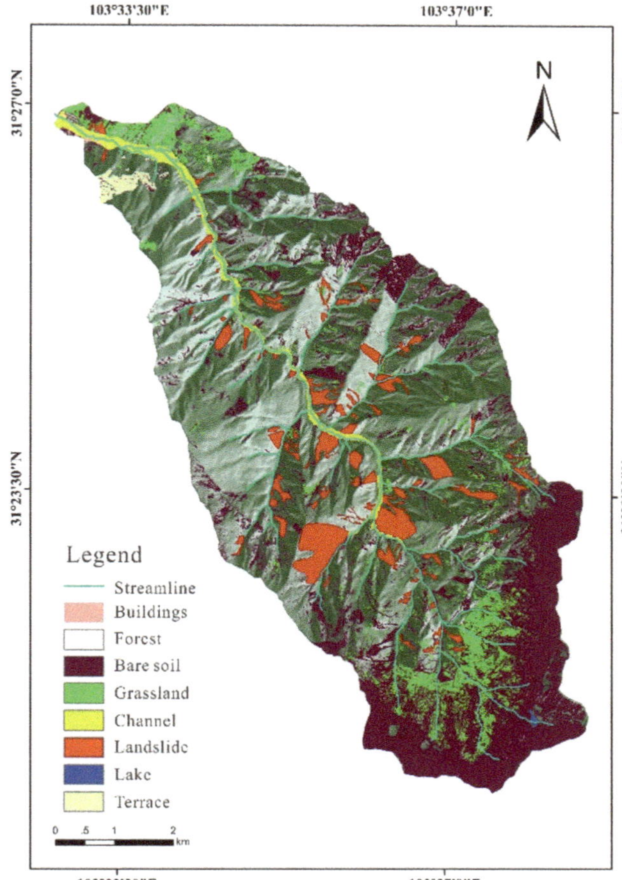

Fig. 5 Land classification based on UAV remote sensing image

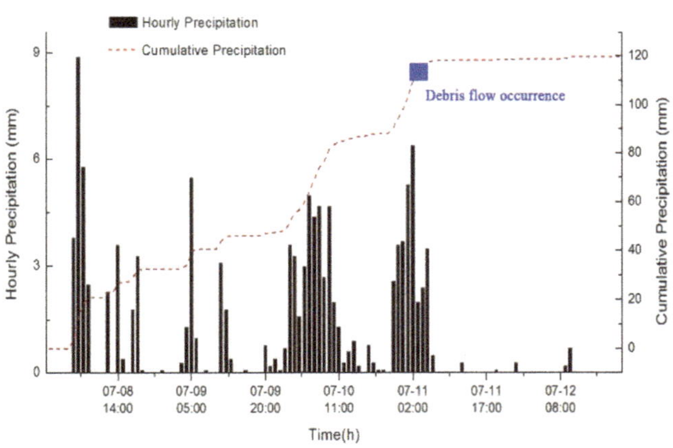

Fig. 4 Rainfall intensity prior and after the event in Qipan catchment (The location of Weizhou meteorological station is 31°08′27.69″N, 103°29′12.92″E)

Six kinds of land types in Qipan, including woodland, grassland, farmland, bare soil, debris flow deposit soil and landslide deposit area, are chosen to conduct in-situ tests as the same as Fatehnia et al. (2016), and the test results are shown in Table 1.

Result and Discussion

Peak Discharge Calculation

After inputting the parameter values, the hydrograph of debris flows between 22:00 on July 10 and 03:00 on July 11, 2013 was simulated via the debris-flow peak discharge model (Fig. 6). According to field survey after the event, the peak discharge was 1745 m^3/s and the average flow depth was approximately 5 m at the cross section K1. Compared with survey results, the peak discharge at K1 section by the numerical simulation was 1836.5 m^3/s, and it occurred at approximately 2:40 AM and lasted for nearly 33 min.

Verification

The flow peak discharges along the main channel measured by Hu and Huang (2017) are compared with the simulated results (Table 2). 14 cross-sections shown in Fig. 6 are selected to verify the simulation results.

Table 1 Final stable infiltration rate (K_s) of different soil types

Types	Forestland	Grassland	Farmland	Soil	Channel	Landslide
K_s (mm/s)	0.0083	0.025	0.0283	0.014	0.0417	0.0508
n	0.2	0.05	0.04	0.055	0.055	0.08

Note the final stable infiltration rate (K_s) is used instead of *fm i,j* in Eq. (3) due to the infiltration rate varies with time in the in-situ tests

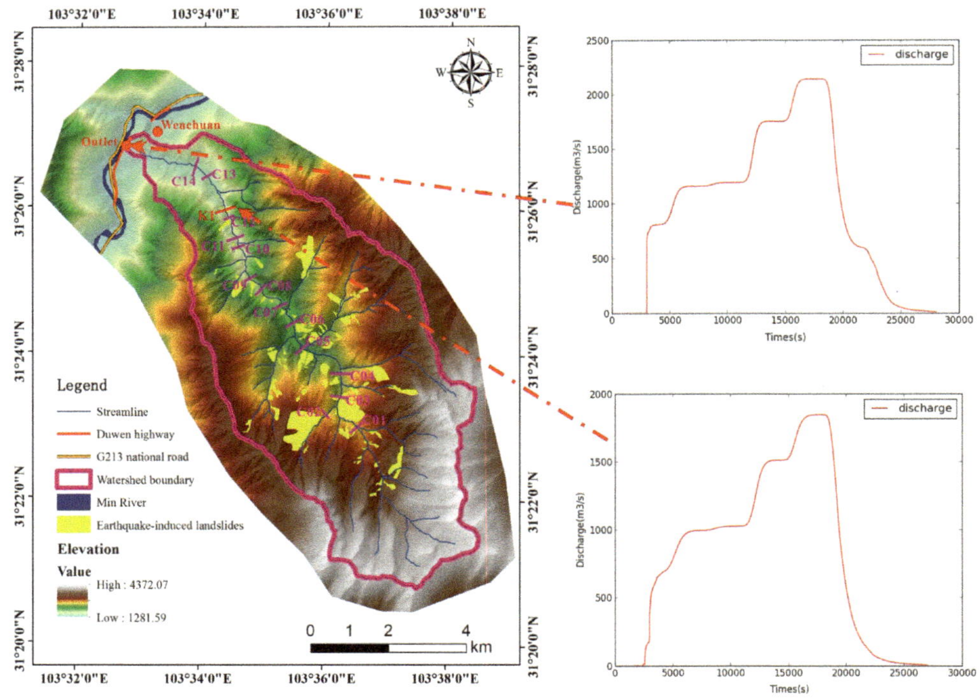

Fig. 6 Simulated hydrographs of debris flows at two different cross sections. (The left image is the map of filled DEM; C01-C14 represent different cross-sections)

Moreover, the total inundation area for the simulation is 644,099 m^2 while the measured area is 630,713 m^2. The simulation depth at the K1 section is approximately 6.6 m while the measured depth is 4–7 m. In addition, the simulation depths at the other two places are 4.0 m and 5.9 m respectively while the measured depths are 4–6 m and 4–7 m, respectively (Fig. 7).

Although the impact of earthquake-induced landslides on debris-flow process is simplified as a conversion coefficient, the overall simulation results are satisfactory with an error range of 3% to 20%. (Table 2). Comparisons of the debris-flow inundation area and flow depth also show that the proposed model has good performance.

Table 2 Comparison of the peak discharges between field measurement and numerical simulation

No. section	Q_C (m^3/s)	Q_P (m^3/s)	Error(%)	No. section	Q_C (m^3/s)	Q_P (m^3/s)	Error(%)
C01	535.8	521.2	2.7	C08	1389.7	1524.9	9.7
C02	59.4	72.3	21.7	C09	1669.6	1798.3	7.7
C03	985.8	774.3	21.5	C10	1997.5	1651.0	17.3
C04	1218.3	1039.6	7.5	C11	1830.3	1710.4	6.6
C05	1068.2	1153.2	8.0	C12	2597.5	2344.7	9.7
C06	1011.7	1236.7	22.2	C13	2242.4	1887.4	15.8
C07	1124.6	1158.6	3.0	C14	1520.4	1784.3	17.4

Notes Q_C is the peak discharge measured by Hu and Huang (2017). Q_P is the simulated peak discharge

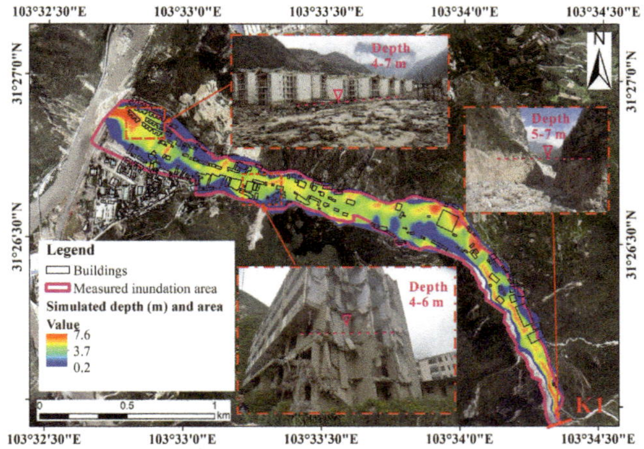

Fig. 7 Comparison of the simulation inundation area and measured inundation area of the event (the aerial photo was provided by the Geomatics Center of Sichuan Province, which was taken shortly after the event.)

Discussion

The newly proposed function β of Eq. (8) plays a critical role in calculating the peak discharge of debris flow in earthquake-affected region. Hu et al. (2010) proposed a nonlinear modified flood method to calculate the debris-flow peak discharge in Wenchuan earthquake region. They lumped the effects of landslides over the whole basin, rather than refined the effect at each landslide areas of the basin as in Eq. (8). Moreover, the blockage coefficient used in the method by Hu and Huang (2017) is determined by expert experience. By contrast, the value of β is dependent on topographical characteristics ($\theta_g x\ i,j$, $d_v x\ i,j$) of the landslide itself.

Comparison of the debris flow inundation areas as shown in Fig. 7 suggests that the methodology with enough precision data in this paper can be applied in delineating hazard zones. Tang et al. and Hungr (2005) both proposed a risk zoning method based on the depth of debris flow. Based their method, the inundation area in Fig. 7 can be delineated into four risk zones: extremely high ($h > 3$ m), high ($1 < h < 3$ m), medium ($0.5 < h < 1$ m), and low ($h < 0.5$ m) risk zone (Tang et al. 1994), or three zones: high ($h > 1$ m), medium ($0.2 < h < 1$ m), and low ($h < 0.2$ m) (Hungr 2005). Therefore, the model is also useful for engineers to employ corresponding prevention and control measures according to different risk zones.

Conclusions

A physically-based distributed debris-flow peak discharge calculation model was developed to model debris-flow routing and estimate its peak discharge in earthquake-affected areas. The cell-based implement of the model divides a catchment into non-erosion, hydraulic erosion, and gravitational erosion zone. A coefficient of β quantified the conversion ratio of landslides into debris flow is introduced in the model, reflecting the impact of earthquake-induced landslides on the debris-flow peak discharge.

The model is applied to simulate the discharge process of debris flows at Qipan catchment. According to field survey, the peak discharge of the event in 2013 was sharply magnified due to the earthquake-induced landslides. The errors between the simulated and field survey peak discharges ranged from 3 to 20%. The methodology also can provide two-dimensional inundation data of debris flows through one-dimensional computation. Comparison of the simulated inundation area (644,099 m^2) with the measured area (630,713 m^2) shows that the model's accuracy is enough to apply to the hazard mapping of debris flows.

Acknowledgements This work has been supported by the National Natural Science Foundation of China (Grant No. 41790434), the National Basic Research Program of China (973 Program) (Grant No. 2015CB452704), the National Natural Science Foundation of China (Grant number 41601011), and the 135 Strategic Program of the Institute of Mountain Hazards and Environment, CAS (SDS-135-1704).

References

Chaudhry MH (1993) Open channel flow. Prentice Hall, Englewood Cliffs, NJ

Chen NS, Yue ZQ, Cui P, Li ZL (2007) A rational method for estimating maximum discharge of a landslide-induced debris flow: a case study from southwestern China. Geomorphology 84(1–2):44–58

Chen XQ, Cui P, Zhao WY (2009) Optimal timing for the control of debris flow in Wenchuan earthquake area. J Sichuan Univ (Eng Sci Ed) 41(3):125–130 ((in Chinese))

Cui P, Chen XQ, Zhu YY, Su FH, Wei FQ, Han YS, Liu HJ, Zhuang JQ (2009) The Wenchuan earthquake (May 12, 2008), Sichuan province, China, and resulting geohazards. Nat Hazards 56 (1):19–36

Cui P, Hu K, Zhuang J, Yang Y, Zhang J (2011) Prediction of debris-flow danger area by combining hydrological and inundation simulation methods. J Mt Sci 8:1–9

Dai FC, Xu C, Yao X, Xu L, Tu XB, Gong QM (2011) Spatial distribution of landslides triggered by the 2008 Ms 8.0 Wenchuan earthquake China. J Asian Earth Sci 40:883–895

Fatehnia M, Tawfiq K, Ye M (2016) Estimation of saturated hydraulic conductivity from double-ring infiltrometer measurements. Eur J Soil Sci 67:135–147

Foody GM (2002) Status of land cover classification accuracy assessment. Remote Sens Environ 80(1):185–201

Hu KH, Cui P, You Y, Zhuang JQ, Chen XQ (2010) Nonlinear modified flood method for calculationg the debris-flow peak discharge in Wenchuan earthquake region. J Sichuan Univ (Eng Sci Ed) 42(5):52–57 ((in Chinese))

Hu T, Huang RQ (2017) A catastrophic debris flow in the Wenchuan earthquake area, July 2013: characteristics, formation, and risk reduction. J Mt Sci 14(1):15–30

Hungr O (2005) Debris flow Hazards and related phenomena. Classification and terminology. In: Jakob M, Hungr O (eds) Springer, Heidelberg

Hungr O, McDougall S (2009) Two numerical models for landslide dynamic analysis. Comput Geosci 35(5):978–992

Iverson RM, Schilling SP, Vallance JW (1998) Objective delineation of lahar-inundation hazard zones. GAS Bulletin 110:972–984

Jain MK, Kothyari UC, Raju KGR (2004) A GIS based distributed rainfall-runoff model. J Hydrol 299(1–2):107–135. https://doi.org/10.1016/j.jhydrol.2004.04.024

Jin X, Hao ZC, Zhang JL, Wang JH (2008) Distributed soil erosion model with the effect of gravitational erosion. Advan Water Sci 19 (2):257–263

Kang ZC (1985) An analysis of maximum discharge of viscous debris flow at Jiangjia Gully of dongchuan in Yunnan. Memoirs of Lanzhou institute of glaciology and cryopedology of Chinese academy of sciences (no. 4), Beijing: Science Press, pp 119–123 (in Chinese)

Lambe TW, Whitman SE (1969) Soil mechanics. Wiley, New York

Montes S (1998) Hydraulics of open channel flows. ASCE Press, Reston

Montgomery DR, Dietrich WE (1994) A physically based model for the topographic control on shallow landsliding. Water Resour Res 30(4):1153–1171

O'Callaghan JF, Mark DM (1984) The extraction of drainage networks from digital elevation data. Comput Vision Graph 28:323–344

Rickenmann D, Laigle D, McArdell B, Hübl J (2006) Comparison of 2D debris-flow simulation models with field events. Comput Geosci 10(2):241–264

Tsai TL, Yang JC (2006) Modeling of rainfall-triggered shallow landslide. Environ Geol 50(4):525–534

Tang C, Zhu J, Ding J, Cui XF, Chen L, Zhang JS (2011) Catastrophic debris flows triggered by a 14 August 2010 rainfall at the epicentre of the Wenchuan earthquake. Landslides 8(4):485–497

Williams JR, Berndt HD (1997) Sediment yield prediction based on Watershed hydrology. Agric Eng 20(6):1100–1104

Wichmann V, Becht M (2003) Modelling of geomorphic processes in an alpine catchment. In: Martin D (ed) 7th international conference on geo computation. University of Southampton, Southampton

Xu ZX (1985) The prevention and control of debris flow of qipan Gully in Wenchuan county. Sichuan Province Mt Res 3(3):166–172

Xu XZ, Liu ZY, Xiao PQ, Guo WZ, Zhang HW, Zhao C, Yan Q (2015) Gravity erosion on the steep loess slope: behavior, trigger and sensitivity. CATENA 135:231–239

Zeng C, Cui P, Ge YG, Zhang JQ, Lei Y, Yan Y (2014) Characteristics and mechanism of buildings damaged by debris flows on 11 July, 2013 in Qipangou of Wenchuan, Sichuan. J Earth Sci Environ 36 (2):81–91 ((in Chinese))

Assessment of Rainfall-Induced Landslides in Tomioka City, Gunma Prefecture, Japan (Oct 2019) Based on a Simple Prediction Model

Akino Watanabe, Thang V. Nguyen, and Akihiko Wakai

Abstract

Landslide is a natural disaster that happens quite often, especially in the rainy season, in monsoon weather countries. High intensity rainfall in short period of time is one of the direct factors that causes shallow landslides on natural slopes. Typhoon Hagibis (October 2019) is one of the biggest storms in 60 years that Japan has suffered. It caused heavy devastation, both economically and humanly. One of the tragic consequences was the landslide that occurred on 12 October 2019 at Tomioka City, Gunma Prefecture, which caused 3 deaths, 5 injuries and destroyed many houses. The prediction of landslide to give early warning and timely evacuation for residents according to the real time rainfall data is essential. This paper will assess the slope stability in the area of this landslide. It is based on a simple prediction model for the rising of underground water in shallow depth of the slope which subjected heavy rainfall. Some of the model's parameters are assumed in reference to a post-landslide survey. The results show that there is considerable rise of the groundwater level in the slope. Consequently, the factor of safety has decreased sharply at the time of the collapse. Additionally, various factors such as topography, crop fields, vegetation and reduced shear strength of weathered pumice soil layer above the bedrock are also thought to be triggers of such landslide. It is expected to perform better with real input parameters as well as extended modelling of other affecting factors in the future.

Keywords

Seepage flow • Groundwater level • Typhoon • Rainfall-induced landslides

Introduction

Rainfall is a major triggering factor causing shallow landslides on the slopes, as it has been highlighted in many studies (Rahardjo et al. 2001; Cascini et al. 2010; Yang et al. 2015). Water infiltration into the soil will cause an increase in moisture content and a decrease in matric suction. On the other hand, infiltration of rainfall results in the rise of groundwater level (GL), and the increase in water pressure. Both the increase in the water pressure and the decrease in the matric suction cause the decrease in the shear strength of soil (Cai and Ugai 2004).

In general, the model of shallow landslides caused by rainfall has also been studied by many authors (Ng and Shi 1998; Cai and Ugai 2004; Cascini et al. 2010). The evaluation models all involve rate of rainwater infiltrates into the soil which is dependent on the matric suction and the volumetric water content (Lumb 1975; Zhang et al. 2004). Therefore, modelling seepage flow in the soil is a complex process that often requires complicated simulation algorithms and long computation times.

Shallow landslides due to heavy rains usually have a thickness of 3–5 m and often occur at relatively high speeds (Huang et al. 2015), so it is difficult to evacuate for residents without early warnings based on real rainfall data. A landslide occurred in Tomioka city, Gunma Prefecture, Japan on 12 October 2019 due to Typhoon Hagibis is rainfall-induced shallow landslide with the depth of about 3 m and the length of about 20 m.

A. Watanabe (✉) · T. V. Nguyen · A. Wakai
Department of Environmental Engineering Science, Gunma University, 1-5-1, Kiryu, Gunma, 376-8515, Tenjin, Japan
e-mail: t191c051@gunma-u.ac.jp

T. V. Nguyen
e-mail: t192c602@gunma-u.ac.jp

A. Wakai
e-mail: wakai@gunma-u.ac.jp

T. V. Nguyen
Department of Civil Engineer, Thuyloi University, 175 Tayson, Dongda, Hanoi 100000, Vietnam

F. Guzzetti et al. (eds.), *Understanding and Reducing Landslide Disaster Risk*,
ICL Contribution to Landslide Disaster Risk Reduction,
https://doi.org/10.1007/978-3-030-60227-7_40

A simple method for prediction of shallow groundwater level rising in natural slopes in time history was proposed based on the parametric study with the finite element analysis under the assumption of semi-infinite homogeneous slope (Wakai et al. 2019). It becomes possible to predict the slope failure in real time.

The objective of this paper is to use such method and assumed parameters based on a landslide survey to assess the factor of safety of the slope in a study area around the location of the landslide in Tomioka city, and clarify assumptions about what happened.

The Simple Prediction Model

The model is proposed to predict the changing of shallow water table in natural slope, which is assumed to have uniform standard soil layer (medium sand or coarse sand). The fluctuation of groundwater level is determined by the amount of vertical rainwater infiltration from the ground surface and the horizontal permeability in the saturated soil layers.

a. **Vertical seepage of rainfall infiltration**

The simulation of one-dimensional vertical infiltration in horizontal ground layers with uniform soil property has been conducted and the simplified calculation model that can describe the fluctuation of the groundwater level has been calculated by VGFLOW model (Cai and Ugai 2004). The parameters of medium sand (standard case study) have been adopted as the material properties from ground surface to the depth of GL = −2.0 m, and the soil deeper than GL = −2.0 m has been assumed as impermeable layer for a base. The initial groundwater level is approximately GL = −1.11 m, an example of rainfall (constant rainfall intensity = 10 mm/hr) is used for calculation. The fluctuation of groundwater level and changing of averaged degree of saturation in unsaturated layer by elapsed time are recorded as in Fig. 1 (Wakai et al. 2019).

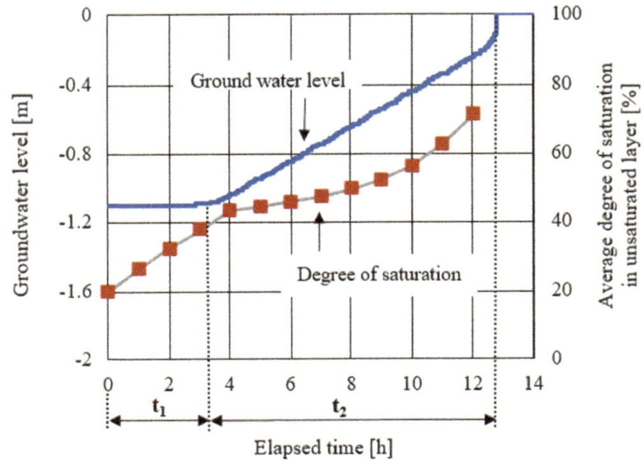

Fig. 1 Time histories of ground water level and degree of saturation (standard case)

The diagram shows that the water table remains the same during t_1 from the beginning of rain, and it only starts to increase with constant velocity during time t_2 until it reaches the ground surface. As a simple method to predict the increase in the water table, it is assumed that the groundwater will only begin to rise when the average of volumetric water content in unsaturated layer exceeds a constant value θ_{cp}. Diagram describing the concept for the proposed model is shown in Fig. 2.

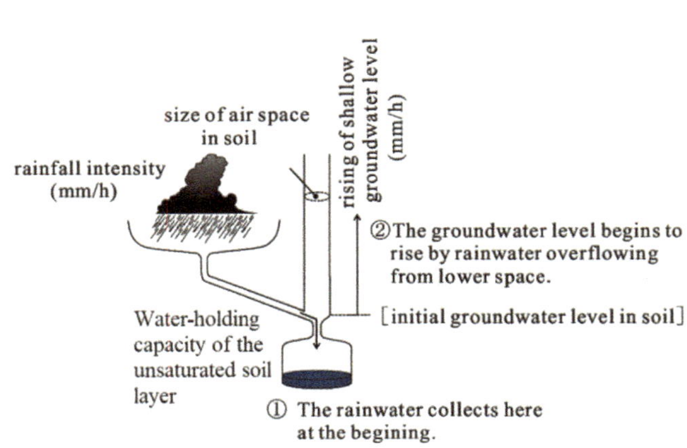

Fig. 2 Concept of proposed method

The limited values of volumetric water content θ_{cp} for medium sands and coarse sands are listed in Table 1. Thus, when θ_{cp} of each position on the slope is determined based on its slope angle and sand case, it is possible to calculate the following values at that location: The duration time t_1 from the beginning of rain to moment when the water table starts rising by Eq. (1); the rising velocity of the groundwater level in the second period v_{wl} by Eq. (2); and the time t_2 for water table reach to the ground surface by Eq. (3).

$$t_1 = \frac{n\left(S_r^* - \tilde{S}_{r0}\right)h}{100\tilde{I}} = \frac{h}{\tilde{I}}\left(\theta_{cp} - n\frac{\tilde{S}_{r0}}{100}\right) \quad (1)$$

$$v_{wl} = \frac{\alpha_v I}{n\left(1 - S_r^*/100\right)} \quad (2)$$

$$t_2 = \frac{h}{v_{wl}} = \frac{n\left(1 - S_r^*/100\right)h}{\alpha_v I} \quad (3)$$

The proposed model will be more accurate by using adjustment factor $\alpha_v = 2.1$ and following modified values of rainfall intensity I, initial saturation degree S_{r0}.

Table 1 Proposed θ_{cp} values considering slope inclination dependency

Angle value of slope	θ_{cp}[−]	
	Medium sands	Coarse sands
0°	0.054	0.036
15°	0.056	0.038
30°	0.058	0.040
45°	0.060	0.042

$$\tilde{I} = \frac{I}{\log_{10}(I)} \tag{4}$$

$$\tilde{S}_{r0} = \frac{S_{r0}}{\log_{10}(S_{r0}^2)} \tag{5}$$

where $\boldsymbol{S_{r0}}$ is the initial saturation degree [%], $\boldsymbol{S_r^*}$ is the critical degree of saturation corresponding to the start of the water level rising [%], n is the porosity of soil [−], $\boldsymbol{h}$ is the initial thickness of the unsaturated layer [m], $\boldsymbol{I}$ is the rainfall intensity [m/h].

b. **Horizontal flow of groundwater**

The horizontal saturated flow in unconfined aquifer is modeled based on the Darcy Eq. (6). This differential equation is computed by discretization method in time and space.

$$\frac{\partial}{\partial x}\left(T_{xx}\frac{\partial \Phi}{\partial x}\right) + \frac{\partial}{\partial y}\left(T_{yy}\frac{\partial \Phi}{\partial y}\right) = n_e\frac{\partial \Phi}{\partial t} \tag{6}$$

where T_{xx} and T_{yy} are transmissivity in x and y directions, n_e is the effective porosity and Φ is the total head of water.

Landslides in Tomioka City, Gunma, Japan

Typhoon Hagibis was a powerful tropical storm that was considered to be the most devastating typhoon in the last decades. It made landfall on Honshu (main island of Japan) on 12 October 2019, bringing record-breaking rainfall in many prefectures.

Meanwhile, around 16:30 on 12 October 2019, the eastern slope of Takumi Village (36.243°N, 138.901°E), Tomioka City, Gunma Prefecture collapsed in two places (Fig. 3), both of which were about 20 m long and about 3 m deep at the top of landslide. It caused 3 deaths, 5 injuries and damaged many properties in the vicinity. This landslide was difficult to predict because it occurred in a place other than the mountainsides. Furthermore, its location was a gentle slope of about 20°, slightly less than 30°, which is the designated requirement for the landslide disaster warning area (collapse of steep slopes) of the landslide disaster prevention law.

Shallow landslides are directly related to weathered soil layer above bedrock, as discussed in previous studies (Dahal et al. 2008; Chigira et al. 2011). The geology of the slope where landslide occurred is depicted in Fig. 4. The total thickness of un-weathered layers is about 2700 mm, including the top soil layer of 1500 mm, followed by the loam layer of 900 mm, and the thin pumice layer of about 300 mm. Sandwiched between the upper un-weathered layer and the lower bedrock layer is a white weathered pumice layer about 200 mm thin, almost parallel to the slope. This weathered pumice layer is estimated to be the slip surface of landslides.

According to the data from a rain gauge station of Gunma prefecture Sediment Control Division, about 3.5 km away from the landslide location, the amount of rainfall measured in 24 h until 17:00 on 12 October 2019 was 388 mm. The figure is greater than one out of three of annual average rainfall (1100 mm) in that place. The maximum amount per hour was almost 50 mm; the precipitation data is presented in Fig. 5.

Fig. 3 The site of the landslide occurrence (Gunma prefecture sediment control division)

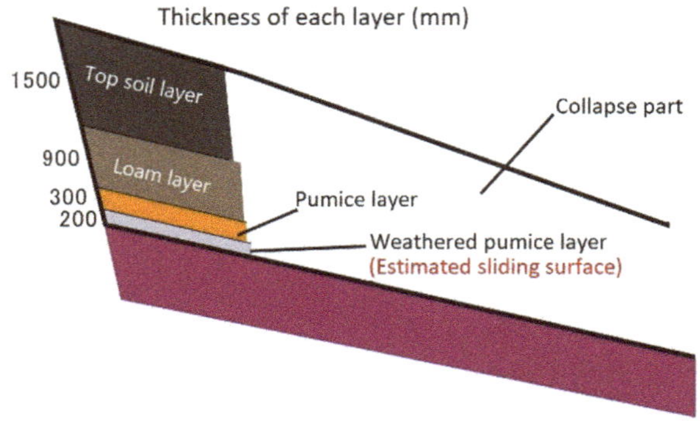

Fig. 4 Geology and thickness of soil layers

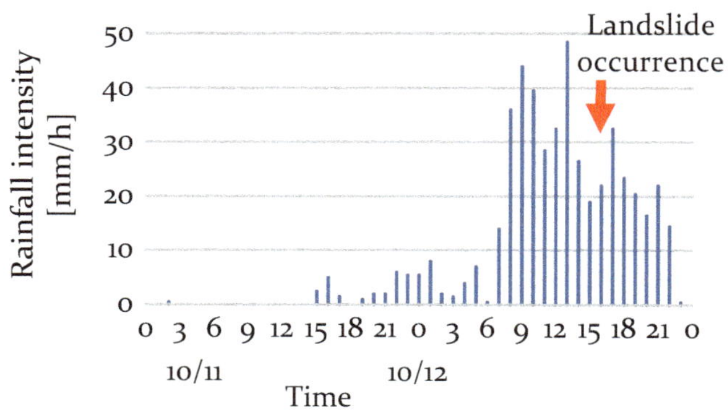

Fig. 5 Actual rainfall data (Gunma prefecture sediment control division)

Numerical Simulation

a. Input parameters of the model

The analysed area is 380 × 355 m^2 including the slope failure area; the computational mesh is 5 × 5 m^2 and the time interval of analysis is 1 min. The elevation of study area is referenced from Geospatial Information Authority of Japan. The slope angle value of each grid element in the area is assigned from the steepest direction of its eight neighbor pixels. The limited value of volumetric water content θ_{cp} of each 5 m × 5 m grid element is interpolated based on slope angle value and sand type from Table 1.

The un-weathered layers are homogeneously simulated as medium sand or coarse sand with the typical value of porosity n. The initial saturation degree S_{r0} is determined from an investigation. Basically, the other properties of the soil layers on the slope remain undefined, so some input parameters are assumed for calculation as Table 2. The estimated sliding layer (the white weathered pumice layer) also has not been tested in term of soil properties. Therefore, it is assumed as a sand layer having cohesion $c^{'} = 0$ and internal friction angle $\phi^{'} = 35°$ for the calculation of factor of safety of the slope.

b. Factor of Safety of slope

The factor of safety of rainfall-subjected slope depends on the groundwater level; the geometric characteristics of slope and the soil properties of sliding soil layer. The factor of safety F_s is given by Eq. (7):

$$F_s = \frac{\tau_f}{\tau} = \frac{c^{'} + [\gamma_t.h_1 + (\gamma_{sat} - \gamma_w)h_2]\cos^2\theta.\tan\varphi^{'}}{(\gamma_t.h_1 + \gamma_{sat}.h_2).\sin\theta.\cos\theta} \quad (7)$$

where τ is the sliding force [kN/m2], τ_f is the shear resistance force [kN/m2], $c^{'}$ is the cohesion [kN/m2], γ_t is the wet unit weight of soil [kN/m^3], γ_{sat} is the saturated unit weight of soil [kN/m^3], γ_w is the unit weight of water [kN/m^3], h_1 is the depth from ground surface to groundwater level [m], h_2 is the depth from groundwater level to sliding surface [m], θ and $\phi^{'}$ are the gradient of slope and the angle of internal friction of soil, respectively [°].

c. Results and discussion

The groundwater level of the study area using medium sand and coarse sand at three moments are displayed in Figs. 6 and 7, respectively. It can be clearly realized that the dispersion of groundwater level in the coarse sand case is clearer than that in the medium sand case. This is explained by the fact that the limited volumetric water content values of the coarse sand are smaller than those of medium sand, so the water table in coarse sand tends to start rising earlier. In addition, the permeability coefficient of coarse sand ($k = 10^{-3}$) is much higher than that of medium sand ($k = 10^{-4}$); thanks to this, the horizontal redistribution of groundwater also occurs faster in coarse sand.

Table 2 Assumed properties of soil layers

Properties	Medium sand	Coarse sand
Permeability factor K [m/s]	10^{-4}	10^{-3}
Porosity n	0.3	0.3
Initial saturated degree S_{r0} [%]	60	60
Layer Thickness [m]	2.7	2.7

Fig. 6 Distributions of groundwater level in study area—medium sand case

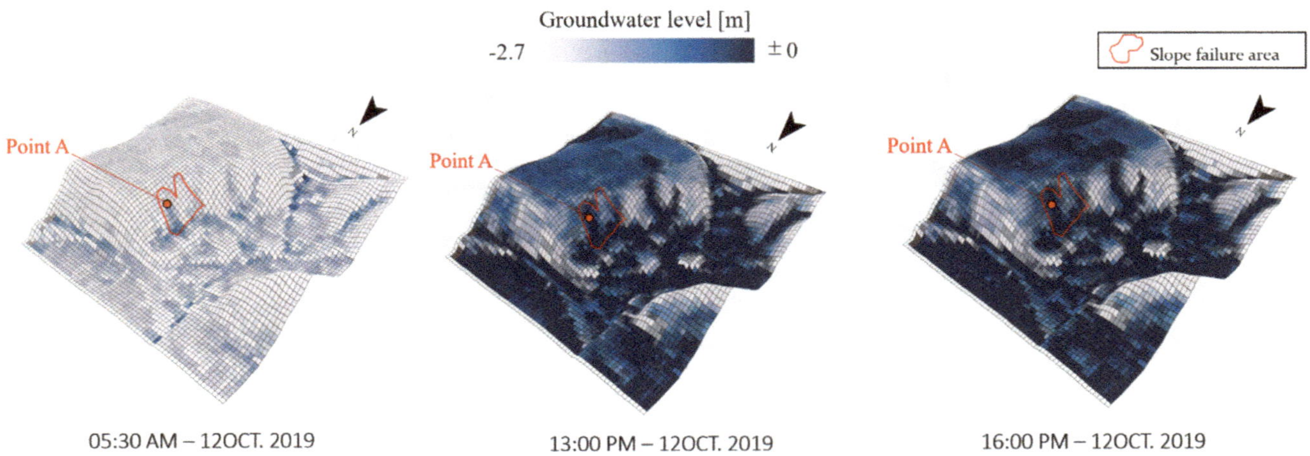

Fig. 7 Distributions of groundwater level in study area—coarse sand case

At 5:30 AM on 12 October, there is a slight difference in water table of all grid elements in both sand cases. This is because the infiltration water in the previous time is still retained mostly in the upper unsaturated soil layer and does not make groundwater level rising in all elements. Then, when the upper unsaturated soil layer can no longer receive additional water, the rainwater is provided to make the water table rise, which can be seen in the pictures at 13:00 PM of that day. At the time of 16:00 PM, the groundwater level reaches the ground surface in almost all grid elements of the area in medium sand case, whereas in coarse sand case it only occurs in some points and is still fairly dispersed. This is due to the larger horizontal permeability flow in coarse sand case.

Figure 8a is the graph of water table by time at one location (point A) in the failure area of both sand cases. It is shown that in coarse sand the groundwater level starts rising at 15:00, much earlier than in medium sand at 21:30 on 11 October. However, the time at which it reaches the ground surface is later, 16:40 on 12 October compared to 13:30 in medium sand.

The factor of safety of slope Fs by time at point A is shown in Fig. 8b. There is a significant decrease in its value, from 1.62 to 0.87 in medium sand case and from 1.62 to 0.88 in coarse sand case from the moment the groundwater level begins rising until it reaches ground surface.

Conclusion

The paper applies a simple model to assess the stability of the eastern slopes of Takumi Village during the heavy rain time of Typhoon Hagibis. Accordingly, the groundwater level had risen and reached the ground surface quite rapidly. As a result, it had reduced the factor of safety of slope considerably before the landslide occurred.

Additionally, there are still other factors which have yet to be modelled that can be considered to affect this landslide.

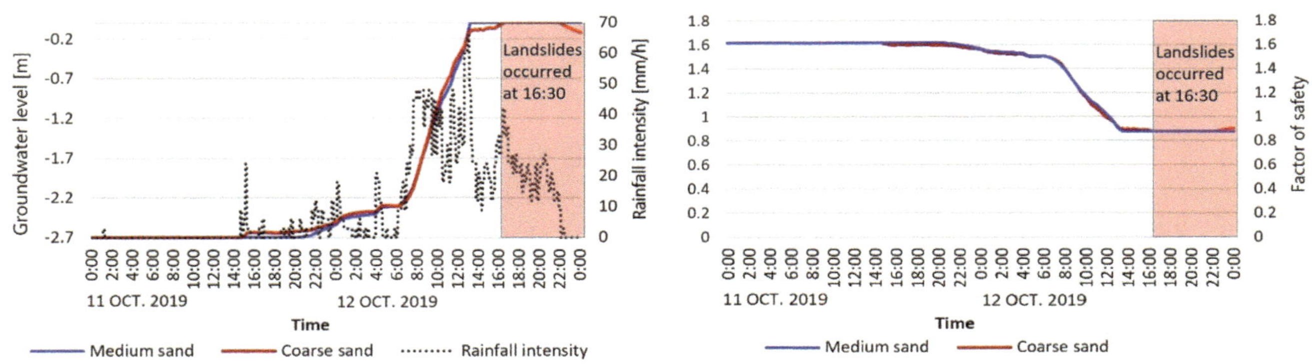

a. Groundwater level b. Factor of safety

Fig. 8 Groundwater level and factor of safety of slope at point A

Firstly, the thin weathered pumice soil layer has low permeability and extremely sensitive soft strength compared to the above layers. The more groundwater accumulates above it, the higher the uplift water pressure and the driving force. Gradually, as the shear deformation develops at this weathered pumice layer, its shear resistance decreases quickly making it unable to withstand the driving forces formed by the weight of the above soil mass. Another factor is a large crop field at the top of the slope (Fig. 3).). It is highly possible that the rainfall intensity is much higher than the infiltration rate of soil for the majority of the time, which lead to the excess water being increasingly accumulated on the ground surface. Inevitably, it then flows down to the lower slope surface. In general, surface water tends to selectively downward in locations with easy-flow terrain surfaces, such as recessed sections, less rugged slopes or less obstructed vegetation. Furthermore, since the surface water velocity is heavily influenced by slope angle and surface roughness, it will definitely increase rapidly after moving from the field surface to the slope surface. Consequently, the erosion of the soil surface at the location of the flow quickly follows. This makes rainwater and surface water penetrate directly to lower layers. Simultaneously, the underground water of the upper field also seeps down the slope over time, causing the increase of the water table at the slope faster.

Acknowledgements This research is supported by JST SICORP (e-ASIA Joint Research Program) Grant Number JPMJSC18E3, Japan. The authors wish to acknowledge the e-ASIA members. The authors also gratefully acknowledge Gunma prefecture Sediment Control Division (rainfall data used for analysis and picture of slope failure) and Geospatial Information Authority of Japan (Digital Elevation Model data). We appreciate students of geotechnical engineering laboratory, Mr. Kitazume and Mr. Bando for supporting experiments. We are grateful to the reviewers for their comments on an earlier version of this manuscript.

References

Cai F, Ugai K (2004) Numerical analysis of rainfall effects on slope stability. Int J Geomech 4(2):69–78

Cascini L, Cuomo S, Pastor M, Sorbino G (2010) Modeling of rainfall-induced shallow landslides of the flow-type. J Geotech Geoenviron Eng 136(1):85–98

Chigira M, Mohamad Z, Sian LC, Komoo I (2011) Landslides in weathered granitic rocks in Japan and Malaysia. Bull Geol Soc Malays 57:1–6

Dahal RK, Hasegawa S, Nonomura A, Yamanaka M, Masuda T, Nishino K (2008) Failure characteristics of rainfall-induced shallow landslides in granitic terrains of Shikoku Island of Japan. Environ Geol 56(7):1295–1310

Huang J, Ju NP, Liao YJ, Liu DD (2015) Determination of rainfall thresholds for shallow landslides by a probabilistic and empirical method. Nat Hazards Earth Sys Sci 15:2715–2723

Lumb PB (1975) Slope failures in Hong Kong. Q. J Eng Geologist 8:31–65

Ng CWW, Shi Q (1998) A numerical investigation of the stability of unsaturated soil slopes subjected to transient seepage. Comput Geotech 22(1):1–28

Rahardjo H, Li XW, Toll DG, Leong EC (2001) The effect of antecedent rainfall on slope stability. Geotech Geol Eng 19:371–399

Wakai A, Hori K, Watanabe A, Cai F, Fukazu H, Goto S, Kimura T (2019) A simple prediction model for shallow groundwater level rise in natural slopes based on finite element solutions. J Jpn Landslide Soc 56:227–239 ((in Japanese))

Yang H, Wang F, Miyajima M (2015) Investigation of shallow landslides triggered by heavy rainfall during typhoon Wipha (2013), Izu Oshima Island, Japan. Geoenvironmental Disasters. pp 2–15

Zhang LL, Fredlund DG, Zhang LM, Tang WH (2004) Numerical study of soil conditions under which matric suction can be maintained. Can Geotech J 41:569–582

WLF 2020 Kyoto Japan

Rainfall-Induced Lahar Occurrences Shortly After Eruptions and Its Initiation Processes in Japan

Takashi Koi, Yasuhiro Fujisawa, and Nobuo Anyoji

Abstract

This study examined the eruption magnitude and timing of lahar occurrences in recent eruptions in Japan. The volcanic deposition status and rainfall in time series were investigated for lahar frequent occurrences shortly after the eruption. The purpose of this study was to clarify the lahar initiation process and characteristics triggering rainfall during periods of approximately one month from the main eruption. Of the nine eruptions in Japan in recent years, lahars frequently occurred due to airborne tephra shortly after the eruption in three cases: the Usu 1977 eruption, Unzen 1990 eruption, and Miyakejima 2000 eruption. For each eruption case, the eruption transition and the occurrence/non-occurrence of lahar are summarized. The primary lahar was initiated in the range of 11–26 mm of hourly rainfall, i.e. "triggered rainfall". After the primary lahar, subsequent lahars were generated with smaller rainfall of less than 10 mm per hour due to the accumulation of volcanic falls and changes in the hydrological regime associated with the preceding rainfall. The existence of "triggered rainfall" is due to the change of unstable sediment that produce initiation conditions. Furthermore, compared to debris flows generated by factors outside the volcanic eruption, the rainfall related to lahar in relatively short durations or at a lower rainfall intensity is also characterized.

Keywords

Lahar • Debris flow • Rainfall intensity • Volcanic falls • Volcanic hazards • Disaster mitigation • Japan

Introduction

Volcanic eruptions cause debris flow and mudflow called lahar. Lahars occur mainly due to snowmelt around craters, the bursting of crater lakes, and post-eruption rainfall. Lahar outbreaks can cause massive damage to downstream settlements. More than 20,000 people were killed by snow-melting lahars of the Nevado del Ruiz volcano in Columbia in 1985 (e.g. Lowe et al. 1986). In Japan, 144 deaths due to snow-melting lahar occurred at Mt. Tokachi-dake, Hokkaido in 1926 (Uesawa 2014). On the other hand, lahars due to rainfall may occur continuously with each rainfall for several years after the eruption (Lavigne and Thouret 2002; Scott et al. 2005).

In Japan, where there are more than 100 active volcanoes in the Asian monsoon region, lahars caused by rainfall after an eruption are one of the most serious volcanic disaster concerns. The Unzen eruption of 1990 caused subsequent lahars due to rainfalls frequently occurring from May to June 1991, causing damage such as the outflow of bridges (Ikeya and Ishikawa 1993). Volcanic fall due to eruptions have been pointed out to change the topography and hydrological environment of the catchment, increasing runoff and sediment transport (Major et al. 1980; Gonda et al. 2014). However, the process of lahar occurrence due to rainfall after eruptions is poorly understood as there are few eruption cases and the information on lahar occurrence time and the characteristics that trigger such rainfall is feeble.

During the 2011 eruption of Shinmoedake, Kyushu, Japan, a large amount of pyroclastic materials fell due to the subplinian eruption. Evacuation information from the local government was issued frequently when it rained due to

T. Koi (✉)
Centre for Natural Hazards Research, Hokkaido University, Kita 9, Nishi 9, Kita-ku, Sappro, Hokkaido, 0608589, Japan
e-mail: koi@cen.agr.hokudai.ac.jp

Y. Fujisawa
Sabo and Landslide Technical Center, 2-7-5, Hirakawa-cho, Chiyoda, Tokyo, 1020093, Japan
e-mail: y-fujisawa@stc.or.jp

N. Anyoji
Sabo and Landslide Technical Center, 2-7-5, Hirakawa-cho, Chiyoda, Tokyo, 1020093, Japan
e-mail: nobanyoji@stc.or.jp

F. Guzzetti et al. (eds.), *Understanding and Reducing Landslide Disaster Risk*,
ICL Contribution to Landslide Disaster Risk Reduction,
https://doi.org/10.1007/978-3-030-60227-7_41

lahar concerns; as a result, lahar that reaching the downstream settlement did not occur. To understand the rainfall conditions that cause lahar during the brief period shortly after the eruption, it is particularly essential for disaster mitigation efforts to consider the evacuation and alerting of downstream residents and provide adequate evacuation information.

In this study, we examined the eruption magnitude and timing of lahar occurrence in recent eruptions in Japan. The volcanic deposition status and rainfall in time series were investigated for frequent occurrences lahar after the eruption. The purpose of this study was to clarify the lahar initiation process and characteristics triggering rainfall during periods of approximately one month from the main eruption. Debris flow and mudflow generated after volcanic fall are collectively called lahar in this study.

Materials and Methods

The eruption and lahar occurrences were summarized in nine cases (seven volcanoes located in Japan) in which lahars occurred after eruptions in recent years. The relationship between the eruption transition and occurrence/non-occurrence of lahar during rainfall events was compared in chronological order. For information on the eruption scale and volcanic phenomena related to sediment movement, we referred mainly to data published on the web site of the National Institute of Advanced Industrial Science and Technology (AIST 2020) and other published papers (Tajima et al. 2015; Maeno et al. 2016).

The occurrence of lahar is summarized by referring to past literature, the date and time of the occurrence of lahars after eruptions, the hourly and continuous rainfall just before the occurrence of lahar, and the hourly maximum rainfall when no lahar occurred. Various lahar occurrence detection methods are used such as lahar-detective sensor disconnection, camera monitoring, notification information, etc., however, in this study, the differences of the occurrence detection methods are not considered and the lahar occurrence time is unified. Unless otherwise specified, hourly rainfall and continuous rainfall refer to the rainfall just before the occurrence of lahars. Lahars whose time of occurrence is unknown are excluded in this study. For the rainfall, the observation value of the rain gauge station located the closest to the target stream was used as-is and 1-h rainfall (60 min of rainfall if there is 10 min rainfall) that can be uniformly compared for each eruption case is used as the rainfall index.

Sakurajima, where intermittently erupt in recent years and lahars frequently occur, was excluded in this study as determining the relationship between a specific eruption event and lahar-generated rainfall is difficult. In addition, if pyroclastic flow occurs, a large amount of fine-granulated pyroclastic material is supplied into the catchment. As it is necessary to consider it separately, this study focus only on the effect of volcanic fall (ash and pumice) associated with the eruption and not on the effect of pyroclastic flow.

Occurrence of Frequent Rainfall-Induce Lahars in Japan

The eruption status and lahar occurrence status of the nine cases were organized (Table 1). Of the nine cases, five had lahars due to rain shortly after the eruption (approximately less than one month after the main eruption). The remaining four cases had lahar occurrences approximately more than four months after the main eruption even despite multipole rainfall. Of the five lahars that occurred shortly after the main eruption, lahars occurred frequently in three cases (the Usu 1977, Unzen 1990, and Miyakejima 2000 eruptions). In the remaining two cases (the Kuchinoerabu-jima 2013 and Ontake 2014 eruptions), lahars occurred only once. Although pyroclastic flow occurred at the Unzen since May 24, lahars occurred before the pyroclastic flow occurred.

All three cases in which lahars occurred frequently shortly after the main eruption were magmatic eruptions, and the amount of volcanic fall were approximately from 10^5 to 10^6 m^3 order, which was larger than the other eruption cases; this is true for all cases except for the 2011 Shinmoedake eruption, which was also a magmatic eruption whose amount of volcanic fall (25×10^6 m^3) was the second largest amount of the nine eruption cases (Table 1). Lahars, however, occurred five months after the main eruption and did not occur frequently. The eruption of the Shinmoedake 2011 eruption was mainly composed of pumice and had a small amount of fine-grained volcanic ash; therefore, lahar did not occur because the infiltration related to lahar generation did not decrease (Koi et al. 2013). Therefore, this suggests that the difference in the timing and frequency of lahar occurrences after the eruption as a predisposition may be affected not only by the magnitude of the eruption but also by the properties of the ejecta.

Time-Series Relationship Among Eruptions, Rainfalls and Lahar Occurrences

For three cases (the Usu 1977, Unzen 1990, and Miyakejima 2000 eruptions), that had lahars occur frequently shortly after the eruption, the time series of eruptions and lahar occurrence/non-occurrence rainfall were arranged (Fig. 1) and described as follows.

Table 1 Occurrences of volcanic eruptions and subsequent rainfall-induced lahars in Japan. Lava F/D indicates lava flow and lava dome, and Pyro F/S indicates pyroclastic flow and pyroclastic surge

	Volcano (Region)	Main eruption date (start date)	Amount of volcanic ejecta ($\times 10^6$ m^3)				Date of the primary lahar	Lahar type
			Total	Lava F/D	Fall	Pyro F/S		
1	Usu (Hokkaido)	07–14/08/1977	100	–	80<	?	16/8/1977	Debris flow Frequent
2	Unzen (Kyushu)	9/4/1991 (17/11/1990–)	300	100	0.9	200	15/05/1991	Debris flow Frequent
3	Hokkaido-Komagatake (Hokkaido)	31/02/1996	0.1	–	0.1	–	01/07/1996	Mud flow Sporadic
		25/10/1998	0.05	–	0.05	–	01/07/1999	
		04/09/2000	0.1	–	0.1	–	16/07/2001	
4	Miyakejima (Izu Islands)	08/07/2000	16	–	13	3	16/07/2000	Debris flow Frequent
5	Shinmoedake (Kyushu)	26–27/01/2011 (19/01/2011)	40	15	25	?	25/06/2011	Mud flow Sporadic
6	Ontake (Chubu)	27/09/2014	0.7–1.3	<10% of the total is Pyro F, and the rest is Fall. No Lave F/D			05/10/2014	Debris flow Once
7	Kuchinoerabu-jima (Kyushu)	29/05/2015	?	–	0.5	?	03/06/2015	Debris flow Once

The Usu 1977 Eruption

On August 7, 8, 9, 13, and 14 of 1977, five large-scale magmatic eruptions occurred and thick volcanic ash was deposited around the crater. Before the eruption on August 7, there were four rainfall events with an hourly rainfall of 1 mm to 12 mm. On August 12 after the eruption started, there was rainfall (maximum hourly rainfall of 3 mm) with no lahar occurrence.

The primary lahars occurred at multiple streams during the rainfall on August 16 (hourly rainfall 11 mm, continuous rainfall 48 mm), causing flooding to farmland. Later, on September 11 and 14, lahars occurred with hourly rainfall of 6 mm (continuous rainfall of 25 mm with flood damage) and 8 mm (continuous rainfall of 32 mm without flood damage), respectively. Although lahar has also been recorded on August 20 and September 3, there is no rainfall record and the amount of rainfall just before the occurrence is unknown.

The Unzen 1990 Eruption

An eruption began on November 17, 1990. A small eruption occurred on February 12, 1991, and the main eruption on April 9 caused tephra fall. After the eruption on May 24, pyroclastic flow occurred frequently. From the eruptions on April 9 to May 14, there were five rain events (maximum hourly rainfall of 6–11 mm), but no lahars occurred.

Before the occurrence of pyroclastic flow, the primary lahar occurred at 1:00 am on May 15 with an hourly rainfall of 14 mm (continuous rainfall of 20 mm without flood damage). Then, at 6:00 am on the same day, a lahar occurred at an hourly rainfall of 25 mm (continuous rainfall of 77 mm), causing damage to bridges. Thereafter, one lahar occurred on May 19 with an hourly rainfall of 9 mm (continuous rainfall 32 mm with facility damage); three lahars without flood damage occurred on May 20 and two lahars without flood damage on May 21.

The Miyakejima 2000 Eruption

The main eruption occurred on July 8, 2000, and there were subsequent eruptions on July 14 and 15. After the main eruption on July 8, there was a maximum rainfall of 27 mm for the event on July 14 and four rainfall events (maximum rainfall of 2–7 mm) by July 25 with no lahar occurrences.

The primary lahar occurred on July 26 in three streams on the island. The hourly rainfall at this time was 12 mm (continuous rainfall 43 mm, without flood damage), 22 mm (continuous rainfall 31 mm, with flood damage), and 26 mm (continuous rainfall 28 mm, with flood damage), respectively. Lahar occurred in three rain events since the primary lahar. The rainfall for each is as follows: on August 10, an hourly rainfall 1–2 mm (continuous rainfall 1–2 mm, with flood damage); on August 12, an hourly rainfall 3–8 mm (continuous rainfall 32–33 mm, with flood damage); and on August 18, an hourly rainfall 1 mm (continuous rainfall 1 mm, no flood damage).

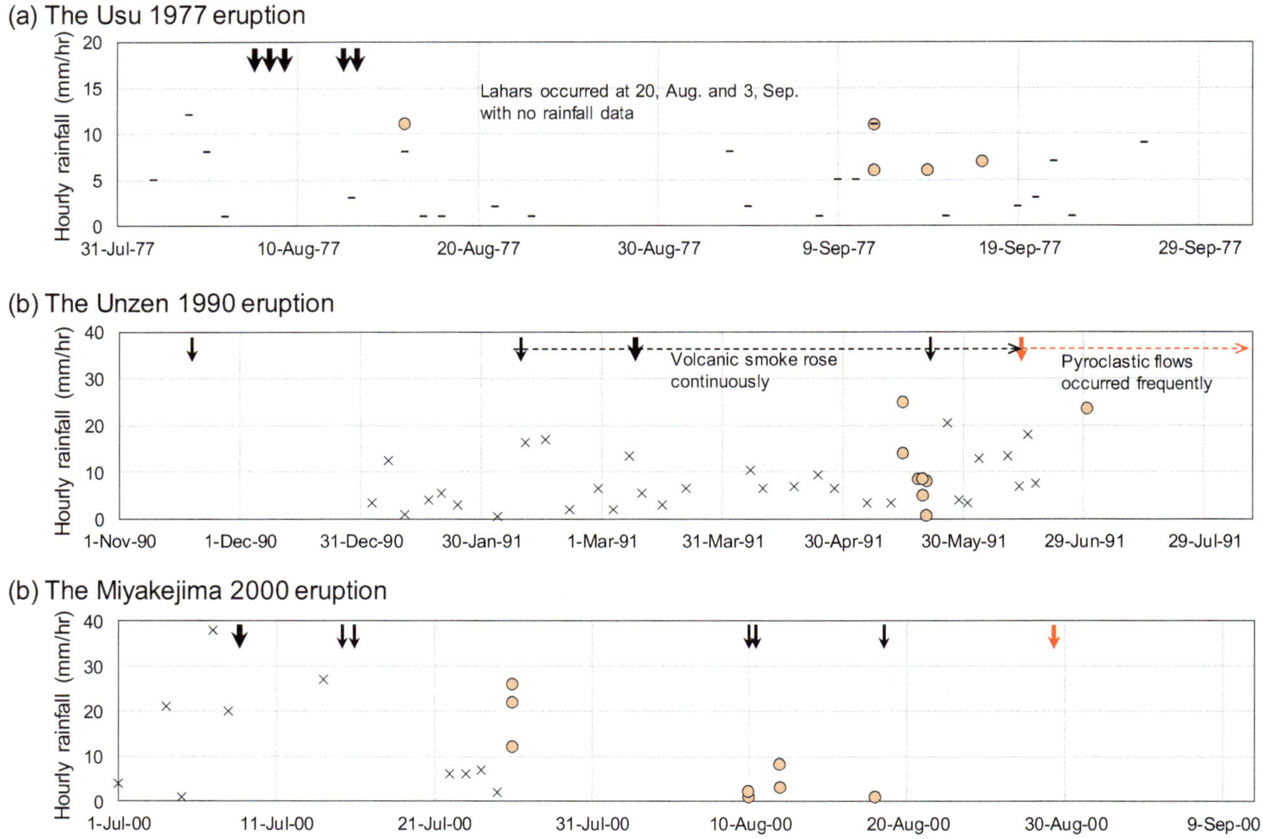

Fig. 1 Time-series relationship among eruptions, hourly rainfalls and lahars occurrence/non-occurrence. Bold black arrows indicate main eruption date; slim black arrows indicate the other eruption and red arrows indicate eruption date with pyroclastic flows. Circles represent lahar occurrence and crosses represent no lahar occurrence. In the Usu 1977 eruption, because only the rainfall at the Lahar occurrences was recorded, the rainfall at another station with rainfall observation data is shown for reference; marks of black bars

Initiation Processes of Rainfall-Induced Lahars

The rainfall that caused the primary lahars ranged from an hourly rainfall of 11 mm (the Usu 1977 eruption) to 26 mm (the Miyakejima 2000 eruption). Lahars did not occur before the eruptions or in the early stages of the eruptions, even though there was similar rainfall, which induced the primary lahar. The accumulation of volcanic falls accompanying the eruptions changed the hydrological environment and the lahar occurrence was considered. After the accumulation of volcanic falls, the primary lahar had the largest rainfall up to that point; however, after the primary lahar occurred, subsequent lahars induced by small magnitude rainfall of less than 10 mm (hourly rainfall).

At Mount Merapi, Indonesia, the average rainfall of lahar occurrence is reported to be 40 mm/2 h (Lavigne et al. 2000); the primary lahar occurring in Japan has similar rainfall intensity. In contrast, the subsequent lahars were caused by smaller rainfall. At the Unzen 1990 eruption, after lahar occurred for the first time on May 15, 1991, an amount of rainfall, smaller than the rainfall that caused the primary lahar, that caused subsequent lahars was reported (Ikeya and Ishikawa 1993). In this study, similar trends were observed in the Usu 1977 eruption and the 2000 eruption of Miyakejima.

When volcanic ash accumulates on hillside slopes, infiltration decreases, and surface erosion is considered to be dominant (Ikeya et al. 1995); and rills and gullies develop over time on the mountain slopes where volcanic products are deposited (Anyoji 2004). In particular, surface erosion progresses rapidly just after the eruption (Collins and Dunne 1986). Moreover, after the Shinmoedake 2011 eruption during the years following the typhoon passage, bed-material load discharge increased during small-scale flooding over the subsequent years due to the movement of unstable sediment (pyroclastic materials) from the slope to around the river channel (Koi et al. 2018).

From above, the initiation processes of frequent rainfall-induced lahar occurrences shortly after the eruptions (Fig. 2) are assumed as follows; the lahars were caused not only by volcanic deposition associated with the eruptions and changes in the hydrological regime associated with the rainfall events experienced before the primary lahar but also

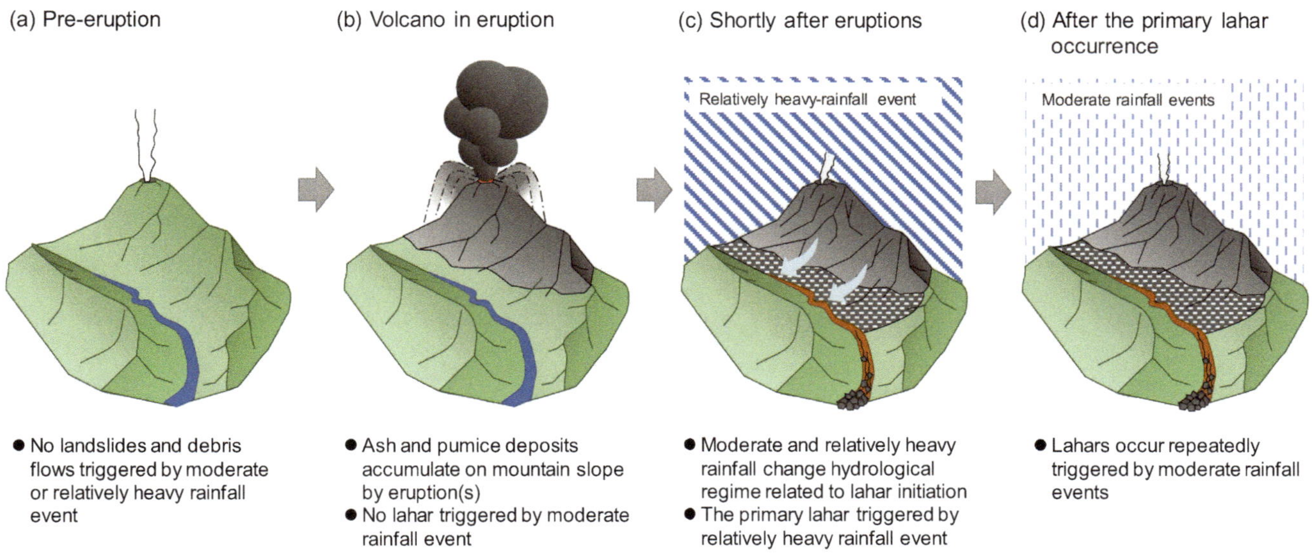

Fig. 2 Schematic illustrations of the initiation processes of frequent rainfall-induced lahar occurrences shortly after the eruptions. Relatively heavy and moderate rainfall event indicate 11–26 mm/hr and <10 mm/hr in this study

by the existence of "triggered rainfall" that caused the primary lahar. After the primary lahar, the subsequent lahar was likely to be generated even during low-magnitude rainfall (moderate rainfall events), because the sediment is likely to move as the sediment accumulates near the river channel.

Characteristics of Rainfall-Induced Lahars

Of the three eruptions, the occurrence times of 22 cases of the lahars that occurred during the period of approximately one month from the main eruption were detected. To understand the characteristics of the rainfall that cause lahars shortly after the eruption, we compared them with the published rainfall intensity-duration (I-D) threshold (Fig. 3). The global I-D threshold for shallow landslides and debris flows was firstly proposed by Caine (1980) and updated by Guzzetti et al. (2008). The event rainfall duration in 22 lahar occurrence cases of 3 eruption events ranges from 0.167 h to 67 h; the rainfall intensity ranges from 0.6 mm/hr to 12.83 mm/hr.

The 22 cases generally plot within the global I-D threshold proposed by Guzetti et al. (2008). In the case of debris flow, it can be seen that the rainfall intensity is much lower than the regional I-D threshold of Wieczorek et al. (2000). Lahars were also generated with lower rainfall intensities than after a wildfire or large-scale earthquake, in which debris flows are relatively likely to occur (Cannon and Gartner 2005; Jan and Chen 2005). In addition, it can be seen that lahars are generated due to rainfall in a short period of rainfall compared with the shallow landslides standard in Japan (Saito et al. 2010). This suggests that lahars, which occur shortly after the eruption, tend to occur during light-intensity and short-duration rainfall events compared to debris flow and landslide caused by factors outside the volcanic eruption.

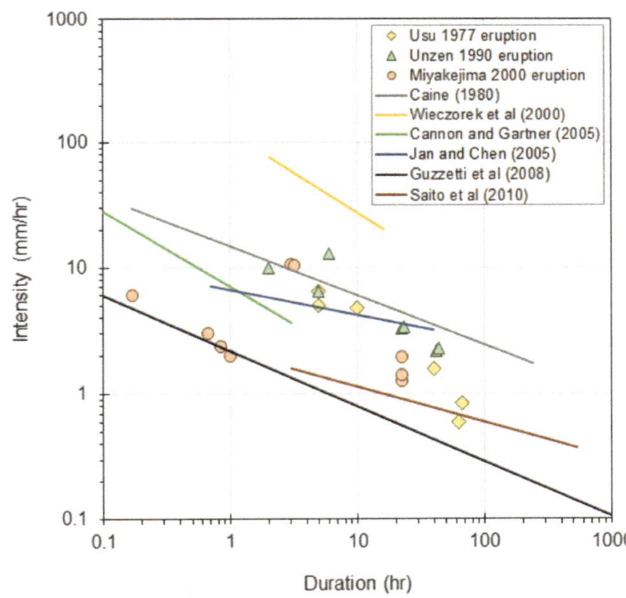

Fig. 3 Relationship between rainfall intensity and duration of frequent rainfall-induced lahars, and the published rainfall I-D thresholds. Marks indicate rainfall conditions of the rainfall-induced lahars and lines indicate the published rainfall I-D thresholds

Conclusions

In Japan, lahars frequently occurred due to volcanic falls shortly after eruption in three cases: the Usu 1977, Unzen 1990, and Miyakejima 2000 eruptions. For each eruption case, the eruption transition and occurrence/non-occurrence of lahars are summarized; the primary lahar was initiated in the range of 11–26 mm of hourly rainfall, i.e. "triggered rainfall". After the primary lahar, the subsequent lahars were generated with a smaller rainfall of less than 10 mm per hour due to the accumulation of volcanic falls and change in the hydrological regime associated with the preceding rainfall. These results suggested that the time-series rainfall condition was due to the change of unstable sediment and lahar initiation conditions, and thus, the prediction and notice of "triggered rainfall" is important shortly after eruptions. Furthermore, compared to debris flows generated by factors outside the volcanic eruption, the rainfall related to lahars in a relatively short duration or at a lower rainfall intensity is also characterized. As there is little information yet on rainfall that causes frequent lahars shortly after eruptions, to try to acquire data during future eruptions is essential for disaster mitigation.

References

AIST (2020) Volcanoes of Japan. URL: https://gbank.gsj.jp/volcano/index.htm [Accessed 11 March 2020]

Anyoji N (2004) Study on prediction of sediment discharge in the active volcano, based on geomorphic analysis of gully erosion. Res Bull Hokkaido Univ Fores 61(2):11–71 ((in Japanese with English summary))

Caine N (1980) The rainfall intensity-duration control of shallow landslides and debris flows. Geogr Ann 62A:23–27

Cannon SH, Gartner JE (2005) Debris-flow hazards and related phenomena. In: Jakb M, Hungr O (eds) Springer. (ISBN_978-3-540-27129-1). p 739

Collins BD, Dunne T (1986) Erosion of tephra from the 1980 eruption of Mount St. Helens. Geol Soc Am Bull 97:896–905

Gonda Y, Legono D, Sukatja B, Santosa UM (2014) Debris flows and flash floods in the Putih river after the 2010 eruption of Mt. Merapi, Indonesia. Int J Erosion Control Eng 7(2):63–68

Guzzetti F, Peruccacci S, Rossi M, Stark C (2008) The rainfall intensity-duration control of shallow landslides and debris flows: an update. Landslides. 5:3–17

Ikeya H, Hendrayanto KK, Mizuyama T (1995) Observed changes in the infiltration rates of pyroclastic deposits resurliting form volcanic activities. J Jpn Soc Erosion Control Eng 48(2):22–26 ((in Japanese with English abstract))

Ikeya H, Ishikawa Y (1993) Characteristics of pyroclastic flows and debris flows accompanying the Mt. Unzen-Fugendake eruption. Sed Probl: Strat Monit, Prediction, Control, IAHS Publ 217:57–66

Jan C, Chen C (2005) Debris-flow hazards and related phenomena. In: Jakb M, Hungr O (eds) Springer. (ISBN_978-3-540-27129-1). p 739

Koi T, Fujisawa Y, Fujita K, Anyoji N, Takiguchi S, Sugiyama M (2013) Characteristics of juvenile pyroclastic materials affected on runoff and sediment discharge—a case study of the 2011 eruption of the Shinmoedake. J Jpn Soc Erosion Control Eng 65(6):37–45 ((in Japanese with English abstract))

Koi T, Egashira I, Suzuki T, Anyoji N (2018) Sediment discharge from a forested-mountain watershed affected by volcanic ashfall during the 2011 eruptions of Shinmoe-dake. J Jpn Soc Erosion Control Eng 70(6):24–33 ((in Japanese with English abstract))

Lavigne F, Thouret JC, Voight B, Suwa H, Sumaryono A (2000) Lahars at merapi volcano, central Java: An overview. J Volcanol Geoth Res 100:423–456

Lavigne F, Thouret JC (2002) Sediment transportation and deposition by rain-triggered lahars at merapi volcano, central java, Indonesia. Geomorphology 49:45–69

Lowe DR, Williams SN, Leigh H, Connor CB, Gemmell JM (1986) Lahars initiated by the 13 November 1985 eruption of nevad del Ruiz, Colombia. Nature 324:51–53

Major JJ, Crisafulli CM, Frenzen P, Bishop J, After the disaster: The hydrogeomorphic, ecological, and biological responses to the 1980 eruption of Mount St. Helens, Washington. *The Geological Society of America Field Guide, 15*, 111–134

Maeno F, Nakada S, Oikawa T, Yoshimoto M, Komori J, Ishizuka Y, Takeshita Y, Shimano T, Kaneko T, Nagai M (2016) Reconstruction of a phreatic eruption on 27 September 2014 at Ontake volcano, central Japan, based on proximal pyroclastic density current and fallout deposits. Earth, Planets and Space, p 68

Saito H, Nakayama D, Matsuyama H (2010) Relationship between the initiation of shallow landslide and rainfall intensity-duration thresholds in Japan. Geomorphology 118:167–175

Scott KM, Vaance JW, Kerle N, Macías JL, Strauch W, Devoli G (2005) Catastrophic precipitation-triggered lahar at Casita volcano, Nicaragua: occurrence, bulking and transformation. Earth Surf Proc Land 30:59–79

Tajima Y, Geshi N, Nakagawa S, Nagai M (2015) Ash fall volume estimation by EAI method to Kuchinoerabujima eruption at May 29th, 2015. Volcanological society of Japan, Fall Meeting. p 14

Uesawa S (2014) A study of the Taisho lahar generated by the 1926 eruption of Tokachidake Volcano, central Hokkaido, Japan, and implications for the generation of cohesive lahars. J Volcanol Geoth Res 270:23–34

Wierczorek GF, Morgan BA, Campbell RH (2000) Debris-flow hazards in the Blue Ridge of central Virginia. Environ Eng Geosci 6(1):3–23

Spatiotemporal Assessment of Geological Hazard Safety Along Railway Engineering Using a Novel Method: A Case Study of the Sichuan-Tibet Railway, China

2020 Kyoto Japan

Jiaying Li and Wei-Dong Wang

Abstract

The Sichuan-Tibet Railway in China is the significance for the long-term development of "The Belt and Road". However, railway engineering reshapes the landscape it passes through and consequently influences the geological hazard safety. Because of a paucity of previous studies involving this factor, in the present paper, we make a novel development in our previous study that a novel index system, notably from the spatiotemporal aspect. Meanwhile, the effect of railway construction is proposed for geological hazard assessment. Different from regional or site assessment, the concept of the line element is introduced, and the geological hazard safety along Sichuan-Tibet Railway is comprehensively evaluated using matter-element extension model (MEEM), gray correlation model (GCM), and support vector machine (SVM). Meanwhile, the receiver operating characteristic curve (ROC) analysis demonstrates that the method performs better than the method without considering the construction factor or temporal factor. The finding highlights that the construction factor and temporal factor play an essential role in the assessment of geological hazard safety along the railway.

Keywords

Geological hazard • Sichuan-tibet railway • Safety assessment • Construction factor • Spatiotemporal factor

J. Li · W.-D. Wang (✉)
School of Civil Engineering, Central South University, 68 Shaoshan Road, Changsha, 410075, Hunan, China
e-mail: 147745@163.com

J. Li
e-mail: 184801058@csu.edu.cn

F. Guzzetti et al. (eds.), *Understanding and Reducing Landslide Disaster Risk*, ICL Contribution to Landslide Disaster Risk Reduction,
https://doi.org/10.1007/978-3-030-60227-7_42

Introduction

Many available models, such as logistic regression models (LRM) and fuzzy evaluation model (FEM), have been proposed to generate susceptibility maps of geological hazards currently (Pradhan and Lee 2010). Establishing an optimized evaluation system of the geological hazard safety needs not only a rational evaluation model but also an opposite evaluation index system. The system in previous studies always takes natural environmental factors, such as elevation, lithology, slope and etc., as dominant factors for susceptibility assessment (Pourghasemi et al. 2018).

Geological hazard assessment along the railway engineering also needs to consider construction factors. Despite this, very few studies have investigated the assessment of geological hazards along a railway or highway (Lee et al. 2018). However, the construction of the railway changes the geological characterization, which is an irreversible process and has a great effect on the original stability factor and geological hazard susceptibility (Xiong et al. 2018; Quang et al. 2018).

On the other hand, because of the seasonality of the climate and rainfall, the temporal factor has important implications for analyzing the assessment of geological hazard safety comprehensively. Although some studies have reported the risk assessment of geological hazards in a tunneling project (considered as a line) (Haghshenas et al. 2017), the study only considers the engineering characteristics without the natural environmental factors and temporal factor. A systematic understanding considering spatial factor and temporal factor up to now is still lacking.

Study Area

The Sichuan-Tibet Railway (STR) is significant to Chinese long-term stability and the development of "The Belt and Road". The length of STR is 1620 km and consists of three

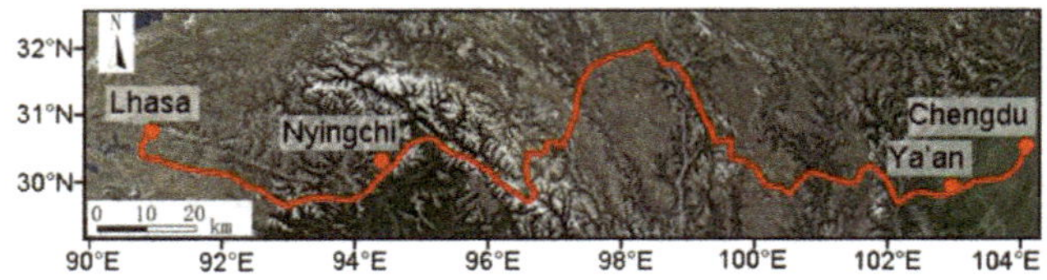

Fig. 1 Map of the Sichuan–Tibet railway

major sections: the Chengya railway, YaNying railway, and LaNying railway (Fig. 1). The study area is the Chengya railway. The total length of the railway section in operation is 41.48 km. It is the first section of the STR design, which is a significant foundation for the STR.

The beginning of the Chengya railway is located on the Chengdu Plain. The terrain of the railway is affected by the Tibetan Plateau where the western part is much higher than the eastern part, and the average annual rainfall differs greatly between the eastern part of the railway and the western part. Meanwhile, part of the railway is located in the Himalaya earthquake zone, which suffers from permafrost, avalanches, landslides, and earthquakes (Wang et al. 2019). The rainfall in the study area is very abundant, which is a typical rainy area.

Methodology

In the present study, the concept of the line element is proposed to analyze geological hazards along a railway. The study railway is divided into 462 line elements, with the average length of an element being 100 m, and the line elements are taken as units in the safety assessment. Meanwhile, the methodologies in this study are as shown in Fig. 2.

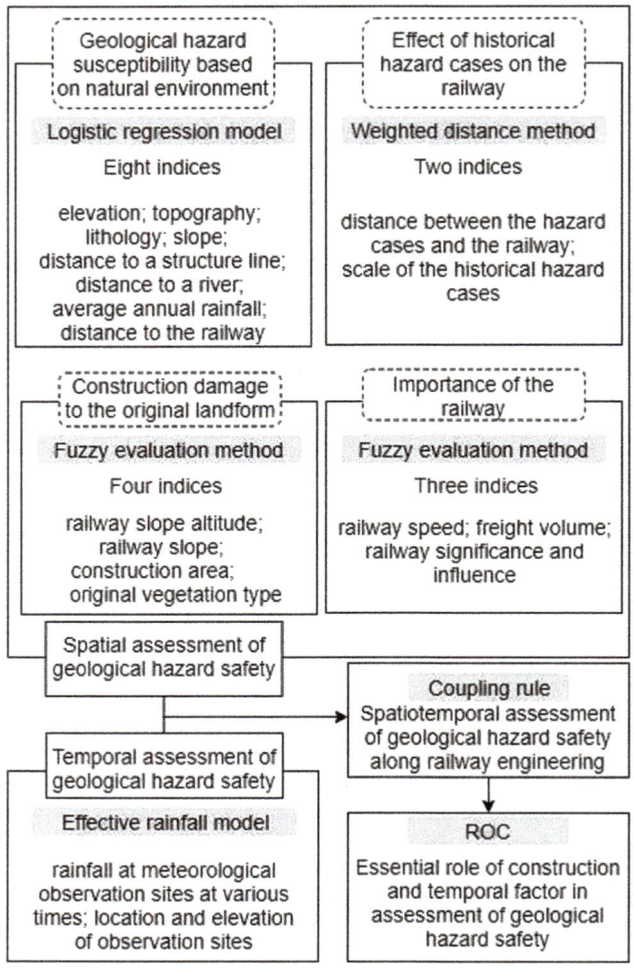

Fig. 2 Flowchart of research applied in this study

Results

Spatial Assessment of Geological Hazard Safety

The present study chooses four index-categories, with seventeen indices, as the spatial assessment index system of geological hazard safety. The index-categories quantified using LRM, weighted distance model (WDM), and FEM. The index system is then used to generate the safety level map employing the matter-element extension model (MEEM), gray correlation model (GCM), and support vector machine (SVM).

Natural Environment

The independent variables are the geological characterization indices and the dependent variable is the probability of the geological hazard occurring. By transforming the universal formula of LRM, the probability Z is obtained:

$$Z = B_0 + B_1 I_{1j} + B_2 I_{2j} + \ldots + B_n I_{nj} \tag{1}$$

where I_{ij} is the geological characterization index; B_i is the LRM coefficient; $Z = \ln(P/(1-P))$; and P is the probability of the geological hazard occurring.

LRM coefficients and regression formula are obtained by LRM in Statistical Product and Service Solutions (SPSS). The probability Z is obtained:

$$\begin{aligned} Z = & -5.107 + 2.805 I_{Slope} + 1.515 I_{Topo} \\ & + 0.785 I_{Lith} + 1.482 I_{Tect} - 2.411 I_{Rail} \\ & + 1.993 I_{River} + 2.251 I_{Rain} - 0.585 I_{Elev} \end{aligned} \tag{2}$$

Using the layer of the Chengya railway, the level map (Fig. 3a) of geological hazard safety based on the natural environment is obtained.

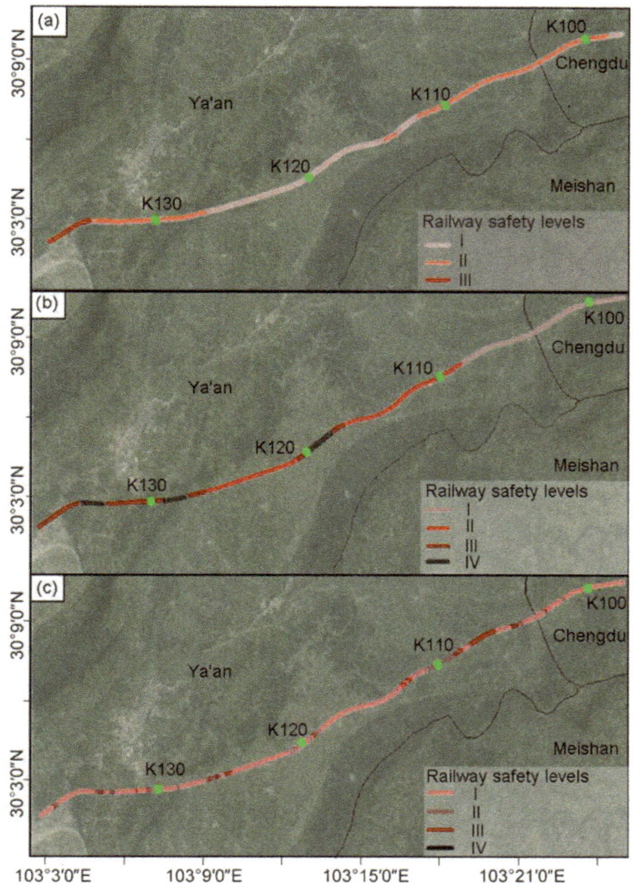

Fig. 3 Level maps of assessment of geological hazard safety based on **a** Natural environment, **b** Effect of historical hazard cases on the railway, and **c** construction damage to the original landform

Effect of Historical Hazard Cases on the Railway

The effect of historical hazard cases on the railway is evaluated using WDM. The WDM introduces the mathematical concept of the fitting degree (Lee and Yu 2012; Liu et al. 2019). Each railway element is considered as a regression line, and historical hazard cases are considered as individual points scattered around the regression lines.

The fitting degree of each railway element is obtained as:

$$u_i(x, y) = 1/e_i(x, y) \tag{3}$$

where $e_i(x, y)$ is the euclidean distance between the individual point and regression line.

The scale of the historical hazard cases is considered as the area of influence of historical hazard cases. The effect of historical hazard cases on the railway can be obtained as:

$$C_i = \sum_{i=1}^{n} a_i u_i(x, y) \tag{4}$$

where C_i is the effect of historical hazard cases on the railway; a_i is the area of a historical hazard case; i is the historical hazard case; j is the railway element. The level map (Fig. 3b) of geological hazard safety based on the effect is therefore obtained.

Construction Damage to the Original Landform

The weights of the indices are determined using the trapezoidal fuzzy number, and the level of index-categories is obtained using the hundred-mark system. The fuzzy evaluation B is calculated as:

$$B = WoR \tag{5}$$

where W is the weight subset; o is the fuzzy operator; and R is the fuzzy evaluation vector.

The construction damage to the original landform is evaluated using FEM. Three experts were invited to evaluate the weights of the indices, and the weight subset is obtained as:

$$W = (0.27, 0.29, 0.22, 0.22) \tag{6}$$

The map (Fig. 3c) of the safety level of the construction damage to the original landform is therefore obtained.

Importance of the Railway

The importance of the railway is divided into four classes using a FEM. The weights of the indices are determined by the trapezoidal fuzzy number. Three experts were invited to evaluate the weights of the three indices, and weights are obtained as:

$$W = (0.31, 0.36, 0.33) \tag{7}$$

The fuzzy evaluation is obtained as:

$$B = (0.31, 0.36, 0.33) \cdot \begin{bmatrix} 74.44 \\ 95.0 \\ 95.0 \end{bmatrix} = 88.68 \tag{8}$$

Spatial Assessment Using MEEM, GCM, and SVM

MEEM, GCM, and SVM are used to obtain the safety levels along the railway.

The steps of MEEM using Visual Basic for Applications (VBA) are as follows: (a) Determination of the classical domain. (b) Determination of joint domain. (c) Determination of the matter-elements to be evaluated. (d) Determination of the levels of the matter-elements.

GCM is a method to determine the degree of variable correlation. The four index-categories of geological hazards susceptibility in the railway are scaled to a range of 0–1 using the GCM (Liu 2011). The correlation coefficients (ξ) are therefore unified reflecting the proximity of the comparison and the reference series. The GCM parameter γ_i is obtained as:

$$\gamma_i = \sum^{n} \lambda_j \xi(j) = \sum^{n} \frac{\lambda_j}{1 + |X_0 - X_i|} \quad (9)$$

where the λ_j is the weight of the index-categories j; $\xi_i(j)$ is the correlation coefficients; X_i is the evaluation index series of the railway line element i; and X_0 is the reference series ($X_0 = (1, 1, 1, 1)$).

The steps of SVM using Matrix Laboratory (MATLAB) are as follows: (a) The raw data of the index-categories are quantified in the range of 0–10 by linear interpolation. (b) Evaluation results of the railway safety are obtained using the MEEM and GCM. (c) The radial basis function is then chosen as the kernel function. The kernel function parameter and penalty factor are obtained through a grid search and cross-validation. (d) The test samples are evaluated using the SVM.

In general, the spatial assessment maps (Fig. 4) of geological hazard safety are obtained.

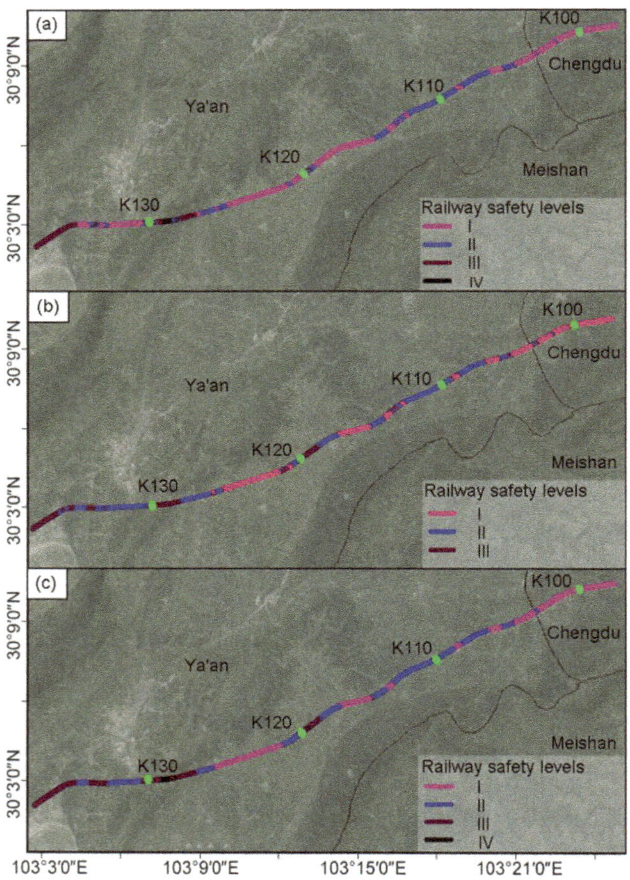

Fig. 4 Level maps of spatial safety assessment obtained using **a** MEEM, **b** GCM, and **c** SVM

Temporal Assessment of Geological Hazard Safety

Taking the rainfall as a significant factor, a temporal assessment of geological hazard safety is obtained using an effective rainfall model (ERM). The effective rainfall refers to the rainfall which stays in the rock and contributes to the development of geological hazards. Based on the hydrological data, antecedent precipitation index (API) is proposed to the study on the prediction of geological hazard. The API is therefore obtained as:

$$R_c = \sum_{i=0}^{n} K^i R_i \quad (10)$$

where R_c is the effective rainfall; R_i is the rainfall of the previous i th day; and K is the attenuation coefficient.

Therefore, the steps of temporal assessment of geological hazard safety are as follow: (a) Determine the time for studying and collect the rainfall data from various meteorological observation stations in the study area. (b) Obtain effective rainfall using the effective rainfall model. (c) Determine the map of effective rainfall using spatial interpolation in ArcGIS. (d) Obtain the temporal assessment of safety along the railway by overlaying the railway map.

In the present study, taking June 8 in 2015 as the example, the rainfalls from June 3 to June 8 in 2015 are selected to study (Table 1). The effective rainfall data of 146 stations in Sichuan is therefore obtained. Based on effective rainfall data and elevation of various stations, the level map of temporal safety assessment in Sichuan is obtained by using Kriging Interpolation. Therefore, the level map of assessment of geological hazard safety is presented (Fig. 5).

Spatiotemporal Assessment of Geological Hazard Safety

By coupling the spatial maps of geological hazard safety with the temporal map of geological hazard safety, the spatiotemporal maps are obtained. The discrimination rule (Table 2) of geological hazard safety in the present study is used widely by the Ministry of Land and Resources of China.

Based on the discrimination rule, the level maps of spatiotemporal assessment of geological hazard safety are presented by using the temporal map and the spatial maps obtained by MEEM, GCM, and SVM (Fig. 6).

Table 1 Rainfall of various stations

Rainfall	R_5	R_4	R_3	R_2	R_1	R_0
Bazhong	54.5	32.5	75.9	34.8	60.4	50.2
Chengdu	29.7	28.0	50.8	59.2	55.7	36.4
Dazhou	50.1	23.4	29.4	10.3	8.8	34.0
Deyang	21.5	9.5	12.9	9.4	6.5	15.0
Ganzi	7.6	12.5	45.7	60.3	58.4	22.0
Guangan	26.7	15.8	35.4	5.8	30.8	23.1
Guangyuan	8.3	10.4	3.5	9.4	2.3	8.1
Leshan	21.9	48.4	10.8	8.5	75.6	26.3
Liangshan	22.3	57.5	30.6	53.8	48.2	37.7
Luzhou	75.0	51.5	22.3	61.9	43.4	56.1
Meishan	11.5	3.6	18.3	20.4	10.6	11.4
Mianyang	22.8	31.6	43.1	60.1	53.7	33.2
Nanchong	50.3	11.4	12.4	23.4	23.4	28.4
Neijiang	38.1	46.8	47.4	36.7	13.7	42.4
Abeizang	13.3	32.4	39.5	54.0	39.5	28.4
Panzhuhua	15.3	22.4	34.5	10.5	34.8	20.8
Suining	33.4	22.5	21.2	47.1	44.2	29.1
Yaan	6.9	11.6	12.5	24.0	9.1	11.1
Yibing	22.8	24.3	28.4	29.5	19.4	25.0
Zigong	65.0	25.6	36.1	35.0	63.3	44.4
Ziyang	22.7	55.1	39.7	34.0	50.7	36.9

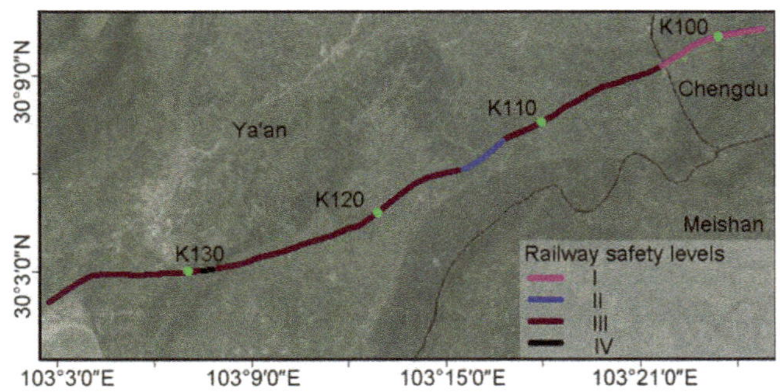

Fig. 5 Level map of temporal safety assessment

Table 2 Discrimination rule of geological hazard safety

Spatiotemporal assessment level		Temporal assessment level			
		I	II	III	IV
Spatial assessment level	I	I	I	I	IV
	II	I	II	III	IV
	III	I	III	IV	V
	IV	IV	IV	V	V

As shown in Fig. 6, although the assessment results obtained using the three models are different, there are many similarities. Only a few railway elements have different safety levels. Even if very few results are different, the different evaluation is mostly at the adjacent level. The average proportion of the railway elements at level I, II, and III is 83.56%, which do not need to be protected. The average proportion of elements at levels IV and V is 16.44%. These elements require protection. On the basis of these results, it is concluded that the design of the railway takes full account of the geological hazards so that the railway keeps away from the geological hazards.

Discussion

The correlation of the results obtained using the three models is determined through correlation analysis (Table 3). According to the literature (Nijmeijer et al. 2001), the correlation coefficient (r) is divided into five classes, namely very low correlation ($0 \leq r < 0.2$), low correlation ($0.2 \leq r < 0.4$), medium correlation ($0.4 \leq r < 0.7$), high correlation ($0.7 \leq r < 0.9$), and very high correlation ($0.9 \leq r < 1$).

Table 3 shows that the coefficients of correlation between the three models are 0.762, 0.807, and 0.902. The results demonstrate the correlation coefficient between GCM and SVM is the highest. Meanwhile, by comparing the results

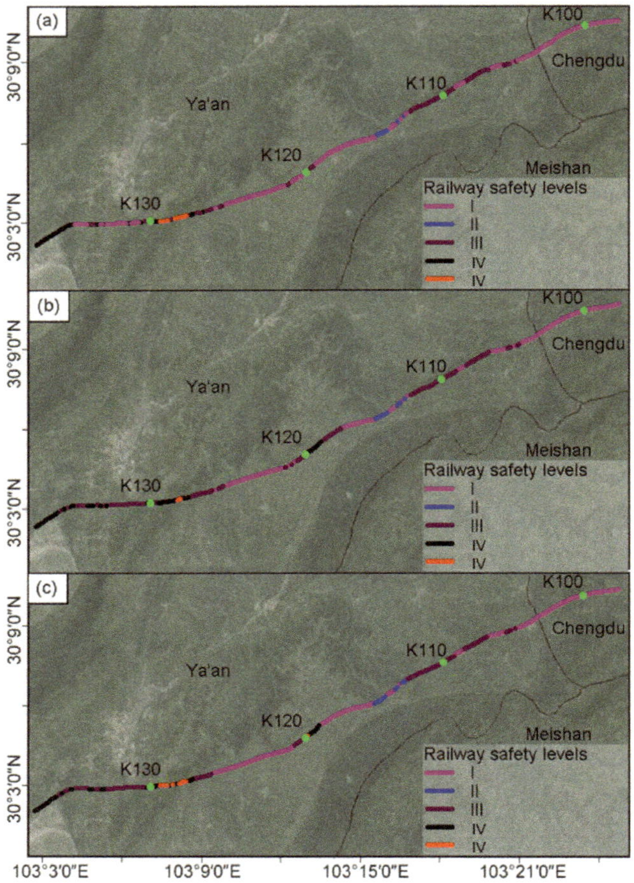

Fig. 6 Level maps of spatiotemporal assessment using temporal map and the spatial maps obtained by **a** MEEM, **b** GCM, and **c** SVM

Table 3 Correlation coefficients between the three models

Model	MEEM	GCM	SVM
MEEM	1	0.762	0.807
GCM	0.762	1	0.902
SVM	0.807	0.902	1

obtained using the three models with the historical hazard cases, ROC curves (Fig. 7) are obtained in SPSS (Ozdemir and Altural 2013). ROC fitting analysis determines the applicability of the three models to study railway.

The area under the ROC curve (AUC) is a measure of the performance in a diagnostic test because it reflects the test performance at all possible cut-off levels (Westin 2001). The larger the AUC is and the smaller the standard deviation is, the more applicable the model is. As can be seen from Fig. 7, the GCM is thus more applicable in evaluating the railway safety levels in this case.

On the other hand, railway safety levels ignoring the construction factor and temporal factors are analyzed using GCM. The ROC (Fig. 8) is used to compare the above

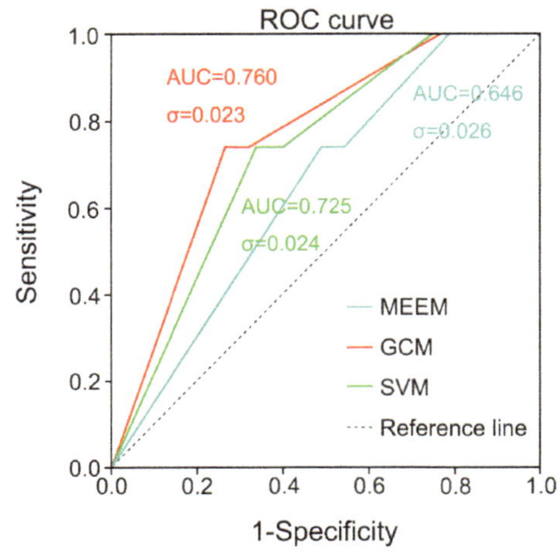

Fig. 7 ROC curves of the three models

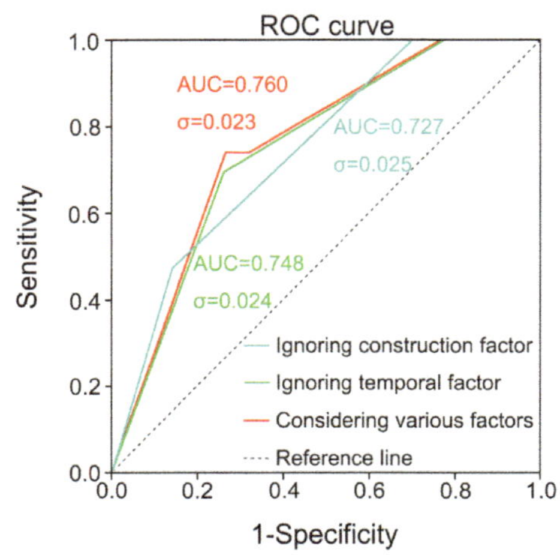

Fig. 8 ROC curves of whether assessment results consider the construction factor and temporal factor

assessment results obtained using GCM. What can be clearly seen above is that the construction factor and temporal factor are significant to geological hazard assessment of railway engineering.

Conclusion

In the present study, taking the characteristics of geological hazards and railway engineering into account, a new evaluation index system is proposed. Railway safety levels are evaluated using the MEEM, GCM, and SVM to indicate that the GCM is more applicable in evaluating the safety in this case. Meanwhile, the assessment results ignoring construction factor and temporal factor are compared with the spatiotemporal assessment result by using ROC fitting analysis comprehensively. The results reveal that the construction factor and temporal factor are significant to geological hazard assessment of railway engineering. The optimized factors thus help us to establish a greater degree of accuracy in the assessment of geological hazard safety along the railway engineering.

Acknowledgements The study was financially supported by the National Natural Science Foundation of China (Grant No. 51478483). The financial supports are gratefully acknowledged.

References

Haghshenas SS, Mikaeil R, Sirati Moghadam P, Haghshenas AS (2017) A new model for evaluating the geological risk based on geomechanical properties-case study: the second part of emamzade hashem tunnel. Electron J Geotech Eng 22(01):309–320

Lee CF, Huang WK, Chang YL, Chi SY, Liao WC (2018) Regional landslide susceptibility assessment using multi-stage remote sensing data along the coastal range highway in northeastern Taiwan. Geomorphology 300:113–127

Lee PH, Yu PLH (2012) Mixtures of weighted distance-based models for ranking data with applications in political studies. Comput Stat Data Anal 56(8):2486–2500

Liu D (2011) E-commerce system security assessment based on grey relational analysis comprehensive evaluation. Int J Digit Content Technol Appl 5(10):279–284

Liu D, Qi X, Fu Q, Li M, Zhu W, Zhang L, Faiz MA, Khan MI, Li T, Cui S (2019) A resilience evaluation method for a combined regional agricultural water and soil resource system based on Weighted Mahalanobis distance and a Gray-TOPSIS model. J Cleaner Prod 229:667–679

Nijmeijer R, Haas A, Dost R, Budde P (2001) ILWIS 3.0 user's guide. Eschede. Int Inst Aerosp Surv Earth Sci

Ozdemir A, Altural T (2013) A comparative study of frequency ratio, weights of evidence and logistic regression methods for landslide susceptibility mapping: Sultan Mountains, SW Turkey. J Asian Earth Sci 64:180–197

Pourghasemi HR, Yansari ZT, Panagos P, Pradhan B (2018) Analysis and evaluation of landslide susceptibility: a review on articles published during 2005–2016 (periods of 2005–2012 and 2013–2016). Arab J Geosci 11:193

Pradhan B, Lee S (2010) Delineation of landslide hazard areas on Penang Island, Malaysia, by using frequency ratio, logistic regression, and artificial neural network models. Environ Earth Sci 60 (5):1037–1054

Quang LH, Loi DH, Sassa K, Takara K, Ochiai H, Dang K, Abe S, Asano S, Ha DN (2018) Susceptibility assessment of the precursor stage of a landslide threatening Haivan railway station. Vietnam Landslides 15(2):309–325

Wang W, Li J, Qu X, Han Z, Liu P (2019) Prediction on landslide displacement using a new combination model: a case study of Qinglong landslide in China. Nat Hazards 3:1121–1139

Westin LK (2001) Receiver operating characteristic (ROC) analysis: Valuating discriminance effects among decision support systems. Dept Comput Sci Umeå Univ 21(13):1849–1865

Xiong Z, Guo J, Xia Y, Lu H, Wang M, Shi S (2018) A 3D multi-scale geology modeling method for tunnel engineering risk assessment. Tunn Undergr Space Technol 73:71–81

Slope Stability and Landslide Hazard in Volubilis Archaeological Site (Morocco)

2020 Kyoto Japan

Mohamed Rouai, Abdelilah Dekayir, and Khaoula Qarqori

Abstract

Volubilis is the most important and famous Roman archaeological site in Morocco. Despite it represents the well preserved archaeological city in Morocco, the site shows a serious marks of soil instability affecting some ancient buildings and mosaics. From geological point of view, Volubilis is located mainly on Miocene marls at mid slope of Zerhoune hill. A field study showed obvious geotechnical instability consisting meanly of a superficial gravitational creep. This landslide phenomenon affects here and there the precious Roman mosaics, and threatens to ruin ancient house walls, especially in northern city. Spatial statistics have led to describe and quantify the inclination of walls as instability index of soil movement, and to map vulnerable zones. The Miocene marls mineralogical composition, determined by X-Ray Diffraction analysis show that the clay fraction is dominated by the Smectite mineral which has high swelling potential. In order to realize an advanced diagnosis of soil instability, geotechnical characterization of Miocene marls has been conducted to assess their mechanical behaviour as expansive soils.

Keywords

Slope stability • Landslide hazard • Geotechnical behaviour • Volubilis • Morocco

M. Rouai (✉) · A. Dekayir
Faculty of Sciences of Meknes, Earth Sciences Department, Moulay Ismail University, Geotech team, 50000 Meknes, Morocco
e-mail: mohamed.rouai@gmail.com

A. Dekayir
e-mail: Dekayir@yahoo.fr

K. Qarqori
Faculty of Sciences and Techniques, Moulay Ismail University, 52000 Errachidia, Morocco
e-mail: khaoula.qarqori@laposte.net

Introduction

Many archaeological sites in the World are prone to natural hazards. Landslides are one of major risk threatening Cultural Heritage in general (Sassa et al. 2003).

Our aim in this paper is to present a preliminary study case on geotechnical instability linked to superficial landslide in the archaeological site of Volubilis (Northern Morocco).

Volubilis Archaeological Site

Volubilis is the most famous Roman archaeological site in Morocco (Figs. 1 and 2), located in the northern part of the country at 26 km from the historical city of Meknes.

Founded in the third century BC, Volubilis holds fine monuments and beautiful Roman mosaics, and is classified as world cultural heritage by Unesco.

Setting and Geology

Volubilis is located 400 m height on the south-western hillside of Jbel Zerhoun mount (1050 m), with 10% average slope, the most part of the city lies on marls of Miocene age, overlying calcareous and marly Jurassic series (Faugeres 1978), Fig. 3.

Detailed geological information is presented in Fig. 4. The associated Digital Elevation Model (DEM) describes the morphology of the area, the location and the gentle slope of the archaeological site.

A NE-SW geological transect on the area (Fig. 5) shows that the Roman site is located mainly on Miocene marls showing some part a geomorphologic disorders. The south-western part of Volubilis site lies on quaternary sandy alluvia.

F. Guzzetti et al. (eds.), *Understanding and Reducing Landslide Disaster Risk*,
ICL Contribution to Landslide Disaster Risk Reduction,
https://doi.org/10.1007/978-3-030-60227-7_43

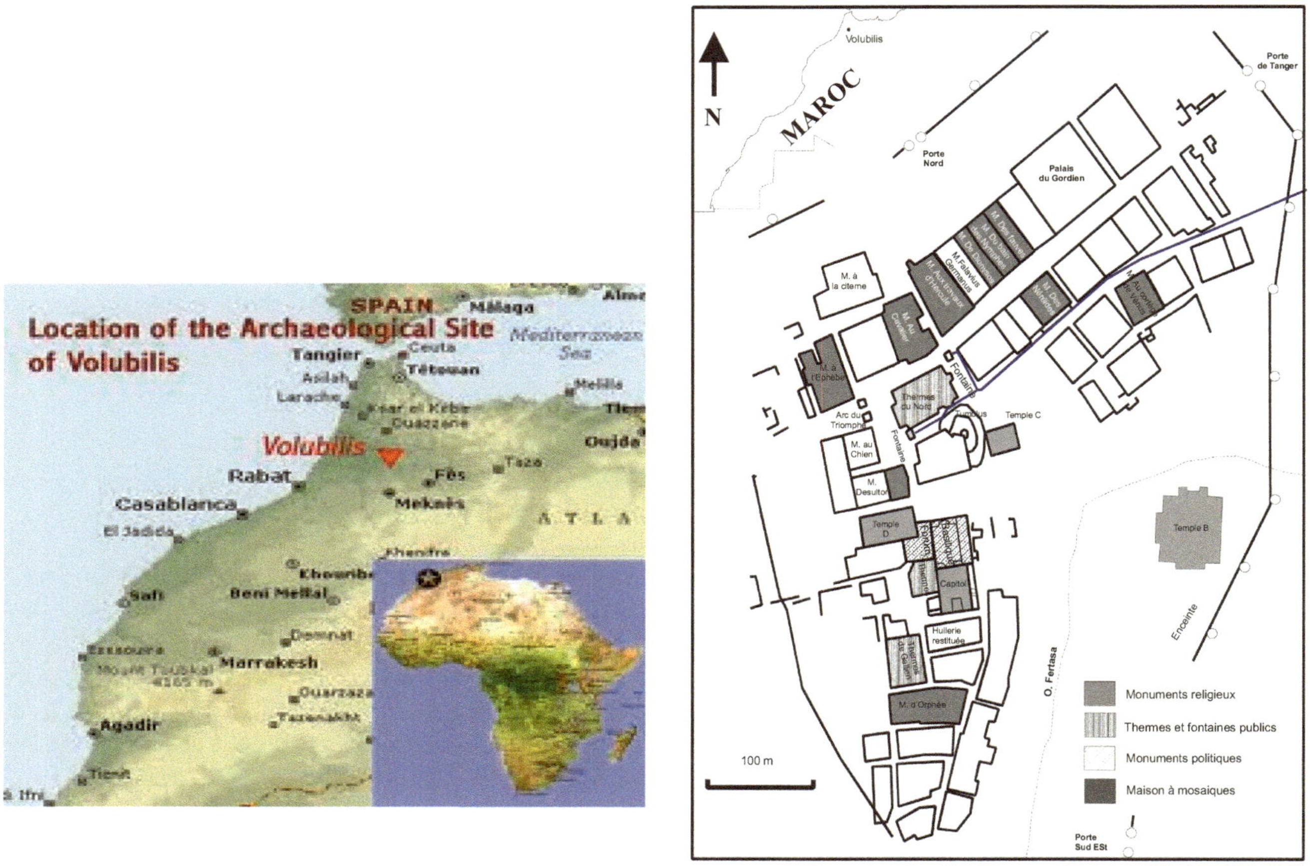

Fig. 1 Volubilis site location and map

Fig.2 Examples of monuments and mosaics in Volubilis

From geo-structural context, the area belongs to the Prerif Ridges which are elongated crests associated with folds and reverse faults developed by the southward propagation of the Alpine deformation of Rif belt.

The archaeological site lies on Volubilis depression; on the north-eastern syncline flank leaning against Zerhoune thrust nappe. Jbel Zerhoune, is a south-vergent fold formed mainly by Jurassic limestones and deformed in its southern part by reverse faults (Ruano et al. 2006).

No geological faults have been reported on the near vicinity of the archaeological site, at least at the surface.

Field Survey

Volubilis represents the well preserved archaeological city in Morocco; but unfortunately it shows serious marks of soil instability affecting some ancient buildings and mosaics.

Fig. 3 Geological scenery showing Volubilis site

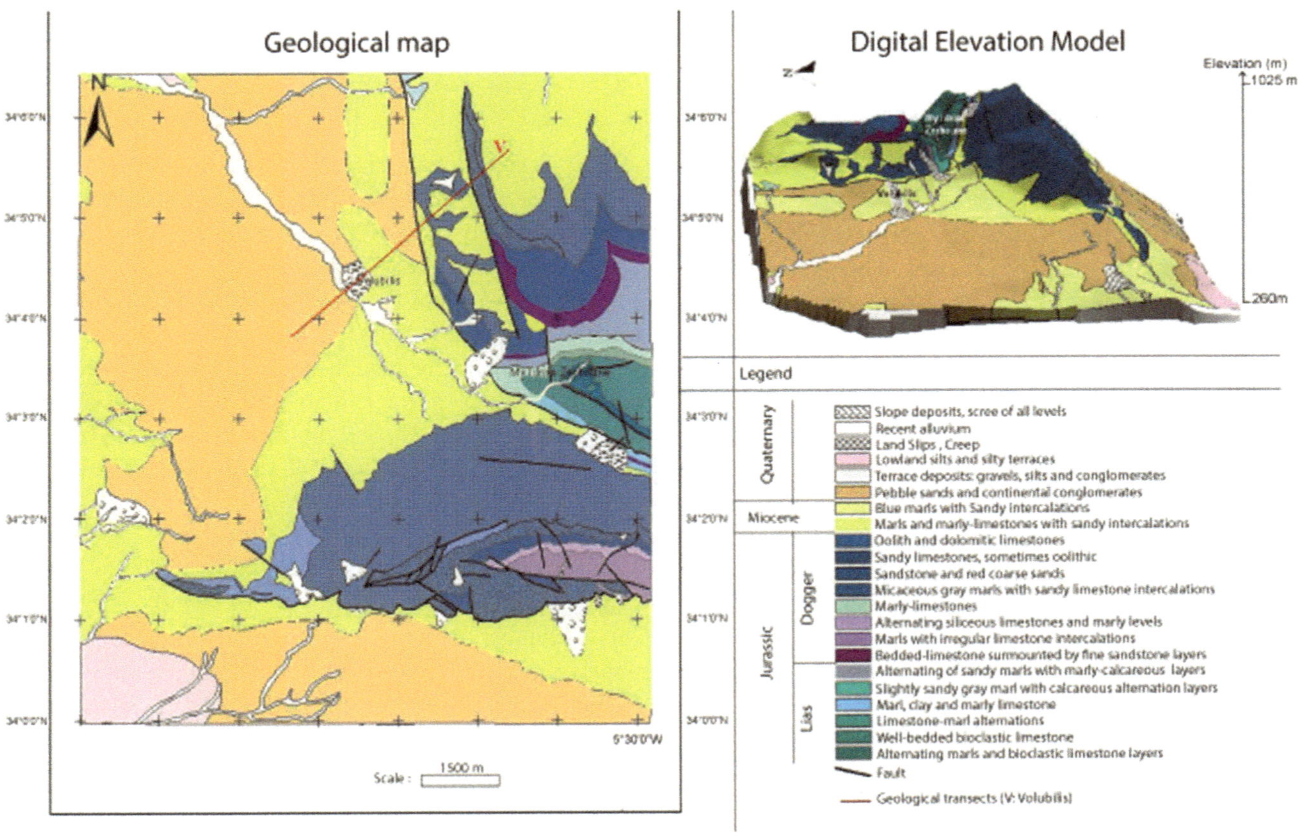

Fig. 4 Geological setting and DEM

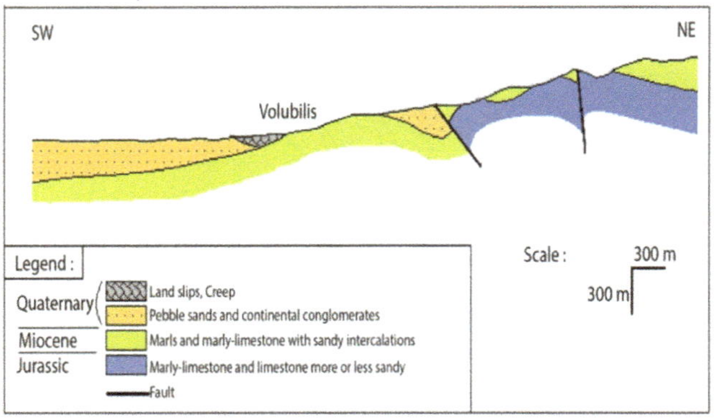

Fig. 5 Geological cross-section on Volubilis area

Fig. 6 Example of wall inclination

Fig. 7 Direction of wall inclination (arrows)

A field study showed obvious soil instability, at least at the surface (Fig. 6). It consists meanly of a superficial gravitational creep: a slow down-slope movement of marls under the effect of gravity and water. This landslide phenomenon affects here and there the precious Roman mosaics, and threatens to ruin ancient house walls, especially in northern city.

Slope Instability Analysis

The slope instability diagnosis is performed starting from superficial indices of soil movements, essentially by inspecting tilted walls.

Walls Inclination

In the north-eastern quarter of Volubilis, measurements of ancient house walls inclination using a compass and dip-meter device have been carried out, leading to determine the direction and magnitude of soil displacements (Fig. 7).

The statistical analysis of 143 tilted walls, 86% are inclined towards the SW, 7.5% and 6.5% towards SE and NW respectively (Fig. 8), confirming that the dominant movement follows the topographic slope.

Analyzing walls of the most prevailing direction NW-SE and inclined towards SW (Fig. 9).

The maximum inclination angle (α) varies between 5° and 40°, with 15° average. Statistics indicate that 40.6% of walls have $\alpha < 10°$; 37.4% with inclination $\alpha \in]10°, 20°]$; 19.5% with $\alpha \in]20°, 30°]$, 2.5% have $\alpha \in]30, 40°]$.

Inclination Mapping

This preliminary study has led to draw quantitative cartography of marly soil movement, mapping both the direction and intensity of displacement, and to determine the most vulnerable zones and walls under the threat of fall over (Figs. 10 and 11). This vulnerability map, based mainly on wall tilt angle spatial distribution contouring, could be well improved taking into account soil geotechnical parameters among others.

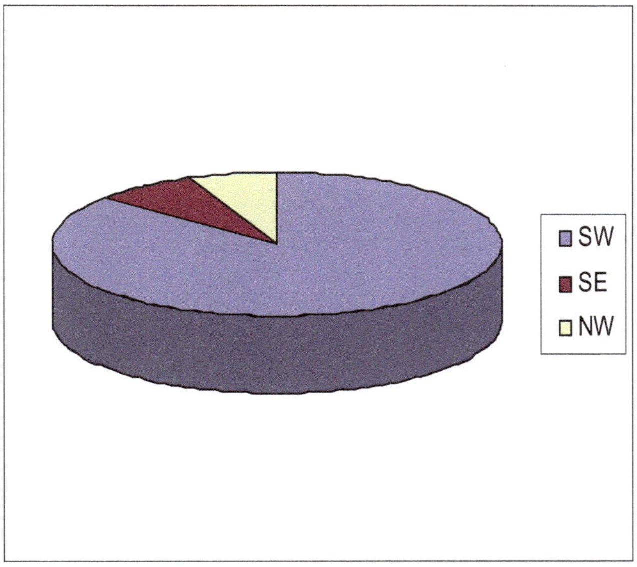

Fig. 8 Statistics of wall inclinations

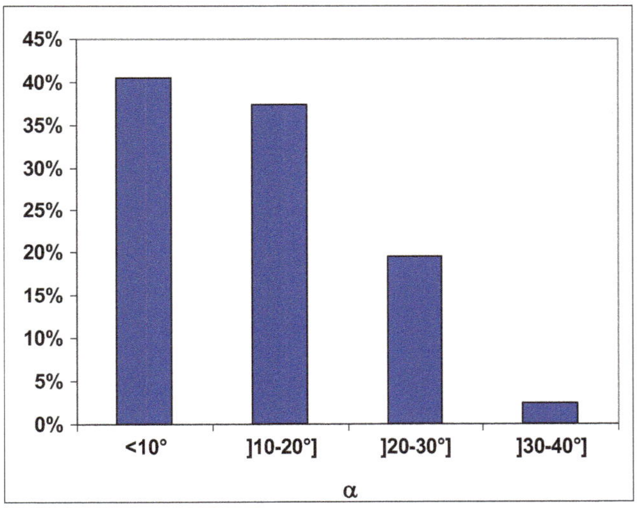

Fig. 9 Inclination degree histogram

It is worthwhile to note the heterogeneous distribution of vulnerable zones with a gradient up-slope.

Soil Characteristics

Marls Mineralogy

The marls mineralogical composition, determined by X-ray diffraction analysis (Fig. 12), shows that the clay fraction is dominated by the Smectite mineral which has high expansion potential and behaviour, confirmed by preliminary geotechnical tests.

Atterberg Limits

An average Atterberg limits of sampled fine-grained marls soil are presented on Table 1, showing the Liquid Limit (LL), the Plastic Limit (PL) and the Plasticity Index PI = (LL − PL). These results confirm that the swelling potential calculated by PI is high.

It is worth to note that the average dry density of marl samples $\gamma d = 1.7$ g/cm^3, and the water content reaches 22%. The Methylene Blue Values (MBV) is about 5.2%.

Oedometric Tests

A representative oedometric curve of tested samples is presented in Fig. 13. Results confirm the high plasticity of Miocene marls. The calculated index of swelling (0.084) is high. Thus the expansive nature of this soil is demonstrated.

Conclusion and Perspectives

This study confirms the presence of superficial gravitational creep: a slow down-slope movement of Miocene marls under the effect of gravity and water.

The developed methodology seems to be very useful to characterize and to map soil instabilities and the prone area to ruin.

A deep rotational landslide has not been identified from field survey, but more geophysical prospecting techniques (Electrical Resistivity Tomography, etc.) are needed.

Part of Volubilis archaeological site is located on Quaternary alluvium, so it is very important to complete advanced geotechnical study and mapping of the whole archaeological site.

It is of high importance to realize an advanced diagnosis of soil instability, to follow its evolution and to assess the exposed risk. More quantitative risk assessment and vulnerability mapping are needed.

Quantitative maps under a GIS for the whole site will be of great importance for the monitoring of the physical problem and the proposed solutions for remediation in order to mitigate risk.

It is urgent to implement the suitable methods for landslide monitoring (GPS, Interferometry, theodolite, inclinometers, piezometry, pore pressure, etc.) and also to propose the more adapted remediation measures (micro-draining, vegetation remediation, mechanical wall consolidation, etc.), especially for the protection and conservation of Roman Mosaics houses and important monuments of this universal cultural heritage.

Fig. 10 Wall inclination degree map

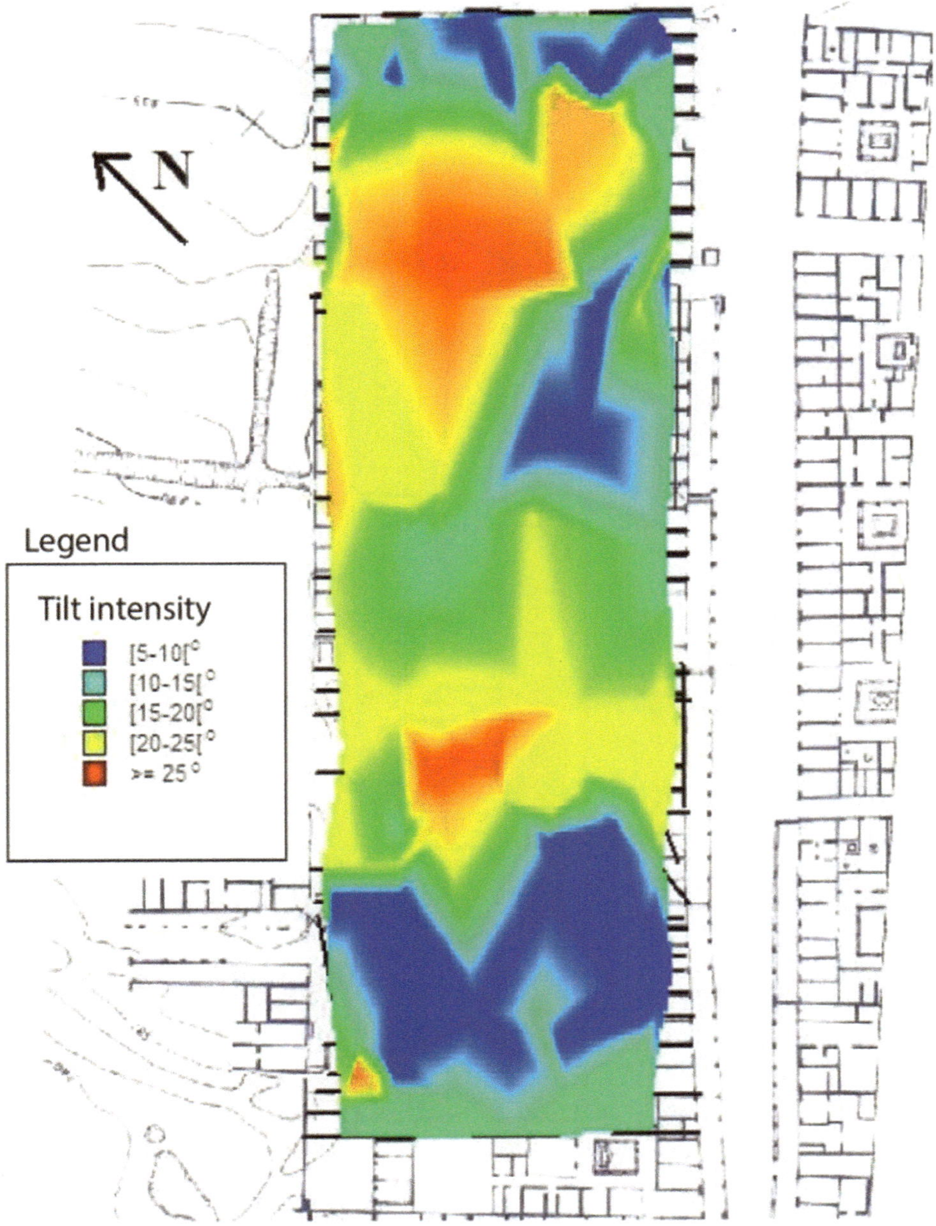

Fig. 11 Vulnerable zones mapping

Fig. 12 XRD of clay mineralogy

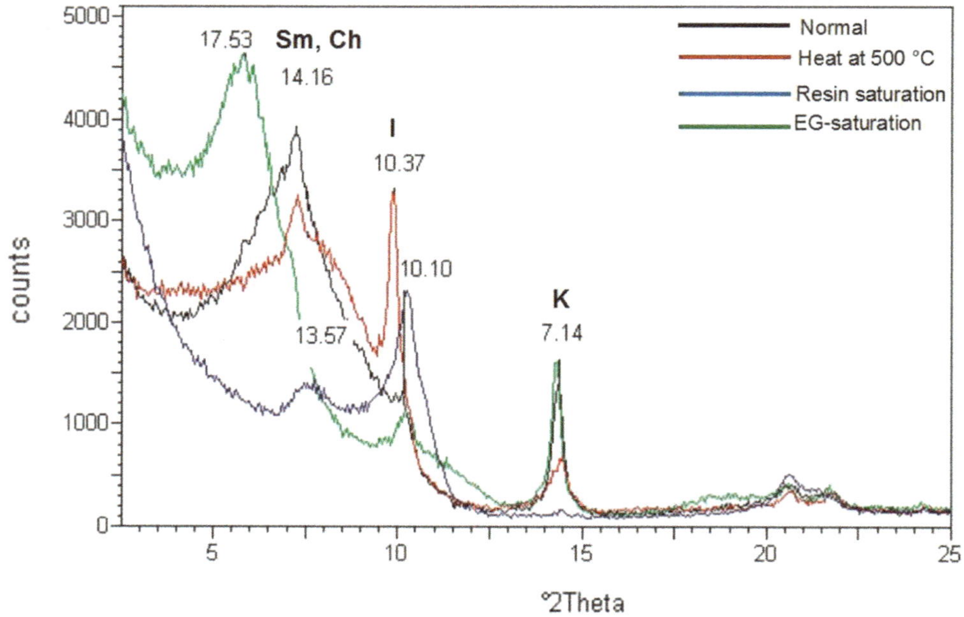

Table 1 Atterberg limits as water content (%)

Atterberg limits	LL	PL	LI
Results (%)	56	27	29

Fig. 13 Oedometric curve

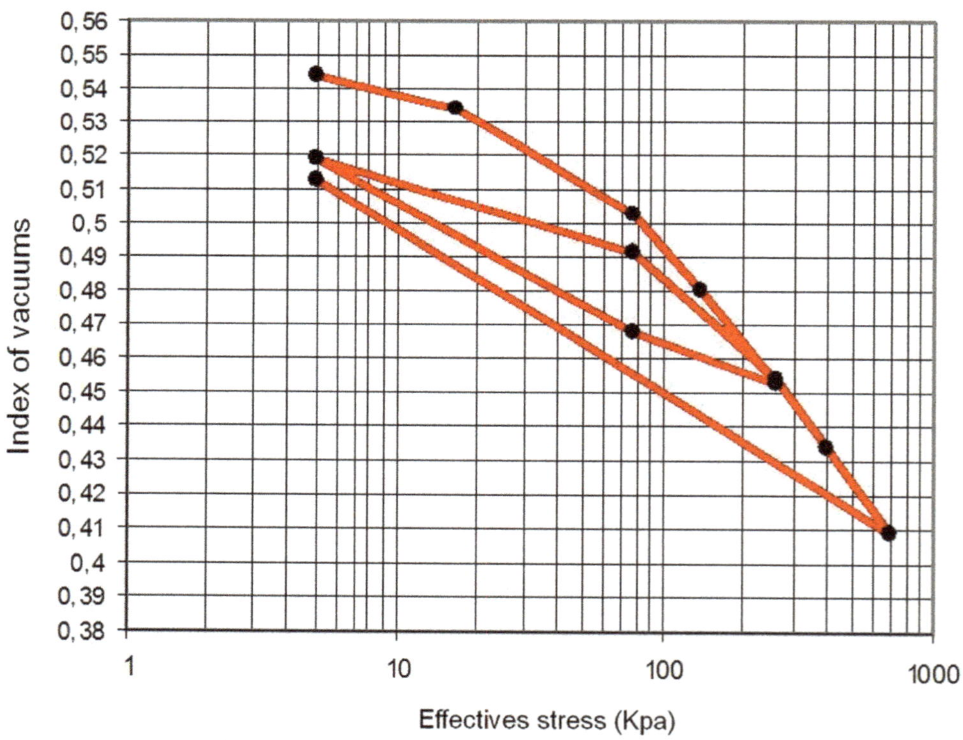

Acknowledgements Authors are very grateful to the Moroccan Ministry of Culture trough Volubilis archaeological site administration and keepers. We would like to thank sincerely Mrs. A. Charki, R. Bouzidi and M. Atki and all persons who helped us during this study.

References

Faugères JC (1978) Les rides sud-rifaines. Evolution sédimentaire et structurale d'un bassin atlantico-mésogéen de la marge africaine. Thèse de Doctorat d'Etat, Univ. Bordeaux 1, France, 480p

Ruano P, Bargach K, Galindo-Zaldivar J, Chalouan A, Ahmamou M (2006) Recent palaeostesses from striated pebbles related to fold development in a mountain front: the Prerif Ridges (Rif Cordiliera, Morocco). Geological Society, London, Special Publications, vol 262, pp 87–99

Sassa K, Fukuoka H, Shuzui H, Hoshino M (2003) Landslide risk evaluation in Macchu Picchu. World Heritage. In: Proceedings of Unesco/IGCP symposium on landslide risk mitigation and protection of Cultural Heritage, Kyoto, Japan, pp 1–20

Part V

Landslide Data and Information for Disaster Mitigation

Slope Hazard and Risk Mapping Project (PBRC)—An Overview of Disaster Risk Reduction Initiative

WLF 2020 Kyoto Japan

Zamri Ramli and Ferdaus Ahmad

Abstract

Extreme climate, rapid urbanisation, and environmental degradation substantially increase our exposure and vulnerability to natural hazards and disaster risk in a developing tropical country. Therefore, it is crucial to address the issues and possible solutions related to the natural hazards in an objective and quantitative manner. The Slope Hazard and Risk Mapping Project (Projek Penghasilan Peta Bahaya dan Risiko Cerun, PBRC) is the first Malaysia initiative that tackles the issues of area-based landslide hazard and risk with the use of multi-sensory Light Detection and Ranging (LiDAR) technology, supported by many expertise and disciplinary partners. This project is one of the action plans formulated under the National Slope Master Plan (NSMP) 2009–2023 to reduce risk and build the resilience of nations and communities to disasters. It involves comprehensive methodology framework and operational needs driven by modern and advanced geospatial technology. The model has been formulated and evaluated with the complexity of risk scenarios in this knowledge driven project. Slope instability problems in the urban, mountainous, and tectonic landscapes are considered, and their spatial information is crucial for regional landslide assessment. This project established standard procedures with optimal parameterisation for susceptibility, hazard, and risk assessment in the selected regions. It also put forth the critical and practical framework, ranging from updating landslide inventory to mitigating landslide risk as an attempt to support the establishment of a comprehensive disaster risk management in Malaysia. The PBRC project is a great example of producing hazard and risk maps that follow international standards and meet the need of multi- sectoral agencies. It is a practical and evidence-based guidance to support implementation, ensure engagement by stakeholders, and strengthen accountability in disaster risk reduction. The developed methods can be used to assess tropical landslide geomorphology in an objective, reproducible, and quantitative manner.

Keywords

Disaster risk reduction • Slope hazard and risk mapping • LiDAR

Introduction

Slope instability occurs in many parts of urban and rural area of Malaysia and often affected housing, roads, and other infrastructures. Slope related disasters may not only limit to loss of life but also huge economic losses due to property damage and other consequence effects. As the importance of addressing slope hazard and associated risk in tropical area become more crucial, Malaysia has taken a significant move and step forward to look for the best methodological framework and operational need to manage the risk.

The project on the Slope and Hazard Risk Mapping or also known as the *Projek Penghasilan Peta Bahaya dan Risiko Cerun* (PBRC) is the national initiative by the government to assess and evaluate critical slope and associate geological process with support of modern mapping technology, coupled with expert knowledge. This project requires technical and professional expertise from several fields such as engineering geology, geotechnical engineering, land survey, remote sensing, and geospatial expertise. In

Z. Ramli (✉)
Department of Mineral and Geoscience Malaysia, Aras 9, Menara PjH, No. 2, Jalan Tun Abdul Razak, Presint 2, 62100 Putrajaya, Malaysia
e-mail: zamri@jmg.gov.my

F. Ahmad
Technical Services Division, Department of Mineral and Geoscience Malaysia, Jalan Sultan Azlan Shah, 31400 Ipoh, Perak, Malaysia
e-mail: ferdaus@jmg.gov.my

F. Guzzetti et al. (eds.), *Understanding and Reducing Landslide Disaster Risk*, ICL Contribution to Landslide Disaster Risk Reduction,
https://doi.org/10.1007/978-3-030-60227-7_44

order to achieve the Disaster Risk Reduction (DRR) aim, the objectives of the PBRC project are as follows:

1. To produce slope hazard and risk information in the selected areas for the local authorities in Malaysia for a better slope management.
2. To provide conceptual proposal and cost estimation for mitigating and strengthening high risk slope.
3. To assist government agencies and local authorities in land-use planning towards a sustainable land management.
4. To develop geospatial infrastructure for national geological terrain and slope information.

Project areas had been selected based on landslide-prone areas (Fig. 1) identified in National Slope Master Plan (2009–2023). A National Slope Master Plan is a national policy, strategy, and action plan for reducing risk from landslides nationwide.

Methodology

Assessment of hazard and associated slope risks in tropical environment required an appropriate methodological framework and operational flow for a better and comprehensive disaster risk management. The selection of methodological framework and operational elements are based on the local conditions and should fulfil the project's objectives need.

Advanced and Modern Geospatial Technique

Topographic laser scanning or also known as light detection and ranging (LiDAR) for data acquisitions is mainly used for landslide investigation to create accurate and precise high resolution digital elevation models (HRDEM) in raster grids or triangulated irregular networks (TINs), which are 2.5D representations of the topography, or in true 3D point clouds

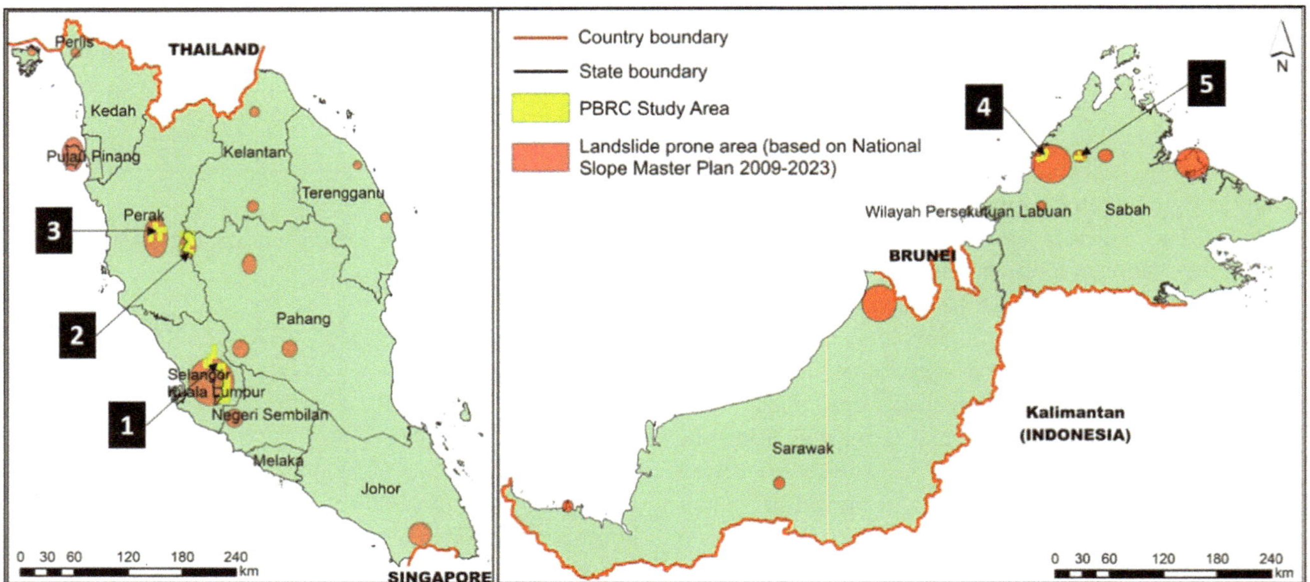

NO.	SLOPE HAZARD AND RISK MAPPING PROJECT (PBRC) AREA	GEOENVIRONMENT	AREA (km^2)
1.	Klang Valley	Urbanised, populated, environmental degradation	650
2.	Cameron Highlands, Pahang	Mountainous region, tourism area, anthropogenic activity	475
3.	Ipoh, Perak	Geoheritage city, settlement expansion	200
4.	Kota Kinabalu, Sabah	Urban-suburban, rapid urbanization	155
5.	Kundasang, Sabah	World Heritage Site, tectonically active region	70
	TOTAL AREA		**1550**

Fig. 1 Slope hazard and risk mapping project area

with a high density of information (Jaboyedoff et al. 2012). This is a relatively new technique in Malaysia and significant to this project. The technique stands capable of providing crucial data for solving an old geological problem (tropical landslide geomorphology), and its hazard and risk mapping is an effective preventive measure. Two techniques of LiDAR acquisitions are being employed: Airbone Laser Scanning Technique (ALS) and Terrestrial Laser Scanning Technique (TLS).

Field Data Collection

This is to verify data captured by TLS and ALS and to collect additional information on site utilising different proformas. There are four types of proforma as listed:

i. Primary Slope Proforma: Covering information on geomorphological condition of the slope, and risk and consequences factor.
ii. Incident Proforma: Encapsulate details of landslide morphology and its activity as well as damages caused by landslide.
iii. Geology Proforma: Covers the evidence of geodynamic features, type of discontinuities, and potential failures on site.
iv. Structural Support Proforma: Summarise building and foundation types; retaining structures and evidence of distress on them; and slope information.

Database Development and Preparation

To prepare Landslide Inventory Maps, LiDAR -derived geomorphological factor Maps, LiDAR -derived geological factor Maps, LiDAR -derived hydro-topographic factor Maps, LiDAR -derived anthropogenic factor Maps.

Landslide Susceptibility Assessment

Landslide susceptibility map was developed based on the analysis of spatial landslide distribution with several geo-environmental parameters. The fundamental concept of bivariate statistical analysis is to determine the relationship between landslide distribution and landslide controlling factors (Guzzetti et al. 1999). This method is also known as landslide index method (*Wi*), in which the weighted value for a parameter class is defined as the natural logarithm of the landslide density in the class, divided by the landslide density in the entire map (van Westen 1997). Wi method is based on the following formula:

$$Wi = ln\frac{Densclass}{Densmap}$$

where *Wi* denotes the weight given to a certain parameter class; *Densclass* denotes the landslide density within the parameter class *Densmap* denotes the landslide density within the entire map. The natural logarithm is used to determine the influence of a certain parameter class in landslide development within the entire map. Negative values of *Wi* indicates that the presence of that particular parameter class has less effect to the landslide development, whereas positive values indicates a relevant relationship between the presence of such parameter class with landslide development (Yin and Yan 1988; Zêzere 2002).

Hazard and Risk Assessment

Landslide hazard is a function of susceptibility (spatial tendency to landslide activity) and temporal frequency of landslide triggers, and its assessment. In landslide hazard assessment, several factors that influence the propensity to landslide activity (susceptibility factors), as well as landslide triggering factors (precipitation, earthquakes, human activity) are taken into consideration. Susceptibility factors are divided into four main groups which consist of geological, geomorphological, hydro topographical and anthropogenic factors, whereas for triggering factors only rainfall and earthquakes are considered. The assessments depending on many scenarios, modelling schemes and parametrization subjected to site conditions and environment.

Generally, risk can be defined as a likelihood of specified adverse consequences arising from an event, circumstance or action. It concerned with the likelihood and scale of adverse consequences that are the product of the interaction of hazard and vulnerability (Lee and Jones 2004). Due to data availability limitation, for risk assessment it only considers number of element at risk and the vulnerability of individual structures such as buildings, pylons, roads, utilities and costing information of individual structures obtained from various stakeholders.

Geotechnical Risk Assessment

Assessment of critical slope especially man-made slope, by utilizing/modification existing slope hazard and risk ranking system and subsequently proposed mitigation measures and estimate cost of construction.

In principal, there are ten operational needs to manage the methodology as shown in the conceptual framework of slope hazard and risk mapping (Fig. 2).

Outputs and Applications

Derivative Maps and Technical Reports

The outputs and their applications obtained from the project are summarized in Tables 1 and 2. These documents and information can be used by relevant government agencies or decision maker for land-use planning and evaluation of development proposal projects. Furthermore, it may provide

Fig. 2 Conceptual framework for slope hazard and risk mapping project

Table 1 Slope hazard and risk mapping output and application

Derivative maps	Information gained	Relatives importance to risk management
Landslide Inventory Map	Landslide characteristics (type, material, activity, magnitude), date of occurrence, triggering factor, and distribution	Identification landslide mitigation option based on landslide characteristics, Primary data to produce subsequence Maps (Susceptibility, Hazard, Vulnerability, and Risk Maps)
Susceptibility Map	Hierarchy of landslide susceptibility areas, landslide causal factors	Identification of area susceptible to landslide and most probable causal factors, parameters for early warning system
Hazard Map	Hierarchy of landslide hazard area, extension, return period	Identification of probable extent of landslide hazard area, repetition of extreme landslide incident, parameters for early warning system
Vulnerability Map	Vulnerability of element at risk (infrastructure, building, road, and human)	Identification of types of construction materials/design less vulnerable to hazard, relocate most vulnerable peoples to non-landslide area
Risk Map	Probable negative impact (loss of life/casualties/economic loss) on element at risk due to landslide incident	Identification of suitable (selective) development, Insurance coverage, Cost and Benefit analysis

Table 2 Slope hazard and risk mapping reports and application

Reports	Information gained	Relatives importance to risk management
Landslide Risk Management	Risk and hazard condition, inventory of slopes and element at risk	Prioritization of repair, strengthening works, expenditure, and avoidance of disaster
Geotechnical Risk Assessment	Method of geotechnical analysis and assessment	As tool for risk and hazard ranking
Guideline for Mitigation	Procedures of risk and hazard reduction and mitigation	Effectiveness of mitigation measures
Geotechnical Risk Analysis and Mitigation Measures Proposal	Potential failure modes and probable methods of repair and strengthening works	Prioritization of repair and strengthening and expenditure

conceptual recommendations and cost estimation for mitigation and remedial measures of critical slopes. All these outputs are useful in designing a comprehensive risk management framework conceptual.

National Geospatial Terrain and Slope Information System (NaTSIS)

In parallel to the fourth objective of the project, the development of National Geospatial Terrain and Slope Information System (NaTSIS) have been initiated and completed in 2017.

The system overview is shown in Fig. 3. Generally, the system acts as a geospatial databank, which can be shared among respective government agencies for spatial information on landslides, terrain, and slopes.

Discussion

One of the PBRC study area, Kota Kinabalu, Sabah (area 4 in Fig. 1), which covers an area of about 155 km^2 is currently one of the most rapidly undergone urbanization area in Malaysia and has experienced various scales of landslides due to increasing demands of physical development and human activities.

It is underlain by Tertiary sedimentary rocks known as Crocker Formation while recent alluvium deposits formed during the Quaternary period is largely covering the coastal area. The airborne LiDAR survey was carried out in 2015 prior to landslide hazard and risk mapping in 2016. The digital terrain model developed from LiDAR data was used to facilitate the field mapping as well as to execute slope hazard and risk assessment.

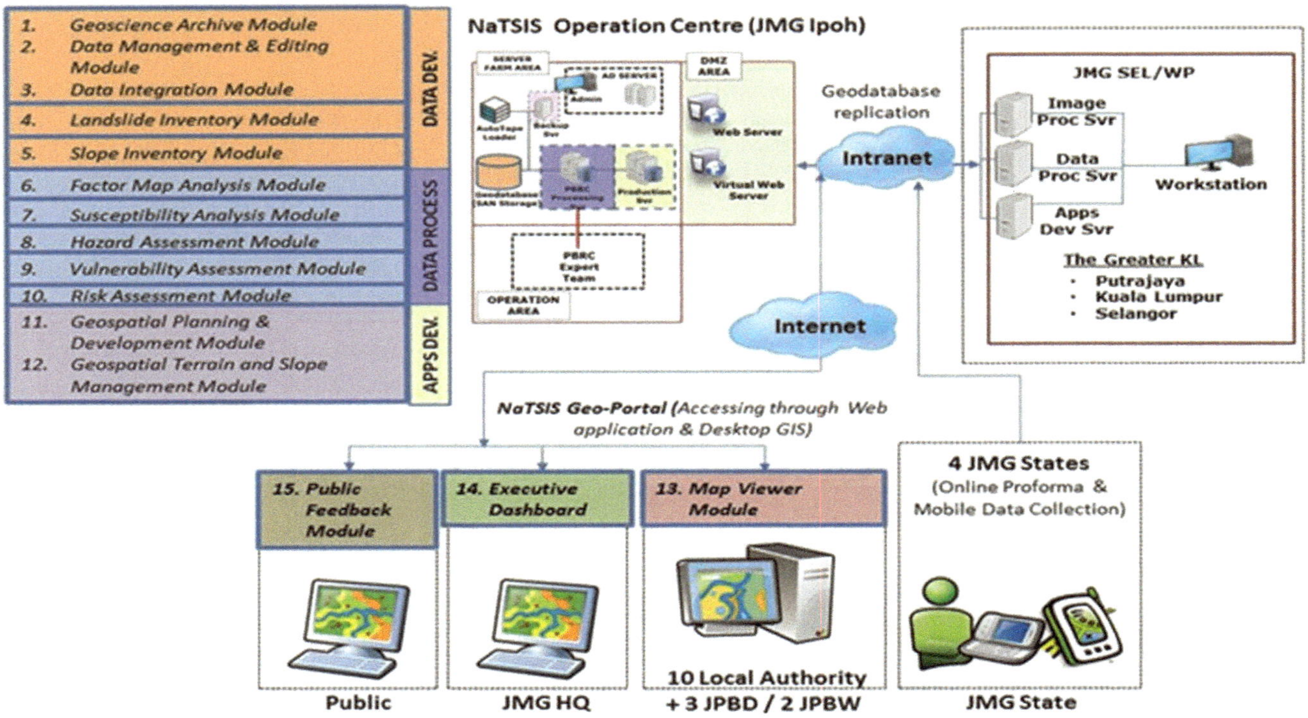

Fig. 3 Overview of National Geospatial Terrain and Slope Information System (NaTSIS)

A total of 1126 inventories have been made which consists of 868 slopes and 258 landslides. The inventorized data are employed for susceptibility, hazard and risk, geotechnical, and critical slope assessment. From hazard analysis, about 6.6% is categorized as very high hazard slopes, whereas 38.2% is considered high hazard slopes, followed by 29.0% is moderate hazard, 9.9% is low hazard, and the remaining 16.3% is classified as very low hazard slopes (Fig. 4). The vulnerability map was produced based on two types of element at risk which is buildings and roads. The risk analysis shows that 0.2% of element at risk is categorized as very high risk, 17.0% is under high risk, 5.4% is in moderate risk, 12.6% is categorized as low risk, and the remaining 64.8% is very low risk.

By incorporating hazard and risk analyses, and expert judgement, there are 35 slopes identified as critical slopes. For Kota Kinabalu, all the critical slopes have been categorized into three main category which is Critical Slopes Required Immediate Strengthening Works (Category 1), Critical Slopes Need Strengthening Works (Category 2) and Critical Slopes Required Monitoring (Category 3). From the analysis and site evaluation, 7 slopes are under Category 1, 9 slopes belong to Category 2, and 19 slopes go to Category 3.

Conclusions

The implemented methodology and the project output fulfilled the designing comprehensive disaster risk reduction framework in Malaysia through the following:

1. Provide an efficient and effective slope management system which include slope mitigation measures taking into account prioritization of repair, strengthening works, expenditure, and avoidance of disaster.
2. Expedite delivery of advisory services on geohazard risk, development proposal, and land- use zonings evaluations for stake-holders. This will assist local authority planning committee to make decisions and approval within time frame.
3. Alleviates awareness and preparedness among the local authorities and its community thus to minimise loss of life and damage of properties.
4. Geospatial data acquisition, storage, access, and sharing among agencies which act as reference centre for geological terrain and slopes information to promote data sharing amongst agencies.

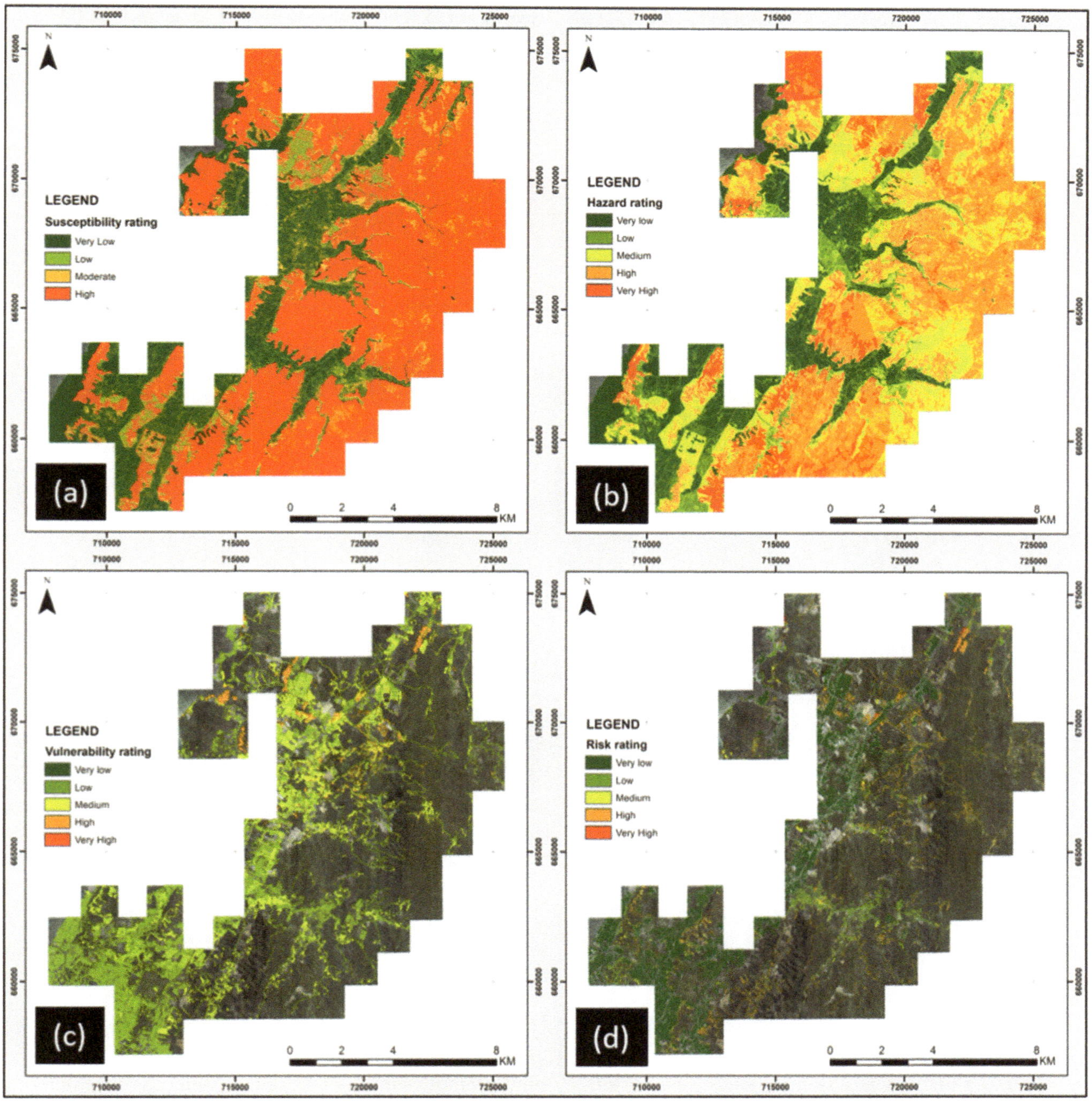

Fig. 4 **a** Landslide Susceptibility Map, **b** Landslide Hazard Map, **c** Landslide Vulnerability Map and **d** Landslide Risk Map of Kota Kinabalu, Sabah

Nevertheless, there is still much left to be done. The PBRC project is a great example of producing hazard and risk maps that follow international standards and meet the need of multi- sectoral agencies. The developed methods can be used to assess tropical landslide geomorphology in an objective, reproducible, and quantitative manner, which can be extended to other slope areas in Malaysia.

Acknowledgements Thanks are due to the Government of Malaysia as this project could not have been attempted without the financial support provided by them. We are also grateful to Department of Mineral and Geoscience of Malaysia for the support and assistance provided by the staff. Finally, thanks are also due to PBRC team members for their technical assistance and guidance throughout the course of this project.

References

Guzzetti F, Carrara A, Cardinali M, Reichenbach P (1999) Landslide hazard evaluation: a review of current techniques and their application in a multi-scale study, Central Italy. Geomorphology 31(1):181–216. https://doi.org/10.1016/S0169-555X(99)00078-1

Jaboyedoff M, Oppikofer T, Abellán A, Derron M-H, Loye A, Metzger R, Pedrazzini A (2012) Use of LIDAR in landslide investigations: a review. Nat Hazards 61(1):5–28. https://doi.org/10.1007/s11069-010-9634-2

Lee EM, Jones DKC (2004) Landslide risk assessment. Thomas Telford Publishing

van Westen CJ (1997) Statistical landslide hazard analysis. ILWIS 2.1 for windows application guide. ITC Publication, Enschede, The Netherlands

Yin KL, Yan TZ (1988) Statistical prediction model for slope instability of metamorphosed rocks. Paper presented at the 5th international symposium on landslides, Lausanne, Switzerland

Zêzere J (2002) Landslide susceptibility assessment considering landslide typology. A case study in the area north of Lisbon (Portugal). Nat Hazards Earth Syst Sci 2. https://doi.org/10.5194/nhess-2-73-2002

Risk-Informed Land Use Planning for Landslide Disaster Risk Reduction: A Case Study of Cameron Highlands, Pahang, Malaysia

2020 Kyoto Japan

Mohd Farid Abdul Kadir, Khamarrul Azahari Razak, Ferdaus Ahmad, and Dzul Khaimi Khailani

Abstract

The 2030 Agenda for Sustainable Development Goals (SDG) has recognized the urgent need to reduce the risk of disasters in a changing climate. With complexity and interaction of human, economic and political systems, risk becomes increasingly systemic. Therefore, it requires concerted and urgent effort to reduce it in an integrated way. This paper describes a new insight into risk-informed development planning in Cameron Highlands (Pahang, Malaysia), the most populated and economically developed mountainous area in Malaysia. We compiled landslide inventories dated since 1961 coupling with the historical records, and national landslide hazard risk mapping project resulted in 1842 landslides. We found that several landslide hotspots contain more than 10 landslides/km^2. Most of the study area was marked as very high to high hazard (38.96%). The GIS-based spatio-temporal assessment indicated about 10.0% and 0.4% of land use changes for the period 2003–2015 and 2015–2018, respectively. Spatial analysis of landslide hazard indicated a positive correlation with land use changes. This study review of 100-year trends in land use changes (1930–1960, 1960–2003, 2003–2015, 2015–2030) indicates that anthropogenic activities dominantly affected by physical development and agriculture activity contributed significantly towards environmental degradation in a populated highland area. Since the urbanization rate in Malaysia is expected to increase up to 85% with 46.1 million people in 2040, integration of Disaster Risk Reduction (DRR) and resilience is critical in all sectors. As a conclusion, this research provides a significant input for developing local DRR and resilience strategies, as one of seven global targets of the Sendai Framework for Disaster Risk Reduction 2015–2030.

Keywords

Landslide inventory • LiDAR technology • Hazard assessment • Land use changes • Risk-informed sustainable development

Introduction

Land use changes are one of the issues in the sustainable development and environmental degradation (Guan et al. 2011; Halmy et al. 2015; Zheng et al. 2015). It is the result of interaction between socio-economic factors, population pressure, land use type and climate change (Ayele et al. 2016).

Land use changes have affected and will continue to affect the vulnerable communities and even so with climate change. Land use changes caused by improper management of agricultural, urban and forest land could lead to environmental problems such as landslides and floods (Reis 2008; Yaakup et al. 2000). The mountainous region is an environmentally sensitive area for any development process, especially in the rugged terrain with steeper slopes. Improper

M. F. A. Kadir (✉) · F. Ahmad
Department of Mineral and Geoscience Malaysia, Technical Service Division, 31400 Ipoh, Malaysia
e-mail: mhd.farid@jmg.gov.my

F. Ahmad
e-mail: ferdaus@jmg.gov.my

K. A. Razak
Razak Faculty of Technology and Informatics & Disaster Preparedness and Prevention Center (DPPC), Malaysia-Japan International Institute of Technology (MJIIT), Universiti Teknologi Malaysia, 54100 Kuala Lumpur, Malaysia
e-mail: khamarrul.kl@utm.my

D. K. Khailani
Federal Department of Town and Country Planning Malaysia, Wisma Tun Sambanthan, 50000 Kuala Lumpur, Malaysia
e-mail: dzulkhaimi@townplan.gov.my

F. Guzzetti et al. (eds.), *Understanding and Reducing Landslide Disaster Risk*, ICL Contribution to Landslide Disaster Risk Reduction,
https://doi.org/10.1007/978-3-030-60227-7_45

land use management would lead to mass wasting movement, such as a landslide. In 2009, the Federal Department of Town and Country Planning (PLANMalaysia) established the Guideline for Development Planning at Hilly Area and Highland in order to manage development activity in the hilly terrain and highlands in Malaysia. The connection of land use and disaster has been studied by numerous researchers and institutions. Rapid changes of land use or land cover can lead to land degradation and environmental problems due to anthropogenic factors such as agricultural activity, deforestation and urbanization (Rendana et al. 2015). Land use changes study has far-reaching significance for decision makers to feed the output into land planning, forest management, ecosystem conservation as well as mitigating disaster issue such as landslide (Ayele et al. 2016). This study aims to provide insight into the risk-informed development planning in the populated mountainous region in Cameron Highlands, Pahang in Malaysia. Cameron Highlands is one of the landslide prone areas highlighted in the National Slope Master Plan. The first recorded landslide in Cameron Highlands was the landslide at Ringlet on May 11th, 1961 (Public Work Department 2009). This tragedy claimed 16 lives and approximately 30 people were rescued. Due to the demand of land for agricultural activity, the number of illegal land clearings is increasing. Poorly planned agricultural activity and development has contributed to the instability of the slopes leading to severe erosion processes and landslides. This paper reviews 100 years of land use trends in Cameron Highlands.

Study Area

Cameron Highlands, Pahang is one of the tourist hotspots in Malaysia. It is known as the largest hilltop resort in Malaysia. The cool mountainous climate is suitable for cultivating fresh vegetables, fruits, blooms, and teas. It occupies an area of 712.18 km^2 and lies between latitudes 4° 18′44″ N–4°36′35″ N and longitudes 101°19′02″ E–101°36′ 05″ E as presented in Fig. 1. The average temperature is about 18 °C, hence it is suitable for sub-tropical agricultural activity.

As a popular tourist destination in Malaysia, the number of tourists increases every year since 2009 except in 2012 and 2014, as a result of flood and landslide disasters in Cameron Highlands (Cameron Highlands District Council 2015). Based on the 2010 census, the area recorded a population of 36,978 people which represents a 2.7% increment with respect to the population record of 2000. This population rate is higher compared to the state's population rate (2.0% a year) and the national population rate (2.53% a year). Based on 2010 records, 47.6% of total job opportunities derived from the primary sector, especially in the agricultural sector. Geographically, Cameron Highlands is located on the highland's "spine" of Peninsular Malaysia. Most of the Cameron Highlands areas (74%) are located at elevations beyond 1000 m above sea level, characterised by steep slopes and undulating terrains. Geologically, Cameron Highlands is underlain by two main lithologies known as granite and schist. Granite is forming parts of several large

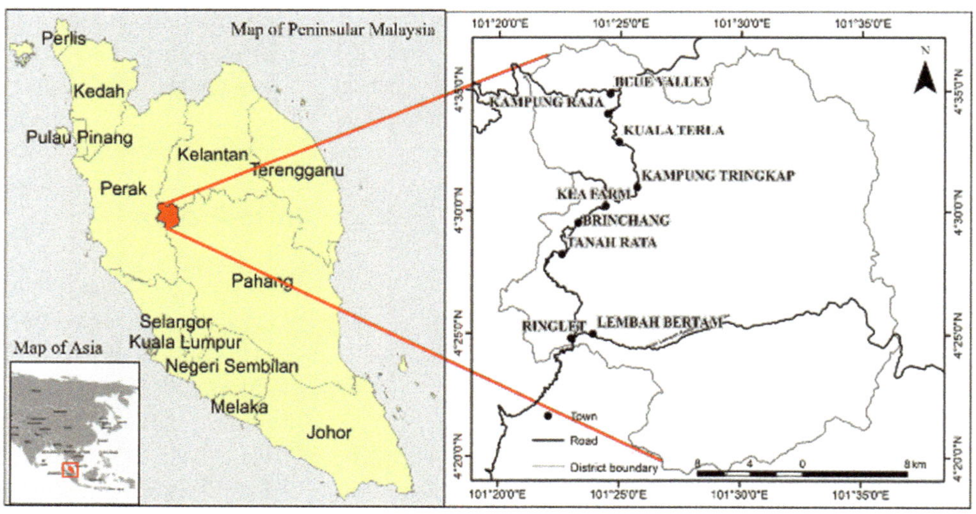

Fig. 1 Study area in Cameron Highlands (Pahang, Malaysia)

batholiths, while the meta-sediments occur as roof pendants. The depth of weathering varies but is greatest (>50 m thick) below ridge crests and thins towards valley floors where fresh bedrock may outcrop. The schists have also been weathered, though the depth of weathering is much less than granite (Raj 2002).

Methodology

Landslide Inventory Mapping and Causal Factor Analysis

Landslide inventory has contributed significantly to the quality of landslide hazard and risk analysis. A comprehensive landslide inventory map was produced by combining historical landslide record, LiDAR-derived inventory, field-derived inventory and expert-knowledge input. It is worth to mention that this is the first comprehensive study carried out in Malaysia to map landslides including those beneath tropical vegetation, rugged terrain and inaccessible areas by utilising high density airborne LiDAR data.

Two series of airborne LiDAR data acquisition campaigns were carried out in 2014 and 2017 under the National Landslide Hazard and Risk Mapping Project (PBRC) managed by the Department of Mineral and Geoscience Malaysia (JMG) (Department of Mineral and Geoscience Malaysia 2018; Mohamad et al. 2015). Historical landslide inventory was obtained from various sources, namely from JMG archives, National Slope Master Plan 2009–2023, newspaper reports, technical papers and research articles. This LiDAR-derived inventory mainly utilised high-resolution 0.25 m LiDAR-derived Digital Terrain Model and 0.07 m orthophoto. LiDAR-derived datasets such as orthophoto, hillshade, color composite and slope angle images are generated to visualize landslides in 2D and 3D modelling views (Fig. 2). Landslides and manmade slope failure signs such as hummocky topography, tension crack, and crown-shaped scarp are delineated in ArcGIS software.

Each landslide and slope failure are characterized based on type, activity, and process in attribute tables and systematically stored in a geodatabase. Then, the quality of an interpreted landslide was critically checked and verified with the field mapping by professional geologists. The landslide body was identified and digitised in a polygon form within GIS environment. In this research, every landslide location was marked in a point form representing the landslide crown. Landslide density analyses are conducted using a point density tool. Density analysis is done by dividing the study area into 1 km $\times$ 1 km grid cells. Each cell was classified as either containing no landslide, 1 landslide/km^2, 2–10 landslides/km^2 or more than 10 landslides/km^2 (Calvello and Piciullo 2015). This information gives an overview of the region that is susceptible to landslides and it is possible to correlate landslide occurrence with landslide causal factors such as geology, geomorphology, hydrology and anthropogenic.

Spatio-temporal Land Use Changes and Hazard Zonation

Spatial analysis was carried out by utilising land use maps (2003, 2015 and 2018) produced by PLANMalaysia. The analysis was conducted to obtain the distribution of each land use class. The percentage of overall land use changes was calculated by dividing the area of changes with the area for the initial year. Positive values suggest an increase in the land use area, whereas negative values imply a decrease in the extent.

Land use changes analysis of each land use type was conducted to obtain the conversion of one particular land use type to another land use type in one period of time. Land use conversion matrix for each period was generated using the intersection tool in ArcGIS 10.4 software. The output of the intersection process was compiled in a matrix table using Microsoft Excel. Results from land use changes analysis produced land use conversion matrix. The larger number indicates the area that experiences the most changes in surface activity from one particular land use type to another. These changes could give a positive or negative impact on the environment. For instance, conversion of forest to agriculture land probably exposes soil surface against rainfall, allows rapid infiltration and lead to the instability of the slope or increases the potential of erosion process.

The landslide hazard classification was obtained using the statistical analysis as a result of combining landslide susceptibility map and the temporal probability of rainfall produced by the Department of Mineral and Geoscience Malaysia in 2016 and 2018. The landslide hazard rating was categorised into 5 classes known as very low, low, medium, high and very high (Department of Mineral and Geoscience Malaysia 2016, 2018).

Impact of Land Use Changes on Landslide

Deforestation and land clearing are common examples of land use change. Glade (2003), Gökceoglu and Aksoy (1996) and Karsli et al. (2008) emphasised the positive impact of vegetation to influence the slope stability.

This study investigates the impact of land use changes, as a result of extreme anthropogenic activities, on landslide occurrence in Cameron Highlands, Pahang. Besides the remotely sensed data analysis, this study arranged a series of field investigations to observe current site conditions and

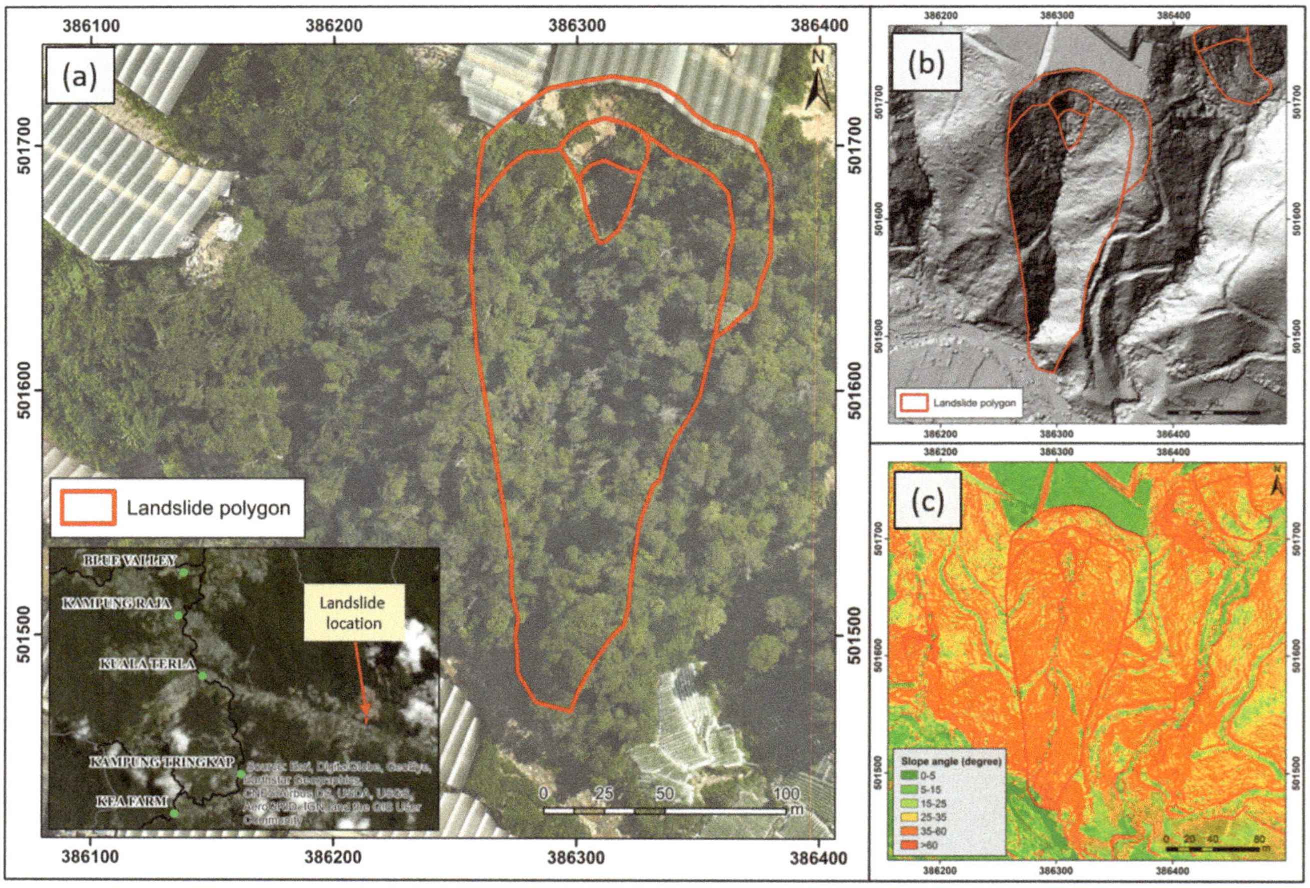

Fig. 2 Airborne LiDAR derived dataset for landslide mapping. **a** Relict landslide in the vegetated terrain (see polygon on the orthophoto obtained from the LiDAR campaign), **b** LiDAR-derived hillshade and **c** slope gradient map

visits to relict landslides location. The classified land use maps in a vector dataset (polygon) were compared in two periods: 2003–2015 and 2015–2018. In this research, land use changes were overlaid with the landslide points to obtain the correlation between landslide occurrence and land use change. Areas that were classified as very high landslide hazard were overlaid with an areas characterized by land use changes to highlight the impact of land use changes towards the present level of hazard and future probability of losses. The overall research process is simplified in the operational framework as Fig. 3.

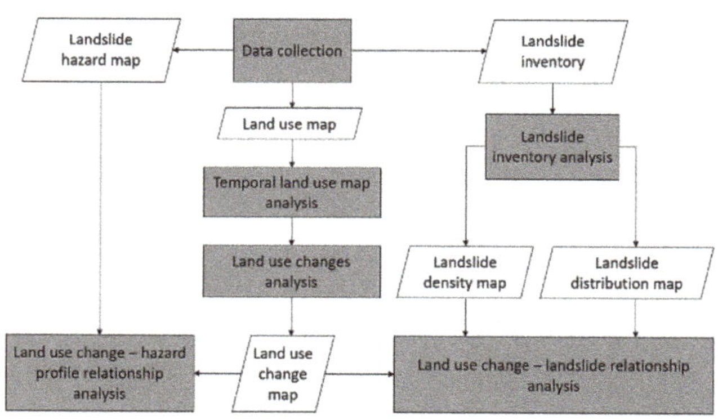

Fig. 3 Processes (rectangle) and outputs (parallelogram) carried out in this research

Result and Discussion

Landslide Inventory Map

Based on the historical data, we recorded 29 landslides since 1961. A total of 1842 landslides were identified within PBRC study area in 2014/2015 and 2017/2018 by using visual and interpretation method.

The landslide density map was generated based on the distribution of historical landslide and LiDAR-derived landslide inventory. Figure 4 shows that most of the landslide locations are concentrated adjacent to the main road where human activities took place. Several landslide hotspots that contain more than 10 landslides/km^2 are identified in Kuala Terla, Kea Farm, Tringkap, Brinchang, Tanah Rata, Ringlet and Lembah Bertam. Landslide location is also concentrated at the central part of Cameron Highlands.

Fig. 4 Distribution of historical landslides and landslide density map of Cameron Highlands

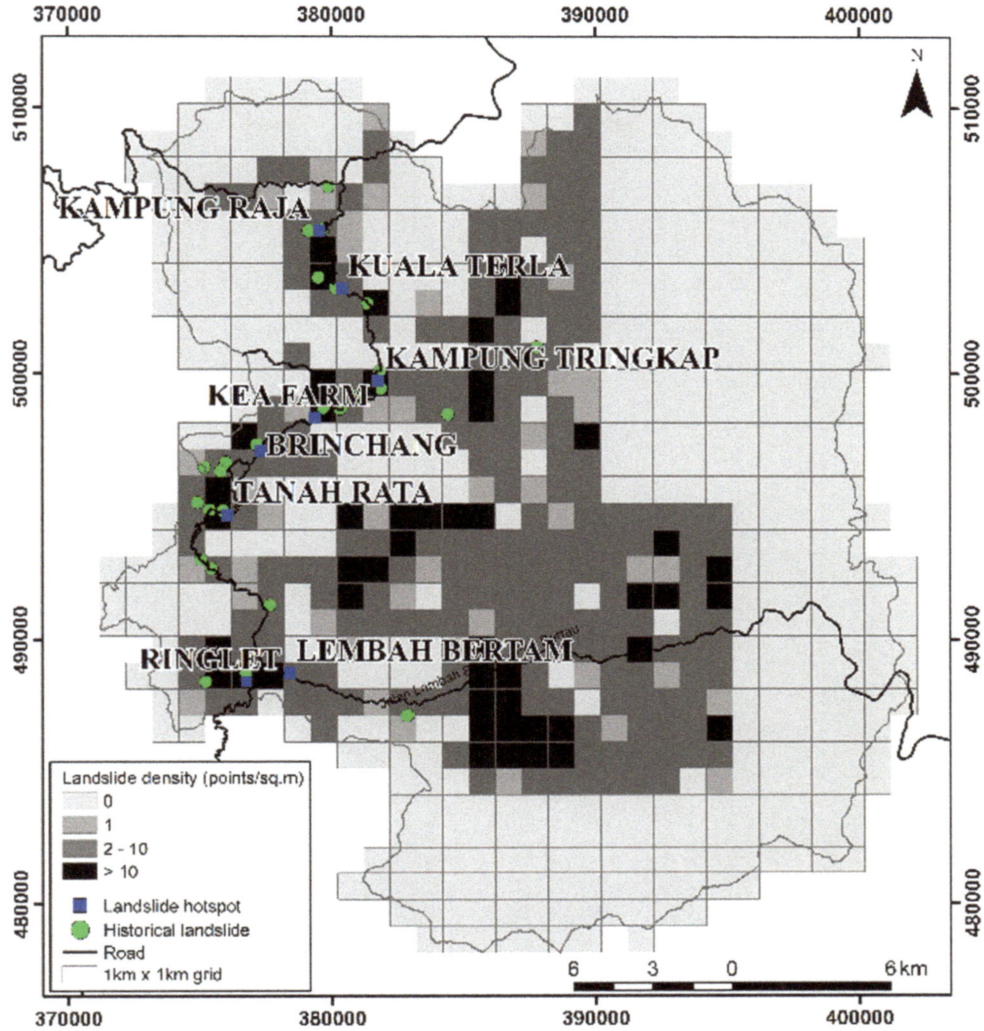

Land Use Changes and Hazard Zonation

Forest areas in Cameron Highlands dominated three periods of study in 2003, 2015 and 2018, making up 90.0%, 87.7% and 87.6% of the study landscape, respectively.

In 2003, the agriculture land was the second largest land use cover (8.4%) followed by transportation (0.6%), residential (0.5%), institution and public amenity (0.2%) and commercial (0.1%). After 12 years, agriculture land remained the second largest land use type where it covered 10.6% of the total land area. The residential, water body, transportation and institution and public amenity accounted 0.4%, 0.4%, 0.3% and 0.3%, respectively. In 2018, there were no significant changes that accounted for land use distribution. Forest continued to be the dominant land use type, followed by agriculture land (10.7%), residential (0.5%), water body (0.4%) transportation (0.3%) and institution and public amenity (0.2%). The reduction of the forest areas represents a loss of 15.8 km^2 for the first period and 0.8 km^2 for the second period. On the contrary, agriculture land increased by 15.3 km^2 from 2003 to 2015 and further increased by 0.3 km^2 from 2015 to 2018. The land use change matrix analysis from 2003 to 2015 shows that about 67.7 km^2 (10.0%) of the Cameron Highlands area experienced land use changes as shown in Table 1. The conversion of 35.2 km^2 of forest to agriculture land is the largest land use change recorded in 12 years. From 2015 to 2018, the land use change matrix analysis encounters only 2.5 km^2 (0.4%) of the total area of Cameron Highlands experiencing land use changes. Most of the land use condition in 2015 has remained unchanged.

Landslide hazard analysis resulted in a 3.44% of the study area classified as very low hazard. About 36.45% and 21.15% were classified as low and medium, respectively. Most of the study area was marked as high hazard with the percentage of 37.61% and a very small area was marked as very high hazard, which is about 1.35% (Department of Mineral and Geoscience Malaysia 2016, 2018).

Table 1 Land use change matrix in square kilometer and percentage per classes (in parenthesis) from 2003 to 2015

	To / From	Land Use 2015												
		Agr.	Com.	For.	Ind.	Infr.	Inst.	Mix.	Open.	Res.	Trans.	Vac.	Wat.	Grand total
Land Use 2003	Agr.	34.67 (61.24)	0.01 (0.02)	21.22 (37.49)	0.00 (0.00)	0.01 (0.01)	0.37 (0.65)	0.00 (0.00)	0.01 (0.01)	0.04 (0.07)	0.14 (0.25)	0.04 (0.08)	0.10 (0.17)	**56.61**
	Com.	0.10 (13.34)	0.31 (40.21)	0.08 (9.75)	0.00 (0.01)	0.01 (1.31)	0.01 (0.76)	0.04 (5.57)	0.00 (0.34)	0.06 (7.70)	0.08 (10.08)	0.07 (9.54)	0.01 (1.40)	**0.78**
	For.	35.23 (5.78)	0.09 (0.01)	570.33 (93.50)	0.03 (0.01)	0.34 (0.06)	0.14 (0.02)	0.01 (0.00)	0.05 (0.01)	0.66 (0.11)	0.75 (0.12)	0.15 (0.03)	2.16 (0.35)	**609.96**
	Ind.	0.00 (0.00)	0.00 (0.00)	0.00 (2.22)	0.06 (97.78)	0.00 (0.00)	0.00 (0.00)	0.00 (0.00)	0.00 (0.00)	0.00 (0.00)	0.00 (0.00)	0.00 (0.00)	0.00 (0.00)	**0.06**
	Infr.	0.05 (27.44)	0.00 (0.47)	0.04 (25.54)	0.00 (0.00)	0.05 (29.87)	0.02 (9.35)	0.00 (0.00)	0.00 (0.18)	0.00 (0.12)	0.00 (0.91)	0.00 (0.00)	0.01 (6.12)	**0.17**
	Inst.	0.01 (0.89)	0.00 (0.21)	0.48 (29.94)	0.00 (0.00)	0.00 (0.00)	1.07 (66.96)	0.00 (0.00)	0.00 (0.29)	0.01 (0.84)	0.01 (0.81)	0.00 (0.04)	0.00 (0.01)	**1.60**
	Open.	0.01 (2.31)	0.01 (2.92)	0.01 (3.08)	0.00 (0.00)	0.00 (0.01)	0.00 (0.89)	0.00 (0.00)	0.41 (87.40)	0.00 (0.59)	0.00 (1.04)	0.01 (1.58)	0.00 (0.17)	**0.47**
	Res.	0.26 (6.97)	0.09 (2.55)	0.67 (17.93)	0.00 (0.07)	0.01 (0.25)	0.08 (2.03)	0.00 (0.00)	0.05 (1.45)	2.03 (54.52)	0.36 (9.64)	0.16 (4.41)	0.01 (0.18)	**3.72**
	Trans.	1.56 (39.42)	0.04 (0.91)	1.27 (32.14)	0.00 (0.06)	0.00 (0.06)	0.02 (0.57)	0.00 (0.02)	0.06 (1.53)	0.07 (1.87)	0.75 (19.02)	0.01 (0.37)	0.16 (4.03)	**3.95**
	Wat.	0.00 (0.23)	0.00 (0.00)	0.03 (10.14)	0.00 (0.00)	0.00 (0.00)	0.00 (0.00)	0.00 (0.00)	0.00 (0.00)	0.00 (0.00)	0.00 (0.14)	0.00 (0.00)	0.23 (89.48)	**0.25**
	Grand total	**71.90**	**0.56**	**594.13**	**0.10**	**0.42**	**1.71**	**0.05**	**0.59**	**2.88**	**2.11**	**0.46**	**2.67**	**677.58**

Agr.—Agriculture; Com.—Commercial; For.—Forest; Ind.—Industrial; Infr.—Infrastructure and utility; Inst.—Institution and public amenity; Mix.—Mixed development; Open.—Open space and recreation; Res. —Residential; Trans.—Transportation; Vac.—Vacant land; Wat.—Water body

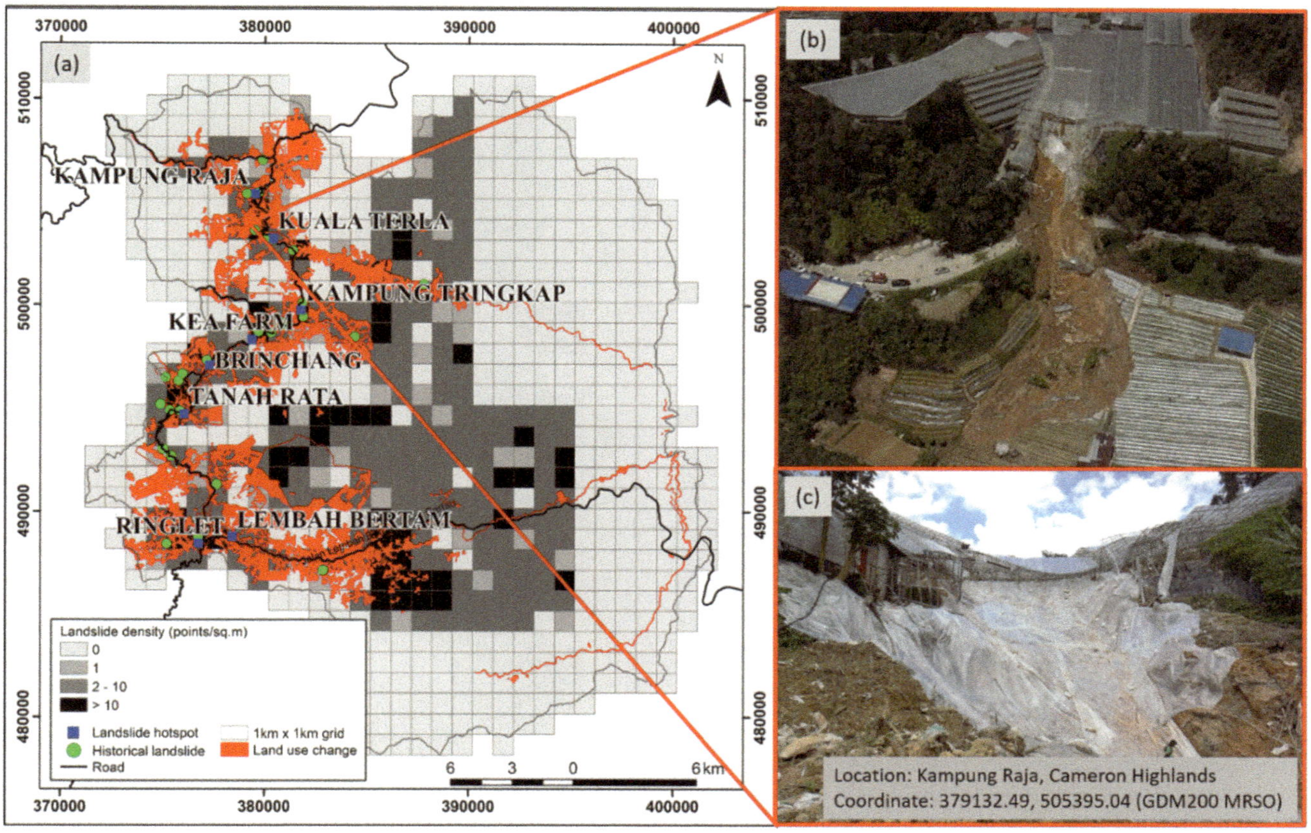

Fig. 5 Land use changes in 2003–2015 period overlaid with **a** the historical landslide inventory and landslide density; **b** plan view of landslide area in Kampung Raja occured on April 23rd, 2019 (Picture taken by: Abdul Kadir, M.F.); **c** disrupted agriculture activity at the crown of landslide in Kampung Raja (Picture taken by: Abdul Kadir, M.F.)

Impact of Land Use Changes on Landslide Occurrence

GIS-based intersection analysis for land use changes and historical landslide inventory between 2003–2018 shows a significant correlation. Out of 16 historical landslides within this period, 5 landslides (31.3%) fall in the areas with changes of land use and all landslides occurred in the period of 2003–2015.

Figure 5 depicts that the landslide density distribution is concentrated in areas where changes in land use have taken place. For the period of 2015–2018, there are 8 landslide occurrences, but none of these landslides are related to land use changes, since the overall land use change is minimum (0.4%). Nonetheless, a number of landslides within this 3-year period needs attention. Even though land use changes seem not to have a direct impact on the slope stability, landslides in this period could be the effect of excessive land use change during the previous period, especially the conversion of forest to the agriculture land. There are 6 out of 8 landslides in 2015–2018 period that is related to land use change in 2003–2015 period. Mismanagement of agriculture activity and illegal land clearing led to erosion and landslide occurrence. Figure 5 also shows that a landslide occurring in Kampung Raja on April 23rd, 2019 may be associated with agricultural activity. Cultivation was performed on the slope with a slope gradient of about 20°. Plastic covers are used to protect plants from extreme weather and increase productivity. However, improper water management, such as rainwater draining directly onto the slope, could rapidly affect the slope stability. This landslide falls within an area that underwent changes in land use during the period of 2003–2015. It could be related to the consequences of conversion of a land use type within the aforementioned period.

Figure 6 shows the intersection of land use change and the very high hazard areas of the landslide hazard map. In the western part of Cameron Highlands, where most of the built-up area and human activity are located, landslide hazard area is related to land use change. Spatial analysis shows about 62% of the very high hazard areas in the landslide hazard map were intersected with land use change areas within the built-up area. In the central and eastern part of Cameron Highlands, there is no significant change in land use, even though very high landslide hazard areas can be seen.

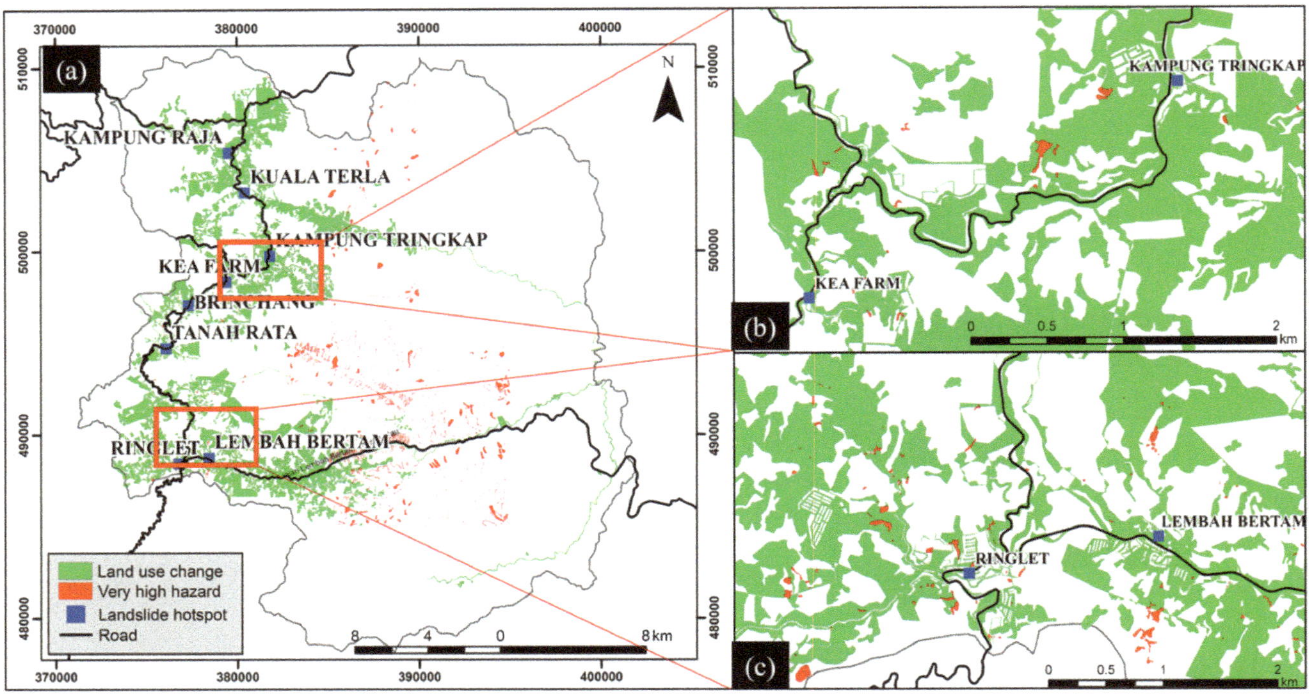

Fig. 6 **a** Intersection of land use change for the period 2003–2018 with landslide hazard map of Cameron Highlands; **b** Focused image at Kea Farm-Tringkap areas; **c** Focused image at Ringlet-Lembah Bertam areas indicates the intersection of land use change areas (green) and very high hazard areas in the landslide hazard map (red)

Mainstreaming DRR into Development Planning

Cameron Highlands is classified as the Environmentally Sensitive Area and also part of the Special Management Area in the Main Range where agriculture and urbanisation have been permitted under strict management control to ensure continuity of the environment quality and the safety of people. The suitable development area is limited in Cameron Highlands. It is important to mainstream Disaster Risk Reduction (DRR) initiatives into development planning to ensure sustainable development and restore Cameron Highlands as the greatest highland hotspot in Malaysia.

Land use planning can be a tool to resolve a conflict between the need for country development with the awareness of protecting the environment and safety of people. Understanding the current and potential risk could lead to better design of strategy and plan for risk-sensitive development planning. Geological information is vital to understand the potential geological hazard and it is required at all levels of development planning (Bell 2006). Landslide hazard and risk assessment are used to aid the respective authority in the decision making process by providing and highlighting important geological information. The spatial analysis on the relationship of land use changes and landslide occurrence in Cameron Highlands shows a positive correlation between both elements. Thus, it is good to mention that landslide hazard and risk information are crucial to be adopted in the land use planning process to avoid new emerging risk and minimise the existing risk.

Figure 7 shows timeline data of significant anthropogenic events and landslide records in Cameron Highlands. Even though it began with tea plantation activity around 1929, Cameron Highlands had started to develop as a residential and highland resort in 1960, after the emergency order was stopped. Cameron Highlands, nowadays, has developed to become the main vegetable producer region after it was firstly introduced in the 1970s. The development of agriculture for vegetables and flowers production also has increased since 1982, while many cultivations were done on the steep slope (Ko et al. 1987). Landslide events became more frequent in recent years which could be associated with the anthropogenic factors in the changing climate.

The preparation of a local plan document seems not to contribute much to reduce the number of landslides, even though its purpose is to ensure the proper and systematic land use development in the Cameron Highlands district. A comprehensive local plan considering landslide disaster risk may enhance the capacity of the city to manage landslides. National Security Council (NSC), in charge of leading special operation called Ops Lestari (formerly known as

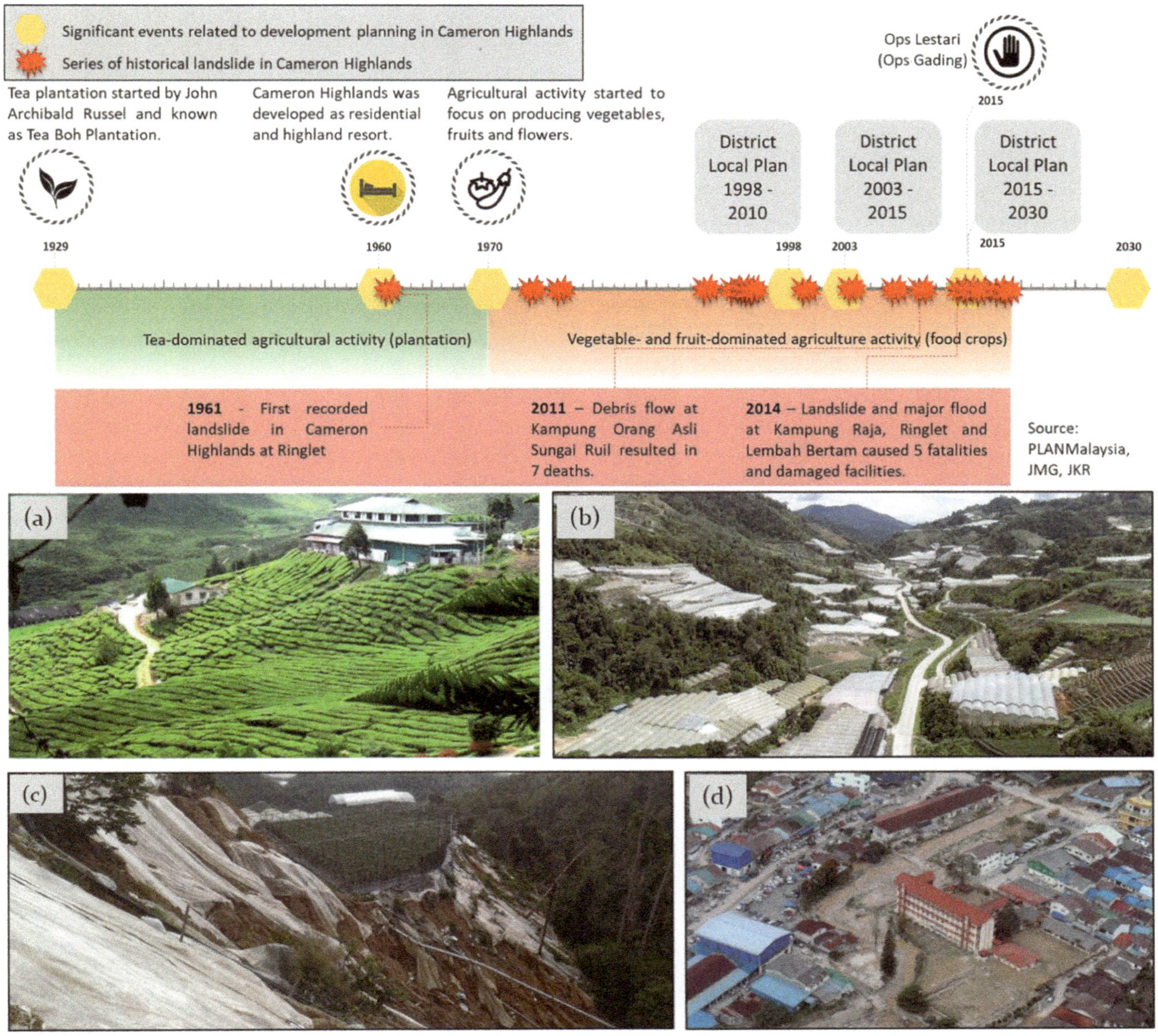

Fig. 7 100-year timeline data of anthropogenic factors and series of historical landslides in Cameron Highlands. **a** Tea plantation in Tanah Rata. Picture retrieved from https://www.cameronhighlandsinfo.com/attraction/Cameron_Valley_Bharat_Tea_Plantations/; **b** Vegetable- and fruit-dominated agriculture in Lembah Bertam. Picture taken by Abdul Kadir, M.F.; **c** Agriculture-related landslide in Terisu, Kuala Terla. Picture taken by Abdul Kadir, M.F.; **d** Debris flow and mud flood in Lembah Bertam. Picture retrieved from https://www.bharian.com.my/node/16037

Ops Gading), has controlled illegal agricultural activity and evaluated the suitability of existing agricultural land since 2014. The frequency and magnitude of the landslide disaster may affect the tourism industry. In 2014, the number of tourists who visited Cameron Highlands has reduced (18.9%) from the previous year record due to massive landslide and debris flow tragedy. Certainly, understanding disaster risk is needed to provide an insight into local risk profiles and serve an important scientific judgement for the stakeholders in the decision making process.

Conclusion and Recommendation

As a conclusion, this study provides a significant input for developing local DRR and resilience strategies, as one of seven global targets of the Sendai Framework for Disaster Risk Reduction 2015–2030. We assessed land use changes, and reviewed a 100-year trend. We found only 10% and 0.4% of the land in Cameron Highlands experienced changes within the 2003–2015 and 2015–2018 periods, respectively. Conversion of forest into agricultural land is the most significant change to meet the local and regional demand for agricultural products.

Landslide is one of the common disaster issues in Cameron Highlands. A mountainous region with hilly and steep slope made Cameron Highlands limited for vast development. Human intervention on the slope terrain often contributed to land degradation. Historical landslide records are very limited in Cameron Highlands and only concentrated at the populated area. 3D visual analysis using LiDAR data helps to identify relict landslides even under a thick canopy of a tropical forest. Coupling with the field landslide identification, this study shows a significant concentration of landslides (more than 10 landslides/km^2) at Kuala Terla, Kea Farm, Tringkap, Brinchang, Tanah Rata, Ringlet and Lembah Bertam. Spatial and temporal analysis of land use changes with landslide inventory and hazard zonation shows a positive relationship. Approximately 31.3% of historical landslides and 62% of very high landslide hazard areas fall within areas that underwent changes.

Several recommendations have been identified to enhance the concept of landslide risk-informed land use planning for disaster risk reduction. Up-to-date landslide risk information is a tool to promote risk-sensitive development. Proper documentation of landslide events is vital to get better results in conducting spatial and temporal analyses on the relationship of landslide occurrence and land use change. In Cameron Highlands, where landslides are recurrent, it is important to produce multi-temporal landslide maps to understand the unique characteristic of landslide occurrence. The temporal landslide analysis is also vital for landslide hazard and risk assessment (Guzzetti et al. 2012). High accuracy and detailed scale of Landslide Hazard and Risk Assessment (LHRA) are needed at various scales of development. Adopting LiDAR technology would greatly improve bare-surface model even under dense vegetation. A detailed digital terrain model allows better interpretation of geomorphological and relict landslides (Razak et al. 2013). Currently, detailed landslide assessment conducted by the JMG Malaysia is only limited to the PBRC area. Absence of risk assessment in the development planning process will lead to the misleading development strategy where new settlement and building are allowed in the areas exposed to future disaster. Updated landslide hazard risk maps are vital to ensure risk-based decision to be made, taking into account the current condition at the site and considering future scenarios of climate change and land use changes.

One of the seven global targets in the Sendai Framework for Disaster Risk Reduction 2015–2030 is to substantially increase the number of countries with the national and local disaster risk reduction strategies by 2020 (Target E). It is recommended to allow the National Disaster Management Agency (NADMA), assisted by technical agencies at the state and district level, to co-design the national and local DRR and resilience strategies. This research provides valuable input in terms of understanding the importance of risk-based land use planning in the agriculture-based region. Risk-driven land use management, sustainable agriculture management, risk communication, and development policy are recommended strategies to achieve sustainable development in Malaysia. With the strong commitment by the local authorities in promoting risk-informed development planning, we are able to reduce current risk, prevent future disaster and strengthen societal resilience.

The Global 2030 Agenda for SDG has promoted risk-informed sustainable development, and our urgent action to reduce impacts of climate. As a core development strategy, 10 of the 17 SDGs with 25 targets related to DRR are identified. With complexity and interaction of human, economic and political systems, risk becomes increasingly systemic. Therefore, it requires concerted and urgent effort to reduce it in an integrated way. It is timely for us to better cope with our capacity and assess our increased vulnerability to climate risk. A transdisciplinary approach is a way forward to identify the prevention of new risk, the reduction of existing risk, and the strengthening of resilience, as central to global efforts in realizing sustainable development pathways in the twenty-first century (Matsuura and Razak 2019).

Acknowledgements A very special thanks to JMG, PLANMalaysia, Universiti Teknologi Malaysia and Cameron Highlands Local Council for providing valuable dataset and input for this research. Thanks to financial supports from Disaster Preparedness and Prevention Center (DPPC) Malaysia-Japan International Institute of Technology (MJIIT) UTM Kuala Lumpur and research grants, PY/2019/03529.

References

Ayele GT, Demessie SS, Mengistu KT, Tilahun SA, Melesse AM (2016) Multitemporal land use/land cover change detection for the Batena Watershed, Rift Valley Lakes Basin, Ethiopia. In Melesse AM, Abtew W (eds) Landscape dynamics, soils and hydrological processes in varied climates. Springer International Publishing, Cham, pp 51–72. https://doi.org/10.1007/978-3-319-18787-7_4

Bell FG (2006) Engineering geology, 2nd edn

Calvello M, Piciullo L (2015) Assessing the performance of regional landslide early warning models: the EDuMaP method. Natural Hazards and Earth System Sciences Discussions, vol 3. https://doi.org/10.5194/nhessd-3-6021-2015

Cameron Highlands District Council (2015) Draf Rancangan Tempatan Daerah Cameron Highlands 2030 (Penggantian)

Department of Mineral and Geoscience Malaysia (2016) Slope hazard and risk mapping of Ipoh, Perak and Cameron Highlands, Pahang Area

Department of Mineral and Geoscience Malaysia (2018) Landslide hazard and risk assessment report of Cameron Highlands, Pahang

Glade T (2003) Landslide occurrence as a response to land use change: a review of evidence from New Zealand. CATENA 51(3):297–314. https://doi.org/10.1016/S0341-8162(02)00170-4

Gökceoglu C, Aksoy H (1996) Landslide susceptibility mapping of the slopes in the residual soils of the Mengen region (Turkey) by deterministic stability analyses and image processing techniques. Eng Geol 44(1):147–161. https://doi.org/10.1016/S0013-7952(97)81260-4

Guan D, Li H, Inohae T, Su W, Nagaie T, Hokao K (2011) Modeling urban land use change by the integration of cellular automaton and Markov model. Ecol Model 222(20):3761–3772. https://doi.org/10.1016/j.ecolmodel.2011.09.009

Guzzetti F, Mondini AC, Cardinali M, Fiorucci F, Santangelo M, Chang K-T (2012) Landslide inventory maps: new tools for an old problem. Earth Sci Rev 112(1):42–66. https://doi.org/10.1016/j.earscirev.2012.02.001

Halmy MWA, Gessler PE, Hicke JA, Salem BB (2015) Land use/land cover change detection and prediction in the north-western coastal desert of Egypt using Markov-CA. Appl Geogr 63:101–112. https://doi.org/10.1016/j.apgeog.2015.06.015

Karsli F, Atasoy M, Yalcin A, Reis S, Demir O, Gokceoglu C (2008) Effects of land-use changes on landslides in a landslide-prone area (Ardesen, Rize, NE Turkey). Environ Monit Assess 156(1):241. https://doi.org/10.1007/s10661-008-0481-5

Ko WW, Syed AR, Shukor NM, Safruddin H, Azhar JA (1987) Agriculture in the Cameron Highlands. Paper presented at the International Conference on Steepland Agriculture in the Humid Tropics, Kuala Lumpur

Matsuura S, Razak KA (2019) Exploring transdisciplinary approaches to facilitate disaster risk reduction. Disaster Prev Manag Int J 28 (6):809–822. https://doi.org/10.1108/DPM-09-2019-0289

Mohamad Z, Azahari Razak K, Ahmad F, Manap MA, Ramli Z, Ahmad A, Mohamed Z (2015) Slope hazard and risk assessment in the tropics: Malaysia' experience. Paper presented at the EGU General Assembly Conference Abstracts. https://ui.adsabs.harvard.edu/abs/2015EGUGA..17.7746M

Public Work Department (2009) National Slope Master Plan 2009–2023

Raj JK (2002) Land use changes, soil erosion and decreased base flow of rivers at Cameron Highlands, Peninsular Malaysia. Paper presented at the Geological Society of Malaysia Annual Geological Conference 2002, Kota Bharu, Kelantan, Malaysia. https://gsm.org.my/products/702001–100750-PDF.pdf. 10.7186/bgsm45200201

Razak KA, Santangelo M, Van Westen CJ, Straatsma MW, de Jong SM (2013) Generating an optimal DTM from airborne laser scanning data for landslide mapping in a tropical forest environment. Geomorphology 190:112–125. https://doi.org/10.1016/j.geomorph.2013.02.021

Reis S (2008) Analyzing land use/land cover changes using remote sensing and GIS in Rize North-East Turkey. Sensors 8(10):6188–6202. https://doi.org/10.3390/s8106188

Rendana M, Rahim SA, Idris WMR, Lihan T, Rahman ZA (2015) CA-Markov for predicting land use changes in tropical catchment area: a case study in Cameron Highland, Malaysia. J Appl Sci 15:689–695. https://doi.org/10.3923/jas.2015.689.695

Yaakup A, Ibrahim M, Sulaiman SM, Sosi Z (2000) Assessment of urban development and landuse changes' impact on the environment: a case study of urban development in Klang Valley region, Malaysia. Paper presented at the SENVAR, Surabaya, Indonesia. eprints.utm.my/id/eprint/631/2/ASSESSMENT_OF_URBAN_DEVELOPMENT%282000%29Ahris_Yaakup.pdf

Zheng HW, Shen GQ, Wang H, Hong J (2015) Simulating land use change in urban renewal areas: a case study in Hong Kong. Habitat Int 46:23–34. https://doi.org/10.1016/j.habitatint.2014.10.008

Landslides in Steep-Slope Agricultural Landscapes

Paolo Tarolli, Anton Pijl, and Sara Cucchiaro

Abstract

Agricultural landscapes cover a significant part of the Earth. In floodplains, we can find large areas dedicated to intensive agriculture. However, also on hills and mountains, agricultural activity can be relevant from the socio-economic point of view. Nowadays, such areas are increasingly under threat because of global environmental changes. Widespread growing rainfall aggressiveness due to climate change, in addition to land abandonment, lack of structural maintenance, and in some cases unsuitable agronomic practices are exposing steep-slope agricultural landscapes to increased hazard of landslides. A suitable hazard assessment and zonation of these phenomena would help better management of such agricultural landscapes. The purpose of this article is to provide an overview of this relevant problem focusing on (i) the contribution of remote sensing technologies (e.g., LiDAR and UAV photogrammetry) in mapping the investigated processes, and (ii) discussing advances and limitations of susceptibility modelling.

Keywords

Landslide • Remote sensing • Modelling • Landscape • Agriculture

P. Tarolli (✉) · A. Pijl · S. Cucchiaro
Department of Land, Environment, Agriculture and Forestry, University of Padova, viale dell'Università 16, 35020 Agripolis Legnaro, PD, Italy
e-mail: paolo.tarolli@unipd.it

S. Cucchiaro
Department of Agricultural, Food, Enviromental and Animal Sciences, Università of Udine, via delle Scienze 206, 33100 Udine, Italy

F. Guzzetti et al. (eds.), *Understanding and Reducing Landslide Disaster Risk*, ICL Contribution to Landslide Disaster Risk Reduction,
https://doi.org/10.1007/978-3-030-60227-7_46

Background

Agricultural land use is responsible for an unprecedented transformation of natural environments worldwide, with vast and long-term impacts on geomorphology and soil properties (Baartman et al. 2012). Cultivated areas typically involve reduced soil cover and cohesion, which particularly in steep-slope environments greatly affect soil erosion and slope instability (Koulouri and Giourga 2007; Prosdocimi et al. 2016; Tarolli and Straffelini 2020). Several researchers around the world have studied the increased landslide susceptibility of cultivated hillslopes, and several factors related to agricultural practices (i.e. not considering climate or slope steepness) have been discussed. Agricultural transformation affects soil stability due to the removal of permanent deep-rooted vegetative cover (Perotto-Baldiviezo et al. 2004), while the natural soil structure is affected due to land levelling (DeGraff and Canuti 1988; Ramos et al. 2007). In addition, an unsuitable or degraded terracing or drainage system can further aggravate landslide hazard (Tarolli et al. 2014). This is often related to the lack of maintenance as a result of land abandonment and loss of labour, as widely reported for Mediterranean (Arnáez et al. 2011; Cevasco et al. 2014; Tarolli et al. 2014) and Asian steep-slope agricultural areas (Gerrard and Gardner 2002; Raj Khanal and Watanabe 2006). Other factors include the cultivation and subsequent reactivation of dormant landslides (Sugawara 2013), and the construction of agricultural roads for machinery (Tarolli et al. 2015). The latter is able to divert runoff and create concentrated patterns of water flow, which are often related to the initiation of landslides (as illustrated in Fig. 1).

The high landslide hazard in agricultural areas can have considerable impact on production and human safety. Nonetheless, reliable inventories are missing for many marginalised steep rural areas around the world. Modern developments in remote sensing, computer technologies and models may help contributing to this. In this work, we

Fig. 1 Landslides that occurred after an intense rainfall event in a typical steep-slope vineyard terraced landscape. Landslide crowns are highlighted with a white dashed line (photograph modified from Tarolli et al. (2020))

discuss the opportunities and challenges of remote sensing techniques, digital terrain analysis and landslide modelling, based on literature and few original examples.

Remote Sensing

Techniques

Landslide inventory maps are generally limited in terms of spatial coverage and time period (Guzzetti et al. 2012), which can partly be attributed to the intensive mapping methods used in the past (Galli et al. 2008). However, there is a strong potential to address this gap by the use of modern remote sensing techniques, which allow more accurate topographic analysis by use of faster and cheaper surveys (Tarolli 2014). The most recent platform that proved highly successful is the use of an Unmanned Aerial Vehicle (UAV, sometimes referred to as UAS or RPAS) and the parallel development of Structure from Motion (SfM) photogrammetry technique (Giordan et al. 2018), which allows high-accuracy surveying by low-cost applications of a simple drone mounted with a non-metric camera (Fig. 2). This platform rapidly gained popularity for mapping topographic features, as it is flexible to deploy in varying conditions (including difficult-to-access sites) and is able to capture complex geomorphologic features (Eltner et al. 2016; Cucchiaro et al. 2018). The typical coverage of light-weight UAVs is in the order of tens of hectares and is optimal for

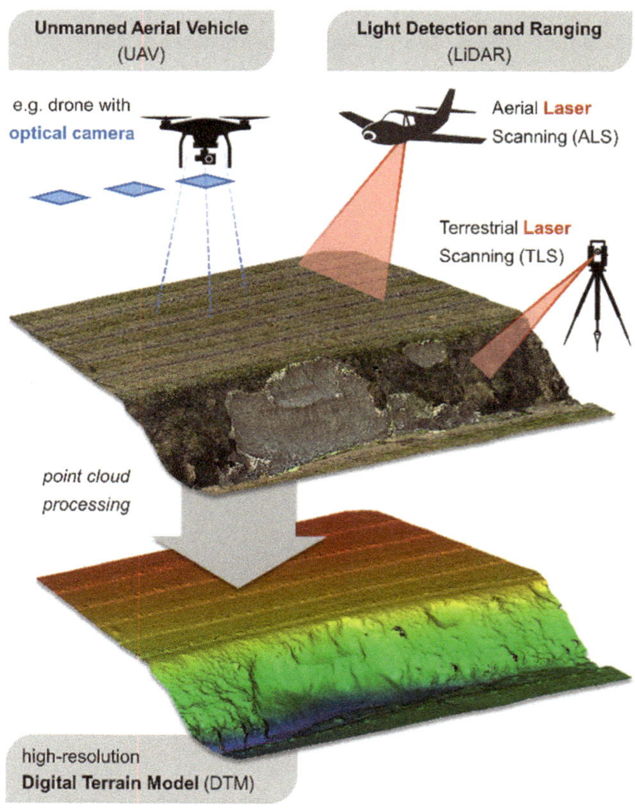

Fig. 2 Schematic illustration of the creation of a high-resolution Digital Terrain Model (DTM) by remote sensing techniques of Unmanned Aerial Vehicles (UAVs) or Light Detection and Ranging (LiDAR) instruments (figure modified from Tarolli et al. (2020))

agricultural conditions (Colomina and Molina 2014; Pijl et al. 2020). Indeed, specific examples of UAV-based analysis of agricultural slope failure can be found, e.g. for monitoring mass-movements (Lucieer et al. 2014; Turner et al. 2015), or for the detection and modelling of terrace failures (Pijl et al. 2019).

Another modern remote sensing technique that is commonly used for high-resolution topographic mapping is Light Detection and Ranging (LiDAR) from a terrestrial or aerial platform (Fig. 2). It is probably one of the most popular technique in landslide research, by providing highly-detailed topographic information for the detection, monitoring or modelling of slope failure (Jaboyedoff et al. 2012; Scaioni et al. 2014). Although LiDAR involves considerably higher deployment costs than UAV-based surveying (Guzzetti et al. 2012), it offers a major advantage over the latter due to its ability to retrieve multi-layered return signals and resulting ground detection in densely vegetated areas (Tarolli 2014). As such, it is probably the most established platform for geomorphologic research, and countless examples can be found of LiDAR-based research of agricultural landslides, e.g. in Italian terraced landscapes (Preti et al. 2018a; Tarolli et al. 2014, 2015; Giordan et al. 2017; Brandolini et al. 2018).

Digital Terrain Analysis

The improving spatial accuracy of topographic data (e.g. from UAV or LiDAR source) also allows higher accuracy in the detection of landslides and their geomorphology. Basic 2-D digital terrain analysis can be used to support visual detection, while facilitating more advanced purposes such as volume estimation, multi-temporal monitoring and susceptibility modelling (as discussed in the following chapter). Several authors have tested the use of terrain derivatives such as slope, surface roughness or landform curvature for the automatic delineation of geomorphologic features using a statistical threshold (McKean and Roering 2004; Booth et al. 2009; Tarolli et al. 2012). In this work, we illustrate this concept using an original topographic data sample, consisting of a road-induced landslide (Fig. 3). The shaded 0.05-m DTM clearly shows how the landslide crown is captured, appearing as a dark edge at the road side. This crown presents a disruption with respect to the surrounding geomorphology, as can be identified using the surface roughness index (Fig. 3), here calculated according to Cavalli et al. (2008) using a 31-cell kernel. Similarly, high values of maximum landform curvature indicate convex terrain elements (according to Evans 1979 using a 31-cell kernel); and the statistic threshold approach by Sofia et al. (2014) has been be used to automatically extract the landslide crown (Fig. 3). This approach could rapidly be applied over large extents, hence becoming a powerful tool for automatic geomorphologic feature extraction.

Modelling

For estimating the susceptibility, or spatial probability of slope failures occurrence, numerous models are available (Guzzetti et al. 2005), but the number of publications on landslide hazard assessment in steep agricultural slopes is still rather modest. In this anthropic context, further variables must be considered than the classic ones used to better describe the landslide triggering processes. Here, slope failure hazard is not only dependent on meteorological events and geotechnical land attributes, but also agricultural practices and human activities can strongly affect the occurrence of instability phenomena (Shrestha et al. 2004). For example, Perotto-Baldiviezo et al. (2004) developed and tested a spatial model for predicting the spatial distribution of landslide hazard in steep areas of Honduras considering four model variables as slope, aspect, stream proximity and land cover type. As the slope increased, the percentage of land affected by landslides, increased sharply on cropland when the soils were saturated. This indicated that agricultural activity and the associated removal of deep-rooted permanent vegetation increased the landslide hazard on steep sites. Jaiswal et al. (2010) also proposed a probabilistic landslide model to quantify hazard of first-time slope failure on natural slopes and tea plantations, underlining how agricultural and constructional activities make an area more susceptible to landslides. At the local scale, numerical simulations must consider the presence of widespread agricultural practices such as terraces that heavily influence the hydrological processes (Gallart et al. 1994; Preti et al. 2018a) and can trigger superficial mass-movements when they collapse. Camera et al. (2014) analyzed the processes that can lead to failure. They used a numerical modelling approach of groundwater movement and related stability analysis, to provide pore water pressure distributions, which are generated by different rainfall amounts, as parameters for a stress-strain analysis that can directly determine the influence of various rainfall parameters on dry-stone wall stability. Penna et al. (2014) evaluated the predictive power of the quasi-dynamic shallow landslide model QD-SLaM (Tarolli et al. 2008) to simulate shallow landslide locations in a small-scale steep-slope watershed. The study area represented a typical anthropogenic Mediterranean landscape, with terraces (partly abandoned), roads and a village. The applied landslide model did not incorporate the description of road-related or terrace-related failures, thus highlighting its limits in the correct interpretation of hydrological processes in an anthropogenic context. The model predictive power was shown to be DTM-resolution dependent. The use

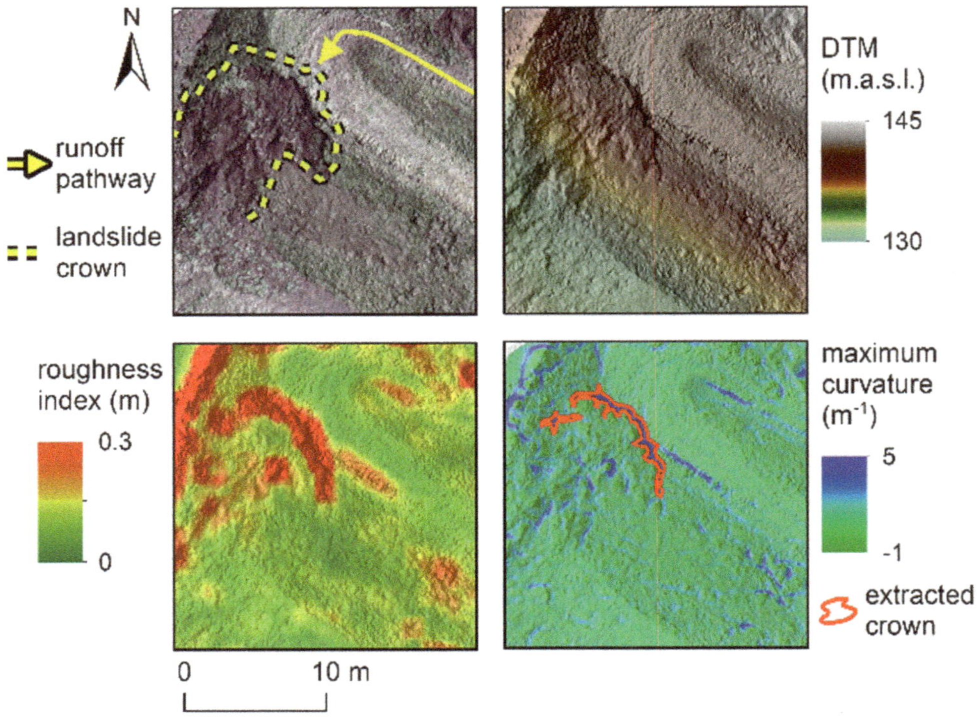

Fig. 3 Example of a landslide induced by a runoff pathway on a curving agricultural road (top-left, yellow lines). Additionally, illustrated are the shaded high-resolution DTM derived by UAV-SfM workflow (top-right), and two terrain derivatives: the roughness index (bottom-left) and the maximum landform curvature (bottom-right). The latter shows the automatically extracted landslide crown using a threshold approach (red outline). (figure modified from Tarolli et al. (2020), based on 0.05-m resolution sample dataset)

of a coarser resolution had a smoothing effect on terrain attributes, and therefore on predictive model performance. The authors concluded that to realize the full potential of high-resolution topography thus including anthropogenic geomorphic features, more extensive work is needed to specifically identify the extent of the artificial structures and their impact on shallow landsliding processes. Tarolli et al. (2015) used LiDAR elevation data for a detailed hydro-geomorphological analysis of terraced vineyards. The geomorphic Relative Path Impact Index (RPII) was tested in two vineyards to identify terrace-induced and road-induced erosion. Using such an index, the authors then simulated different scenarios of soil conservation measures, establishing the optimal solution to reduce erosion. The results highlighted the effectiveness of high-resolution topography in the analysis of erosion at the local scale of terraced vineyards when surface water flow is the main factor triggering the instabilities. An example of the predictive power of RPII in the recognition of potential slope failure in a steep-slope terraced site is shown in Fig. 4, where we considered the case study of Fig. 1.

Preti et al. (2018b) proposed a more specific model able to describe hydrological processes in terraced landscapes. In detail, they analyzed the destabilizing pressures acting on the retaining dry-stone walls in the most critical portion of each terrace. The results showed good capability of the model to predict the distribution and intensity of stress on the dry-stone wall over time and space. This stress was related to the combined earth pressure and hydrostatic pressure (water accumulation), without the occurrence of soil saturation. A better understanding of the main hydrological processes that govern surface and subsurface water flow pathways and that are responsible for terrace failure is essential for appropriate water resource management and rural landscape maintenance in terraced areas (Preti et al. 2018a). Such insights could support landowners and land planners in managing these complex and fragile environments.

Final Remarks

Climate change (e.g., the increase of rainfall intensity) and changing societal trends (e.g., land abandonment) are aggravating land degradation in agricultural landscapes, resulting in increased mass movements that should not be neglected. Not only does this affect agricultural production, it also poses a risk to local communities of people. Several approaches for the spatial analysis of such processes are

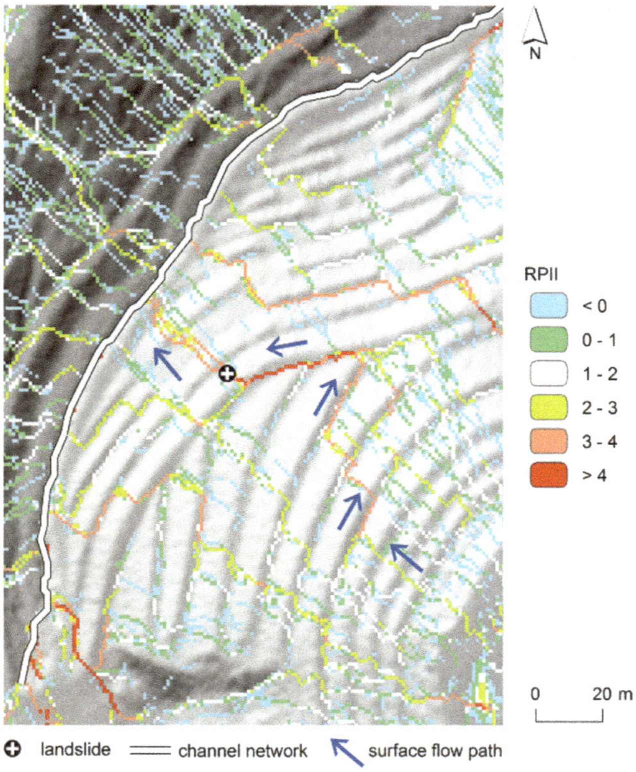

Fig. 4 Relative Path Impact Index (RPII) calculated for the case study shown in Fig. 1 using a 1-m LiDAR-derived DTM. Surface flow modifications induced by an agricultural road and consequent landslide (warm colours) match well with the location of the observed landslide (plus symbol), thus confirming the predictive power of the index for the detection of slope failure in this context (spatial data analysis original work by the authors)

available, however high-resolution topography derived by modern remote sensing techniques (e.g., SfM photogrammetry using UAV images, or airborne or terrestrial LiDAR data) is highly recommended for the detection of the characteristic local-scale geomorphic features usually visible with sub-meter DTM grid cell size. In addition, it is also recommended to improve the understanding of the main hydrological processes that govern surface and subsurface water flow pathways. Existing modelling approaches seem to be not optimal for a satisfactory understanding of such processes. Indeed, to meet the full potential of high-resolution topography, they should be designed to include the anthropogenic geomorphic features and model their impact on the physical processes or vice versa. More recently, few authors developed some advances along this line. This will absolutely be a future challenge for scientific studies in this field. Novel insights from research could support land owners and land planners in managing these complex and fragile landscapes, in order to preserve cultural heritage, ecosystem services, and food safety while maintaining the economic and environmental sustainability.

Acknowledgements This work is supported by ViTE “Vineyard Terraced landscapes: understanding the Environmental constraints to improve sustainable managements”, a project funded by TESAF department of the University of Padova (Italy).

References

Arnáez, J, Lasanta, T, Erreas, MP, Ortigosa, L (2011) Land abandonment, landscape evolution, and soil erosion in a Spanish Mediterranean mountain region: the case of Camero Viejo. Land Degrad Dev 22:537–550. https://doi.org/10.1002/ldr.1032

Baartman JEM, Temme AJAM, Schoorl JM et al (2012) Did tillage erosion play a role in millennial scale landscape development? Earth Surf Process Landforms 37:1615–1626. https://doi.org/10.1002/esp.3262

Booth AM, Roering JJ, Perron JT (2009) Automated landslide mapping using spectral analysis and high-resolution topographic data: Puget Sound lowlands, Washington, and Portland Hills. Geomorphology, Oregon. https://doi.org/10.1016/j.geomorph.2009.02.027

Brandolini P, Cevasco A, Capolongo D et al (2018) Response of terraced slopes to a very intense rainfall event and relationships with land abandonment: a case study from Cinque Terre (Italy). L Degrad Dev 29:630–642. https://doi.org/10.1002/ldr.2672

Camera C, Apuani T, Masetti M (2014) Mechanisms of failure on terraced slopes: The Valtellina case (northern Italy). Landslides 11:43–54. https://doi.org/10.1007/s10346-012-0371-3

Cavalli M, Tarolli P, Marchi L, Dalla Fontana G (2008) The effectiveness of airborne LiDAR data in the recognition of channel bed morphology. CATENA 73:249–260. https://doi.org/10.1016/j.catena.2007.11.001

Cevasco A, Pepe G, Brandolini P (2014) The influences of geological and land use settings on shallow landslides triggered by an intense rainfall event in a coastal terraced environment. Bull Eng Geol Environ 73:859–875. https://doi.org/10.1007/s10064-013-0544-x

Colomina I, Molina P (2014) Unmanned aerial systems for photogrammetry and remote sensing: a review. ISPRS J Photogramm Remote Sens 92:79–97.https://doi.org/10.1016/j.isprsjprs.2014.02.013

Cucchiaro S, Cavalli M, Vericat D et al (2018) Monitoring topographic changes through 4D-structure-from-motion photogrammetry: application to a debris-flow channel. Environ Earth Sci 77:632. https://doi.org/10.1007/s12665-018-7817-4

De Graff JV, Canuti P (1988) Using isopleth mapping to evaluate landslide activity in relation to agricultural practices. Bull Int Assoc Eng Geol 38:62–71

Eltner A, Kaiser A, Castillo C et al (2016) Image-based surface reconstruction in geomorphometry-merits, limits and developments. Earth Surf Dyn 4:359–389. https://doi.org/10.5194/esurf-4-359-2016

Evans IS (1979) An integrated system of terrain analysis and slope mapping. Final report on grant DA-ERO-591–73-G0040. University of Durham, England

Gallart F, Llorens P, Latron J (1994) Studying the role of old agricultural terraces on runoff generation in a small Mediterranean mountainous basin. J Hydrol 159:291–303. https://doi.org/10.1016/0022-1694(94)90262-3

Galli M, Ardizzone F, Cardinali M et al (2008) Comparing landslide inventory maps. Geomorphology 94:268–289. https://doi.org/10.1016/J.GEOMORPH.2006.09.023

Gerrard J, Gardner R (2002) Relationships between landsliding and land use in the Likhu Khola Drainage Basin, Middle Hills, Nepal. Mt Res Dev 22:48–55. https://doi.org/10.1659/0276-4741(2002)022[0048:rblalu]2.0.co;2

Giordan D, Cignetti M, Baldo M, Godone D (2017) Relationship between man-made environment and slope stability: the case of 2014 rainfall events in the terraced landscape of the Liguria region (northwestern Italy). Geomatics, Nat Hazards Risk 8:1833–1852. https://doi.org/10.1080/19475705.2017.1391129

Giordan D, Hayakawa Y, Nex F et al (2018) Review article: the use of remotely piloted aircraft systems (RPASs) for natural hazards monitoring and management. Nat Hazards Earth Syst Sci 18:1079–1096. https://doi.org/10.5194/nhess-18-1079-2018

Guzzetti F, Mondini AC, Cardinali M et al (2012) Landslide inventory maps: new tools for an old problem. Earth-Sci Rev 112:42–66. https://doi.org/10.1016/j.earscirev.2012.02.001

Guzzetti F, Reichenbach P, Cardinali M et al (2005) Probabilistic landslide hazard assessment at the basin scale. Geomorphology 72:272–299. https://doi.org/10.1016/j.geomorph.2005.06.002

Jaboyedoff M, Oppikofer T, Abellán A et al (2012) Use of LIDAR in landslide investigations: a review. Nat Hazards 61:5–28. https://doi.org/10.1007/s11069-010-9634-2

Jaiswal P, van Westen CJ, Jetten V (2010) Quantitative landslide hazard assessment along a transportation corridor in southern India. Eng Geol 116:236–250. https://doi.org/10.1016/j.enggeo.2010.09.005

Koulouri M, Giourga C (2007) Land abandonment and slope gradient as key factors of soil erosion in Mediterranean terraced lands. CATENA 69:274–281. https://doi.org/10.1016/j.catena.2006.07.001

Lucieer A, de Jong SM, Turner D (2014) Mapping landslide displacements using Structure from Motion (SfM) and image correlation of multi-temporal UAV photography. Prog Phys Geogr Earth Environ 38:97–116. https://doi.org/10.1177/0309133313515293

McKean J, Roering J (2004) Objective landslide detection and surface morphology mapping using high-resolution airborne laser altimetry. Geomorphology. https://doi.org/10.1016/S0169-555X(03)00164-8

Penna D, Borga M, Aronica GT, Brigandì G, Tarolli P (2014) The influence of grid resolution on the prediction of natural and road-related shallow landslides. Hydrol Earth Syst Sci 18:2127–2139. https://doi.org/10.5194/hess-18-2127-2014

Perotto-Baldiviezo HL, Thurow TL, Smith CT et al (2004) GIS-based spatial analysis and modeling for landslide hazard assessment in steeplands, southern Honduras. Agric Ecosyst Environ 103:165–176. https://doi.org/10.1016/j.agee.2003.10.011

Pijl A, Tosoni M, Roder G, Sofia G, Tarolli P (2019) Design of terrace drainage networks using UAV-based high-resolution topographic data. Water 11:814. https://doi.org/10.3390/w11040814

Pijl A, Bailly JS, Feurer D, El Maaoui MA, Boussema MR, Tarolli P (2020) TERRA: terrain extraction from elevation rasters through repetitive anisotropic filtering. Int J Appl Earth Obs Geoinf 84:101977. https://doi.org/10.1016/j.jag.2019.101977

Preti F, Guastini E, Penna D et al (2018a) Conceptualization of water flow pathways in agricultural terraced landscapes. L Degrad Dev 29:651–662. https://doi.org/10.1002/ldr.2764

Preti F, Errico A, Caruso M et al (2018b) Dry-stone wall terrace monitoring and modelling. L Degrad Dev 29:1806–1818. https://doi.org/10.1002/ldr.2926

Prosdocimi M, Cerdà A, Tarolli P (2016) Soil water erosion on Mediterranean vineyards: a review. CATENA 141:1–21. https://doi.org/10.1016/j.catena.2016.02.010

Raj Khanal N, Watanabe T (2006) Abandonment of agricultural land and its consequences. Mt Res Dev 26:32–40. https://doi.org/10.1659/0276-4741(2006)026[0032:aoalai]2.0.co;2

Ramos MC, Cots-Folch R, Martínez-Casasnovas JA (2007) Sustainability of modern land terracing for vineyard plantation in a Mediterranean mountain environment—the case of the Priorat region (NE Spain). Geomorphology 86:1–11. https://doi.org/10.1016/j.geomorph.2006.08.004

Scaioni M, Longoni L, Melillo V, Papini M (2014) Remote sensing for landslide investigations: an overview of recent achievements and perspectives. Remote Sens 6:1–53. https://doi.org/10.3390/rs60x000x

Shrestha DP, Zinck JA, Van Ranst E (2004) Modelling land degradation in the Nepalese Himalaya. CATENA 57:135–156. https://doi.org/10.1016/j.catena.2003.11.003

Sofia G, Dalla Fontana G, Tarolli P (2014) High-resolution topography and anthropogenic feature extraction: testing geomorphometric parameters in floodplains. Hydrol Process 28:2046–2061. https://doi.org/10.1002/hyp.9727

Sugawara J (2013) Landslides in tea plantation fields in Shizuoka, Japan. Int J Geomate, 495–500. https://doi.org/10.21660/2013.7.21154

Tarolli P (2014) High-resolution topography for understanding earth surface processes: opportunities and challenges. Geomorphology 216:295–312. https://doi.org/10.1016/j.geomorph.2014.03.008

Tarolli P, Borga M, Dalla Fontana G (2008) Analyzing the influence of upslope bedrock outcrops on shallow landsliding. Geomorphology 93:186–200. https://doi.org/10.1016/j.geomorph.2007.02.017

Tarolli P, Preti F, Romano N (2014) Terraced landscapes: from an old best practice to a potential hazard for soil degradation due to land abandonment. Anthropocene 6:10–25. https://doi.org/10.1016/j.ancene.2014.03.002

Tarolli P, Sofia G, Calligaro S et al (2015) Vineyards in terraced landscapes new opportunities from Lidar data. L Degrad Dev 26:92–102. https://doi.org/10.1002/ldr.2311

Tarolli P, Sofia G, Dalla Fontana G (2012) Geomorphic features extraction from high-resolution topography: landslide crowns and bank erosion. Nat Hazards 61:65–83. https://doi.org/10.1007/s11069-010-9695-2

Tarolli P, Straffelini E (2020) Agriculture in hilly and mountainous landscapes: threats, monitoring and sustainable management. Geogr Sustain. https://doi.org/10.1016/j.geosus.2020.03.003

Tarolli P, Pijl A, Cucchiaro S, Wei W (2020) Slope instabilities in steep cultivation systems: process classification and opportunities from remote sensing. L Degrad Dev (in print). https://doi.org/10.1002/ldr.3798

Turner D, Lucieer A, de Jong S et al (2015) Time series analysis of landslide dynamics using an unmanned aerial vehicle (UAV). Remote Sens 7:1736–1757. https://doi.org/10.3390/rs70201736

From Satellite Images to Field Survey: A Complete Scheme of Landslide InSAR Monitoring

2020 Kyoto Japan

Matteo Del Soldato, Lorenzo Solari, Davide Festa, Pierluigi Confuorto, Silvia Bianchini, and Nicola Casagli

Abstract

The free availability of Sentinel-1 satellite radar images, combined with the wide-area coverage, the huge benefits-cost ratio, the short revisiting time and the non-invasiveness allows a continuous monitoring of ground displacements. In Italy, Sentinel-1-based continuous monitoring activities are operating in three regions and a possible chain for a practical use of satellite images for local authorities and civil protection operators is presented. A continuous monitoring is set up through the systematic collection of imagery and the repeated processing of each new pair acquired. The Persistent Scatterers (PS) resulting from such chain are subsequently analyzed by means of a data-mining algorithm for highlighting the points with relevant trend variations in the time series. In this way, a huge amount of data and wide areas may be monitored and temporal changes due to various phenomena could be assessed in a short time. The highlighted points, taking advantage of the surrounding PS data, are thus interpreted and classified according to a list of possible causes (e.g. accelerations of landslide or subsidence phenomena). The most relevant cases are notified to local authorities as potential active hazardous phenomena to be verified by means of field survey. A supporting scheme for the operative chain is here proposed and two case studies are presented. This operating workflow is intended to represent a standard procedure which could be applied on a national scale, aiming at the real-time monitoring of hazardous scenarios, and may be used as a guide for civil protection procedures for the risk reduction in large areas.

Keywords

Sentinel-1 • InSAR • Monitoring • Landslides • Preliminary risk assessment • Field survey

Introduction

Rapid urbanization and climate changes in the last decades are among the main factors causing a relevant transformation of the environment with geohazards impacts (Schuster and Highland 2001). The main implications caused by geo-hydrogeological disasters, e.g. landslides and subsidence, concern economic and social fabrics (Hu et al. 2002; Herrera et al. 2018) producing direct and indirect impacts on population as well as on structures and infrastructures (Kjekstad and Highland 2009; Del Soldato et al. 2017).

The increase of hazard consequences can be addressed by incrementing the Civil Protection monitoring activities and enlarging the cooperation between the authorities in charge and the scientific community. Valid support for monitoring

The original version of this chapter was revised: Chapter author's incorrect family name has been updated from "Del Soldato Matteo" to "Matteo Del Soldato" . The correction to this chapter is available at https://doi.org/10.1007/978-3-030-60227-7_57

M. Del Soldato (✉) · D. Festa · P. Confuorto · S. Bianchini · N. Casagli
Earth Science Department, University of Florence, Via G. La Pira 4, 50121 Firenze, Italy
e-mail: matteo.delsoldato@unifi.it

D. Festa
e-mail: davide.festa@unifi.it

P. Confuorto
e-mail: pierluigi.confuorto@unifi.it

S. Bianchini
e-mail: silvia.bianchini@unifi.it

N. Casagli
e-mail: nicola.casagli@unifi.it

L. Solari
Centre Tecnològic de Telecomunicacions de Catalunya (CTTC), Division of Geomatics. Av. Gauss 7, 08860 Castelldefels, Spain
e-mail: lorenzo.solari@cttc.es

F. Guzzetti et al. (eds.), *Understanding and Reducing Landslide Disaster Risk*, ICL Contribution to Landslide Disaster Risk Reduction, https://doi.org/10.1007/978-3-030-60227-7_47

activities has been given by the first SAR (Synthetic Aperture Radar) missions. The SAR data allowed measuring slow-moving phenomena since the early 90s thanks to the Interferometric SAR (InSAR) techniques. Moreover, the rapid technological evolution of InSAR algorithm combined with the large-area coverage, the millimetric precision, the high spatial and temporal resolution (mainly with the latest constellations) (Plank 2014) and the good cost/benefits ratio compared to conventional monitoring methods (Tomas et al. 2014) allows detecting and monitoring several ground deformational scenarios (Joyce et al. 2014).

These techniques aim to extract information of displacement taking advantage from an interferogram containing the phase difference between two passages of the satellite over the same area. Thanks to the use of several algorithms (Crosetto et al. 2016) on large SAR imagery (usually at least 35 images) it is possible to derive temporal and spatial evolution of ground deformation by identifying several PS points (Permanent Scatterers) over the area of interest. Two main approaches are present in literature, mainly related to two types of reflectors (Crosetto et al. 2016): (i) Point-like Scatterers Interferometry (PS-InSAR), featured by the constant response overtime of the radar from strong reflecting object (Ferretti et al. 2001); (ii) Distributed Scatterers (DS) or SBAS-like (Small BAseline Subset), characterized by different small scattering objects over areas with coherent backscattering signal (Berardino et al. 2002). The final output of all these Multi-Temporal InSAR (MTInSAR) approaches is a ground deformation map composed of thousands of PS points having information on the displacement over time and an average annual value of velocity.

The Sentinel-1 constellation (Sentinel-1A & 1B), launched in 2014 by ESA (European Space Agency), gave the possibility to the scientific community to exploit the data for a large spectrum of potential applications, thanks to a systematic and regular acquisition plan of SAR data. The shorter revisiting time of Sentinel-1 (6 days), with respect to the previous ESA systems, i.e. ERS1/2 and ENVISAT (35 and 24 days, respectively), enables continuous monitoring of ground deformation occurring on Earth surface. In addition, the C-band is well-suited for monitoring urban and peri-urban areas, being capable of analyzing large areas with a ground resolution up to 20 m. Considering all the above-mentioned aspects, along with technological advancements of data processing (i.e. cloud computing and parallel processing), systematic continuous monitoring is nowadays a solid reality. A first example is the Tuscany Region, where systematical satellite-based monitoring of natural phenomena at regional scale is active since 2016 (Raspini et al. 2018). Such a system revealed to be useful also for updating the slow-moving areas (Montalti et al. 2019). In the following years, other Italian regions followed the same example of the Tuscany region, e.g. the Valle d'Aosta for both updating the mapping of the slow-moving areas (Solari et al. 2019) and implementing a continuous monitoring system.

In this work, an automatic data-mining algorithm is applied for identifying situations requiring a priority of analysis, based on: (i) the processing of the SAR images; (ii) detection of PS data with a relevant trend change (Raspini et al. 2018); (iii) identification and delivery of SAR anomalous data to the hydro-geomorphological risk management actors, in charge of deciding to proceed with a field survey (Del Soldato et al. 2019a).

Two examples from two different Italian Regions are here presented, to better explain how the SAR data, combined with a field check and validation, may represent a great support for geohazard risk management at regional and local scale.

Methodology

The workflow chain here proposed is based on 4 different steps (Fig. 1), from the collection of satellite images to the field survey. The first step is about the processing of the Sentinel-1 data and data-mining of the derived PS points that

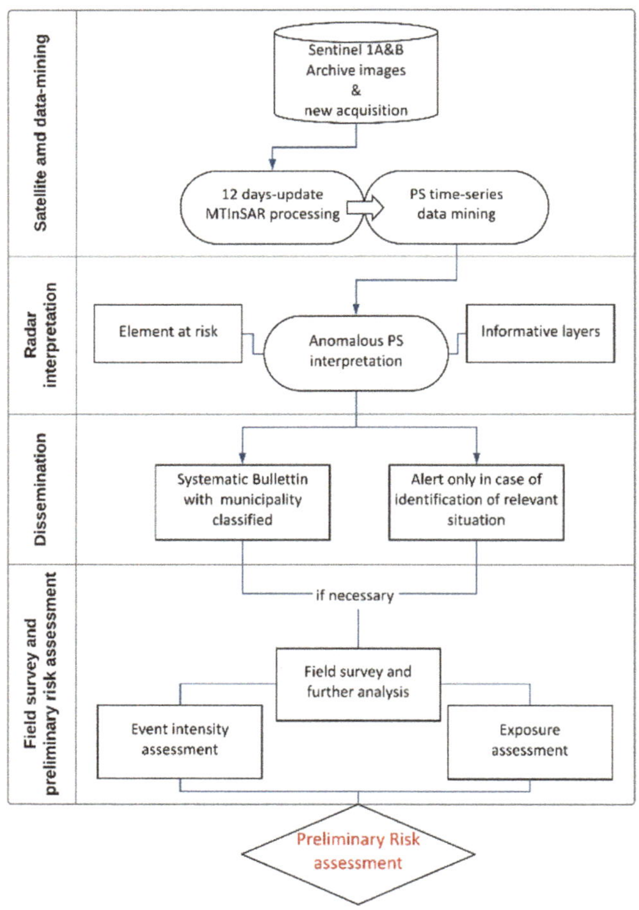

Fig. 1 Summary of the proposed workflow chain from satellite images to field survey

should be radar-interpreted by experts for assigning possible causes (second step). The results have to be shared with the hydrogeological risk managing actors (third step) for deciding if proceed with a field survey in the highlighted areas (fourth step).

Satellite Data Processing and Data-Mining

First of all, the complete stack of Sentinel-1 images, both in ascending and descending orbit, has to be collected and the archive continuously updated every time at each acquisition (6 days), on the Sentinel-Scientific Data Hub (https://scihub.copernicus.eu).

The images have to be processed by a MT-InSAR technique in order to have ground deformation maps continuously updated over all the investigated area. The resulting PS points are subsequently filtered by a data-mining algorithm ables to identify the trend variation with determined parameters. The operator has to select properly the change in velocity (Δv) in a time span (Δt) in order to identify the relevant PS to recognized significant situations, which may vary according to the area or to the targeted phenomena. For in-depth information about this approach refers to Raspini et al. (2018).

Radar-Interpretation

The radar-interpretation is fundamental to recognize if the highlighted PSs by data-mining are real, i.e. not affected by processing or environmental noise, and to assign them a possible cause.

For an accurate radar-interpretation of both ascending and descending ground deformation, several informative layers and databases can be involved:

- Archive of past interferometric dataset, e.g. ERS1/2, ENVISAT and, where available, COSMO-SkyMed (Costantini et al. 2017);
- Digital Elevation Models and derived maps, e.g. Aspect and Slope;
- Geological and lithologica maps;
- Corine Land Cover (CLC) map;
- Landslide inventory;
- Other databases, e.g. subsidence, mining, gothermal or quarries areas;
- Multi-temporal aerial and orthophotos optical imgery.

All these information, combined with the expertise of the operators, can lead to recognize possible causes of the anomaly in trend changes. The main classes to categorize the PS points with relevant trend variations could be identify in: *slope instability*, *aereal subsidence*, *local subsidence*, *uplift*, *quarries*, *geothermal activities* and *mining activities*. In addition, a class of *noise* and one of *undefined* could be useful for highlighted PS time series characterized by intrinsic errors of the techniques, a possible high atmospheric influence or temporal snow cover, as well as not clear signal in time-series, respectively.

Dissemination to Hydrogeological Risk Management Actors

Based on the radar-interpreted anomalous PS points, two different ways could be adopted to draw the attention to the hydrogeological risk managing actors:

(i) Send a direct alert in case of identification of one or more relevant and persistent cases on which draw the attention;
(ii) Prepare and deliver a monitoring bulletting every update of the ground deformation map including anomalous PS. In the bulletin the municipality of the area under monitoring should be coloured according to the presence or absence of anomalous PS, their persistency and relevance. For specific cases, a detailed scheme with ground deformation map and sampled time-series of the area under investigation should be prepared.

The regional officers, based on the delivered disclosures and on other information, if available, from instruments installed, could state the necessity or not to proceed with a field survey in the area of interest.

Field Survey and Preliminary Risk Assessment

Following the notification, regional officers could decide to proceed with a joint field survey with Civil Protection

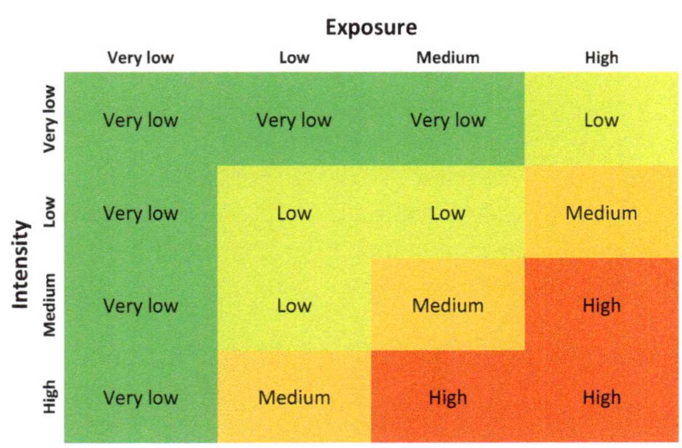

Fig. 2 Matrix of preliminary risk assessment based on the combination of four levels of intensity and exposure

Table 1 List of parameters for intensity and exposure estimation

Intensity	Exposure
Anomalous PS density	Number of structures
Anomalous PS velocity	Type of structures
Anomalous PS distribution	Use of facilities
PS density	Number of inhabitants
PS velocity	Number of infrastructures
PS distribution	Type of infrastructures
Building crack dimension	Geothermal area (Y/N)
Ground fissure	Hydraulic work (Y/N)
Geomorphological markers	Quarry site (Y/N)

personnel, if necessary, and SAR-interpreting operators. The field survey has the purpose to verify the situation in the field, the induced-damage on structures and infrastructures and try to recognize the source of the ground deformation, as well as to identify a possible solution for avoiding further consequences. In addition, combining the Intensity of the phenomenon and the Exposed element at risk, a preliminary risk can be quickly assessed (Fig. 2).

The Intensity is a combination of distribution, density and magnitude of the anomalous PS and the surrounding PSs, along with the damaging level recognized on structures and infrastructures during the field survey.

The Exposure is derived by the number and type of edifices or infrastructures in the area under investigation, as well as their use destination (e.g. industrial, private, communication route).

A summary of the relevant parameters for the estimation of Intensity and Exposure is in Table 1.

Once estimated the preliminary risks, actions for mitigating further consequences can be adopted and their severity varies according to the Intensity, the Exposures and the estimated preliminary Risk.

Practical Examples

In this section, two examples of application of the procedure are presented. The two cases show two different phenomena, i.e. a landslide in Tuscany and one in Valle d'Aosta Regions, occurred in two different environments.

In the two examples, a Δv (differential of velocity) of 10 mm/year in a Δt (time-span) of 150 days were set in the data mining algorithm to detect the anomalous trends.

Bosmatto Landslide, Valle d'Aosta Region

The Bosmatto landslide, affecting the village of Bosmatto in the Gressoney-Saint-Jean municipality, is in the alpine Valle d'Aosta Region, in the North-West of Italy. The landslide is a well-known moving area classified as complex (Luino 2005) and composed by two parts involving the upper debris and the bedrock. In 2000, several edifices were hit by a rainfall-triggered debris flow transporting blocks with estimated volume up to 100 m^3 (Luino 2005).

Since March 2017 continuous monitoring of ground deformation over the Valle d'Aosta region is active. The PS data, available only in descending orbit due to the geometrical distorition due to the Line of Sight (LoS) of the satellite, confirm the state of activity of the area surrounding Bosmatto, as active, with high deformation rates, up to 50 mm/year in the upper sector of the landslide, gradually decreasing to 15 mm/year in the central portion (Solari et al. 2019). The trend of the time-series of the Bosmatto landslide until June 2018 shows a quite linear behavior. From the end of June the data-mining algorithm pointed out PS points showing a relevant acceleration (Fig. 3). For the Valle d'Aosta Region no bullettin is delivered at every update, but only alerts in case of significant and persistent situations have to be reported.

According to the procedure, in early July 2018, an alert to the regional officers was sent. After the note, the officers of the Valle d'Aosta compared PS measurements with other instruments installed in the area (e.g. GPS) asking a systematic monitoring by PS data and a continuous update in case of evolution of the situation, i.e. other relevant trend variation or back to linear trend. No field survey has been done during the alert phase.

Zeri Landslide, Toscana Region

The municipality of Zeri (Massa-Carrara province, North Tuscany) is affected by various active and dormant landslides, involving several edifices and infrastructures. The activity of some of these landslides and the induced damage on structures are known since the middle '90s (Federici et al. 2001; Baldi et al. 2008) and are confirmed by the velocity analysis of recent PS data (Del Soldato et al. 2019b; Tomàs et al. 2019). During the continuous monitoring of the Tuscany Region in May 2018 some significant and persistent

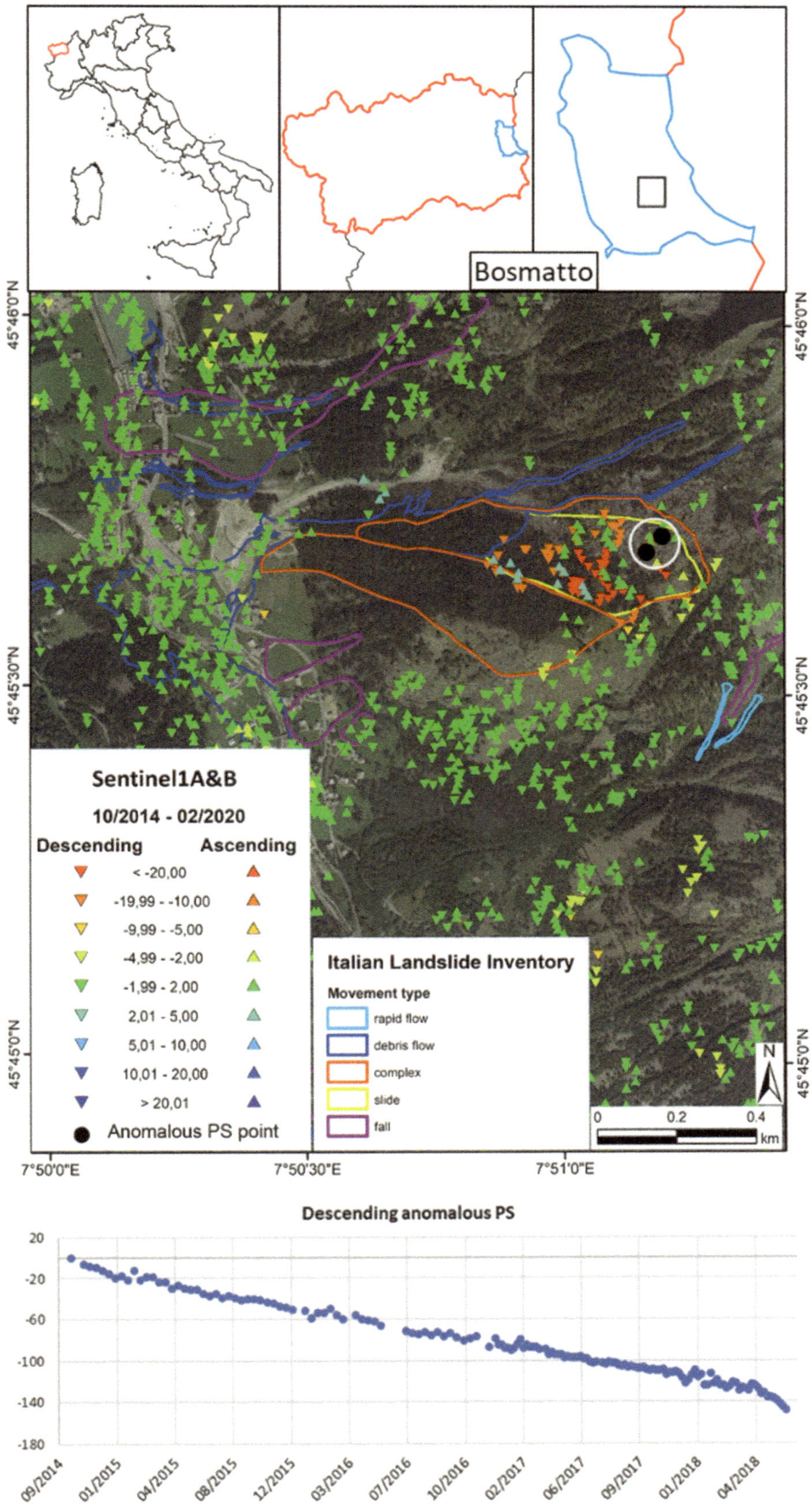

Fig. 3 Localization of gressoney-saint-jean municipality and focus on the area with PS anomalous points in June 2018. In the lower part the time-series of the anomalous PS data in June 2018 is visible (white circle in the ground deformation map)

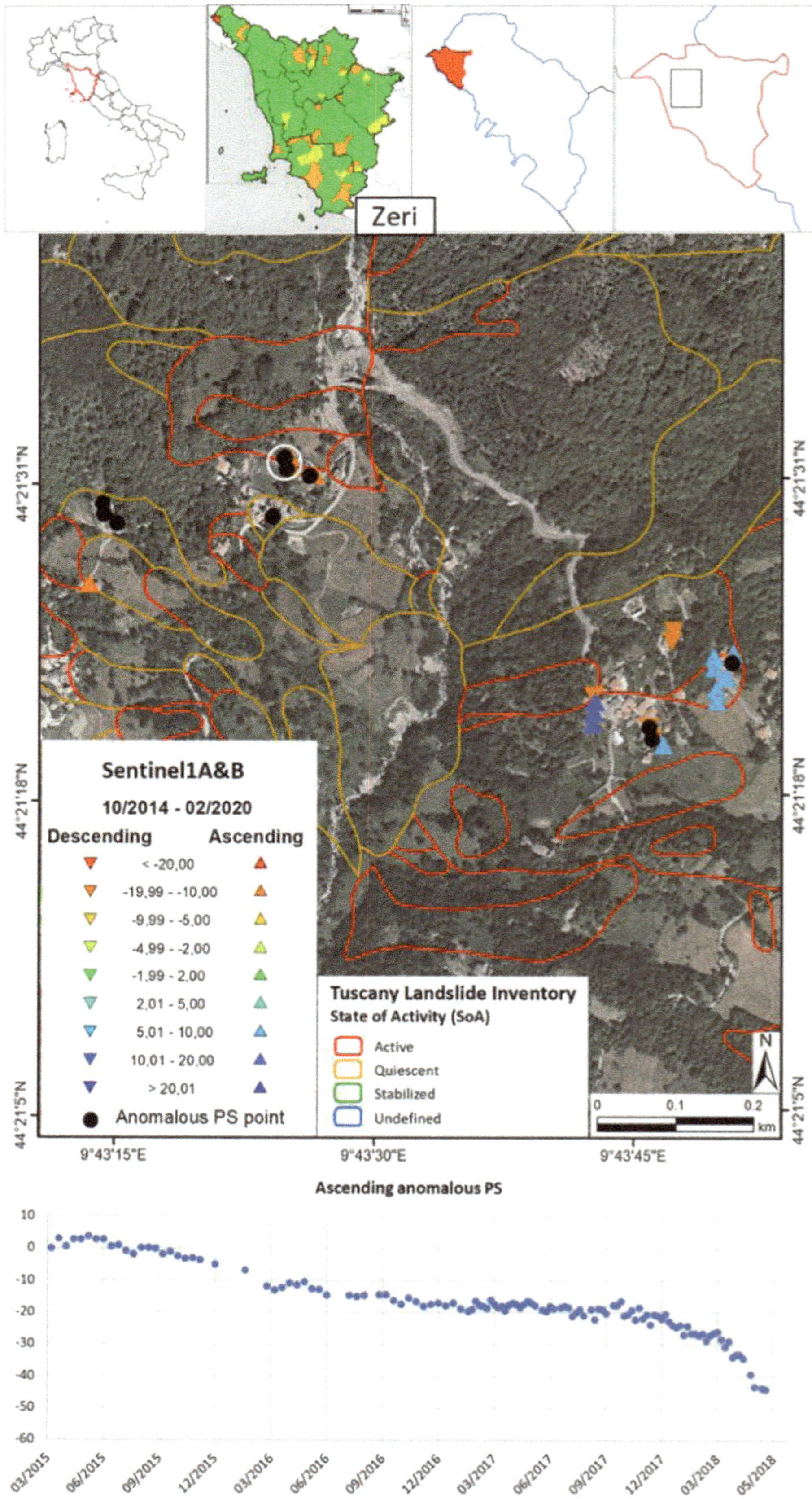

Fig. 4 Localization of Zeri municipality and of the area with PS anomalous point in May 2018. At the top also the delivered bulletin with the municipality classified (green—no anomalies; yellow—one anomaly; orange—at least one persistent anomaly; red—at least one persistent and significant anomaly) is shown, while at the bottom the time-series of the anomalous PS data in May 2018 (white circle in the ground deformation map)

anomalous time-series of PS points were highlighted and classified as caused by landslide movement (lower part in Fig. 4).

For the Tuscany Region, differently for the Valle d'Aosta Region, a monitoring bulletin is delivered every new update (commonly each 12 days) in which every municipality is coloured according to the presence, the relevance and persistency of anomaly PS time-series (Del Soldato et al. 2019a). In the bulletin at the end of May 2018 delivered to the regional hydro-geomorphological risk management actors, the municipality of Zeri was pointed out (Fig. 4).

Consequently to the bulletin, the officers organized a field surveys in few days in order to investigate the induced damage on structures and infrastructures. Some damage on the edifices close to the PS points with trend changes were recorded. Vertical and diagonal cracks, disjunctions of external stairs and some cracks in fencing walls or roads were mapped and their appearance and evolution were stated by the inhabitants. According to the scheme proposed by Del Soldato et al. (2019a) the Intensity and the Exposure were classified in E3 (Medium) and I2 (Low), respectively. According to the matrix (Fig. 2), the preliminary risk was assessed as Low and only systematic field surveys and annual/monthly ground-based manual monitoring tools, in addition to the continuous InSAR monitoring, were suggested.

Conclusion and Discussion

The scientific community benefits of the systematic availability of Sentinel-1 SAR images every six days, being possible to continuously monitor wide areas with low costs and high precision.

The technological and technical advancement has allowed processing large imagery dataset in short time and automatically detecting, under set parameters, trend changes in time-series by data-mining. This process permits to highlight particular situations over wide areas, in few days, taking into account the radar-interpretation operated by experts, and the delivery to officers operating in the hydro-geomorphological risk management. These outcomes can guide the end-users in defining site-specific strategies for the Civil Protection procedures of prevention and for ear-marking funds in a more conscious way in order to focus on very significant phenomena.

The entire workflow chain requires less then 10 days to be completed considering the download of the Sentinel-1 images, with a delay approximatively of few hours with respect to the shot, the processing and the data mining, the radar-interpretation by experts, the dissemination and the field survey, if required.

Two examples on which the same approach was applied were here proposed to show the potentiality of the use of satellite SAR data in different environments for identifying landslide-prone areas with changing trend, both in apennine and alpine landscapes, i.e. Zeri (Tuscany Region) and Bosmatto (Valle d'Aosta Region).

Acknowledgements The monitoring activity was founded and supported by the agreements "Monitoring ground deformation in Tuscany Region with satellite radar data" and "Monitoring ground deformation in Valle d'Aosta Region with satellite radar data" between the Civil Protection Centre of the University of Florence, the Italian Department of Civil Protection and Tuscany Region, and the Valle d'Aosta Region, respectively.

References

Baldi P, Cenni N, Fabris M, Zanutta A (2008) Kinematics of a landslide derived from archival photogrammetry and GPS data. Geomorphology 102(3–4):435–444

Berardino P, Fornaro G, Lanari R, Sansosti E (2002) A new algorithm for surface deformation monitoring based on small baseline differential SAR interferograms. IEEE Trans Geosci Remote Sens 40(11):2375–2383

Costantini M, Ferretti A, Minati F, Falco S, Trillo F et al (2017) Analysis of surface deformations over the whole Italian territory by interferometric processing of ERS, envisat and COSMO-SkyMed radar data. Remote Sens Environ 202:250–275

Crosetto M, Monserrat O, Iglesias R, Crippa B (2010) Persistent scatterer interferometry. Photogram Eng Remote Sens 76(9):1061–1069

Crosetto M, Monserrat O, Cuevas-González M, Devanthéry N, Crippa B (2016) Persistent scatterer interferometry: a review. ISPRS J. Photogrammetry Remote Sens 115:78–89

Del Soldato M, Bianchini S, Calcaterra D, De Vita P, Martire DD et al (2017) A new approach for landslide-induced damage assessment. Geomatics, Nat Hazards Risk 8(2):1524–1537

Del Soldato M, Solari L, Raspini F, Bianchini S, Ciampalini A et al (2019) Monitoring ground instabilities using SAR satellite data: a practical approach. ISPRS Int J Geo-Inf 8(7):307

Del Soldato M, Solari L, Poggi F, Raspini F, Tomás R et al (2019) Landslide-induced damage probability estimation coupling InSAR and field survey data by fragility curves. Remote Sens 11(12):1486

Federici PR, Puccinelli A, Chelli A, D'Amato Avanzi GA, Ribolini A, Verani M (2001) La grande frana di Patigno di Zeri (Massa-Carrara)

Ferretti A, Prati C, Rocca F (2001) Permanent scatterers in sar interferometry. IEEE Trans Geosci Remote Sens 39:8–20

Herrera G, Mateos RM, García-Davalillo JC, Grandjean G, Poyiadji E et al (2018) Landslide databases in the geological surveys of Europe. Landslides 15(2):359–379

Hu R, Wang S, Lee C, Li M (2002) Characteristics and trends of land subsidence in Tanggu, Tianjin, China. Bull Eng Geol Env 61 (3):213–225

Joyce K, Samsonov S, Levick SR, Engelbrecht J, Belliss S (2014) Mapping and monitoring geological hazards using optical, LiDAR, and synthetic aperture RADAR image data. Nat Hazards 73:137–163

Kjekstad O, Highland L (2009) Economic and social impacts of landslides. In: Landslides–disaster risk reduction. Springer, Berlin, Heidelberg. pp 573–587

Luino F (2005) Sequence of instability processes triggered by heavy rainfall in the Northern Italy. Geomorphology 66:13–39. https://doi.org/10.1016/j.geomorph.2004.09.010

Montalti R, Solari L, Bianchini S, Del Soldato M, Raspini F, Casagli N (2019) A Sentinel-1-based clustering analysis for geo-hazards mitigation at regional scale: a case study in central Italy. Geomatics, Nat Hazards Risk 10(1):2257–2275

Plank S (2014) Rapid damage assessment by means of multi-temporal SAR-A comprehensive review and outlook to Sentinel-1. Remote Sens 6(6):4870–4906

Raspini F, Bianchini S, Ciampalini A, Del Soldato M, Solari L et al (2018) Continuous, semi-automatic monitoring of ground deformation using sentinel-1 satellites. Sci Rep 8(1):1–11

Schuster RL, Highland L (2001) Socioeconomic and environmental impacts of landslides in the western hemisphere. Denver (CO): US Department of the Interior, US Geological Survey. pp 1–50

Solari L, Del Soldato M, Montalti R, Bianchini S, Raspini F et al (2019) A sentinel-1 based hot-spot analysis: landslide mapping in north-western Italy. Int J Remote Sens 40(20):7898–7921

Tomás R, Pagán JI, Navarro JA, Cano M, Pastor JL et al (2019) Semi-automatic identification and pre-screening of geological–geotechnical deformational processes using persistent scatterer interferometry datasets. Remote Sens 11(14):1675

Tomás R, Romero R, Mulas J, Marturià JJ, Mallorquí JJ et al (2014) Radar interferometry techniques for the study of ground subsidence phenomena: a review of practical issues through cases in Spain. Environ Earth Sci 71(1):163–181

2020 Kyoto Japan

Slope Disaster Risk Reduction Map as a Communication Tool for Community Based DRR in Japan and Vietnam

Toyohiko Miyagi, Nguyen Kim Thanh, Dinh Van Tien, Le Hong Luong, and Quang Van Viet

Abstract

The need to reduce the damage of local residents who are direct victims of natural disasters has long been pointed out. In recent years, disasters of an unexpectedly large scale occur frequently, and it is not possible to protect local communities and human lives simply by taking disaster prevention measures. In order to mitigate disasters, it is necessary to build voluntary disaster mitigation measures by residents. However, local residents have not accurately acknowledged and understood disaster risks existing in the region. Through this report, we have discussed the significance of the process of creating a communication base DRR map jointly by the community and engineers in order to realize "local residents' understanding of regional disaster risks". Specifically, we have introduced how a slope disaster risk map should be, the flexibility and ease of acquiring necessary data in the present age, the actual work progress and results, etc., using examples from Japan and Vietnam.

T. Miyagi (✉) · L. H. Luong
Advantechnology Co. Ltd., 1-4-8 Kakyoin Aobaku, Sendai, 980-0013, Japan
e-mail: c1934009@mail.tohoku-gakuin.ac.jp

L. H. Luong
e-mail: lehongluong@gmail.com

N. K. Thanh (✉) · D. V. Tien · L. H. Luong
Institute of Transport Science and Technology, 1252 Lang Street, Lang Thuong, Dong Da, Ha Noi, Vietnam
e-mail: nguyenkimthanh.itst@gmail.com

D. V. Tien
e-mail: dvtien.gbn@gmail.com

N. K. Thanh
Tohoku Gakuin University, Sendai, Japan

Q. V. Viet
Department for Agriculture and Rural Development (DARD), Lao Cai, Vietnam

Keywords

Landslide mapping • Slope DRR • Local resident • Communication base mapping • Japan • Vietnam

Background to Making Useful Maps in the Area

Sendai Bosai Frame Work 2015–30 was proposed at the 3rdWCDRR held in Sendai in 2014. It states that upfront investment for preventing disasters is more reasonable and can be achieved at a lower cost than disaster prevention measures after a disaster occurs. Therefore, disaster prevention, in other words, disaster mitigation, has become the mainstream of disaster prevention. However, in order to achieve disaster mitigation, "evacuation" must be considered. Participation of local residents is essential for evacuation. In the Sendai Action Plan, 4 out of 7 items are closely related to disaster risk reduction of the community. In order for the region to acquire disaster mitigation abilities, its inhabitants must understand the disaster risks of the region and obtain know-how to avoid it. Maps of appropriate accuracy should be created to help build an evacuation system that leads to avoidance of risks. Therefore, it is necessary to create maps of slope disasters suitable for a time of disaster mitigation.

This report presents examples of mapping of several types of slope disaster risks at different scales. Presented is the possibility of realizing acknowledgement and understanding of slope disaster risk by central administrative organization to local communities through a series of mapping. Corresponding data is required for mapping slope disaster vulnerability. There has been a great difference in data availability between Japan and Vietnam. In recent years, data of various spatial resolutions has become available worldwide, and large-scale images and digital three-dimensional information have become available freely,

F. Guzzetti et al. (eds.), *Understanding and Reducing Landslide Disaster Risk*, ICL Contribution to Landslide Disaster Risk Reduction, https://doi.org/10.1007/978-3-030-60227-7_48

at low costs and at the appropriate times and places. Needless to say, in order to realize disaster prevention, we need to understand the accuracy of map creation, how and what data is used, what kind of map needs to be created. Concrete maps must be proposed to appropriate people and a system should be built to reach a common understanding among engineers, governments and local residents.

There are two viewpoints for viewing the slope (terrain). The first viewpoint is a perspective which views the terrain as "continuity". A terrain refers to a so-called seamless film that covers the entire surface of the earth. For example, it is based on this perspective that contour lines can represent the same height with a single line. In addition, the terrain can be comprehended and expressed by various numerical indicators such as direction, inclination, and curvature.

The other is a perspective that views the slope as a "Unit". The slope is created in various ways to form, maintain and destroy it, which spreads out as spatially distinctive mosaics. So-called slope names such as Alluvial fan, Valley bottom, Landslide topography, and Volcano are defined from this perspective. A slope grasped from this perspective is defined by its own shape, material, process and age. Landform classification maps and landslide topography maps are created from this perspective.

Hazard maps and disaster maps are created for the purpose of comprehending the site potential of various actions (landslides, debris flow, and surface failure) that cause slope disasters. A characteristic mapping is proposed from the above two perspectives. They can be broadly divided into two types of maps.

Susceptibility map: The slope of the target area is gridded with an appropriate mesh size, and its form and disaster-related factors are indexed (digitized) and displayed. The aim is to combine this with the actual results of landslide disasters and optimize the model by increasing its accuracy for predicting future disasters. This is from the perspective of seeing the terrain as a continuum, regardless of consciousness. This technique is being advanced globally. In general, susceptibility (potential for disaster occurrence) is expressed as a level division for each grid.

Landform classification map: Created from the perspective of grasping the terrain of the target area as a mosaic of terrain units. This method generally involves transferring the three-dimensional photo interpretation of an aerial photograph to a topographic map. Interpretation is based on the premise that terrain areas such as landslides and general

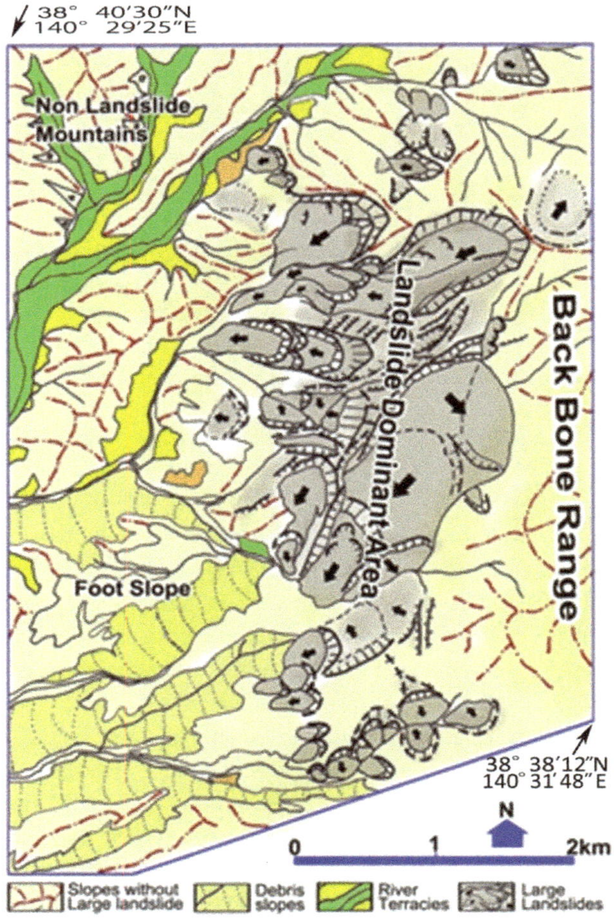

Fig. 1 Sample of the landform classification map at a part of Japan

slopes can be observed, grasped, analyzed, and illustrated. Based on accurate terrain classification, they can be used as basic data for hazard/safe area classification and disaster prevention planning. Figure 1 illustrates mapping by stereoscopic interpretation. At present, technology for automated interpretation has not been established.

The susceptibility map is a technique for screening the target area uniformly and comprehending the risk level. The classification map is a technique to clearly demarcate risky and non-risk areas in the target area. As shown in Fig. 1, landslide topographic and non-landslide areas are widely mixed in many areas. Since a large-scale landslide area has the possibility of fluctuating integrally beyond the grid size, it is not reasonable to divide this area with the same mesh size as the others. Therefore, the mapping of the local

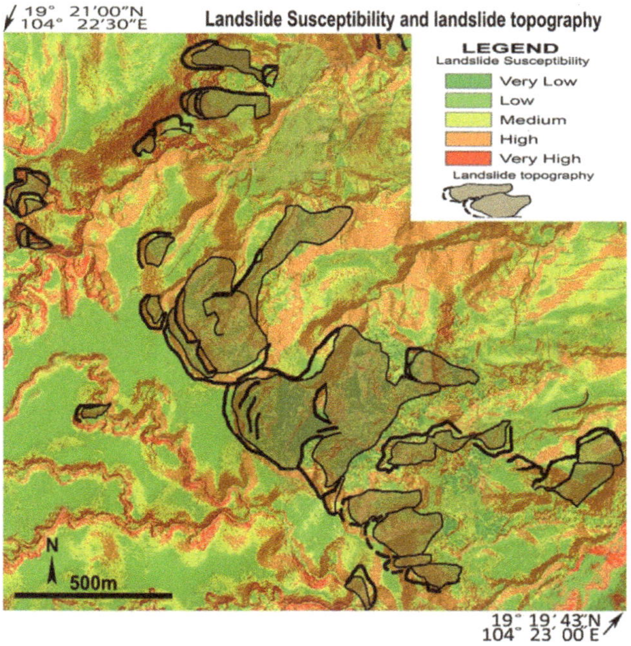

Fig. 2 Combination of the susceptibility and landslide topographies

disaster entity should not be completed by only one type of map. An integrated disaster prevention map can be achieved by the two complementing each other. An example is shown in Fig. 2.

From the Perspective of Local Residents

Whether in Japan or Vietnam, it is necessary to create a system to acknowledge disaster risks and manage both residents and danger points in order to avoid disasters and achieve evacuation in times of emergency to reduce disasters. In fact, in the Sendai Framework 2015–30, among the three of seven items below suggest the tasks of the local community:

- All society joint and prevention and reduction
- Capacity building the local com./Community.
- Before measure is low cost then the after one.

However, Vietnam's provincial officials have pointed out many difficulties in achieving disaster prevention at the regional level (Fig. 3). It is necessary to understand how to deal with disasters separately for disaster prevention and mitigation (Fig. 4), and **"in disaster mitigation, evacuation and avoidance actions must be achieved by the local residents' will."** From this perspective, it is thought that **"a proposal of a regional-level slope disaster map that is useful for the residents themselves to grasp the locational nature of the disaster specifically and to consider countermeasures"** is required.

A Paradigm Shift in Mapping Technology

Understanding the causes of disasters such as landslides has always been a major issue in disaster prevention, and it has been extremely important to determine potentially dangerous slopes. However, the technology for accurately comprehending it has not quite been completed. On the other hand, sensing technology and mapping technology themselves have developed in various ways, while in recent years, a rapid and revolutionary technological innovation is progressing. Figure 5 illustrates various sensing technologies, from satellite imagery to human exploration. There are many difficulties in using aircraft surveying and satellite sensing for local residents' familiar disaster prevention. These include cost, technology, and voluntary data acquisition. On the other hand, the use of UAVs has the flexibility of low cost, easy technology, shooting time, scale, etc. For example, it is extremely effective when detailed information is required in spite of small-scale issues such as living and production space. In the present age, the use of UAVs to acquire regional information, use of its software for mapping (for example, SfM) and easy use of GIS have been realized, making it possible to map and acquire information according to the purpose anywhere in the world. By using these, it became possible for local residents to create maps that manage regional disaster risks and disasters.

Fig. 3 Disaster prevention plan in community and the difficulties in Vietnam

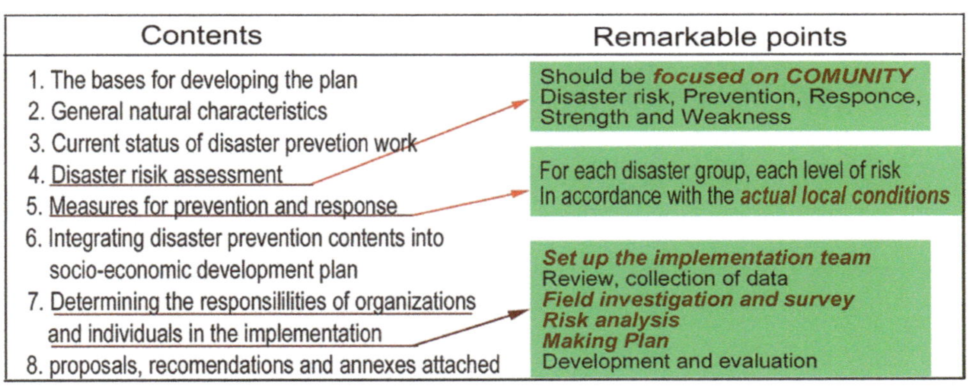

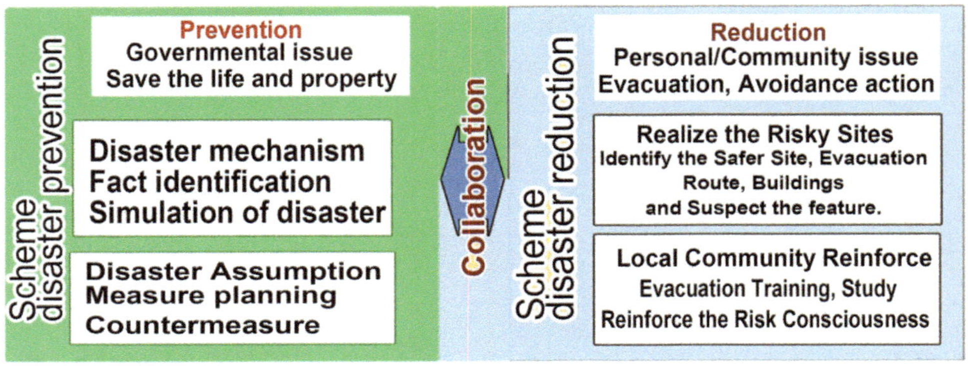

Fig. 4 Structures of disaster prevention and disaster reduction (After Miyagi 2016)

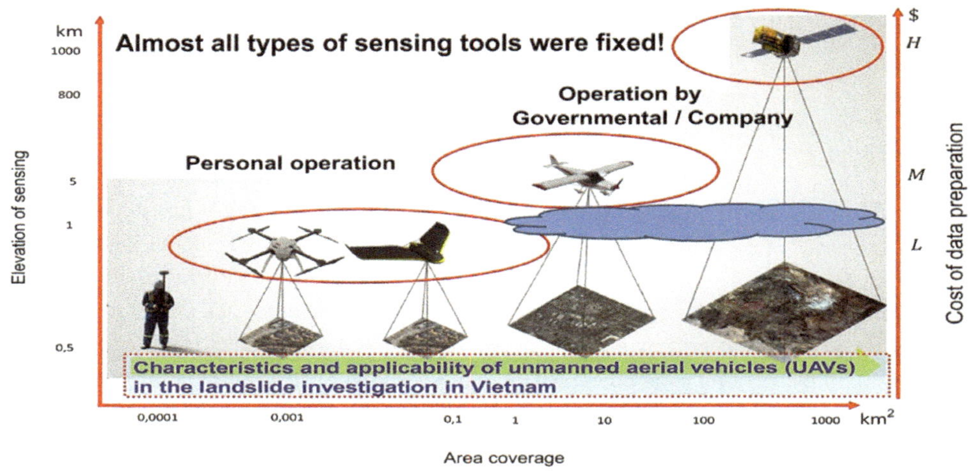

Fig. 5 Variety of the sensing tools, cost etc. (Modified Nguyen et al. 2020)

Communication Based DRR Mapping

Each region has its own unique natural characteristics as well as history and culture of life cultivated as a community. Without these understandings, disaster vulnerability mapping for local residents to match local issues cannot be accomplished. In Japan, most municipalities have created and released various hazard maps based on national guidelines. However, it has been pointed out that many residents in the areas affected by the recent great disaster have "**signified that they did not see or did not know about the hazard map**", which had great significance as forecast data. This means that "realization of the risk reduction shown in Fig. 4 requires decision-making by local residents themselves, such as avoiding disasters and evacuating." Also, given the reality that "regions have their own characteristics and regional characteristics", the necessity of maps that can be used by all means is implied. The authors propose Communication Base DRR Mapping as a form of maps which can be used. In Japan, the East Japan Great Earthquake and the Tsunami 2011 occurred, causing severe damage. Among them, the mapping and evacuation drills that the authors conducted with the local residents showed great effects (Miyagi 2016).

Immediately after the disaster, pledges were received from residents such as, "Our disaster prevention map did not help, but it saved our lives!" or "The tsunami did not come exactly as the teacher said, but our lives were saved because we had studied. Thank you!" Realization of local residents' appropriate disaster avoidance behavior may not be due to the map creation itself, but the process of making it which helped them understand the map and characteristics of the area indicated by the map. Needless to say, a disaster is a situation that residents want to avoid. The process of building a map with the goal of understanding local disaster risks is what is truly important. Figure 6 shows the map creation and a map of the tsunami and the places where victims were found in the autonomous region. In community base DRR mapping, it is essential that the scale and display method be "understandable and understandable" according to the region. In addition, it should be easy to revise in response to changes in the region. Figure 7 shows a map revised and distributed to each house three months before the disaster. Currently, map creation has begun in a certain area in Vietnam (Fig. 8). Map making has just begun.

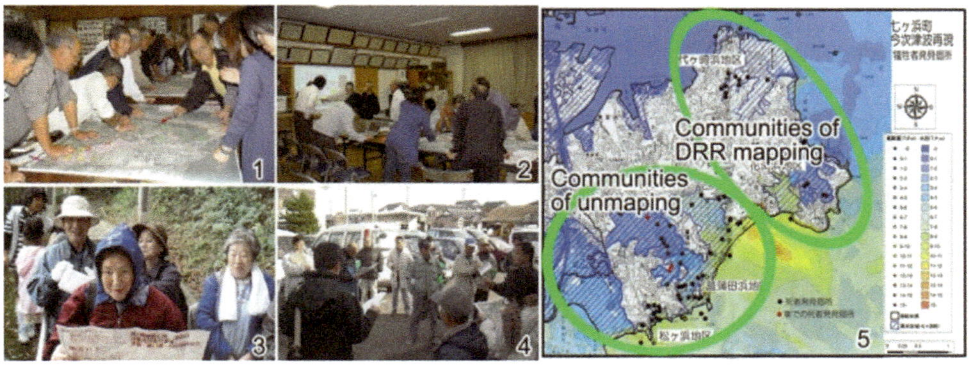

Fig. 6 Processes involved in the community based DRR mapping and the result in case of Shichigahama, Japan. 1. Discussion with community about map. 2. Reconstruction the evacuation route and the site in GIS platform. 3&4. Evacuation drill. 5. Result as the human victim in the town from comunity without DRR mapping

Fig. 7 Communication base disaster risk reduction map in a community, Shichigahama town, Northeastern Japan

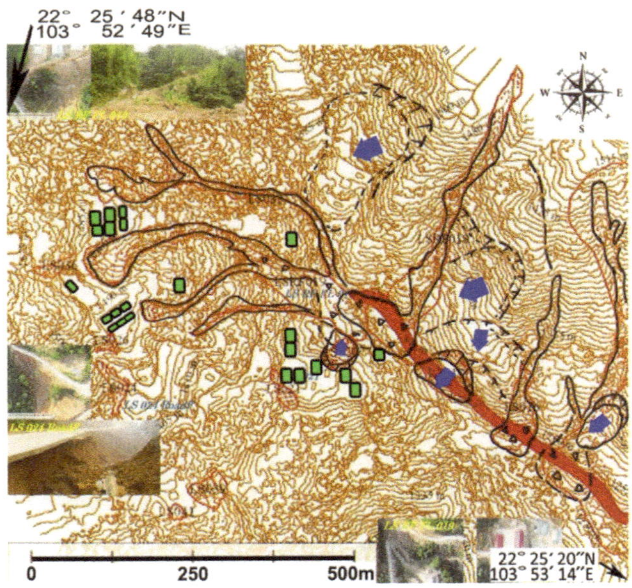

Fig. 8 First step of communication base disaster reduction map in a part of Lao Cai province, VIETNAM

Opinion exchange and regional surveys will be repeated, a provisional map will be created.

Summary

The need to reduce the damage of local residents who are direct victims of natural disasters has long been pointed out. In recent years, disasters of an unexpectedly large scale occur frequently, and it is not possible to protect local communities and human lives simply by taking disaster prevention measures. In order to mitigate disasters, it is necessary to build voluntary disaster mitigation measures by residents. However, local residents have not accurately acknowledged and understood disaster risks existing in the region. Through this report, we have discussed the significance of the process of creating a communication base DRR map jointly by the community and engineers in order to realize "local residents' understanding of regional disaster risks'. Specifically, we have introduced how a slope disaster risk map should be, the flexibility and ease of acquiring necessary data in the present age, the actual work progress and results, etc., using examples from Japan and Vietnam.

Reference

Miyagi T (2016) Drill of hazard mapping as the practice of physical geography and disaster prevention study in the East Japan great earthquake 2011. Crustal disaster reduction and the academic/education society, Japan scientific council series no. 22, pp 88–99

Nguyen KT, Miyagi T, Shinobu S, Tien Dv, Luong LH, Ha DN (2020) Developing recognition and simple mapping by UAV for local resident in mountainous area in Vietnam. (Submitting to WLF 5)

Part VI

Landslide Vulnerability of People, Communities and the Built Environment

People Vulnerability to Landslide: Risky Behaviours and Dangerous Conditions by Gender and Age

WLF 2020 Kyoto Japan

Paola Salvati, Mauro Rossi, Cinzia Bianchi, and Fausto Guzzetti

Abstract

Landslides are frequent and destructive geo-hydrological events that cause harm to people every year worldwide. We used a catalogue on 1039 landslide fatalities, occurred in Italy in the 50-year period 1970–2019, to determine the dependence of the fatalities on gender, age and circumstance of death. The updated version of the Italian landslide fatalities catalogue includes information on the exact time of the occurrence of the fatal landslide events and the circumstances of death. Possible relations between time of the day and fatalities occurrence are analysed. Males landslides fatalities occurred frequently outdoor, along roads mainly involving drivers or passengers travelling in vehicles, indicating a specific dangerous death contingence preferentially occurring in daylight. Conversely, female landslide fatalities occurred more frequently indoor. To consider the demographic and socio-cultural changes over time, we performed a temporal analysis splitting the catalogue into three overlapping subsets 30-year each. In the three time periods, we estimated the expected fatalities by gender and age, using national census data and a multinomial distribution. Such estimates were compared with the observed landslide fatalities distribution. We identified the age categories over or under represented when the observed fatalities were respectively higher or lower than the modelled expected deaths. The analysis shows that for all the periods landslide male fatalities compared to the female ones are significantly higher than those expected by census data, indicating both a diverse propensity towards the risk taking and a different degree of exposure between males and females.

P. Salvati (✉) · M. Rossi · C. Bianchi · F. Guzzetti
Istituto di Ricerca per la Protezione Idrogeologica (CNR-IRPI), Consiglio Nazionale delle Ricerche, via Madonna Alta 126, 06128 Perugia, Italy
e-mail: paola.salvati@irpi.cnr.it

Keywords

Loss of life • Landslide fatalities • Landslide mortality • Statistical approach • Gender and age analysis

Introduction

Landslides cause significant societal and economic damage and large number of fatalities worldwide. Froude and Petley (2018) recorded 55, 997 landslide fatalities in 4,862 distinct landslide events in the period from January 2004 to December 2016 at global scale. Different catalogues of landslide fatalities are also available at national and regional level. This information is used to analyse the geographical distribution of fatal landslides (Sepulveda and Petley 2015; Petley 2012), to assess landslide mortality rates (Guzzetti 2000; Salvati et al. 2016), the societal landslide risk at national scale (Rossi et al. 2019), and to identify the circumstances of landslide fatalities (Wong et al. 1997; Pereira et al. 2015; Badoux et al. 2016; Salvati et al. 2018). Age and gender are known to influence human vulnerability to natural hazards (Sharif et al. 2015; FitzGerald et al. 2010; Ashley and Ashley 2008). The identification of possible vulnerable groups of people in terms of gender and age and the recognition of the circumstances in which people lost their life are fundamental to improve people safety and to increase community resilience to geo-hydrological hazards, including landslides (Davies et al. 2015; Weichselgartner and Pigeon 2015). A number of studies in the scientific literature have attempted to analyse fatalities by gender and age, though statistical approaches comparing quantitatively the expected number of fatalities with the demographic data distributions are rare. In 2018 Salvati et al. used a detailed historical catalogue to quantify the expected number of landslide fatalities, stratified by gender and age on the base of the census data. In the attempt to update previous published statistics, we first updated the catalogue of landslide

F. Guzzetti et al. (eds.), *Understanding and Reducing Landslide Disaster Risk*, ICL Contribution to Landslide Disaster Risk Reduction, https://doi.org/10.1007/978-3-030-60227-7_49

fatalities in the 50-year period 1970–2019 and then we searched for additional information regarding possible relation between the time of the day of landslides and the fatalities occurrence.

Data on Landslide Fatalities in Italy

For Italy, comprehensive information on the human consequences of landslides, including the dead, injured people, missing persons, evacuated and homeless people, is available from 68 AD (Salvati et al. 2016, 2018). For this work, we searched additional data and we looked to the times of the day of fatal landslides occurrence and to the circumstances of the fatal events occurred in the 50-year period 1970–2019, including information on 1,039 persons killed by 283 landslides at 277 sites (Fig. 1). In this period all the 20 Italian regions suffered fatal landslides. In northern Italy, those are mostly concentrated in the mountainous areas of the internal Alps and along the mountainous coast of Liguria region. In southern Italy the most affected areas are in Campania region and along the coast of the Peloritan mountains in Sicily and in the southern coast of Calabria.

We looked specifically to the times of occurrence classifying them in by day or by night. Additional details were found to describe the causes and the circumstances of the fatal events in the 50-year period 1970–2019 and the age and gender of the fatalities. We searched different sources of information including written reports, interviews to eyewitnesses, newspaper articles and official government reports and documents. We examined web sites, social networks and blogs, and we collected and analysed photographs and videos. As it was first mentioned by Salvati et al. 2018, when the landslide events have caused large numbers of fatalities the identification of the circumstances of death of the single fatalities remains often impossible. This occurred for the Stava catastrophic mudflow event of 19 July 1985 that killed 268 persons (Tosatti 2003) for which the complexity of the rescue and recovery operations and the difficulties in the legal identification of the deaths, made impossible the reconstruction of the circumstance of death.

Gender and Age

In the updated version of the catalogue of landslide fatalities, we determined the gender (from the name) of 1,039 landslide fatalities (96.6%), and the age for 1,027 fatalities (93.5%) occurred in Italy in the period 1970 and 2019. Overall, males account for 56.7% and females for 43.3% of the landslide fatalities for which the gender is known in the catalogue (Table 1).

In our catalogue, differences exist in the age of the landslide fatalities, and between females and males and are shown in Fig. 2. The average age of the fatalities was 42.4, being higher for the female fatalities (43.6) respect to the males once (41.4). Except for children (9-year-old and younger) and the elderly (70-year-old and older), landslides killed more males than females, with maxima in the age category 30–39 for males and 60–69 for females. Old females (80-year-old and older) are considerably more vulnerable to landslides than old males.

Times of the Day and Places

The updated version of the Italian landslide fatalities catalogue includes information on the exact times of the occurrences of the fatal landslide events. This information remained the most difficult to find, particularly for the events with one or two fatalities and for those occurred between the 1970 and 1990. Conversely, it was easier for the more recent landslides and for those that caused a large number of fatalities, thanks to the higher availability of sources of information. Finally, we were able to identify the exact times of 194 (68.5%) fatal landslide events and of 921 (88.6%) fatalities. We used this data to find possible relation between the light condition, the places (indoor or outdoor) and the gender of landslide fatalities. For different geo-hydrological hazard (i.e. flood and flash flood), similar studies found evidences that in dark conditions the percentage of fatalities is slightly higher (Petrucci et al. 2019; Terti et al. 2017). We first classified the times in two categories, respectively by night and by day, on the bases of the hour of the fatalities occurrence and considering the daytime variation during the year.

Then, for each fatality, we related the times of occurrence with the place (outdoor and indoor) were people lost their life and with the gender and age. We elaborated the analyses for different categories of data and we prepared mosaic plots in R (https://www.r-project.org/) for visualizing the co-occurrence of multivariate categorical data. Differences between gender and the other variables emerged as shown in Fig. 3, where pink scale colours represent females and the blue ones males.

Visual inspection of plot A shows that the majority of males lost their life outdoor, while females indoor and that both males and females lost their lives during the daylight hours. This is in contrast with previous finding for flood fatalities. In plot B it is evident that females, differently from

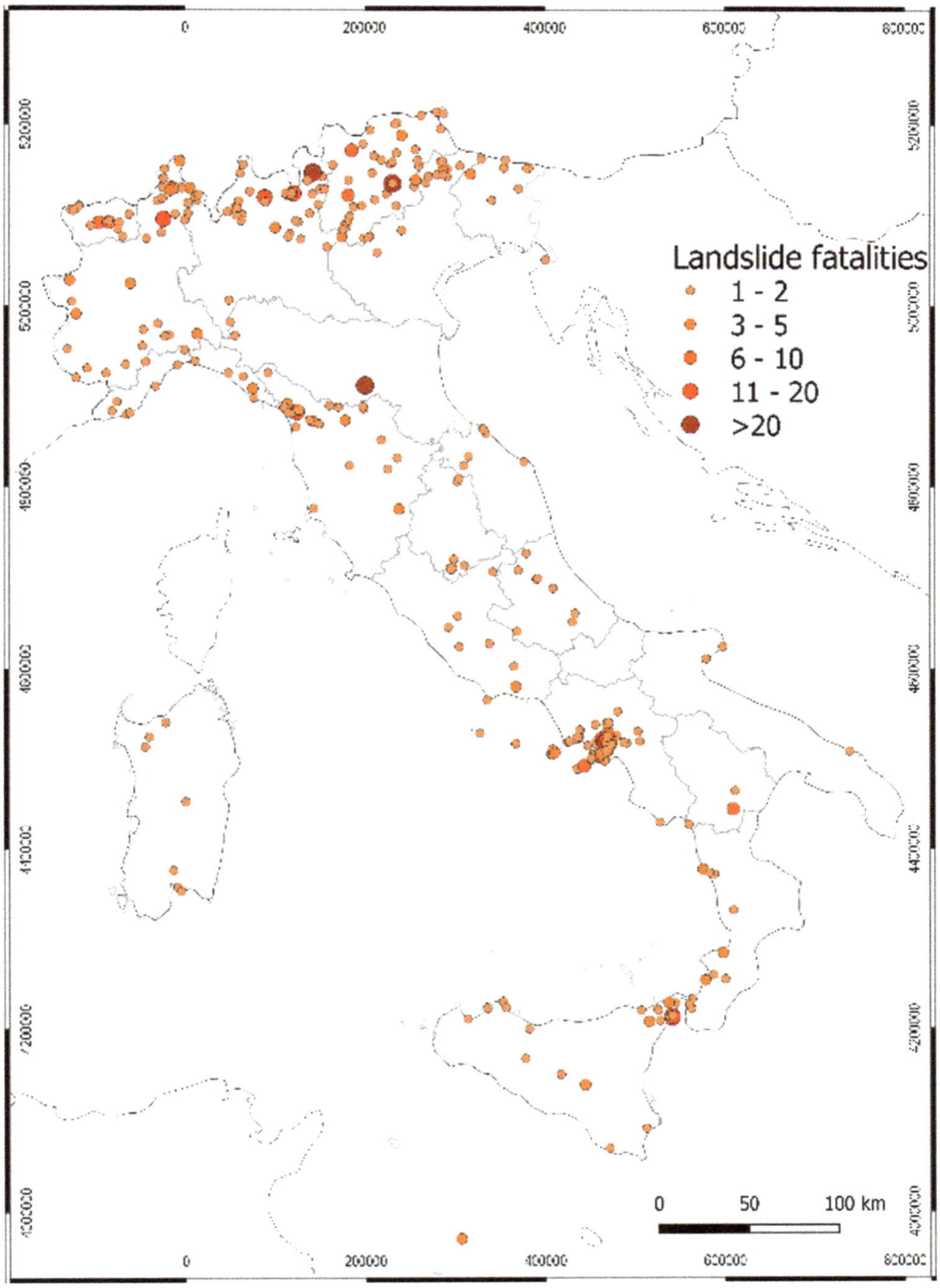

Fig. 1 Map of the landslide fatalities sites in Italy in the period 1970–2019

Table 1 Number and percentage of fatalities caused by landslides in Italy in the period 1970–2019

Col Head 1	#	%
Fatalities in the national catalogue	1098	
Fatalities with known gender	1039	94.6
Males	589	56.7
Females	450	43.3
Fatalities with known age	1027	93.5
Males	583	56.8
Females	444	43.2

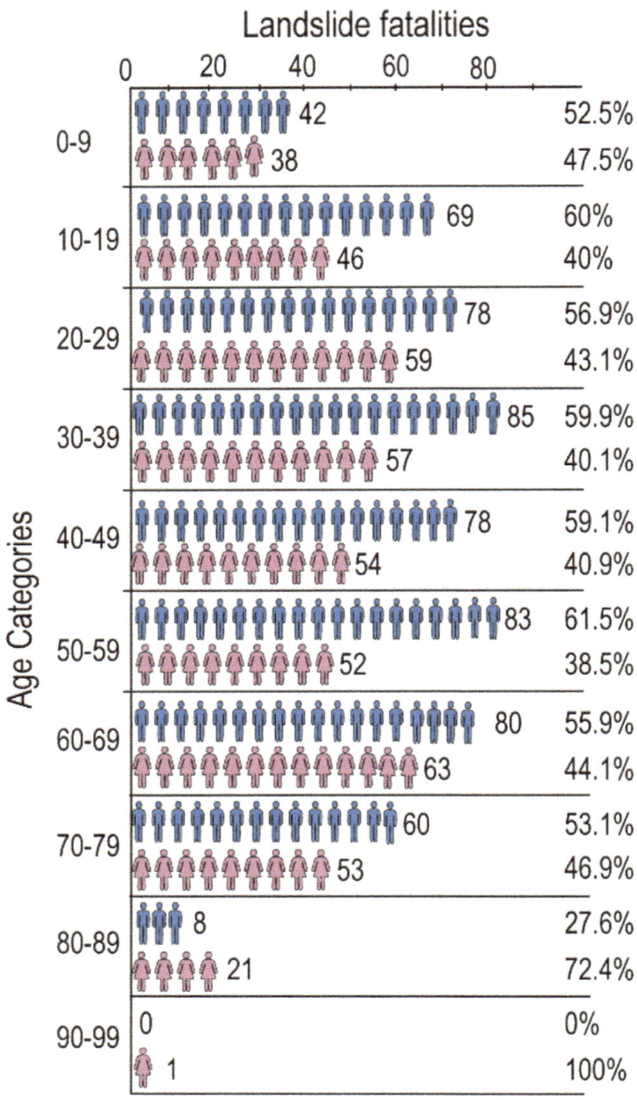

Fig. 2 Statistics for landslide fatalities by gender and age in the period 1970–2019

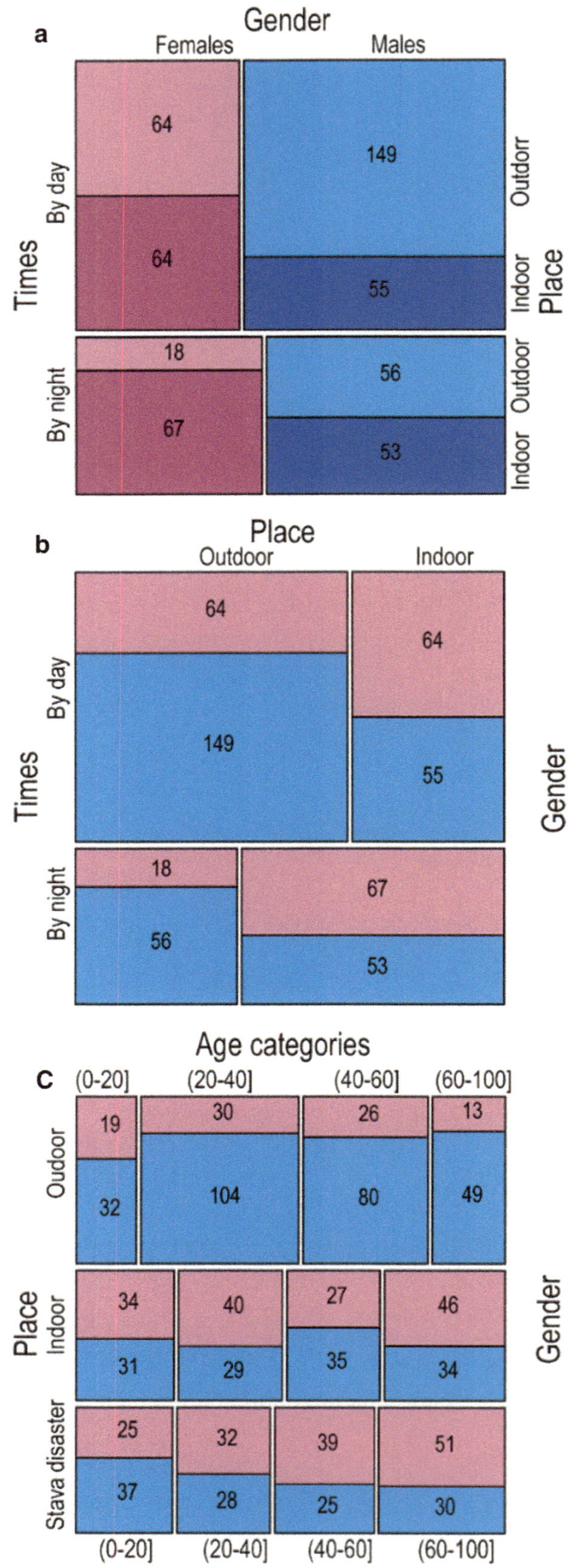

Fig. 3 **a-b** Statistics for landslide fatalities by gender, times (by day or by night), place (indoor or outdoor). **c** Statistics by age categories, gender and place (indoor or outdoor or Stava). Stava is the place of occurrence of the catastrophic mudflow event of 19 July 1985

males, resulted more vulnerable when they were indoor during night and that, for them, the darkness condition could be more dangerous. In plot C data are analysed by age categories and data are related by gender and place of occurrence. Differently from the 10 age categories of Fig. 2, in which we give the statistics by age, in plot C we grouped data using four larger age categories and we divided the information on places in three classes with an additional class for the Stava mudflow disaster. During the very rapid Stava mudflow (up to 90 kmh^{-1}) buildings and infrastructures were destroyed and the high velocity of the mudflow did not allow people to escape the very rapid moving landslide.

For this catastrophic landslide event, the distinction between indoor and outdoor fatalities was impossible, while official documents made available the age and gender of all the victims. Analysis of plot C show that the males are more numerous outdoor when they are younger (between 20 and 40) while females are more numerous indoor when they are older (between 60 and 100). During the Stava event older women and young males in the 0–20 resulted the most vulnerable age categories.

Circumstances

To better understand the most hazardous situations and the places where it is more dangerous to be during a landslide event, we looked for information on the exact places and situations. For a number of the fatalities for which the place of death was known (641–61.7%), we were able to determine the type of place (624–60.1%), the local condition (354, 34.1%), and the dynamic of the fatal event (617–59.4%).

Analysis of Table 2 allows to recognize that people inside their home or in buildings (277–26.7%) primarily lost their life due to the partial or total collapse of the buildings (234–22.5%) caused by the landslides, or secondarily, dragged by earth, debris or mud outside their house (32–3%). Inside buildings lost their lives 148 females and 129 males. Landslides killed people also along roads (167–16.1%), and where mostly debris flows and rockfall that caught people in vehicles (e.g. cars, vans, buses, tracks, excavators) or that hit pedestrians along the road (37). Landslide fatalities along roads were predominantly males (74.8%) and subordinately females (25.2%). Landslide-induced train derailments produced 65 fatalities in seven cases, chiefly males (72.3%). Landslides killed people at work (50) involved in outdoor

Table 2 Places, conditions and dynamics of fatalities caused by landslides in Italy in the period 1970–2019

Circumstances	Landslide fatalities			
Where	#	%	M	F
Indoor	277	26.7	129	148
Outdoor	364	35	271	93
No data	398	38.3	189	209
Type of place				
Public or private building	277	26.7	271	93
Proximity of house	14	1.3	9	5
Road/Highway	167	16.1	125	42
Railway	65	6.3	47	18
Countryside or Mountain	82	7.9	66	16
Beach	18	1.7	9	9
No data	415	39.9	205	211
Conditions				
By foot	153	14.7	112	41
By car	114	11	86	28
By work vehicle	22	2.1	20	2
By train	65	6.3	47	18
No data	685	65.9	324	361
Dynamic				
Collapse of building	234	22.5	110	124
Dragged indoor	32	3	18	14
Dragged outdoor	92	8.8	70	22
Dragged by car	65	6.3	47	18
Dragged down in a slope	20	1.9	17	3
Hit	92	8.8	66	26
Buried	16	1.5	14	2
Involved in derailment	65	6.3	47	18
No data	422	40.6	203	219

activities (43), the majority males (96.4%). The most common dynamics for landslide fatalities were persons involved in the partial or total collapse of a building (234), dragged by rapid debris flows or mudflows (209), or hit by rocks (92). A limited number of the fatalities (16) were buried by landslide deposits (Table 2). Determining the circumstance of death was difficult and we successfully identify the 60.1% of the precise types of place, the 60.6% of dynamics but less than half of the conditions (34.1%).

Gender and Age Analysis

When analysing landslide hazard it is important to consider that the number of the deaths is a function of multiple factors that include the intensity and magnitude of the landslide (e.g. volume, the velocity) (Guzzetti 2000; Guzzetti et al. 2005) and the type and distribution of the population at risk (e.g. population density, vulnerability of building and people). The latter can vary geographically (Cutter et al. 2003) and temporally, as a consequence of demographic changes. In the 50-year period 1970–2019, the Italian population has increased of about 6 million, from about 54 million in 1971 to the current 60 million, recording a stop in growth from about 10 years. In this period culture and society have transformed the family and social roles of males and females. It is proper to investigate if these changes have affected the impact of geo-hydrological hazards on the population and in particular by genders and age. For the purpose, we have analysed the landslide fatalities variation during time. To investigate possible variations with gender and age, we split the available data into three overlapping 30-year data-subsets (1970 and 1999, 1980 and 2009, 1999 and 2019). We then applied the statistical approach applied by Salvati et al. 2018 (Eq. 1). We generated for each gender and age category, and for each data subset, a set of 10,000 samples drawing randomly from a multinomial distribution, with each sample of the same size (number) of fatalities as in the specific gender and age category. The probability mass function (i.e. equivalent to the probability density function for discrete data) of the multinomial distribution is given by:

$$f(x_1,\ldots,x_k;p_1,\ldots,p_k) = \frac{\Gamma(\sum_i x_i + 1)}{\prod_i \Gamma(x_i+1)} \prod_{i=1}^{k} p_i^{x_i} \quad (1)$$

where $x_1,\ldots,x_k$ is the number of males and females in each of the k age categories, and $p_1,\ldots,p_k$ the corresponding probability. We then prepared, for each 30-year data subset, box plots to show the expected distribution of the fatalities in each gender and age category, and we overlaid the observed number of the landslide fatalities (Fig. 4). Comparison of the expected fatalities, based on the census data (the box plots), with the observed landslide fatalities (blue dots for males

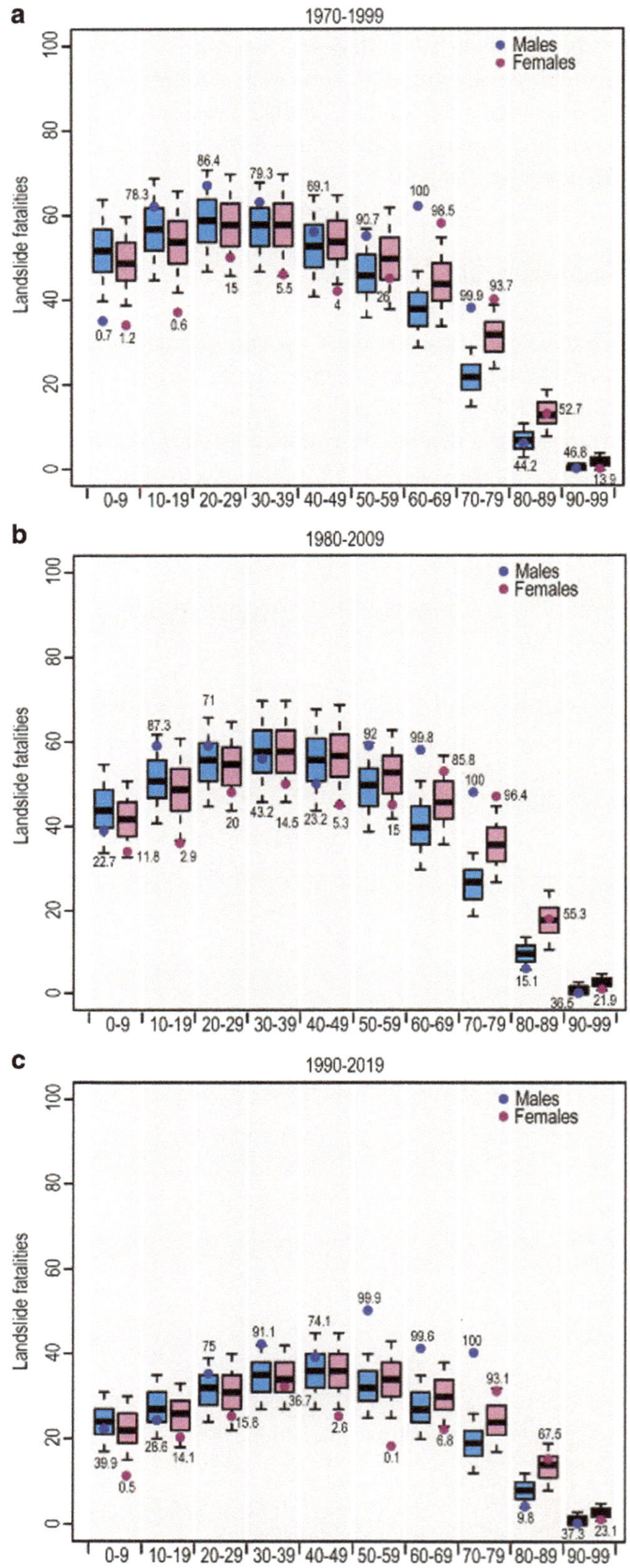

Fig. 4 Dots show the observed number of fatalities for females in pink and males in blue and age categories in Italy in the two 30-year periods 1970–1999 **a**, 1980–2009 **b**, 1990–2019 **c**. Box plots show the expected distribution of the fatalities in each gender and age category for the three periods

and pink dots for females) allowed us to identify the gender and age categories over or under represented (statistically higher or lower than those expected by census data). In Fig. 4, for each observed landslides fatalities data (points) we reported the corresponding percentile value respect those modelled by census data.

Inspection of Fig. 4 allows noting the largest difference between the three census data-subsets consisting in the decrease in the number of children (0–9) in the period 1980–2009 (plot B) and, more evident, in the period 1990–2019 (plot C) and the variation of the age category where population is more numerous, varying from 20–29 (plot A) to 40–49 (plot C) age category, for both genders. We note significant differences between the observed landslide fatalities (points) and those expected by census (box plots). For both genders, children (0–9) are under-represented, probably as a result of the care taken to protect them during hazardous events. The female fatalities were underrepresented up to the age of 59, in the recent period 1990–2019 the female fatalities were underrepresented up to the age of 69. Other differences concern the young boys (10–19). A decrease in the observed number of fatalities in the period 1990–2019 was registered, indicating a general increase in the safety of young boys, or a different risk exposure of young boys to natural hazard.

Discussion

Our gender analysis revealed that, in the 50-year observation period 1970–2019, in Italy landslides have killed more males (589–56.7%) than females (450, 43.3%) (Table 1) and the results are in accordance with previous findings obtained for Italy (Salvati et al 2018). 20-year-old males or older are over-represented up to the age of 79. The over-representation is statistically significant (Fig. 4), denoting difference in gender propensity towards risk taking, and the different exposure to landslides of the two genders. The low gender ratio for landslide fatalities M/F = 1.3, indicate that difference between male and female landslide fatalities is limited, and we attribute this evidence to the physical characteristics of the fatal landslides in Italy, which are mainly rapid to very rapid slope failures, including rock falls, debris flows, rapid mudflows, and soil slips (Guzzetti et al. 2005). Our result is in general agreement with Hungr (2007) who argued that rapid and very rapid landslides stand out in terms of human vulnerability. The majority of the landslide fatalities occurred outdoor (53.0%) (Table 2). The sources indicate that, most frequently, people were caught by surprise by the landslide event, and were apparently unaware of the imminent risk, showing no familiarity of the individual with the local environment, the landscape, and the type of the hazardous event and its potentially fatal consequences. The updated version of the Italian landslide fatalities catalogue includes information on the exact times of the occurrences of the fatal landslide events. We find that the majority of the fatalities occurred by day. This indicates a major landslide risk exposure during daylight since people are engaged in their occupation outdoor and they move along roads by cars or other work vehicle to reach their offices.

This confirms that landslide risk to the population varies depending on the number of people exposed, the time of the day and the place (Rossi et al. 2019).

Conclusion

We used information on landslides with human consequences in Italy in the 50-year period between 1970 and 2019 to determine new statistics by gender and age. We identified the times of occurrence of fatal landslide events and we found that the majority of the fatalities occurred by day since people are engaged in their work activities or is moving along roads, increasing their exposition to landslide risk. It does not seem that the dark conditions worsen the danger of landslides (Fig. 3). We performed a statistical analysis that allow to identify and compare the most vulnerable gender and age categories. Our gender analysis revealed that landslides have killed more males than females, in most of the age categories (Fig. 4). The statistical analysis we applied to evaluate if the observed number of fatalities are over or under represented compared with the expected deaths by census, revealed a systematic over-representation of male fatalities in the range of age from 20 to 79-year and the trend was independent by the demographic variations and the cultural and economic changes occurred in Italy in the observation period (Fig. 4). Our results confirm a different propensity towards risk taking of females and males, and a different degree of exposure to landslide of the two genders.

References

Ashley ST, Ashley WS (2008) Flood fatalities in the United States. J Appl Meteor Climatol 47:805–818

Badoux A, Andres N, Techel F, Hegg C (2016) Natural hazard fatalities in Switzerland from 1946 to 2015. Nat Hazards Earth Syst Sci 16:2747–2768

Cutter S, Boruff B, Shirley W (2003) Social vulnerability to environmental hazards. Soc Sci Quart 84(2):242

Davies T, Beaven S, Conradson D, Densmore A, Gaillard JC, Johnston D, et al (2015) Towards disaster resilience: A scenario-based approach to co-producing and integrating hazard and risk knowledge. Int J Disaster

FitzGerald G, Du W, Jamal A, Clark M, Hou XY (2010) Flood fatalities in contemporary Australia (1997–2008). Emerg Med Australas 22:180–186

Froude MJ, Petley DN (2018) Global fatal landslide occurrence from 2004 to 2016. Nat Hazards Earth Syst Sci 18:2161–2181

Guzzetti F (2000) Landslide fatalities and the evaluation of landslide risk in Italy. Eng Geol 58:89–107

Guzzetti F, Stark CP, Salvati P (2005) Evaluation of flood and landslide risk to the population of Italy. Environ Manage 36(1):15–36. https://doi.org/10.1007/s00267-003-0257-1

Hungr O (2007) Dynamics of rapid landslides. In: Sassa K, Fukuoka H, Wang F, Wang G (eds) Progress in landslide science. Springer, Berlin, pp 47–57

Pereira S, Zêzere JL, Quaresma I, Santos PP, Santos M (2015) Mortality patterns of hydro-geomorphologic disasters. Risk Anal 36 (6):1188–1210

Petley D (2012) Global patterns of loss of life from landslides. Geology 40(10):927–930. https://doi.org/10.1130/G33217.1

Petrucci O, Aceto L, Bianchi C, Bigot V, Brázdil R, Pereira S, Kahraman A, Kılıç Ö, Kotroni V, Llasat MC, Llasat-Botija M, Papagiannaki K, Pasqua AA, Řehoř J, Geli JR, Salvati P, Vinet F, Luis Zêzere J (2019) Flood Fatalities in Europe, 1980–2018: variability, features, and lessons to learn. Water 11:1682. https://doi.org/10.3390/w11081682

Rossi M, Guzzetti F, Salvati P, Donnini M, Napolitano E, Bianchi C (2019) A predictive model of societal landslide risk in Italy. Earth Sci Rev 196:102849

Salvati P, Bianchi C, Rossi M, Guzzetti F (2016) Landslide risk to the population of Italy and its geographical and temporal variation. In: Chavez M, Ghil M, Urrutia J (eds) Extreme events: observations, modeling and economics. Wiley-Blackwell, Hoboken, pp 177–194

Salvati P, Petrucci O, Rossi M, Pasqua AA, Bianchi C, Guzzetti F (2018) Gender, age and circumstances analysis of flood and landslide fatalities in Italy. Sci Total Environ 610–611:867–879. https://doi.org/10.1016/j.scitotenv.2017.08.064

Sepulveda SA, Petley DN (2015) Regional trends and controlling factors of fatal landslides in Latin America and the Caribbean. Nat Hazards Earth Syst Sci 15:1821–1833

Sharif H, Jackson T, Hossain M, Zane D (2015) Analysis of flood fatalities in Texas. Nat Hazards Rev 16(1)

Terti G, Ruin I, Anquetin S, Gourley JJ (February 2017) A situation-based analysis of flash flood fatalities in the United States. Am Meteorol Soc

Tosatti G (ed) (2003) A review of scientific contributions on the Stava Valley disaster (eastern Italian Alps), 19th July 1985. Pitagora, Bologna

Weichselgartner J, Pigeon P (2015) The role of knowledge in disaster risk reduction. Int J Disaster Risk Sci 6:107–116. https://doi.org/10.1007/s13753-015-0052-7

Wong HN, Ho KK, Chan YC (1997) Assessment of consequences of landslides. In: Cruden DM, Fell R (eds) Landslide risk assessment. Balkema, Rottherdam, pp 111–149

2020 Kyoto Japan

Using Mixed-Methods to Understand Community Vulnerability to Debris Flows in Montecito, CA

Erica Akemi Goto, Summer Gray, Edward Keller, and Keith C. Clarke

Abstract

More communities around the world are living in hazard-prone areas. These communities are susceptible to natural hazards and negative impacts, which can be mitigated by adopting Disaster Risk Reduction measures that take into account an assessment of vulnerability. However, vulnerability studies can be limited and often do not take a spatio-temporal approach. In this study, we applied a spatio-temporal approach to the case of Montecito, CA, in order to better understand the variables that make a community vulnerable and to identify those who are the most vulnerable. In 2018, Montecito was heavily impacted by debris-flows resulting in material and human losses. Our results showed that informal workers, residents of voluntary evacuation areas, and renters were most vulnerable, and the main aspect that contributed to their vulnerability was a lack of understanding of debris-flow risk. Moreover, local authorities failed to use historical data to communicate and educate the community before the event to reduce the overall community vulnerability and rebuilding process is changing the community vulnerability.

Keywords

Montecito • Debris-flows • Vulnerability • Mixed-methods • Spatio-temporal

Introduction

> "Even when they said the rains were coming, you don't expect a mountain to come down." (Respondent 04 2018)

Natural hazard events impact communities negatively around the world. The increase in negative impacts in these communities is linked to the increase of population and, consequently, more people living in at-risk areas. There are multiple ways to assess risk, but the framework we are using to understand risk is the one that combines hazardous events and vulnerability (Maskrey 1989; Wisner et al. 2003). Hazardous events will continue to be hazardous, but a better assessment of vulnerability can prevent disasters by building stronger and more resilient communities.

Vulnerability assessment is part of Disaster Risk Reduction (DRR) measures. A good vulnerability assessment should consider all members of the community, the local dynamic, and particularities of geography. Broad quantitative vulnerability assessments consider the spatial extent of the community but do not consider the local socio-economic heterogeneity. Using the same variables to assess vulnerability nationwide can help researchers understand general vulnerability patterns, but not specific details, which can include cultural or local dynamics. General assessments have no tools to include the marginalized populations within a community, and these are usually the most vulnerable (Wisner 1998). For instance, the homeless during the Tokyo earthquake (Wisner 1998) or informal workers during a wildfire (Thomas fire) in Ventura and Santa Barbara Counties, CA. Not including the marginalized populations continues to maintain them as marginalized, and when they survive a natural disaster impact, they do not recover their

E. A. Goto (✉)
Department of Geography, UCSB, Ellison Hall, Santa Barbara, 93106, USA
e-mail: ericagoto@gmail.com

S. Gray
UCSB, Environmental Studies, Santa Barbara, 93106, USA
e-mail: summer_gray@ucsb.edu

E. Keller
UCSB, Earth Science, Santa Barbara, 93106, USA
e-mail: keller@ucsb.edu

K. C. Clarke
Department of Geography, UCSB, Santa Barbara, 93106, USA
e-mail: kcclarke@ucsb.edu

F. Guzzetti et al. (eds.), *Understanding and Reducing Landslide Disaster Risk*, ICL Contribution to Landslide Disaster Risk Reduction,
https://doi.org/10.1007/978-3-030-60227-7_50

livelihoods (Wisner & Luce 1993), and become even more vulnerable. Therefore, quantitative and qualitative assessments targeting the local community can tell a better story about community vulnerability.

In this study, we do not aim to assess vulnerability but rather to demonstrate how a temporal-spatial approach that considers the type of natural hazard combined with a mixed-methods study can enhance understanding of community vulnerability. Considering Montecito and the 2018 debris-flows, we aim to answer the following questions: (1) Who were the most vulnerable people? (2) What were the characteristics of the community that made it more vulnerable? We used a parallel mixed-method approach to explore in-depth the particularities of the local community and develop a case study by using a survey, one-on-one interviews, analysis of local news, and records of community meetings.

Background

Vulnerability Assessment

The focus (dimension) of vulnerability assessment varies in disaster research. The main dimensions are social, economic, environmental, and institutional (Birkmann 2013). The fragmentation of vulnerability into these dimensions might help its assessment but might also favor one dimension over another. For example, when the focus is just on the social dimension, it might wrongly assume vulnerability as synonymous with poverty, but they are not the same thing (Chambers 1989). However, it is important to recognize the difficulty of assessing vulnerability and that none of the variables used in each dimension will perfectly explain the vulnerability of the community (Dwyer et al. 2004).

The Spatial Component of Vulnerability

Vulnerability assessment uses local, regional (e.g. Cardozo and Monteiro 2019), or national scales (Cutter et al. 2003), and can assess one type of natural hazard (e.g. Kaynia et al. 2008; Cardozo and Monteiro 2019) or many (Cutter et al. 2003; Flanagan et al. 2011; Hummell et al. 2016; Frigerio et al. 2018). In this study, we understand the spatial component as the local scale, which allows for a complex understanding of the community and its vulnerability towards a specific natural hazard (in this study, debris-flows). Moreover, the impact of a natural hazard in a community is not dependent just on the magnitude of the event but also on the damage caused by the conditions and dynamics to the different social groups inside of the community (Birkmann 2013). Communities are unique, and generalization of the vulnerability assessment can exclude important variables or exclude the most vulnerable people. Other than that, vulnerability assessment should consider the type of natural hazard. For instance, preparedness and education measures are unique for each type of natural hazard. Earthquake education is not the same as for tsunamis or landslides; preparedness and mitigation measures for floods are not the same as for earthquakes, or safe homes for landslides are not the same as for earthquakes. Furthermore, vulnerability is not static; it can change over time, and a community can become more or less vulnerable. Maybe more elderly are living in the community because a retirement community was built in the prior year, or the municipality has trained residents for a specific natural hazard, and they became more prepared.

The Temporal Component of Vulnerability

Natural disasters have three main phases: before, during, and after. Vulnerability assessment should be done in the before phase, and take into consideration the conditions of the community before the disaster (before phase) that can contribute to negative impacts in the during and after phases. The after phase (Fig. 1) has three main phases: short-term, intermediate, and long term (Homeland Security 2016).

This temporal assessment of vulnerability helps to identify who are the people that are part of the community, the local dynamics, and the local characteristics that make the community more or less vulnerable. However, the temporal component of vulnerability should be assessed together with the spatial component and the vulnerability dimensions previously described.

Methods

Figure 2 illustrates the methodology workflow of this study, and each of the phases is described in this section.

Literature Review

A literature review was fundamental to define the research question and to design survey questions and the semi-structured one-on-one interview guidelines.

Quantitative Survey Design

Based on our literature review, we asked target questions that would provide information about community vulnerability, which includes demographic, evacuation, and risk perception questions.

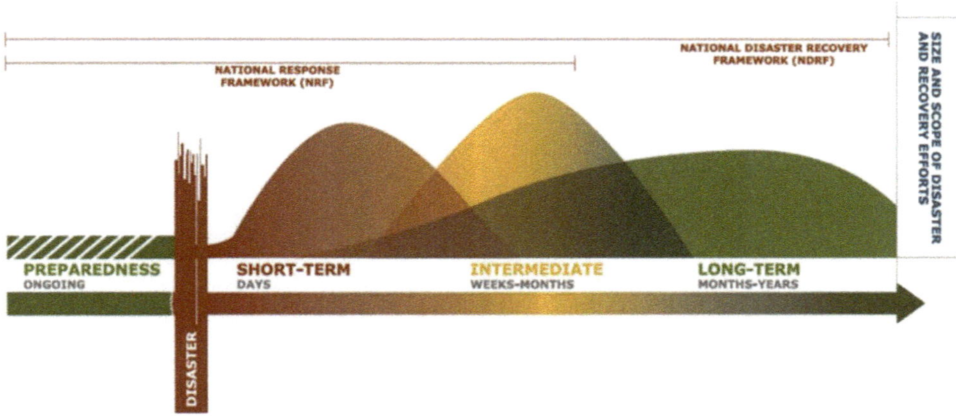

Fig. 1 Phases of disaster used by FEMA as part of the National Disaster Recovery Framework. During the After phase, you see the short-term, intermediate, and long-term (*Source* Homeland Security 2016)

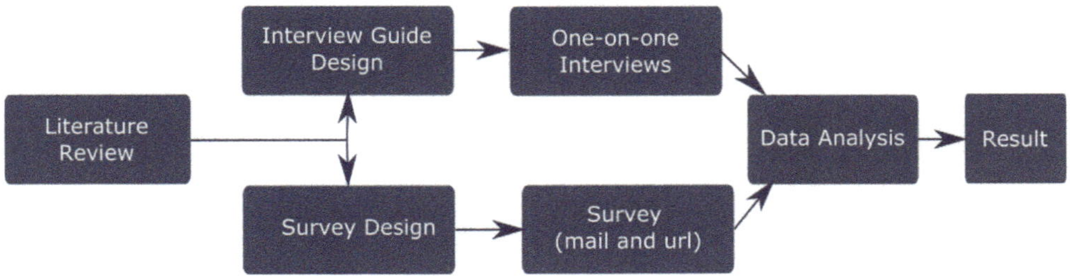

Fig. 2 Methodology workflow. Parallel mixed-methods were used in this study

We targeted residents of Montecito by including all buildings in the postal zip code 93108. Montecito has 32% of its residents older than 60 years (CENSUS 2010). Therefore, we decided to mail postcards with a hyperlink to the online survey and paper survey with a pre-paid envelope. We sent 4,798 postcards and paper surveys, and we got a 11% response rate. Since questions were voluntary, some questions had fewer answers than others. Therefore, the total number of answers varies from one question to another.

Qualitative Interview Design

Personal interviews allow for a richer understanding of the vulnerability of the community since they are not as rigid as surveys. We used a semi-structured technique, covering main topics of interest. However, we provided interviewees with some flexibility to talk about their reasoning, opinions, feelings, values, and any topic they thought was relevant.

We did 22 in depth interviews starting three months after the disaster. These interviews included residents and residents' relatives, members of grassroots organizations, and officers of government organizations that were directed affected by or worked with people affected by the debris-flows.

The first interviewees were residents that had their property impacted by the event. After these first interviewees, we reached out to more interviewees using snowball sampling. As we began receiving surveys, we contacted more residents who said they were interested in participating in an individual interview.

Data Analysis

We applied descriptive statistics to the survey data to summarize and describe our sample of the Montecito population, and used inferential statistics to draw conclusions using R software. Moreover, we used a phenomenological approach to explore meaning-making among people involved with the debris-flows, and coded the interviews and used themes in NVivo software. Themes emerged from our semi-structured survey, together with recurring topics brought up by different interviewees.

Study Area: Montecito, CA

In January 2018, a concentrated series of debris-flows heavily impacted the community of Montecito, CA (Fig. 3). They occurred as a result of wildfire (Thomas Fire) followed by intense rainfall in the catchments of the Santa Ynez Mountains (Kean et al. 2019). These catchments are steep and short and combine as they reach the Pacific Ocean. The debris-flows killed 23 people, injured at least 167, and damaged at least 400 homes (Kean et al. 2019). Residents, workers, and businesses in the community were impacted. The area was covered by mud and debris for at least two

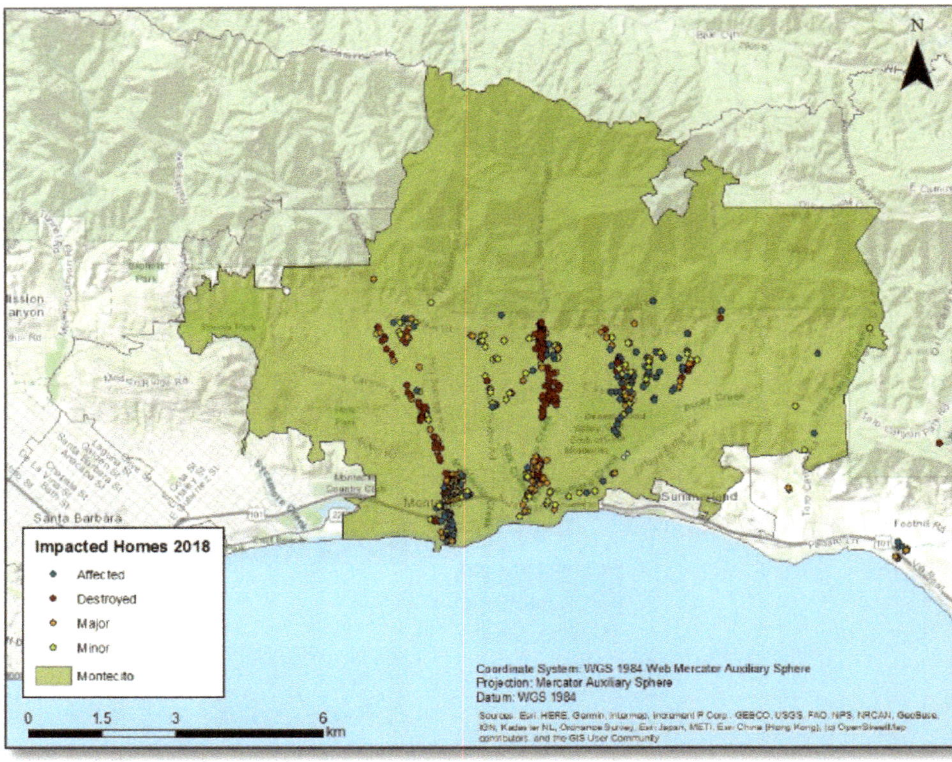

Fig. 3 Study area and buildings impacted by the 2018 debris-flows

weeks, the main highway was closed for ten days, water was considered not safe for drinking, and utilities (electricity and gas) were shut down. Beyond these direct impacts, there were many indirect impacts. For instance, emotional and psychological issues caused by the loss of family members and/or neighbors, and loss of income due to the closure of the highway not allowing people to work.

Results and Discussion

The Spatial Component of Vulnerability

The Montecito community is a combination of residents, workers, and businesses. Overall, Montecito residents are wealthy, have little ethnic diversity, and are over sixty. Most residents (66%) make more than $150,000/year, but in the same community, some (4%) make less than $25,000/year. Furthermore, most residents (63%) hire at least one worker at their property. These workers are gardeners, housekeepers, caretakers, babysitters, and home care for the elderly. These workers are commonly immigrants, including documented or undocumented. Considering that one-third of the human losses during debris-flows were from immigrant families and that only 9% of the Montecito residents are non-white, we can state that people of color and workers were more vulnerable than white residents. Moreover, after the debris-flows, workers were financially impacted because they could not work in the community for around two to four weeks. Part of the community and access to the community was closed due to debris and mud on the roads, and houses had no electricity.

Montecito has around 9% of renters. Commonly, these renters are families with school-aged children who attend Montecito Union School. The school is known to provide a good education and has a small student/teacher ratio. These families moved to Montecito to provide their children with a public education. Therefore, these residents have lived in the area for some years but usually do not know the history of past hazard events in the community.

Non-profit organizations and Santa Barbara County were very organized and helpful after the debris-flows. These organizations played different roles, from psychological health to cleaning the mud. Some of these organizations were Montecito Union School District, Cottage Health with the Post-Disaster Healing Process, Bucket Brigade, Montecito Association, Habitat for Humanity, Montecito Center for Preparedness, Recovery, and Rebuilding, and churches.

Interviewees frequently mentioned the amazing work done by the Bucket Brigade. Their volunteers helped clean the mud off many homes. Moreover, the volunteers created a hopeful environment as mentioned by one interviewee: "*Not just that they did it, but that they were there, they were comforting, they were happy.*" (Respondent 09 2018) The volunteers would go to a home, clean it, and leave.

Santa Barbara County frequently organized community meetings in Montecito to help residents to deal with the aftermath. A center was open in the community, the Montecito Center for Preparedness, Recovery, and Rebuilding, to offer support to impacted people. Interviewees highlighted the resources Santa Barbara County offered and how helpful the Montecito Center was, especially by locating aid organizations on the same site. The survey showed that 55% of residents attended at least one community meeting planned by the county. Additionally, meetings were broadcast online live and uploaded online, so people did not necessarily need to attend the meeting to know what was discussed.

The Temporal Component of Vulnerability

The temporal dimension of the vulnerability of the 2018 debris-flows in Montecito can be divided into before, during, and after.

During the Before Phase, residents had evacuation fatigue due to the Thomas Fire. There were multiple evacuations orders the month before the debris-flow evacuation order, as described by one of the interviewees "*we were evacuated six times.*" (Respondent 20 2018) Moreover, some residents have pets, which makes many evacuations very complicated.

The evacuation order and map (68.5%) were clear for residents. But, the volume of information released by different sources about evacuation was overwhelming. Residents that have lived longer knew that creeks could flood or that La Conchita, in Ventura County, had a major landslide in the past, but majority of them did not hear about debris-flows before and had no knowledge of past debris-flows in Montecito or Santa Barbara County. Local, regional, and national public authorities, on the other hand, knew that there was a 60–80% likelihood for debris-flows following the Thomas Fire combined with 15-minutes intense rainfall (USGS 2017). For example, Montecito Union School District decided to close school the day of the debris-flow based on the convincing argument given by public authorities that a debris-flow was very likely to occur, and the school could be isolated if the community was impacted.

The January debris-flow evacuation order used the same line as the Thomas Fire voluntary and mandatory; an east-west for a north-south hazard. Debris-flow risk is very different from fire risk, and the impact of the debris-flows confirmed that. Homes in the voluntary zone were destroyed, and 19 of the people that died lived in the voluntary evacuation zone, not in the mandatory (Hayden 2018). Mandatory evacuation along the channel corridors would have saved lives. It was missing an understanding of debris-flows hazard by the ones deciding the evacuation zones.

In the During Phase, the lack of knowledge of debris-flows increased residents' vulnerability. During the event, some residents decided to drive to escape the mud they saw coming toward their homes. The interviewees were safer at their homes (small impact at their homes) but put themselves at risk. These residents were driving during the storm, without power, with mud and debris coming downhill. One interviewee mentioned a family that was trying to escape and was swept away inside their car by the debris-flows, while their home did not get mud inside.

In the After Phase, a lack of information about what was going on in the community for people that did not evacuate increased their risk. A couple of residents that did not have their houses impacted got stuck in the mud the day after the event when they went to the pharmacy to get medication. Another example was an older woman whose house was not impacted was walking around the neighborhood, which could end up with her getting stuck in the mud.

In the initial process of the aftermath, there was much attention to the people that were impacted, but after some time, the impacted people were left by themselves to deal with rebuilding, getting permits, dealing with insurance, contractors, and so on, which can be mentally challenging and can be frustrating as an interviewee brought to our attention. It also brought tension between neighbors. To protect themselves from future debris-flows, some residents decided to build high walls or rebuild their homes on higher elevation ground. However, these changes in the landscape might impact other homes since they change the pattern of water flow. And now, homes that were safe, might not be safe anymore.

> "We had someone approach us when we soon after we bought this house [where they are living now] and asked if we were interested in selling [the house that was destroyed], and we had a couple of offers …" (Respondent 18 2018)

Properties that had homes destroyed were being bought by real estate agencies, as described by the interviewee. These homes are in at-risk areas, and people impacted want to move out, and they are willing to sell their homes cheaper than the market value. These real estate agencies buy these properties, rebuild and sell or rent for new families that did not experience the natural disaster. Another resident, when commenting about the rebuilding of their home stated: "*From an ethical perspective, I don't think that we should be allowed to rebuild. We're in a flood zone, and that's not gonna change.*" (Respondent 19 2018).

Conclusion

In conclusion, we discussed the spatial and temporal dimensions of vulnerability and why it is important to point out who are the vulnerable people of the community and the main aspects that made the community more vulnerable.

Considering the social and economic dimensions, homeowners that were in the voluntary evacuation zone, renters, and workers were the most vulnerable people in the community. The main characteristic that made the community more vulnerable was the lack of knowledge about debris-flows and the previous events in the region. Due to this lack of knowledge, some residents increased their risk unnecessarily by driving in an unsafe environment when they were actually safe at home. Moreover, the institutional dimensional of vulnerability was high for the whole community. Public authorities did not educate the community about past debris-flows and what they are, but also made a poor choice by choosing the same evacuation line as the Thomas Fire. On the other hand, authorities have been very supportive of residents in the after phase, creating Montecito Center for Preparedness, Recovery and Rebuilding and hosting many community meetings in the aftermath, which contributed to reducing the negative impact for homeowners.

These findings can help public authorities in their DRR measures and Disaster Risk Management. Moreover, the present study can provide valuable information for a future assessment of the spatial distribution of debris-flows vulnerability and consequent debris-flows risk in Montecito.

References

Birkmann J (2013) Measuring vulnerability to promote disaster-resilient societies and to enhance adaptation: conceptual frameworks and definitions. In: Birkmann J (ed) Measuring vulnerability to natural hazards: towards disaster resilient societies. Institute for Environment and Human Security, UNU—EHS

Cardozo CP, Monteiro AM (2019) Assessing Vulnerability to natural hazards in nova friburgo, Rio De Janeiro Mountain Region, Brazil. Revista de Estudios Latinoamericanos Sobre Reducción Del Riesgo de Desastres REDER 3(2):71–83

CENSUS (2010) DP-1 Profile of General Population and Housing Characteristics: 2010. https://factfinder.census.gov

Chambers R (1989) Vulnerability, coping and policy. Inst Dev Stud Bull 2(20):1–7

Cutter SL, Boruff BJ, Shirley WL (2003) Social vulnerability to environmental hazards. Soc Sci Q 84(2):242–261

de L Hummell BM, Cutter SL, Emrich CT (2016) Social vulnerability to natural hazards in Brazil. Int J Disaster Risk Sci 7(2):111–122

Dwyer A, Zoppou C, Nielsen O (2004) Quantifying social vulnerability: a methodology for identifying those at risk to natural hazards. Spat Inf Nation 101

Flanagan BE, Gregory EW, Hallisey EJ, Heitgerd JL, Lewis (2011) A social vulnerability index for disaster management a social vulnerability index for disaster management. J Homel Secur Emerg Manage 8(1)

Frigerio I, Carnelli F, Cabinio M, De Amicis M (2018) Spatiotemporal pattern of social vulnerability in Italy. Int J Disaster Risk Sci 9:249–262

Guzzetti F, Carrara A, Cardinali M, Reichenbach P (1999) Landslide hazard evaluation: a review of current techniques and their application in a multi-scale study, central Italy. Geomorphology 31(1–4):181–216. https://doi.org/10.1016/S0169-555X(99)00078-1

Hayden T (2018, June 14). Public survey exposes montecito debris flow communication failures. Santa Barbara Independent. Retrieved from https://www.independent.com/2018/06/14/public-survey-exposes-montecito-debris-flow-communication-failures/

Homeland Security (2016) National disaster recovery framework, 2nd Edn

Kaynia AM, Papathoma-Köhle M, Neuhäuser B, Ratzinger K, Wenzel H, Medina-Cetina Z (2008) Probabilistic assessment of vulnerability to landslide: Application to the village of Lichtenstein, Baden-Württemberg, Germany. Eng Geol 101(1–2):33–48. https://doi.org/10.1016/j.enggeo.2008.03.008

Kean JW, Staley DM, Lancaster JT, Rengers FK, Swanson BJ, Coe JA, Hernandez JL, Sigman AJ, Allstadt KE, Lindsay DN (2019) Inundation, flow dynamics, and damage in the 9 January 2018 Montecito debris-flow event, California, USA: opportunities and challenges for post-wildfire risk assessment. Geosphere 15 (4):1140–1163. https://doi.org/10.1130/ges02048.1

Maskrey A (1989) Disaster mitigation: a community based approach. Oxfam Publishing

Respondent 04 (2018, August) Personal interview

Respondent 09 (2018, September) Personal interview

Respondent 18 (2018, October). Personal interview

Respondent 19 (2018, November) Personal interview

Respondent 20 (2018, November) Personal interview

USGS (2017) Thomas fire (Ventura and Santa Barbara County). Retrieved from https://landslides.usgs.gov/hazards/postfire_debrisflow/detail.php?objectid=178

Wisner B (1998) Marginality and vulnerability. Appl Geogr 18(1): 25–33

Wisner B, Luce HR (1993) Disaster vulnerability: scale, power and daily life. GeoJournal 30(2):127–140

Wisner B, Blaikie P, Cannon T, Davis I (2003) At risk: natural hazards, people's vulnerability, and disasters. Human Ecology (Second, vol 24)

Innovation in Analysis and Forecasting of Vulnerability to Slow-Moving Landslides

Dario Peduto, Gianfranco Nicodemo, Nicoletta Nappo, and Giovanni Gullà

Abstract

The analysis of consequence induced by landslides to the built environment is a key step in the landslide risk management process. For this purpose, a thorough knowledge of both the landslide mechanisms and the behaviour/characteristics of the exposed elements turn out to be crucial. Currently, scientists and technicians have access to unprecedented big data sources that provide a variety of useful information on landslide-affected areas. With reference to slow-moving landslides, this paper presents some case studies of an ongoing research aimed at defining and testing original procedures pursuing the exploitation of innovative monitoring/surveying techniques for the analysis and forecasting of vulnerability of buildings and roads.

Keywords

Slow-moving landslides • Vulnerability • Damage • Remote sensing • DInSAR

D. Peduto (✉) · G. Nicodemo
Department of Civil Engineering, University of Salerno, Via Giovanni Paolo II, 132, 84084 Fisciano (SA), Italy
e-mail: dpeduto@unisa.it

G. Nicodemo
e-mail: gnicodemo@unisa.it

N. Nappo
Department of Science and High Technology, University of Insubria, Varese, Italy
e-mail: nnappo@uninsubria.it

G. Gullà
Research Institute for Geo-Hydrological Protection, National Research Council of Italy (IRPI—CNR), Rende, CS, Italy
e-mail: giovanni.gulla@irpi.cnr.it

F. Guzzetti et al. (eds.), *Understanding and Reducing Landslide Disaster Risk*, ICL Contribution to Landslide Disaster Risk Reduction,
https://doi.org/10.1007/978-3-030-60227-7_51

Introduction

Landslides are one of the natural hazards yearly causing highest impact worldwide in terms of consequences on either people (i.e. injury and life losses) or the built-up environment (i.e. damages of different severity levels to structures and infrastructure).

The analysis/forecasting of consequences induced by a landslide represents a key step for evaluating the risk to which people and/or (infra)structures interacting with a landslide are (or can be) exposed (Fell et al. 2008). Focusing on slow-moving landslides, consequence analysis involves identifying/quantifying the structures (i.e. buildings) and/or the infrastructures (i.e. roads) at risk as well as estimating their vulnerability (Peduto et al. 2017; Argyroudis et al. 2019). This latter depends on both the intensity of the landslide (e.g. the kinematics and related displacements) and peculiar features of exposed facilities (structural and foundation typology, geometry, material properties, state of maintenance, etc.). Therefore, its estimation is a difficult task that entails a thorough knowledge of both the landslide mechanisms (Mavrouli et al. 2014; Peduto et al. 2018) and the behaviour of the exposed elements (Ferlisi et al. 2019) as well as the monitoring of landslide intensity parameters (Peduto et al. 2017).

Nowadays, the technological innovation in monitoring systems (both ground-based or remotely operating) offers the concrete possibility to get access to unprecedented rich datasets that the landslide community can profitably use for risk analysis purposes. In this regard, the present paper aims to provide an example of how multi-source monitoring/surveying conventional/innovative data can complement with each other to enhance vulnerability analysis and/or forecasting capabilities in the field of landslide consequence analysis.

Proposed Approach

The proposed approach to investigate the vulnerability of structures and infrastructures in landslide-affected urban areas includes three phases (Fig. 1). The first phase (Phase I) pursues the detection of the elements (i.e. aggregate/single buildings and/or road/railway networks, including bridges, embankments, etc.) exposed to landslide risk. According to the scale of analysis, quality and resolution of the used data (Ferlisi et al. 2019), they are identified by overlapping base and thematic maps (e.g. built-up area map, DEM, landslide inventory map) with available observations (e.g. optical imagery) and/or monitoring (i.e. provided by conventional or innovative field/remote monitoring techniques) data (Peduto et al. 2018; Greenwood et al. 2019).

In Phase II, a damage severity level (DL) as well as a value of the intensity parameter (IP) are assigned to every identified exposed element. In particular, the DL is classified based on the analysis of field and/or remote surveys adopting a proper damage ranking. The selected IP (e.g. velocity, settlement or related parameters) is computed on the base of the quality and resolution of available monitoring data. In Phase III, one can combine both recorded DL and retrieved magnitude of IP to analyse the vulnerability of the exposed elements. For this purpose, data are treated based on the aim of the study (i.e. in relation to the scale of analysis), the type (i.e. empirical or numerical) of the analysis, the quantity and quality of available data, via deterministic (i.e. intended to investigate the established relationships between the cause (IP) and effect (DL)) or probabilistic (i.e. aimed at generating consequence forecasting tools such as fragility and/or vulnerability curves) approaches (Ferlisi et al. 2019).

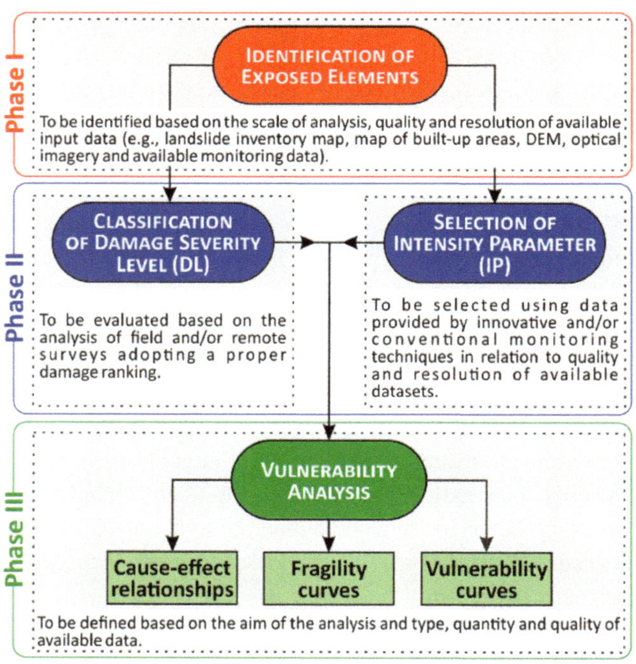

Fig. 1 Flowchart of the proposed approach

Results

Two well-documented case studies in Calabria region (southern Italy) were analysed. As for buildings, the urban centre of the Lungro town, which has been extensively suffering from several slope instabilities that caused damages of different severity to the built-up area, was selected as test site. The state road SS 177 connecting the municipalities of Longobucco (from km 51 + 400) and Paludi (to km 68 + 100), which is affected by several ground displacements that over time have caused frequent closures and cutting of the connections between many municipalities along its track, was chosen as test infrastructure.

Lungro case study

The urban area of Lungro (Fig. 2a), located at 650 m a.s.l. in the northern sector of the Calabria region (southern Italy), presents a complex geological context in which the Lungro-Verbicaro Unit dating back to the Middle Trias made up of metapelites and metacarbonates prevails (Gullà et al. 2017). The urban centre is widely affected by slow-moving landslides with either active or dormant state based on geomorphological criteria (Fig. 2a). Consequently, over the years, several damages (Fig. 2a) were recorded to both masonry buildings—mainly located in the historic centre and its surroundings (Carmine and Lafcantino), made up of 2–3 floors with pebbles, or erratic/irregular stones on shallow foundations—and reinforced concrete buildings up to 5–6 floors mainly located in the San Leonardo area, built since the early 1950–1960s (Peduto et al. 2017, 2018). Previous studies on the area (Gullà et al. 2017) typified the inventoried slow-moving landslides in six categories (Fig. 2b), through the application of the "*aPosIn*" procedure that relies on the combination of geological information, geomorphological criteria and both geotechnical and remote displacement monitoring data. These categories differ for geometric and kinematic characteristics, involved soils, and type of movement. In the area, in addition to deep and surface displacement monitoring data deriving from conventional (inclinometers and GPS) techniques (Gullà et al. 2017; Peduto et al. 2018), innovative monitoring data provided by the interferometric processing of spaceborne synthetic aperture radar images (DInSAR) are available. These latter, consists of high- and very high-resolution DInSAR data, respectively acquired on ascending orbit by ENVISAT (period August 2003 to January 2010) and COSMO-SkyMed (period October 2012–April 2014) radar sensors, which were processed according to the SAR tomographic analysis (Fornaro et al. 2009). Figure 2b shows the spatial distribution of the DInSAR coherent pixel velocities derived from the processing of COSMO-SkyMed images that are particularly effective for monitoring built-up areas (Peduto et al. 2017, Nicodemo et al. 2018).

Following the described approach (Fig. 1), in Phase I, single buildings exposed to landslide risk were identified by combining the available input data. Then, in Phase II, both DL and IP were estimated. The former was assigned to each single building (Fig. 2a) based on the data collected via ad hoc predisposed building fact-sheets (Ferlisi et al. 2015;

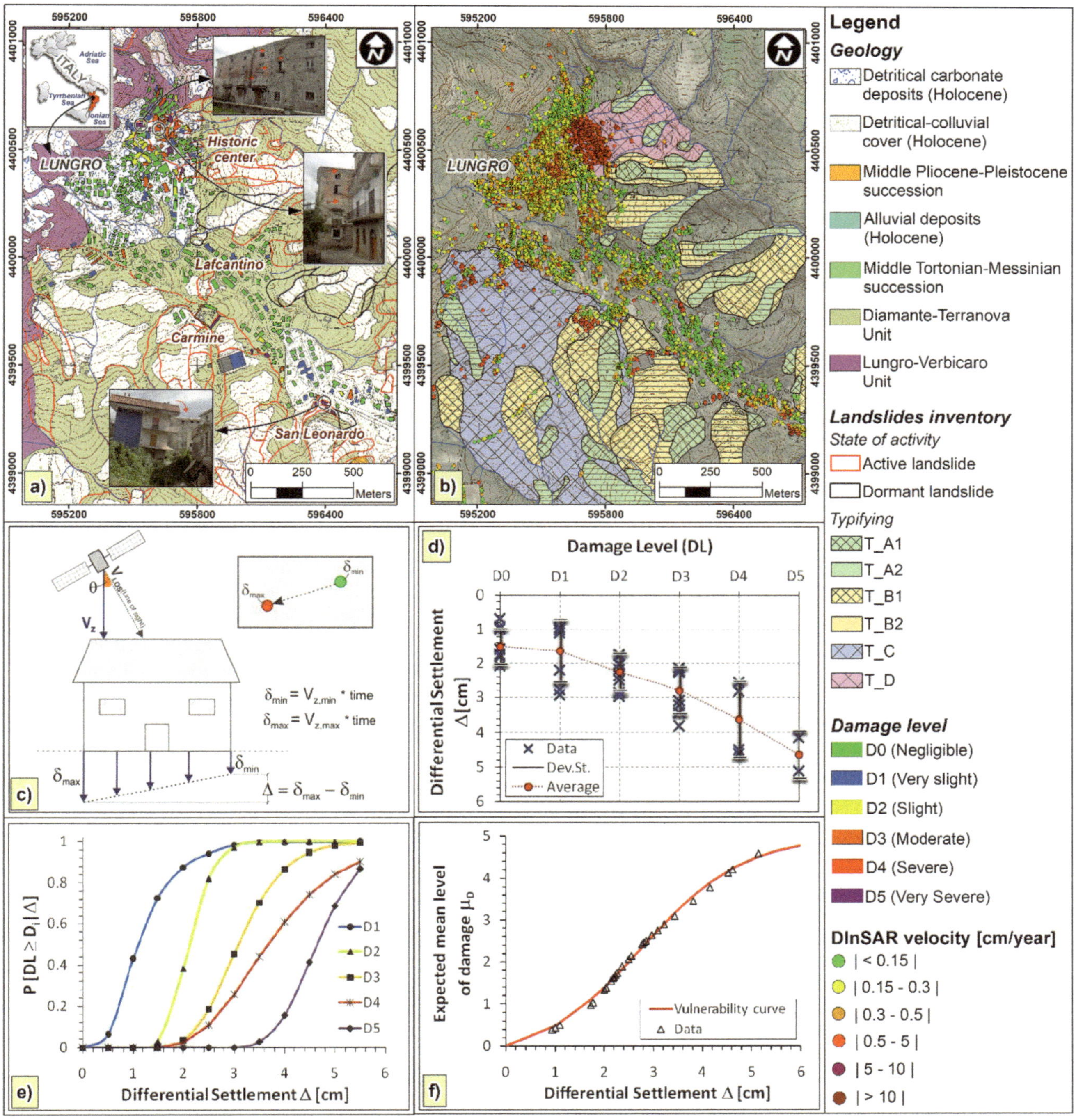

Fig. 2 Lungro case study: **a** geological and landslide inventory maps along with their state of activity (modified from Gullà et al. 2017) and building damage map with some examples of crack patterns exhibited by both reinforced concrete and masonry buildings (modified from Peduto et al. 2017); **b** map of typified landslides with spatial distribution of DInSAR data acquired by COSMO-SkyMed satellite (ascending orbit, 2012–2014) (modified from Gullà et al. 2017 and Peduto et al. 2018); **c** sketch of the DInSAR-derived differential settlement (Δ) of a building; **d** damage level (DL) vs. differential settlements (Δ); **e** empirical fragility curves and **f** vulnerability curve for masonry buildings in Lungro area (modified from Peduto et al. 2017)

Peduto et. 2017) during an extensive in situ damage survey carried out in October 2015 (including the whole urban area over a total of 540 buildings). The recorded DLs were categorized by adapting the classification system proposed by Burland et al. (1977) on the basis of the visual interpretation of crack patterns exhibited by the building façades considering six categories (D0 = negligible; D1 = very slight; D2 = slight; D3 = moderate; D4 = severe; D5 = very severe). The IP was assumed as the differential settlement (Δ) computed using the DInSAR data on 49 exposed single buildings (12 of reinforced concrete and 37 of masonry structure) out of 111 total buildings interacting with slow-moving landslides; for this purpose only those buildings covered by at least two coherent DInSAR benchmarks in both Envisat and COSMO-SkyMed datasets were considered.

The DInSAR-derived differential settlement (Δ) was computed as the maximum difference (Fig. 2c) of the cumulative settlements (δ, derived by multiplying the DInSAR velocity along the vertical direction for the observation period of both available SAR datasets) recorded by the DInSAR benchmarks within the building perimeter (Peduto et al. 2017).

In the last phase (Phase III), referring to 37 masonry buildings composing a homogeneous sample (i.e. masonry 2–3 floored structures on shallow foundations) and exhibiting all the different damage severity levels (i.e. from D1 to D5), the building vulnerability was investigated. First, DInSAR-derived IP were merged with the recorded DLs to retrieve the relationship (Fig. 2d) between cause (Δ) and effect (DL) showing a generally increasing trend of DL when Δ increases. Then, according to the procedure described by Peduto et al. (2017, 2018), assuming the log-normal distribution function as probabilistic model, empirical fragility curves providing the conditional probability for a randomly selected building at risk to be in, or exceed, a certain DL when the IP (Δ) equals a given value were generated (Fig. 2e). Finally, using the tangent hyperbolic function (Lagomarsino and Giovinazzi, 2006) as regression model to fit the empirical data, the empirical vulnerability curve (Fig. 2f), which relates the expected mean level of damage severity (μ_D) to a given building to an IP (Δ) value, was derived.

State Road SS 177 case study

The State Road SS 177 from km 51 + 400 to 68 + 100 (Fig. 3) is located in the north-eastern sector of the Calabria region; it connects the municipalities of Longobucco, Cropalati and Paludi. The road was built on steep slopes mainly made up of the following lithological units (Fig. 3a): flysh-type sequences (Mesozoic-Tertiary), varicoloured shales (Cretaceous-Oligocene), and very low to low-grade metamorphic rocks at places with ophiolites (Paleozoic-Mesozoic) (Nappo et al. 2019). The analysed 17 km of the SS 177 intersect 32 slow-moving landslides that differ for type and state activity (Fig. 3a and 3b). In this area, high and very-high resolution DInSAR data from Envisat (February 2003–July 2010) and COSMO-SkyMed (May 2011–April 2014) ascending orbits were collected. Figure 3b shows the spatial distribution of COSMO-SkyMed Line of Sight (LOS) velocities. During the field survey performed in 2018, 245 damaged sections were identified along the SS 177 (Fig. 3a). In Phase I, by overlapping the road map with other thematic maps (i.e. landslide inventory and DInSAR data), 135 damaged sections resulted exposed to slow-moving landslides. Based on both Google Street View imagery and field reconnaissance, a damage severity level (DL) was associated to each damaged section of the road according to the following classes: D1—deformation of the road pavement without the occurrence of cracks; D2—cracks on the road pavement; D3—dislocation of the road pavement (Nappo et al. 2019). Like it was done for buildings in Phase II, the differential displacement ($\bar{\Delta}$) was used as intensity parameter (IP) to investigate the relationship between exposed roads and slow-moving landslides. To compute the IP, two areas of maximum (δ_{max}) and minimum cumulative displacement (δ_{min}) were identified around the damaged road section, here considered as points, by drawing an inner and an outer buffer with 10 m and 20 m diameters respectively (Fig. 3c). Following Nappo et al. (2019) the cumulative displacements (i.e. δ_{max}, δ_{min}) were retrieved by selecting the maximum/minimum value of the DInSAR-derived cumulative displacement computed in the inner/outer buffer by multiplying the modulus of the velocity projected along the steepest slope direction for the observation period of the satellite dataset. The differential displacement $\bar{\Delta}$ was then obtained as the difference between δ_{max} and δ_{min} (Nappo et al. 2019), and the distribution of its values represented in Fig. 3d.

In Phase III, the percentage distribution of DLs was computed for three different ranges of $\bar{\Delta}$ (Fig. 3e) chosen from 0-20 mm to > 40 mm according to the observed values. Figures 3e and 3f show that as the cause ($\bar{\Delta}$) increases, the number of road sections exhibiting low and medium DL (i.e. D1 and D2) decreases, whereas those with high DL (i.e. D3) increase.

Discussion and Conclusions

The presented case studies showed that technological innovation of multi-source monitoring/survey data can provide useful contribution to procedures aimed at landslide risk analyses. For instance, Google Street View was proved to be

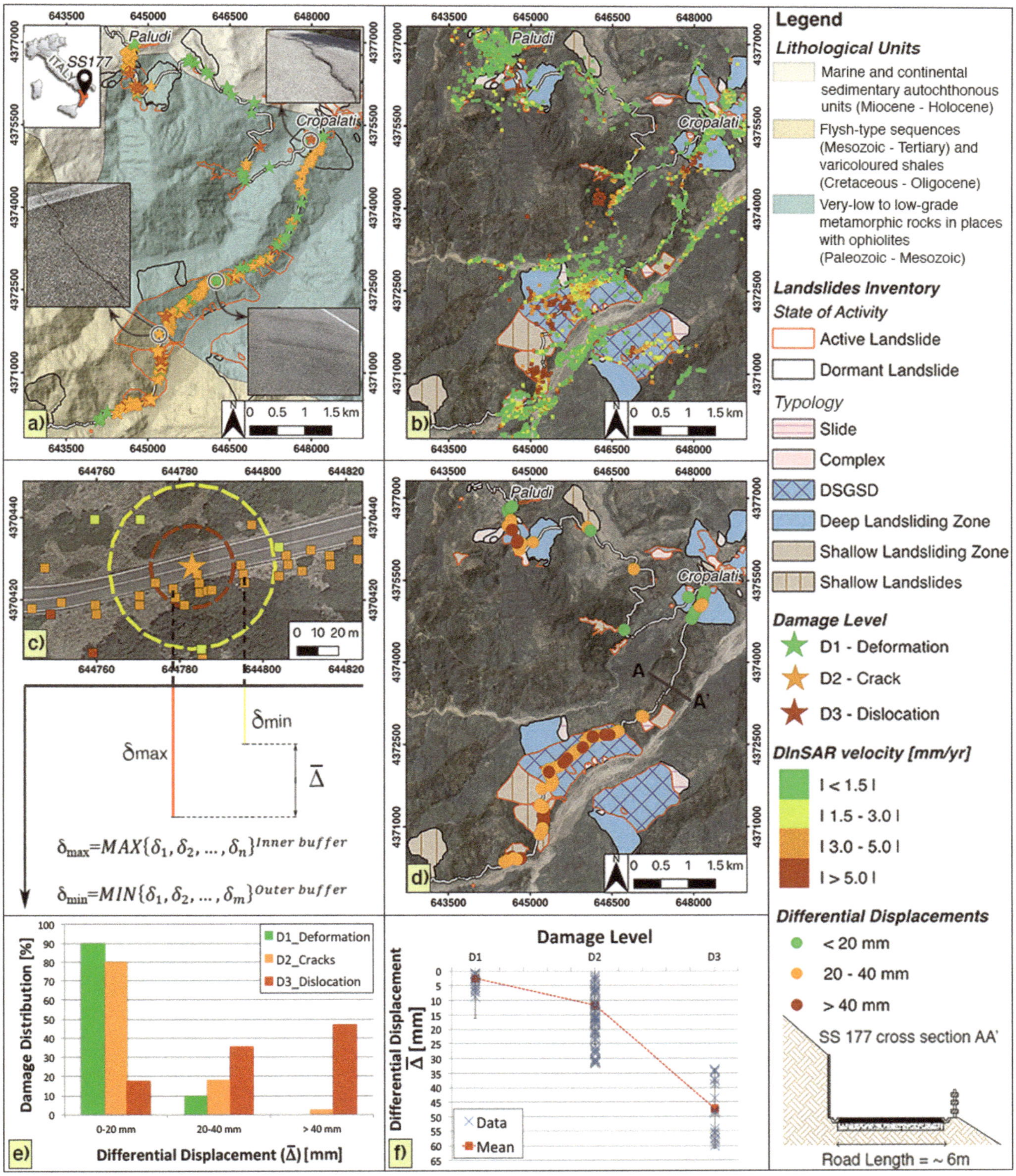

Fig. 3 Road SS 177 case study: **a** geological sketch (modified from Gullà et al. 2008; Sorriso-Valvo 1993), inventory map and state of activity of landslides, map of damaged road sections with severity levels; **b** map of landslides interacting with the investigated road and spatial distribution of DInSAR data acquired by COSMO-SkyMed satellite (ascending orbit, 2011–2014); **c** sketch of DInSAR-derived differential displacement ($\bar{\Delta}$) for a damaged road section; **d** distribution of the $\bar{\Delta}$ ranges along the road stretches interacting with landslides; **e** percentage distribution of damage levels versus differential displacements ($\bar{\Delta}$); **f** damage levels versus differential displacements ($\bar{\Delta}$) (modified from Nappo et al. 2019)

a useful tool for rapid, easy and cost-effective road damage inspections; whereas DInSAR data confirmed to be a well-established complement of conventional displacement monitoring techniques. Both the abovementioned datasets, which are broadly available in archives worldwide, can foster deterministic/probabilistic vulnerability analyses of buildings/roads exposed to slow-moving landslides.

Once further validated, the presented approach could be part of an iterative procedure to help building/road maintenance prioritization and scheduling as well as risk management in slow-moving landslide-affected urban area and extended to other geo-hazards. Nevertheless, the exportability to other areas will require updated landslide inventories, local geotechnical and building/road characteristics to be taken into account to typify representative landslide-damaged buildings/roads in homogeneous geo-lithological contexts.

Acknowledgements The research was carried out partly within the Agreement between the Dept. of Civil Eng. (DICIV) of the Univ. of Salerno (Italy) and the CNR-IRPI (Section of Cosenza) on "*Attività di ricerca mirate ad approfondire le conoscenze sulla tipizzazione di movimenti franosi a cinematica lenta e sulla vulnerabilità di strutture e infrastrutture che con essi interagiscono anche mediante l'impiego di tecniche innovative di monitoraggio satellitare e da terra*". Olga Mavrouli and Cees van Westen of the Faculty of Geo-Information Science and Earth Observation (ITC) of the University of Twente (The Netherlands) are gratefully acknowledged for the analysis of roads. This work is also part of the activity of the ICL-IPL project No 248: "*Innovation in slow-moving landslide risk assessment of roads and urban sites by combining multi-sensor multi-source monitoring data*". The processed DInSAR data were provided by the Italian Ministry of Environment (MATTM) within the Italian Piano Straordinario di Telerilevamento (PST). Cosmo-SkyMed data were provided by Italian Space Agency under license no. 0000155.

References

Argyroudis SA, Mitoulis SA, Winter MG, Kaynia AM (2019) Fragility of transport assets exposed to multiple hazards: state-of-the-art review toward infrastructural resilience. Reliab Eng Syst Saf 191:106567. https://doi.org/10.1016/j.ress.2019.106567

Burland JB, Broms BB, de Mello VFB (1977) Behaviour of foundations and structures. Proceedings of the 9th Int Conf on Soil Mechanics and Foundation Engineering, Tokyo. SOA Report, vol 2, pp 495–546

Fell R, Corominas J, Bonnard Ch, Cascini L, Leroi E, Savage WZ (2008) Guidelines for landslide susceptibility, hazard and risk zoning for land-use planning. Eng Geol 102(3–4):99–111. https://doi.org/10.1016/j.enggeo.2008.03.014

Ferlisi S, Gullà G, Nicodemo G, Peduto D (2019) A multi-scale methodological approach for slow-moving landslide risk mitigation in urban areas, southern Italy. Euro-Mediterr J Environ Integr 4:20. https://doi.org/10.1007/s41207-019-0110-4

Ferlisi S, Peduto D, Gullà G, Nicodemo G, Borrelli L, Fornaro G (2015) The use of DInSAR data for the analysis of building damage induced by slow-moving landslides. Engineering Geology for Society and Territory—Landslide Processes, © Springer International Publishing, vol 2. pp. 1835–1839. https://doi.org/10.1007/978-3-319-090573_325

Fornaro G, Reale D, Serafino F (2009) Four dimensional SAR imaging for height estimation and monitoring of single and double scatterers. IEEE Trans Geosci Remote Sens Lett 47(1):224–237

Greenwood WW, Lynch JP, Zekkos D (2019) Applications of UAVs in civil infrastructure. J Infrastruct Syst 25(2):04019002

Gullà G, Antronico L, Iaquinta P, Terranova O (2008) Susceptibility and triggering scenarios at a regional scale for shallow landslides. Geomorphology 99:39–58

Gullà G, Peduto D, Borrelli L, Antronico L, Fornaro G (2017) Geometric and kinematic characterization of landslides affecting urban areas: the Lungro case study (Calabria, southern Italy). Landslides 14(1):171–188. https://doi.org/10.1007/s10346-015-0676-0

Lagomarsino S, Giovinazzi S (2006) Macroseismic and mechanical models for the vulnerability and damage assessment of current buildings. Bull Earthq Eng 4(4):415–443. https://doi.org/10.1007/s10518-006-9024-z

Mavrouli O, Fotopoulou S, Pitilakis K, Zuccaro G, Corominas J, Santo A, Cacace F, De Gregorio D, Di Crescenzo G, Foerster E, Ulrich T (2014) Vulnerability assessment for reinforced concrete buildings exposed to landslides. Bull Eng Geol Environ 73:265–289. https://doi.org/10.1007/s10064-014-0573-0

Nappo N, Peduto D, Mavrouli O, van Westen C, Gullà G (2019) Slow-moving landslide interacting with the road network: analysis of damage using ancillary data, in situ surveys and multi-source monitoring data. Eng Geol 260, https://doi.org/10.1016/j.enggeo.2019.105244

Nicodemo G, Peduto D, Ferlisi S, Gullà G, Reale D, Fornaro G (2018) DInSAR data integration in vulnerability analysis of buildings exposed to slow-moving landslides. Proceedings of IEEE International Geoscience and Remote Sensing Symposium (IGARSS 2018), 22–27 July, 2018—Valentia, Spain. pp. 6111–6114. https://doi.org/10.1109/igarss.2018.8518808

Peduto D, Ferlisi S, Nicodemo G, Reale D, Gullà G (2017) Empirical fragility and vulnerability curves for buildings exposed to slow-moving landslides at medium and large scales. Landslides. 14(6):1993–2007. https://doi.org/10.1007/s10346-017-0826-7

Peduto D, Nicodemo G, Caraffa M, Gullà G (2018) Quantitative analysis of consequences induced by slow-moving landslides to masonry buildings: a case study. Landslides. 15(10):2017–2030. https://doi.org/10.1007/s10346-018-1014-0

Sorriso-Valvo M (1993) The geomorphology of Calabria. A Sketch Geogr Fis Din Quat 16:75–80

Sentinel-1 PSI Data for the Evaluation of Landslide Geohazard and Impact

Silvia Bianchini, Lorenzo Solari, Anna Barra, Oriol Monserrat, Michele Crosetto, and Filippo Catani

Abstract

In this work we exploited Sentinel-1 satellite radar images processed by means of Persistent Scatterers Interferometry (PSI) techniques for the evaluation of landslide geohazard and impact on a mountainous region. In particular, we used PSI data as starting point in a working chain whose final goal is the estimation of the potential worth of loss of the structures involved by slope instability phenomena. We applied this approach on a test area in the Valle d'Aosta Region (North Italy) where more than fifty percent of the territory is above 2000 m a.s.l. and extensively affected by landslides. Firstly, PSI Sentinel-1 data permitted to scan the territory and to highlight the areas characterized by the highest ground motion rates, namely Active Deformation Areas (ADA). These areas were used to derive the intensity of potential landslides in terms of magnitude. Then, for the different elements at risk (EAR) we estimated both the vulnerability, by referring to values already proposed in literature for similar working scale, and the exposure, by considering the current real estate market values of the EAR in the area. We finally derived color-scale maps showing landslide intensity and values of potential loss expressed in quantitative terms (Euros for square meters). This operational methodology can provide useful indications and outputs for landslide risk management at regional scale. Considering the present availability of Sentinel-1 SAR images with 6-days revisiting time, this procedure can represent an example of satellite InSAR monitoring as supporting tool for Civil Protection activities and geohazard mitigation long-term strategies.

Keywords

Satellite interferometry • Sentinel-1 • Landslide • Potential worth of loss

Introduction

Landslide risk assessment is an efficient approach to mitigate geohazards and reduce potential disasters. Nevertheless, landslide risk and its components (hazard and intensity of phenomena, vulnerability and exposure of elements at risk) are still a difficult issue to address. This is especially true over large areas, where specific data about landslide occurrence and kinematics, as well as values of vulnerability and exposure are usually quite challenging to collect (Liu & Miao 2018; Corominas et al. 2014).

In the last decade, landslide risk studies have made relevant progress and many qualitative or quantitative methods have been proposed and tested (e.g. Dai et al. 2002; van Western et al. 2006). In this work, we assessed the impact of potential landslides on built-up areas and road networks by estimating intensity, vulnerability, exposure and potential loss on a test area in Valle D'Aosta region (northwestern Italy). The area is characterized by steep slopes and widespread landslides.

S. Bianchini (✉) · F. Catani
Earth Sciences Department, University of Firenze, Via La Pira 4, 50121 Florence, Italy
e-mail: silvia.bianchini@unifi.it

F. Catani
e-mail: filippo.catani@unifi.it

L. Solari · A. Barra · O. Monserrat · M. Crosetto
Centre Tecnològic de Telecomunicacions de Catalunya Geomatics Division, 08860 Castelldefels, Spain
e-mail: lorenzo.solari@cttc.cat

A. Barra
e-mail: anna.barra@cttc.cat

O. Monserrat
e-mail: oriol.monserrat@cttc.cat

M. Crosetto
e-mail: michele.crosetto@cttc.cat

F. Guzzetti et al. (eds.), *Understanding and Reducing Landslide Disaster Risk*,
ICL Contribution to Landslide Disaster Risk Reduction,
https://doi.org/10.1007/978-3-030-60227-7_52

In particular, we exploited space-borne interferometric products as starting point of an operational procedure, based on the so-called ADA (Active Deformation Areas) (Barra et al. 2017; Solari et al. 2018) derived from Persistent Scatterers Interferometry (PSI) data. PSI products were derived from Sentinel-1 radar images with the aim to provide landslide intensity as magnitude of potential slow-moving landslides (Bianchini et al. 2017). The final goal is to quantify by an economic point of view the potential worth of loss suffered by buildings or roads where the fastest motions are detected (i.e. where ADA are found).

The outputs of this work are simplified colour-scale maps indicating those structures and infrastructures with a greater probability to suffer for the impact of a geohazard and those ones affected by the dynamic of an active geohazard.

Study Area

Geomorphological Setting

The test area of this work is the eastern portion of Valle d'Aosta, a small alpine region in north-western Italy (Fig. 1). The area is mainly mountainous and characterized by four N-S oriented valleys and slopes with peaks that reach up 4000 m a.s.l. The geological/geomorphological context, the high topographic relief as well as local climatic conditions control landslide processes. Debris flows, rock-slides, complex and rotational landslides as well as Deep-Seated Gravitational Slopes Deformations (DSGSD), which cover the 14% of the regional territory, are the most widespread typologies (Trigila et al. 2010; Giordan et al. 2017).

The area of interest is characterized by 972 mapped landslides and 101 DSGSD included in the IFFI catalogue (Inventario dei Fenomeni Franosi in Italia—Italian landslide inventory—Trigila et al. 2010) (Fig. 1).

In the area, landslides are potential threat for the population and responsible for great economic losses. The Mont de La Saxe rockslide is an example; its activity could generate socio-economic consequences estimated in one billion of euros, involving 1200 inhabitants. Fatal events in the area were registered in October 2000, when a series of debris flows killed 17 people, or in August 2018 when two landslides caused two fatalities in the Val Ferret in Mont Blanc Massif (Ratto et al. 2003).

Input Data

The Valle d'Aosta Region is characterized by sparse urban areas, with the largest cities, including Aosta, located along the main NW-SE oriented valley (Dora Baltea Valley). In this context, InSAR data are a valuable support for landslide investigations, offering a good compromise between costs and spatial-temporal resolution and coverage.

Available satellite radar data over the AOI consist of 153 SAR images acquired in C-band by the Sentinel-1 satellite in

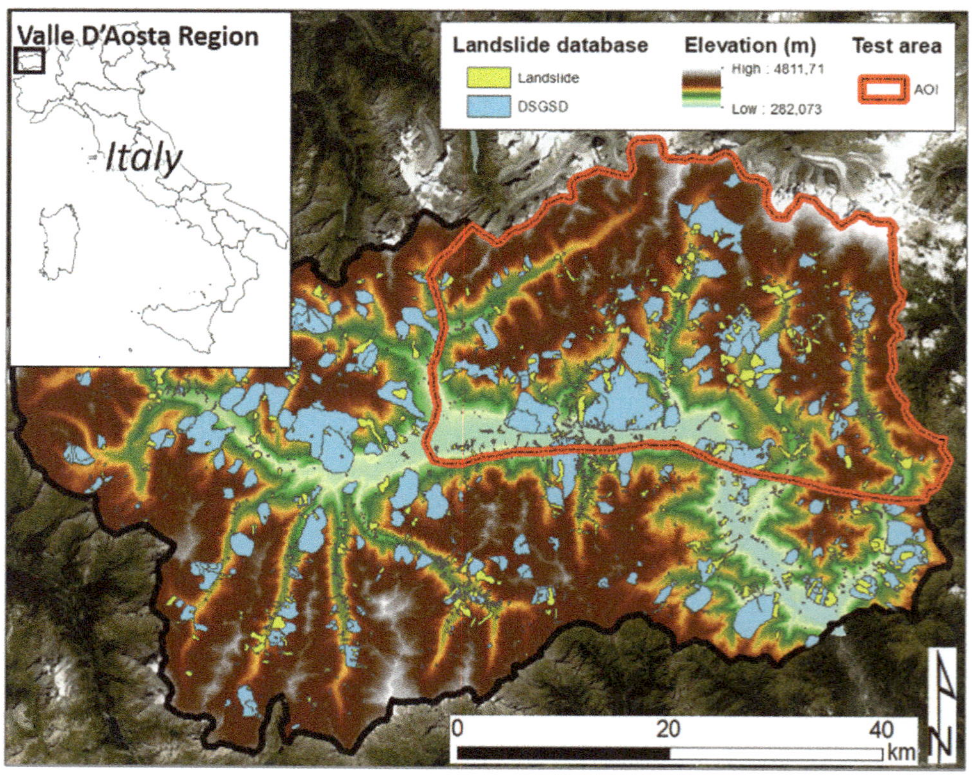

Fig. 1 Location of the Area Of Interest (AOI)

the period January 2015—August 2018 in descending orbit. Radar images were processed by means of a dedicated PSI approach (Barra et al. 2017). The spatial distribution of PSI LOS (Line Of Sight) mean yearly velocities is shown in Fig. 2a. Some wide moving areas related to deep-seated landslides are recognizable.

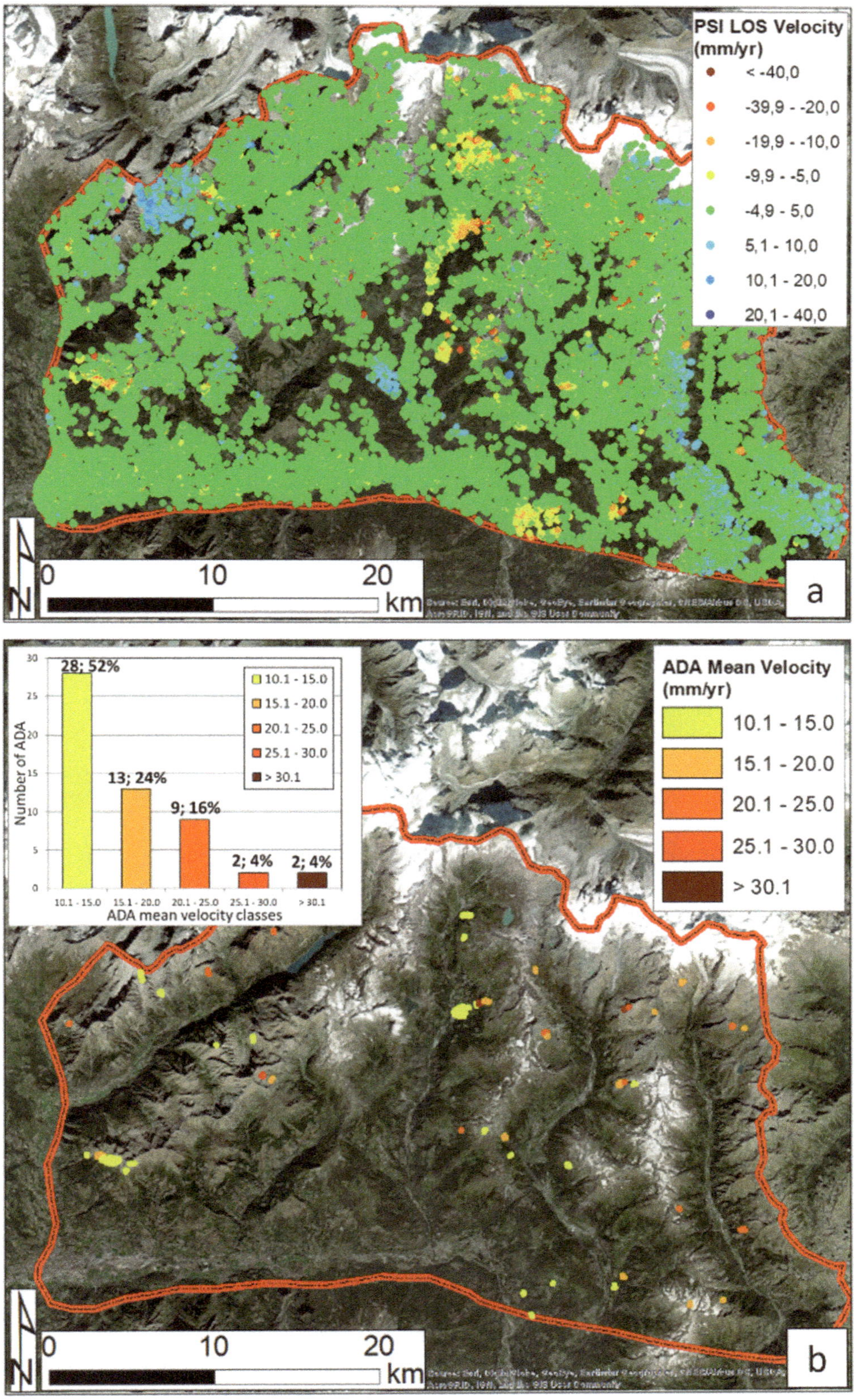

Fig. 2 **a** Sentinel-1 PSI descending data; **b** ADA database classified according to mean velocity

The stability threshold of the velocity map was fixed at ± 5 mm/year, value equal to the standard deviation value (σ) of the dataset.

The ADA were extracted from the PSI points population by following a semi-automatic procedure already proposed and tested by Barra et al. (2018) and Solari et al. (2018).

The ADA are clusters of moving PS selected on the basis of a velocity threshold (set as 2σ where σ is the standard deviation value of the LOS velocity of the PSI dataset) and of simple spatial aggregating parameters, i.e. minimum number of PS that an ADA must contain (set as 5 PSI), and clustering radius built around every moving PS (set as 28 m, which is almost two times Sentinel-1 ground resolution).

A total number of 54 ADA clusters were detected in the AOI. In Fig. 2b ADA are classified on the basis of mean velocity value, obtained by averaging the velocities of all the PS points that form each moving area. The ADA are well-distributed on the territory, mostly on W-facing slopes given the satellite descending geometry. The majority of ADA records average velocities higher than 15 mm/yr (52%), while few ones show velocities higher than 25 mm/yr (8%) (Fig. 2b).

Step-by-Step Methodology

The methodology aims to provide the potential worth of loss (D) due to possible mass movements on the territory starting from a regional scale deformation map obtained through PSI data and ADA-derived moving areas (Fig. 3).

The landslide intensity is assessed using ADA in a dual form depending on two conditions: i) if the ADA directly overlaps an urban area (intersection with one or more buildings/roads), then it is taken as direct estimation of landslide magnitude and the landslide intensity depends on the average velocity of the ADA (ADA-related intensity); ii) if the ADA does not intersect any urban fabric, but it is found within a possible unstable debris area with presence of detrital cover, then the ADA is taken as possible source area of an unstable debris mass that could evolve into a mass movement (i.e. debris flow). In this second case, a run-out basin-scale model -Gravitational Process Path, GPP (Wichmann, 2017)—is used to evaluate the possible landslide evolution, in terms of landslide path, spatial distribution and thickness of the accumulation zone and the intensity is derived from GPP outputs (Model-related intensity). GPP is available as tool of the open source system SAGA GIS.

Once we have derived vulnerability (V), which depends on intensity and elements at risk (EAR) typology, and exposure (E), determined by considering the cadastral database and market monetary value, the potential worth of loss (D) is computed as product between V and E. The final output contains a preliminary quantitative evaluation of landslide impact and loss degree.

Results

Evaluation of Landslide Intensity

For the landslide intensity evaluation, ADA clusters were used as direct estimation of landslide magnitude ("ADA-related intensity") or as an effort to detect potential source areas of moving debris masses ("Model-related intensity").

(1) **ADA-related intensity**

This approach relies on the overlapping between ADA and urban area/single building/road (Solari et al. 2020). Landslide intensity is here conceived as direct expression of landslide velocity (derived from LOS deformation rates of the PSI data) and thus defined by the mean velocity of the ADA, following the classification:

- intensity 1: average velocity lower than 16 mm/yr;
- intensity 2: average velocity between 16 and 32 mm/yr;
- intensity 3: average velocity higher than 32 mm/yr.

The first threshold (16 mm/yr) is representative for the passage between extremely slow and very slow landslides, as assumed by Cruden and Varnes (1996), while the second one (32 mm/yr) is obtained by doubling up the first value.

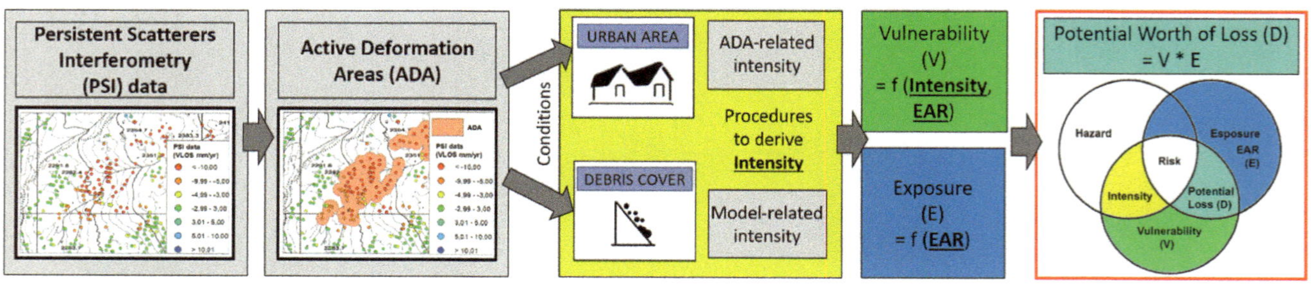

Fig. 3 Flowchart of the procedure for deriving intensity, EAR exposure and potential worth of loss

(2) **Model-related intensity**

In the Model-related approach the ADA are both a proxy for ground moving areas and an input for the GPP model, acting as the main factor for the source areas definition. We assumed that a debris mass in correspondence of an ADA, already in motion (as testified by the PS data), could evolve into a debris flow if external triggering factors are present (i.e. an intense rainfall period). Therefore, if an ADA is found within a possible unstable debris area, defined on the basis of geological analysis and orthophoto interpretation, the GPP run-out model is used to evaluate the possible motion along the slope, spatial distribution and thickness of the accumulation zone.

A 2 m DEM was used as input for the model. The source areas were defined within each ADA and selected considering the distribution of moving points and the local morphology. For soil/debris thickness, we used the results derived within the area of interest by Salvatici et al. (2018) using the Geomorphologically Indexed Soil Thickness model (Catani et al. 2010). The height of material deposited at every cell was used as a proxy for landslide intensity. The three intensity classes are:

- intensity 1: height of material lower than 1.25 m;
- intensity 2: height of material between 1.26 and 2 m;
- intensity 3: height of material higher than 2.01 m.

The values were chosen by following the vulnerability functions derived by Papathoma-Kohle et al. (2012) considering debris flow events occurred in South Tyrol in Austria. The three intensity classes are equal to a loss degree of 30%, 60% and higher than 60%, respectively.

Evaluation of Vulnerability and Exposure of the Elements at Risk

We defined Vulnerability (V) as a parameter that varies between 0 (no damage) and 1 (complete loss). V is a function of landslide intensity. Each value of vulnerability was defined on the type of element at risk by following a data driven approach and considering the possible interactions between elements and landslide. For deriving the polygons/lines of EAR, a cadastral map at 1:2000 scale of the whole Valle D'Aosta Region was used.

The values of vulnerability assigned to each EAR type were chosen by considering previous literature works at similar scale (Catani et al. 2005; Bianchini et al. 2017) and they are reported in Table 1.

The exposure (E) is a monetary value that was assigned to each EAR type by extracting information from various bibliographic resources (market value, construction cost, renovation cost). For private houses, E (€/m^2) was determined by following the market value, as defined in the OMI

Table 1 Vulnerability and exposure values for the area of interest: V, vulnerability; I, intensity; E, exposure

EARtype	V for I = 1	V for I = 2	V for I = 3	Exposure E (€/m^2)
Barn	0.2	0.4	0.6	80
Camping	0.4	0.6	0.8	2600
Greenhouse	0.2	0.4	0.6	50
Hotel	0.15	0.3	0.5	1550–4600
Industry	0.1	0.2	0.5	740–1000
Local road	0.6	0.8	1	50
Motorway	0.4	0.6	0.8	350
Municipal road	0.6	0.8	1	150
Office/service	0.1	0.3	0.6	1175–2600
Private house	0.2	0.35	0.6	1075–4350
Provincial road	0.6	0.8	1	180
School complex	0.3	0.5	0.7	4000
Shed	0.2	0.4	0.6	540–890
Restaurant	0.2	0.35	0.6	865–2100
Shopping mall	0.2	0.35	0.6	1400–2200
Sport facilities	0.3	0.5	0.7	15–120
Stable	0.15	0.4	0.6	120
Regional road	0.4	0.6	0.8	250
Warehouse	0.2	0.4	0.6	680–1000

(Osservatorio del Mercato Immobiliare—Real estate market observatory) managed by the Italian revenue agency (OMI 2018). In Table 1 the range values of E refer to the minimum and maximum value among the 35 municipalities for the AOI, while single value are unique values that are valid for all the 35 municipalities. For road categories, the E value was assigned on the basis of the construction/renovation cost for square meter.

The values of vulnerability of Table 1 are common for both the ADA-related and Model-related approaches used to evaluate the 3 landslide intensity classes I.

Evaluation of Potential Worth of Loss

The potential worth of loss (D) is function of the elements at risk and vulnerability for a given intensity hazardous event. We computed D in quantitative terms (Euros for square meters) deriving it as the product between vulnerability and exposure. If one or both vulnerability or exposure are null, the potential loss is zero (Catani et al. 2005).

In Fig. 4 an example of application of the ADA-related approach is shown. The site is located in the Valtournenche municipality (Chaloz hamlet) along a west-facing slope in which several landslides are already known and mapped (Fig. 4a). In particular, a large DSGSD involves the entire flank from 2900 m a.s.l. to 1500 m a.s.l. altitude (Giordan et al. 2017). Within the lower part of the deep-seated landslide, four ADAs are detected with average LOS velocities between 11 and 16 mm/year, resulting in the intensity class 1. Several hundred buildings (especially private houses warehouses and hotels) as well as some local and provincial roads are located within ADA boundaries and thus exposed to potential landslide risk.

Considering the intensity class and the type of EAR, the vulnerability ranges 0.1–0.2 for buildings and ranges 0.4–0.6 for roads. After assigning E values, the potential worth of loss varies between 21 and 490 €/m^2 (Fig. 4b).

In Fig. 5 an example of Model-related application is shown. The site is located in the Montjovet municipality (Tron hamlet). One DSSGD and two complex landslides are mapped in the landslide inventory of the area. One ADA is detected in the upper part of the complex landslides, where debris deposits are present.

Starting from two source areas (traced in dashed lines and labelled as S1 and S2 in Fig. 5) within the ADA, the GPP model simulated the run-out of possible debris flows on the slope. The input thickness of the material deposits were 2-2.4 m and 0.6-1 m for S1 and S2 areas, respectively. The flows originated from these two source areas converge and create a fan-like deposit with variable thickness between 0.2 and 1.6 m, thus the resulting intensity class is 1. Possible simulated mass movements intersect two local roads as shown in Fig. 5b.

In this case, by considering an exposure value of 50 €/m^2 and a vulnerability of 0.6, the output potential damage and worth of loss is 30 €/m^2 (Fig. 5b).

Discussion and Conclusions

In this work, we proposed a procedure which starts from Sentinel-1 PSI data and leads to the estimation of potential loss experienced by one or more elements at risk (building and roads) on areas with a certain level of landslide intensity.

Given the working scale (i.e. regional, basin scale) some approximations needed to be made. The maximum level of detail about economic exposure value of buildings is obtainable from the real estate market. No further information regarding people occupancy or residency type (e.g. inhabited all-year round or rented for holiday) is collectable at this scale in a reasonable time. Furthermore, the use of the ADA database is practical and useful for rapid and semi-automatic detection of the areas characterized by the highest ground motions rates at wide scale where attention must be focused. Nevertheless, the exploitation of ADA must be carefully carried out: the ADA interpretation permits to outline the dynamics and morphological setting of a site, but it does not specifically include the recognition of landslide type or the real spatial extension of the phenomenon.

In conclusion, the outcomes of this work provide preliminary space-borne indications of the most moving areas at regional scale. This information is useful for Civil Protection entities and authorities in charge of risk management to easily know which are the areas that are moving faster, and which is their potential impact on built-up areas.

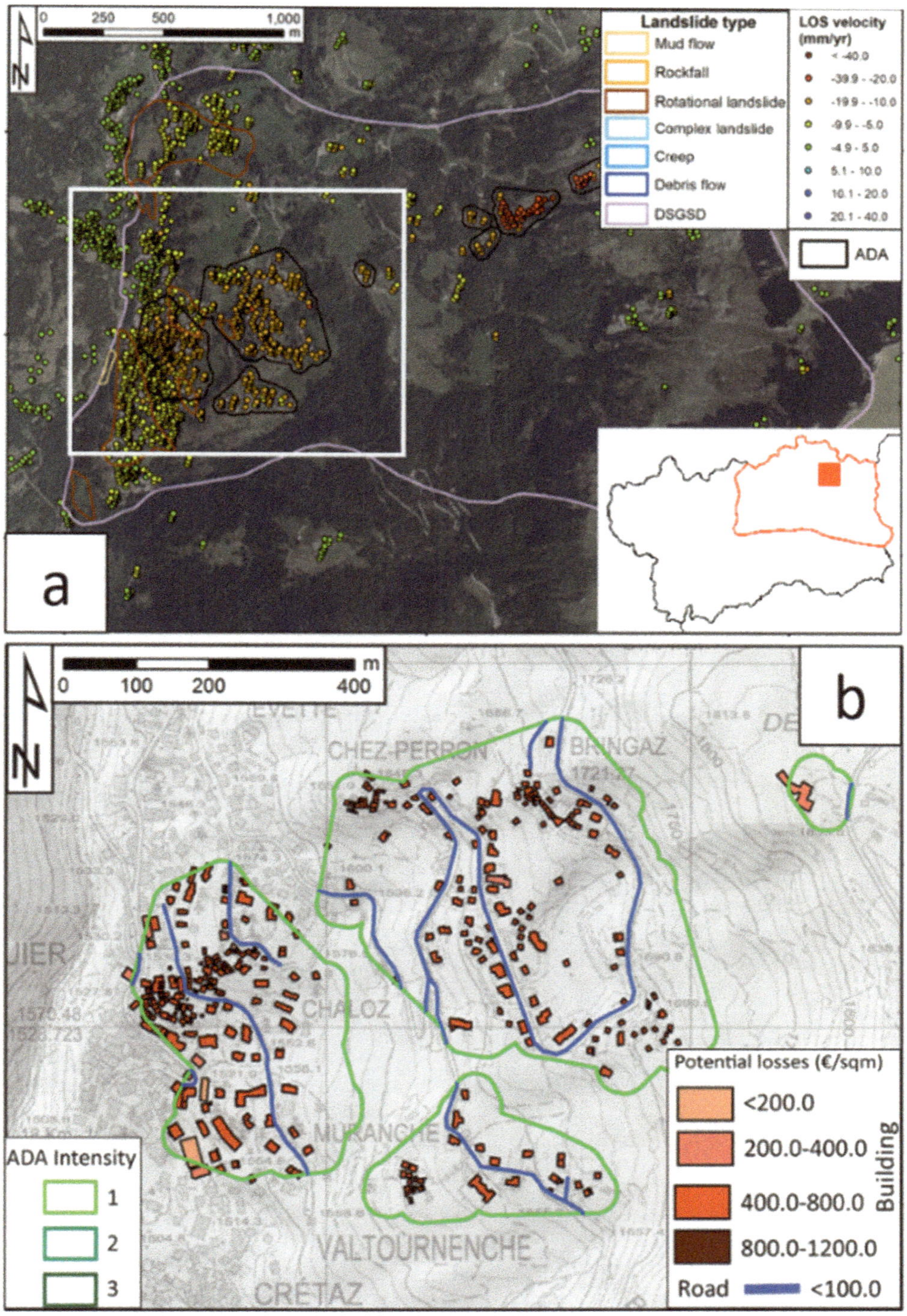

Fig. 4 **a** PSI data and landslide database; **b** Involved EAR classified in potential loss on Chaloz hamlet

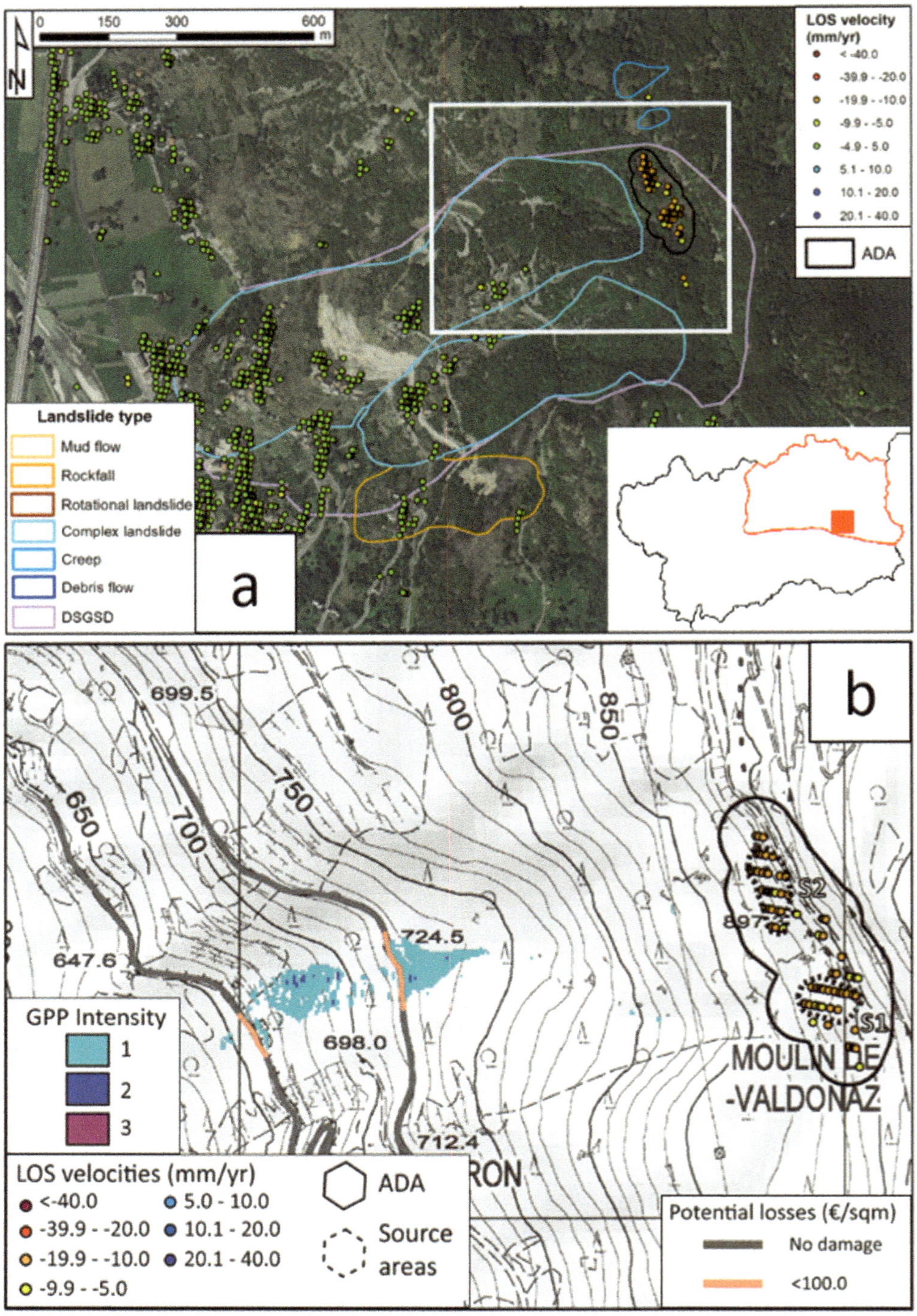

Fig. 5 **a** PSI data and landslide database; **b** Involved EAR classified in potential loss on Tron hamlet

Acknowledgements The work has been conceived and developed in the framework of the "U-Geohaz—Geohazard impact assessment for urban areas" project, co-funded by the European Commission, Directorate-General Humanitarian Aid and Civil Protection (ECHO).

References

Barra A, Solari L, Béjar-Pizarro M, Monserrat O, Bianchini S, Herrera G et al (2017) A methodology to detect and update active deformation areas based on sentinel-1 SAR images. Remote Sens 9 (10):1002

Bianchini S, Solari L, Casagli N (2017) A gis-based procedure for landslide intensity evaluation and specific risk analysis supported by persistent scatterers interferometry (PSI). Remote Sens 9(11):1093

Catani F, Casagli N, Ermini L, Righini G, Menduni G (2005) Landslide hazard and risk mapping at catchment scale in the Arno River basin. Landslides 2(4):329–342

Catani F, Segoni S, Falorni G (2010) An empirical geomorphology-based approach to the spatial prediction of soil thickness at catchment scale. Water Resour Res 46(5)

Corominas J, van Westen C, Frattini P, Cascini L, Malet JP et al (2014) Recommendations for the quantitative analysis of landslide risk. Bull Eng Geol Env 773(2):209–263

Cruden DM, Varnes DJ (1996) Landslides: investigation and mitigation. Landslide types and processes. Transportation research board special report 247

Dai FC, Lee CF, Ngai YY (2002) Landslide risk assessment and management: an overview. Eng Geol 64(1):65–87

Giordan D, Cignetti M, Bertolo D (2017) The use of morpho-structural domains for the characterization of deep-seated gravitational slope deformations in Valle d'Aosta. Springer, Cham, pp 59–68

Liu X, Miao C (2018) Large-scale assessment of landslide hazard, vulnerability and risk in China. Geomatics Nat Hazards Risk 9 (1):1037–1052

OMI database (2018). Agency of Revenues https://wwwt.agenziaentrate.gov.it

Papathoma-Köhle M, Totschnig R, Keiler M, Glade T (2012) A new vulnerability function for debris flow-The importance of physical vulnerability assessment in alpine areas. Internationale Forschungsgesellschaft. pp 1033–1043

Ratto S, Bonetto F, Comoglio C (2003) The October 2000 flooding in Valle d'Aosta (Italy): event description and land planning measures for the risk mitigation. Int J River Basin Manage 1:105–116

Salvatici T, Tofani V, Rossi G, D'Ambrosio M, Stefanelli CT et al (2018) Application of a physically based model to forecast shallow landslides at a regional scale. Nat Hazards Earth Syst Sci 18 (7):1919–1935

Solari L, Barra A, Herrera G, Bianchini S, Monserrat O et al (2018) Fast detection of ground motions on vulnerable elements using Sentinel-1 InSAR data. Geomatics Nat Hazards Risk 9(1):152–174

Solari L, Bianchini S, Franceschini R, Barra A, Monserrat O et al (2020) Satellite interferometric data for landslide intensity evaluation in mountainous regions. Int J Appl Earth Obs Geoinf 87:1–16

Trigila A, Iadanza C, Spizzichino D (2010) Quality assessment of the Italian Landslide Inventory using GIS processing. Landslides 7 (4):455–470

Van Westen CJ, Van Asch TW, Soeters R (2006) Landslide hazard and risk zonation—why is it still so difficult? Bull Eng Geol Environ 65 (2):167–184

Wichmann V (2017) The gravitational process path (GPP) model-GIS-based simulation framework for gravitational processes. Geosci Model Dev 10(9)

On the Use of UAVs for Landslide Exposure of Households: La Gloria Neighbourhood, Teziutlán, Puebla

2020 Kyoto Japan

Ricardo J. Garnica-Peña, Galdino García-Marroquin, and Irasema Alcántara-Ayala

Abstract

Owing to landslide susceptibility of pyroclastic deposits, coupled with high vulnerability and exposure, the Sierra Norte de Puebla region, in Mexico, is characterised by significant levels of landslide disaster risk. This study aimed at evaluating landslide exposure of households in La Gloria neighbourhood, Teziutlán, Puebla, an area that was severely affected by the disaster of October 1999. The methodological approach comprised three stages: (1) definition of the parameters used for developing the landslide exposure index; (2) aerial survey using a drone; and, (3) elaboration of the landslide exposure map. According to the produced landslide exposure map, 57% of the households are in areas of high level of exposure, whereas 32 and 11% are in moderate and low exposure zones, correspondingly.

Keywords

Landslide • Exposure • UAV • Households • Disaster risk • Teziutlán

Introduction

In Mexico, disasters associated with rainfall induced landsliding are of major relevance due to high vulnerability levels of population and exposure. In October 1999, hundreds of rainfall-induced landslides affected considerably the municipality of Teziutlán, in the Sierra Norte de Puebla (SNP). Damages were severe in varies neighbourhoods, among them, La Gloria, situated in the urban area of the municipality capital (Fig. 1). Although no fatal victims were registered, considerable effects on housing were accounted for. Owing to high levels of vulnerability, exposure and landslide susceptibility, the potential occurrence of disasters in Teziutlán is evident.

Different cartographic tools, and scientific and technological developments have been used for hazard characterisation and disaster risk assessment, including instrumentation, monitoring and remote sensing. As such, Unmanned Aerial Vehicles (UAVs) have long been used in various fields to gain knowledge of territories; recently their application has been also directed to landslide disaster risk research, risk reduction and disaster response based on both social and physical approaches (Niethammer et al. 2012; Erenoglu 2016; Thiebes et al. 2016; Casagli et al. 2017; Peppa et al. 2017; Chen et al. 2018; Busa et al. 2019).

Along this vein, this study aimed at analysing landslide exposure of households comprised within La Gloria neighbourhood, at local scale, by using UAVs or drones. This included field characterisation and the generation of high-accuracy cartography.

Teziutlán, Puebla: General Context

The municipality of Teziutlán is in the northeast of the state of Puebla in The Sierra Norte de Puebla (SNP), a region where the physiography of the country combines two main provinces; the Volcanic Trans-Mexican Belt and the Sierra Madre Oriental (Fig. 1). It comprises 32 localities and in 2015 had a population of 97,590 inhabitants. Population density was estimated as 1,054 inhabitants per km^2 (INEGI 2015a).

R. J. Garnica-Peña (✉) · I. Alcántara-Ayala
National Autonomous University of Mexico (UNAM), Institute of Geography, Mexico City, 04510, Mexico
e-mail: garnica@igg.unam.mx

I. Alcántara-Ayala
e-mail: ialcantara@igg.unam.mx

G. García-Marroquin
Faculty of Philosophy and Letters, National Autonomous University of Mexico (UNAM), Mexico City, 04510, Mexico
e-mail: garyrocker32@hotmail.com

F. Guzzetti et al. (eds.), *Understanding and Reducing Landslide Disaster Risk*, ICL Contribution to Landslide Disaster Risk Reduction, https://doi.org/10.1007/978-3-030-60227-7_53

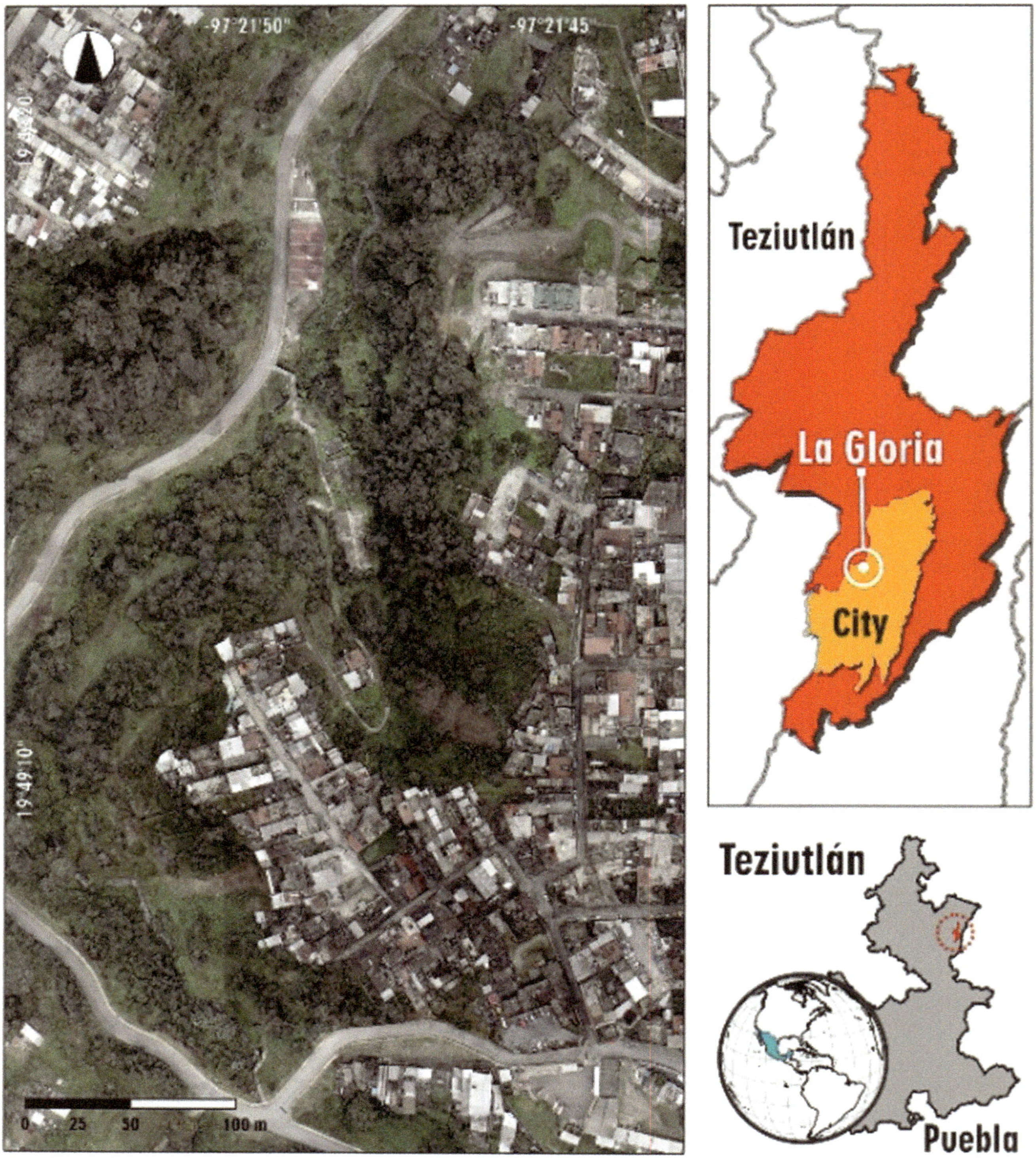

Fig. 1 Location of La Gloria neighbourhood, Teziutlán

The SNP comprises high mountain ranges and steep hills with variable altitude ranging from 1000 to 3000 ma.s.l. (Alcántara-Ayala et al. 2017). In Teziutlán there are sedimentary and metamorphic rocks of different ages; the oldest metamorphic rocks are shales of the Chilis formation (SGM 2011). However, it is the Quaternary volcanic material, particularly the presence of pyroclastic deposits generated from the Caldera de Los Humeros, overlaying the sedimentary and metamorphic rocks, the most important lithological unit highly susceptible to landsliding (Fig. 2).

The municipality is in a transition zone from the temperate climates of the Sierra Norte to the warmer areas of the

Fig. 2 Aerial view of La Gloria neighbourhood, Teziutlán

plain of the Gulf of Mexico. There are three types of climate. In the northern part, the semi-warm humid climate with rains all year round covers most of the area (INEGI 2015b). The central portion is characterised by humid temperate climate with rainfall all year around, while the southern part, humid temperate climate with precipitation in summer, covers a small portion of the territory (García 1998).

Landslide Exposure: Methodology

According to the UNISDR (2017), exposure is the situation of people, infrastructure, housing, production capacities and other tangible human assets located in hazard-prone areas". In the case of landslides, exposure is strongly linked not only to areas susceptible to landslides, but to housing conditions. Garnica-Peña and Alcántara-Ayala (2017) mention that urban growth and lack of planning are among the main drivers of landslide exposure clearly expressed in the unplanned household spatial distribution and existing conditions in Francia neighbourhood (Teziutlán city) and these two factors also are relevant in the case of La Gloria neighbourhood (Fig. 3).

The involved methodology included three main steps: (1) definition of the parameters used to elaborate the exposure index applied for housing conditions in La Gloria neighbourhood; (2) aerial survey using UAVs; and (3) the development of the landslide exposure map.

Definition of the parameters

Parameters used to prepare the exposure index included information on quality and materials of the households (walls, floor and ceiling), conditions of maintenance of buildings and access to basic services (drinking water, electricity, sewerage). Field evaluation was carried out by using Mapit GIS Pro. This allowed high-accuracy GPS records in addition to the corresponding information about characterization of the 298 households comprised within the neighbourhood to be included on a database, on which also photographs were integrated. Evaluation was carried out in the field on house to house basis.

Aerial survey using UAVs

The aerial survey using a DJI Phantom 4 Pro drone included flight route cabinet planning and aerial survey and image

Fig. 3 Poor housing and unmanaged settlement construction on unstable slopes in La Gloria neighbourhood increase exposure conditions to landslides and other hazards

processing. Flight route planning was done using the MapPilot).

Flight routes were planned with North-South orientation, and East-West with a total of 96.2 hectares of coverage at a height of 150 m. This involved 791 images with a resolution of 4.1 cm. The overlay used was 80 × 80 between each photograph (Fig. 4).

In order to get an optimum coverage of the area, it was necessary to implement, in addition to the E-O and N-S routes, a specific project for the oblique shots of the surface (Table 1. This project was divided into 3 sub-projects of 2 levels each. Oblique photographs were taken by using the FVP Camera (Fig. 4).

Images obtained from the UAV aerial survey were processed using the ContextCapture software. The program allowed to assemble the 3D model of the study area. Subsequently, a Raster mosaic, a Digital Elevation Model (MDE) and the map of landslide exposure indicators were produced in ArcGis.

Landslide exposure map

The digital surface model and point clouds derived from the aerial survey were used to create a digital terrain model (DTM) by removing buildings and any other element different from the terrain. This procedure was carried out using the LAS dataset tools of the ESRI ArcGIS 10.6 program. The DTM was afterwards utilised to calculate the topographic wetness index (TWI), shape of the slopes (concave, straight and convex), and the slope of the terrain (slope). This information was combined with the inventory and landslides susceptibility map of the municipality of Teziutlán prepared by Murillo-García and Alcántara-Ayala (2017) and Murillo-García et al. (2019), respectively. Data concerning field evaluation of the households was also included to develop a landslide exposure index. This included six criteria: level of landslide susceptibility, topographic wetness index, hillslope shape, proximity to existing landslides, number of built household storeys, as well as the condition or level of deterioration of the facade.

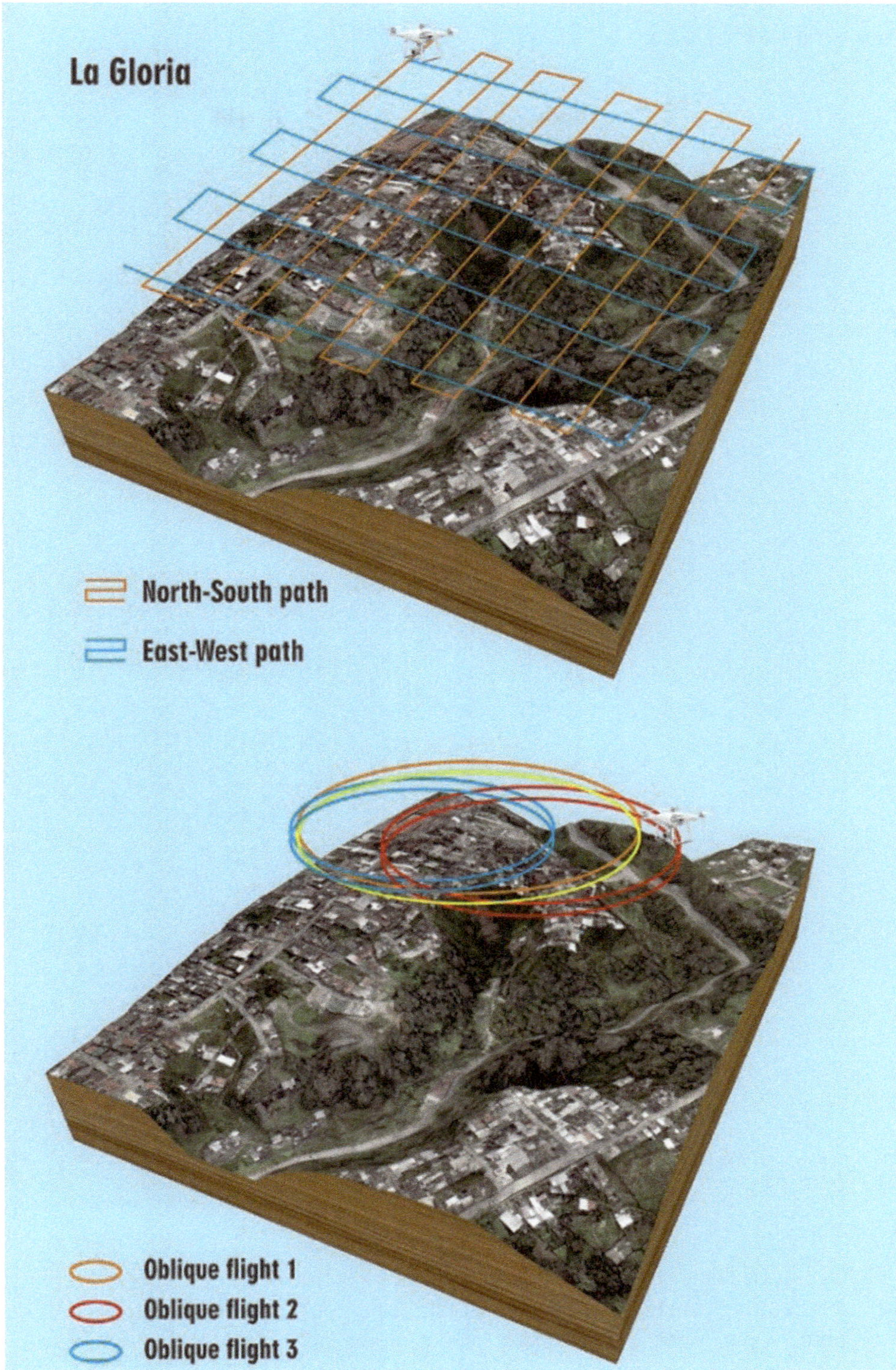

Fig. 4 Grid and oblique surveys

Table 1 Characteristics of the scheduled flights

Criteria	Oblique flight 1		Oblique flight 2		Oblique flight 3	
Flight height above ground level (m)	**50**	**75**	**50**	**80**	**50**	**80**
Radius (m)	**156**	**156**	**130**	**130**	**120**	**120**
Images	**80**	**80**	**80**	**80**	**80**	**80**
Angle (°)	**15**	**24**	**21**	**32**	**22**	**33**

Fig. 5 Landslide exposure in La Gloria neighbourhood

Before the elaboration of the final landslide exposure map, several weighting tests were carried out. Landslide susceptibility, proximity to landslides and number of household storeys were among the most significant factors (Figs. 5 and 6). Three classes were established for the landslide exposure map; low, moderate and high.

Results

The predominant land use in La Gloria neighbourhood is of housing type (96%) and the remaining 4% housing with commerce. In terms of the characterisation of existing dwellings, 64% of households have between two and three

Fig. 6 Households in La Gloria neighbourhood

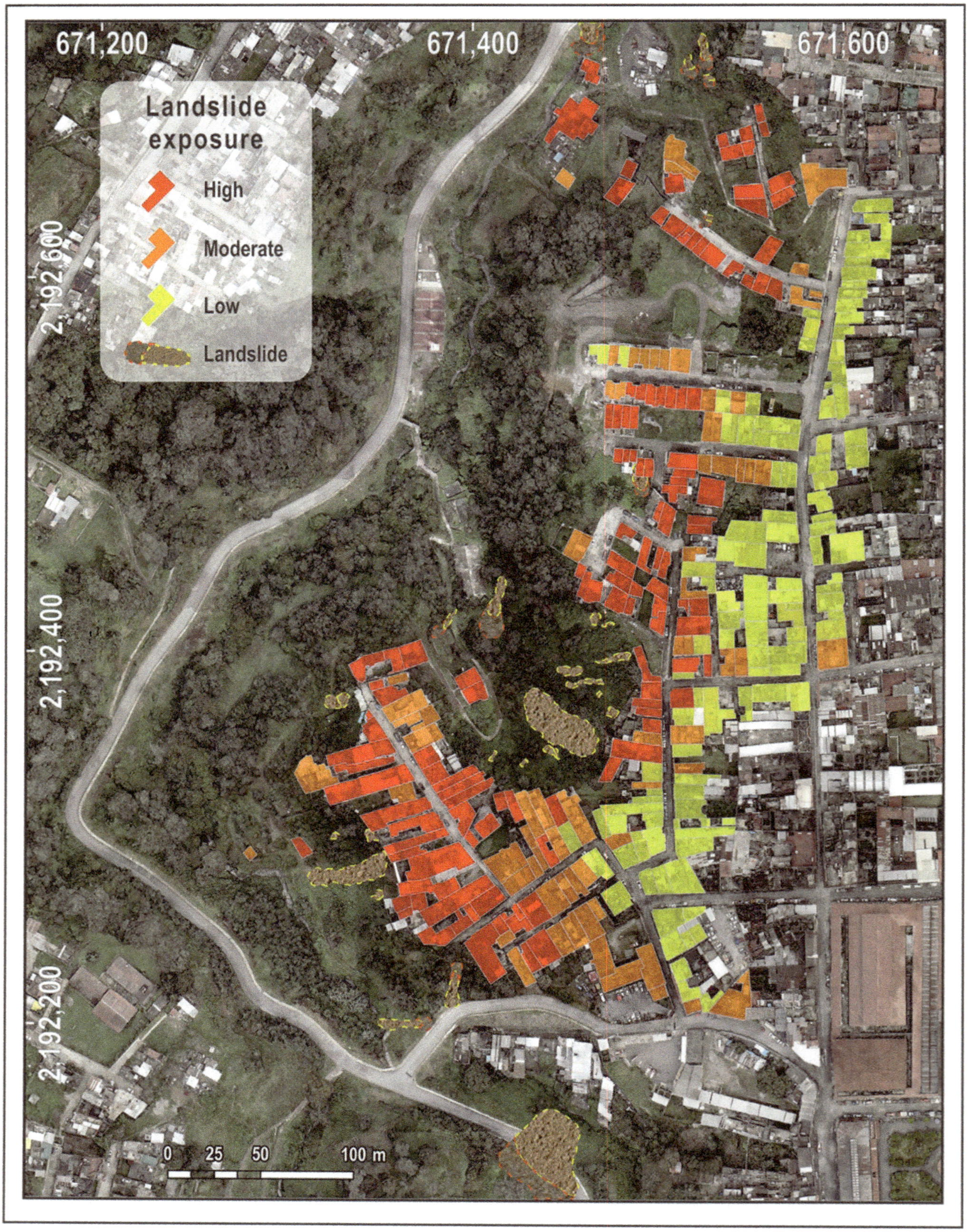

Fig. 7 Landslide exposure map of La Gloria neighbourhood, Teziutlán, Puebla, México

storeys built, and 53% of the total properties possess a good building structural condition, 45% are characterised by moderate condition, while the outstanding 2% of the houses is very deteriorated.

According to the produced map, 57% of the households are in areas of high level of exposure, whereas 32% and 11% are in moderate and low exposure zones, respectively. Those considered as of low exposure are near downtown Teziutlán (Fig. 7). Owing to lack of population data at local scale, it was not possible to estimate how many people live in those areas exposed to landslides. Nonetheless, raising landslide disaster risk awareness is relevant for all inhabitants of the neighbourhood. As for the proximity to landslides, it was possible to estimate that 42% of households is located within 0–25 m from landslides, followed by 37.5% of houses situated 25–50 m away, and the remaining 19.79% where the distance is more than 50 m.

Concluding Remarks

Owing to the social and physical context of the municipality, landslide disaster risk in Teziutlán will continue to be of major concerns in the years to come. Disaster reduction would require efforts to decrease vulnerability and exposure. However, twenty years after the disaster of 1999, there has not been a sound transformation regarding policy making and practice in DRR. As such, scientific studies become more valuable as they intend to rasise awareness and preparedness of the population and local authorities.

The performed evaluation expresses clearly the positive aspects with respect to the use of UAVs for landslide exposure mapping. UAVs can be regarded as an accessible alternative, effective and low-cost tool that allows the analysis of exposure conditions of dwellings.

Urban growth, unplanned land-use and lack of territorial management in an area highly susceptible to landslides, are among the main drivers of landslide exposure in La Gloria neighbourhood.

Consequently, there is an evident need to delineate and implement regulation frameworks regarding the use of the land, especially in areas with greater exposure. This ought to contemplate that real estate companies should provide hazard studies and risk assessments to demonstrate viability of future constructions so that no more inhabitants are placed at risk. What is more, strong relationships between Civil Protection and land-use authorities must be enhanced, from an integrated disaster risk management (IDRiM) perspective.

It is considered necessary to carry out this type of studies in other localities of Teziutlán, both urban and rural, where population growth results in the occupation of areas susceptible to landslides derived from regional or local migration. Nonetheless, what also remains as a challenge is the availability of economic and human resources, in the understanding that in addition to financial support to develop strategic initiatives, disaster risk management needs to be carried out by people with the appropriate skills and committed to disaster risk reduction.

The participation of local communities becomes also quite relevant not only for coordinated response, but in terms of their collaboration in local initiatives on which their necessities are reflected and the constrains they face are considered. Landslide disaster risk awareness and preparedness should also need to be sustain through time to guarantee efficient efforts to reduce disasters.

Acknowledgements Thanks are due to members of La Gloria neighbourhood and authorities from the Civil Protection Unit of Teziutlán. Financial support was kindly provided by UNAM-DGAPA, through the project PAPIIT IN300818.

References

Alcántara-Ayala I, Garnica-Peña RJ, Coll-Hurtado A, Gutiérrez de MacGregor MT (Coords.) (2017) Inestabilidad de laderas en Teziutlán, Puebla. Factores inductores del riesgo de desastres, Instituto de Geografía, Universidad Nacional Autónoma de México, p 223. In Spanish

Busa J, Rusnak M, Kusnirak D, Greif V, Bednarik M, Putiska R, Dostal I, Sladek J, Rusnakova D (2019) Urban landslide monitoring by combined use of multiple methodologies-a case study on Sv. Anton town, Slovakia. Physical Geography. https://doi.org/10.1080/02723646.2019.1630232

Casagli N, Frodella W, Morelli S, Tofani V, Ciampalini A, Intrieri E, Raspini F, Rossi G, Tanteri L, Lu P (2017) Spaceborne, UAV and ground-based remote sensing techniques for landslide mapping, monitoring and early warning. Geoenviron Disasters 4(1). https://doi.org/10.1186/s40677-017-0073-1

Chen ML, Lv PF, Zhang SL, Chen XZ, Zhou JW (2018) Time evolution and spatial accumulation of progressive failure for Xinhua slope in the Dagangshan reservoir, Southwest China. Landslides 15 (3):565–580. https://doi.org/10.1007/s10346-018-0946-8

Erenoglu RC (2016) Generating a digital elevation model using unmanned aerial system for a deep-seated rotational landslide on forest cover and vegetation. Austrian J Forest Sci 133(1):47–61

García E (1998) Climas de México. Comisión Nacional para el Conocimiento y Uso de la Biodiversidad (CONABIO). In Spanish

Garnica-Peña RJ, Alcántara-Ayala I (2017) Landslide synchronic evaluation by using unmanned aerial vehicles (UAV): some insights on disaster risk in Teziutlán, Puebla, México. In: Mikos M, Tiwari B, Yin Y, Sassa K (eds) Advancing Culture of Living with Landslides, vol 2. Advances in Landslide Science, 209–218, Springer. https://doi.org/10.1007/978-3-319-53498-5_24

INEGI (2015a) Encuesta Intercensal 2015. http://www.inegi.org.mx

INEGI (2015b) Unidades climatológicas (Climatologic units) http://www.inegi.org.mx/geo/ contenidos/recnat/clima/infoescala.aspx. In Spanish

Murillo-García F, Alcántara-Ayala I (2017) Landslide inventory, Teziutlán municipality, Puebla, México (1942–2015). J Maps 13 (2):767–776

Murillo-García F, Steger S, Alcántara-Ayala I (2019) Landslide susceptibility: a statistically based assessment on a depositional pyroclastic ramp. J Mt Sci 16(3):561–580

Niethammer U, James MR, Rothmund S, Travelletti J, Joswig M (2012) UAV-based remote sensing of the Super-Sauze landslide: Evaluation and results. Eng Geol 128:2–11. https://doi.org/10.1016/j.enggeo.2011.03.012

Peppa MV, Mills JP, Moore P, Miller PE, Chambers JE (2017) Brief communication: landslide motion from cross correlation of UAV-derived morphological attributes. Natural Hazards and Earth System Sciences 17(12):2143–2150. https://doi.org/10.5194/nhess-17-2143-2017

SGM (2011) Carta Geológico-Minera: Teziutlán E14-B15 (Geological-Mining map: Teziutlán E14-B15). Servicio Geológico Mexicano, Spanish

Thiebes B, Tomelleri E, Mejia-Aguilar A, Rabanser M, Schlogel R, Mulas M, Corsini A (2016) Assessment of the 2006 to 2015 Corvara landslide evolution using a UAV-derived DSM and orthophoto. Landslides Eng Slopes: Experience, Theory Pract 1–3:1897–1902

UNISDR (2017) Report of the open-ended intergovernmental expert working group on indicators and terminology relating to disaster risk reduction. UNISDR, Geneva

Ordinal Logistic Regression to Automatic Classify Shallow Landslide Risk Level in Sao Paulo City, Brazil

2020 Kyoto Japan

Erica Akemi Goto and Keith C. Clarke

Abstract

Landslides are natural events that can cause disasters and catastrophes and threaten human lives and assets. Moreover, landslide disasters are more common in developing countries than developed countries. In these countries, more people live in risk-prone areas. Disaster Risk Reduction measures are fundamental to mitigate the risk. In Brazil, public authorities have been mapping these risk areas with inventory-based assessment and classifies the risk into four levels. Pairs of technicians do field inventory and decide on the risk level based on their personal and technical experience. However, the risk areas need to be constantly updated and the inventory-based assessments are not standardized and so carry the technicians' subjectivity. In this study, we quantify the inventory-based methodology for shallow landslides by developing corresponding mathematical equations that can automatically classify the risk level based on input data regardless of the technicians' decision. Using a dataset from Sao Paulo city risk mapping and Ordinal Logistic Regression (OLR), we quantify the field methodology by first dividing the dataset into training and test sets, secondly creating equations using training set and accounting for ordinal-fashion, proportionality, and p-value, and lastly, we compared the predictive power of the equations using the test set. Based on the results, we concluded the model 2 has the best predictive power (0.73). In conclusion, this methodology, by more rapidly updating the sector risk levels, standardizes the methodology, diminishes the subjectivity of the mapping, and so can contribute to the prevention of future shallow landslides.

Keywords

Ordinal logistic regression • Landslide • Risk • Sao Paulo • Assessment

Introduction

Landslides are natural events that threaten human lives and assets. Landslide disasters and catastrophes are more common in developing countries, as also are deaths and loss of property (Alcántara-Ayala 2002). For example, in 1987, 1,000 people were killed due to the Reventador landslides in Ecuador; and in 1994, 271 people were killed and 1,700 were missing due to the Paez landslides in Colombia (USGS 2019). In these countries, a larger number of people live in at-risk areas due to the rapid increase of population, unplanned urbanization, and poverty.

Brazil is a developing country that has faced many landslide catastrophes. In the Southeastern states, these landslide events frequently occur during the rainy season (summer) and past landslides have resulted in major disasters, like those in Rio de Janeiro state in 2011 with 947 people killed (Dourado et al. 2012). Rio de Janeiro's landslide risk areas are a combination of the geography of the hazard events (landslides) and the vulnerability of the local residents. As a consequence of the residents' financial situation, they often occupied public lands, commonly in flood-plain or hillslope areas, and self-engineered the construction of their houses (Maricato 2008).

Neither the local nor the federal government can remove all the residents that live in at-risk areas in the short term. As a preventative measure, it is important to monitor these areas to anticipate and prevent human casualties. By mapping the

E. A. Goto (✉)
Department of Geography, UCSB, Ellison Hall 1723, Santa Barbara, 93106, USA
e-mail: ericagoto@gmail.com

K. C. Clarke
Department of Geography, UCSB, Santa Barbara, 93106, USA
e-mail: kcclarke@ucsb.edu

F. Guzzetti et al. (eds.), *Understanding and Reducing Landslide Disaster Risk*, ICL Contribution to Landslide Disaster Risk Reduction,
https://doi.org/10.1007/978-3-030-60227-7_54

landslide-prone areas and their risk levels, local governments can closely monitor the risk conditions and allocate resources or prioritize removals of residents living in high and very high risk level sectors. Moreover, these risk areas need continuous updates since they are constantly changing.

The Brazilian government has developed an inventory-based landslide risk mapping methodology to map landslide and river-bank erosion (Carvalho et al. 2007). In this study, we call this method the Brazilian Government Methodology (BGM). In the BGM, pairs of technicians do a field inventory and map the risk areas by observing local characteristics. Based on their evaluation and expertise, and the BGM recommendations, they decide upon the overall risk level. The methodology is not standardized and does not apply mathematical computation to the risk level classification, allowing for classification variability and subjectivity.

In this study, we propose using ordinal logistic regression (OLR) to quantify the BGM for shallow landslides. The methodology allows the mathematical computation of classifier weights and standardization of the risk level classification, diminishing the subjectivity of field technician decisions. The developed application can help municipalities update and validate their mapped risk areas.

Background

Risk Areas in Brazil

In Brazil, many people live in at-risk areas (IBGE 2018). In the southeast, major cities like Sao Paulo and Rio de Janeiro have 6 and 7% of the population living in these risk areas, respectively (IBGE 2018). Smaller cities also have a large number of people living in at-risk areas, such as Ribeirão das Neves (60.5%) and São João de Meriti (18.8%) (IGBE 2018). Knowing the location, the type of risk, and the level of risk is important to prevent landslide disasters. By mapping these areas, local municipalities can manage and monitor the risk sectors. Also, they can allocate financial resources to remove or prioritize reallocation of residents in the very high risk sectors.

Risk, in this study, is a combination of hazard, exposure, and vulnerability of the elements at risk (UN-SPIDER 2019). In developing countries, such as Brazil, the vulnerability of the residents contributes to the large number of people living in at-risk areas. Vulnerability, in the field of disaster reduction, is "the conditions determined by physical, social, economic and environmental factors or processes which increase the susceptibility of an individual, a community, assets or systems to the impacts of hazards" (United Nations 2015). These vulnerability conditions are related to the root causes: dynamic pressures, fragile livelihoods and unsafe locations (Marchezini and Wisner 2017).

Disaster risk reduction (DRR) plays an important role to prevent and mitigate disasters. Mapping landslide risk areas is part of DRR, and it contributes to knowledge of the type, location, and level of risk, which is fundamental to closely monitor risk, and to help local governments make decisions, such as about dwelling removal or building infrastructure to reduce the risk.

Landslide Assessment Methodologies

We can assess landslides with four main types of methodologies: (i) inventory-based; (ii) deterministic; (iii) heuristic; and (iv) statistical (Das 2011; Van Westen et al. 2006). The assessment can rely on one methodology or combine more than one. Decisions about which type of methodology to use is related to the data available, time, and the site scale. In our study, we are using a type of statistical methodology (ordinal logistic regression) based on the data available (landslide risk mapping by risk level in Sao Paulo city).

Ordinal Logistic Regression (OLR) is a statistical methodology that is frequently used in psychological, medical, and educational studies, in which the response variable (y or dependent variable) has an order and the explanatory variables (x or independent variables) can be discrete (categorical, numerical, or ordinal), and/or contiguous. Examples of ordinal response variables are restaurant satisfaction (low, moderate, high) or academic grades (F, E, D, C, B, A). OLR is not frequently used to assess landslides since the response variable commonly is not ordinal. However, the BGM classifies the landslide risk areas in an ordinal fashion with four risk levels (low, moderate, high, and very high), which makes OLR the most appropriate methodology for this study.

We used the MASS package in the open source R statistical software and the polr equation to estimate the ordinal logistic regression model. The method is also called the ordered logistic regression model, proportional odds model, cumulative link model, ordered logit model, and ordered probit model (Christensen 2018). Equation 1 illustrates polr () computation of the OLR.

$$\log it[P(y \leq j)] = \propto_j - \beta_1 \chi_1 - \ldots - \beta_n \chi_n, \quad j = 1, 2, \ldots, c-1 \tag{1}$$

(Agresti and Finlay 2009).

Method

The flowchart (Fig. 1) illustrates the steps of the methods used in this study. We used RStudio to do all the analysis in this section.

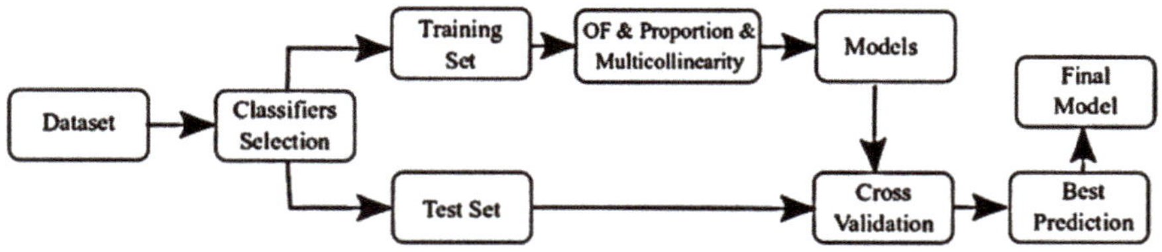

Fig. 1 Flowchart of the study

Dataset

The dataset is from a 2010 Sao Paulo landslide and river bank erosion mapping in which the BGM was used. The mapping was done by the Technological Institute of Research (IPT) and Sao Paulo Municipality during 2009 and 2010 and covered the city of Sao Paulo by sectors. It mapped risk level of the sector considering any type of landslides and/or river bank erosion. We selected those sectors with shallow landslides and excluded all the other sectors. The original dataset has 1,789 risk sectors and it was reduced to 754 sectors. The new dataset was divided into two: 70% training (N = 530) and 30% test (N = 224) sets. We used the training set to create the OLR models and the test set to predict the model's performance (Fig. 2).

Classifiers Selection and Model

To select the classifiers for the model we followed three main steps. First, we divided all the possible classifiers into 4 classes: not relevant, relevant but not consistent, relevant, and outcome. We reduced our 97 initial variables to 32. Secondly, we tested the 32 variables for their ordinal fashion by plotting each possible variable with the risk levels. Based on this test, we divided the variables into three groups: ordinal fashion (OF), moderate ordinal fashion (MOF), and non-ordinal fashion (NOF). Figure 2 illustrates the test with 8 variables, where conc_rainfall is OF, ground_veg is MOF, and deforestation is NOF.

Third, we checked for proportion across the options of the explanatory predictor variables. We divided the variables into two categories: good proportion (GP) and bad proportion (BP). Choices (i.e. True or False) of explanatory predictors with less than 5% in at least half of the choices is classified as BP. GP, on the other hand, has at least 5% of what is expected by each choice. Based on ordinal fashion and proportion, we tested two equations and used p-value to reduce the number of variables (Table 1).

Comparison of Models and Cross Validation

We compared the models by cross validation. We used the test set to predict and compare the four models using the training set. We predicted the models ten times and compared the predicted value, mean, standard deviation, and range. Based on that, we point out the model with the best overall predictive power.

The Application

We developed an application (app) that automatically computes the risk level of shallow landslides. It uses the equation with the best predictive power to do the task. It was

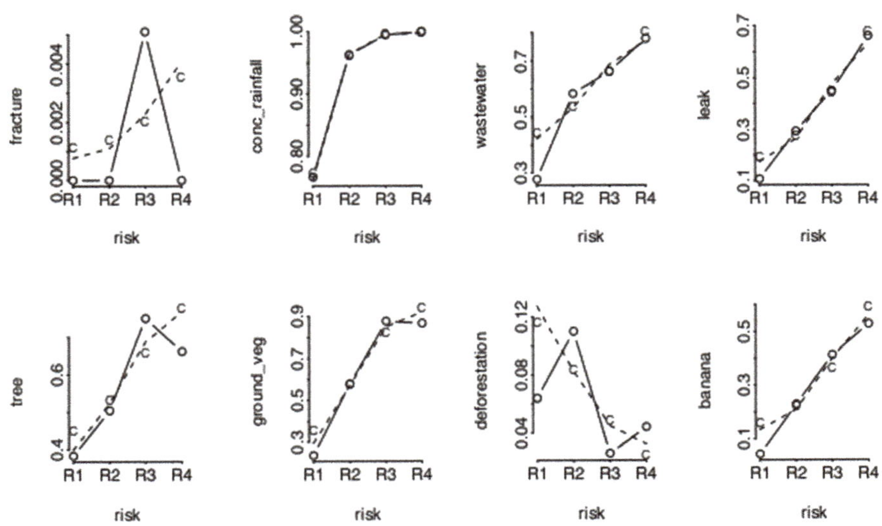

Fig. 2 Test for Ordinal Fashion. Continuous lines connect circles of the computed values and dashed lines connect the estimated expected values. Variables conc_rainfall, wastewater, leak, and banana (banana tree) are in the OF category. Variables tree and ground_veg are in the MOF category. Variables fracture and deforestation are in the NOF

Fig. 3 (1) Mixed material dwellings combine different materials and the dwellings are very precarious. (2) Brick dwellings are the ones with most of the dwellings made with bricks. They are more resistant than wood and mixed material dwellings

Table 1 Model, number of predictable variables, and aspects used to reduce number of variables

Model	N. Variables	Aspects Considered
01	16	OF, MOF, GP, p-value $\leq$ 0.10
02	13	OF, MOF, GP, p-value $\leq$ 0.05
03	20	OF, GP, p-value $\leq$ 0.10
04	13	OF, GP, p-value $\leq$ 0.10

developed in RStudio and mimics the BGM field evaluation paper form on a digital device. In one tab of the digital platform technicians check features observed in the field, in the other tab, table with data collected is created and sector risk level is computed.

Result and Discussion

We used MASS package in Rstudio to compute the classifier coefficients. The majority of our classifiers were binary (TRUE and FALSE). Table 2 displays value of classifiers TRUE since classifier FALSE is the dummy variable and equal to zero.

Scars is the classifier, in all the models, with the highest coefficient value and it does not differ much between models (see Table 2). It represents the presence of prior landslide scars in the sector, which means that in the past a landslide occurred in the sector. Past landslide indicates the possibility of reactivation of the same landslide (Highland & Brobrowsky 2008). Classifiers that represent signs of instability have positive coefficient values (see crack, leaning_wall, downward_floor, and tilted). Signs of instability specify movement on the terrain and suggest that a landslide could be initiated (Carvalho et al. 2007). Therefore, the classifier coefficient value should be positive since its presence in the sector contributes to a high risk level.

Classifiers brick, wood, and mixed represent the type of dwelling. Dwellings made of brick are more resistant to landslides than wood and mixed material dwellings (Carvalho et al. 2007). Coefficient value of classifiers brick and wood follow what would be expected, but not the classifier mixed. Mixed material dwellings are more precarious than wood since they combine pieces of cardboard boxes, plastic, wood, and any other material available, and mixed material and wood dwellings are more precarious than brick (see Fig. 3). Risk areas are usually located in impoverished neighborhoods and residents commonly build their initial dwellings with any materials available on site (mixed material dwellings), later upgrading to wood dwellings, and eventually to brick as their financial situation improves. One of the reasons mixed material dwellings have lower coefficient values than wood dwellings could be related to the fact there are less mixed material dwellings in a sector or mixed material could be misclassified as wood (since it usually has some pieces of wood). Combining wood and mixed material would be one way to solve this problem and should yield a more consistent coefficient value for this classifier.

Landfill (landfill classifier), garbage (garbage classifier), and construction material (construction classifier) can be

Table 2 Classifiers and their specific weight (α) and odds ratio (OR) for each model

Classifier	Model 1		Model2		Model 3		Model 4	
	α	OR	α	OR	α	OR	α	OR
Brick	−1.6	0.20	−1.6	0.20	−1.7	0.18	−1.8	0.17
Wood	0.84	2.32	0.94	2.56	1.02	2.77	1.04	2.83
Mixed					0.09	1.09		
EN	0.39	1.48			0.7	2.01	0.65	1.92
TC_mature_soil	0.54	1.72			0.58	1.79	0.65	1.92
TC_saprolite_soil					0.25	1.28		
Spring					−0.22	0.80		
Landfill					0.1	1.11		
Garbage					0.03	1.03		
Construction	0.61	1.84	0.69	1.99	0.46	1.58		
Leak	−0.6	0.54	−0.6	0.57	−0.4	0.66	−0.4	0.68
Conc_rainfall_water	1.01	2.75			1.63	5.10	1.75	5.75
Wastewater	0.39	1.48			0.6	1.82	0.64	1.90
Drainage.L	1.31	3.71	1.4	4.06				
Drainage.Q	−0.03	0.97	−0.1	0.95				1.00
Crack	2.27	9.68	2.26	9.58	2.18	8.85	2.23	9.30
Leaning_wall	2.21	9.12	2.26	9.58	2.03	7.61	2.06	7.85
Scars	4.2	66.7	4.21	67.4	4.25	70.11	4.31	74.4
Downward_floor	1.1	3.00	1.07	2.92	1.28	3.60	1.26	3.53
Tilted	0.75	2.12			0.85	2.34	0.87	2.39
AngleD (60 ≤ < 90)					0.24	1.27		
AngleE (= 90)					0.32	1.38		
Banana					0.39	1.48	0.48	1.62
Ground_veg			0.92	2.51				
TC_unstable_material	−0.9	0.41						
R1|R2	**−1.5**	**0.21**	**−1.6**	**0.19**	**−0.7**	**0.49**	**−1.1**	**0.33**
R2|R3	**3.23**	**25.3**	**3.03**	**20.7**	**3.56**	**35.16**	**3.08**	**21.8**
R3|R4	**8.77**	**6438**	**8.59**	**5378**	**9.02**	**8267**	**8.5**	**4915**

OR is interpreted as the same within risk levels and it is computed by exponentiating the coef. For instance, in Model 1, OR = 0.2 for wood. The odds of having higher level of risk is 0.2 times for sectors with wood houses (woodTRUE) when compared with sectors without wood houses (woodFALSE) if all other variables in the model are held constant. Empy cells represents coefficents that were not used in the model
EN is natural slope, TC_mature_soil is cut slope with mature soil, TC_sprolite_soil is cut slope with saprolite soil, spring is presence of spring in the slope, drainage.L isprecarious drainage system, drainage.Q is no drainage system, TC_unstrable_structure is cut slope with unstable material, R1 is low risk level, R2 is moderate risk level, R3 is high risk level, R4 is very high risk level

found in the risk sectors and they contribute to increase the level of risk. Garbage and construction material on the slope (see Fig. 4) can move downslope and dwellings built on top of landfills or disturbed soil are less stable (Carvalho et al. 2007). The quantity and composition of the materials in each sector are different (see Fig. 4). They have positive coefficient values in the models, contributing to an increase of the risk level. Since garbage and construction material are usually found in the same site, combining both into one single classifier could improve models. Another aspect to consider, in the field inventory form, there is the option to choose if there are garbage, landfill, and construction material on the sector, but whether the amount is small, moderate, or high, the choice is the same. A Likert scale would improve the data collection and contribute to better models.

Shallow landslides in these urban risk sectors in Brazil are triggered by rainfall (Tominaga et al. 2009). Increase of the soil water content in the slope can contribute to shallow landslide initiation. In these risk sectors, self-engineered dwellings have precarious pipelines that can leak, and

Fig. 4 Slope with garbage and construction material. Notice the difference in amount of material in each part of the hill

sewage is left to flow onto the slope (see Fig. 5). Some sectors have dwellings with a septic tank. Overtime, as these informal sectors become more formal, municipalities attempt improvements in public utilities, like drainage system and electricity. However, these improvements take time and it depends on the municipality's financial resources and priorities. Classifiers that contribute to increasing the water content on the terrain should have a positive coefficient value since they contribute to increasing the risk level. Wastewar and conc_rainfall_water have positive coefficient values, however Leak has negative coefficient value in all the models, but it would be expected to be positive. A negative coefficient value of the classifier could be related to the difficulty to quantifying the leakage in the sectors since on the field evaluation form you have just the options of presence or absence of leakages in the sector. For instance, a sector with just one dwelling with a leakage has the same weight as a sector with all dwellings with leakages.

Vegetation classifiers in all the models have coefficient values as would be expected. Banana tree (banana) and ground vegetation (ground_veg) have positive classifiers values, and tree (tree), has a negative value. Banana trees are known to increase water content in the soil, which can increase the risk level since the soil is more saturated (Carvalho et al. 2007). Ground vegetation can increase risk level since the soil becomes more protection than with the original vegetation. However, it would be expected that banana trees in the terrain would contribute more for a high risk level than ground vegetation, but in the models ground vegetation contributes more toward a high risk level than banana trees. Aspects that are difficult to account for the classifiers related to vegetation is the quantity of vegetation, its location (near the dwellings or not), and the combination of one or more vegetation types.

We ran the models ten times to compare the predictive power of each of them. Model 2 has the best prediction (Mean = 0.729, SD = 0.024, Range = 0.071). Still, the mean prediction power of model 1 and 4 is very similar (0.720, 0.721, respectively). Based on our results, models with smaller numbers of classifers performed better than those with larger numbers.

Fig. 5 Precarious wastewater pipes and pipes to conduct water, some of these pipes have water leakeage

Conclusion

Updating risk level more rapidly can prevent disasters in shallow landslides risk areas, allowing a local municipality to better monitor the area and to relocate financial resources to sectors with high risk levels. Automatically computing shallow landslide risk level using an application (app) allows the standardization of the BGM, rapid updates to the mapped areas, and a reduction in human subjectivity in assessing the landslide risks.

In this study, when comparing the models, those with a smaller number of classifiers had better performance and were simpler. Model 2 had the best mean predictive power when compared with three other models. The mean predictive power was 0.73. The total number of predictors variables are 14 and it accounts for OF, MOF, GP, and p-value $\leq$ 0.05. To improve the model for shallow landslides, we recommend that the BGM field evaluation form adopt a Likert scale for some classifiers, like leaks, presence of garbage, and construction material. This would allow the differentiation of sectors with small, moderate, and high leak or garbage, and not treat the small and high amounts as the same. Combining classifiers that are similar and found in the same sector, like wood and mixed material dwellings, or classifiers that are found in same site, like garbage and construction material, could improve the model.

Acknowledgements The Brazilian Ministry of Science and Technology (MCT) through CAPES by providing doctor fellowship Science without Borders.

References

Agresti A, Finley B (2009) Statistical methods for social sciences.4th edn. Upper Saddle River, New Jersey, Pearson Prentice Hall

Alcantara-Ayala I (2002) Geomorphology, natural hazards, vulnerability and prevention of natural disasters in developing countries. Geomorphology 47 (September 2016), 107–124

Carvalho CS, de Macedo ES, Ogura AT (2007) Mapeamento de Riscos em Encostas e Margem de Rios. Ministério das Cidades & Instituto de Pesquisas Tecnológicas, Brasilia

Christensen RHB (2018) Cumulative link models for ordinal regression with the R package ordinal. J Stat Softw, 1–40

Das IC (2011) Spatial statistical modelling for assessing landslide hazard and vulnerabiity. Enschede, The Netherlands: ITC—Faculty of Geo-information science and earth observation—University of Twente

Dourado F, Arraes TC, Silva MF (2012 Dec) O Megadesastre da Região Serrana do Rio de Janeiro - as causas do evento, os mecanismos dos movimentos de massa e a distribuição espacial dos investimentos de reconstrução no pós-desastre. Anu Inst Anu Inst Geocienc 35(2):43–54

Highland LM, Bobrowsky P (2008) The landslide handbook—a guide to understanding landslides. Reston, Virginia, U.S. Geological Survey Circular 1325, p 129

IBGE—Instituto Brasileiro de Geografia e Estatística (2018) População em áreas de risco no Brasil. Rio de Janeiro: IBGE, 2018. p 91. ISBN 978-85-240-4468-7

Marchezini V, Wisner B, Londe LR, Saito SM (2017) Reduction of vulnerability to disasters: from knowledge to action. RiMa Editora, São Carlos

Maricato E (2008) Brasil, cidades: alternativas para a crise urbana. Vozes, Petropolis, RJ

Tominaga LK, Santoro J, Amaral R (2009) Desastres naturais: conhecer para prevenir. Instituto Geologico, Sao Paulo

UN-SPIDER—United Nations Office for Outer Space Affairs (2019 June) Knowledge Portal. Retrieved from UN-SPIDER: http://www.un-spider.org/risks-and-disasters/disaster-risk-ma-gement

United Nations (2015) Report of the open-ended intergovernmental expert working group on indicators and terminology relating to disaster risk reduction

USGS (2019 June) Catastrophic landslides of the 20th Century—Worldwide

Van Westen CJ, Van Asch TW, Soeters R (2006) Landslide hazard and risk zonation—why is it still so difficult? Bull Eng Geol Env 65:167–184

Site-Specific Risk Assessment of Buildings Exposed to Rock Fall in India—a Case Study

2020 Kyoto Japan

Aditi Singh, Debi P. Kanungo, and Pravin Kr. Singh

Abstract

Rock falls are the most common type of fast- moving landslide where rock fragments fall (bounces or slides or rolls) freely from a cliff face having little or no shear displacement, thereby posing serious threat to life and infrastructure. In order to prevent these accidental happenings, assessment of site-specific risk becomes essential. The specific risk (R_S) is useful in assessing risk due to site-specific landslide in a particular area at a particular intensity. In the present study, landslide risk (rock fall risk in this case) has been assessed spatially as a function of physical vulnerability (PV) and proximity (P_{rox}) of the buildings from site-specific rock fall zone and drainage channels associated with them by using a semi-quantitative approach in a GIS platform. The site-specific risk exposed to rock falls has further been classified into three classes to determine the severity of risk level as low, medium and high. These site-specific risk classes will be useful in taking appropriate protection measures such as bolts, nets, removal of unstable blocks, simple light fences, buttress walls. The proposed methodology has been presented as a case study in a rock fall prone region of Chamoli District, Garhwal Himalayas of India.

Keywords

Rock fall • Site-Specific risk • Physical vulnerability

Introduction

Rock falls are landslides composed of detachable rocks of varying sizes which falls, rolls, bounces, slides freely from a cliff under the effect of gravity (Blahut et al. 2013; Youssef et al. 2015; Varnes 1984; Arbanas et al. 2012; Ferrari et al. 2013; Leine et al. 2014). The detachment of rocks is basically due to the presence of discontinuity and weathering along the surface. Some other triggering factors resulting in rock falls are high intensity rainfalls leading to pore water pressure, surface runoff & erosion, earthquakes, thermal expansion–contraction of rocks, frequent freeze–thaw cycles, etc. (Asteriou et al. 2012; Wyllie 2014; Sabatakakis et al. 2015). Despite small magnitude of events, the falling rocks can be extremely dangerous due to their high velocities reaching up to tens of meters per second and the potential of damaging the surrounding areas of slopes. In order to protect the surrounding areas comprising of life and infrastructure risk assessment becomes necessary, so that appropriate measures (i.e. land-planning & structural measures) can be taken.

Rock fall risk can be assessed using qualitative or quantitative methods, depending upon the input data, adopted procedures and final outputs (Ferrari et al. 2016). Qualitative methods are based on hazards, elements at risk and their vulnerabilities using qualitative descriptors, such as scoring schemes, weighted indices, ranked attributes, ranking matrices and rating systems and classifications (i.e. low, moderate & high) (Cruden and Fell 1997). On the other hand, Quantitative methods encompass numerical values or ranges of values by using numerical modelling (such as UDEC, FLO-2D, FLO-3D, etc.) instead of qualitative terms. However, due to lack of sufficient knowledge on

A. Singh (✉)
Department of Civil Engineering, Sarala Birla University, Ranchi, 835103, Jharkhand, India
e-mail: aditi.peace@gmail.com

D. P. Kanungo
Geotechnical Engineering Division, CSIR—Central Building Research Institute (CBRI), Roorkee, 247667, Uttarakhand, India
e-mail: debi.kanungo@gmail.com

P. Kr.Singh
Department of Mechanical Engineering, AMITY University, Ranchi, 834001, Jharkhand, India
e-mail: pravinsingh.phd@gmail.com

F. Guzzetti et al. (eds.), *Understanding and Reducing Landslide Disaster Risk*,
ICL Contribution to Landslide Disaster Risk Reduction,
https://doi.org/10.1007/978-3-030-60227-7_55

vulnerability of elements at risk, the intensity of rock fall events, site-specific nature of rock falls, the assessment of risk is complex (Wang et al. 2014).

Several landslide risk assessment methodologies, overviews and guidelines can be found in the literature (Dai et al. 2002; Catani et al. 2005; AGS 2007, Singh et al. 2017). However, due to scarcity of data in terms of rock fall inventory, frequency and intensity of rock fall and differences in geological, topographical, climatic and economic conditions, these guidelines cannot be directly implemented in India. Therefore, a methodology has been proposed keeping in view the above discussed conditions.

In the present study, risk assessment exposed to rock fall has been assessed spatially as a function of physical vulnerability (PV) and proximity (P_{rox}) of the buildings from site-specific landslide and drainage channels associated with them by using a semi-quantitative approach in a GIS platform.

Study Area

The Chamoli district which is located in the north-eastern part of Uttarakhand state has been selected as the study area. Chamoli District covers an area of 8030 km^2 (https://chamoli.gov.in). Being the district headquarters and second largest district of Uttarakhand state, Chamoli Township is a densely populated region. Alaknanda river flowing towards south-east is the main source of drainage in the Gopeshwar Township of Chamoli District.

The rock fall prone region in the study area is located along the Gopeshwar - Mandal road from latitude 30° 25'22.99"N to 30°25'11.95"N and longitude 79°18'46.51"E to 79°19'5.83" E. Figure 1 presents a comparison of the images of the rock fall zone acquired from field survey and ArcGIS. The affected rock fall zone is approximately 650 m of length along the road. Further, covering around 360 m from the initiation zone up to road level and approximately 300 m below the road level up to the built-up region. According to locals, rock fall occur more frequently during rainy seasons. Field observation showed blocks of different sizes on the roadside. The boulder size in the field varied from 0.2 kg to 6 kg. Large quantity of blocky boulders has also travelled down towards the built-up area (Ward Number 2, Gopeshwar Gaon); however, due to dense forest downhill the intensity of the rocks is reduced, and the travelled rock boulders get accumulated in the forest area (Fig. 2). Therefore, the forest in this region acts as a barrier, thereby providing natural protection to the houses and people living in it.

Methodology

The site-specific risk (R_s) due to a particular rock fall event can be obtained by determining the physical vulnerability (PV) of the buildings and their proximity from the specific rock fall zone and the drainage channels associated with it. This can be done by developing multiple buffer rings at a distance of 0–50 m, 50–100 m, 100–150 m and $\geq$ 150 m using analyst tool in ArcGIS around the specific rock fall zone and the drainage channel feeding from them. Then the building blocks present in the study area are overlaid as a single data layer on the buffer layer in ArcGIS. The detailed methodology for the assessment of site-specific risk (R_s) in the study area is shown in the form of a flow chart in Fig. 3.

Further, depending upon the position of the building in the rock fall proximity zone (i.e. different buffer zones), the physical vulnerability (PV) of the buildings will decrease as per the criteria given in flowchart (Fig. 3).

Fig. 1 Comparison of images acquired from **a** field survey and **b** base map (ArcGIS) showing rock fall zone in the study area

Fig. 2 Rock fall zone as viewed from downhill showing **a** the rock fall path and **b** accumulated rocks in forest area

Thus, the specific risk (R_s) can be obtained using Eq. (1).

$$R_S = PV_i - P_{rox} \quad (1)$$

where, R_s: Specific risk; PV_i: Physical vulnerability due to a particular or varying intensity such as I_{RL}, I_{RM} and I_{RH} (low, moderate and high intensities respectively) for rock fall; P_{rox}: Proximity of the buildings from the rock fall region.

Further, to understand the acceptability of risk indicating low, moderate and high risks the buildings are classed into three categories as: (i) risk class I ($R_s \leq 0.5$) (ii) risk class II ($0.5 < R_s \leq 0.9$) and (iii) risk class III ($0.9 < Rs \leq 1$) respectively.

Physical Vulnerability (PV)

In the present study PV is assessed based on the rock fall intensity and resistance of the buildings (Singh et al. 2019a, b, c) and empirically by Li et al. (2010) expressed as

$$\boldsymbol{PV} = \boldsymbol{f}(\boldsymbol{I}, \boldsymbol{R}) \begin{cases} 2\frac{I^2}{R^2} & \frac{I}{R} \leq 0.5 \\ 1 - \frac{2(R-I)^2}{R^2} & 0.5 < \frac{I}{R} \leq 1.0 \\ 1.0 & \frac{I}{R} > 1 \end{cases} \quad (2)$$

where, I refer to the rock fall intensity and R refers to the resistance offered by buildings.

In the present study rock fall intensity (I) is classified as low (I_{RL}= 0.2), medium (I_{RM}= 0.6) & high (I_{RH}= 1.0) as per Singh et al. (2019c). Further resistance of the buildings is taken by encompassing fifteen structural and non-structural indicators, namely: year of construction, roof design, structural typology, surrounding wall of the building, number of floors, vertical configuration, horizontal configuration, presence of windows & doors $\geq$ 1 m from each other, distance between two houses $\geq$ 1 m from each other, building position from steep slope, surrounding vegetation, quality of construction, state of maintenance, type of construction and retrofitted buildings (Singh et al. 2019b).

Proximity of the Buildings from the Rock Fall Prone Region

Since drainage channels put significant stress on the hilly terrains, the stability of slope decreases. Therefore, most of the landslides in hilly areas occur due to the erosional activity associated with drainage. Nevertheless, if the rock fall zone is present near such source of drainage channel, the rock fall driven debris flow may also occur along the drainage channel, thereby affecting the buildings in the proximity of those drainage channels. Several studies have also

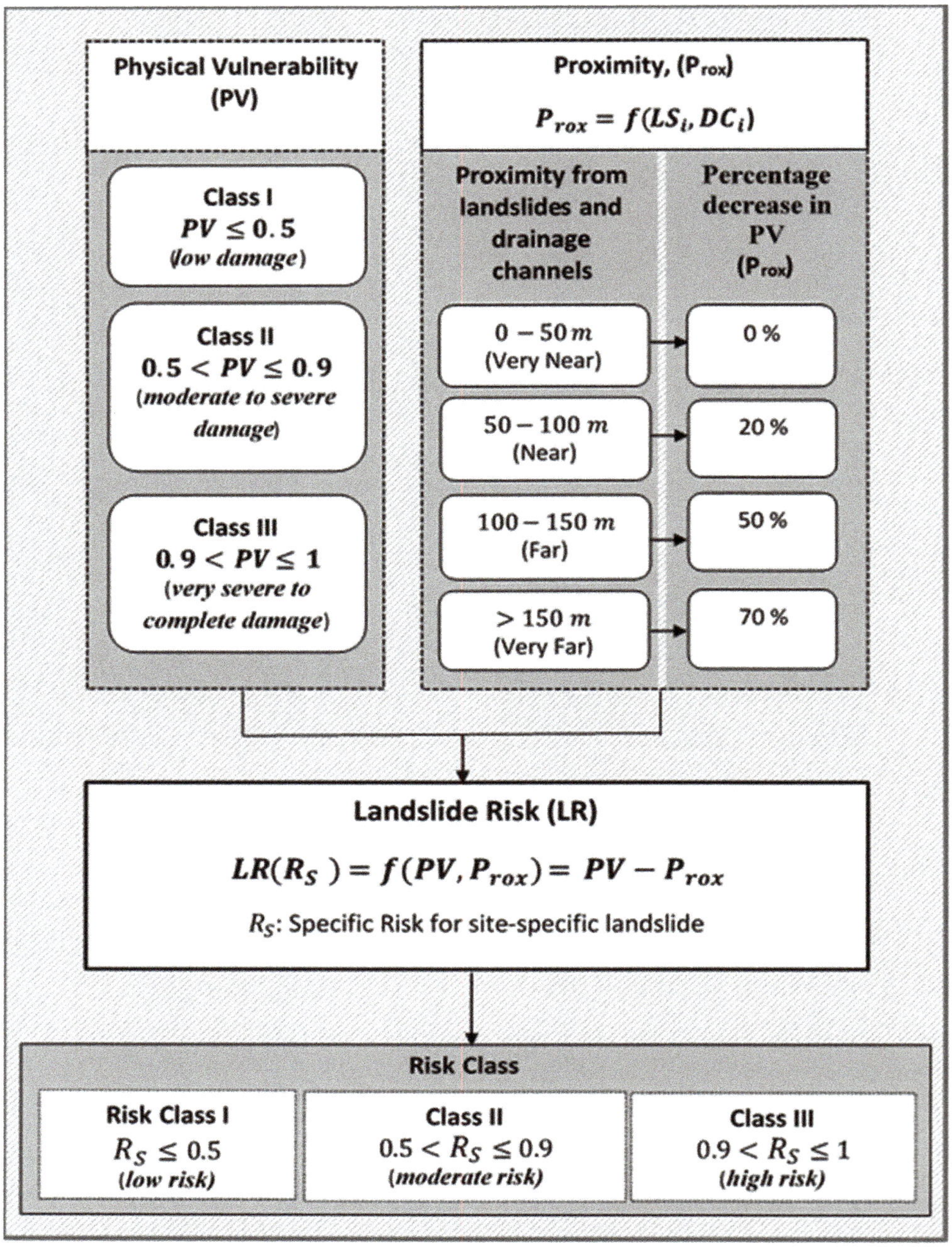

Fig. 3 Methodological flowchart to assess landslide risk (LR) of buildings exposed to landslides

proven that proximity of slopes to river networks is an important factor controlling the occurrence of landslides (Gokceoglu and Aksoy 1996; Pourghasemi et al. 2012; Park et al. 2013). Such related phenomenon is most commonly used for the assessment of landslide susceptibility such as proximity to streams, proximity to river networks, distance to rivers, proximity of slope to drainages distance to rivers, drainage, drainage buffer etc. However, for the assessment of landslide risk via proximity of drainage channel to buildings, no concrete literature is yet available. Singh et al. (2019b) stated that impact of proximity of buildings from the landslides will vary depending upon the its intensity and distance; if the landslide intensity is high and very high, the impact will be high on the buildings in near proximity and the impact will reduce with the increase in the distance of the buildings from the landslide.

Therefore, proximity of buildings in this study is taken as the function of a particular rock fall zone and the drainage channel from that particular zone and expressed in Eq. 3.

$$P_{rox} = f(LS_i - DC_i) \tag{3}$$

where, P_{rox}: Proximity of the buildings from drainage channel; LS_i: site-specific landslide, i.e. rock fall in this case; and DC_i: Drainage channel feeding from a particular site-specific landslide.

It is very obvious that the buildings with same physical vulnerability (PV) may pose different level of risk, due to its proximity with the drainage channel. Singh et al. (2019b) suggested the percentage impact of buildings to landslide at different intensity based on the field conditions. Since visual observations play a vital role in landslide studies (Bhandrai 2006), the percentage of impact of buildings at different landslide intensity as presented by Singh et al. (2019b) has been modified in this study as shown in Fig. 3.

Case Study

In case of rock fall, the physical vulnerability (PV) of the buildings is exposed to high rock fall intensity (i.e. $I_{RH} = 1$) (Singh et al. 2019c). Further, the proximity of the buildings from the rock fall is obtained by creating multiple buffer rings of rock fall zone and the drainage channels associated with it. And then the buildings of ward number 2 are overlaid on it in ArcGIS (Fig. 4).

The specific risk (*Rs*) of all the buildings in ward number 2 is calculated due to the rock fall using Eq. (1) and is shown in Table 1. From the assessment, it is found that 74%, 12% and 14% of buildings lie in risk class I, II and III respectively as shown in Fig. 4.

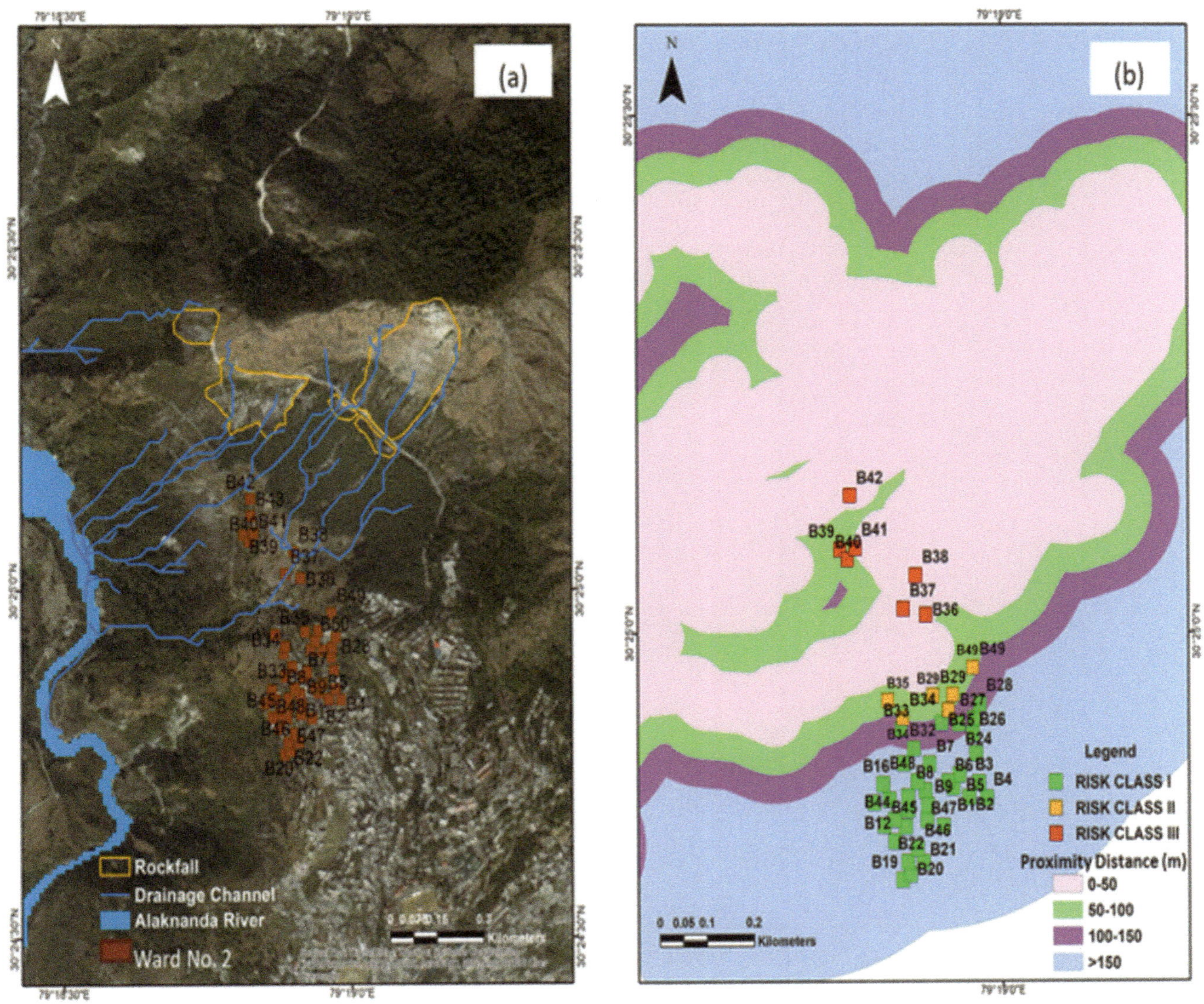

Fig. 4 **a** Ward number 2 exposed to rock fall and the drainage channel feeding from it; **b** Classification of buildings in different risk classes due to proximity of buildings from rock fall

Table 1 Specific risk (Rs) of buildings present in ward number 2 exposed to rock fall

Building ID	Resistance, (*R*)	Rock fall intensity, (*I*)	PV at $I_{RH} = 1$	Proximity (P_{rox})	% decrease in PV	Specific risk (R_S)	Risk class
B1	0.50	High ($I_{RH} = 1$)	1	>150 m	70	0.3	I
B2	0.62			>150 m	70	0.3	I
B3	0.57			>150 m	70	0.3	I
B4	0.63			>150 m	70	0.3	I
B5	0.47			>150 m	70	0.3	I
B6	0.64			>150 m	70	0.3	I
B7	0.55			>150 m	70	0.3	I
B8	0.65			>150 m	70	0.3	I
B9	0.61			>150 m	70	0.3	I
B10	0.65			>150 m	70	0.3	I
B11	0.64			>150 m	70	0.3	I
B12	0.47			>150 m	70	0.3	I
B13	0.61			>150 m	70	0.3	I
B14	0.56			>150 m	70	0.3	I
B15	0.68			>150 m	70	0.3	I
B16	0.69			>150 m	70	0.3	I
B17	0.69			>150 m	70	0.3	I
B18	0.47			>150 m	70	0.3	I
B19	0.45			>150 m	70	0.3	I
B20	0.45			>150 m	70	0.3	I
B21	0.47			>150 m	70	0.3	I
B22	0.63			>150 m	70	0.3	I
B23	0.64			>150 m	70	0.3	I
B24	0.57			100–150 m	50	0.5	I
B25	0.6			100–150 m	50	0.5	I
B26	0.64			>150 m	70	0.3	I
B27	0.58			100–150 m	50	0.5	I
B28	0.55			100–150 m	50	0.5	I
B29	0.45			50–100 m	20	0.8	II
B30	0.63			50–100 m	20	0.8	II
B31	0.47			>150 m	70	0.3	I
B32	0.47			100–150 m	50	0.5	I
B33	0.47			100–150 m	50	0.5	I
B34	0.50			50–100 m	20	0.8	II
B35	0.57			50–100 m	20	0.8	II
B36	0.55			0–50 m	0	1	III
B37	0.55			0–50 m	0	1	III
B38	0.61			0–50 m	0	1	III
B39	0.47			50–100 m	20	1	III

(continued)

Table 1 (continued)

Building ID	Resistance, (R)	Rock fall intensity, (I)	PV at I_{RH} = 1	Proximity (P_{rox})	% decrease in PV	Specific risk (R_S)	Risk class
B40/B41	0.50/0.56			50–100 m/0–50 m	20/0	1	III
B42	0.55			0–50 m	0	1	III
B43	0.47			>150 m	70	0.3	I
B44	0.57			>150 m	70	0.3	I
B45	0.54			>150 m	70	0.3	I
B46	0.50			>150 m	70	0.3	I
B47	0.55			>150 m	70	0.3	I
B48	0.47			>150 m	70	0.3	I
B49	0.57			50–100 m	20	0.8	II
B50	0.55			50–100 m	20	0.8	II

Discussion and Conclusions

This paper describes and discusses a semi-quantitative methodology to assess site-specific risk of buildings exposed to rock fall in India. In this study, site-specific risk exposed to rock fall is assessed as the function of a particular rock fall zone and the drainage channel from that particular region. This assessment is carried out in ArcGIS platform. In the present study it was observed that drainage played a crucial role in triggering the rock fall. It is quite evident from the field check that the rock fall path and drainage channels produced on GIS platform follows a same path. Therefore, it can be concluded that the drainage feature in GIS alone can highlight the probable risk zones of a particular area.

The methodology introduced in this paper is crucial to understand the level of risks in a particular area, in order to take appropriate control measures such as bolts, nets, removal of unstable blocks, simple light fences, buttress walls. The developed methodology is easy to apply and economically feasible in real time in a data scarce scenario of India. Further, the methodology can be used in other countries as well, where the kinetic energy of rock fall is known. The level of risk (i.e. low risk, moderate risk and high risk) is designed to help administrators and practitioners to carry out preliminary site-specific risk assessment. Therefore, it is expected that the developed methodology will be useful in assessing the site-specific risk in India and other developing countries affected by rock falls.

References

AGS (Australian geomechanics society sub-committee on landslide risk management) (2007). A national landslide risk management framework for Australia, Aust Geomech J 42(1). www.australiangeomechanics.org

Arbanas Ž, Grošić M, Udovič D, Mihalić S (2012) Rockfall hazard analyses and rockfall protection along the Adriatic coast of Croatia. J Civil Eng Architect 6(3):344–355

Asteriou P, Saroglou H, Tsiambaos G (2012) Geotechnical and kinematic parameters affecting the coefficients of restitution for rock fall analysis. Int J Rock Mech Min Sci 54:103–113

Bhandari RK (2006) The Indian landslide scenario, strategic issues and action points. First India disaster management congress, New Delhi 29–30, Session A2, Keynote address, 1–18

Blahut J, Klimeš J, Vařilová Z (2013) Quantitative rockfall hazard and risk analysis in selected municipalities of the ČeskéŠvýcarsko National Park, Northwestern Czechia. Geografie 118(3), 205–220

Catani F, Casagli N, Ermini L, Righini G, Menduni G (2005) Landslide hazard and risk mapping at catchment scale. Landslides 2:329–342. https://doi.org/10.1007/s10346-005-0021-0

Cruden DM, Fell R (1997) Landslide risk assessment. Balkema, Rotterdam

Dai FC, Lee CF, Ngai YY (2002) Landslide risk assessment and management: an overview. Eng Geol 64(1):65–87

Ferrari F, Giani GP, Apuani T (2013) Why can rockfall normal restitution coefficient be higher than one? Rendiconti online SocietàGeologicaItaliana, 122

Ferrari F, Giacomini A, Thoeni K (2016) Qualitative Rockfall Hazard Assessment: A Comprehensive Review of Current Practices. Rock Mech Rock Eng 49:2865–2922. https://doi.org/10.1007/s00603-016-0918-z

Gokceoglu C, Aksoy H (1996) Landslide susceptibility mapping of the slopes in the residual soils of the Mengen region (Turkey) by deterministic stability analyses and image processing techniques. Eng Geol 44(1–4):147–161

Leine R, Schweizer A, Christen M, Glover J, Bartelt P, Gerber W (2013) Simulation of rockfall trajectories with consideration of rock shape. Multibody Sys Dyn 32(2):1–31

Li Z, Nadim F, Huang H, Uzielli M, Lacasse S (2010) Quantitative vulnerability estimation for scenario-based landslide hazards. Landslides 7(2):125–134

Park S, Choi C, Kim B, Kim J (2013) Landslide susceptibility mapping using frequency ratio, analytic hierarchy process, logistic regression, and artificial neural network methods at the Inje area. Korea Environ Earth Sci 68:1443–1464

Pourghasemi HR, Pradhan B, Gokceoglu C, Moezzi KD (2012) Landslide susceptibility mapping using a spatial multi criteria evaluation model at Haraz Watershed, Iran. In: Terrigenous mass movements, Springer-Berlin Heidelberg, pp 23–49

Sabatakakis N, Depountis N, Vagenas N, (2015) Evaluation of rockfall restitution coefficients. In: Engineering geology for society and territory, vol 2. Berlin, Springer, pp 2023–2026

Singh A, Pal S, Kanungo DP, Pareek N, (2017) An overview of recent developments in landslide vulnerability assessment—presentation of a new conceptual framework. In: Mikos M, Tiwari B, Yin Y, Sassa K (eds) Advancing culture of living with landslides. WLF 2017, vol 2. pp 795–802. https://doi.org/10.1007/978-3-319-53498-5_91

Singh A, Kanungo DP, Pal S (2019a) A modified approach for semi-quantitative estimation of physical vulnerability of buildings exposed to different landslide intensity scenarios. Georisk 13(1):66–81. https://doi.org/10.1080/17499518.2018.1501076

Singh A, Kanungo DP, Pal S (2019b) Physical vulnerability assessment of buildings exposed to landslides in India. Nat Hazards 96:753–790. https://doi.org/10.1007/s11069-018-03568-y

Singh A, Pal S, Kanungo DP (2019c) Site-specific vulnerability assessment of buildings exposed to rockfalls. In: Chattopadhyay J et al (eds) Renewable energy and its innovative technologies. Springer Nature Singapore Pte Ltd. 2019. https://doi.org/10.1007/978-981-13-2116-0_1

Varnes DJ (1984) Landslide hazard zonation: a review of principles and practice, p 64

Wang X, Frattini P, Crosta GB, Zhang L, Agliardi F, Lari S, Yang Z (2014) Uncertainty assessment in quantitative rockfall risk assessment. Landslides 11(4):711–722. https://doi.org/10.1007/s10346-013-0447-8

Wyllie DC (2014) Calibration of rock fall modeling parameters. Int J Rock Mech Min Sci 67:170–180

Youssef AM, Pradhan B, Al-Kathery M, Bathrellos GD, Skilodimou HD (2015) Assessment of rockfall hazard at Al-Noor Mountain, Makkah city (Saudi Arabia) using spatio-temporal remote sensing data and field investigation. J Afr Earth Sci 101:309–321

Cutting-Edge Technologies Aiming for Better Outcomes of Landslide Disaster Mitigation

WLF 2020 Kyoto Japan

Kazuo Konagai

The International Consortium on Landslides (ICL) and The Global Promotion Committee of the International Programme on Landslides (GPC/IPL) have been responsible for organizing the World Landslide Forums (WLFs) every three years since 2008. Ever since the 1st WLF, the forums have long been the arena for landslide researchers and practitioners to exchange up-to-date information of recent devastations caused by landslides, cutting-edge technologies for landslide disaster mitigations and early warnings etc. to establish synergies among all participants worldwide.

Though the upcoming WLF5 has officially been postponed by one year to 2–6 November 2021 due to the global disruption caused by the coronavirus pandemic, the WLF5 will be all the more important with the Kyoto Landslide Commitment 2020 (KLC2020) to be launched as planned in the final online signatory meeting on 5 November 2020; the KLC 2020 is intended to be our action goals as the further advanced successor of the 'Sendai Landslide Partnerships 2015–2025 for Global Promotion of Understanding and Reducing Landslide Disaster Risk' in line with some of 17 Sustainable Development Goals (SDGs), particularly SDG 11, "Make cities and human settlements inclusive, safe, resilient and sustainable," of the United Nations.

For these important goals, the ICL has been inviting sponsorship from industries, businesses, and government agencies; all leading players in landslide science and technologies. They have been supporting a variety of the ICL/IPL activities such as publishing the International full-color journal "Landslides (Journal of the International Consortium on Landslides), full-color books for WLFs, exhibiting their cutting-edge technologies in WLFs, etc. Here follow short introductions of their activities with their names, addresses and contact information.

Marui & Co. Ltd.

1-9-17 Goryo, Daito City, Osaka 574-0064, Japan
URL: http://marui-group.co.jp/en/index.html
Contact: hp-mail@marui-group.co.jp

Marui & Co. Ltd. celebrates its 100th anniversary in 2020. Marui, as one of the leading manufacturers of testing apparatuses in Japan, has been constantly striving to further improve its service since its foundation in 1920, thus contributing to the sustainable development of our nation and society. Our main products cover a wide variety of destructive and non-destructive testing apparatuses in the fields of geotechnical engineering, concrete engineering (mortar, aggregates, etc.), and ceramic engineering. Of special note is that Marui has been helping manufacture ring-shear apparatuses half-century long based on the leading-edge idea of Dr. Kyoji Sassa, Professor Emeritus at the Kyoto University. Marui has delivered total seven ring-shear apparatuses to the Disaster Prevention Research Institute, Kyoto University, and 2 to the International Consortium on Landslides. Also the apparatuses were exported to the United States of America, China, Croatia and Vietnam.

Marui & Co. Ltd. takes great pleasure in developing, manufacturing, and providing new products of high value sharing the delight of achievement with our customers, and thus contributing to the social development. The whole staff of Marui & Co. Ltd. are determined to devote ceaseless effort to keep its organization optimized for its speedy and high-quality services, by the motto "Creativity and Revolution", and strive hard to take a step further, as a leading manufacturer of testing apparatuses, to answer our customer's expectations for the twenty-second century to come.

K. Konagai (✉)
Organizing Committee of the Fifth World Landslide Forum, International Consortium on Landslides, Kyoto, 606-8226, Japan
e-mail: konagai@iclhq.org

F. Guzzetti et al. (eds.), *Understanding and Reducing Landslide Disaster Risk*, ICL Contribution to Landslide Disaster Risk Reduction,
https://doi.org/10.1007/978-3-030-60227-7_56

Nippon Koei Co. Ltd.

5-4 Kojimachi, Chiyoda-ku, Tokyo 102-8539, Japan
URL: https://www.n-koei.co.jp/english/
Contact: https://www.n-koei.co.jp/english/contact/input

Nippon Koei Co. Ltd. and its group companies conduct many projects to support the growth of developing countries in Asia, Africa, the Middle and Near East, Latin America and other regions. Examples of their efforts include environmental measures to combat global warming, development of regional transportation infrastructure to support the rapid growth of emerging economies, and reconstruction assistance for regions affected by conflict and/or natural disasters.

OSASI Technos, Inc.

65-3 Hongu-cho, Kochi City, Kochi 780-0945, Japan
URL: http://www.osasi.co.jp/en/
Contact: cs@osasi.co.jp

OSASI Technos, Inc. has been making its best efforts to develop its cutting-edge technologies for landslide early warning. Its unique compact and lightweight sensors making up the Landslide Early Warning System enable long-term monitoring of unstable landslide mass movements, precipitations, porewater pressure buildups, etc. in a remote mountainous area where commercial power is often unavailable. OSASI Technos, Inc. is also proud of its advanced technology to transfer observed data even in areas with poor telecom environments as proven in the successful implementations in South Asia.

All stuff members of OSASI Technos work together for mitigation of landslide disasters worldwide.

Godai Corporation

1–35 Kuroda, Kanazawa City, Ishikawa Prefecture, Japan
URL: https://soft.godai.co.jp/En/Soft/Product/Products/LS-RAPID/
Contact: pp-sales@godai.co.jp

Ever since its foundation in 1965, Godai Kaihatsu Co. Ltd. a civil engineering consulting firm, has long been providing a variety of software and measures particularly for natural disaster mitigation. With its rich expertise in both civil engineering and information technology (IT), the company has its primary goal to address real world needs of disaster mitigation. All the staff of Godai Kaihatsu Co. Ltd. feel it more than happy that their cutting-edge technologies help mitigate natural disasters.

Japan Conservation Engineers & Co. Ltd.

3-18-5 Toranomon, Minato-ku, Tokyo 1050001, Japan
URL: https://www.jce.co.jp/en/
Contact: go_info@jce.jp

Japan Conservation Engineers & Co. Ltd. (JCE) is a general consulting firm working on landslide prevention research and consulting. JCE provides various disaster prevention technologies for debris flows, landslides, slope failures, rockfalls, etc. In addition, JCE is proud of its expertise having been conducting surveys and consulting works on coastal erosions and tsunami countermeasures for about 20 years. JCE contributes to the world through its activities in the realm of both structural and non-structural measures to build a resilient society.

OYO Corporation

7 Kanda-Mitoshiro-cho, Chiyoda-ku, Tokyo 101-8486, Japan
URL: https://www.oyo.co.jp/english/
Contact: https://www.oyo.co.jp/english/contacts/

OYO Corporation, the top geological survey company in Japan established in Tokyo in 1957, is well known as one of leading companies providing cutting-edge technologies and measures for natural disasters such as landslides, earthquakes, tsunamis, and floods. Not just developing and selling measuring instruments related to disaster prevention, OYO also delivers a market-leading services in 3D ground/geological modeling and 3D exploration technologies.

Kokusai Kogyo Co. Ltd.

2 Rokubancho, Chiyoda-ku, Tokyo 102-0085, Japan
URL: https://www.kkc.co.jp/english/index.html
Contact: overseas@kk-grp.jp

Kokusai Kogyo Co. Ltd. as a leading company of geospatial information technologies, has long been providing public services with its comprehensive expertise to address real world needs and cutting-edge measurement technologies. Kokusai Kogyo Co. Ltd. helps rebuild “Green Communities,” which has been of our great concern in terms of “environment and energy,” “disaster risk reduction” and "asset management". Kokusai Kogyo Co. Ltd. offers the advanced and comprehensive analyses of geospatial information for developing new government policies, maintaining and operating social infrastructures safe and secure, and implementing low-carbon measures in cities.

Influenced by the recent global climate change, extreme rainfall events have become more frequent worldwide and resultant hydro-meteorological hazards are creating more deaths and devastations particularly in many developing countries where effective advanced countermeasures are not readily available. Kokusai Kogyo Co. Ltd. is proud of its achievements in establishing resilient infrastructure systems and implementing effective monitoring/early warning systems in developing countries, which have long been helping reduce the risks from natural hazards.

Geobrugg AG

Aachstrasse 11, 8590 Romanshorn, Switzerland
URL: www.geobrugg.com
Contact: info@geobrugg.com

Swiss company Geobrugg is the global leader in the supply of high-tensile steel wire safety nets and meshes – with production facilities on four continents, as well as branches and partners in over 50 countries. True to the philosophy "Safety is our nature" the company develops and manufactures protection systems made of high-tensile steel wire. These systems protect against natural hazards such as rockfall, landslides, debris flow and avalanches. They ensure safety in mining and tunneling, as well as on motorsport tracks and stop other impacts from falling or flying objects. More than 65 years of experience and close collaboration with research institutes and universities make Geobrugg a pioneer in these fields.

Ellegi Srl

Via Petrarca, 55 I-22070 Rovello Porro (CO) Italy
URL: http://www.lisalab.com/engl/?seze=1
Contact: info@lisalab.com

Ellegi srl provides worldwide monitoring services and produces Ground Based synthetic aperture radar (GBInsAR) for remote measurement of displacements and deformations on natural hazards and manmade buildings using its own designed and patented LiSALab system.

Its activities started in 2003 as a spin off project to exploit commercially the Ground Based Linear Synthetic Aperture Radars technology developed by European Commission's Ispra Joint Research Centre and based on the results of more than 10 years of research. Since then Ellegi has industrialized and developed the core technology of the LiSALab system and latest LiSAmobile system represents the 5th generation of development.

In 2003 it was the first commercial company in the world to provide GBInSAR measurements of natural hazards and structure.

Ellegi srl offers:

- Displacement fields measurement, control and monitoring of the deformation caused by natural hazards, like landslides, rockslides, sinkhole, volcanic deformation in every operative condition, including emergencies,
- Structural strain fields measurement, control, monitoring and diagnosis of the deformation affecting buildings, bridges, viaducts, dams.
- GBInSAR monitoring systems, installation, management and maintenance in order to provide information about natural hazards or anthropic activity, that can generate or cause slopes failures or buildings instabilities.

In all the above-mentioned activities Ellegi srl uses the GBInSAR LiSALab technology that represents a real "break-through".

Chuo Kaihatsu Corporation

3-13-5 Nishi-waseda, Shinjuku-ku, Tokyo 169-8612, Japan
URL: https://www.ckcnet.co.jp/global/
Contact: https://www.ckcnet.co.jp/contactus/

Chuo Kaihatsu Corporation (CKC) was founded in 1946, and has been aiming to become the "Only One" consultant for our customers. We engage in the hands-on work that will "Remain with the earth, Remain in people's hearts, and Lead to a prosperous future". We focus on road, river and dam engineering to flesh out industrial infrastructures specifically by means of geophysical/geotechnical/geological investigations, civil engineering surveys and project implementations. In recent years, we make significant efforts on earthquake disaster mitigation, sediment disaster prevention/mitigation and ICT information services. Many achievements of ours have already contributed to mitigation of natural disasters such as landslides, earthquakes and slope failures in Japan, Asia and the Pacific Region.

IDS GeoRadar S.R.L

Via Augusto Righi, 6, 6A, 8, Loc. Ospedaletto, Pisa, Italy, 56121
URL: https://idsgeoradar.com/
Contact: info@idsgeoradar.com

IDS GeoRadar, part of Hexagon, provides products and solutions, based on radar technology, for monitoring applications including landslides, rockfalls, complex structures, mining and civil engineering. The company is a leading provider of Ground Penetrating Radar (GPR) and Interferometric Radar solutions worldwide. IDS GeoRadar is committed to delivering best-in-class performance solutions and to the pursuit of product excellence, through the creation of application-specific, innovative and cost-efficient systems for a wide range of applications.

METER Group, Inc

2365 NE Hopkins Court, Pullman, WA 99163, USA
URL: metergroup.com/wlf5
Contact: bryan.wacker@metergroup.com

METER Group provides accurate, rugged, and dependable instrumentation to monitor moisture in all its phases within an unstable slope. METER specializes in instrumentation for near real-time monitoring of incoming moisture in the form of rain and weather. In addition, we provide real-time below-surface monitoring of existing moisture conditions like moisture content and soil suction which show how the soil profile is filling with water to saturation, including the transition to positive pore water pressure.

The ZL6 advanced cloud data logger works together with ZENTRA Cloud data software to simplify and speed up data collection, management, visualization, and alerting. Our well-published instrumentation is used worldwide in universities, research and testing labs, government agencies, and industrial applications.

For almost four decades, scientists and engineers have relied on our instrumentation to understand critical moisture parameters. We've even partnered with NASA to measure soil (regolith) moisture on Mars. Wherever you measure, and whatever you're measuring, rely on METER for accuracy, affordability, and simplicity that will make your job easier.

Asia Air Survey Co. Ltd.

Shinyuri 21 BLDG 3F, 1-2-2 Manpukuji, Asao-Ku, Kawasaki, Kanagawa 215-0004, Japan
URL: https://www.ajiko.co.jp/en/
Contact: service@ajiko.co.jp

Asia Air Survey (AAS), as one of the leading engineering and consulting companies, has long been providing disaster prevention and mitigation services for over 65 years, particularly in the fields of landslide, debris flow, erosion control, etc. AAS is proud of being the inventor of Red Relief Image Map (RRIM), which is a cutting-edge 3D terrain visualization method allowing great geomorphological details to be visualized in one glance, thus has been used in various facets of disaster prevention and mitigation.

Kiso-Jiban Consultants Co. Ltd.

Kinshicho Prime Tower 12 Floor, 1-5-7 Kameido, Koto-ku, Tokyo 36-8577, Japan
URL: https://www.kisojiban.com/
Contact: kisojiban-contactus@kiso.co.jp

Kiso-Jiban Consultants, established in 1953, is an engineering consulting firm especially well known in the field of geotechnical engineering. The areas of its comprehensive services are listed below:

- Geological and Geotechnical Survey
- Geotechnical Analysis and Design
- Disaster Prevention and Management
- GIS (Geographic Information Systems)
- Soil and Rock Laboratory Tests
- Instrumentation and Monitoring
- Geophysical Exploration and Logging
- Distribution of Geosynthetics Products.

Much-talked-about new service is Kiso-SAR System allowing accurate estimation of both extent and rate of landslide movements based upon a comprehensive interpretation of InSAR results from geotechnical and landslide engineering viewpoint (see the one-page introduction of Kiso-Jiban Consultants Co. Ltd.). With Kiso-SAR system, the following pieces of important geotechnical information can be provided:

(1) Extent of a deforming landslide mass (and the rate of its movement
(2) Consolidation buildup in soft clay underlying a fill
(3) Deformation buildups induced by slope cutting.

Okuyama Boring Co. Ltd.

10-39 Shimei-cho, Yokote City, Akita 013-0046, Japan
URL: https://okuyama.co.jp/en/
Contact: info@okuyama.co.jp

Okuyama Boring Co. Ltd. is proud of its achievements in various projects to help solve many landslide problems. The company has been offering services in geological surveys and analyses, developing rational countermeasures against various geotechnical problems as well as safe workflow diagrams, and providing necessary pieces of advice for ensuring safety during landslide countermeasure works. For this purpose, Okuyama Boring Co. Ltd. works on monitoring, observations, field surveys, numerical analyses, countermeasure works, etc. of landslides.

Kawasaki Geological Engineering Co. Ltd.

Mita-Kawasaki Bldg, 2-11-15 Mita, Minato-ku, Tokyo108-8337, Japan
URL: http://www.kge.co.jp/
Contact: post-master@kge.co.jp

Kawasaki Geological Engineering Co. Ltd. as one of the leading members of SAAM Research Group, has proactively been involved in developing "Sustainable Asset Anchor Maintenance (SAAM, hereafter) System," enabling easy maintenance of ground anchors. Its unique jack, weighing about half the weight of a conventional jack, together with a newly developed jig, can be applied to any type of anchor even with a short extra length, thus allowing for in situ lift-off tests on these anchors. The SAAM system also has an optional weight meter that can be installed after performing a lift-off test.

Nissaku Co. Ltd.

4-199-3 Sakuragi-cho, Omiya-ku, Saitama 330-0854, Japan
URL: https://www.nissaku.co.jp/
Contact: survey@nissaku.co.jp

Nissaku Co. Ltd. founded in 1912 as a well drilling company, provides services for far-flung fields of not only groundwater exploitation but also measures for landslides. Having its rich expertise in these fields, Nissaku Co. Ltd. offers general reliable one-stop technical services including designs, investigations, analyses, constructions, and maintenances.

Full-color presentations from the above seventeen exhibitors focusing on their landslide technologies are shown on the following pages. Their cutting-edge technologies have of course been instrumental in the progress that we have made in landslide risk-reduction worldwide, and we want to exert even greater effort to aim high given the KLC 2020 as our new action goals. The International Consortium on Landslides seeks volunteers willing to support our activities introducing their brand-new technologies for landslide disaster mitigation in our international journal "Landslides," full color books for WLFs, exhibitions at WLFs, etc. If you are interested in being engaged in supporting ICL activities, please contact the ICL secretariat <secretariat@iclhq.org>.

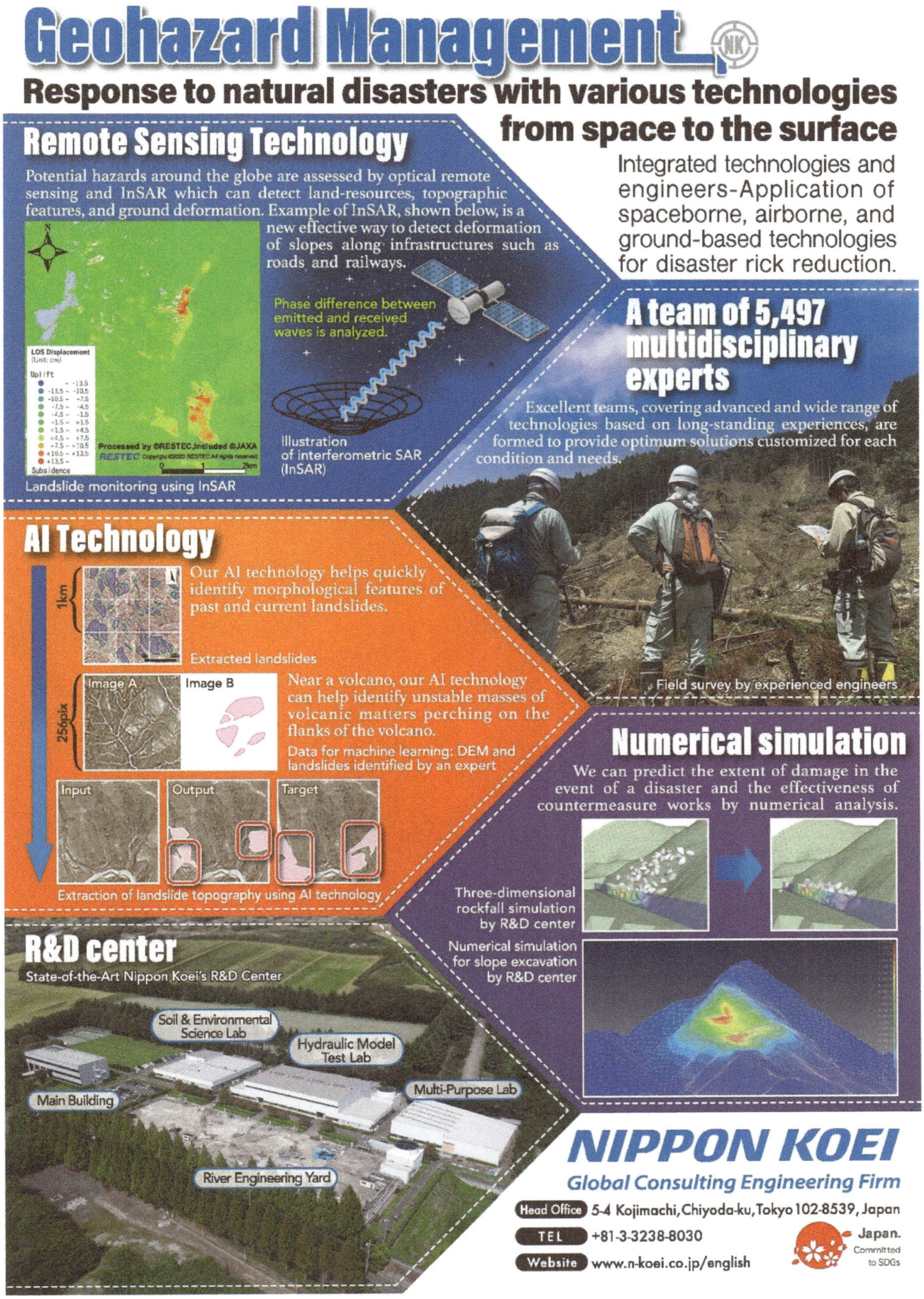
Geohazard Management
Response to natural disasters with various technologies from space to the surface
Integrated technologies and engineers-Application of spaceborne, airborne, and ground-based technologies for disaster rick reduction.
Remote Sensing Technology
Potential hazards around the globe are assessed by optical remote sensing and InSAR which can detect land-resources, topographic features, and ground deformation. Example of InSAR, shown below, is a new effective way to detect deformation of slopes along infrastructures such as roads and railways.
Phase difference between emitted and received waves is analyzed.
Illustration of interferometric SAR (InSAR)
Landslide monitoring using InSAR
A team of 5,497 multidisciplinary experts
Excellent teams, covering advanced and wide range of technologies based on long-standing experiences, are formed to provide optimum solutions customized for each condition and needs.
Field survey by experienced engineers
AI Technology
Our AI technology helps quickly identify morphological features of past and current landslides.
1km
Extracted landslides
Image A
Image B
256pix
Near a volcano, our AI technology can help identify unstable masses of volcanic matters perching on the flanks of the volcano.
Data for machine learning: DEM and landslides identified by an expert
Input
Output
Target
Extraction of landslide topography using AI technology
Numerical simulation
We can predict the extent of damage in the event of a disaster and the effectiveness of countermeasure works by numerical analysis.
Three-dimensional rockfall simulation by R&D center
Numerical simulation for slope excavation by R&D center
R&D center
State-of-the-Art Nippon Koei's R&D Center
Soil & Environmental Science Lab
Hydraulic Model Test Lab
Main Building
Multi-Purpose Lab
River Engineering Yard
NIPPON KOEI
Global Consulting Engineering Firm
Head Office 5-4 Kojimachi, Chiyoda-ku, Tokyo 102-8539, Japan
TEL +81-3-3238-8030
Website www.n-koei.co.jp/english
Japan. Committed to SDGs

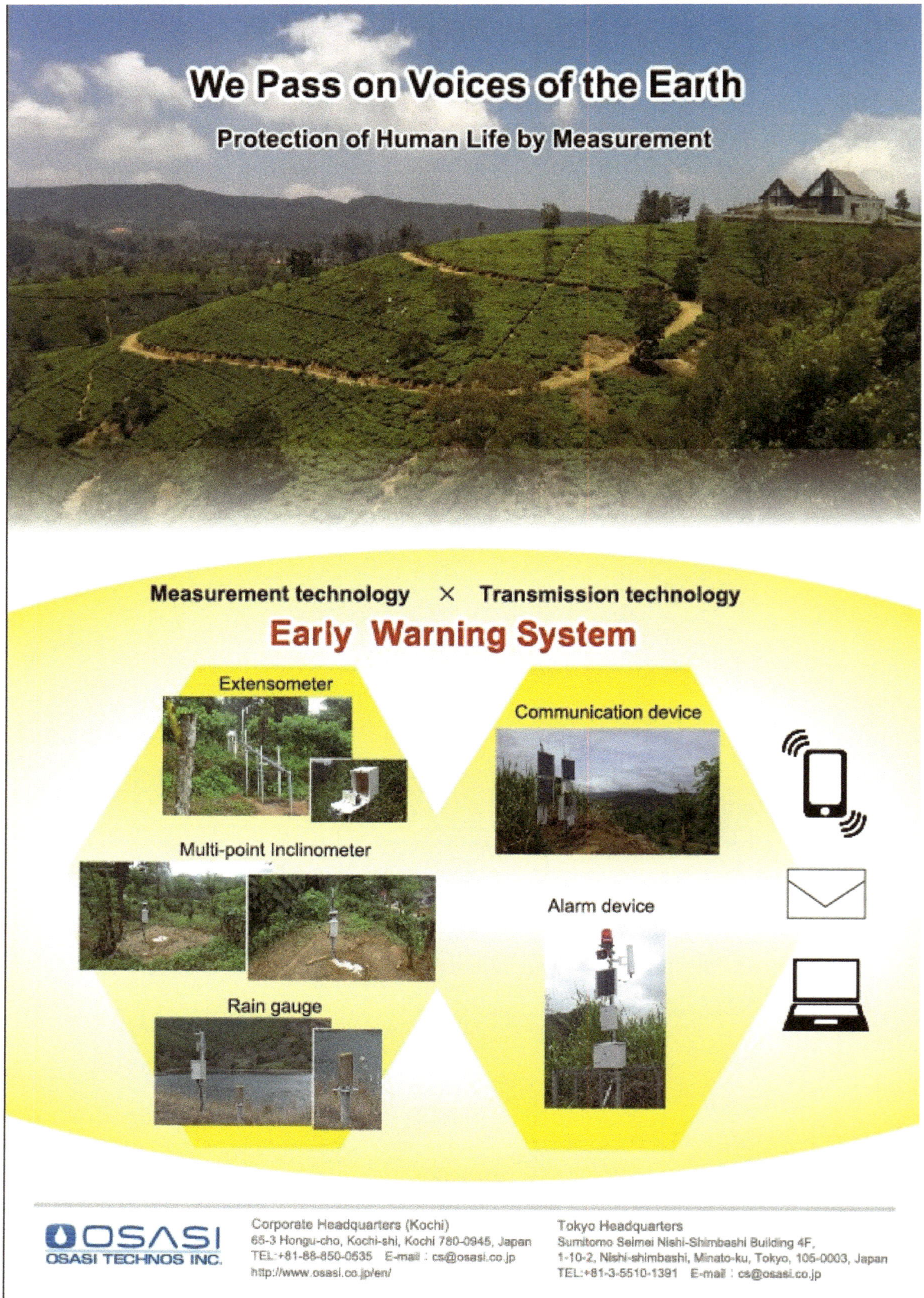
We Pass on Voices of the Earth
Protection of Human Life by Measurement
Measurement technology × Transmission technology
Early Warning System
Extensometer
Communication device
Multi-point Inclinometer
Alarm device
Rain gauge
OSASI
OSASI TECHNOS INC.
Corporate Headquarters (Kochi)
65-3 Hongu-cho, Kochi-shi, Kochi 780-0945, Japan
TEL:+81-88-850-0535 E-mail : cs@osasi.co.jp
http://www.osasi.co.jp/en/
Tokyo Headquarters
Sumitomo Seimei Nishi-Shimbashi Building 4F,
1-10-2, Nishi-shimbashi, Minato-ku, Tokyo, 105-0003, Japan
TEL:+81-3-5510-1391 E-mail : cs@osasi.co.jp

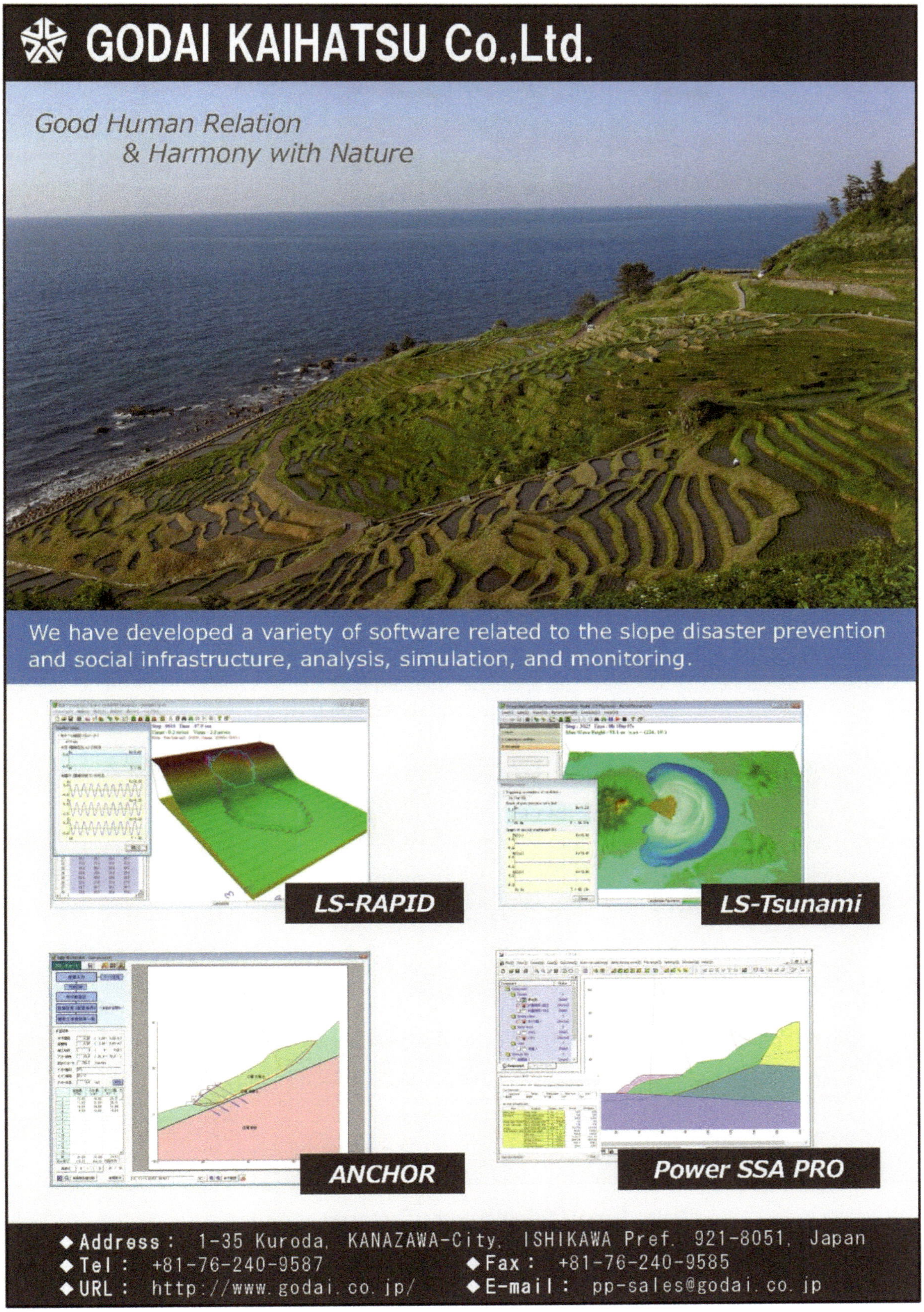
GODAI KAIHATSU Co.,Ltd.
Good Human Relation
& Harmony with Nature
We have developed a variety of software related to the slope disaster prevention and social infrastructure, analysis, simulation, and monitoring.
LS-RAPID
LS-Tsunami
ANCHOR
Power SSA PRO
◆Address : 1-35 Kuroda, KANAZAWA-City, ISHIKAWA Pref. 921-8051, Japan
◆Tel : +81-76-240-9587
◆Fax : +81-76-240-9585
◆URL : http://www.godai.co.jp/
◆E-mail : pp-sales@godai.co.jp

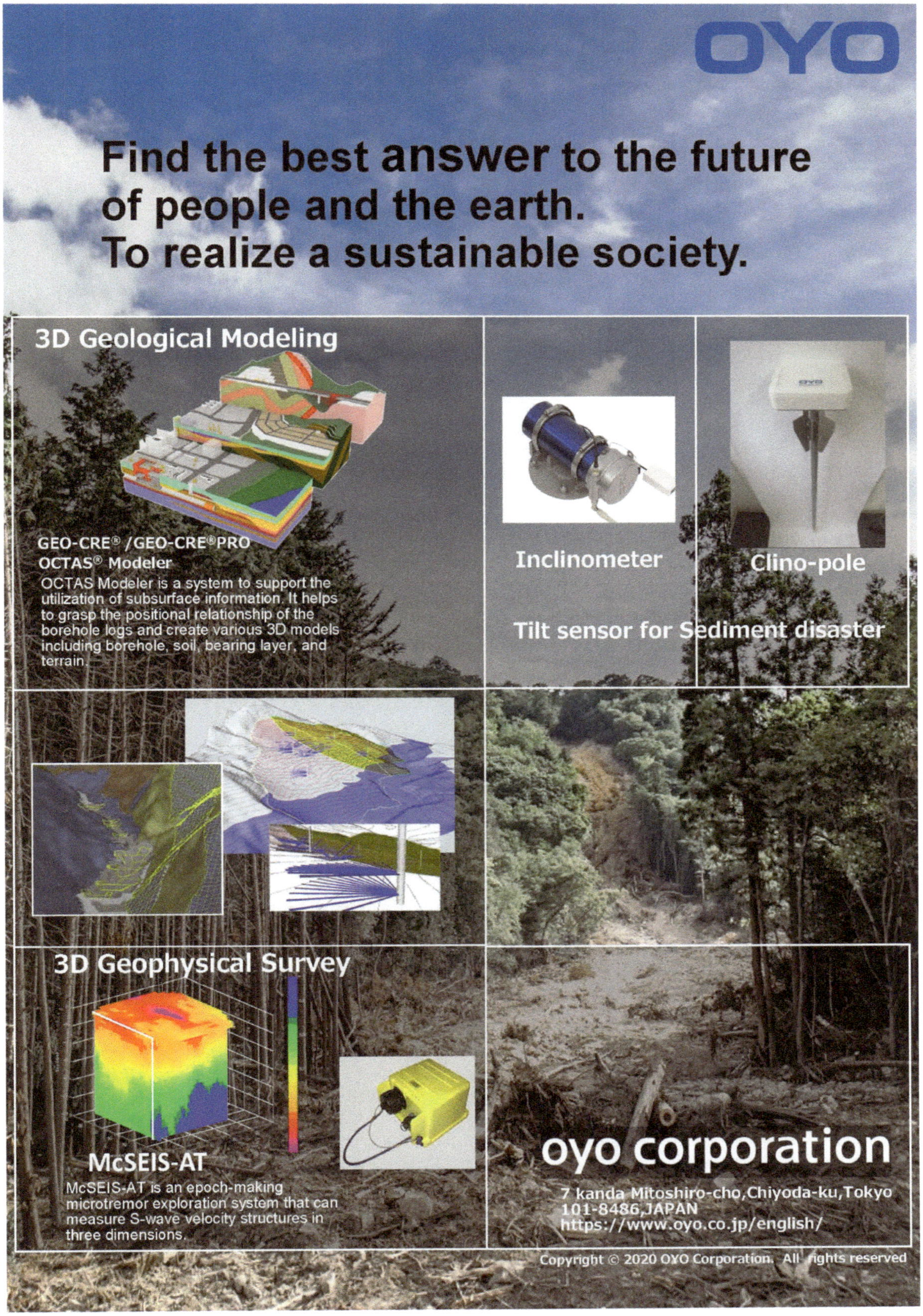
OYO
Find the best answer to the future of people and the earth.
To realize a sustainable society.
3D Geological Modeling
GEO-CRE® /GEO-CRE®PRO
OCTAS® Modeler
OCTAS Modeler is a system to support the utilization of subsurface information. It helps to grasp the positional relationship of the borehole logs and create various 3D models including borehole, soil, bearing layer, and terrain.
Inclinometer
Clino-pole
Tilt sensor for Sediment disaster
3D Geophysical Survey
McSEIS-AT
McSEIS-AT is an epoch-making microtremor exploration system that can measure S-wave velocity structures in three dimensions.
oyo corporation
7 kanda Mitoshiro-cho,Chiyoda-ku,Tokyo
101-8486,JAPAN
https://www.oyo.co.jp/english/
Copyright © 2020 OYO Corporation. All rights reserved

Free dimensioning tool
www.mygeobrugg.com
Ring net barriers made of high-tensile steel wire
THE ECONOMICAL SOLUTION
TO DEBRIS FLOW

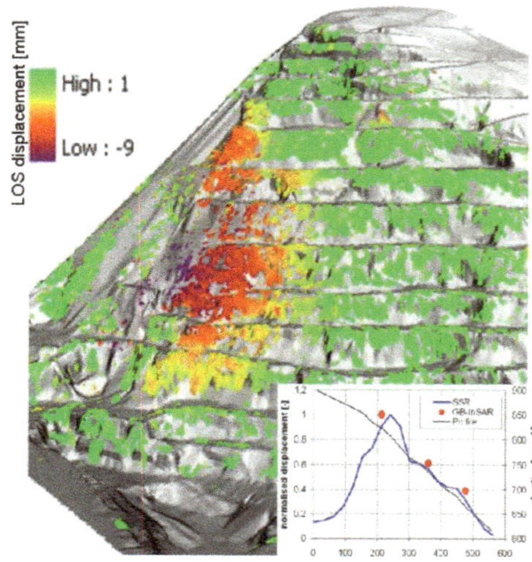

LOS displacement [mm]
High : 1
Low : -9

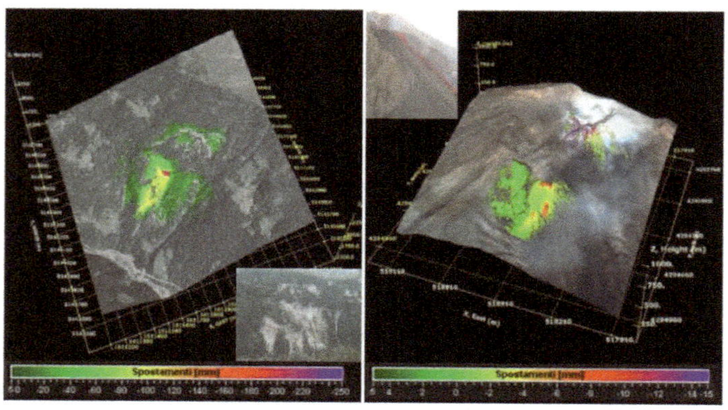

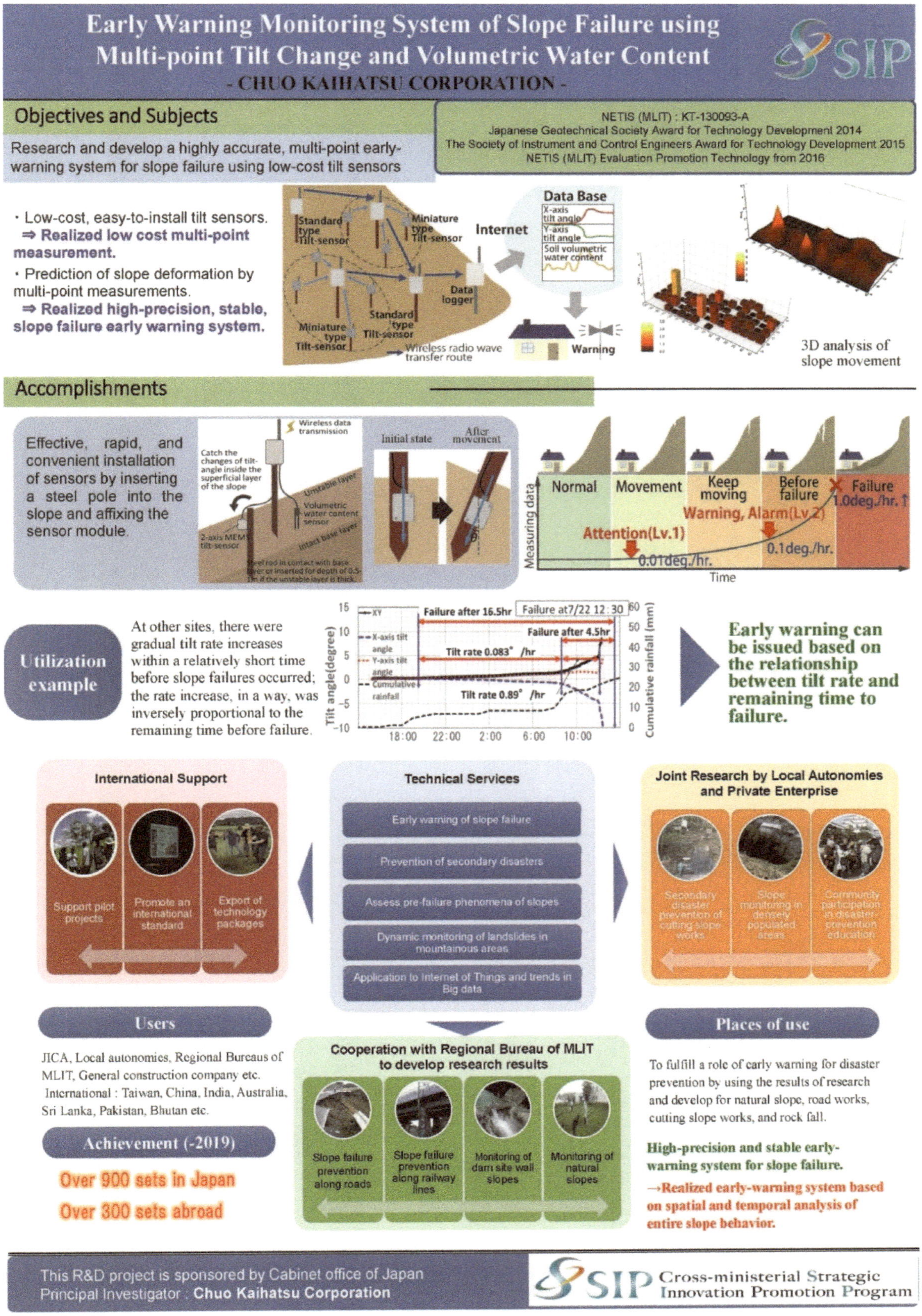
Early Warning Monitoring System of Slope Failure using
Multi-point Tilt Change and Volumetric Water Content
- CHUO KAIHATSU CORPORATION -
SIP
Objectives and Subjects
Research and develop a highly accurate, multi-point early-warning system for slope failure using low-cost tilt sensors
NETIS (MLIT) : KT-130093-A
Japanese Geotechnical Society Award for Technology Development 2014
The Society of Instrument and Control Engineers Award for Technology Development 2015
NETIS (MLIT) Evaluation Promotion Technology from 2016
・Low-cost, easy-to-install tilt sensors.
⇒ Realized low cost multi-point measurement.
・Prediction of slope deformation by multi-point measurements.
⇒ Realized high-precision, stable, slope failure early warning system.
Standard type Tilt-sensor
Miniature type Tilt-sensor
Data logger
Internet
Data Base
X-axis tilt angle
Y-axis tilt angle
Soil volumetric water content
Warning
Wireless radio wave transfer route
3D analysis of slope movement
Accomplishments
Effective, rapid, and convenient installation of sensors by inserting a steel pole into the slope and affixing the sensor module.
Wireless data transmission
Catch the changes of tilt-angle inside the superficial layer of the slope
Unstable layer
Volumetric water content sensor
2-axis MEMS tilt-sensor
Intact base layer
Initial state
After movement
Normal
Movement
Keep moving
Before failure
Failure
1.0deg./hr.↑
Warning, Alarm(Lv.2)
Attention(Lv.1)
0.01deg./hr.
0.1deg./hr.
Measuring data
Time
Utilization example
At other sites, there were gradual tilt rate increases within a relatively short time before slope failures occurred; the rate increase, in a way, was inversely proportional to the remaining time before failure.
Failure after 16.5hr
Failure at7/22 12:30
Failure after 4.5hr
Tilt rate 0.083° /hr
Tilt rate 0.89° /hr
XY
X-axis tilt angle
Y-axis tilt angle
Cumulative rainfall
Tilt angle(degree)
Cumulative rainfall (mm)
18:00 22:00 2:00 6:00 10:00
Early warning can be issued based on the relationship between tilt rate and remaining time to failure.
International Support
Support pilot projects
Promote an international standard
Export of technology packages
Technical Services
Early warning of slope failure
Prevention of secondary disasters
Assess pre-failure phenomena of slopes
Dynamic monitoring of landslides in mountainous areas
Application to Internet of Things and trends in Big data
Joint Research by Local Autonomies and Private Enterprise
Secondary disaster prevention of cutting slope works
Slope monitoring in densely populated areas
Community participation in disaster-prevention education
Users
JICA, Local autonomies, Regional Bureaus of MLIT, General construction company etc.
International : Taiwan, China, India, Australia, Sri Lanka, Pakistan, Bhutan etc.
Achievement (-2019)
Over 900 sets in Japan
Over 300 sets abroad
Cooperation with Regional Bureau of MLIT to develop research results
Slope failure prevention along roads
Slope failure prevention along railway lines
Monitoring of dam site wall slopes
Monitoring of natural slopes
Places of use
To fulfill a role of early warning for disaster prevention by using the results of research and develop for natural slope, road works, cutting slope works, and rock fall.
High-precision and stable early-warning system for slope failure.
→Realized early-warning system based on spatial and temporal analysis of entire slope behavior.
This R&D project is sponsored by Cabinet office of Japan
Principal Investigator : Chuo Kaihatsu Corporation
SIP Cross-ministerial Strategic Innovation Promotion Program

IDS GeoRadar

IBIS FM - Natural Hazard Monitoring Solution

GB-InSAR System with Advanced Atmospheric Correction

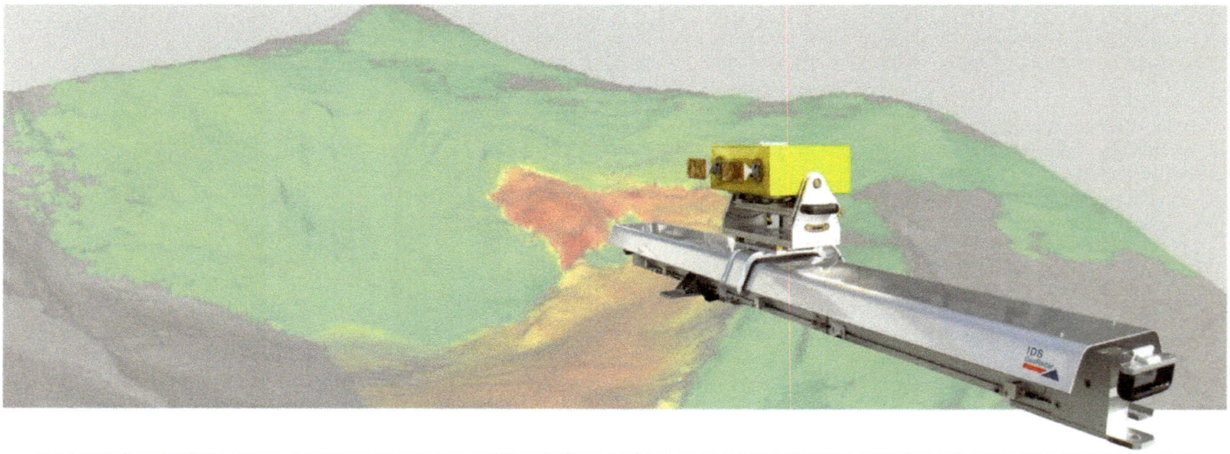

IDS GeoRadar is nowadays the leader company in the Radar monitoring solutions applied to landslide monitoring, thanks to the combination of the highest performance radar technology with the most advanced data processing algorithm for the removal of atmospheric artifacts integral part of Guardian Software.

The unique IBIS-FM radar system accurately monitors multiple scales of displacements in real time, from early detection of slow movements to fast accelerations associated with slope collapse. The great operative range, up to 4500 m, allows to safely deploying the system in comfortably accessible areas, without exposing people and equipment to hazardous zones.

Fully automatic

The atmospheric correction procedure does not require any input from the user (does not need any stable area selection); it is a completely automatic software processing, not requiring advanced know-how to the user. The atmospheric compensation algorithm is based on hundreds of thousands of pixels automatically selected by the software and updated at each radar scan. It permits to achieve the best performances even in case of extreme or unstable atmospheric conditions.

Advanced 3D atmospheric effects modelling

IDS GeoRadar algorithm for atmospheric correction offers unique capability able to model and remove complex atmospheric effects in extensive areas overcoming the limits of the standard linear model algorithms, providing very clean displacement maps in the most extreme atmospheric conditions environment.

Slow and fast movement detection

IDS GeoRadar algorithm is able to distinguish very slow movement together with the faster ones providing the capability to measure in real time movements across four orders of magnitudes: from very fast movements (up to 150 mm/hour) to extremely low displacements (few mm/month).

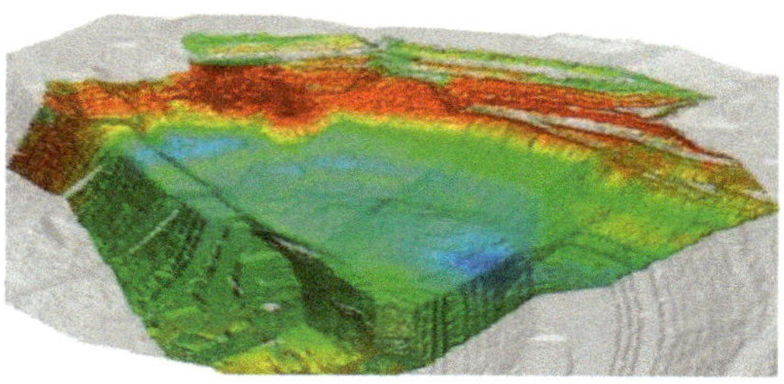

Displacement map obtained with standard processing and standard atmospheric correction

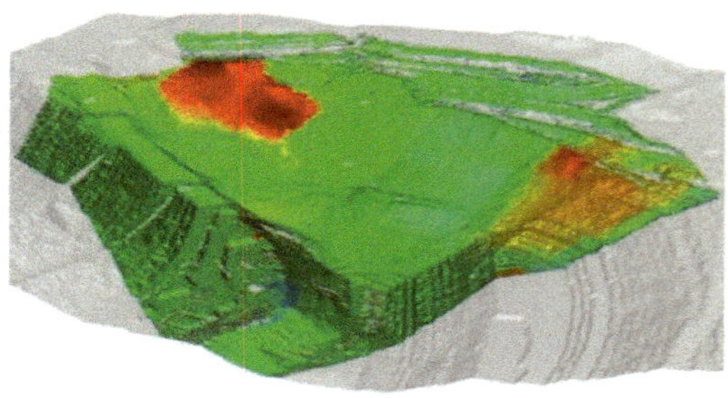

Displacement map obtained with IDS GeoRadar Guardian atmospheric correction and processing

IDS GeoRadar Srl
Part of Hexagon
Via Augusto Righi, 6, 6A, 8 - 56121 Ospedaletto, Pisa, Italy
Tel: +39 050 89 34 100
info@idsgeoradar.com
www.idsgeoradar.com

METER
ENVIRONMENT
LANDSLIDE PREDICTION AND ALERTS
POSITIVE & NEGATIVE PORE WATER
PRESSURE MONITORING
DISCOVER NOW
For more information visit
metergroup.com/wlf5

Kiso-Jiban Consultants Co., Ltd.

Monitoring Service for Landslide by Kiso-SAR System

What's SAR ?

SAR (Synthetic Aperture Radar) is a technique that utilizes interference of radio waves for precise determination of distance.
InSAR (interferometric SAR), the phase of the received backscattered signal from two images of the same scene are used to measure path length differences with an accuracy of a few millimeters.

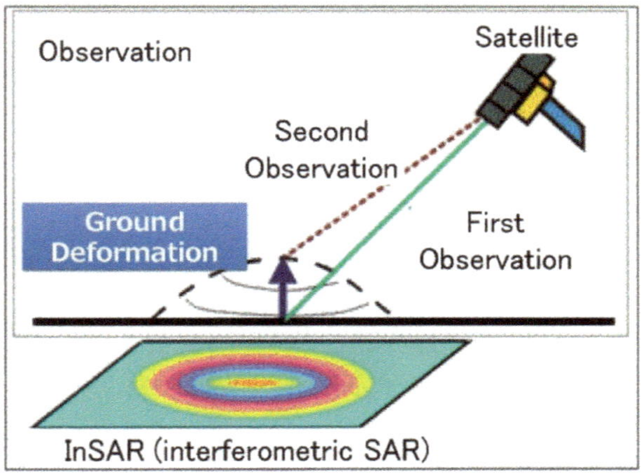

Ground Deformation Estimation in Landslides

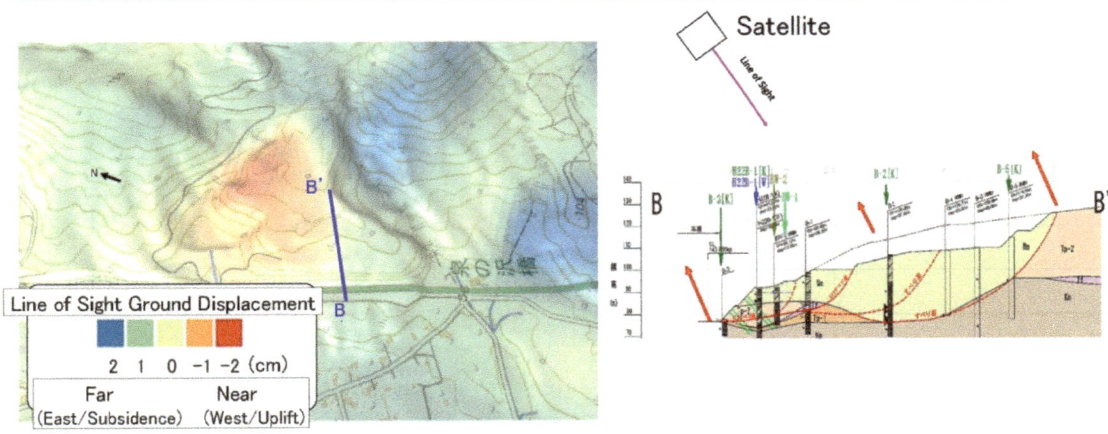

Ground Deformation of Landslide Observed by Kiso-SAR System

Kiso-Jiban Consultants Co., Ltd.
Kinshicho Prime Tower 12 Floor,1-5-7
Kameido,Koto-ku,Tokyo 136-8577,Japan
Tel.:+81-3-6861-8800

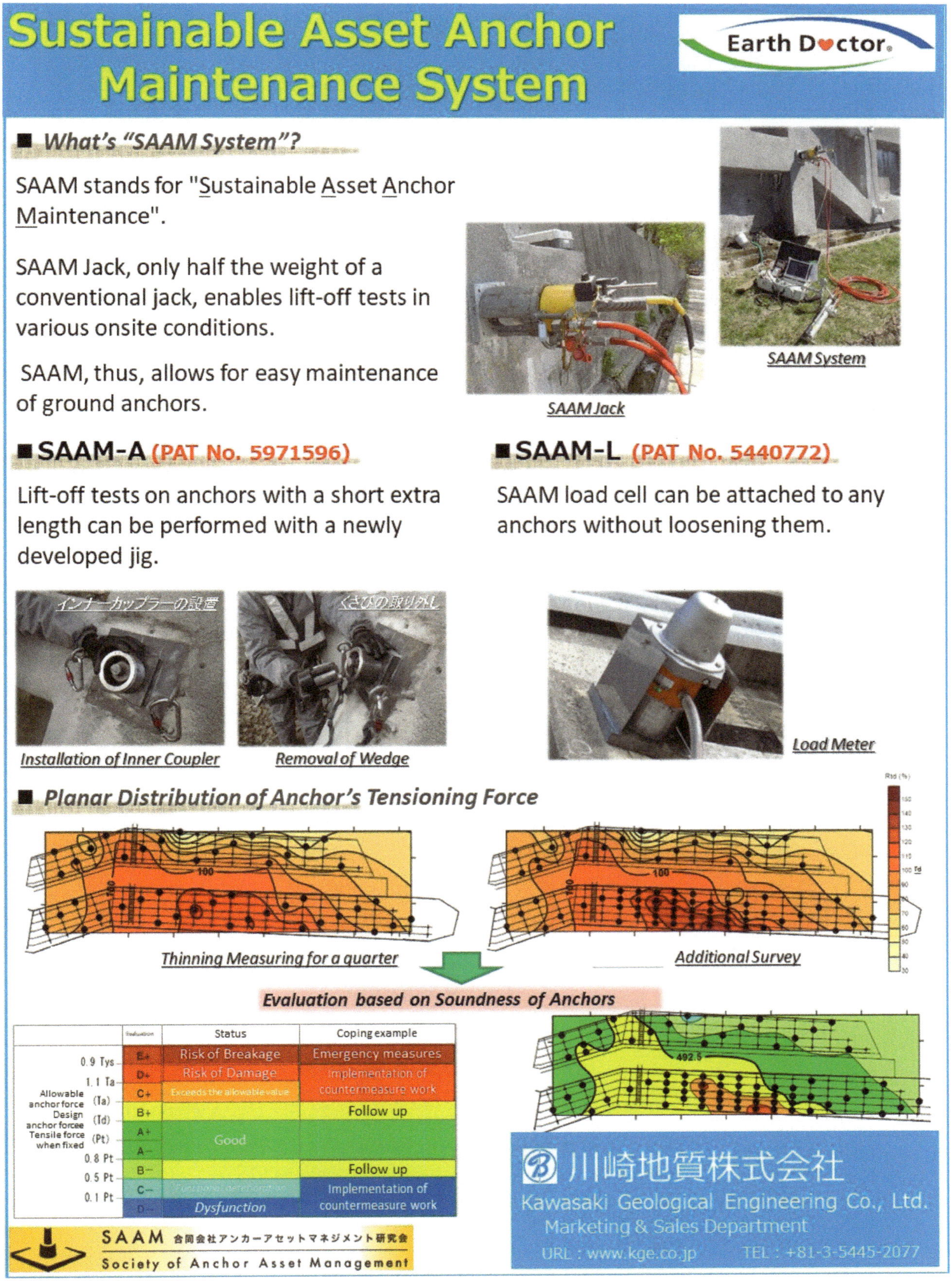
Sustainable Asset Anchor Maintenance System
Earth Doctor
■ What's "SAAM System"?
SAAM stands for "Sustainable Asset Anchor Maintenance".
SAAM Jack, only half the weight of a conventional jack, enables lift-off tests in various onsite conditions.
SAAM, thus, allows for easy maintenance of ground anchors.
SAAM Jack
SAAM System
■ SAAM-A (PAT No. 5971596)
Lift-off tests on anchors with a short extra length can be performed with a newly developed jig.
インナーカップラーの設置
くさびの取り外し
Installation of Inner Coupler
Removal of Wedge
■ SAAM-L (PAT No. 5440772)
SAAM load cell can be attached to any anchors without loosening them.
Load Meter
■ Planar Distribution of Anchor's Tensioning Force
Thinning Measuring for a quarter
Additional Survey
Evaluation based on Soundness of Anchors
Status
Coping example
0.9 Tys
1.1 Ta
Allowable anchor force (Ta)
Design anchor forcee (Td)
Tensile force when fixed (Pt)
0.8 Pt
0.5 Pt
0.1 Pt
E+
D+
C+
B+
A+
A−
B−
C−
D−
Risk of Breakage
Risk of Damage
Exceeds the allowable value
Good
Dysfunction
Emergency measures
Implementation of countermeasure work
Follow up
Follow up
Implementation of countermeasure work
492.5
川崎地質株式会社
Kawasaki Geological Engineering Co., Ltd.
Marketing & Sales Department
URL : www.kge.co.jp
TEL : +81-3-5445-2077
SAAM 合同会社アンカーアセットマネジメント研究会
Society of Anchor Asset Management

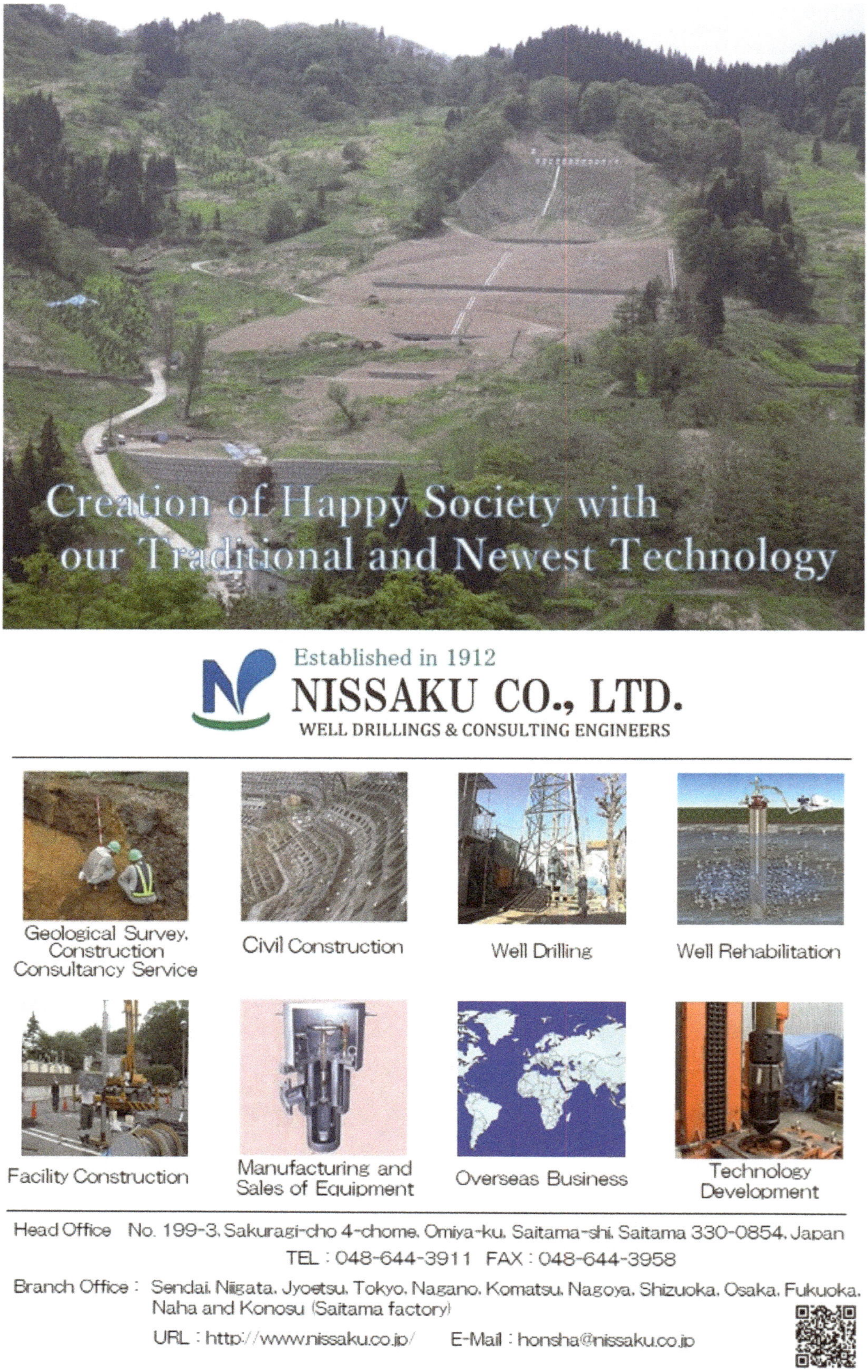
Creation of Happy Society with
our Traditional and Newest Technology
Established in 1912
NISSAKU CO., LTD.
WELL DRILLINGS & CONSULTING ENGINEERS
Geological Survey, Construction Consultancy Service
Civil Construction
Well Drilling
Well Rehabilitation
Facility Construction
Manufacturing and Sales of Equipment
Overseas Business
Technology Development
Head Office No. 199-3, Sakuragi-cho 4-chome, Omiya-ku, Saitama-shi, Saitama 330-0854, Japan
TEL : 048-644-3911 FAX : 048-644-3958
Branch Office : Sendai, Niigata, Jyoetsu, Tokyo, Nagano, Komatsu, Nagoya, Shizuoka, Osaka, Fukuoka, Naha and Konosu (Saitama factory)
URL : http://www.nissaku.co.jp/ E-Mail : honsha@nissaku.co.jp

Correction to: From Satellite Images to Field Survey: A Complete Scheme of Landslide InSAR Monitoring

Matteo Del Soldato, Lorenzo Solari, Davide Festa, Pierluigi Confuorto, Silvia Bianchini, and Nicola Casagli

Correction to:
Chapter "From Satellite Images to Field Survey: A Complete Scheme of Landslide InSAR Monitoring" in: F. Guzzetti et al. (eds.), Understanding and Reducing Landslide Disaster Risk, ICL Contribution to Landslide Disaster Risk Reduction, https://doi.org/10.1007/978-3-030-60227-7_47

The book was inadvertently published with chapter author's incorrect family name. This information has been updated from "Del Soldato Matteo" to "Matteo Del Soldato" in the initially published version of chapter "From Satellite Images to Field Survey: A Complete Scheme of Landslide InSAR Monitoring". The correction chapter and the book have been updated.

The updated version of this chapter can be found at https://doi.org/10.1007/978-3-030-60227-7_47

F. Guzzetti et al. (eds.), *Understanding and Reducing Landslide Disaster Risk*,
ICL Contribution to Landslide Disaster Risk Reduction,
https://doi.org/10.1007/978-3-030-60227-7_57

International Consortium on Landslides

International Consortium on Landslides

An international non-government and non-profit scientific organization
promoting landslide research and capacity building for the benefit of society and the environment

President: Peter Bobrowsky (Geological Survey of Canada)
Vice Presidents: Matjaz Mikos (University of Ljubljana, Slovenia), Dwikorita Karnawati (Agency for Meteorology, Climatology, and Geophysics, Indonesia), Nicola Casagli (University of Florence, Italy), Binod Tiwari (California State University, USA), Zeljko Arbanas (University of Rijeka, Croatia)
Executive Director: Kaoru Takara (Kyoto University, Japan), Treasurer: Kyoji Sassa (Prof. Emeritus, Kyoto University, Japan)

ICL Full Members:

Geotechnical Engineering Office, Hong Kong Special Administrative Region, China
UNESCO Chair for the Prevention and the Sustainable Management of Geo-hydrological Hazards - University of Florence, Italy
Korea Institute of Geoscience and Mineral Resources (KIGAM)
University of Ljubljana, Faculty of Civil and Geodetic Engineering (ULFGG), Slovenia

Albania Geological Survey / The Geotechnical Society of Bosnia and Herzegovina / Center for Scientific Support in Disasters – Federal University of Parana, Brazil/ Geological Survey of Canada / University of Alberta, Canada / Northeast Forestry University, Institute of Cold Regions Science and Engineering, China / China University of Geosciences / Chinese Academy of Sciences, Institute of Mountain Hazards and Environment / Tongji University, College of Surveying and Geo-Informatics, China / The Hong Kong University of Science and Technology, China / Shanghai Jiao Tong University, China / The University of Hong Kong, China / Universidad Nacional de Colombia / Croatian Landslide Group (Faculty of Civil Engineering, University of Rijeka and Faculty of Mining, Geology and Petroleum Engineering, University of Zagreb) / City of Zagreb, Emergency Management Office, Croatia / Charles University, Faculty of Science, Czech Republic / Institute of Rock Structure and Mechanics, Department of Engineering Geology, Czech Republic / Brown Coal Research Institute, Czech Republic / Cairo University, Egypt / Technische Universitat Darmstadt, Institute and Laboratory of Geotechnics, Germany / National Environmental Agency, Department of Geology, Georgia / Universidad Nacional Autonoma de Honduras (UNAH), Honduras / Amrita Vishwa Vidyapeetham, Amrita University / Vellore Institute of Technology, India / National Institute of Disaster Management, India / Agency for Meteorology, Climatology, and Geophysics of the Republic of Indonesia (BMKG Indonesia) / University of Gadjah Mada, Center for Disaster Mitigation and Technological Innovation (GAMA-InaTEK), Indonesia / Parahyangan Catholic University, Indonesia / Building & Housing Research Center, Iran / Italian Institute for Environmental Protection and Research (ISPRA) - Dept. Geological Survey, Italy / University of Calabria, DIMES, CAMILAB, Italy / Istituto de Ricerca per la Protezione Idrogeologica (IRPI), CNR, Italy / DIA–Universita degli Studi di Parma, Italy / University of Torino, Dept of Earth Science , Italy / Centro di Ricerca CERI - Sapienza Università di Roma, Italy / Kyoto University, Disaster Prevention Research Institute, Japan / Japan Landslide Society / Korean Society of Forest Engineering / National Institute of Forest Science, Korea / Korea Infrastructure Safety & Technology Corporation / Korea Institute of Civil Engineering and Building Technology / Slope Engineering Branch, Public Works Department of Malaysia / Institute of Geography, National Autonomous University of Mexico (UNAM) / International Centre for Integrated Mountain Development (ICIMOD), Nepal / University of Nigeria, Department of Geology, Nigeria / Moscow State University, Department of Engineering and Ecological Geology, Russia / JSC "Hydroproject Institute", Russia / University of Belgrade, Faculty of Mining and Geology, Serbia / Comenius University, Faculty of Natural Sciences, Department of Engineering Geology, Slovakia / Geological Survey of Slovenia / University of Ljubljana, Faculty of Natural Sciences and Engineering (ULNTF), Slovenia / Central Engineering Consultancy Bureau (CECB), Sri Lanka / National Building Research Organization, Sri Lanka / Landslide group in National Central University from Graduate Institute of Applied Geology, Department of Civil Engineering, Center for Environmental Studies, Chinese Taipei / National Taiwan University, Department of Civil Engineering, Chinese Taipei / Asian Disaster Preparedness Center, Thailand / Ministry of Agriculture and Cooperative, Land Development Department, Thailand / Institute of Telecommunication and Global Information Space, Ukraine / California State University, Fullerton & Tribhuvan University, Institute of Engineering, USA & Nepal / Institute of Transport Science and Technology, Vietnam / Vietnam Institute of Geosciences and Mineral Resources (VIGMR).

ICL Associates

State Key Laboratory of Geohazard Prevention and Geoenvironment Protection (Chengdu University of Technology), China / Czech Geological Survey, Czech Republic / Department of Earth and Environmental Sciences, University Aldo Moro, Bari, Italy / Department of Sciences and Technologies, University of Sannio, Italy / Department of Earth and Environmental Sciences – University of Pavia, Italy / Geotechnical Engineering Group (GEG), University of Salerno, Italy / Niigata University, Research Institute for Natural Hazards and Disaster Recovery, Japan / Ehime University Center for Disaster Management Informatics Research, Japan / Tian-Shan Geological Society, Kyrgyzstan / Institute of Environmental Geoscience RAS (IEG RAS), Russia / Russian State Geological Prospecting University n.a. Sergo Ordzhonikidze (MGRI-RSGPU) / TEMPOS, environmental civil engineering Ltd., Slovenia / Institute of Earth Sciences – Faculty of Geoscience and Environment, University of Lausanne, Switzerland / Middle East Technical University (METU), Turkey / North Dakota State University, USA

ICL Secretariat:

Secretary General: Kyoji Sassa
International Consortium on Landslides, 138-1 Tanaka Asukai-cho, Sakyo-ku, Kyoto 606-8226, Japan
Web: http://icl.iplhq.org/, E-mail: secretariat@iclhq.org
Tel: +81-75-723-0640, Fax: +81-75-950-0910

F. Guzzetti et al. (eds.), *Understanding and Reducing Landslide Disaster Risk*,
ICL Contribution to Landslide Disaster Risk Reduction,
https://doi.org/10.1007/978-3-030-60227-7